ADVANCED MODERN
CONTROL SYSTEM
THEORY AND DESIGN

ADVANCED MODERN CONTROL SYSTEM THEORY AND DESIGN

Stanley M. Shinners

Program Manager
Lockheed Martin Federal Systems, Inc.
Great Neck, New York

and

Adjunct Professor of Electrical Engineering
The Cooper Union for the Advancement of Science and Art
New York, New York

A WILEY-INTERSCIENCE PUBLICATION
JOHN WILEY & SONS, INC.
New York / Chichester / Weinheim / Brisbane / Singapore / Toronto

The author and publisher of Advanced Modern Control System Theory and Design have used their best efforts in preparing this book and the accompanying Advanced Modern Control System Theory and Design Toolbox of software. These efforts have included the research, development, checking, testing, and evaluation of the theories, analysis, applications, and computer programs to determine their correctness, accuracy, and effectiveness. The software contained in the Advanced Modern Control System Theory and Design Toolbox have been supplied to The MathWorks, Inc for distribution, upon request, on an "as is" basis. The accuracy of the solutions contained may be affected by The MathWorks' upgrades to their software. The author and publisher make no warranty of any kind, expressed or implied, with regard to the documentation of the programs contained in this book. Although this material have been carefully checked, the author and publisher shall not be liable in any event for incidental or consequential damages resulting in connection with, or arising out of, the use, performance, or furnishing of this material.

This book is printed on acid-free paper. ∞

Copyright © 1998 by John Wiley & Sons, Inc. All rights reserved.

Published simultaneously in Canada.

No part of this publication may be reproduced, stored in a retrieval system or transmitted in any form or by any means, electronic, mechanical, photocopying, recording, scanning or otherwise, except as permitted under Sections 107 or 108 of the 1976 United States Copyright Act, without either the prior written permission of the Publisher, or authorization through payment of the appropriate per-copy fee to the Copyright Clearance Center, 222 Rosewood Drive, Danvers, MA 01923, (978) 750-8400, fax (978) 750-4744. Requests to the Publisher for permission should be addressed to the Permissions Department, John Wiley & Sons, Inc., 605 Third Avenue, New York, NY 10158-0012, (212) 850-6011, fax (212) 850-6008, E-Mail: PERMREQ @ WILEY.COM.

Library of Congress Cataloging-in-Publication Data is available.

Shinners, Stanley M.
    Advanced modern control system theory and design / Stanley M. Shinners
      p. cm.
    Includes index.
    ISBN 0-471-31857-4

To my wife, Doris,
and to
Sharon, Stuart, Jessica, Jonathan,
Walter, Laurie, Jason, Rebecca
and
Daniel

CONTENTS

Preface   xi

1 Introduction   1

     1.1. Introduction / 1
     1.2. Goal of Advanced Modern Control System Theory and Design / 2
     1.3. Control System Performance Objectives / 2
     1.4. The Procedure for Designing a Control System / 5
     1.5. Outline of Advanced Modern Control System Theory and Design / 7
     1.6. Advanced Modern Control System Theory and Design Toolbox / 9
     1.7. Illustrative Problems and Solutions / 9
     Problems / 9
     References / 31

2 Linear Control-System Compensation and Design   33

     2.1. Introduction / 33
     2.2. Cascade-Compensation Techniques / 36
     2.3. Minor-Loop Feedback-Compensation Techniques / 44
     2.4. Proportional-Plus-Integral-Plus Derivative (PID) Compensators / 46
     2.5. Example for the Design of a Second-Order Control System / 48
     2.6. Compensation and Design using the Bode-Diagram Method / 51
     2.7. Approximate Methods for Preliminary Compensation and Design using the Bode Diagram / 67
     2.8. Compensation and Design using the Nichols Chart / 74
     2.9. Compensation and Design using the Root-Locus Method / 75
     2.10. Tradeoffs of using Various Cascade-Compensation Methods and Minor-Loop Feedback / 93
     2.11. Illustrative Problems and Solutions / 96
     Problems / 100
     References / 117

3 Modern Control-System Design using State-Space, Pole Placement, Ackermann's Formula, Estimation, Robust Control, and H^∞ Techniques   118

     3.1. Introduction / 118

- 3.2. Pole-Placement Design using Linear-State-Variable Feedback / 119
- 3.3. Controller Design using Pole Placement and Linear-State-Variable-Feedback Techniques / 124
- 3.4. Controllability / 131
- 3.5. Observability / 135
- 3.6. Ackermann's Formula for Design using Pole Placement / 137
- 3.7. Estimator Design in Conjunction with the Pole Placement Approach using Linear-State-Variable Feedback / 143
- 3.8. Combined Compensator Design Including a Controller and an Estimator for a Regulator System / 146
- 3.9. Extension of Combined Compensator Design Including a Controller and an Estimator for Systems Containing a Reference Input / 154
- 3.10. Robust Control Systems / 156
- 3.11. An Introduction to H^∞ Control Concepts / 162
- 3.12. Foundations of H^∞ Control Theory / 169
- 3.13. Linear Algebraic Aspects of Control-System Design Computations / 175
- 3.14. Illustrative Problems and Solutions / 177
  Problems / 198
  References / 207

4 Digital Control-System Analysis and Design   209

- 4.1. Introduction / 209
- 4.2. Characteristics of Sampling / 211
- 4.3. Data Extrapolators / 215
- 4.4. z-Transform Theory / 221
- 4.5. z-Transform Block-Diagram Algebra / 236
- 4.6. Characteristic Response of a Sampler and Zero-Order Hold Combination / 242
- 4.7. Stability Analysis Using the Nyquist Diagram / 245
- 4.8. Stability Determination Using Mathematical Tests / 252
- 4.9. Stability Analysis and Design Using the Bode Diagram / 256
- 4.10. Stability Analysis and Design Using the Root-Locus Diagram / 262
- 4.11. Bode and Root-Locus Diagrams for Discrete-Time Systems using MATLAB / 272
- 4.12. Ragazzini's Method / 273
- 4.13. The Digitization Process and the Design of Digital Filters / 285
- 4.14. Summary / 289
- 4.15. Illustrative Problems and Solutions / 290
  Problems / 311
  References / 324

5 Nonlinear Control-System Design   325

5.1. Introduction / 325
5.2. Nonlinear Differential Equations / 326
5.3. Properties of Linear Systems that are not Valid for Nonlinear Systems / 327
5.4. Unique Characteristics of Nonlinear Systems / 328
5.5. Methods Available for Analyzing Nonlinear Systems / 329
5.6. Linearizing Approximations / 330
5.7. Describing-Function Concept / 334
5.8. Derivation of Describing Functions for Common Nonlinearities / 336
5.9. Use of the Describing Function to Predict Oscillations / 352
5.10. Compensation and Design of Nonlinear Control Systems Using Describing Functions / 357
5.11. Describing-Functions Analysis and Design Using MATLAB / 361
5.12. Digital Computer Programs for Performing the Describing Function Analysis / 363
5.13. Piecewise-Linear Approximations / 369
5.14. State-Variable Analysis: The Phase Plane / 371
5.15. Construction of the Phase Portrait / 372
5.16. Characteristics of the Phase Portrait / 381
5.17. Phase Plane for Systems Containing External Forcing Functions / 389
5.18. Design of Nonlinear Feedback Control Systems Using the State-Variable Phase-Plane Method / 392
5.19. Digital Computer Program for Obtaining the Phase Plane / 397
5.20. Liapunov's Stability Criteria / 399
5.21. Popov's Method / 415
5.22. Generalized Circle Criterion / 422
5.23. Guidelines for Selecting the "Best" Nonlinear Control System Method(s) Presented for Analysis and Design / 425
5.24. Illustrative Problems and Solutions / 428
Problems / 435
References / 454

6 Introduction to Optimal Control Theory and Its Applications   457

6.1. Introduction / 457
6.2. Characteristics of the Optimal Control Problem / 458
6.3. Calculus of Variations / 462
6.4. Dynamic Programming / 467
6.5. Pontryagin's Maximum Principle / 473
6.6. Application of the Maximum Principle to the Space Attitude-Control Problem / 477
6.7. Application of the Maximum Principle to the Lunar Soft-Landing Problem / 488

6.8. Illustrative Problems and Solutions / 492
Problems / 496
References / 502

7 Control-System Design Examples: Complete Case Studies 505

7.1. Introduction / 505
7.2. Outline of Procedure for Designing a Control System / 506
7.3. Example 1: Design for the Positioning System of a Tracking Radar using Linear and Nonlinear Techniques Jointly / 507
7.4. Example 2: Design of the Angular Control System for a Robot's Joint / 516
7.5. Example 3: Design of the Controller and Full-order Estimator for a Space Satellite's Attitude-Control System with Pole Placement Using Linear-State-Variable Feedback / 522
7.6. Example 4: Design of a Sampled-Data Control System for Controlling the Temperature of a Liquid in a Tank / 528
7.7. Example 5: Design of a Robust Control System for Controlling the Flaps of a Hydrofoil / 537
Problems / 542
References / 548

A Tutorial for the Effective Use of MATLAB 549

B Characteristic Response of Second-Order Control Systems 566

C Static Accuracy 574

Answers to Selected Problems 586

Index 601

PREFACE

The goal of *Advanced Modern Control System Theory and Design* is to present a unified treatment of advanced control system design techniques, and to illustrate how to apply the theory presented to realistic design problems. It assumes that the reader (student or practicing control system engineer) has completed an introductory control system course which has used any of the control system books which cover the subject at an introductory level. It can serve as an excellent book at universities for a follow-up advanced course (in the senior-year or at the graduate level) to an introductory course. It alleviates the need for a professor to request that their students purchase several control system books to cover various advanced subjects. This book contains a unified presentation of the advanced subjects that are of current interest for a follow-up course to an introductory course. In addition, it can serve as an excellent book for practicing control system engineers who need to learn more advanced control system subjects which are required to perform their tasks.

The book stands alone and does not assume that the reader has studied previously from my companion book *Modern Control System Theory and Design, Second Edition*. The book was written assuming that the reader has studied previously from any introductory control system book.

Throughout this book reference is made to various items in the companion volume, *Modern Control System Theory and Design, Second Edition*, to assist the reader in referencing introductory control system concepts. These references are indicated by a double dagger (‡) after the relevant item number.

The contents of this book correspond to advanced control system courses that I have taught at universities and in industry and reflects my industrial experience coupled with extensive teaching experience in both university and industrial programs.

The book has been written assuming that the reader will learn the subject in conjunction with a PC. Several very good control system software packages are available. MATLAB™ (MATLAB is a trademark of The MathWorks, Inc.) was selected for this book, and it is integrated into the text. MATLAB is now considered a control-system industry standard. All computational problems are solved with MATLAB.

In addition to the MATLAB programs integrated in the text, free MATLAB software and solutions to the computational problems in the book are available from The MathWorks. These can be downloaded from The MathWorks anonymous File Transfer Protocol (FTP) server at ftp://ftp.mathworks.com/pub/books/advshinners. The *Advanced Modern Control System Theory and Design* Toolbox software additionally contains the following: (a) a Tutorial File explaining the MATLAB fundamentals, notations, and the use of the software; (b) a Demonstration M-file which gives an overview of the various utilities; (c) a Synopsis File that reviews and highlights features of each chapter. Together, the integrated learning package and the MATLAB software are self-contained, so it is not necessary to purchase additional books/materials to learn how to use MATLAB, the *Advanced Modern Control System Theory and Design* Toolbox, or the various applications within this book. In this manner, the user of this book has available a toolbox of features/utilites created to enhance The MathWorks' Control System Toolbox, optionally utilize The MathWorks' Simulink, Nonlinear and Symbolic Toolboxes, as well as supply the computer-generated solutions of the problems in the book. The *Advanced Modern Control System Theory and Design* Toolbox runs equally well with The Student Edition of MATLAB as it does with the professional version of MATLAB (with or without The MathWorks' Control System Toolbox, Simulink, the Nonlinear Toolbox, or the Symbolic Toolbox). The *Advanced Modern Control System Theory and Design Toolbox* is designed to run with MATLAB for Windows (and all predecessor versions), and the Student Edition of MATLAB for Windows (and all predecessor versions). Accordingly, the resultant presentation of this book is an integration of computer-aided software engineering (CASE) and computer-aided design (CAD) techniques with the control-system analysis and design methods illustrated.

For those who do not have MATLAB, a set of working digital-computer programs containing their logic flow diagrams, listings, and representative output based on the programs' application to practical design problems is also contained.

Advanced Modern Control System Theory and Design contains the important advanced control system subjects that a professor requires in a follow-up course to an introductory control system course in one book. It is also an important and useful reference for the practicing control system engineer who only had an introductory control system course as an undergraduate, and who now has to be concerned with advanced modern control design methods (such as robust control) to achieve challenging modern design goals incorporating digital control, nonlinearities, and the other advanced techniques for a design project.

Advanced Modern Control System Theory and Design contains the following important subjects presented in dedicated chapters:

- *Linear-Control System Compensation and Design* presents linear control systems design methods using the Bode-diagram method, Nichols Chart, and the root-locus method.
- *Modern Control-System Design Using State-Space, Pole Placement, Ackermann's Formula, Estimation, Robust Control, and H^∞ Techniques* presents the modern approaches for designing linear control systems.
- *Digital Control-System Analysis and Design* extends the continuous concepts presented to discrete systems.
- *Nonlinear Control-System Design* extends the linear concepts presented in this volume to nonlinear systems.
- *Introduction to Optimal Control Theory and its Applications* presents such important topics as dynamic programming and the maximum principle, and applies it to the space attitude control problem and the lunar soft-landing problem.

- *Control-System Design Examples: Complete Case Studies* presents complete case studies of five control system designs which illustrate practical design projects.

Major Enhancement for Learning

Advanced Modern Control System Theory and Design provides dedicated sections at the end of each chapter (but before problems at ends of chapters) containing illustrative problems and their solutions which cover material in their respective chapter. This is a very important major enhancement for learning by students, and by practicing control system engineers using this book for self study.

Primary Features of Advanced Modern Control System Theory and Design

1. *Unifies and blends the conventional and modern approaches.* It uses the conventional block-diagram-transfer function and state-variable methods in parallel throughout the book. Modern control-system design techniques using the state-space approach, pole placement, estimation, robust control, and H^∞ are presented in addition to the classical Nyquist, Bode, Nichols, and root-locus methods. A chapter on digital controls extends the continuous concepts presented to discrete systems. Nonlinear control systems, and modern optimal control theory (including such important topics as dynamic programming and the maximum principle) are presented in this book. This interesting presentation also shows how we can extend our capabilities for analyzing and designing linear control systems to nonlinear control systems. It also introduces students to the optimal control concept (and its important applications) which adds another dimension to their capabilities. In addition, the modern state-space, pole placement, Ackermann's formula for pole placement, estimation, robust control, the H^∞ method, nonlinear and optimal control chapters present ideas for design projects at the undergraduate level and ideas for specialization and future research at the graduate level. The concluding chapter presents the complete case studies of five design examples which illustrate the undergraduate type design projects.

2. *Emphasizes design.* The development of theoretical topics is coupled with clear applications of the theory in engineering design. Recognizing that control theory is interdisciplinary and cuts across all specialized engineering fields, I have presented modern illustrations and practical problems from the fields of robotics, space-vehicle systems, aircraft, submarines, hydrofoils, servomechanisms, economics, management, biomedical engineering, and nuclear reactor systems. This should prove to be of interest to electrical, mechanical, aerospace, system, chemical, nuclear, biochemical, and industrial engineers.

3. *Presents a complete set of working digital computer programs and summarizes commercially available software packages available for control system analysis in addition to the many solutions presented in the text using MATLAB.* Although computer techniques are extremely valuable to the student and the practicing engineer, I believe they must first understand the details of the techniques being applied so that computer results can be properly interpreted and judged for their analytical reasonableness. Therefore, I present basic theory while applying it to design problems with and without computer solutions.

4. *Covers a variety of topics of recent importance.* In addition to covering modern state-space design techniques including pole placement, Ackermann's formula for pole placement, estimation, robust control, and the H^∞ method for design, the presentation also includes dynamic programming and the maximum principle, Liapunov's stability criteria, Popov's method, the generalized circle criterion, controllability, observability, and linear-state-variable feedback (including the design of controllers, estimators, and compensators).

5. *Adapts to the student's training, ability, and to various curricula.*

6. *Contains problems (with answers to approximately one-third).* This is helpful to the student and also to the practicing control-system engineer using the book for self-study. The available MATLAB computer-generated solutions will guide the reader in understanding the approach used for solving the problems.

7. *Offers an accompanying Solutions Manual (available to professors) containing detailed solutions to the remaining problems whose answers are not found in this book.*

By means of these features, *Advanced Modern Control Systems Theory and Design* is a computer-oriented, state-of-the-art book that is comprehensive and unique and fills a rather large gap in the existing literature on advanced control-system design.

Chapter Organization

After presenting an introduction to the subject in Chapter 1, the design of linear control systems is presented in Chapter 2. Single-degree and two-degrees of-freedom compensation techniques using cascade-compensation and minor-loop feedback compensation techniques are developed and applied. Phase-lag, phase-lead, and phase-lag–lead networks are illustrated, in addition to proportional-plus-integral-plus-derivative (PID) compensators.

Chapter 3 is dedicated to presenting modern control-system design using state-space methods, including pole-placement design using linear-state-variable-feedback, Ackermann's formula for design using pole placement, and estimator design in conjunction with the pole placement approach using linear-state-variable feedback. The concepts of controllability and observability are introduced. Robust control is developed and applied to a variety of problems. The modern H^∞ control concepts are introduced, developed, and applied. The chapter concludes with a presentation of linear algebraic aspects of control-system design computations, and several illustrative problems and solutions in a dedicated section.

An extension of the continuous methods of analysis and design to digital control-system analysis and design is presented in Chapter 4. The characteristics of sampling, data extrapolators, z-transform theory, and stability analyses using mathematical tests, the Nyquist diagam, Bode diagram (in the w-plane), root locus, and Ragazzini's method are presented. MATLAB programs for obtaining the Bode and root-locus diagrams of discrete time systems are developed and applied. The digitization process and the design of digital filters are presented.

Chapter 5 discusses the theory of designing nonlinear systems along with practical examples. Included are linearizing approximations, the describing function, piece-wise-linear approximations, the state-variable analysis technique (the phase plane), Liapunov's stability criteria, Popov's stability criterion, the generalized circle criterion, and the use of simulation. MATLAB programs for the analysis and design of nonlinear control systems using the describing function method are developed and applied. In addition, non-MATLAB working digital computer programs are also presented and used for applying the describing-function and state variable analyses using the phase-plane method. Guidelines for selecting the appropriate nonlinear control-system method(s) to use for a particular problem are presented. Chapter 6 introduces optimal control theory and presents its interesting applications. The following material is covered at an introductory level: calculus of variations, Bellman's dynamic programming, and Pontryagin's maximum principle. The maximum principle is applied to the space attitude-control and the lunar soft-landing problems.

The book concludes with Chapter 7 on complete case studies for designing five different control systems. These include the application of continuous linear control-system design, nonlinear control-system design, pole placement using linear-state-variable-feedback design, digital control-system design techniques, and the design of a robust control system are applied to practical problems. The book also contains Appendix A which is a tutorial for the effective use of MATLAB. This appendix is designed for self-study by the reader who is unfamiliar with MATLAB or needs a refresher, and is coordinated with this text. The book concludes with Appendix B on the characteristic response of second-order systems, and Appendix C on static accuracy.

The Learning Package

The following educational materials comprise the learning package for *Advanced Modern Control System Theory and Design*, and supplement and enhance this textbook:

1. *Advanced Modern Control System Theory and Design Toolbox*, which is available free to the reader from The MathWorks Inc. anonymous FTP server at ftp://ftp.mathworks.com/pub/books/advshinners. This software contains MATLAB solutions to the computational problems in this book. In addition, the *Advanced Modern Control System Theory and Design* Toolbox software contains the following features:

   - Tutorial File explaining the MATLAB fundamentals, notations, and the use of the software.
   - Demonstration M-File which gives an overview of the various utilities.
   - Synopsis File that reviews and highlights features of each chapter.

   The *Advanced Modern Control System Theory and Design* Toolbox runs equally well with the professional and Student Edition of MATLAB. By virtue of the *Advanced Modern Control System Theory and Design* Toolbox, and the fact that MATLAB is also integrated into this textbook throughout, it is not necessary for the reader to buy a supplementary book to learn how to use MATLAB.

2. *Solutions Manual for Advanced Modern Control System Theory and Design* by Stanley M. Shinners contains the detailed solutions to approximately 66% of

the problems at the ends of each chapter. The *Solutions Manual* is available only to qualifying faculty members.

ACKNOWLEDGMENTS

I am most grateful to my wife Doris for her encouragement, understanding, patience, and word-processing assistance throughout this project. In addition, I express thanks and appreciation to my parents for their efforts, encouragement, and inspiration. I thank the editor of my book at John Wiley & Sons, Mr. George Telecki, who managed the smooth production of the book, and to Lisa Van Horn of John Wiley & Sons for her professional production of this book.

STANLEY M. SHINNERS

Jericho, New York
May 1998

1

INTRODUCTION

1.1. INTRODUCTION

The desire of people to control nature's forces successfully has been the catalyst for progress throughout history. Our goal has been to control these forces in order to help perform physical tasks which were beyond our own capabilities. During the dynamic and highly motivated 20th century, the control-system engineer has transformed many of our hopes and dreams into reality.

Control-system engineers have made very important contributions to our advancements in the 20th century, and they are building the foundation for greater advancements as we approach the 21st century. As we look back, control-system engineers have made contributions to robotics; space-vehicle systems, including the successful accomplishment of the lunar soft landing; aircraft autopilots and controls; control systems for ships and submarines; guidance systems for intercontinental missiles; and automatic control systems for hydrofoils, surface-effect ships, high-speed rail systems; and, most recently, control systems for the magnetic-levitation rail systems. The future lies in our imagination, creativity, and ability to transform ideas into working automatic control systems that we can build to work reliably and accurately, and which can be manufactured at a profit and on schedule. It is hoped that this book contributes to the attainments of our future goals and dreams in the advancement of automatic control systems.

The control of systems is an interdisciplinary subject and cuts across all specialized engineering fields. This book recognizes this fact and presents illustrations of control systems from the fields of electrical, mechanical, aeronautical, chemical, nuclear, economics, management, bioengineering, and other related fields. The versatile subject of automatic control ranks today as one of the most promising fields, and its growth potential appears unlimited.

1.2. GOAL OF ADVANCED MODERN CONTROL SYSTEM THEORY AND DESIGN

The goal of *Advanced Modern Control System Theory and Design* is to present a unified treatment of advanced control system design techniques, and to illustrate how to apply the theory presented to realistic design problems. It assumes that the reader (student or practicing control system engineer) has completed an introductory control system course which has used any of the control system books which cover the subject at an introductory level. It can serve as an excellent book at universities for a follow-up advanced course (in the senior-year or at the graduate level) to an introductory course. It alleviates the need for a professor to request that their students purchase several control system books to cover various advanced subjects, because this book contains a unified presentation of the advanced subjects that are of current interest. In addition, it can serve as an excellent book for practicing control system engineers who need to learn more advanced control system subjects, such as digital control, robust control, nonlinearities, and other advanced techniques which are required to perform their tasks.

I have written an accompanying companion introductory book to *Advanced Modern Control System Theory and Design*, entitled *Modern Control System Theory and Design, Second Edition*. [1] However, this book does not assume that the reader has studied previously from my companion book. This book was written assuming that the reader has studied previously from any introductory control system book, [1–5] or has attained a basic understanding of control systems from experience.

Throughout this book reference is made to various items in the companion volume, *Modern Control System Theory and Design, Second Edition*, to assist the reader in referencing introductory control system concepts. These references are indicated by a double dagger (‡) after the relevant item number.

This book has been written assuming that the reader will learn the subject in conjunction with a computer. Several good control system software packages are available. MATLAB^{TM} (MATLAB is a trademark of The MathWorks, Inc.) was selected for this book, and it is integrated into this text. [6] MATLAB is now considered a control-system industry standard. Free software is available to the reader of this book with the *Advanced Modern Control System Theory and Design Toolbox* which complements this book and contains MATLAB solutions to computational problems. It can be retrieved free from The MathWorks anonymous File Transfer Protocol (FTP) server at ftp://ftp.mathworks.com/pub/books/advshinners. All computational problems in this book are solved with MATLAB. The *Advanced Modern Control System Theory and Design Toolbox* runs equally well with The Student Edition of MATLAB as it does with the professional version of MATLAB. Appendix A is included which provides a tutorial for the effective use of MATLAB for readers who are unfamiliar with or need a refresher of MATLAB, and it is coordinated with the text.

1.3. CONTROL SYSTEM PERFORMANCE OBJECTIVES

The performance of a control system is generally specified in terms of stability, sensitivity, accuracy, transient response, and residual noise jitter. The exact specifications are usually dictated by the required system performance. [1–5] Certain characteristics are more important in some systems than in others.

Stability

A fundamental requirement of a control system is that it be stable. The system must maintain its stability when the control system is subjected to commands at its input, extraneous inputs anywhere within the feedback loop, power-supply variations, and changes in the parameters of the elements comprising the control system due to aging and environmental conditions.

If a control system has zero initial conditions, then for every bounded input, the output is bounded and the system is stable. This is popularly referred to by control system engineers as the bounded input-bounded output (BIBO) stability. This book illustrates how to design and stabilize linear continuous time-invariant control systems, that is, systems for which the principle of superposition is valid and which may be described by ordinary linear differential equations with constant coefficients.

The total response of a control system, c(t), is composed of the natural or homogeneous response due to initial conditions, and the forced response due to the external input:

$$c(t)_{total} = c(t)_{homogenous} + c(t)_{forced} \tag{1.1}$$

A linear, time-invariant control system is stable if the natural or homogeneous response approaches zero, and it is unstable if the natural or homogeneous response grows without bound as time approaches infinity. If the linear, time-invariant system neither grows nor decays, and an oscillation of constant magnitude results, then this represents simple harmonic motion. Such a system is considered to be an unstable control system practically.

The design of linear, time-invariant, control system stability is presented in Chapters 1, 2 and 3. The design of digital control system stability is presented in Chapter 4. The design of linear time-variable and that of nonlinear control system stability are presented in Chapter 5. Case studies of five control system designs are presented in Chapter 7 which applies stability designs to linear, digital, nonlinear, and robust control systems.

Sensitivity

Sensitivity is a measure of the dependence of a control system's characteristic on those of a particular element. The differential sensitivity, S, of a control system's closed-loop transfer function $H(s)$ with respect to the characteristics of a given element $K(s)$ is defined as

$$S_K^H(s) = \frac{d \ln H(s)}{d \ln K(s)} = \frac{\% \text{change in } H(s)}{\% \text{change in } K(s)}, \tag{1.2}$$

4 INTRODUCTION

where $H(s)$ is defined in terms of the control system input, $R(s)$, and output, $C(s)$, as

$$H(s) = C(s)/R(s).$$

A more meaningful definition can be obtained by rewriting Eq. (1.2) as

$$S_K^H(s) = \frac{dH(s)/H(s)}{dK(s)/K(s)}. \quad (1.3)$$

Equation (1.2) states that the differential sensitivity of $H(s)$ with respect to $K(s)$ is the percentage change in $H(s)$ divided by that percentage change in $K(s)$ that has caused the change in $H(s)$ to occur. This definition is valid only for small changes. It is important to note that sensitivity is a function of frequency and an ideal system has zero sensitivity with respect to any parameter.

Sensitivity of control systems containing disturbances and measurement noise is presented in Chapter 1 and developed further in Chapter 3 as part of the presentation on H^∞ control concepts. This approach is a new control system design technique developed in the 1980s which combines both the frequency- and time-domain approaches to provide a unified answer for the design of control systems. Sensitivity is applied in this presentation to both single-input single-output, and multiple-input multiple-output control systems.

Accuracy

Accuracy is another very important characteristic of a control system. The control system designer always strives to design the control system to minimize error due to anticipated reference inputs and disturbances.

The specification which a control system engineer strives to achieve always contains control system accuracy. Accuracy is usually specified in terms of the steady-state response to reference inputs containing some kind of combination of position, velocity, acceleration, and changes of higher-order derivatives. (See Appendix C.) The control system characteristics define the steady-state error constants such as the position constant, velocity constant, and acceleration constant. [1–5] From these error constants, and knowledge of the reference inputs, the static accuracy of control systems can be determined. This is illustrated in this book with the design of linear control systems in Chapters 2 and 3, and with the illustrative case studies in Chapter 7.

Transient Response

In addition to stability, sensitivity, and accuracy, the control system engineer is also very concerned with the transient response of a control system. Transient response characteristics such as maximum percent overshoot, time-to-peak, settling time, and rise time are defined on the basis of the response of a second-order control system to a unit step-input. (See Appendix B.) For the case of control systems which are greater than second-order, a good approximation can be made if one pair of complex-conjugate roots dominates. This is illustrated in Chapter 2 on designing linear control systems, and it is further illustrated in Chapter 7 on the design of control systems with complete case studies.

Other Practical Considerations

The control system engineer must also be concerned with several other practical aspects before becoming able to state intelligently and completely the expected system performance. These include considerations of feedback system bandwidth, nonlinearities, size, weight, power consumption, and economics.

Feedback system bandwidth is usually defined as the frequency at which the open-loop magnitude equals unity. Sometimes it is defined where it is $-3\,\text{dB}$. The bandwidth of a system is indicated by its particular application. Usually, the control engineer is interested in designing the system to respond to a certain spectrum of input signal frequencies and to suppress all inputs above a certain frequency. It is important to emphasize that we should not arbitrarily design for a large bandwidth. Although large bandwidths usually result in large error constants, with small resulting system error, they also result in a system that responds to extraneous noise inputs and has considerable jitter due to the noise. The desirable approach is to design the feedback system bandwidth to be just large enough to pass the desired input-signal frequency spectrum and then attenuate all higher-frequency signals [1]. Feedback system bandwidth considerations are discussed in detail in Chapters 1, 2, 3 and 7, where methods for obtaining the frequency response are presented.

Nonlinearities are other factors that affect the performance of a control system. Primary concern is with backlash, stiction (starting friction) and Coulomb friction. Backlash is the amount of free motion of one gear while its mating gear is held fast. Stiction is the frictional force that prevents motion until the driving force exceeds some minimum value. Coulomb friction is a constant frictional drag that opposes motion, but has a magnitude that is independent of velocity. Each of these nonlinearities has an effect on performance. More will be said regarding nonlinearities in Chapter 5.

Other factors of concern are size, weight, power consumption, and economics. The system must conform to certain specifications of size and weight. These are very important factors that usually dictate the design of the system. For example, these specifications may decide the type of power drive to be used. Power is another very important consideration. The system must usually perform within a certain allowable power limitation. Size, weight, and power consumption are usually very critical items for airborne and space applications. Last, but not least, is the question of economics. A basic fact of life is that most engineers work for organizations whose primary purpose is to make profit. Systems must therefore be designed as inexpensively as possible within the framework of good performance. A generally useful rule of thumb is that minimum-bandwidth systems will consume the least power and be the most economical.

1.4. THE PROCEDURE FOR DESIGNING A CONTROL SYSTEM

The design procedure for designing a control system is an orderly sequence of steps. Good engineering design is interdisciplinary and requires that the engineer first thoroughly understand the customer's requirements, the defined control system specifications, the environment that the control system will operate in, the available power, the schedule that it must be built in, and the available budget to do the job.

6 INTRODUCTION

Other considerations are reliability and maintainability which may dictate the kind of motor to use (i.e., electric motor or hydraulic motor).

Due to the availability of a large number of techniques to solve the great variety of control-system problems present, the element of experience is very important to the approach used for the solution of a specific problem. An outline of a logical step-by-step procedure for designing a control system from its conception through the final hardware stage is illustrated in Figure 1.1 and described as follows:

1. Obtain a complete understanding of the job requirements with respect to
    (a) a general description of the problem;

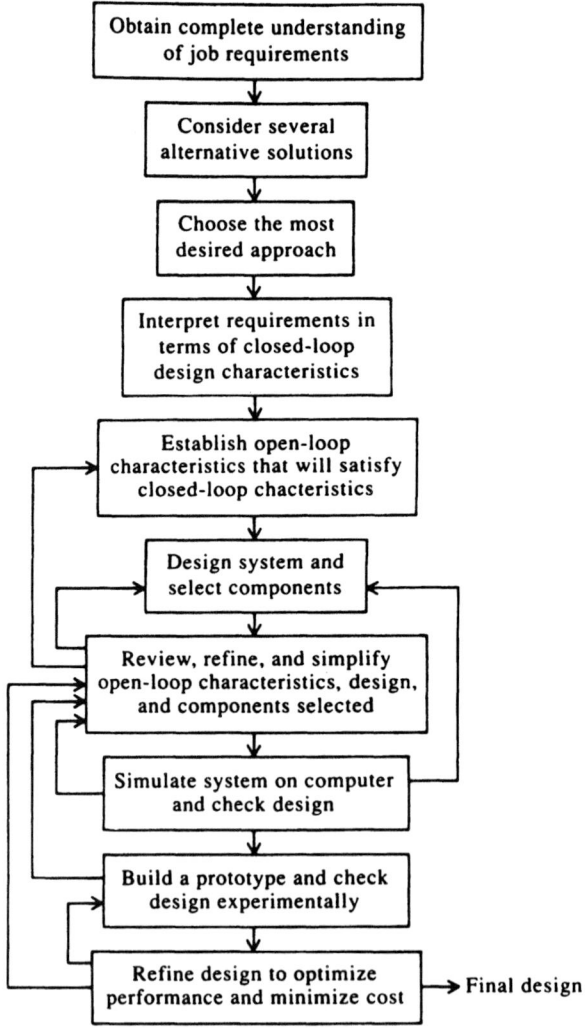

Figure 1.1 Procedure for designing a control system.

(b) the overall control-system performance and accuracy with respect to the steady-state and transient phases;

(c) identification of the transfer function of the controlled process;

(d) miscellaneous requirements as to reliability, schedule, cost, maintainability, size, weight, and available power.

2. Consider several alternative solutions, including electric and hydraulic power servo drives, the use of continuous control or digital control, etc.
3. Choose the most desired approach based on the specifications, requirements, and elements fixed by the customer.
4. Interpret these requirements in terms of such closed-loop design characteristics as frequency and transient response.
5. Establish the approximate open-loop characteristics that will satisfy the closed-loop requirements.
6. Design the system and select the sensors, actuators, amplifier, and stabilization required (analog or digital) in order to satisfy step 5.
7. Review, refine, and simplify steps 5 and 6.
8. Simulate the system on a computer, including its linear and nonlinear characteristics, to check the design. Make any necessary changes to the design.
9. Build a prototype, and check the design experimentally. Make any necessary changes to the design.
10. Refine the design in order to optimize performance and minimize cost.

Observe from this approach that the procedure is an iterative one, and is itself a feedback process, as illustrated in Figure 1.1.

1.5. OUTLINE OF ADVANCED MODERN CONTROL SYSTEM THEORY AND DESIGN

Advanced Modern Control System Theory and Design contains the following important subjects presented in dedicated chapters:

Chapter 1: *Introduction* provides the foundation for the following chapters. Besides introducing this book's goals, providing a summary of a control system's performance objectives, presenting the procedure for designing a control system, providing a synopsis of this book's chapters in this section, and discussing the features of the accompanying software *Advanced Modern Control System Theory and Design Toolbox* in the following section, a key feature of this chapter is the last section entitled *Illustrative Problems and Solutions*. This book provides dedicated sections at the end of each chapter containing illustrative problems and their solutions which cover material in their respective chapter. This is a very important major enhancement for learning by students, and by practicing control system engineers using this book for self study. This introductory chapter uses this section on *Illustrative Problems and Solutions* to review introductory control system analysis methods which will serve as the basis for some of the design problems illustrated in succeeding chapters. [1] MATLAB programs for obtaining the Nyquist diagram, Bode diagram, Nichols Chart, and the root-locus diagram are developed and applied.

Chapter 2: *Linear-Control System Compensation and Design* presents linear control system design methods using the Bode-diagram method, Nichols Chart, and the root-locus method. Designs are developed which meet stability and accuracy specifications. Single-degree and two-degrees-of-freedom compensation techniques using cascade-compensation and minor-loop feedback compensation techniques are developed and applied. Phase-lag, phase-lead, and phase-lag-lead networks are illustrated, in addition to proportional-plus-integral-plus-derivative (PID) compensators.

Chapter 3: *Modern Control-System Design Using State-Space, Pole Placement, Ackermann's Formula, Estimation, Robust Control*, and H^∞ *Techniques* presents these modern approaches for designing linear control systems. The concepts of controllability and observability are also introduced. These modern control system techniques are developed and applied to the design of a variety of control system problems. The chapter concludes with a presentation of linear algebraic aspects of control-system design computations, and several illustrative problems and solutions.

Chapter 4: *Digital Control-System Analysis and Design* extends the continuous concepts presented in Chapters 1–3 to discrete systems. The characteristics of sampling, data extrapolators, z-transform theory, and stability analyses using mathematical tests, the Nyquist diagram, Bode diagram (in the w-plane), root locus, and Ragazzini's method are presented. MATLAB programs for obtaining the Bode and root-locus diagrams of discrete time systems are developed and applied. The digitization process and the design of digital filters are presented.

Chapter 5: *Nonlinear Control-System Design* extends the linear concepts presented in this book to nonlinear systems. Included are linearizing approximations, the describing function, piecewise-linear approximations, the state-variable analysis technique (the phase plane), Liapunov's stability criteria, Popov's stability criterion, the generalized circle criterion, and the use of simulation. MATLAB programs for the analysis and design of nonlinear control systems using the describing function are developed and applied. In addition, non-MATLAB working digital computer programs are also presented and used for applying the describing-function and state variable analyses using the phase-plane method. Guidelines for selecting the appropriate nonlinear control-system method(s) to use for a particular problem are presented.

Chapter 6: *Introduction to Optimal Control Theory and its Applications* presents such important topics as dynamic programming and the maximum principle, and applies it to the space attitude control problem and the lunar soft-landing problem. The following material is covered at an introductory level: calculus of variations, Bellman's dynamic programming, and Pontryagin's maximum principle. The maximum principle is applied to the space attitude-control and the lunar soft-landing problems.

Chapter 7: *Control-System Design Examples: Complete Case Studies* presents complete case studies of five control system designs which illustrate practical design projects. These include the application of linear control-system design, nonlinear control-system design, pole placement using linear-state-variable-feedback design, digital control-system design techniques, and the design of a robust control system are applied to practical problems.

Apppendix A: *Tutorial for the effective use of MATLAB* presents the information necessary for the effective use of MATLAB which is used throughout this book.

Appendix B: Reviews the *Characteristic response of second-order control systems.*
Appendix C: Reviews *Static accuracy.*

1.6. ADVANCED MODERN CONTROL SYSTEM THEORY AND DESIGN TOOLBOX

Control system engineers are very fortunate today with the availability of software that greatly enhances the speed and capability for performing control system analysis, simulation, and design. Using computers, the control system engineer can easily, on-line, make changes and quickly determine whether a new design is acceptable. Alternative solutions can easily be attempted to determine which provides the best design. The effects of nonlinearities and noise on control system performance can quickly and easily be determined.

Advanced Modern Control System Theory and Design has been written assuming that the reader will learn the subject in conjunction with a computer. MATLAB was selected for the software for this book, and is integrated into this book. [6] For those who don't know MATLAB, Appendix A provides a tutorial for self-study of MATLAB. Working digital-computer programs containing their logic flow diagrams, listings, and output based on the programs' application to problems are also contained for those who don't have MATLAB.

An important feature of this book is that most of the solutions shown in this book were generated using the commercially available software packaged called MATLAB which is available from the MathWorks, Inc. The *Advanced Modern Control System Theory and Design (AMCSTD) Toolbox*, and the M-files that were used to develop these solutions can be retrieved free from The MathWorks, Inc. anonymous FTP server at ftp://ftp.mathworks.com/pub/books/advshinners. These M-files are the *AMCSTD Toolbox* and were used to develop the graphical figures and problem solutions in this book. In this manner, the user of this book has available a toolbox of features/utilities created to enhance The Mathworks' Control System Toolbox, and the computer-generated solutions to the problems in this book. The *Advanced Modern Control System Theory and Design Toolbox* runs equally well with *The Student Edition of MATLAB* as well as with or without the professional versions of MATLAB (with or without The MathWorks' Control System, Simulink, Nonlinear, and the Symbolic Toolboxes).

1.7. ILLUSTRATIVE PROBLEMS AND SOLUTIONS [1]

This section provides a set of illustrative problems and their solutions to review introductory control system analysis methods. These problems will serve as the basis for some of the design problems illustrated in succeeding chapters.

I1.1. The control system of Figure I1.1 contains position and rate feedback:

(a) Determine the closed-loop transfer function $C(s)/R(s)$.

10 INTRODUCTION

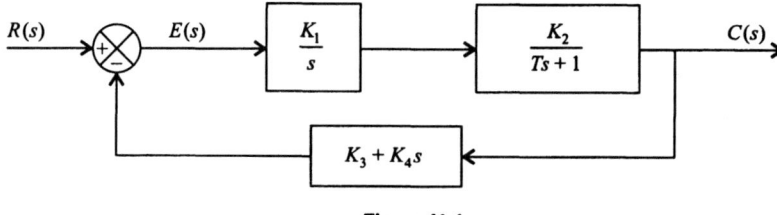

Figure I1.1

(b) Determine the sensitivity of the closed-loop transfer function to K_2 as a function of s.

(c) Determine the sensitivity of the closed-loop transfer function to K_2 at dc.

SOLUTION

(a)

$$H(s) = \frac{C(s)}{R(s)} = \frac{\left(\frac{K_1}{s}\right)\left(\frac{K_2}{Ts+1}\right)}{1 + \left(\frac{K_1}{s}\right)\left(\frac{K_2}{Ts+2}\right)(K_3 + K_4 s)} = \frac{K_1 K_2}{Ts^2 + (K_1 K_2 K_4 + 1)s + K_1 K_2 K_3}.$$

(b)
$$S_{K_2}^H(s) = \frac{\frac{dH}{H}}{\frac{dK_2}{K_2}} = \left(\frac{K_2}{H}\right)\left(\frac{dH}{dK_2}\right)$$

where

$$\frac{dH}{dK_2} = \frac{(K_1 s)(Ts + 1)}{Ts^2 + (K_1 K_2 K_4 + 1)s + K_1 K_2 K_3}.$$

Therefore,

$$S_{K_2}^H(s) = \frac{s(Ts + 1)}{Ts^2 + (K_1 K_2 K_4 + 1)s + K_1 K_2 K_3}.$$

(c)
$$S_{K_2}^H(s)\bigg]_{s=0} = 0.$$

I1.2. The forward-loop transfer function of a unity-feedback control system is given by the following:

$$G(s) = \frac{2}{(s+1)^4}.$$

(a) Sketch the Nyquist diagram, and determine whether the control system is stable.

(b) Determine the intersection of $G(j\omega)H(j\omega)$ and the negative real axis of the $G(j\omega)H(j\omega)$ plane *analytically*, and find the values of the *frequency*, ω, and *location* of the intersection on the negative real axis.

(c) How much can the system gain, 2, be increased before the roots of the characteristics equation cross the $j\omega$ axis and go into the right-half plane?

SOLUTION: (a) System is stable.

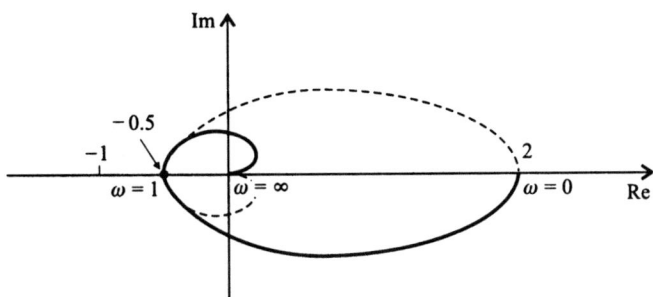

Figure I1.2 Solution.

(b) $$G(j\omega) = \frac{2}{(j\omega + 1)^4} = \frac{2}{\omega^4 - 4\omega^3 j - 6\omega^2 + 4\omega j + 1}.$$

Separating the real and imaginary terms in the denominator, we obtain the following:

$$G(j\omega) = \frac{2}{(\omega^4 - 6\omega^2 + 1) + (4\omega - 4\omega^3)j}.$$

Therefore,

$$G(j\omega) = \frac{2[(\omega^4 - 6\omega^2 + 1) - (4\omega - 4\omega^3)j]}{(\omega^4 - 6\omega^2 + 1)^2 + (4\omega - 4\omega^3)^2}.$$

Setting the imaginary part of the numerator equal to zero, we obtain:

$$-2(4\omega - 4\omega^3) = 0.$$

Therefore,

$$\omega = 0, \quad 1, \quad -1$$

and we are interested in the condition of $\omega = 1$. The case of $\omega = 0$ refers to the intersection of the Nyquist diagram with the positive real axis.

Substituting $\omega = 1$ into $G(j\omega)$, we obtain that the intersection of the Nyquist diagram and the real axis occurs at -0.5.

(c) Since the intersection of the Nyquist diagram and the negative real axis occurs at −0.5, we can increase the gain, 2, to 4 at which point the Nyquist diagram crosses the negative real axis at −1 and the roots of the characteristic equation cross the $j\omega$ axis.

I1.3. The Nyquist diagram of a control system is to be obtained using MATLAB*. If the system is defined by a transfer function, the Control System Toolbox command which determines the Nyquist plot is:

$$\text{nyquist.}$$

The Nyquist diagram can be used with polynomial transfer functions, defined as $G(s) = \text{num}(s)/\text{den}(s)$ where "num" and "den" contain the polynomial coefficients, if invoked with three right-hand arguments as follows

$$[\text{re,im}] = \text{nyquist(num,den,w)}$$

where

re = real part on the Nyquist diagram
im = imaginary part on the Nyquist Diagram
num = row matrix format representing the numerator of the polynomial
den = row matrix format representing the denominator of the polynomial
w = frequency in rad/sec.

The Nyquist diagram will occasionally go to infinity. Without special precautions in the MATLAB program, an erroneous Nyquist plot may occur. We can avoid this with MATLAB by specifying the finite area we want plotted. This will be illustrated next.

Determine the Nyquist diagram of a unity feedback control system whose forward transfer function is given by

$$G(s) = \frac{1}{s(1+s)^2}.$$

using MATLAB.

SOLUTION:

Since the Nyquist diagram goes to infinity, we can avoid this with **MATLAB** by using a manually determined range. For example, we can limit it from −1.8 to 0 on the real axis, and from −2.5 to 2.5 on the imaginary axis by entering the commands

$$v = [-1.8 \quad 0 \quad -2.5 \quad 2.5]$$
$$\text{axis(v)}$$

---
*It is suggested that those readers who are unfamiliar with MATLAB or need a refresher read Appendix A before writing MATLAB programs.

1.7. ILLUSTRATIVE PROBLEMS AND SOLUTIONS

Alternatively, we can use the single command

$$\text{axis}([-1.8 \quad 0 \quad -2.5 \quad 2.5])$$

Another important point to emphasize in this problem is that the numerator is clearly defined as a constant. However, the denominator is given as the product of terms, with the result that we must multiply these terms to get a polynomial in s. We can use the *convolution* command in MATLAB to perform this multiplication as follows. Defining

$$a = s(s+1) = s^2 + s : a = [1 \quad 1 \quad 0]$$
$$b = s + 1 \qquad\qquad : b = [0 \quad 1 \quad 1]$$

and using the MATLAB command

$$c = \text{conv}(a, b)$$

will result in the product of the terms in the denominator. The resulting MATLAB program for multiplying the factors in the denominator is illustrated in Table I1.3-1.

We are now ready to develop the MATLAB program for plotting this Nyquist diagram which is shown in Table I1.3-2.

The resultant Nyquist plot obtained is illustrated in Figure I1.3(a). Observe that the Nyquist diagram illustrated in Figure I1.3(a) is only a partial Nyquist diagram because we limited the real and imaginary axes range with the MATLAB command "axis." If we could let the real and imaginary axes approach infinity with MATLAB, this Nyquist diagram would be the Nyquist diagram shown in Figure I1.3(b). However, enlarging the real and

Table I1.3-1. MATLAB Program for Multiplying Factors

```
a = [1  1  0]
b = [0  1  1]
c = conv(a,b)
c =
     1  2  1  0
```

Table I1.3-2. MATLAB Program for Obtaining the Nyquist Diagram

```
num = [0  0  0  1];
den = [1  2  1  0];
nyquist(num,den)
Warning: Divide by zero
axis([-1.8  0  -2.5  2.5])
grid
title('Nyquist Plot of G(s) * H(s) = 1/[(s * (s + 1)^2]')
```

14 INTRODUCTION

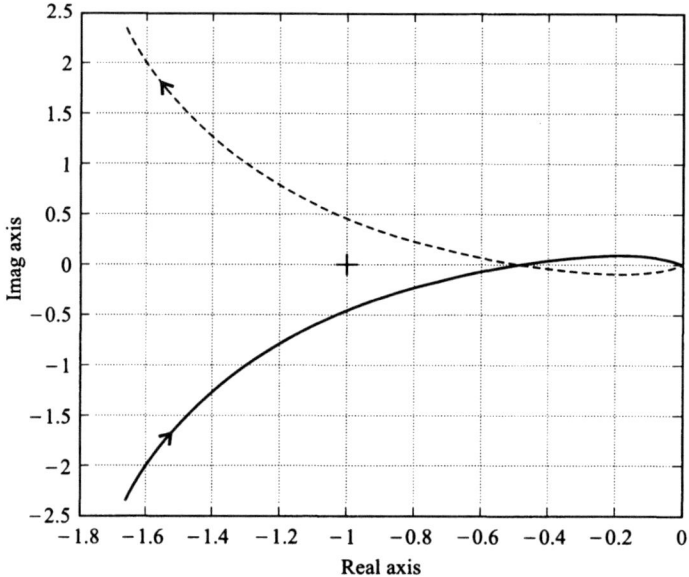

Figure I1.3(a) Nyquist diagram for

$$G(s)H(s) = \frac{1}{s(s+1)^2}.$$

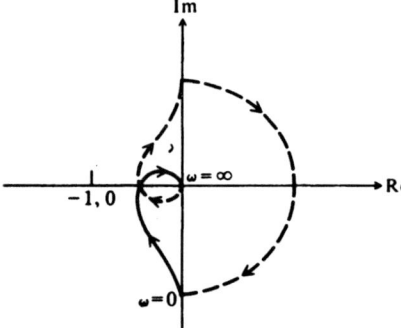

Figure I1.3(b) Complete Nyquist diagram for

$$G(s)H(s) = \frac{1}{s(s+1)^2}.$$

imaginary axes with MATLAB results in losing observability of the portion of the Nyquist diagram closest to the origin which intersects the real axis at -0.5. This is a very important value to know, because it means that the gain could be doubled from 1 to 2 before the Nyquist diagram intersects the $-1, 0$ point.

I1.4. Repeat the Nyquist diagram of Problem I1.3 if the control system is defined in state-space form

$$\dot{x}(t) = Ax(t) + Bu(t) \qquad (I1.4\text{-}1)$$

$$y(t) = Cx(t) + Du(t) \qquad (I1.4\text{-}2)$$

rather than in the transfer function form.

(a) Convert from the transfer function given in Problem I1.3 to the state-space form using MATLAB.

(b) Determine the Nyquist diagram from the state-space form using MATLAB.

SOLUTION:

If the control system is defined in state-space form, the *Control System Toolbox* command

$$\text{nyquist}(A,B,C,D)$$

will produce the Nyquist diagram, where the matrices **A**, **B**, **C** and **D** are defined and discussed in Appendix A. We will show how to obtain the Nyquist diagram for the same control system illustrated earlier in this section, but we will now obtain it from its state-space form.

To convert from the transfer function to the state-space form, the resulting MATLAB program for accomplishing this is shown in the MATLAB Program in Table I1.4-1.

Table I1.4-1. MATLAB Program for Converting from the Transfer Function to the State-Space Form

```
num = [0  0  0  1];
den = [1  2  1  0];
[A, B, C, D] = tf2ss(num,den)
A =
   -2  -1  0
    1   0  0
    0   1  0
B =
    1
    0
    0
C =
    0  0  1
D =
    0
```

16 INTRODUCTION

Therefore, the MATLAB program for determining the Nyquist diagram from the state-space form is given by the MATLAB program in Table I1.4-2.

Table I1.4-2. MATLAB Program for Determining the Nyquist Diagram from the State-Space Form

A = [−2 −1 0; 1 0 0; 0 1 0];
B = [1; 0; 0]
C = [0 0 1];
D = [0];
nyquist(A,B,C,D)
axis([−1.8 0 −2.5 2.5])
grid
title('Nyquist plot of G(s) ∗ H(s) = 1/[s ∗ (s + 1)^2]')

The result is the same Nyquist diagrams shown in Figures I1.3(a) and I1.3(b).

I1.5. The Bode diagram of a control system is to be obtained using MATLAB. There are many examples to practice with for creating Bode diagrams on the *Advanced Modern Control System Theory and Design* (AMCSTD) Toolbox. Two functions exist that assist in Bode diagrams:

1. "bode" returns/plots the Bode response of a system.

2. "margins" is described in the AMCSTD toolbox. This, in my opnion, has an advantage over the professional version of the Control System toolbox's "margin" routine. Margins analytically calculates, with analytic precision, the gain and phase margins and their associated frequencies (versus interpolating a single value from a plot).

When trying to find the proper syntax to call the "bode" utility, either use the help feature or look in the reference manual. I personally prefer the help feature, unless I need an example.

Valid syntax for the "bode" utility, for transfer functions, is:

1. [mag,phase,w] = bode(num,den)

2. [mag,phase,w] = bode(num,den,w)

3. [mag,phase] = bode(num,den,w)

4. bode(num,den,w)

5. bode(num,den)

Use MATLAB to determine the Bode diagram, if the open-loop transfer function of a control system is given by the following:

$$G(s)H(s) = \frac{(s+4)(s+40)}{s^3(s+200)(s+900)}. \tag{I1.5-1}$$

1.7. ILLUSTRATIVE PROBLEMS AND SOLUTIONS

SOLUTION:

In order to ease its transformation to MATLAB notation, we multiply all terms in the numerator and denominator (or use the "conv" command) as follows:

$$G(s)H(s) = \frac{s^2 + 44s + 160}{s^5 + 1100s^4 + 180,000s^3} \quad (I1.5\text{-}2)$$

Therefore, the row matrices for the numerator and denominator are as follows:

$$\text{num} = [0 \quad 0 \quad 0 \quad 1 \quad 44 \quad 160]$$
$$\text{den} = [1 \quad 1100 \quad 180,000 \quad 0 \quad 0 \quad 0].$$

The resulting MATLAB program will first be provided by the listing in Table I1.5-1, and new commands will then be explained. The resulting Bode diagram is shown in Figure I1.5-1.

MATLAB and the AMCSTD Toolbox automatically select the frequency range used in Figure I1.5-1 when using the MATLAB program in Table I1.5-1. If the control-system engineer wants to select a different frequency range, such as from 0.01 to 10,000 rad/sec instead of from 0.1 to 10,000 rad/sec, then we have to use the "log-space" command which is defined as follows:

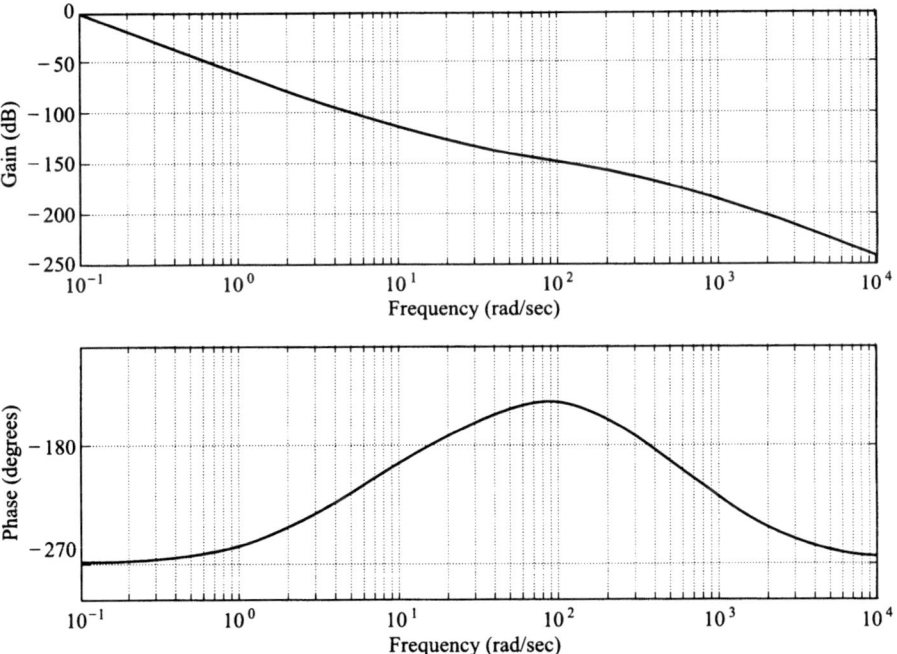

Figure I1.5-1 Bode diagram for control system whose transfer function is defined by Eq. I1.5-1.

18 INTRODUCTION

Table I1.5-1. MATLAB Program for Obtaining Bode Diagram of System Defined in Eq. (6.106)

```
num = [0  0  0  1  44  160];
den = [1  1100  180000  0  0  0];
[mag,ph] = bode(num,den);
grid
title('Bode Diagram of
G(s)H(s) = (s + 4)(s + 40)/s^3(s + 200)(s + 900)')
```

w = logspace (-2,4)   Generates 50 points equally spaced between $w = 10^{-2}$ and 10^4 rad/sec

In addition, we now have to use the MATLAB command "bode(num,den,w)" which is defined as follows:

bode(num,den,w):   Generates the Bode diagram from the user-supplied num, den, and the frequency vector **w** which specifies the frequencies at which the Bode diagram will be calculated.

Using the MATLAB commands "logspace" and "bode(num,den,w)", in the MATLAB Program in Table I1.5-1, is modified as shown in the MATLAB Program in Table I1.5-2.

Table I1.5-2. MATLAB Program for Obtaining Bode Diagram of System Defined in Eq. (I1.5-1) with the Logspace Command

```
num = [0  0  0  1  44  160];
den = [1  1100  180000  0  0  0];
w = logspace(-2,4);
[mag,ph] = bode(num,den,w);
grid
title ('Bode Diagram of
G(s)H(s) = (s + 4)(s + 40)/s^3(s + 200)(s + 900)')
```

The resulting Bode diagram is shown in Figure I1.5-2. Observe that the frequency range of this Bode diagram is 0.01 to 10,000 rad/sec, compared to 0.1 to 10,000 rad/sec in Figure I1.5-1.

The AMCSTD Toolbox command *margins* provide the following:

- gain margin (gm)

- phase margin (pm)

- frequency (rad/sec) where the phase equals $-180°$ (wcg: ω_{cg}, gain crossover frequency)

- frequency (rad/sec) where the gain equals zero dB (wcp: ω_{cp}, phase crossover frequency).

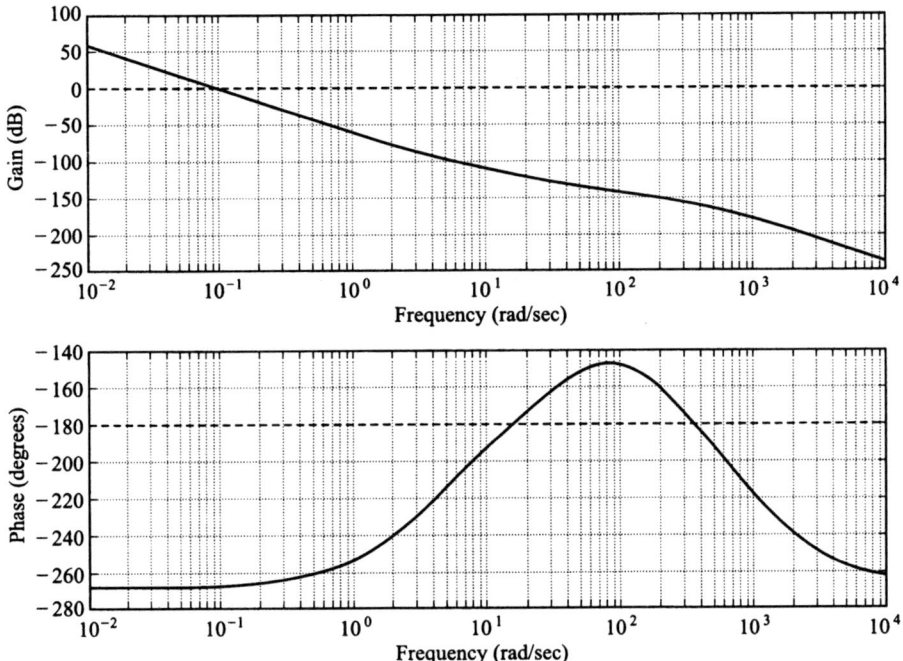

Figure I1.5-2 Bode diagram of

$$G(s)H(s) = \frac{(s+4)(s+40)}{s^3(s+200)(s+900)}$$

with the logspace command added.

Therefore, adding the following AMCSTD Toobox command to the MATLAB Program in Table I1.5-2,

[gm, pm, wcg, wcp] = margins(num,den)

will result in the Bode diagram of Figure I1.5-3 which contains everything previously shown on Figure I1.5-2 plus the phase margin, the gain margin, the frequency where the gain equals zero dB, and the frequencies where the phase is −180°. Observe from Figure I1.5-3 that this control system has two gain margins and one phase margin.

I1.6. Repeat the Bode diagram of Problem I1.5 if the control system is defined in the state-space form defined by Eqs. (I1.4-1) and (I1.4-2).

   (a) Convert from the transfer function given in Problem I1.5 to the state-space form using MATLAB.

   (b) Determine the Bode diagram from the state-space form using MATLAB.

20 INTRODUCTION

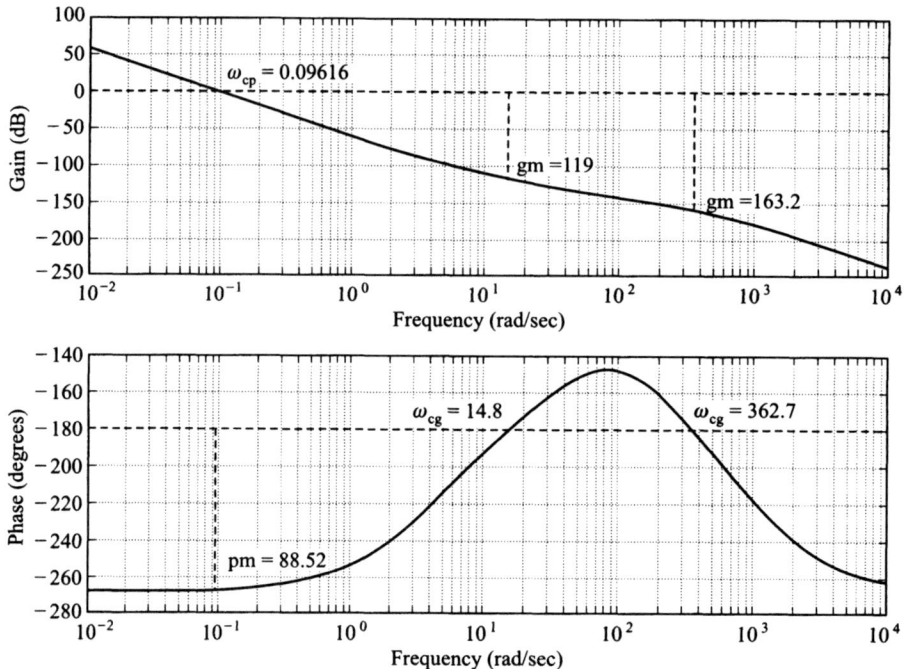

Figure I1.5-3 Bode diagram of

$$G(s)H(s) = \frac{(s+4)(s+40)}{s^3(s+200)(s+900)}$$

with the logspace command added.

SOLUTION:

(a) The MATLAB command for this case is given by

bode(A,B,C,D).

To demonstrate this procedure, let us obtain the state-space form from the transfer function given by Eq. I1.5-2 using the MATLAB command

[A,B,C,D] = tf2ss(num,den)

whose application has been demonstrated earlier in the section on the Nyquist diagram (Illustrative Problem I1.4). The resulting MATLAB program for accomplishing this is shown in the MATLAB program in Table I1.6-1.

Therefore, the resulting MATLAB program for obtaining the Bode diagram of this problem from the state-space formulation is given by the MATLAB program in Table I1.6-2. The resulting Bode diagram is identical to that shown in Figure I1.5-2.

1.7. ILLUSTRATIVE PROBLEMS AND SOLUTIONS

Table I1.6-1. MATLAB Program for Converting from the Transfer Function to the State-Space Form

```
num = [0   0   1   44   160]
den = [1   1100   180000   0   0]
[A, B, C, D] = tf2ss(num,den)
A =
   -1100   -180000   0   0   0
       1         0   0   0   0
       0         1   0   0   0
       0         0   1   0   0
       0         0   0   1   0
B =
   1
   0
   0
   0
   0
C =
   0   0   1   44   160
D =
   0
```

Table I1.6-2. MATLAB Program for Determining the Bode Diagram from the State-Space Form

```
A = [-1100   -180000   0   0   0; 1   0   0   0   0; 0   1   0   0   0;
     0   0   1   0   0; 0   0   0   1   0]
B = [1; 0; 0; 0; 0]
C = [0   0   1   44   160];
D = [0];
w = logspace(-2,4);
[mag, ph] = bode(A,B,C,D,w)
grid
title('Bode Diagram of
G(s)H(s) = (s + 4)(s + 40)/s^3(s + 200)(s + 900)')
```

I1.7. The Nichols chart is a very useful technique for determining stability and the closed-loop frequency response of a control system. Consider the following third-order feedback control system:

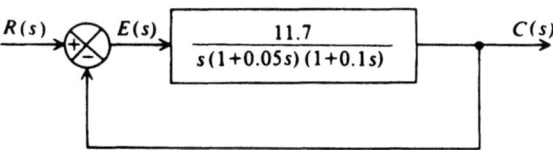

Figure I1.7-1 A third-order feedback control system.

(a) Determine the Nichols chart for this control system using MATLAB.

(b) From the resulting Nichols chart, determine the closed-loop frequency response of this control system.

(c) What are the resulting closed-loop maximum values of $C(j\omega)/R(j\omega)$, M_p, and the frequency where it occurs, ω_p?

SOLUTION:

(a) The MATLAB Control System Toolbox uses the command *nichols* to create the Nichols frequency response plot, and it uses the command *ngrid* to generate the grid lines for a Nichols chart. The AMCSTD Toolbox simplifies the process with the command *nichgrid*.

The definition of the nichgrid command is as follows:

$$\text{function}[x,y] = \text{nichgrid}(g,a,b,n)$$

where

g = grid values for axis (in degrees and dB)
a = angles (in degrees, $-180° < a < 0$)
b = gains (in dB)
n = interpolation value (optional, default = none)
x = phase values
y = magnitude (in dB).

The AMCSTD Toolbox discusses the command nichgrid in great detail.

Let us consider drawing the Nichols chart for a unity-feedback control system whose open-loop transfer function is given by

$$G(s)H(s) = \frac{11.7}{s(1+0.05s)(1+0.1s)}$$

For purposes of MATLAB formulation, we will put this transfer function in the following format:

$$G(s)H(s) = \frac{11.7}{0.005s^3 + 0.15s^2 + s} \qquad (11.7\text{-}1)$$

The resulting MATLAB program is shown in Table 11.7-1 which develops the Nichols chart and also superimposes this transfer function. The resulting Nichols chart obtained from this program is plotted in Figure 11.7-2.

(b) The closed-loop frequency response, obtained from the Nichols chart illustrated in Figure 11.7-2, is illustrated in Figure 11.7-3.

(c) $M_p = 6.758$ dB and $\omega_p = 8.989$ rad/sec.

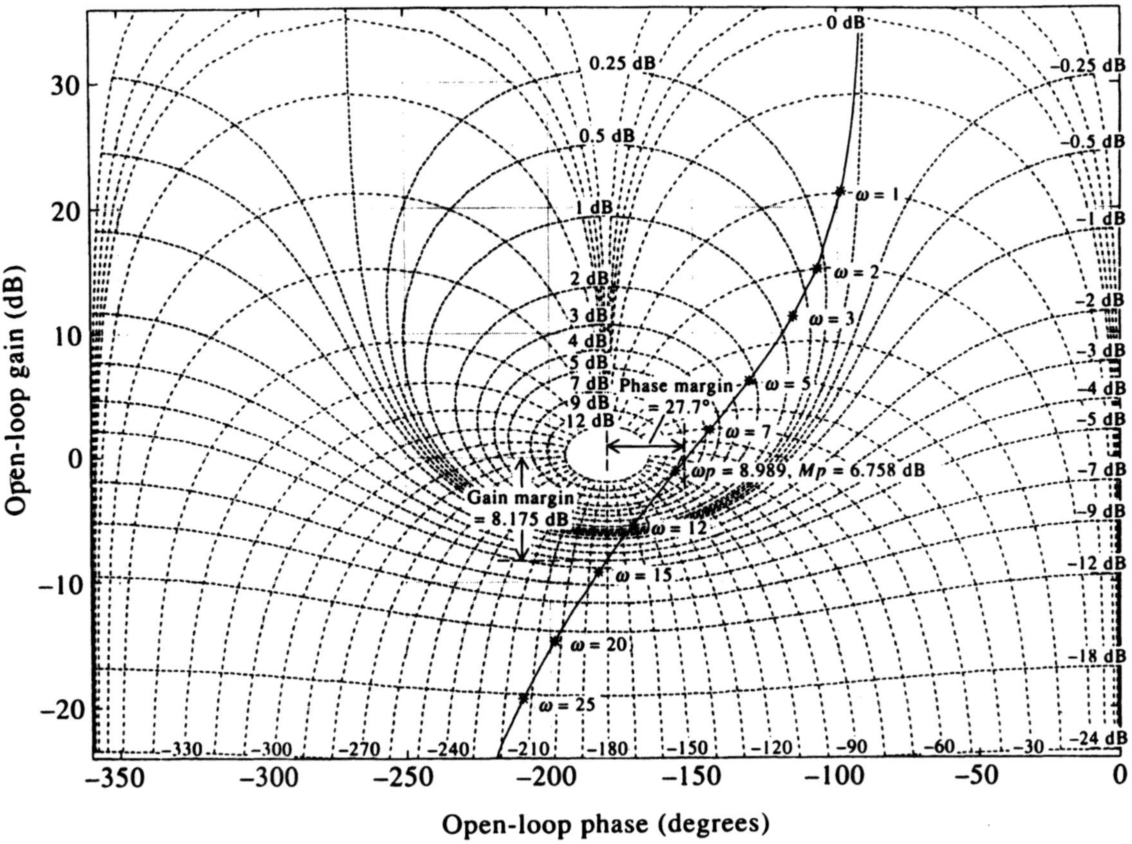

1.7. ILLUSTRATIVE PROBLEMS AND SOLUTIONS 23

Figure I1.7-2 Nichols chart for the system shown in Figure I1.7-1.

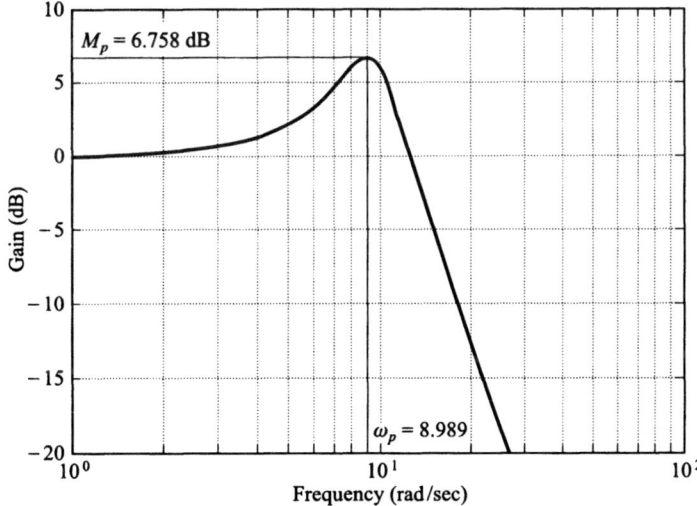

Figure I1.7-3 Closed-loop frequency response of the system shown in Figure I1.7-1 from the Nichols chart of Figure I1.7-2.

24 INTRODUCTION

Table I1.7-1. MATLAB Program to Obtain the Nichols Chart for the System whose Open-Loop Transfer Function is shown in Eq. (I1.7-1)

num = [0 0 0 11.7]
den = [0.005 0.15 1 0]
a = -[.25:.25:1 2 5 10:10:170 179.99];
b = [-24 -18 -12 -9 -7 -5:-1 -.5:.25:.5 1:5 7 9 12];
[x,y] = nichgrid([-360 0 -24 36],a,b,3);
[mag,ph] = bode(num,den,logspace(-1,2));
plot(ph,20 * log10(mag));
W = [1 2 3 5 7 8.989 12 15 20 25];
[mag, ph] = bode(num,den,w);
plot(ph,20 * log10(mag),'* g')
title('Nichols Frequency Response Plot')

I1.8. An engineer is called in to consult on a control system in a piece of equipment in the field. No one can find the design report or test results from the original design of the control system. The engineer, therefore, decides to take a frequency response of the control system by opening the outer feedback loop. The resulting asymptotic gain-frequency characteristic is obtained:

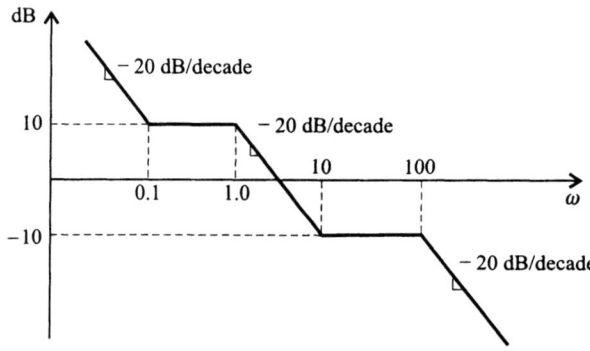

Figure I1.8 Asymptotic gain frequency characteristic.

Assuming that the system has a minimum phase transfer function, determine the transfer functon, $G(s)H(s)$.

SOLUTION: From knowledge of the break points, we know that the transfer function is given by the following transfer function where K is the gain to be determined:

$$G(s)H(s) = \frac{K\left(1 + \frac{s}{0.1}\right)\left(1 + \frac{s}{10}\right)}{s(1 + s)\left(1 + \frac{s}{100}\right)}.$$

The value of K can be obtained from knowledge that at $\omega = 0.1$ rad/sec, the gain is 13 dB (10 dB plus 3 dB due to the difference between the asymptotic gain-frequency characteristic and the actual curve at the break frequency). Therefore, we can approximate the following:

$$20 \log_{10} \left| \frac{K}{0.1j} \right| = 13.$$

Therefore, $K = 0.447$ and the complete transfer function is given by:

$$G(s)H(s) = \frac{0.447(1 + 10s)(1 + 0.1s)}{s(1 + s)(1 + 0.01s)}.$$

I1.9. The root locus of a control system is to be determined using MATLAB.

The MATLAB command used for plotting the root locus is

$$\text{rlocus(num,den)}$$

The *Advanced Modern Control System Theory and Design* (AMCSTD) Toolbox enhances the professional toolbox by not only plotting the root locus with "rlocus" but also calculating almost any point of interest on the root-locus plot directly (the professional version has only "rootfind," a graphical method, to locate these points). The most commonly required value, K_{max}, can now be obtained directly using the AMCSTD Toolbox.

1. "rlocus" is described in Reference 6. This function returns/plots the root-locus response of the system. A single example is given in the text manual.

2. "rlaxis" is described in the AMCSTD Toolbox. This function returns a set of data representing the portion of the root locus that is on the real axis.

3. "rlpoba" is described in the AMCSTD Toolbox. This function returns the value(s) (with their associated gains) of the breakaway/break-in points of the root locus.

4. "rootmag" is described in the AMCSTD Toolbox. This function returns the value(s) (with their associated gains) of any point on the root locus that corresponds to the specified magnitude. In the discrete root locus, a mangitude of 1 corresponds to K_{max}.

5. "rootangl" is described in the AMCSTD Toolbox. This function returns the value(s) (with their associated gains) of any point on the root locus that corresponds to the specified angle. In the continuous root locus, an angle of ± 90 corresponds to K_{max}. Also, desired gains that accomplish specified damping ratios (corresponding to particular

26 INTRODUCTION

angles), can easily be picked off (usually presented in compensation and design using the root locus).

Determine the root locus of a unity negative feedback control system whose forward transfer function is given by:

$$G(s) = \frac{K}{s(s+4)(s+5)}$$

SOLUTION:

In order to ease its transformation to MATLAB notation, we multiply all terms in the denominator (or use in the MATLAB "conv" command):

$$G(s)H(s) = \frac{K}{s^3 + 9s^2 + 20s} \quad (I1.9\text{-}1)$$

Therefore, the row matrices for the numerator and denominator are:

$$\text{num} = [0 \quad 0 \quad 0 \quad 1]$$
$$\text{den} = [1 \quad 9 \quad 20 \quad 0]$$

The resulting MATLAB program to draw this root locus is given in Table I1.9, and the resulting root locus is illustrated in Figure I1.9. We conclude from this root locus diagram that this control system is stable for $0 < K < 180$.

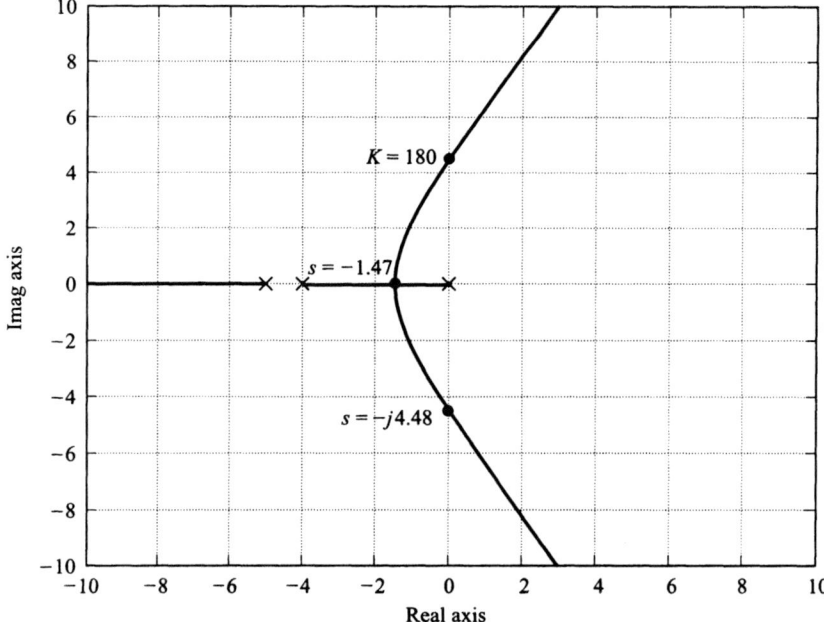

Figure I1.9 Root locus of unity feedback control system defined by Eq. (I1.9-1).

Table I1.9. MATLAB Program for Obtaining the Root Locus for the Unity Feedback Control System Defined by Eq. (I1.9-1)

num = [0 0 0 1]
den = [1 9 20 0]
rlocus(num,den)
grid
[kbreak,sbreak] = rlpoba(num,den)
[kimag,simag] = rootangl(num,den,90)
disp('[kbreak,sbreak] = rlpoba(num,den);')
disp('[kimag,simag] = rootangl(num,den,90);')

I1.10. Repeat Problem I1.9 for a unity negative feedback control system whose forward transfer function is given by

$$G(s) = \frac{K}{(s+1)(s-1)(s+4)^2}$$

SOLUTION:

In order to ease its transformation to MATLAB notation, we multiply all terms in the denominator (or use the MATLAB "conv" command):

$$G(s)H(s) = \frac{K}{s^4 + 8s^3 + 15s^2 - 8s - 16} \qquad (I1.10\text{-}1)$$

Therefore, the row matrices for the numerator and denominator are:

$$\text{num} = [0 \quad 0 \quad 0 \quad 0 \quad 1]$$
$$\text{den} = [1 \quad 8 \quad 15 \quad -8 \quad -16]$$

The resulting MATLAB program to draw this root locus is given in Table I1.10, and the resulting root locus is illustrated in Figure I1.10. We conclude from this root locus that this control system is unstable for $0 < K < \infty$ because at least one root is always in the right-half plane.

Table I1.10. MATLAB Program for Obtaining the Root Locus for the Unity Feedback Control System Defined by Eq. (I1.10-1)

```
num = [0  0  0  0  1]
den = [1  8  15  -8  -16]
rlocus(num,den)
grid
[kbreak,sbreak] = rlpoba(num,den)
disp('[kbreak,sbreak] = rootangl(num,den,90);')
```

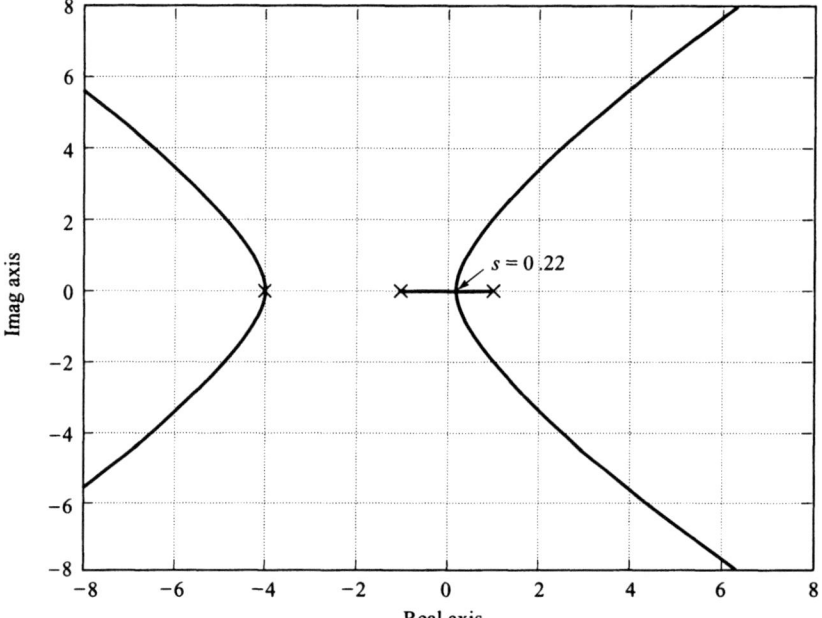

Figure I1.10 Root locus of unity feedback control system defined by Eq. (I1.10-1).

PROBLEMS

1.1. The signal-flow graph representation of the control system in Figure P1.1 is to be analyzed.

(a) Determine the overall transfer function $C(s)/R(s)$.

(b) The selection of $G_4(s)$ is critical. Determine the sensitivity $S_{G_4(s)}^{H(s)}(s)$.

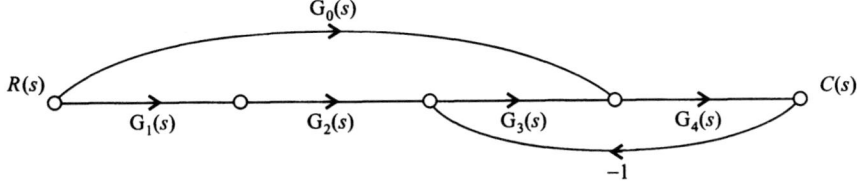

Figure P1.1

1.2. The Nyquist diagram for a control system is as shown in Figure P1.2.

The control system is known to have an open-loop transfer function whose form is given by the following:

$$G(s)H(s) = \frac{K_1}{s^3 + K_2 s^2 + K_3 s}.$$

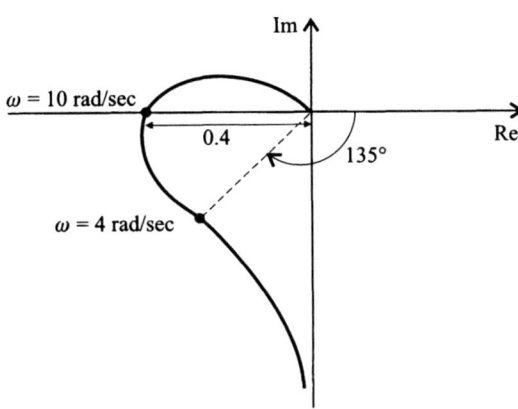

Figure P1.2

Determine the values of K_1, K_2, and K_3.

1.3. We wish to design the control system shown in Figure P1.3 so that the phase margin is 70°.

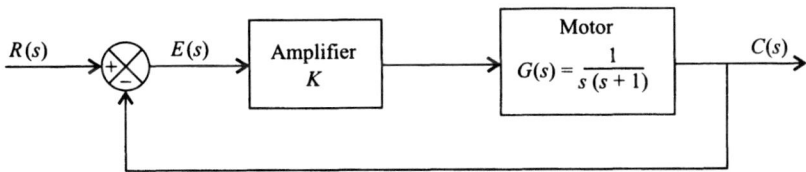

Figure P1.3

Without using semi-log graph paper, determine analytically the value of the amplifier gain K so that the phase margin is 70°.

1.4. An engineer is called in to consult on a control system in a piece of equipment in the field. No one can find the design report or test results from the original design of the control system. The engineer, therefore, decides to take a frequency response of the control system by opening up the outer feedback loop. The resulting asymptotic gain-frequency characteristic is obtained as shown in Figure P1.4.

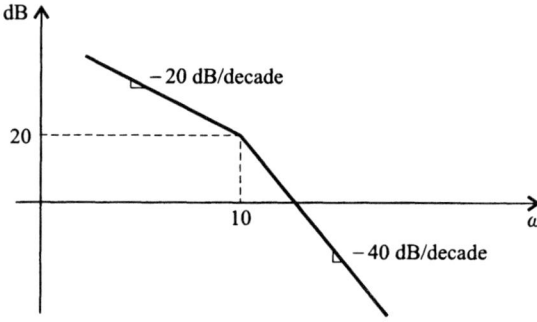

Figure P1.4

Assuming that the system is a minimum-phase transfer function, determine the transfer function, $G(s)H(s)$.

1.5. A feedback control system has the configuration shown in Figure P1.5, where $U(s)$ represents an extraneous signal appearing at the input to the plant.

(a) Assuming that $G_1(s) = 1$, plot the decibel-log frequency diagram and the phase diagram for this system.

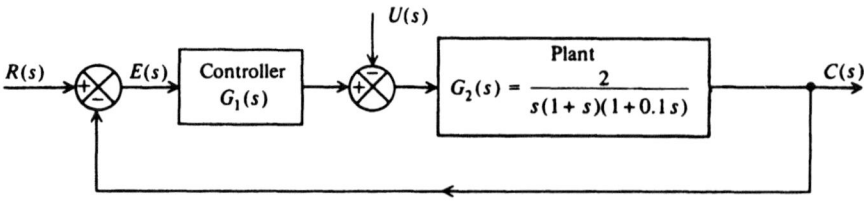

Figure P1.5

(b) It is desired that the steady-state error resulting from an extraneous unit step input signal at $U(s)$ shall be 0.1 unit. Assuming $G_1(s) = K$, determine K to meet this specification.

(c) Determine the gain crossover frequency, phase margin, and gain margin resulting from part (b).

1.6. A second-order servomechanism has a forward transfer function given by

$$G(s) = \frac{16}{s(2+s)}.$$

The feedback transfer function is unity.

(a) Draw the Bode diagram showing the magnitude and the phase characteristics as a function of frequency.

(b) Using the Nichols chart, plot a curve of frequency response of the closed-loop system.

(c) What are ω_p and M_p?

(d) Can this system ever be unstable no matter how large the forward gain is made? Explain.

1.7. (a) Draw the root-locus diagram of a negative-feedback control system whose feedback-path transfer function is unity and whose forward-path transfer function is given by the following:

$$G(s) = \frac{K}{(s+2)^5}.$$

(b) At what value of K does the system become unstable, and where does the root locus intersect the $j\omega$ axis when this occurs?

1.8. Sketch the root locus for the following negative-feedback control system, and determine the range of K for which the system is stable:

$$G(s) = \frac{K(s+6)}{s(s+1)(s+4)}, \quad H(s) = 1.$$

REFERENCES

1. S. M. Shinners, *Modern Control System Theory and Design*, 2nd ed., Wiley, New York, 1998.
2. R. C. Dorf, *Modern Control Systems*, 7th ed., Addison-Wesley, Reading, MA, 1995.
3. K. Ogatta, *Modern Control Engineering*, 3rd ed., Prentice Hall, Englewood Cliffs, NJ, 1997.
4. B. C. Kuo, *Automatic Control Systems*, 7th ed., Prentice Hall, Englewood Cliffs, NJ, 1995.

5. N. S. Nise, *Control Systems Engineering*, 2nd ed., Addison-Wesley, Reading, MA, 1995.
6. *MATLAB^{TM} for MS-DOS Personal Computers Guide*, Control System Toolbox. MathWorks, Inc., Natick, MA, 1997.

2

LINEAR CONTROL-SYSTEM COMPENSATION AND DESIGN

2.1. INTRODUCTION

After the stability of a feedback control system has been analyzed, by using any of the tools presented in Chapter 1, it will often be found that system performance is not satisfactory and needs to be modified. It is necessary to ensure that the open-loop gain is adequate for accuracy, and that the transient response is desirable for the particular application. In order for the system to meet the requirements of stability, accuracy, and transient response, certain types of equipment must be added to the basic feedback control system. We use the term *design* to encompass the entire process of basic system modification in order to meet the specifications of stability, accuracy, and transient response. The term *stabilization* is usually used to indicate the process of achieving the requirements of stability alone; the term *compensation* is usually used to indicate the process of increasing accuracy and speeding up the response.

There are several commonly used configurations for compensating (or stabilizing) a control system. The compensating (or stabilizing) device may be inserted into the system either in cascade with the forward portion of the loop (cascade compensation) as shown in Figure 2.1, or as part of a minor feedback loop (feedback compensation) as shown in Figure 2.2 [1, 2]. The cascade-compensation technique is usually concerned with the addition of phase-lag, phase-lead, and phase-lag–lead passive networks. The feedback-compensation technique is primarily concerned with the addition of rate or acceleration feedback. The type of compensation chosen usually depends on the nonlinearities and the location of the noises in the loop, and economic considerations.

34 LINEAR CONTROL-SYSTEM COMPENSATION AND DESIGN

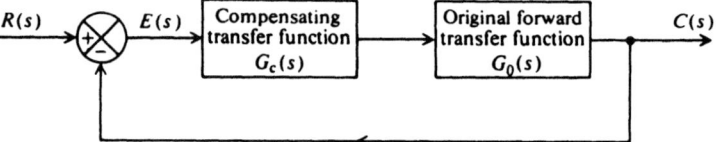

Figure 2.1 Illustration of series or cascade compensation.

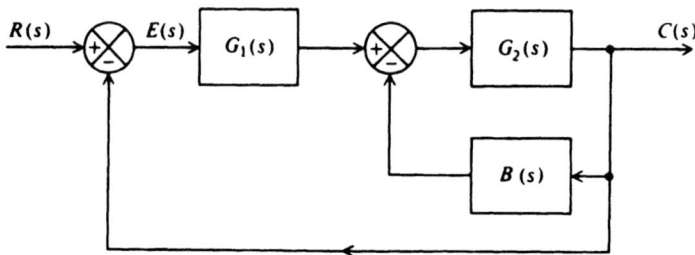

Figure 2.2 Illustration of minor-loop feedback compensation.

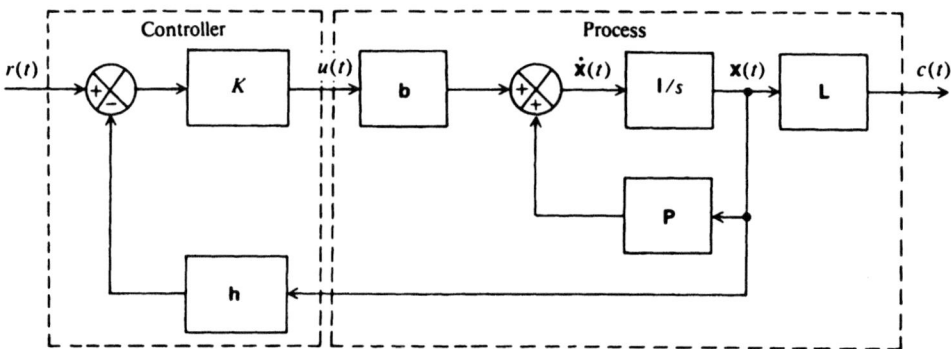

Figure 2.3 Illustration of linear-state-variable-feedback compensation.

Linear-state-variable-feedback compensation, illustrated in Figure 2.3, consists of feeding back the state variables through constant real gains.

Linear-state-variable feedback requires that all state variables be fed back to the controller. For very high-order control systems, it requires a larger number of transducers to sense the state variables for feedback. Therefore, practical applications of this technique can be very costly or impractical. In addition, some state variables may not be directly accessible (e.g., nuclear power plant processes), and this can limit this technique's application even in low-order control systems. For cases where some state variables may not be directly accessible, we use an estimator to estimate the state variables from measurements made of the accessible variables. Figure 2.4 illustrates an estimator system for a regulator system where the reference input $r(t)$, equals zero.

The compensation techniques illustrated in Figures 2.1 through 2.4 all have only one degree of freedom because there is only one controller in each system, even though the controller can have more than one parameter which can be varied. These one-degree-of-freedom controllers have the major disadvantage that the performance criteria which can be achieved is limited. For example, if a control system

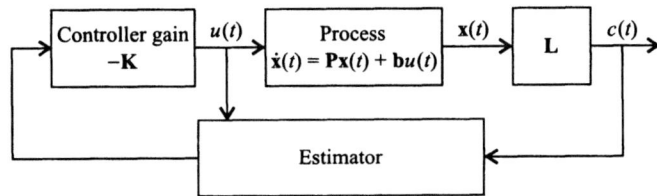

Figure 2.4 Illustration of an estimator system for a regulator system (where the reference input $r(t) = 0$).

is designed to have a desired stability, then it may have poor sensitivity to parameter variations. Let us, therefore, next consider compensation techniques which have two degrees of freedom.

Figure 2.5 illustrates the series-feedback compensation technique in which a series controller and a feedback controller are used. This technique is denoted as having two degrees of freedom compensation because it has two controllers, the series controller $G_c(s)$ and the feedback controller $B(s)$.

Figure 2.6 illustrates forward compensation in which the feedforward controller $G_{c1}(s)$ is in series with the closed-loop control system which also has a controller $G_{c2}(s)$ in the forward part of the control system. Because the controller $G_{c1}(s)$ is not within the feedback loop of the control system, it does not affect the characteristic equation roots of the original control system. Therefore, the poles and zeros of $G_{c1}(s)$ can be designed to cancel (or add) to the poles and zeros of the closed-loop transfer function. This technique is also referred to as having two degrees of freedom compensation because it has two controllers, the series controller $G_{c1}(s)$ and the forward-loop controller $G_{c2}(s)$.

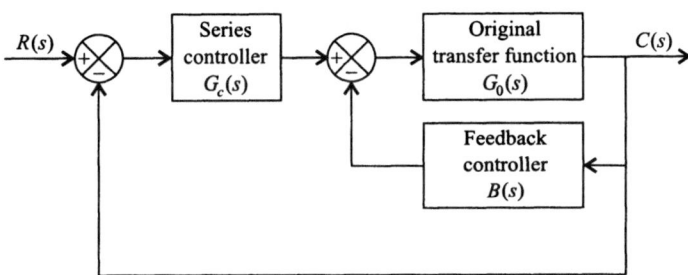

Figure 2.5 Series-feedback compensation technique illustrating two degrees of freedom.

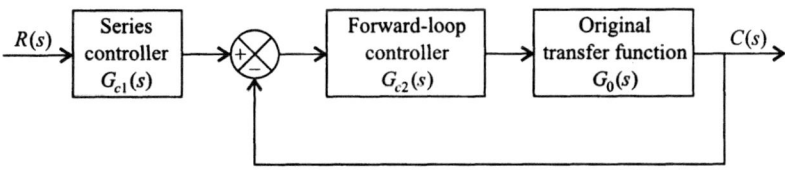

Figure 2.6 Forward compensation with series compensation illustrating two degrees of freedom.

36 LINEAR CONTROL-SYSTEM COMPENSATION AND DESIGN

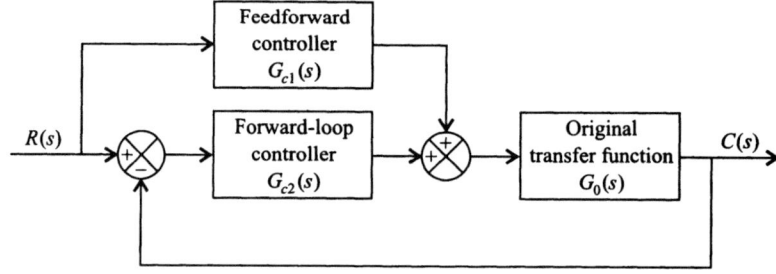

Figure 2.7 Feedforward compensation illustrating two degrees of freedom.

Figure 2.7 illustrates another form of two degrees of freedom compensation system which uses a feedforward controller, $G_{c1}(s)$, which is placed in parallel with the forward path which contains the controller $G_{c2}(s)$.

This chapter focuses attention on the tools presented in Chapter 1 which are of practical and useful interest to the control engineer. We will focus our attention on the application of the techniques to those particular design problems they are most suited to solve. Chapter 3 presents modern control-system design techniques using state-space techniques, Ackermann's formula for pole placement, estimation, robust control, and the H^∞ method for sensitivity reduction. Chapter 6 discusses the design of linear feedback control systems from the point of view of modern optimal control theory. Additional linear design problems are presented in Chapter 7 where actual case studies are outlined.

2.2. CASCADE-COMPENSATION TECHNIQUES

Let us consider the system of Figure 2.1 as our basic starting point in order to analyze the effect of cascade compensation. The compensating transfer function $G_c(s)$ is designed in order to provide additional phase lag, phase lead, or a combination of both at certain frequencies, in order to achieve certain specifications regarding stability and accuracy. We will illustrate and derive the transfer functions for representative compensating passive networks [1–6].

A *phase-lag network* is a device that shifts the phase of the control signal in order that the phase of the output lags the phase of the input over a certain range of frequencies. An electrical network performing this function was illustrated in Table 2.4[‡] as item 4. Its transfer function was

$$\frac{E_{\text{out}}(s)}{E_{\text{in}}(s)} = \frac{1 + R_2 C_2 s}{1 + (R_1 + R_2) C_2 s}. \quad (2.1)$$

This can be written in the following more useful form:

$$\frac{E_{\text{out}}(s)}{E_{\text{in}}(s)} = \frac{1 + Ts}{1 + \alpha Ts}, \quad (2.2)$$

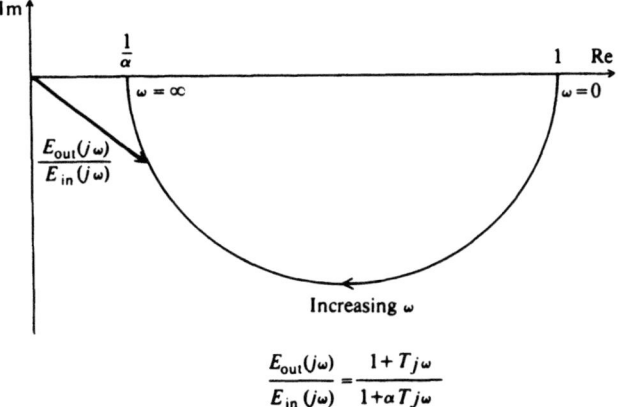

$$\frac{E_{out}(j\omega)}{E_{in}(j\omega)} = \frac{1+Tj\omega}{1+\alpha Tj\omega}$$

Figure 2.8 A complex-plane plot for a phase-lag network.

where

$$T = R_2 C_2, \quad \alpha = 1 + R_1/R_2.$$

Observe that $\alpha T > T$. A complex-plane plot of this network as a function of frequency is shown in Figure 2.8. Notice that the output voltage lags the input in phase angle for all positive frequencies. In addition, observe that the magnitude of $E_{out}(j\omega)/E_{in}(j\omega)$ decreases from unity at $\omega = 0$ to $1/\alpha$ at $\omega = \infty$. The Bode diagram for the phase-lag network is illustrated in Figure 2.9. The frequency at which the maximum phase lag occurs, ω_{max}, and the maximum phase lag, ϕ_{max}, can be easily derived. The results are

$$\omega_{max} = 1/(T\sqrt{\alpha}) \tag{2.3}$$

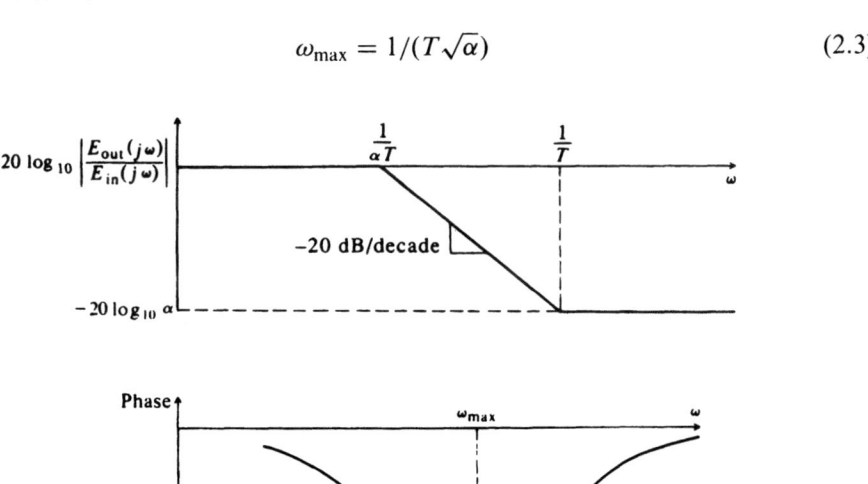

$$\frac{E_{out}(s)}{E_{in}(s)} = \frac{1+Ts}{1+\alpha Ts} \quad \text{where } \alpha T > T$$

Figure 2.9 Bode diagram of a phase-lag network.

and

$$\phi_{\max} = \sin^{-1}\frac{1-\alpha}{1+\alpha}. \tag{2.4}$$

Values of $\phi_{\max}$ for certain values of α, which are useful for design purposes, are listed in Table 2.1.

A *phase-lead network* is a network which shifts the phase of the control signal in order that the phase of the output leads the phase of the input at certain frequencies. An electrical network performing this function was illustrated in Table 2.4[‡] as item 3. Its transfer function was as follows:

$$\frac{E_{\text{out}}(s)}{E_{\text{in}}(s)} = \frac{R_2}{R_1+R_2}\frac{1+R_1C_1s}{1+[R_2/(R_1+R_2)]R_1C_1s}. \tag{2.5}$$

This can be written in the following more useful form:

$$\frac{E_{\text{out}}(s)}{E_{\text{in}}(s)} = \frac{1}{\alpha}\left(\frac{1+\alpha Ts}{1+Ts}\right) \tag{2.6}$$

where

$$T = \frac{R_1R_2}{R_1+R_2}C_1, \quad \alpha = 1 + \frac{R_1}{R_2}.$$

Observe that $\alpha T > T$. A complex-plane plot of this network as a function of frequency is shown in Figure 2.10. Notice that the output voltage leads the input in phase angle for all positive frequencies. In addition, notice that the magnitude of $E_{\text{out}}(j\omega)/E_{\text{in}}(j\omega)$ increases from $1/\alpha$ at $\omega = 0$ to unity at $\omega = \infty$. The Bode diagram for the phase-lead network is illustrated in Figure 2.11. The corresponding values of $\omega_{\max}$ and $\phi_{\max}$ for the phase-lead network are

$$\omega_{\max} = 1/(T\sqrt{\alpha}) \tag{2.7}$$

and

$$\phi_{\max} = \sin^{-1}\frac{\alpha-1}{\alpha+1}. \tag{2.8}$$

Table 2.1. $\phi_{\max}$ as a Function of α

α	$\phi_{\max}$ (degrees)
1	0
2	−19.4
4	−36.9
8	−51.0
10	−55.0

[‡]Throughout the book, this symbol is used to refer to material in the companion volume, *Modern Control System Theory and Design, Second Edition.*

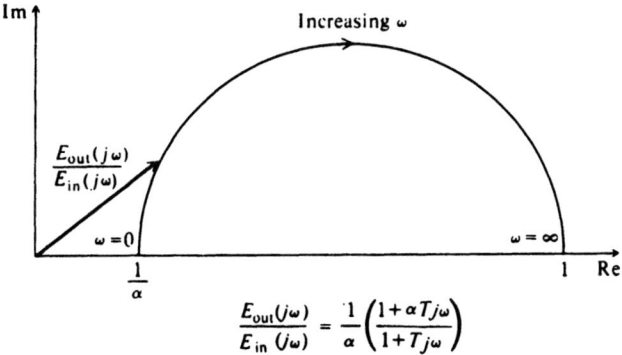

Figure 2.10 A complex-plane plot for a phase-lead network.

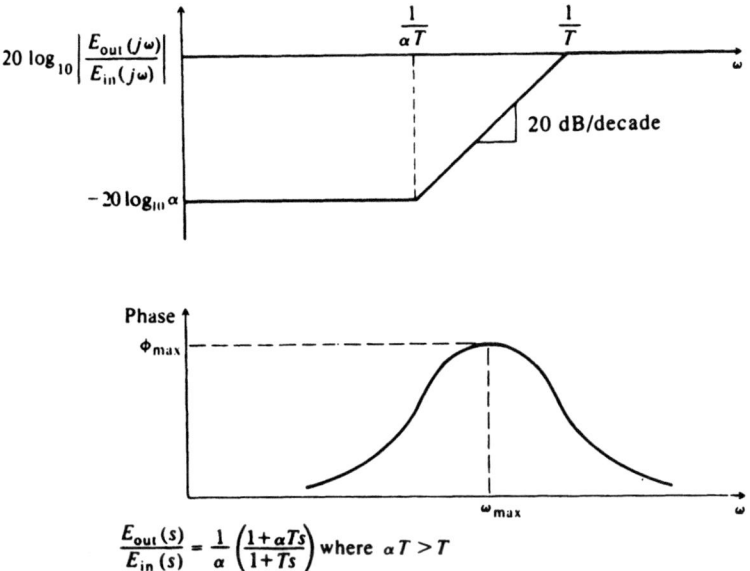

Figure 2.11 Bode diagram of a phase-lead network.

The values shown in Table 2.1 are also true for the phase-lead case except for the sign. An important practical point to emphasize is that the control engineer would not in practice use any ratio of $\alpha > 10$ because the lead network acts as an attenuation, which must be made up for somewhere in the feedback control system, with an amplification whose ratio is α.

A *phase-lag–lead network* is a network that shifts the phase of a control signal in order that the phase of the output lags at low frequencies and leads at high frequencies relative to the input. An electrical network performing this function was illustrated in Table 2.4[‡] as item 5. Its transfer function was as follows:

$$\frac{E_{out}(s)}{E_{in}(s)} = \frac{(1 + R_1 C_1 s)(1 + R_2 C_2 s)}{R_1 R_2 C_1 C_2 s^2 + (R_1 C_1 + R_2 C_2 + R_1 C_2)s + 1}. \tag{2.9}$$

Defining

$$T_1 = R_1C_1, \quad T_2 = R_2C_2, \quad T_{21} = R_1C_2$$

we can rewrite Eq. (2.9) as

$$\frac{E_{\text{out}}(s)}{E_{\text{in}}(s)} = \frac{T_1T_2s^2 + (T_1 + T_2)s + 1}{T_1T_2s^2 + (T_1 + T_2 + T_{21})s + 1}.$$

A complex-plane plot of this network as a function of frequency is shown in Figure 2.12. Notice that the output voltage lags the input in phase angle for low frequencies and leads in phase angle for high frequencies. In addition, notice that the magnitude of $E_{\text{out}}(j\omega)/E_{\text{m}}(j\omega)$ decreases for intermediate frequencies and increases to unity as ω approaches 0 and ∞. A corresponding Bode diagram for the phase-lag-lead network is illustrated in Figure 2.13.

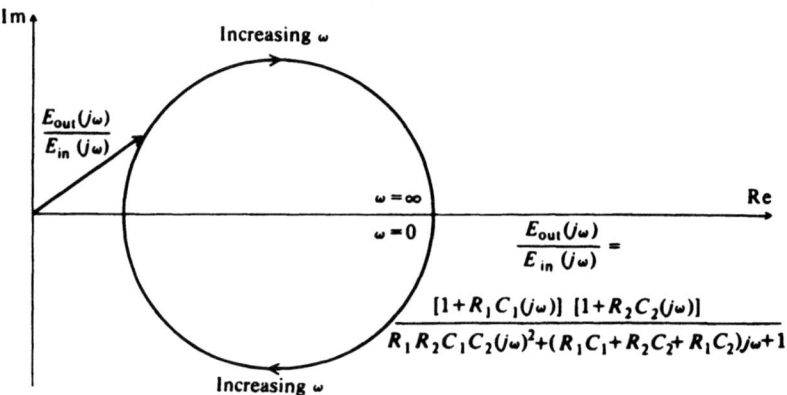

Figure 2.12 A complex-plane plot for a phase-lag-lead network.

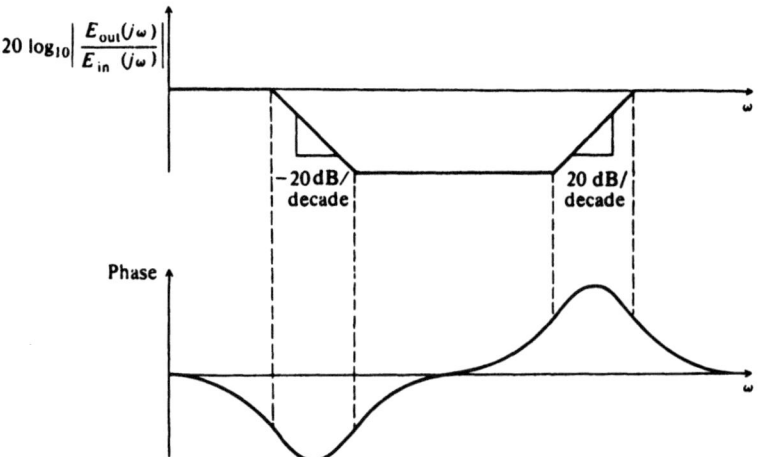

Figure 2.13 Bode diagram of a phase-lag-lead network.

The stabilizing effect of cascaded, phase-shifting networks can easily be demonstrated for a simple second-order system. For example, let us consider the configuration illustrated in Figure 2.1, where the original forward transfer function $G_0(s)$, is given by

$$G_0(s) = \frac{\omega_n^2}{s(s + 2\zeta\omega_n)}. \tag{2.10}$$

If this system were uncompensated [$G_c(s) = 1$], then the transfer function $G_0(s)$ would result in the familiar second-order system response which is discussed at great length in Appendix B. The resulting damping ratio of the system would be given by ζ and its undamped natural frequency by ω_n. Let us now assume that we add a lead network to this system whose transfer function is given by

$$G_c(s) = \frac{1 + \alpha T s}{1 + Ts}. \tag{2.11}$$

It is assumed that the attenuation factor of $1/\alpha$ is compensated for with an amplification increase of α, and the system will maintain the same static error. Let us consider the case where $T \ll \alpha T$ and, therefore, $G_c(s)$ can be approximated by a zero factor (proportional plus derivative control):

$$G_c(s) \approx 1 + \alpha T s, \quad T \ll \alpha T. \tag{2.12}$$

Another way of looking at this approximation is the viewpoint of adding a rate feedback loop. In the following section, it is shown in Figure 2.14a that the addition of rate feedback bs in parallel with unity position feedback is equivalent to adding the pure zero factor term $(1 + bs)$ to the open-loop transfer function, $G(s)H(s)$. Therefore, the approximation in Eq. (2.12) can be viewed as an approximate representation of a phase-lead network, or the exact representation of the addition of rate feedback in parallel with position feedback to the system. In the following analysis, we will use the terminology of Eq. (2.12), which is based on an approximation to the phase-lead network, although it could just as easily represent exactly the addition of rate feedback in parallel with position feedback.

The form of Eq. (2.12) suggests that this lead network (or the addition of rate feedback in parallel with unity position feedback) is equivalent to a proportional plus derivative controller. The resulting system transfer function with the lead network is given by

$$\frac{C(s)}{R(s)} \approx \frac{\omega_n^2(1 + \alpha T s)}{s^2 + (2\zeta\omega_n + \alpha T \omega_n^2)s + \omega_n^2}. \tag{2.13}$$

Comparing the denominators of Eqs. (2.13) and (B.3) of Appendix B, we observe that it is still of second order and ω_n remains the same, but ζ is greater due to the increase in the coefficient of s in the denominator. The equivalent damping ratio with $G_c(s)$ can be obtained as follows:

$$2\zeta\omega_n + \alpha T \omega_n^2 = 2\zeta_{eq}\omega_n \tag{2.14}$$

42 LINEAR CONTROL-SYSTEM COMPENSATION AND DESIGN

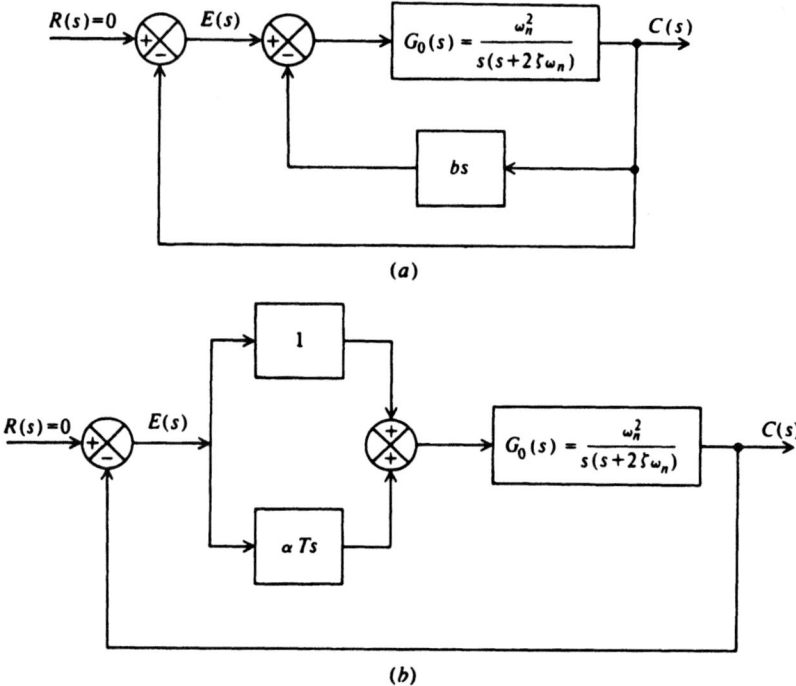

Figure 2.14 The stabilizing effects of the systems illustrated are equivalent. (a) Feedback compensation. (b) Cacade compensation—proportional plus derivative controller

where ζ_{eq} is an equivalent damping ratio with the addition of a zero factor for compensation. Solving for ζ_{eq}, we obtain

$$\zeta_{eq} = \zeta + \alpha T \omega_n/2. \tag{2.15}$$

Therefore, we can conclude that the addition of a zero factor in $G_c(s)$ has increased the damping ratio from ζ to ζ_{eq} by an amount equal to $\alpha T \omega_n/2$. This assumes that T is positive, or the zero of the factor $(1 + \alpha T s)$ is in the left half of the s-plane.

The next question is what is the steady-state error resulting from cascade compensation? To answer this, we must find the steady-state errors resulting from the application of a unit ramp input for the cases of no compensation and compare them with those resulting from cascade compensation. We choose a unit ramp as our input because it is the only input which results in a finite response error for a system with a pole at the origin. The transfer function relating error to input for the system shown in Figure 2.1 is given by

$$\frac{E(s)}{R(s)} = \frac{1}{1 + G_c(s)G_0(s)}. \tag{2.16}$$

Assuming that

$$G_c(s) = 1 \quad \text{(no cascade compensation)},$$
$$G_0(s) = \frac{\omega_n^2}{s(s + 2\zeta\omega_n)},$$

and

$$R(s) = 1/s^2 \quad \text{(a unit ramp input)},$$

we find that

$$E(s) = \frac{s + 2\zeta\omega_n}{s(s^2 + 2\zeta\omega_n s + \omega_n^2)}. \tag{2.17}$$

Applying the final-value theorem to Eq. (2.17), we find the steady-state error to be

$$e_{ss(\text{ramp input})} = \lim_{s \to 0} sE(s) = \frac{2\zeta\omega_n}{\omega_n^2} = \frac{2\zeta}{\omega_n}. \tag{2.18}$$

For the case with cascade compensation, a similar analysis yields the following result:

$$G_c(s) = \frac{1 + \alpha Ts}{1 + Ts}, \quad G_0(s) = \frac{\omega_n^2}{s(s + 2\zeta\omega_n)}, \quad R(s) = \frac{1}{s^2}.$$

Therefore,

$$E(s) = \frac{1}{s} \frac{(s + 2\zeta\omega_n)(1 + Ts)}{[s(s + 2\zeta\omega_n)(1 + Ts) + \omega_n^2(1 + \alpha Ts)]}. \tag{2.19}$$

Applying the final-value theorem to Eq. (2.19), the steady-state error is found to be

$$e_{ss(\text{ramp input})} = \lim_{s \to 0} sE(s) = \frac{2\zeta\omega_n}{\omega_n^2} = \frac{2\zeta}{\omega_n}. \tag{2.20}$$

Comparing the results of Eqs. (2.18) and (2.20), we conclude that the addition of the cascade lead network as given by Eq. (2.11) does not increase or decrease the steady-state response error of the system.

It is important to emphasize that the relationships derived in this analysis apply only to the simple system considered. For example, if a zero factor were contained in the numerator of Eq. (2.10), then these relationships are modified (see Problems 2.5 and 2.6).

If we attempt to extend this analysis of a second-order system to the case of phase-lag compensation, the characteristic equation becomes third order and difficult to factor. For example, if $G_c(s)$ were only to represent the pole factor of the phase-lag network, then

$$G_c(s) = \frac{1}{1 + \alpha Ts}. \tag{2.21}$$

In a similar manner, the closed-loop system transfer function can be found to be given by

$$\frac{C(s)}{R(s)} = \frac{\omega_n^2}{\alpha Ts^3 + (2\zeta\omega_n \alpha T + 1)s^2 + 2\zeta\omega_n s + \omega_n^2}. \tag{2.22}$$

44 LINEAR CONTROL-SYSTEM COMPENSATION AND DESIGN

The factorization of this characteristic equation is not trivial and a similar analysis to that performed for the phase lead-network case is more complex. The root-locus method is an excellent tool which can be used for factorization, and the analysis of this problem for third- and higher-order systems is presented in Sections 1.7 and 2.9.

2.3. MINOR-LOOP FEEDBACK-COMPENSATION TECHNIQUES

Let us consider the general system illustrated in Figure 2.2. The compensating element in this case is the transfer function $B(s)$. In order to have a basis of comparison, we will follow an analysis for minor-loop feedback compensation similar to that performed for the case of phase lead-network cascade compensation.

The minor-loop feedback element $B(s)$ usually represents rate feedback or acceleration feedback. In general, phase-lag, -lead, and/or lag–lead networks may also be cascaded with $B(s)$.

The stabilizing effect of minor-loop feedback compensation can easily be demonstrated for a simple second-order system. We assume that the system illustrated in Figure 2.2 contains simple rate feedback. The specific transfer functions for the system are

$$G_1(s) = 1 \tag{2.23}$$

$$G_2(s) = \frac{\omega_n^2}{s(s + 2\zeta\omega_n)}, \tag{2.24}$$

$$B(s) = bs. \tag{2.25}$$

The system is redrawn with these transfer functions and shown in Figure 2.14a.

Without any rate feedback, the configuration represents a simple second-order system whose damping ratio is ζ and undamped natural frequency is ω_n. The resulting system transfer function with rate-feedback compensation is given by

$$\frac{C(s)}{R(s)} = \frac{\omega_n^2}{s^2 + (2\zeta\omega_n + \omega_n^2 b)s + \omega_n^2}. \tag{2.26}$$

Comparing the denominator of Eq. (2.26) with that of Eq. (B.3), we observe that it is still of second order and ω_n remains the same, but ζ is greater due to the increase in the coefficient of s in the denominator. The equivalent damping ratio with rate feedback added can be obtained by setting the coefficients of the s terms equal to each other, as follows:

$$2\zeta\omega_n + \omega_n^2 b = 2\zeta_{eq}\omega_n. \tag{2.27}$$

where ζ_{eq} = an equivalent damping ratio with rate feedback added. Solving for ζ_{eq} we obtain

$$\zeta_{eq} = \zeta + \frac{\omega_n b}{2}. \tag{2.28}$$

2.3. MINOR-LOOP FEEDBACK-COMPENSATION TECHNIQUES

Therefore, we can conclude that the addition of the minor loop using rate feedback has increased the damping ratio from ζ to ζ_{eq} by an amount equal to $\omega_n b/2$. This assumes that b is positive (negative feedback).

It is important at this point to compare Eqs. (2.15) and (2.28). Note that they are very similar, and they imply that

$$\alpha T = b \tag{2.29}$$

The fact that rate feedback behaves as the approximated phase lead network, as defined by Eq. (2.12) (porportional plus derivative controller), can be easily demonstrated from Figure 2.14a and b. Let us assume that there is zero input to both systems, because we are concerned only with the system poles. Clearly, in both cases, there are two negative-feedback paths in parallel around $G_0(s)$. In the cascade-compensation case, the total feedback around $G_0(s)$ is $1 + \alpha Ts$; in the rate-feedback-compensation case, the total feedback around $G_0(s)$ is $1 + bs$. Therefore, the stabilizing effects of αT and b are equivalent. We discuss proportional plus derivative (PD) controllers fully in the next section, 2.4, on proportional-plus-integral-plus-derivative (PID) compensators.

Let us next determine the steady-state error resulting from the use of minor-loop rate-feedback compensation. We assume that the input to this system is a unit ramp in order to have a finite steady-state response error and a basis for comparison. From our discussion of cascade compensation in Section 2.2 we know from Eq. (2.18) that the resulting steady-state error of this system without any compensation ($b = 0$) is $2\zeta/\omega_n$. For the case of minor-loop rate-feedback compensation, the resulting expression for $E(s)$ is given by

$$E(s) = \frac{1}{s}\left[\frac{s + 2\zeta\omega_n + b\omega_n^2}{s(s + 2\zeta\omega_n + b\omega_n^2) + \omega_n^2}\right]. \tag{2.30}$$

Applying the final-value theorem, the steady-state error is found to be

$$e_{ss(\text{ramp input})} = \lim_{s \to 0} sE(s) = \frac{2\zeta\omega_n + b\omega_n^2}{\omega_n^2} = \frac{2\zeta}{\omega_n} + b. \tag{2.31}$$

Therefore, the steady-state response error of the system with minor-loop rate-feedback compensation has increased by a factor of b. This unfavorable result can easily be remedied by placing a high-pass filter in cascade with the rate device. Such a filter would block the steady-state value of the rate output. This technique is illustrated in Figure 2.15.

As in the preceding section, it is important to emphasize that the relationships derived apply only to the simple system considered. Problems 2.5 and 2.6 illustrate how these relationships change if a zero factor is added to the basic system transfer function considered.

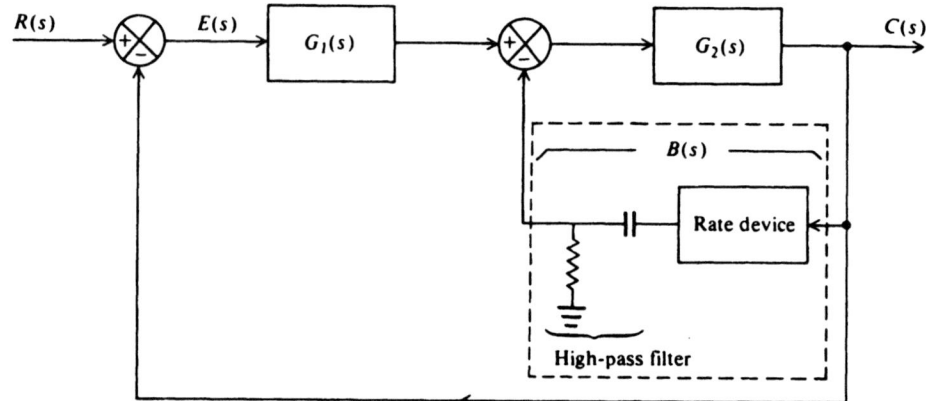

Figure 2.15 Illustration of minor-loop feedback compensation using a rate device in cascade with a high-pass filter

2.4. PROPORTIONAL-PLUS-INTEGRAL-PLUS DERIVATIVE (PID) COMPENSATORS

PID compensators are another form of compensation frequently used in control systems, especially in the industrial process control field. They are very popular in the industrial process control field due to their robust (insensitive) performance over a wide range of operating conditions including plant uncertainty, parameter variation, and external disturbances. Robust control is presented in Section 3.10 of Chapter 3.

Assuming that the input to the PID compensator is $e(t)$ and its output is $u(t)$, the equation defining the operation of the PID compensator is given by:

$$u(t) = K_P e(t) + K_I \int_0^t e(\tau) \, d\tau + K_D \frac{de(t)}{dt}. \qquad (2.32)$$

Figure 2.16 illustrates a block diagram representation of Eq. (2.32). The transfer function of the PID compensator is obtained as follows:

$$U(s) = K_P E(s) + \frac{K_I}{s} E(s) + K_D s E(s) \qquad (2.33)$$

$$U(s) = \left(K_P + \frac{K_I}{s} + K_D s\right) E(s). \qquad (2.34)$$

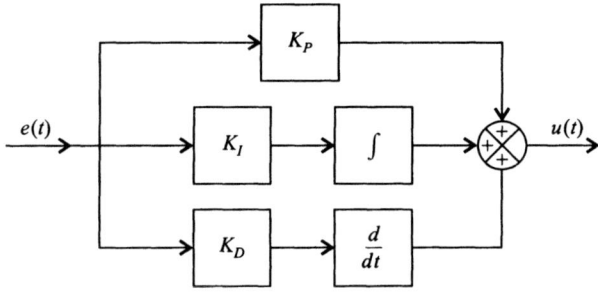

Figure 2.16 PID compensator block-diagram representation of Eq. (2.32).

2.4. PROPORTIONAL-PLUS-INTEGRAL-PLUS DERIVATIVE (PID) COMPENSATORS

The resulting transfer function of the PID compensator, $G_c(s)$, is given by

$$G_c(s) = \frac{U(s)}{E(s)} = K_P + \frac{K_I}{s} + K_D s. \qquad (2.35)$$

Figure 2.17 illustrates a block diagram representation of Eq. (2.35). Observe from Eqs. (2.32) and (2.35) that the PID compensator provides a proportional term, an integration term, a derivative term, as its name implies.

Very often, only a portion of the very general PID compensator is used. For example, if $K_D = 0$, then we have

$$G_c(s) = K_P + \frac{K_I}{s} \qquad (2.36)$$

which is denoted as the proportional plus integral, or PI, compensator. If $K_I = 0$, then we have

$$G_c(s) = K_P + K_D s \qquad (2.37)$$

which is denoted as the proportional plus derivative, or PD, compensator. This was illustrated previously in Figure 2.14.

To consider the design of a PID compensator, let us reconsider the transfer function of $G_c(s)$ given by Eq. (2.35):

$$G_c(s) = K_p + \frac{K_I}{s} + K_D s. \qquad (2.38)$$

This can be rewritten as

$$G_c(s) = \frac{K_D s^2 + K_P s + K_I}{s} \qquad (2.39)$$

or,

$$G_c(s) = \frac{K_D(s^2 + K_1 s + K_2)}{s} \qquad (2.40)$$

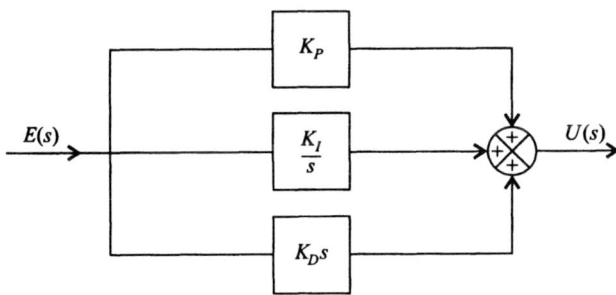

Figure 2.17 PID compensator block diagram representation of Eq. (2.35).

where

$$K_1 = \frac{K_P}{K_D}, \quad K_2 = \frac{K_I}{K_D}.$$

Factoring the numerator of Eq. (2.40), and assuming the quadratic factors as two real zeros, we obtain the following:

$$G_c(s) = \frac{K_D(s + z_a)(s + z_b)}{s}. \tag{2.41}$$

The very pleasing result of Eq. (2.41) is that the PID compensator results in two zero factors which can be located anywhere in the left-hand of the s-plane, in addition to the pole at the origin. We know from the analysis in Sections 2.2 and 2.3 that a zero factor provides a phase lead and aids in the compensation of a control system. In particular, as illustrated in Figure 2.14b, a PD is equivalent to one zero factor (which the rate feedback compensator also provides and is illustrated in Figure 2.14a). With the complete PID compensator, two zero factors are provided for compensation, in addition to the pole at the origin.

Applications of PI, PD, and PDI compensators are illustrated throughout this book for compensating control systems. These techniques are very useful for compensation of control systems in addition to the phase-lag, phase-lead, and phase-lag–lead networks illustrated in Section 2.2, and minor-loop rate feedback illustrated in Section 2.3.

2.5. EXAMPLE FOR THE DESIGN OF A SECOND-ORDER CONTROL SYSTEM

In this section we consider the design and resulting performance of the second-order system by means of cascade and minor-loop rate-feedback techniques. This problem is useful in unifying concepts which are introduced in Chapter 1, and Appendices B and C, together with the design techniques illustrated in this chapter.

Let us consider the second-order system illustrated in Figure 2.18a. We will assume that the original forward-loop transfer function $G_0(s)$ is given by

$$G_0(s) = \frac{14.4}{s(0.1s + 1)}. \tag{2.42}$$

The closed-loop transfer function, $C(s)/R(s)$, is given by

$$\frac{C(s)}{R(s)} = \frac{G_0(s)}{1 + G_0(s)} = \frac{14.4/[s(0.1s + 1)]}{1 + 14.4/[s(0.1s + 1)]} \tag{2.43}$$

or

$$\frac{C(s)}{R(s)} = \frac{144}{s^2 + 10s + 144}. \tag{2.44}$$

2.5. EXAMPLE FOR THE DESIGN OF A SECOND-ORDER CONTROL SYSTEM

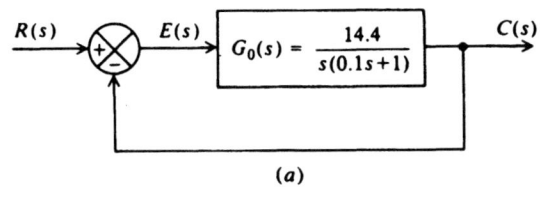

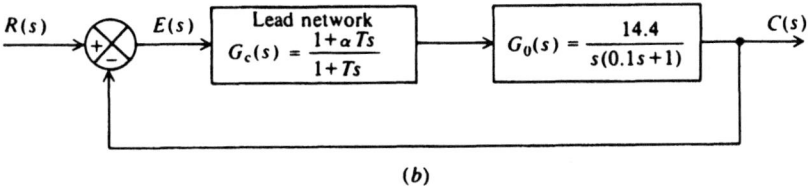

Figure 2.18 Design of a second-order system.

Comparing Eqs. (B.3) and (2.44), we observe that the undamped natural frequency ω_n and damping ratio ζ of the system are given by

$$\omega_n = 12 \text{ rad/sec} \qquad (2.45)$$

and

$$\zeta = 0.417.$$

If the system is subjected to a unit step input, the transient response will have the form shown in Figure B.4 (interpolate between $\zeta = 0.4$ and $\zeta = 0.6$). The maximum percent overshoot can be obtained from Eq. (B.33) and is found to be 23.5%.

Let us assume, for this application, that it is desired to have a critically damped system ($\zeta = 1$). We will demonstrate how this can be achieved using a cascaded network and minor-loop rate feedback.

Figure 2.18b illustrates the form that the system illustrated in Figure 2.18a would have if a cascaded network were used. We attempt to achieve a damping ratio equal to 1 using a phase-lead network, where

$$G_c(s) = \frac{1 + \alpha Ts}{1 + Ts} \qquad (2.46)$$

As was assumed previously, in Section 2.2, we assume that $T \ll \alpha T$, and therefore $G_c(s)$ can be approximated by (proportional plus derivative control)

$$G_c(s) \approx 1 + \alpha Ts. \qquad (2.47)$$

The closed-loop system transfer function for this case was derived in Section 2.2 [see Eq. (2.13)] as

$$\frac{C(s)}{R(s)} \approx \frac{\omega_n^2 + \alpha T \omega_n^2 s}{s^2 + (2\zeta\omega_n + \alpha T \omega_n^2)s + \omega_n^2}. \qquad (2.48)$$

The equivalent damping ratio for this situation was derived in Section 2.2 [see Eq. (2.15)] as

$$\zeta_{eq} = \zeta + \frac{\alpha T \omega_n}{2}. \tag{2.49}$$

The object in this problem is to design $\zeta_{eq} = 1$ for the case where

$$\zeta = 0.417 \quad \text{and} \quad \omega_n = 12. \tag{2.50}$$

Substituting the values into Eq. (2.49), we find that

$$\alpha T = 0.0972.$$

We know that the resulting system will be stable and critically damped, and will have a steady-state response error for a unit step input of zero. The steady-state response error of the system to a unit ramp input was derived [see Eqs. (2.18) and (2.20)] as

$$e_{ss(\text{ramp input})} = \frac{2\zeta}{\omega_n} = \frac{2(0.417)}{12} = 0.0695 \text{ units.} \tag{2.51}$$

Let us next attempt to achieve the same type of performance using minor-loop rate feedback. Figure 2.19 illustrates the form that the system illustrated in Figure 2.18a would have if minor-loop feedback were used. Our goal is to achieve a damping ratio of $\zeta = 1$. The closed-loop system transfer function for this case was derived previously, in Section 2.2 [see Eq. (2.26)] as

$$\frac{C(s)}{R(s)} = \frac{\omega_n^2}{s^2 + (2\zeta\omega_n + \omega_n^2 b)s + \omega_n^2}. \tag{2.52}$$

The equivalent damping ratio for this configuration was derived in Section 2.2 [see Eq. (2.28)] as

$$\zeta_{eq} = \zeta + \omega_n b/2. \tag{2.53}$$

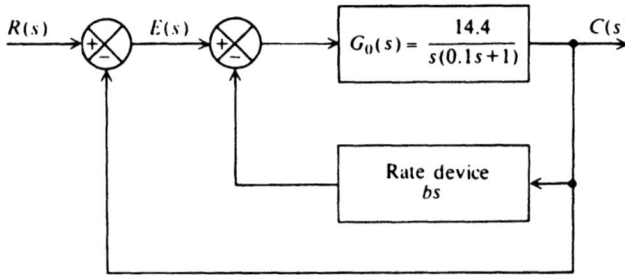

Figure 2.19 Minor-loop feedback added to the system shown in Figure 2.18a.

The object in this problem is to design $\zeta_{eq} = 1$ for the case where

$$\zeta = 0.417 \tag{2.54}$$

and

$$\omega_n = 12. \tag{2.55}$$

Substituting these values into Eq. (2.53), we find that

$$b = 0.0972. \tag{2.56}$$

Notice that b and αT are identical.

We know that the resulting system will be stable and critically damped, and will have a steady-state response error of zero for a unit step input. The steady-state response error of the system to a unit ramp input was derived [see Eq. (2.31)] as

$$e_{ss(\text{ramp input})} = \frac{2\zeta}{\omega_n} + b = 0.0695 + 0.0972 = 0.1667 \text{ units}. \tag{2.57}$$

The increase has been accounted for in Section 2.3; by using a high-pass filter in the feedback path to block a steady-state output from the rate feedback, the steady-state error can be reduced to 0.0695.

2.6. COMPENSATION AND DESIGN USING THE BODE-DIAGRAM METHOD

The techniques necessary to construct and analyze the open-loop frequency response of a feedback control system utilizing the Bode-diagram approach were presented in Section 1.7. This section illustrates how the Bode diagram can be used for designing a feedback-control system in order to meet certain specifications regarding relative stability, transient response, and accuracy. It is important to emphasize that the Bode-diagram approach is used very frequently by the practicing control engineer. Its use is due to the fact that the anticipated theoretical results may be relatively simply checked with actual performance in the laboratory just by opening the feedback loop and obtaining an open-loop frequency response of the system.

Bode's primary contribution to the control art is summarized in two theorems [7]. We introduce the concepts embodied in these theorems first in a qualitative manner, and then the mathematical statements are given.

A. Bode's Theorems

Bode's first theorem essentially states that the slopes of the asymptotic amplitude–log-frequency curve implies a certain corresponding phase shift. For example, in Section 1.7 it was shown that a slope of $20n$ dB/decade (or $6n$ dB/octave) corresponded to a phase shift of $90n°$ for $n = 0, \pm 1, \pm 2, \ldots$. Furthermore, this theorem states that the slope at crossover (where the attenuation–log-frequency curve crosses the 0-dB line) is weighted more heavily toward determining system stability than a slope further removed from this frequency. This results in a rather complex weighting factor which is a measure of relative importance toward determining system stability.

From what has been presented so far, this theorem is intuitively seen to be valid. Gain crossover frequency is one of the two points that is checked to determine the degree of stability when using the Bode diagram. Specifically, the phase shift is measured at this particular frequency in order to determine the phase margin. A feedback system whose slope at gain crossover is -20 dB/decade, and whose other slope sections are relatively far away from crossover in accordance with the relative weighting function, implies a phase shift of approximately $-90°$ in the vicinity of crossover and a corresponding phase margin of about $90°$. This value of phase margin certainly implies a stable system. A system, however, whose slope at crossover is -40 dB/decade, and whose other slope sections are relatively far away from crossover in accordance with the relative weighting function, implies a phase shift of approximately $-180°$ and a corresponding phase margin of about $0°$. This value of phase margin implies a system which is on the verge of being unstable and would probably be so when actually tested. Steeper slopes would indicate negative phase margins and definitely unstable systems. Therefore, one strives to maintain the slope of the amplitude–log-frequency curve in the area of gain crossover at a slope of -20 dB/decade. Notice that the system, illustrated in Figure 6.33‡, has slopes of -60 dB/decade that are relatively far from the gain crossover frequency. Therefore, this system has a fairly respectable phase margin of $56.74°$ by maintaining the 20 dB/decade slope for about an octave below, and about 2 octaves above crossover.

Bode's second theorem essentially states that the amplitude and phase characteristics of linear, minimum-phase-shift systems are uniquely related. When we specify the slope of the amplitude–log-frequency curve over a certain frequency interval, we have also specified the corresponding phase-shift characteristics over that frequency interval. Conversely, if we specify the phase shift over a certain frequency interval, we have also specified the corresponding amplitude–log-frequency characteristic over that frequency interval. The theorem emphasizes the fact that we can specify the amplitude–log-frequency characteristic over a certain interval of frequencies together with the phase-shift–log-frequency characteristics over the remaining frequencies. It should be emphasized that these conclusions apply only if the transfer function is minimum phase.

The second theorem may appear quite trivial at first glance. Its implications however, are quite important. We will make further use of this this theorem when designing feedback control systems using the Bode-diagram approach.

The formal *mathematical statement of Bode's first theorem* is given by the following expression:

$$\phi(\omega_d) = \frac{\pi}{2}\left|\frac{dG}{dn}\right|_d + \frac{1}{\pi}\int_{-\infty}^{\infty}\left[\left|\frac{dG}{dn}\right| - \left|\frac{dG}{dn}\right|_d\right]\ln\coth\left|\frac{n}{2}\right|dn, \qquad (2.58)$$

where $\phi(\omega_d)$ is the phase shift of the system in radians at the desired (e.g., crossover) frequency ω_d, G represents the gain in nepers (1 neper = $\ln|e|$), $n = \ln(\omega/\omega_d)$, $|dG/dn|$ represents the slope of the amplitude–log-frequency curve in nepers per unit change of n (1 neper/unit change of n is equivalent to 20 dB/decade), $|dG/dn|_d$ is the slope of the amplitude–log-frequency curve at the desired (crossover) frequency ω_d, and $\ln\coth|n/2|$ is the weighting function which is plotted in Figure

2.20. The first term of Eq. (2.58) represents the phase shift contributed by the slope of the amplitude–log-frequency curve at the reference frequency ω_d. For example, it yields a phase shift of 90° for every neper per unit of n (20 dB/decade). The second term of Eq. (2.58) is proportional to the integral of the product of the weighting function and the difference in slope of the amplitude–log-frequency curve at a frequency ω as compared to its value at the reference frequency, ω_d. Attention is drawn to the fact that it is the weighting function that determines the phase-shift contribution at ω_d due to the amplitude–log-frequency curve which exists at some frequency ω. Because the second term of Eq. (2.58) is zero for large values of n and where $n = 0$, the value of the integral will be relatively small compared with the first term if the slope of dG/dn is constant over a relatively wide range of frequencies above ω_d. Therefore, under these conditions, the phase shift would be determined primarily by the first term of Eq. (2.58). Following this line of reasoning, the slope of the amplitude–log-frequency curve should be less than -2 nepers per unit of n (-40 dB/decade) over a relatively wide range of frequencies at crossover in order to ensure stability.

The formal *mathematical statement of Bode's second theorem* is given by the following expression:

$$\int_0^{\omega_s} \frac{G\,d\omega}{\sqrt{\omega_s^2 - \omega^2}(\omega^2 - \omega_d^2)} + \int_{\omega_s}^{\infty} \frac{\phi\,d\omega}{\sqrt{\omega^2 - \omega_s^2}(\omega^2 - \omega_d^2)}$$

$$= \begin{cases} \dfrac{\pi}{2} \dfrac{\phi(\omega_d)}{\omega_d \sqrt{\omega_s^2 - \omega_d^2}} & (\text{for } \omega_d < \omega_s) \\[2ex] -\dfrac{\pi}{2} \dfrac{G(\omega_d)}{\omega_d \sqrt{\omega_d^2 - \omega_s^2}} & (\text{for } \omega_d > \omega_s) \end{cases}, \quad (2.59)$$

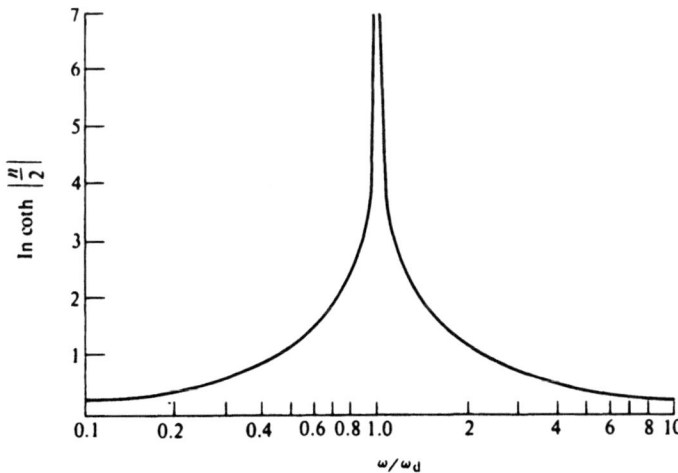

Figure 2.20 Plot of the weighting function used in Bode's first theorem.

where ω_s represents the frequency in radians per second below which the amplitude–log-frequency characteristics is specified and above which the phase characteristic is specified. This theorem emphasizes the interdependence of amplitude and phase shift over the entire range of positive frequencies. In addition, notice that although it is possible to specify amplitude or phase in one range of frequencies, and the other quantity in the remaining frequencies, these quantitites reflect their presence back into the other range of frequencies. Therefore the integration with respect to frequency is performed over the entire range of positive frequencies.

The design of several systems using the Bode-diagram approach is considered next. We shall illustrate a method that determines steady-state accuracy from the Bode diagram as well as meeting relative stability requirements.

B. Example of Phase-Lead and Phase-Lag Network Compensation, and Rate-Feedback Compensation

Let us first consider the third-order system illustrated in Figure 2.21a. Its open-loop transfer function $G_0(s)$ is given by

$$G_0(s) = \frac{80}{s(1 + 0.02s)(1 + 0.05s)}. \quad (2.60)$$

The Bode diagram for the uncompensated system $[G_c(s) = 1]$ is illustrated in Figure 2.22. It shows a gain crossover frequency of 33.78 rad/sec, a phase margin of $-3.402°$, and a gain margin of -1.1422 dB. The frequency where the phase shift equals $-180°$ (phase crossover frequency) is 31.7 rad/sec. This Bode diagram was obtained using MATLAB, and is contained in the M-file that is part of my *Advanced*

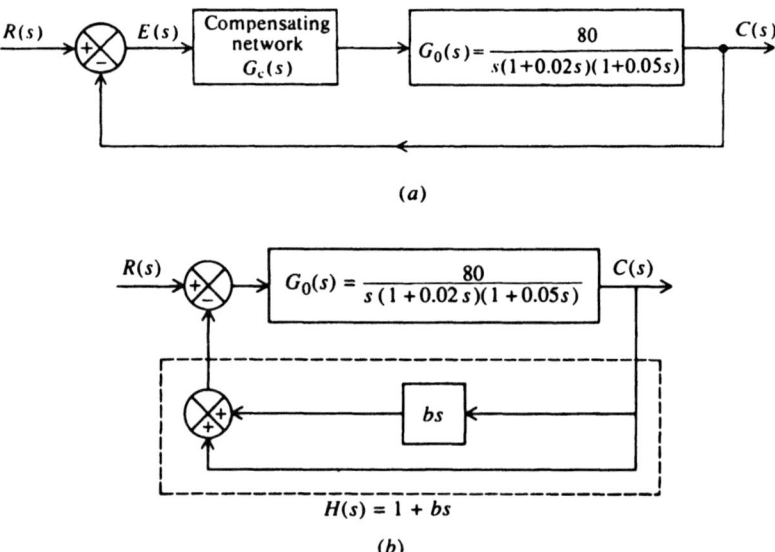

Figure 2.21 A third-order system which is to be compensated using (a) cascade compensation (phase-lag and phase-lead) and (b) rate-feedback compensation.

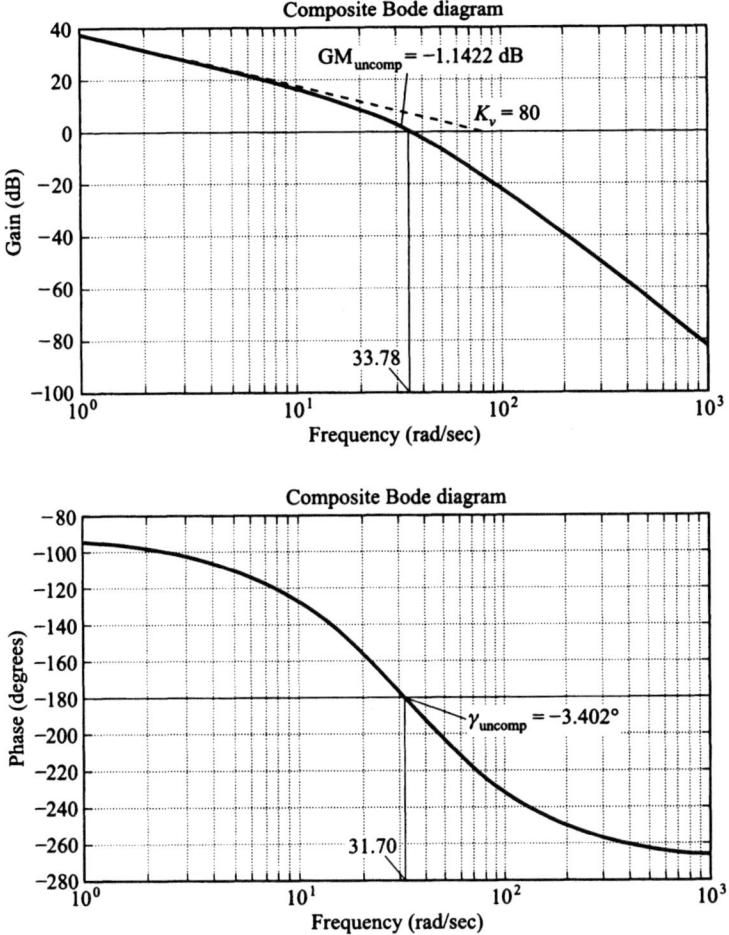

Figure 2.22 Third-order system where the uncompensated transfer function is

$$G_0(s) = \frac{80}{s(1 + 0.02s)(1 + 0.05s)}.$$

Modern Control System Theory and Design (AMCSTD) Toolbox which can be retrieved free from The MathWorks, Inc. anonymous FTP server at ftp://ftp.mathworks.com/pub/books/advshinners.

These results indicate that the uncompensated system is unstable. Let us next attempt to compensate this system with a phase-lag network, a phase-lead network, and the use of rate feedback in parallel with position feedback (which produces a pure zero factor). The specifications for this sytem require a minimum phase margin of 20° and a minimum gain margin of 10 dB. In addition, it is assumed that a sinusoidal disturbance at 1 rad/sec is present, and a gain of at least 35 dB is required at this frequency to nullify its effect.

1. Phase-Lag Network Compensation Let us first consider the phase-lag-network compensation case in Figure 2.21a. Applying Bode's theorems in order to achieve the specified phase and gain margins, we would expect that the -20 dB decade slope in the vicinity of the new crossover frequency should not extend over too wide a range of frequencies because the relative stability that is desired is rather moderate. The phase-lag network is of the form.

$$G_c(s) = \frac{1+Ts}{1+\alpha Ts} \quad (2.61)$$

where $\alpha T > T$. In Section 2.2 we studied the characteristics of the phase-lag network. In particular, Figure 2.9 illustrated the Bode diagram of a general phase-lag network. Notice that this type of network is of such a nature that it attenuated all high-frequency components above $\omega = 1/T$ by a factor of $1/\alpha$. From the Bode-diagram viewpoint, this attenuating characteristic can be used for stabilization purposes by reshaping the uncompensated amplitude characteristics, so that the initial -20 dB/decade is made to cross over the 0-dB line rather than the -40-dB/decade segment. In other words, one would attempt to stabilize this system with a phase-lag network by placing the frequencies $1/\alpha T$ and $1/T$ in the range of frequencies below about 10 rad/sec. It would be desirable that the -20-dB/decade segment start at least around 10 rad/sec cross over the 0-dB line before 20 rad/sec where the amplitude-log-frequency characteristic changes to a slope of -40 dB/decade. In addition, we would not want $1/\alpha T$ to occur at less than 1 rad/sec, because an open-loop gain of 35 dB has been specified at $\omega = 1$ rad/sec. The final phase of the solution is by means of iteration. However, the procedure converges quite rapidly. Usually, two or three iterations should prove sufficient. For the requirements specified, a phase-lag network given by

$$G_c(s) = \frac{1+0.143s}{1+s} \quad (2.62)$$

results in a gain crossover frequency of 11.371 rad/sec, a phase margin of 20.978° and a gain margin of 10.988 dB as shown in Figure 2.23. The frequency where the phase is $-180°$ (phase crossover frequency) occurs at 24.22 rad/sec.

2. Phase-Lead Network Compensation. Let us next consider the phase-lead-network compensation case. The lead network has the form

$$G_c(s) = \frac{1}{\alpha}\left(\frac{1+\alpha Ts}{1+Ts}\right), \quad (2.63)$$

where $\alpha T > T$. In Section 2.2 we studied the characteristics of the phase-lead network. We assume that any low-frequency attenuation, which is due to the value of $1/\alpha$, is made up for by increasing the gain of the feedback control system by a like amount. The Bode diagram of a general phase-lead network was shown in Figure 2.11. From the Bode-diagram viewpoint, this type of characteristic can be used for stabilization purposes by reshaping the uncompensated amplitude characteristics so that a -40-dB/decade slope is made to cross over the 0-dB line along a synthesized -20-dB/decade slope rather than the -40-dB/decade segment. The range of frequencies where one can place the frequencies $1/\alpha T$ and $1/T$ is quite limited in this

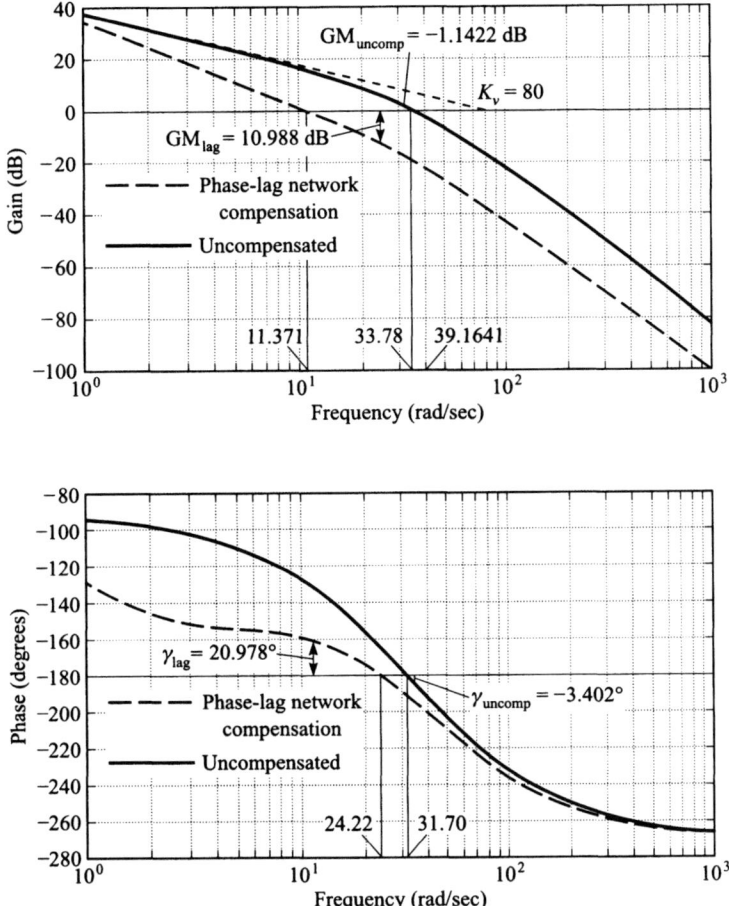

Figure 2.23 Compensation of a third-order system with a phase-lag network of $G_c(s) = (1 + 0.143s)/(1 + s)$ where the uncompensated transfer function is

$$G_0(s) = \frac{80}{s(1 + 0.02s)(1 + 0.05s)}.$$

particular problem. The value of $1/\alpha T$ can be placed between 20 and 33.78 rad/sec. The closer it is to 20 rad/sec, the greater its stabilizing effect. The further away the break $1/T$ is from 33.78 rad/sec, the greater will be the stabilizing effect of the phase-lead network. It can be seen that this particular solution does not modify the open-loop characteristics in the vicinity of 1 rad/sec and the accuracy specification of 35 dB at 1 rad/sec will easily be achieved.

We can obtain the pole and zero terms in Eq. (2.63) using a first-cut design procedure by using Eqs. (2.7) and (2.8) for the phase-lead network and the following reasoning. Equation (2.7) states the frequency where the maximum phase lead occurs, and Eq. (2.8) states the maximum value of the phase lead. In view of the analysis of the previous paragraph, it seems reasonable to place the frequency where maximum phase lead occurs around 47 rad/sec, and to make ϕ_{max} equal to the phase margin desired (a conservative choice of 28° will be selected) plus the magnitude of

58 LINEAR CONTROL-SYSTEM COMPENSATION AND DESIGN

the phase margin obtained of the uncompensated system (because it is negative) at the gain crossover frequency of 33.78 rad/sec (3.402°) plus the additional phase shift that the composite phase shift curve incurs in going from the uncompensated crossover frequency of 33.78 rad/sec to ω_{max} of 47 rad/sec (about 19°) as follows:

$$\omega_{max} = \frac{1}{T\sqrt{\alpha}} = 47 \text{ rad/sec},$$

$$\phi_{max} = \sin^{-1}\frac{(\alpha - 1)}{(\alpha + 1)} = (28° + 3.4° + 19°).$$

Solving these two equations simultaneously, we obtain the phase-lead network

$$G_c(s) = \frac{1 + 0.059s}{1 + 0.00767s}.$$

Because this solution has "blinders" on and does not consider the proximity of other poles and zeros, it usually has to be fine-tuned. Therefore, the following phase-lead network was selected and is illustrated in Figure 2.24:

$$G_c(s) = \frac{1 + 0.04s}{1 + 0.005s}. \tag{2.64}$$

It results in a phase margin of 28.431° and a gain margin of 10.655 dB. This meets the specification requirements. Notice that the resulting phase-lead network has a low-frequency attenuation of 0.005/0.04 which must be made up by increasing the gain by a factor of 0.04/0.005 = 8. Its gain crossover frequency is 47.27 rad/sec, and the phase crossover frequency occurs at 94.06 rad/sec.

If we have a choice of using the phase-lag or phase-lead network, the phase-lag network solution would be preferable because it meets the required specifications with a narrower bandwidth than the phase-lead network case ($\omega_c = 11.371$ rad/sec for the former; $\omega_c = 47.27$ rad/sec for the latter). A feedback control system having a narrower bandwidth will reject a greater amount of noise than one having a wider bandwidth, as well as requiring less power and cost. In addition, the phase-lead network has the disadvantage of requiring a greater amount of amplification within the control system than the phase-lag network. These and other considerations are discussed further in Section 2.10, where tradeoffs of using different kinds of cascade networks and minor-loop feedback compensation methods are compared.

3. Rate Feedback in Parallel with Position Feedback Compensation. Finally, we wish to compensate the system using rate-feedback compensation in parallel with position feedback as shown in Figure 2.21b. The addition of rate feedback provides an $H(s)$ term of

$$H(s) = (1 + bs)$$

which represents a pure zero term that provides a phase lead of

$$\tan^{-1} b\omega = \phi_{max}.$$

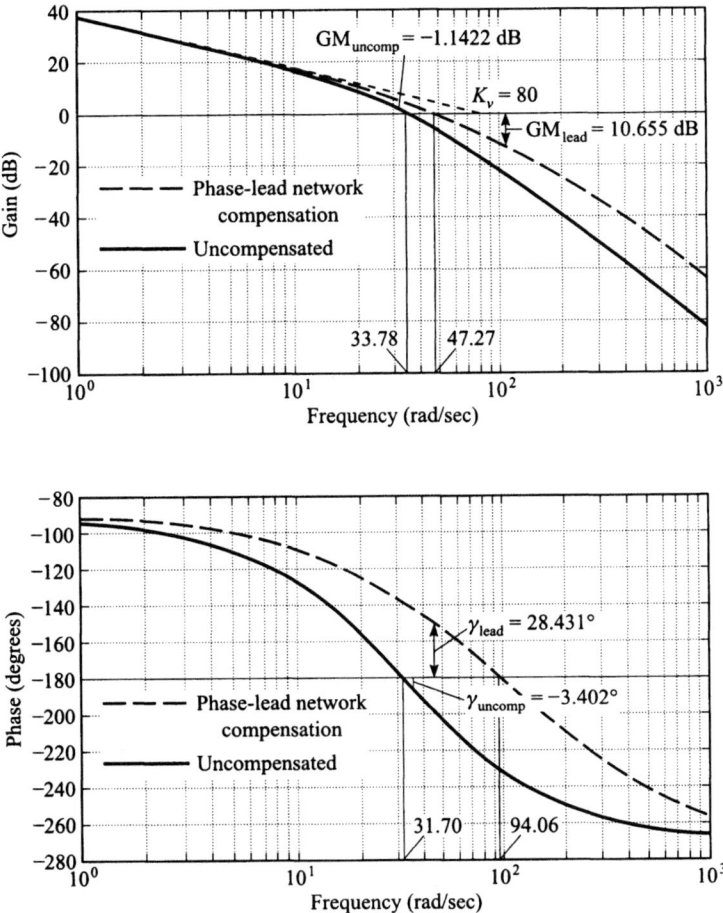

Figure 2.24 Compensation of a third-order system with a phase-lead network $G_c(s) = (1 + 0.04s)/(1 + 0.005s)$ where the uncompensated transfer function is

$$G_0(s) = \frac{80}{s(1 + 0.02s)(1 + 0.05s)}.$$

Remember that a pure zero term is an ideal compensator. Using a similar procedure as for the design of the phase-lead network, let us assume that we desire an approximate 50° phase lead at a frequency of 50 rad/sec. Therefore,

$$b(50) = \tan 50°$$
$$b = 0.0238.$$

The use of rate-feedback compensation is illustrated in Figure 2.25 and indicates a phase margin of 31.9728° at the gain crossover frequency of 39.1641 rad/sec. Its gain margin is infinity, because the phase never is −180° except at $\omega = $ infinity.

Figure 2.26 illustrates the Bode diagram of the uncompensated system, and the Bode diagram of the compensated systems with a phase-lag network, phase-lead

60 LINEAR CONTROL-SYSTEM COMPENSATION AND DESIGN

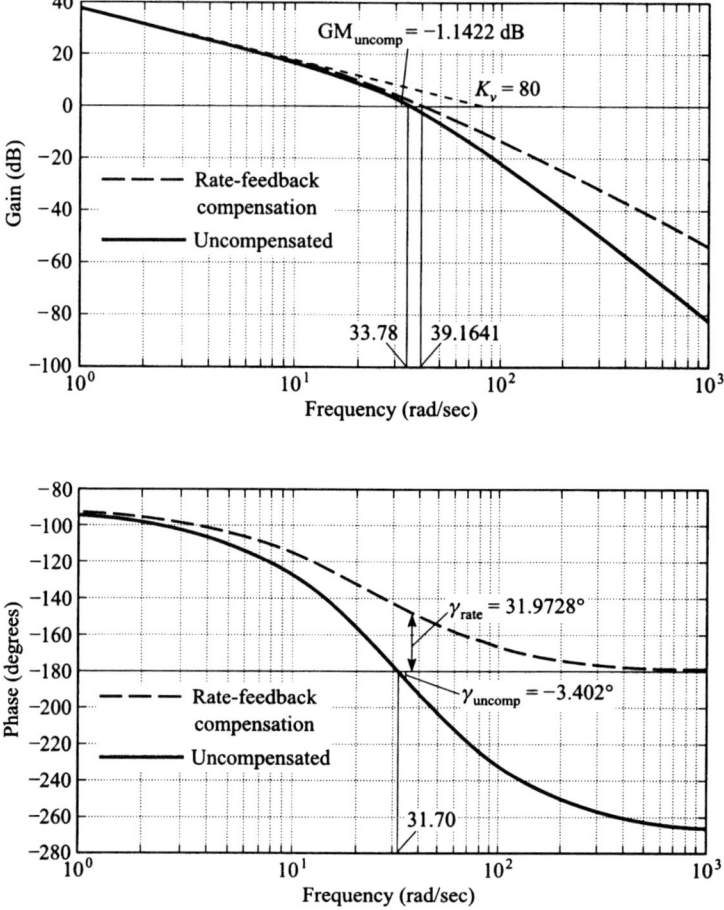

Figure 2.25 Compensation of third-order system with rate feedback ($b = 0.0238$) in parallel with position feedback where the uncompensated transfer function is

$$G_0(s) = \frac{80}{s(1 + 0.02s)(1 + 0.05s)}$$

network, and with rate feedback in parallel with position feedback all superimposed on one Bode diagram.

C. Obtaining Steady-State Error Coefficients from the Bode Diagram

The steady-state error coefficients can be determined from the Bode diagram. The definition and importance of these error coefficients have been discussed in Section 5.4[‡]. For a system having one pure integration, the velocity constant K_v can be obtained by extending the initial -20-dB/decade slope until it intersects the 0-dB line. The frequency at which it intersects this line is equal to the velocity constant. In the discussion in Appendix C, K_v is obtained by letting s approach zero when utilizing the final-value theorem [see Eq. (C.11)]. Therefore, the pole and/or zero terms having the forms $(1 + Ts)$ or $[(Ts)^2 + 2\zeta Ts + 1]$ all approach unity. This per-

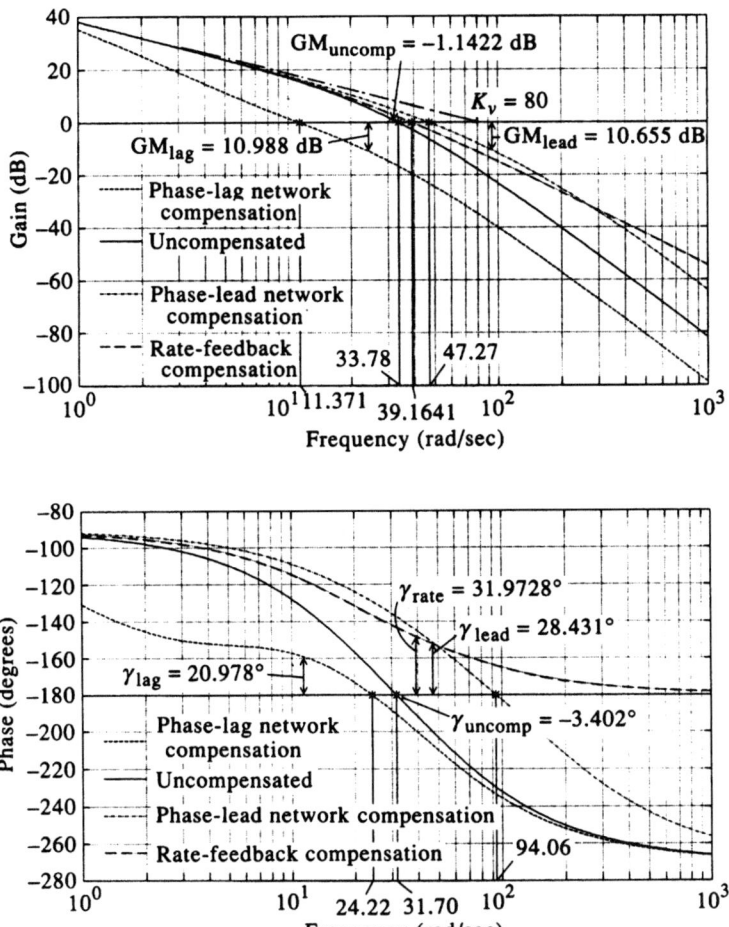

Figure 2.26 Compensation of a third-order system where the uncompensated transfer function is

$$G_0(s) = \frac{80}{s(1+0.02s)(1+0.05s)}.$$

mits one to obtain K_v directly by considering only the initial slope of the Bode diagram. For the Bode diagram shown in Figures 2.22–2.26, the value K_v obtained graphically is 80. As a check, using the definition given by Eq. (C.11), we obtain

$$K_v = \lim_{s \to 0} sG(s)H(s). \quad (2.65)$$

Because

$$G(s) = \frac{80}{s(1+0.02s)(1+0.05s)}, \quad (2.66)$$

$$H(s) = 1, \quad (2.67)$$

then,

$$K_v = \lim_{s \to 0} \frac{s(80)}{s(1+0.02s)(1+0.05s)} = 80. \qquad (2.68)$$

It is also interesting to note that the velocity constant is the same for the uncompensated and compensated systems.

For a system which has a double pole at the origin, the acceleration constant K_a can be obtained in a similar manner. The initial -40-dB/decade slope is extended until it intersects the 0-dB line. The square of the frequency at which it intersects this line is equal to the acceleration constant.

D. Example of Compensating a System Containing a Disturbance Input

The next system we consider is illustrated in Figure 2.27. This consists of a third-order system which has an unwanted external input $U(s)$. The open-loop transfer function for the uncompensated system, $G_0(s)$, is given by

$$G_0(s) = \frac{2.2}{(1+0.1s)(1+0.4s)(1+1.2s)}. \qquad (2.69)$$

It is desired that the steady-state error resulting from an unwanted, external step input signal at $U(s)$, should not exceed 0.1 unit. The compensation device, $G_c(s)$, is to contain amplification which will meet this accuracy requirement together with a phase-lead or phase-lag network which will provide a minimum phase margin of $20°$ and a minimum gain margin of 6 dB.

The value of gain K required for $G_c(s)$ will be computed first. For this calculation, $U(s)$ is assumed to be the input and $E(s)$ is assumed to be the output. The transfer function between these two points is given by

$$\frac{E(s)}{U(s)} = \frac{2.2(1+T_2s)}{(1+T_2s)(1+0.1s)(1+0.4s)(1+1.2s) + 2.2K(1+T_1s)}. \qquad (2.70)$$

Setting $U(s) = 1/s$, we obtain the expression for the Laplace transform of the error, $E(s)$, as

$$E(s) = \frac{2.2(1+T_2s)}{s[(1+T_2s)(1+0.1s)(1+0.4s)(1+1.2s) + 2.2K(1+T_1s)]}. \qquad (2.71)$$

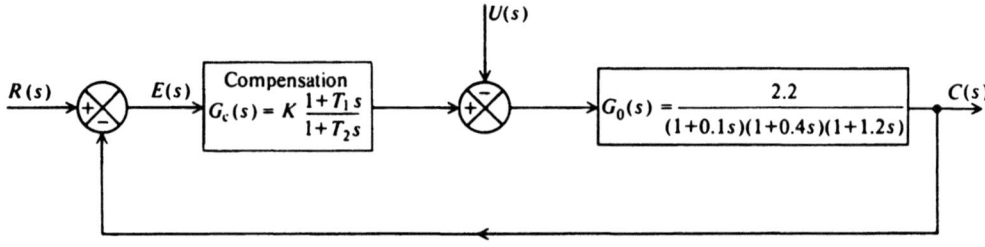

Figure 2.27 A third-order system having an unwanted external input.

2.6. COMPENSATION AND DESIGN USING THE BODE-DIAGRAM METHOD

The required value of K can be obtained by applying the final-value theorem to $E(s)$ and setting the result equal to 0.1 unit. We find

$$K = 9.55. \qquad (2.72)$$

Let us next determine the compensating network required to achieve a phase margin of 20° and a gain margin of 6 dB. The transfer function of the uncompensated system, with $K = 9.55$, is given by

$$KG_0(s) = \frac{21}{(1+0.1s)(1+0.4s)(1+1.2s)}. \qquad (2.73)$$

Its Bode diagram which is drawn in Figure 2.28, indicates a phase margin of 0.7167° (gain crossover frequency is 5.8668 rad/sec) and a gain margin of 0.267 dB (phase

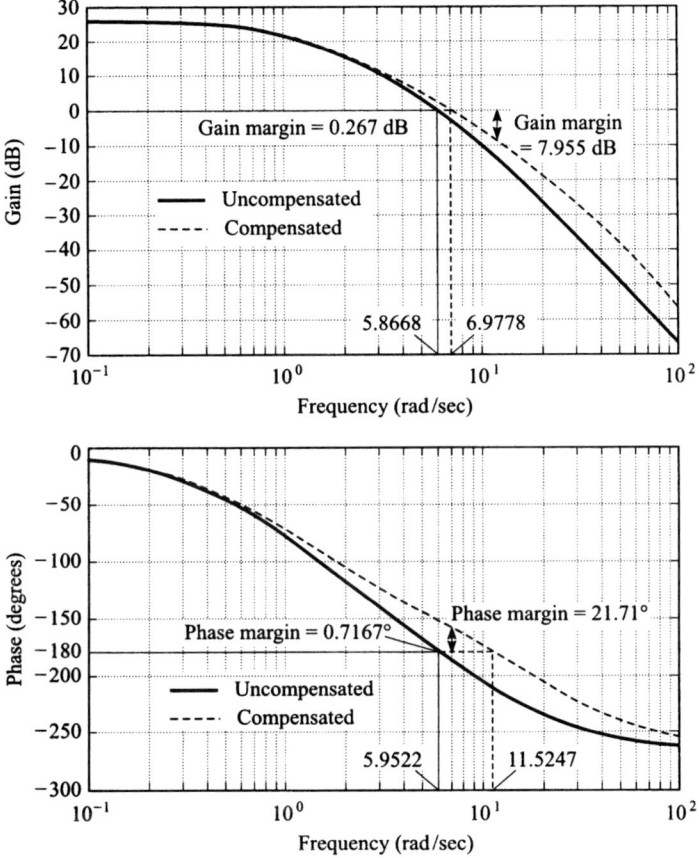

Figure 2.28 Compensation of a third-order system where

$$KG_0(s) = \frac{21}{(1+0.1s)(1+0.4s)(1+1.2s)} \text{ and } G_c(s) = \frac{1+0.167s}{1+0.05s}.$$

Solid line = uncompensated; dashed line = compensated

crossover frequency is 5.9522 rad/sec). We wish to increase the phase margin by about 20°. This Bode diagram was obtained using MATLAB and is contained in the M-file that is part of my AMCSTD Toolbox and can be retrieved free from The MathWorks, Inc. anonymous FTP server at ftp://ftp.mathworks.com/pub/books/advshinners.

As in the previous problem analyzed in this section, we will estimate the values of ω_{max} and ϕ_{max} for Eqs. (2.7) and (2.8), respectively, to obtain a first-cut design. We must be careful when using this approach, because we actually want to have this phase shift at the gain crossover frequency in order to achieve the desired phase margin. However, specifying α, T, and ω_{max} does not ensure that ϕ_{max} will be at the gain crossover frequency. After obtaining this first-cut design, we then fine-tune the results to obtain the specification requirements as in the previous example. For this design and an analysis of the Bode diagram of the uncompensated system shown in Figure 2.28 a value of $\omega_{max} = 8$ rad/sec was selected. The value of ϕ_{max} selected was a conservative value of 34° which includes the difference between the minimum phase margin desired (20°) and the actual phase margin of the uncompensated system (0.7167°) plus the difference in phase shift between the gain crossover frequency of the uncompensated system (5.8668 rad/sec) and 8 rad/sec plus an extra margin to be conservative. The resulting two equations to be solved simultaneously were

$$\omega_{max} = 1/(T\sqrt{\alpha}) = 8,$$
$$\phi_{max} = \sin^{-1}\frac{\alpha - 1}{\alpha + 1} = 34°.$$

The result is a phase-lead network whose transfer function is

$$\frac{1 + 0.2355s}{1 + 0.0663s}.$$

The first-cut design was adjusted a little to better fit the overall Bode diagram and the final phase-lead network selected is as follows:

$$\frac{1 + 0.167s}{1 + 0.05s}. \tag{2.74}$$

This network causes a gain crossover frequency of 6.9778 rad/sec where the phase margin is 21.71°, and a gain margin of 7.955 dB (the phase crossover frequency is at 11.5247 rad/sec). This satisfies the specification. Remember that this network will require an increase in the amplifier gain by 3.34 to make up for the attentuation of $0.05/0.167 = 0.2994$ that it provides.

E. Example for Compensating a Two-Loop System Using Rate Feedback in Cascade with a Filter

The concluding problem we consider using the Bode-diagram approach consists of designing the feedback control system illustrated in Figure 2.29a. For this particular system, we desire that the steady-state error resulting from a velocity input of 110

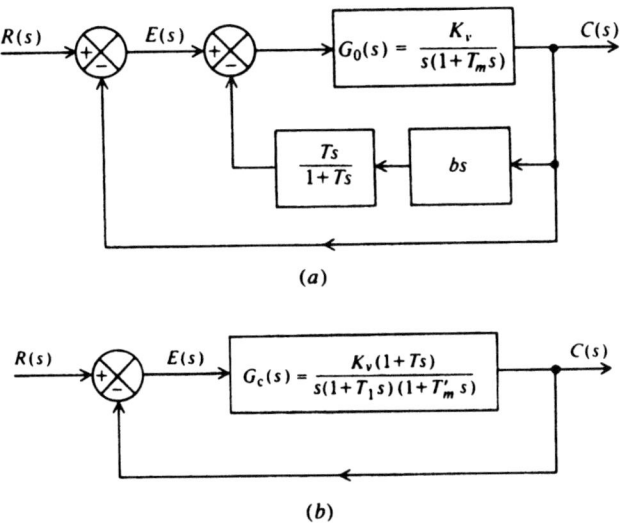

Figure 2.29 (a) Use of rate feedback in cascade with a high-pass filter in order to compensate a feedback control system. (b) Equivalent block diagram for the system shown in (a).

rad/sec be equal to 0.25 rad. The uncompensated open-loop transfer function $G_0(s)$ is given by

$$G_0(s) = K_v/(s(1 + T_m s)), \tag{2.75}$$

where K_v = velocity constant and T_m = motor time constant = 0.025 sec. This transfer function consists of an amplifier, positioning motor, gear train, and load. In order to achieve the required accuracy, K_v must equal

$$K_v = \frac{\omega}{\text{error}} = \frac{110 \text{ rad/sec}}{0.25 \text{ rad}} = 440/\text{sec}. \tag{2.76}$$

We want a phase margin of approximately 55° for this system. This will be achieved by means of minor-loop rate-feedback compensation which is cascaded with a simple RC high-pass filter (phase-lead network) in order not to increase the steady-state response error for velocity inputs. This was discussed in Section 2.3 (see Figure 2.15).

The open-loop frequency response we must plot on the Bode diagram is that obtained with the minor-rate loop closed and the outer position loop opened. Therefore, we are interested in obtaining the equivalent transfer function between $E(s)$ and $C(s)$. This is easily found to be

$$\frac{C(s)}{E(s)} = \frac{K_v(1 + Ts)}{s[(1 + T_m s)(1 + Ts) + K_v bTs]}. \tag{2.77}$$

Expanding the denominator of Eq. (2.77), we obtain the expression

$$s[T_m T s^2 + (T_m + T + K_v bT)s + 1]. \tag{2.78}$$

66 LINEAR CONTROL-SYSTEM COMPENSATION AND DESIGN

This expression can be put into the form

$$s[(1 + T'_m s)(1 + T_1 s)]. \tag{2.79}$$

by defining the time constants T'_m and T_1 as

$$T'_m = \frac{T_m}{T_1} T \tag{2.80}$$

and

$$T_1 = -T'_m + T_m + (1 + K_v b)T. \tag{2.81}$$

Therefore, we may redraw Figure 2.29a as shown in Figure 2.29b. For any set of values for T_m, T, K_v, and b, we can derive T'_m and T_1 by solving the simultaneous equations (2.80) and (2.81). Another approach is to choose T and T_1 from the Bode diagram which meets the specified phase margin and solve for the required rate-feedback constant b.

The procedure we follow when compensating this system is to draw the Bode diagram for the uncompensated system in accordance with Eq. (2.75) and then fit the characteristics of the compensated system in accordance with

$$G_c(s) = \frac{C(s)}{E(s)} = \frac{K_v(1 + Ts)}{s(1 + T'_m s)(1 + T_1 s)} \tag{2.82}$$

until a phase margin of 55° is achieved. The compensated characteristic will then determine T and T_1, from which T'_m and the rate-feedback constant b can be determined. Figure 2.30 illustrates the Bode diagram of the uncompensated and compensated systems. The phase margin of the uncompensated system is 17.1°. Values of

$$T_1 = 0.33, \tag{2.83}$$
$$T = 0.033, \tag{2.84}$$

and

$$T'_m = \frac{T_m}{T_1} T = 0.0025 \quad \text{[from Eq. (2.80)]} \tag{2.85}$$

result in a phase margin of 55.36°. From Eq. (2.81), the corresponding value of rate-feedback constant b is 0.0186. Figure 2.30 was obtained using MATLAB and is contained in the M-file that is part of my AMCSTD Toolbox.

The type of compensation just illustrated is used quite frequently in practice. In order to really understand what is actually happening, it is important to examine the Bode diagram of Figure 2.30 closely. The net effect of the minor-loop rate feedback has been to move the equivalent motor break frequency from $1/T_m$ to $1/T'_m$ by the ratio given in Eq. (2.80). This technique is used quite frequently to compensate power servos. The net effect of the phase-lead network, in the minor-loop feedback path, is to appear as a phase lag, for the equivalent open-loop characteristics of Figure 2.29b. This can be easily understood because we effectively see the reciprocal of the feedback element when looking into a closed-loop system which has an open-

2.7. APPROXIMATE METHODS FOR PRELIMINARY COMPENSATION AND DESIGN USING THE BODE DIAGRAM

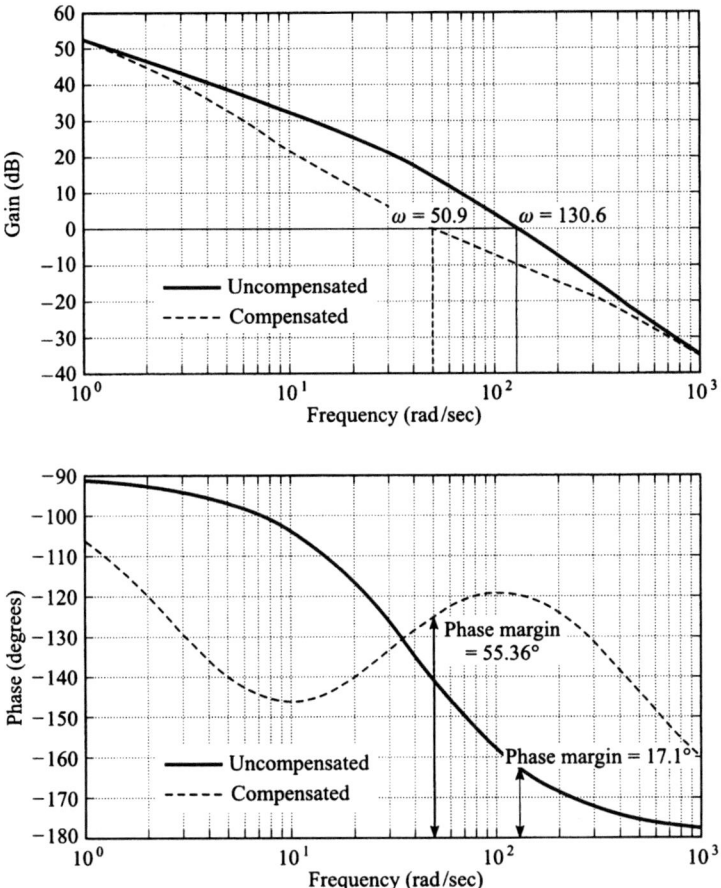

Figure 2.30 Compensation of the system shown in Figure 2.29b.

$$G_0(s) = \frac{440}{s(1 + 0.025s)} \text{ and } G_c(s) = \frac{440(1 + 0.033s)}{s(1 + 0.33s)(1 + 0.0025s)}.$$

loop gain much greater than unity [see Eq. (2.122)‡]. This is a very important fact that can be utilized to approximate the Bode diagram in preliminary designs. This point is now expanded upon in the following section.

The complete case study for the design of an angular control system for a robot's joint is presented in Section 7.4 of Chapter 7. In this problem, the Bode diagram is used for analyzing and designing the control system.

2.7. APPROXIMATE METHODS FOR PRELIMINARY COMPENSATION AND DESIGN USING THE BODE DIAGRAM

Having presented detailed compensation methods, let us next focus our attention on approximate methods for obtaining a first cut at compensation utilizing the Bode diagram [8–10]. Although the procedures presented in this section are approximate, they are generally adequate for the preliminary stage of design. Before the system design is completed, however, the exact magnitude and phase curves should be drawn as indicated previously. The practice of utilizing approximate methods for preliminary design is generally employed as a convenience in obtaining significant time constants, gains, and phase characteristics required for a design.

A. Approximate Closed-Loop Response from the Bode Diagram

In order to develop the concept of this approach, let us consider the feedback system illustrated in Figure 2.31. The closed-loop transfer function of this system is given by

$$\frac{C(s)}{R(s)} = \frac{G(s)}{1 + G(s)H(s)}. \tag{2.86}$$

Let us modify this equation into the following more convenient form:

$$\frac{C(s)}{R(s)} = \frac{1}{H(s)}\left[\frac{G(s)H(s)}{1 + G(s)H(s)}\right]. \tag{2.87}$$

As shown previously, the magnitude (in decibels) of $G(s)H(s)$ can be approximated by straight lines of constant slope when plotted against frequency on a log scale. Therefore, it appears reasonable that the term $1 + G(s)H(s)$ in Eq. (2.87) may also be dealt with in a similar approximate manner as follows:

$$1 + G(s)H(s) \approx 1 \qquad \text{for } |G(s)H(s)| < 1, \tag{2.88}$$
$$1 + G(s)H(s) \approx G(s)H(s) \qquad \text{for } |G(s)H(s)| > 1. \tag{2.89}$$

Substituting these approximations into Eq. (2.87), we find that

$$\frac{C(s)}{R(s)} \approx G(s) \qquad \text{for } |G(s)H(s)| < 1, \tag{2.90}$$

$$\frac{C(s)}{R(s)} \approx \frac{1}{H(s)} \qquad \text{for } |G(s)H(s)| > 1. \tag{2.91}$$

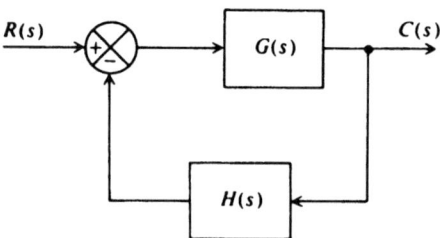

Figure 2.31 A feedback control system.

2.7. APPROXIMATE METHODS FOR PRELIMINARY COMPENSATION AND DESIGN USING THE BODE DIAGRAM

The approximations of Eqs. (2.90) and (2.91) are very useful as a first cut in the preliminary design of a control system. However, it is important to point out that they are approximate relationships, and are subject to error particularly at those frequencies where $G(s)H(s) = 1$. However, the amount by which the approximations are in error can be calculated, and this correction can then be applied to correct the approximate value. The resultant corrected value is then an exact solution.

The technique of applying the approximation of Eqs. (2.90) and (2.91) can best be illustrated through an example. Consider the feedback system of Figure 2.31, where

$$G(s) = \frac{K}{1 + Ts} \quad (2.92)$$

and

$$H(s) = 1. \quad (2.93)$$

This example will illustrate how feedback, around an element containing a time constant, reduces the time constant of that element. Substituting Eqs. (2.92) and (2.93) into Eq. (2.86), the resultant system transfer function is given by

$$\frac{C(s)}{R(s)} = \frac{K/(1+Ts)}{1 + K/(1+Ts)} = \frac{K}{1+K} \frac{1}{1 + Ts/(1+K)}. \quad (2.94)$$

For purposes of illustration, it is assumed that

$$K = 10, \quad (2.95)$$
$$T = 1 \quad (2.96)$$

in this example. For these values, Figure 2.32 illustrates the straight-line approximations for $|C(s)/R(s)|$, $|G(s)|$, and $|G(s)H(s)|$. The dashed curve representing $G(s)$ [and $G(s)H(s)$ as $H(s) = 1$] has a gain of 10 (20 dB) for frequencies lower than $\omega = 1/T = 1$ rad/sec. At frequencies higher than $\omega = 1$ rad/sec, $G(s)$ and $G(s)H(s)$ have an attentuation of 20 dB/decade. At $\omega = \omega_n = K/T = \frac{10}{1} = 10$, $G(s)$ and $G(s)H(s)$ approximately equal 0 dB. Equation (2.91) indicates that for frequencies lower than $\omega_n = 10$, the system transfer function $C(s)/R(s)$ equals $1/H(s) = 1$ and is shown by the solid line at 0 dB. For frequencies greater than ω_n, Eq. (2.90) indicates that $C(s)/R(s)$ is equal to $G(s)$ as shown. Figure 2.32 illustrates very clearly that the use of unity-gain feedback around a single time-constant element results in a reduction in its time constant. This admirable characteristic is achieved at the expense of a loss of gain.

It is important to recognize that the results obtained for the closed-loop transfer function are approximate. For example, Eq. (2.94) indicates that the gain is actually $K/(1 + K) = 10/(1 + 10) = 0.909$ instead of 1 for $\omega < \omega_n$, and the closed-loop time constant is $T/(1 + K) = 1/(1 + 10) = 0.0909$ instead of $T/K = \frac{1}{10} = 0.1$. Note that these errors decrease as the gain K increases. Generally, K is quite large and the error between the approximate and exact curves is very small.

The form of Eq. (2.87) is very well suited for determining the value of $C(s)/R(s)$ when $|G(s)H(s)| > 1$, with the bracketed term representing the difference between the

70 LINEAR CONTROL-SYSTEM COMPENSATION AND DESIGN

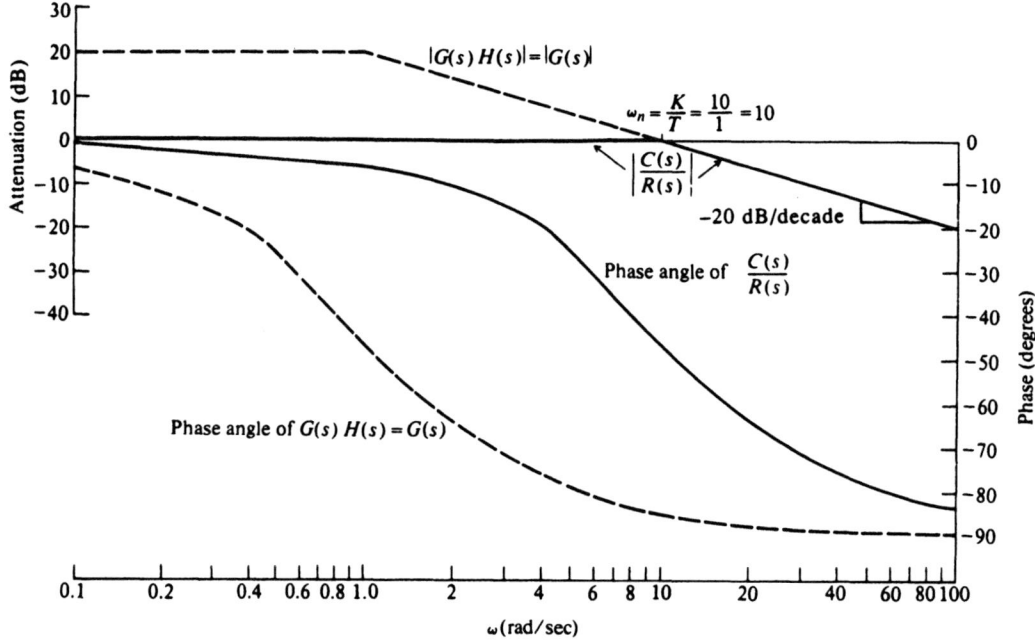

Figure 2.32 Open- and closed-loop gain and phase characteristics for the system of Figure 2.31 where $G(s) = 10/(1 + s)$ and $H(s) = 1$.

approximate and exact curves. In order to determine a general analytic expression for finding the difference between the approximate and exact curves when $|G(s)H(s)| < 1$, let us reconsider Eq. (2.86) and rewrite it as follows:

$$\frac{C(s)}{R(s)} = G(s)\frac{1}{1 + G(s)H(s)}. \quad (2.97)$$

Therefore,

$$\frac{C(s)}{R(s)} = G(s)\left[\frac{1/[G(s)H(s)]}{1 + 1/[G(s)H(s)]}\right]. \quad (2.98)$$

For those frequencies where $|G(s)H(s)| < 1$, the bracketed term of Eq. (2.98) represents the error between the approximate solution

$$\frac{C(s)}{R(s)} \approx G(s) \quad \text{for } |G(s)H(s)| < 1 \quad (2.99)$$

and the exact solution. Another way of looking at this is that the bracketed term represents a correction factor which can be used to correct the results of the approximation given by Eq. (2.90).

B. The Straight-Line Phase-Shift Approximation

For preliminary design purposes, it is usually sufficient to obtain quantitative information regarding phase shift without resorting to the exact but tedious method of

2.7. APPROXIMATE METHODS FOR PRELIMINARY COMPENSATION AND DESIGN USING THE BODE DIAGRAM

calculating the phase. We have found it convenient to represent the amplitude characteristics by a straight-line approximation, and can utilize a similar technique for the phase-shift function. In order to introduce the method, let us consider the phase shift due to the transfer function representing a zero factor given by

$$G(j\omega) = 1 + j\left(\frac{\omega}{\omega_1}\right). \tag{2.100}$$

Figure 2.33a illustrates the exact and straight-line approximation and Figure 2.33b illustrates the exact phase shift and its straight-line approximation. The straight-line approximate phase shift has been constructed by drawing a straight line tangent to the actual curve at $\omega = \omega_1$.

In order to construct these straight-line phase-shift approximations by inspection, the dependence of ω_A and ω_B on ω_1 must be known. This can be obtained by first considering the slope of the actual curve at $\omega = \omega_1$ as follows:

$$\phi = \tan^{-1}\frac{\omega}{\omega_1}. \tag{2.101}$$

The slope at $\omega = \omega_1$ can be obtained by differentiating this expression:

$$\frac{d\phi}{d(\ln \omega)}\bigg]_{\omega=\omega_1} = \frac{d\phi}{d\omega}\frac{d\omega}{d(\ln \omega)}\bigg]_{\omega=\omega_1} = \frac{\omega/\omega_1}{1+(\omega/\omega_1)^2}\bigg]_{\omega=\omega_1} = \frac{1}{2}. \tag{2.102}$$

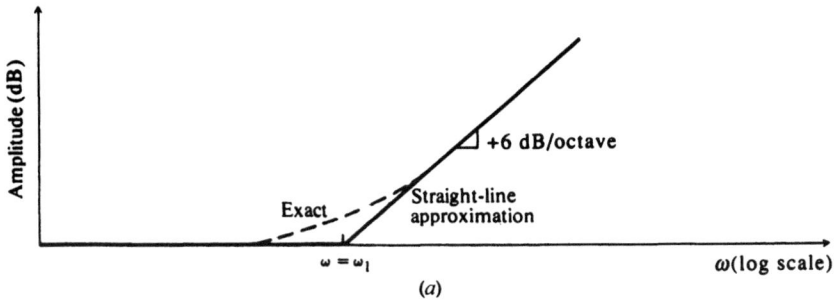

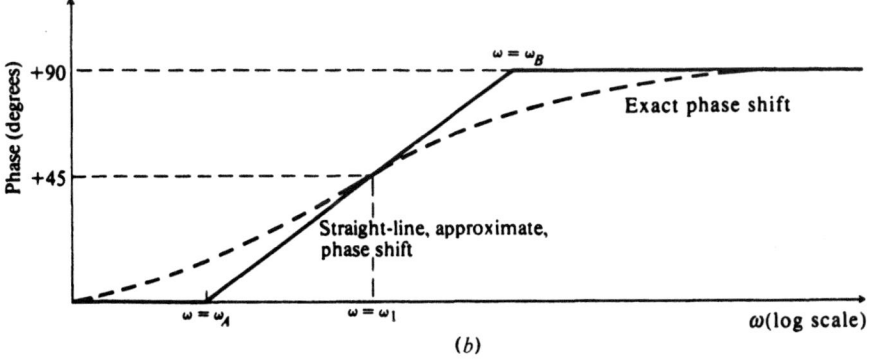

Figure 2.33 Straight-line, approximate, phase-shift methods for a zero factor term.

72 LINEAR CONTROL-SYSTEM COMPENSATION AND DESIGN

Knowing the slope at $\omega = \omega_1$, the intercepts at ω_A and ω_B can be obtained from

$$\frac{\pi/4}{\ln \omega_1 - \ln \omega_A} = \frac{1}{2}, \quad (2.103)$$

which gives

$$\frac{\pi}{4} = \frac{1}{2} \ln \frac{\omega_1}{\omega_A}. \quad (2.104)$$

Therefore,

$$\frac{\omega_1}{\omega_A} = \frac{\omega_B}{\omega_1} = e^{\pi/2} = 4.81. \quad (2.105)$$

It is interesting to note that if a number other than e were chosen for the logarithmic base, the slope in Eq. (2.105) would change, but the frequency ratio given by Eq. (2.105) would remain the same.

Based on this result, complicated approximate straight-line phase-shift curves can be obtained. Figure 2.34 illustrates the application of this approach for a transfer function given by

$$G(s) \frac{(1 + 10s)(1 + 0.01s)^2}{(1 + s)^2(1 + 0.001s)} \quad (2.106)$$

and compares the approximate straight-line phase-shift curve with the exact phase-shift curve. Observe that the errors obtained by using the approximate phase curve are larger than the corresponding errors of the amplitude approximation. However, the use of this approximation greatly aids the control-system engineer in obtaining a first cut preliminary design.

The example just presented contained roots that were all real. What happens if the transfer function contains underdamped quadratic factors? In order to analyze this situation, let us consider the following normalized quadratic phase-lag factor ($\zeta < 1$):

$$G(j\omega) = \frac{1}{(j\omega/\omega_n)^2 + 2\zeta(j\omega/\omega_n) + 1}. \quad (2.107)$$

It was shown in Section 6.7[‡] that the amplitude and phase-shift terms are given respectively by [see Eq. (6.92)[‡]]:

$$-20 \log_{10} \left[\left(\frac{2\zeta\omega}{\omega_n}\right)^2 + \left(1 - \frac{\omega^2}{\omega_n^2}\right)^2 \right]^{1/2}, \quad (2.108)$$

$$\phi = -\tan^{-1} \frac{2\zeta\omega_n\omega}{\omega_n^2 - \omega^2}. \quad (2.109)$$

2.7. APPROXIMATE METHODS FOR PRELIMINARY COMPENSATION AND DESIGN USING THE BODE DIAGRAM

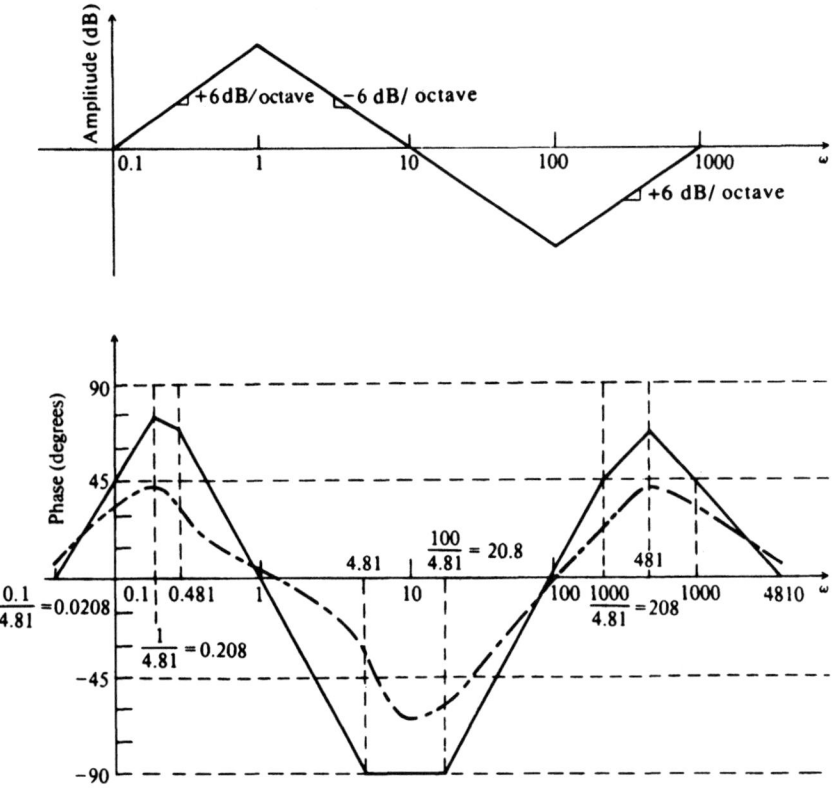

Figure 2.34 Approximate straight-line phase-shift (solid line) and exact phase-shift (short dash–long dash line) characteristics of

$$G(s) = \frac{(1+10s)(1+0.01s)^2}{(1+s)^2(1+0.001s)}.$$

Let us focus attention on the phase-shift term. Differentiation of Eq. (2.109) results in the slope at $\omega = \omega_n$ being given by

$$\text{slope} = \frac{d\phi}{d(\ln \omega)} = \frac{1}{\zeta}. \tag{2.110}$$

As before, the two intercepts ω_A and ω_B can be found as follows:

$$\frac{\pi/2}{\ln \omega_n - \ln \omega_A} = \frac{1}{\zeta}. \tag{2.111}$$

Therefore,

$$\frac{\omega_n}{\omega_A} = \frac{\omega_B}{\omega_n} = e^{(\pi/2)\zeta} = 4.81^\zeta. \tag{2.112}$$

Exact phase characteristics corresponding to Eq. (2.107) and its straight-line approximation, based on the relationship of Eq. (2.112), are illustrated in Figure 2.35. In

74 LINEAR CONTROL-SYSTEM COMPENSATION AND DESIGN

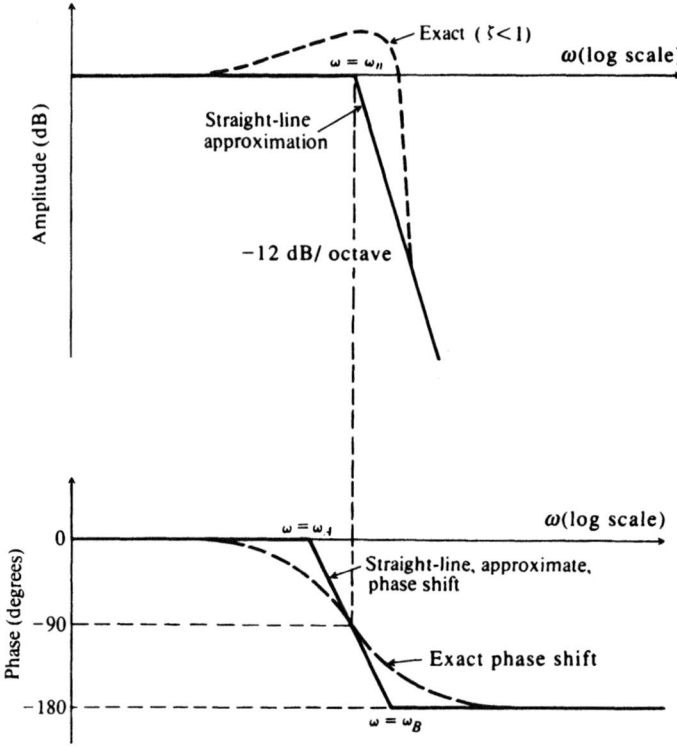

Figure 2.35 Straight-line approximate phase-shift method for a quadratic lag factor ($\zeta < 1$).

addition, the exact amplitude characteristics of Eq. (2.108) and its straight-line approximation are also illustrated. Observe from Eq. (2.112) that as ζ approaches zero, the two frequency ratios decrease and approach unity in the limits. This certainly agrees with the exact phase characteristics of a quadratic phase lag as illustrated in Figure 6.28[‡]. On the other hand, when $\zeta = 1$, the roots become real and Eq. (2.112) reduces to Eq. (2.105).

The reader is again reminded that the straight-line phase-shift approximation is not exact. It should only be used in order to obtain a first cut for preliminary design work. It should also be noted that the errors of the straight-line phase-shift approximation are generally larger than the corresponding errors of the amplitude approximation. For this reason, the phase-shift approximation is not used for final design. In the final design, the actual phase-shift characteristics must be employed as previously illustrated.

2.8. COMPENSATION AND DESIGN USING THE NICHOLS CHART

The Nichols-chart method has been developed in Section 1.7. We demonstrated in that section how one could obtain the closed-loop frequency response of a feedback control system by superimposing the open-loop gain-phase characteristics onto the Nichols chart. Specifically, we obtained the closed-loop frequency response of the system shown in Figure I1.7-1. The intersections of the open-loop gain–phase characteristics and the Nichols closed-loop gain characteristics were shown in Figure I1.7-2. The resulting closed-loop frequency response was illustrated in Figure I1.7-3. It indicated a maximum value of peaking M_p of 6.758 dB (2.2) and the frequency at which it occurred, ω_p, was 8.989 rad/sec. A value of $M_p = 6.758$ dB (2.2) does not represent a good design. This section demonstrates how the control engineer may use the Nichols chart in order to achieve a specified performance.

Let us assume for this problem that an acceptable value of M_p is 1.4 (2.92 dB). This may be achieved by adding a phase-lag or phase-lead network in cascade with the forward-loop transfer function $G(s)$. A phase-lag network is used for this problem although a solution can be found as easily using a phase-lead network. We shall demonstrate that for an M_p of 2.92 dB the object is to modify the gain–phase characteristics on the Nichols chart so that it is just tangent to the 2.92 dB locus and does not enter it. By restricting the gain-phase characteristics to areas external to the $M = 2.92$ dB locus, we will have limited M_p to 2.92 dB, because the interior of this locus represents values of M greater than 2.92 dB.

Studying the characteristics of Figure I1.7-2, we see that relatively large magnitudes of $G(j\omega)$ exist for $\omega < 3$ rad/sec. Therefore, it is not desirable to shift these magnitudes inside the $M_p = 2.92$ dB curve. In addition, it is desirable to attenuate $G(j\omega)$ by a factor of about 3 in the range of frequencies of $\omega = 5 - 12$ rad/sec. A phase-lag network, $(1 + Ts)/(1 + \alpha Ts)$, whose factor α equals 3 will achieve this if $\omega_{max} = 1/(T\sqrt{\alpha})$ is chosen at about 1 rad/sec. Solving for αT and T, we get $\alpha T = 1.74$ and $T = 0.58$.

In order to obtain the gain-phase characteristics of the open-loop system, the Bode diagram is first drawn as indicated in Figure 2.36. Then for each value of ω the magnitude and phase of the open-loop compensated characteristics are plotted into a Nichols chart as shown in Figure 2.37. Because the open-loop gain-phase characteristics are just tangent to the constant-magnitude locus corresponding to $M = 2.976$ dB (1.4), we have achieved our goal. Notice that we have shifted $\omega = 1$ rad/sec by about $-35°$, but this does not increase M_p. Figures 2.36 and 2.37 were obtained using MATLAB, and are contained in the M-files that are part of my AMCSTD Toolbox. It is important to note that the use of MATLAB and the AMCSTD Toolbox permits one to obtain the Nichols chart directly without having to first obtain the Bode diagram.

2.9. COMPENSATION AND DESIGN USING THE ROOT-LOCUS METHOD

The root-locus technique has been developed previously in Sections 1.7, and MATLAB computer programs for obtaining the root locus were also presented in Section 1.7. It is a very helpful tool that the control engineer can use in order to study the variation of gain, system parameters, and effect of compensation. We demonstrated in Section 1.7 the migration of poles in the complex plane as the gain of the system was varied from zero to infinity. We obtained the root locus for several feedback systems including the following in Ilustrative Problem I1.10:

76 LINEAR CONTROL-SYSTEM COMPENSATION AND DESIGN

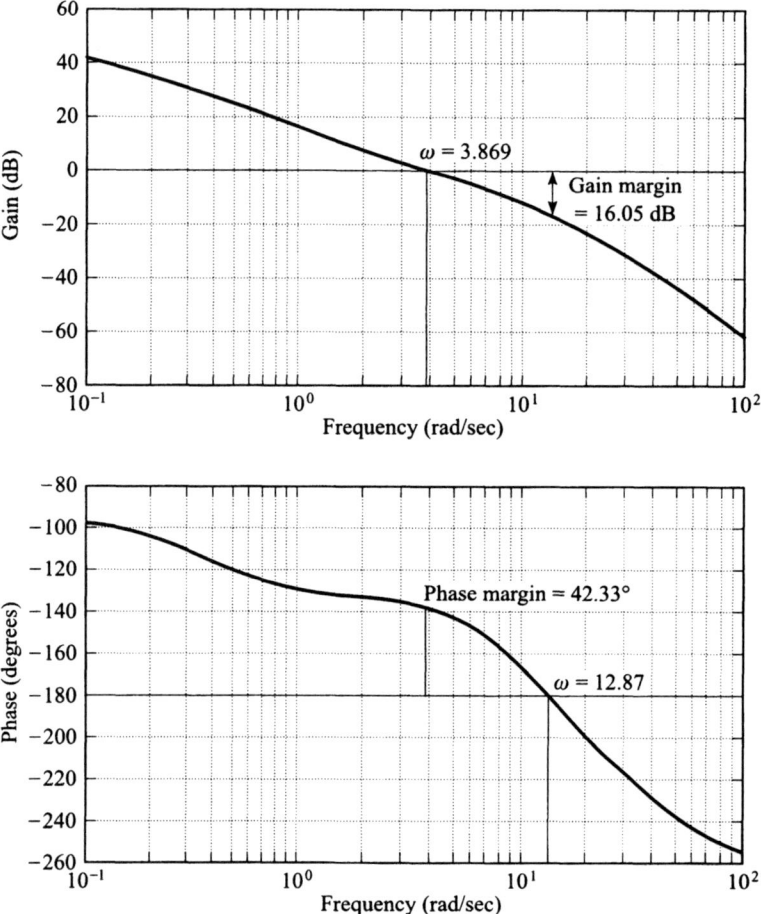

Figure 2.36 Bode diagram for the compensated system of Figure I1.7-1, where

$$G_c(s)G(s)H(s) = \frac{11.7(1 + 0.58s)}{s(1 + 0.05s)(1 + 0.1s)(1 + 1.74s)}.$$

$$G(s)H(s) = \frac{K}{(s + 1)(s - 1)(s + 4)^2}. \tag{2.113}$$

This was illustrated in Figure I1.10. An analysis of the root locus for this system indicated that it was always unstable, because at least one of the roots of the characteristic equation always occurred in the right-half-plane. This section demonstrates how this system may be compensated by means of a phase-lead network, and/or rate feedback in parallel with position feedback. This problem is followed by considering phase-lag-network compensation for the system illustrated in Figure I1.9. In addition, we shall demonstrate how the control engineer may determine the transient response of the compensated systems in order to meet certain specifications.

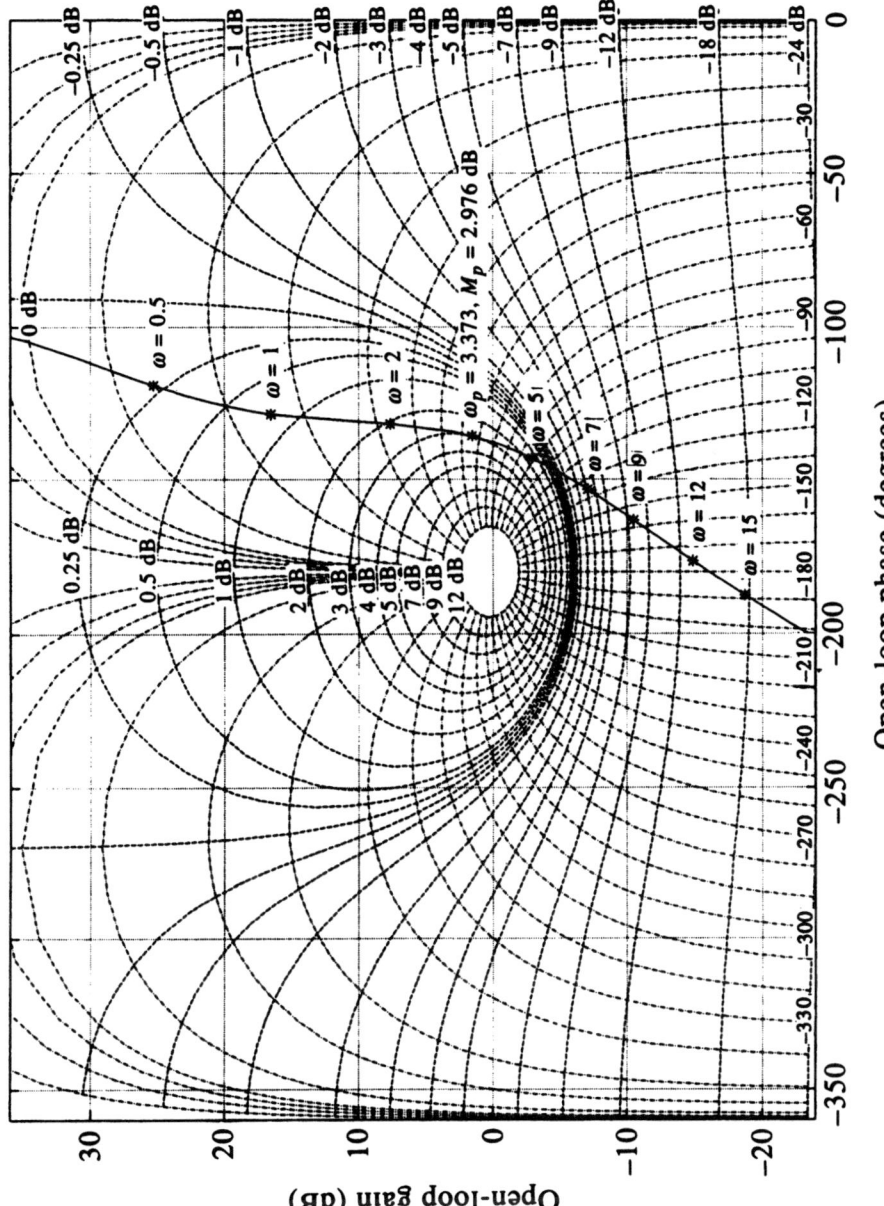

Figure 2.37 Compensation of the system shown in Figure I1.7-1 for $M = 2.976$ dB (1.4).

A. Example of Phase-Lead Network and Rate-Feedback Compensation

Let us attempt to stabilize the system of Figure I1.10 by means of a phase-lead network in cascade with the forward-loop transfer function. The form of its transfer function is given by

$$G_c(s) = \frac{s+\alpha}{s+\beta}. \tag{2.114}$$

We assume that the effect of the pole introduced by the phase-lead network has a negligible effect compared with its zero. Therefore, we assume that the transfer function of the cascaded phase-lead network can be approximated by the zero term only, which is also equivalent to the pure zero obtained with rate feedback in parallel with position feedback (see Figure 2.14a):

$$G_c(s) \approx s+\alpha. \tag{2.115}$$

The resulting value of $G_c(s)G(s)H(s)$, which is to be examined on the root locus, is given by

$$G_c(s)G(s)H(s) = \frac{K(s+\alpha)}{(s+1)(s-1)(s+4)^2}. \tag{2.116}$$

We wish to investigate the effect on stability of a variation in α as follows:

Case A : $\alpha = 0.5$ Case D : $\alpha = 4$
Case B : $\alpha = 1$ Case E : $\alpha = 6$.
Case C : $\alpha = 2$

The resulting root loci for all these cases are presented next. It is important to emphasize at this point that, although we make use of most of the anlaytic tools developed in Section 1.7, we do not use all of them, This omission is due to the fact that some of the analytic techniques developed are too complex to use for higher-order systems. For example, it is very tedious to determine the value of the gain K along the root loci utilizing the relationship given by Eq. (6.133)[†]. Fortunately, this can be obtained much more easily by using MATLAB (Section 1.7). The approach taken in presenting the resulting root loci for this problem is to outline the results of the 12 rules developed previously in Section 6.14[†], and use MATLAB (Section 1.7) wherever it is helpful. In addition, the values of gain obtained using MATLAB which are pertinent for an intelligent evaluation of the problem will be indicated. It is my feeling that this dual approach is the best procedure to use when explaining the root-locus method. This detailed analysis will be applied to all five cases.

Case A. $\alpha = 0.5$. See Figure 2.38 for the root-locus sketch.

Rule 1. There are four separate loci because the characteristic equation,

$$1 + G_c(s)G(s)H(s) = 0,$$

is a fourth-order equation.

2.9. COMPENSATION AND DESIGN USING THE ROOT-LOCUS METHOD 79

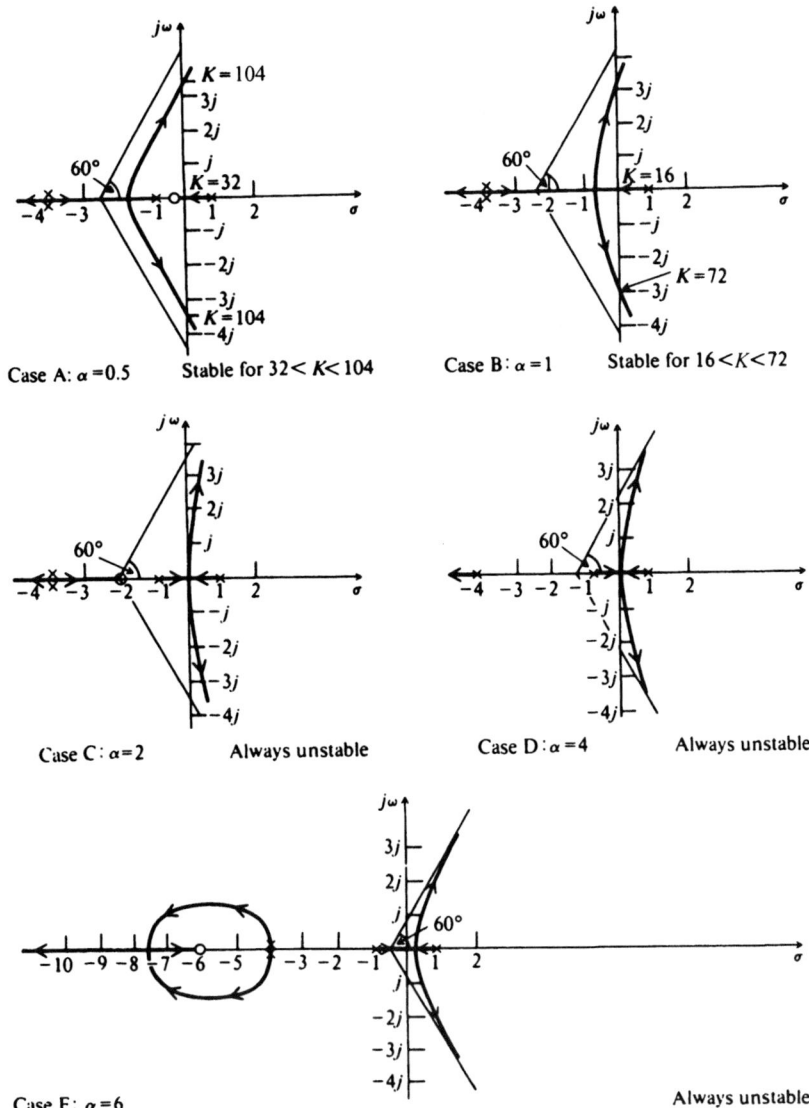

Figure 2.38 Compensation of the root locus shown in Figure I1.10 using a pure zero term where $G_c(s) = (s + \alpha)$.

Rule 2. The root locus starts ($K = 0$) from the poles located at $1, -1$, and a double pole at -4. One branch terminates ($K = \infty$) at the zero located at $\alpha = -0.5$ and three branches terminate at zeros located at infinity.

Rule 3. Complex portions of the root locus occur in complex-conjugate pairs.

Rule 4. The portions of the real axis between 1 to $-\alpha$, -1 to -4, and -4 to ∞ are part of the root locus.

Rule 5. The branches approach infinity as K becomes large at angles given by

$$\alpha_0 = \pm\frac{\pi}{3} = \pm 60°,$$

and

$$\alpha_1 = \pm\frac{3\pi}{3} = \pm 180°.$$

Rule 6. The intersection of the asymptotic lines and the real axis occur at

$$s_r = \frac{-8 - (-0.5)}{4 - 1} = -2.5.$$

Rule 7. Using MATLAB, we found the point of breakaway from the real axis to occur at -1.7549.

Rule 8. Using MATLAB, we found the intersection of the root locus and the imaginary axis to occur at $s = \pm j3.3$, where the gain is 104, and at the origin, where the gain is 32.

Rule 9. This rule does not apply here.

Rule 10. This rule shows that as certain of the loci turn to the right, others turn to the left to ensure that the sum of the roots is a constant.

Rule 11. This rule does not apply here.

Rule 12. The root locus does not cross itself.

The resulting root locus indicates that the system is stable when $32 < K < 104$.

Case B. $\alpha = 1$. See Figure 2.38 for the root-locus sketch. A zero at $s = -1$ cancels the pole at -1. The resulting root-locus sketch indicates that this system is stable when $16 < K < 72$. (The point of breakaway occurs at -0.667.)

Case C. $\alpha = 2$. See Figure 2.38 for the root-locus sketch. The resulting root-locus sketch indicates that the system is always unstable, because at least one of the roots of the characteristic equation always occurs in the right half-plane except for the condition where two poles exist at the origin. (The point of breakaway occurs at the origin.)

Case D. $\alpha = 4$. See Figure 2.38 for the root-locus sketch. A zero at $\alpha = 4$ cancels one of the poles located at -4. the resulting root-locus sketch indicates that the system is always unstable since at least one of the roots of the characteristic equation always occurs in the right half-plane. (The point of breakaway occurs at 0.1196.)

Case E. $\alpha = 6$. See Figure 2.38 for the root-locus sketch. The resulting root-locus sketch indicates that the system is always unstable, because at least one of the roots of the characteristic equation always occurs in the right half-plane. (The points of breakway occur at 0.1156 and −4, and a break-in occurs at −7.0611.)

Conclusions of Cases A through E

Interpretation of Figure 2.38 is quite interesting and revealing. It indicates that the exact location of the zero is very important from a stability viewpoint. Cases A and B were the only configurations which had regions of stability. As a matter of fact, the closer the zero lies to the imaginary axis, the greater is its stabilizing effect. This point is very important. Because Case A resulted in larger values of gain, it would result in a more accurate system and is, therefore, preferred to Case B.

The reader should not get the impression that α could be made extremely small (e.g. 0.0001) to satisfy this guideline because you will be paying a penalty in the size of the capacitors and resistors used in a passive phase-lead network. Remember that the form of the zero is $(s + \alpha)$, which can be modified to $\alpha[(1/\alpha)s + 1]$, where $1/\alpha$ is the time constant of the zero term. Therefore, very small values of α mean that the values of the components (resistors and capacitors) would be too large which is undesirable from a practical viewpoint. If the zero term $(s + \alpha)$ is obtained using rate feedback in cascade with position feedback, as in Figure 2.14a, then you would have to make $1/\alpha = b$ and very small values of α would mean extremely large values of b. This would require an amplifier in cascade with the rate-feedback sensor (e.g., tachometer or rate gyro), which is undesirable due to the added cost and possible noise problems.

B. Determination of the Transient Response

The transient response of the system can also be obtained by reasoning along these lines: The transient performance is often dominated by the pair of complex conjugate poles located closest to the origin. This occurs when the other poles are far to the left of the dominant poles, or the other poles are near a zero. The resulting transient components due to these other poles are small under these conditions and diminish rapidly. For this case, the poles closest to the origin are conventionally referred to as the dominant poles. The relative dominance of the closed-loop poles is found from the ratio of the real parts of the complex-conjugate poles. As a general guideline, reasonable dominance exists if the ratios of the real parts are at least five, and there are no zeros nearby. For such conditions, the closed-loop complex-conjugate poles closest to the origin will dominate the transient response. From the discussion of Appendix B, the expression associated with these complex poles can be given by the following expression [see Eq. (B.3)]:

$$\frac{C(s)}{R(s)} = \frac{\omega_n^2}{s^2 + 2\zeta\omega_n s + \omega_n^2} \qquad (2.117)$$

where $\zeta =$ damping ratio, and $\omega_n =$ undamped natural frequency. We found in Appendix B that the transient response to a unit step input, for $\zeta < 1$, is given by the following expression [see Eq. (B.24)]:

82 LINEAR CONTROL-SYSTEM COMPENSATION AND DESIGN

$$c(t) = 1 - \frac{e^{-\zeta\omega_n t}}{\sqrt{1-\zeta^2}} \sin\left(\omega_n\sqrt{1-\zeta^2}\,t + \alpha\right), \qquad t \geq 0 \qquad (2.118)$$

where

$$\alpha = \cos^{-1}(\zeta).$$

Figure (B.3) illustrated the complex-plane location of these dominant poles. The values derived for the time to the first peak [see Eq. (B.29)] and maximum percent overshoot [see Eq. (B.33)] are specifically for a second-order system whose closed-loop transfer function is given by Eq. (2.117). These quantities change if other closed-loop poles and zeros exist in addition to the dominant complex pair. However, if the ratios of the real parts of the various complex-conjugate poles are greater than five and there are no zeros nearby, then the approximation gives reasonable results. The damping ratio determined in this case using the pair of complex-conjugate roots closest to the imaginary axis is defined as the *relative damping ratio* of the control system.

Expressions for time to the first peak and percent overshoot, which consider other poles and zeros and give more accurate results, can be derived [6]. These expressions assume that

(a) Other poles are far to the left of the dominant poles, so that the amplitude of transients due to these other poles is small.
(b) Poles which are not far to the left of the dominant poles are near a zero so that the transient amplitude due to such poles is small.

The expressions, for unity-feedback systems, are given by

$$[t_p]_{\text{modified}} = \frac{1}{\sqrt{1-\zeta^2}\,\omega_n}\left[\frac{\pi}{2} - \sum \phi_z + \sum \phi_p\right] \qquad (2.119)$$

where

$\sum \phi_z$ = sum of the angles from the zeros of C/R to one of the dominant poles,
$\sum \phi_p$ = sum of the angles from the poles of C/R to one of the dominant poles,

and the maximum percent overshoot

$$= \frac{\begin{bmatrix}\text{product of distances from all}\\\text{poles of } C/R \text{ to origin}\\\text{excluding distances of two}\\\text{dominant poles from origin}\end{bmatrix}\begin{bmatrix}\text{product of distances from}\\\text{all zeros of } C/R \text{ to}\\\text{dominant pole } P_0\end{bmatrix}}{\begin{bmatrix}\text{product of distances from all}\\\text{other poles of } C/R \text{ to}\\\text{dominant pole } P_0 \text{ excluding}\\\text{distance between dominant}\\\text{poles}\end{bmatrix}\begin{bmatrix}\text{product of distances from}\\\text{all zeros of } C/R \text{ to}\\\text{origin}\end{bmatrix}} e^{-\zeta\omega_n t_p} \times 100\%.$$

The expression for maximum percent overshoot can be stated symbolically as

$$\text{max.\% overshoot} = \left[\left(\frac{P_1}{|P_1 - P_0|}\right)\left(\frac{P_2}{|P_2 - P_0|}\right)\left(\frac{P_3}{|P_3 - P_0|}\right)\cdots\right]$$
$$\times \left[\left(\frac{|Z_1 - P_0|}{Z_1}\right)\left(\frac{|Z_2 - P_0|}{Z_2}\right)\left(\frac{|Z_3 - P_0|}{Z_3}\right)\cdots\right] e^{-\zeta\omega_n t_p} \times 100\%, \quad (2.120)$$

where the first set of brackets represents the product of the ratios of the values of s at which poles occur to their absolute distances from the dominant pole. The second set of brackets represents the product of the ratios of the absolute distances of the zeros from the dominant pole and the values of s at which the zeros occur. Let us next apply these expressions in the following design problem.

C. Example of Phase-Lag-Network Compensation and Overall System Performance

The concluding design problem we consider using the root locus consists of employing cascaded phase-lag compensation in order to improve the steady-state performance of a feedback control system. The object is to increase its gain while maintaining a good dynamic response. Specifically, we consider the system whose root locus was illustrated in Figure I1.9. For this system

$$G(s)H(s) = \frac{K}{s(s+4)(s+5)}. \quad (2.121)$$

The root locus of Figure I1.9 indicated that the system was stable when $0 < K < 180$. Let us assume that a daming ratio of 0.707 achieves a desirable dynamic response for this system. In addition, we must maintain a velocity constant K_v of 30 in order to meet specified accuracy requirments. Analyzing this problem, by means of the root locus, we can find the value of K which will give the required damping ratio. For example, the redrawn version of Figure I1.9 shown in Figure 2.39 indicates that a $K = 21.59$ will result in a damping ratio of 0.707 ($\alpha = \cos^{-1} 0.707 = 45°$). This value of gain does not maintain the required velocity constant of 30. The actual value of K_v resulting from $K = 21.59$ is

$$K_v = \lim_{s \to 0} sG(s)H(s) = \lim_{s \to 0} \frac{s(21.59)}{s(s+4)(s+5)} = 1.08/\text{sec}. \quad (2.122)$$

It is therefore clear that we cannot just increase the gain K to a value that produces the required velocity constant, because this would decrease the damping ratio and adversely affect the transient response or cause the system to become unstable. Using the root locus for a solution, we show how these two conflicting factors can be resolved.

In order to achieve the specification requirements, we must increase the gain, but at the same time maintain the dominant complex-conjugate roots of the root locus where the value of $K = 21.59$ is shown in Figure 2.39 so that the damping ratio of 0.707 is maintained. We can accomplish this with a phase-lag network and an increase in the system gain.

84 LINEAR CONTROL-SYSTEM COMPENSATION AND DESIGN

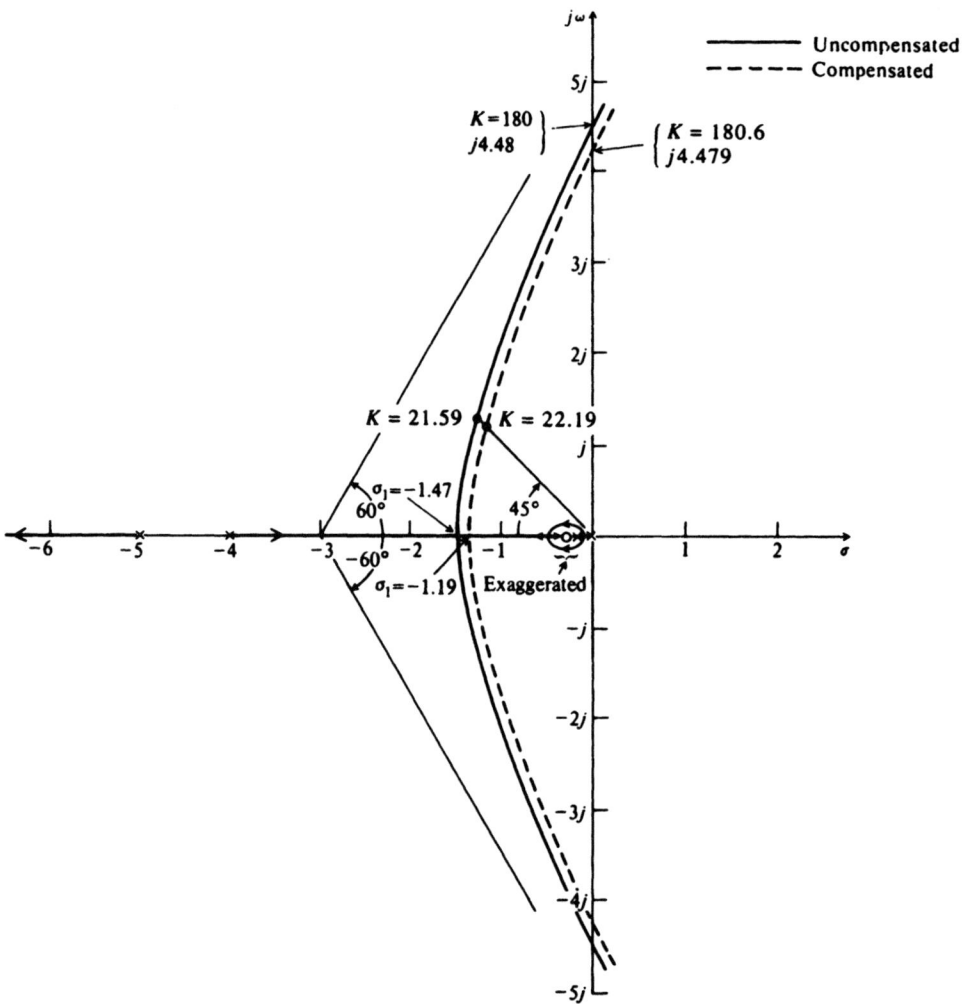

Figure 2.39 Compensation of the root locus shown in Figure I1.9 using a cascaded phase-lag network and an increase in gain.

Let us assume that the combined transfer function representing the increase in gain (n) and the phase-lag network is given by

$$G_c(s) = \frac{s + n\alpha}{s + \alpha}, \qquad (2.123)$$

The form of Eq. (2.123) indicates that this compensator provides a low-frequency gain of n in addition to the phase lag, where n is also the ratio of the break frequencies. The open-loop transfer function of the compensated system is given by

$$G_c(s)G(s)H(s) = \frac{K}{s(s+4)(s+5)} \left(\frac{s + n\alpha}{s + \alpha} \right). \qquad (2.124)$$

In general, the distances of $n\alpha$ and α from the origin in the s-plane are chosen to be small compared with the distances of the other zeros and poles of the uncompensated open-loop transfer function, so that the added pole and zero of the compensator will not contribute significant phase lag in the vicinity of the gain crossover frequency. This result is quite clear from a study of the Bode diagram, which we shall shortly do at the conclusion of this design (see Figure 2.40). Certainly we do not wish to add the phase-lag contribution in the vicinity of the gain crossover frequency. Therefore, the combination of pole and zero will be quite close together on the root locus and very close to the origin. The combination is usually called a dipole.

In order to complete the design, α is chosen close to the origin at 0.01 and n is chosen, using the following derivation, to achieve a $K_v = 30$:

$$K_v = 30 = \lim_{s \to 0} sG_c(s)G(s)H(s). \tag{2.125}$$

Substituting Eq. (2.124) into Eq. (2.125), we obtain

$$30 = \lim_{s \to 0} s\left(\frac{K}{s(s+4)(s+5)}\right)\left(\frac{s+n\alpha}{s+\alpha}\right),$$

or

$$30 = \frac{Kn}{(4)(5)}.$$

Because we desire that $K = 21.59$ from a transient viewpoint, we must have

$$n = \frac{30(4)(5)}{21.59} = 27.79. \tag{2.126}$$

The completed root locus for the compensated system whose open-loop transfer function is given by

$$G_c(s)G(s)H(s) = \frac{K}{s(s+4)(s+5)}\left(\frac{s+0.2779}{s+0.01}\right) \tag{2.127}$$

must now be determined. Because the dipole is added near the origin, the original root locus is not changed significantly, because the two poles and the zero near the origin tend to merge into a single pole.

Let us next determine the new resulting root locus, which is shown as the dashed curve in Figure 2.39, and analyze the effect of the dipole on it. Specifically, we wish to know whether the new root locus will indeed have $K_v = 30$. In addition, we would like to determine the transient response of the compensated system. Each of the 12 rules for constructing the root locus will be considered.

Rule 1. There are four separate branches, because the characteristic equation,

$$1 + G(s)H(s) = 0,$$

is a fourth-order equation.

86 LINEAR CONTROL-SYSTEM COMPENSATION AND DESIGN

Rule 2. The root locus starts ($K = 0$) from the poles located at the origin, -0.01, -4, and -5. One branch terminates ($K = \infty$) at the zero located at -0.2779 and the other three branches terminate at zeros which are located at infinity.

Rule 3. Complex portions of the root locus occur in complex-conjugate pairs.

Rule 4. The portions of the real axis between the origin and -0.01, -0.2779 and -4, and -5 to $-\infty$ are parts of the root locus.

Rule 5. The four branches approach infinity as K becomes large at angles given by

$$\alpha_0 = \pm \frac{\pi}{4-1} = \pm 60°,$$
$$\alpha_1 = \pm \frac{3\pi}{4-1} = \pm 180°.$$

Rule 6. The intersection of the aymptotic lines and the real axis occurs at

$$s_r = \frac{-9.01 - (-0.2779)}{4-1} = \frac{-8.73}{3} = -2.91.$$

Rule 7. The point of breakaway from the real axis can be computed from the following equation:

$$(\beta_1 - \beta_2 - \beta_3 - \beta_4 - \beta_5) = (2n+1)\pi,$$

where

β_1 = angle from the zero at -0.2779 to the point s_1 that is located a small distance δ off the positive real axis,
β_2 = angle from the pole at the origin to the point s_1,
β_3 = angle from the pole at -0.01 to the point s_1,
β_4 = angle from the pole at -4 to the point s_1,
β_5 = angle from the pole at -5 to the point s_1.

The equation of the transition of the root locus from the real axis to a point s_1 which is a small distance δ off the axis is given by

$$\left[\left(\pi - \frac{\delta}{\sigma_1 - 0.2779}\right) - \left(\pi - \frac{\delta}{\sigma_1}\right) - \left(\pi - \frac{\delta}{\sigma_1 - 0.01}\right) - \left(\frac{\delta}{4 - \sigma_1}\right) - \left(\frac{\delta}{5 - \sigma_1}\right)\right] = -\pi.$$

2.9. COMPENSATION AND DESIGN USING THE ROOT-LOCUS METHOD 87

Solving, we obtain $\sigma_1 = -1.19$, which compares with a value of -1.47 obtained using MATLAB for the uncompensated system.

Rule 8. The intersection of the root locus and the imaginary axis can be determined by applying the Routh–Hurwitz stability criterion to the characteristic equation

$$s(s+0.01)(s+4)(s+5) + K(s+0.2779) = 0,$$

which becomes

$$s^4 + 9.01s^3 + 20.09s^2 + (K+0.2)s + 0.2779K = 0.$$

The resulting Routh–Hurwitz array is given by

$$\begin{array}{c|ccc}
s^4 & 1 & 20.09 & 0.2779K \\
s^3 & 9.01 & K+0.2 & \\
s^2 & 20.068 - 0.1111K & 0.2779K & \\
s & \dfrac{-0.111K^2 + 17.55K + 4}{20.068 - 0.111K} & & \\
s^0 & 0.2779K & &
\end{array}$$

An interesting situation occurs in this Routh–Hurwitz array, because the first terms of the third and fourth rows can go to zero for certain values of gain, K. When the equation

$$20.068 - 0.1111K = 0$$

is satisfied, then a possible solution is

$$K_{max} = 180.6.$$

When the equation

$$\frac{-0.111K^2 + 17.55K + 4}{20.068 - 0.111K} = 0$$

is satisfied, then a possible solution is

$$K_{max} = 158.25.$$

Therefore, in order to find which is (or are) valid, let us substitute $s = j\omega$ into the characteristic equation and find out where the root locus crosses the imaginary axis. The results can be separated into a real and imaginary part and written in the following form:

$$\omega^4 - 20.09\omega^2 + 0.2779K + j\omega[-9.01\omega^2 + K + 0.2] = 0. \qquad (2.128)$$

For K to be real and imaginary parts of this equation must each equal zero. Therefore, from the imaginary part:

$$j\omega[-9.01\omega^2 + K + 0.2] = 0,$$

$$\omega = \pm\sqrt{\frac{K + 0.2}{9.01}}.$$

Now, to find the value of K which corresponds to this value of crossing of the imaginary axis, let us substitute this value of ω into the real part of Eq. (2.128). The result is the following equation:

$$K^2 - 180.33K - 36.16 = 0.$$

Therefore, we find that

$$K = 180.6$$

is the only possible real value of gain when the root locus crosses the imaginary axis. The analysis indicates that the maximum value of gain, before the system becomes unstable, is 180.6. In Figure I1.9, we found that the uncompensated system had a maximum allowable gain of 180. Therefore, this result indicates that the dipole has an effect on the maximum allowed gain. The corresponding value of s occuring at the crossing of the imaginary axis is found to be $\pm j4.479$ by substituting K_{max} into the equation for ω. This compares with a value of $s = j4.48$ obtained previously for the uncompensated case [see Figure I1.9].

Rule 9. This rule does not apply to this problem.

Rule 10. This rule shows that as certain of the loci turn to the right, others turn to the left to ensure that the sum of the roots is a constant.

Rule 11. This rule is quite important to us in this case, because we want to determine the value of gain when $\zeta = 0.707$ (the intersection of a line making an angle of $+45°$ with the negative real axis and the dashed curve). For the uncompensated case, we found that $K = 21.59$. The new value can be obtained from the following expression:

$$\left|\frac{K(s + 0.2779)}{s(s + 4)(s + 5)(s + 0.01)}\right| = 1.$$

Measuring the distance from the various poles and zeros to the point of interest, we obtain

$$K = 22.19.$$

Therefore, the gain has increased slightly from 21.59 to 22.19. This will result in a slight increase of K_v from 30 to 30.8.

Rule 12. The root locus does not cross itself.

2.9. COMPENSATION AND DESIGN USING THE ROOT-LOCUS METHOD

To conclude this problem, we can calculate the value of the time to the first peak and the maximum percent overshoot from Eqs. (2.119) and (2.120), respectively. These results can then be compared with the values obtained from Eqs. (B.29) and (B.33), which assumes that the transient response is completely controlled only by the pair of complex-conjugate poles located closest to the origin. The time to the first peak of the compensated system can be calculated from Eq. (2.119). In order to use this equation, the location of the other two roots must be determined. Using MATLAB for determining the root locus, these were found to be located at -6.45 and -0.26. Therefore,

$$t_p = \frac{1}{\left(\sqrt{1-\zeta^2}\omega_n\right)}\left(\frac{\pi}{2} - \sum \phi_z + \sum \phi_p\right)$$

where

$\zeta = 0.707$,

$\omega_n = 1.63$ (distance from the origin of the complex plane to the two dominant poles located at $-1.15 \pm j1.15$—dashed curve of Figure 2.39),

$\sum \phi_z = 141° = 2.47$ rad,

$\sum \phi_p = 136° + 90° + 14° = 240° = 4.19$ rad.

Substituting these values into the equation for t_p, we obtain

$$t_p = \frac{1}{\left(\sqrt{1-0.707^2}\,1.63\right)}(1.57 - 2.47 + 4.19) = 2.86 \text{ sec.}$$

Therefore, the time to the first peak is 2.86 sec. If one simply assumes that the transient response is governed by the pair of complex-conjugate poles located at $1.15 \pm j1.15$ and uses Eq. (4.29)[‡] to determine the time to the first peak, then the following is obtained:

$$t_p = \frac{\pi}{\omega_n\sqrt{1-\zeta^2}} \quad \text{[from Eq. (4.29)}^\ddagger\text{]},$$

$$t_p = \frac{\pi}{1.63\sqrt{1-0.707^2}} = \frac{3.14}{1.15} = 2.73 \text{ sec.}$$

Therefore, we see the slightly improved accuracy obtained using Eq. (2.119).

A similar analysis of the uncompensated system, utilizing Eq. (2.119), results in a time to the first peak of 2.81 sec. For this case, the two complex-conjugate poles are located at $-1.2 \pm j1.2$. The third root can be determined analytically from rule 10:

$$\sum p_i = \sum r_j,$$
$$-4 - 5 = -1.2 + j1.2 - 1.2 - j1.2 + r,$$
$$r = -6.6.$$

90 LINEAR CONTROL-SYSTEM COMPENSATION AND DESIGN

Therefore,

$$t_p = \frac{1}{\sqrt{1-\zeta^2}\omega_n}\left(\frac{\pi}{2} - \sum \phi_z + \sum \phi_p\right),$$

where

$\zeta = 0.707$,

$\omega_n = 1.7$ (distance from the origin of the complex plane to the two dominant poles—solid curve of Figure 2.39),

$\sum \phi_z = 0$,

$\sum \phi_p = 90° + 13° = 103° = 1.8$ rad.

Hence,

$$t_p = \frac{1}{\sqrt{1-0.707^2}(1.7)}(1.57 + 1.8) = 2.81 \text{ sec}.$$

Observe that the dipole compensation increases the time to the first peak slightly.

The maximum percent overshoot of the compensated system can be obtained from Eq. (2.120). To use this equation, we have to determine the location of the other two roots. As mentioned previously, these were found to be at -6.45 and -0.26. Therefore,

$$\text{maximum percent overshoot} = \frac{P_1 P_2}{(|P_1 - P_0|)(|P_2 - P_0|)} \frac{|Z_1 - P_0|}{Z_1} e^{-\zeta\omega_n t_p} \times 100\%$$

$$= \frac{(0.26)(6.45)}{(1.45)(4.9)} \frac{(1.46)}{(0.2779)} e^{-0.707(1.63)2.86} \times 100\%$$

$$= 4.52\%.$$

Therefore, the resulting maximum percent overshoot is only 4.52%. If one simply assumes that the transient response is governed by the pair of complex-conjugate poles located at $-1.15 \pm j1.15$ and uses Eq. (B.33) to determine the maximum percent overshoot, then the following is obtained:

$$\text{maximum percent overshoot} = \left(e^{-\zeta\pi/\sqrt{1-\zeta^2}}\right) \times 100\%$$

$$= \left(e^{-0.707\pi/\sqrt{1-0.707^2}}\right) \times 100\% = 4.3\%.$$

Therefore, we see the slightly increased accuracy obtained using Eq. (2.120).

A similar analysis of the uncompensated system utilizing Eq. (2.120) results in a maximum percent overshoot of 4.48%; for this case, the third root is located at -6.6 as illustrated before:

$$\text{maximum percent overshoot} = \frac{P_1}{|P_1 - P_0|} e^{-\zeta\omega_n t_p} \times 100\%$$

$$= \frac{6.6}{5} e^{-0.707(1.7)(2.81)} \times 100\% = 4.48\%.$$

Observe that the dipole compensation increases the maximum percent overshoot slightly.

The results of the transient analysis indicate that the effect of the dipole is to increase the time to the first peak slightly and to increase the maximum overshoot slightly. Of most importance, the dipole increases the velocity constant greatly, from 1.08 to 30.8.

One should note that the dominant pole concept is very useful here, the results coming quite close to the more accurate calculation. In fact, due to the ever-present uncertainty of the parameters of the actual system, it is rarely, if ever, necessary to carry out the detailed calculations indicated.

D. Comparison of the Root-Locus Method with Bode Diagrams

In our analysis of the system whose transfer function was given by Eq. (2.121) and illustrated in Figure 2.39 using the root-locus method, it was pointed out that this same system's analysis using the Bode diagram would be very interesting. It is always interesting to compare the analysis of a system using more than one of the techniques shown as was discussed in Section 6.20[‡], where 12 commonly used transfer functions were compared from the viewpoints of the Nyquist-diagram, Bode-diagram, Nichols-chart, and root-locus methods. In this problem, it is especially useful because the root-locus solution may be difficult to comprehend initially. What did we really do when we kept the dominant complex-conjugate roots almost fixed, increased the gain and added a phase-lag network? It is very clear if we look at the uncompensated and compensated systems from the viewpoint of the Bode diagram. For purposes of the Bode-diagram analysis, the value of K in Eq. (2.121) is set equal to 21.59, and this equation is modified to the following easier form for a Bode diagram:

$$G(s)H(s) = \frac{1.08}{s(0.25s + 1)(0.2s + 1)}. \qquad (2.129)$$

The Bode diagram of the uncompensated system is shown in Figure 2.40. This Bode diagram was obtained using MATLAB (see Section 1.7) and is contained in the M-file part of my AMCSTD Toolbox disk and can be retrieved free from The MathWorks, Inc. anonymous FTP server at ftp://ftp.mathworks.com/pub/books/advshinners. Notice from this diagram that the system has a gain crossover frequency of 1.025 rad/sec, a phase margin of 64.04°, and a gain margin of 18.42 dB (phase crossover frequency occurs at 4.472 rad/sec). Extension of the initial -20-dB/decade slope intersects the 0-dB line at 1.08 rad/sec indicating a K_v of 1.08 which agrees with Eq. (2.122). (The actual curve intersects the 0-dB line at 1.025 rad/sec.) We now wish to plot the Bode diagram of the compensated system whose transfer function is given by Eq. (2.127). We let $K = 22.19$ and modify Eq. (2.127) into the following equation which is easier to use for the Bode diagram:

$$G_c(s)G(s)H(s) = \frac{30.8(3.60s + 1)}{s(0.25s + 1)(0.2s + 1)(100s + 1)}. \qquad (2.130)$$

92 LINEAR CONTROL-SYSTEM COMPENSATION AND DESIGN

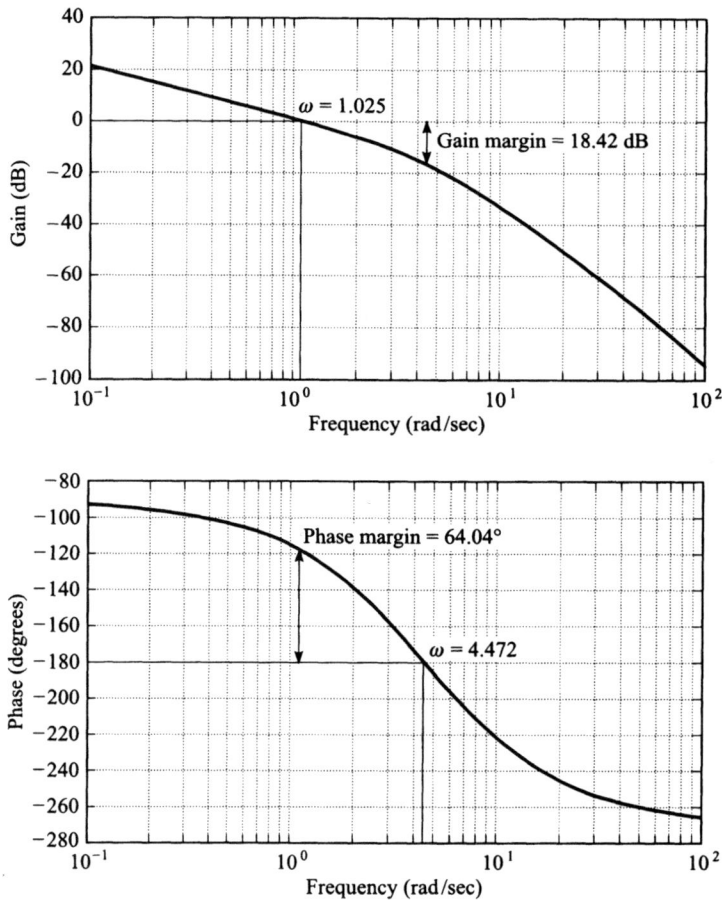

Figure 2.40 Bode diagram for the uncompensated system whose root locus is shown in Figure 2.39.

The resulting Bode diagram for the compensated system, also obtained using MATLAB, whose transfer function is given by Eq. (2.130) is shown in Figure 2.41. It has a gain crossover frequency of 1.138 rad/sec which is close to that of the uncompensated system whose Bode diagram is shown in Figure 2.40 (1.025 rad/sec), because we hardly moved the dominant complex-conjugate poles (see Figure 2.39) in going from the uncompensated to the compensated system. The phase margin of the compensated system is 49.21° (compared to 64.04° of the uncompensated system), and the gain margin of the compensated system is 16.67 dB (compared to 18.42 dB of the uncompensated system). Therefore, the uncompensated system and the compensated system have almost the same relative stability, and about the same gain crossover frequency. From the root-locus viewpoint, this means that the damping factor has hardly changed in going from the uncompensated to compensated systems. (Figure 2.39 shows that the damping factor remains constant at 0.707). However, the velocity constant of the compensated system is 30.8 [see Figure 2.41] compared to the velocity constant of the uncompensated system which is 1.08 [see Eq. (2.122)]. In conclusion, the comparison of the stabilization of this problem using the root locus (Figure 2.39) and the Bode diagram (Figures

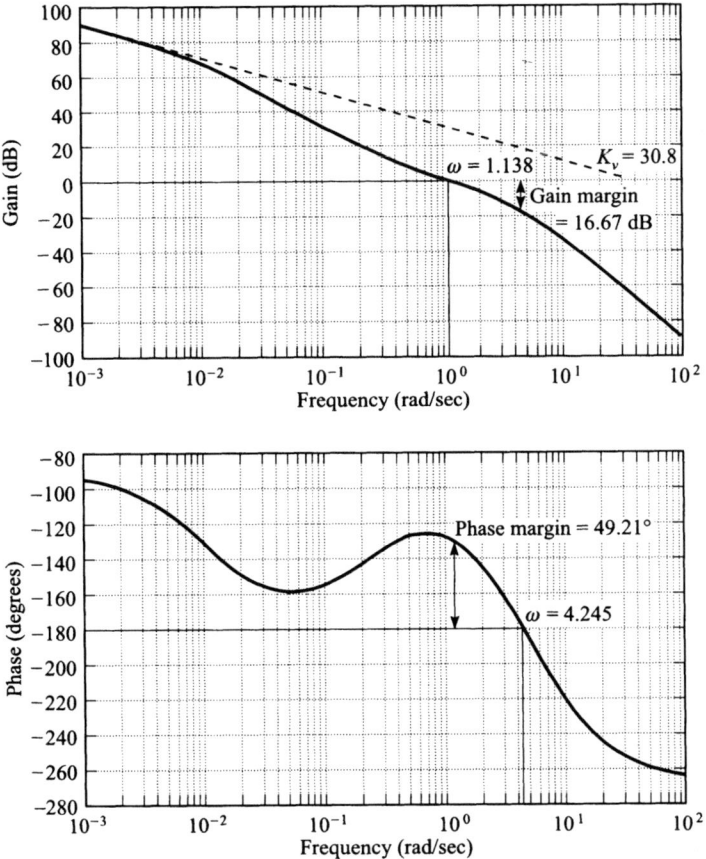

Figure 2.41 Bode diagram for the compensated system whose root locus is shown in Figure 2.39.

2.40 and 2.41) is very interesting as the information shown in the root-locus and Bode diagrams complement each other.

2.10. TRADEOFFS OF USING VARIOUS CASCADE-COMPENSATION METHODS AND MINOR-LOOP FEEDBACK

This chapter has presented the use of phase-lag, phase-lead, and phase-lag–lead cascade compensation networks, and minor-loop feedback, compensation (such as rate feedback). The question arises as to which compensation method should be used: cascade compensation or minor-loop feedback compensation. If the answer turns out to be cascade compensation, the question arises as to whether we should use phase-lag, phase-lead, or phase-lag–lead networks? This section addresses these reasonable questions by discussing the tradeoffs of these different methods of compensation, and guidelines are provided for the practical design of control systems.

I cannot provide a simple answer as to what compensation approach is the "best" method to use. Table 2.2 [11] compares the advantages and disadvantages of phase-

Table 2.2. Comparison of Compensation Methods [11]

	Phase-lag network	Phase-lead network	Phase-lag–lead network	Minor-loop rate feedback
Characteristics	a. Decreases system bandwidth b. Increases low-frequency gain c. Decreases ω_n d. Increases ζ	a. Increases system bandwidth b. Increases high-frequency gain c. Increases ω_n d. Increases ζ	a. Increases low-frequency gain b. Increases high-frequency gain c. Increases ω_n d. Increases ζ	a. Increases system bandwidth b. Increases high-frequency gain c. Increases ω_n d. Increases ζ
Performance advantages	a. Improves stability b. Decreases high-frequency noise c. Reduces steady-state error	a. Improves stability b. Reduces settling time T_s	a. Improves stability b. Reduces steady-state error c. Reduces settling time T_s	a. Improves stability b. Reduces settling time T_s c. Requires small volume d. Relatively independent of environment (altitude and temperature) e. Permits the isolation of undesirable dynamics in one portion of a control system from the complete system

Table 2.2. (continued)

	Phase-lag network	Phase-lead network	Phase-lag–lead network	Minor-loop rate feedback
Performance disadvantages	a. Increases settling time T_s b. Usually requires larger values of resistors and capacitors to achieve desired network time constants	a. Requires additional amplifier gain b. Increases high-frequency noise	a. Increases high-frequency noise b. Usually requires large values of resistors and capacitors to achieve desired network time constants c. May require large values of resistors and capacitors to achieve desired network time constants	a. Increases the steady-state error b. Increases high-frequency noise but less than resulting from a phase-lead network
Relative Cost	Least expensive	Moderately expensive	Relatively inexpensive	Usually the most expensive
Applicability	a. Very applicable when it is desired to increase the steady-state constants b. Cannot be used when the uncompensated phase at low frequencies does not equal the desired phase margin	a. Very useful when a fast transient response is desired	a. Very applicable when it is desired to increase the steady-state constants b. Very useful when a fast transient response is desired	a. Very useful when a fast transient response is desired b. Very useful for hostile environments requiring small space c. Useful for isolating undesirable dynamics

Reprinted from *Control Engineering*, May 1978.
© Copyright Cahners Publishing Company.

lag, phase-lead, and phase-lag–lead networks, and minor-loop rate feedback for compensation. They are compared on the basis of characteristics, performance, applicability, and relative cost. Each of these four approaches will improve system stability if used properly, as shown in Chapters 1 and 2. Therefore, each system must be considered separately in order to determine the most effective, optimum, or "best" approach.

If the system specification requires the need for a low-noise ("quiet") control system having a small steady-state error, then the phase-lag–network approach is the most desirable from the viewpoints of performance and cost. The phase-lead network is desirable for systems requiring a very fast response, and if the resulting increase in high-frequency noise is tolerable. The phase-lag–lead network has essentially the advantages in performance of both the phase-lag and phase-lead networks. It can result in control systems which are very fast, on a relative basis, and they can result in very small steady-state errors.

Minor-loop feedback compensation using rate-feedback from tachometers and rate gyros has the advantage that it results in a pure zero, when connected in parallel with the position feedback, and represents ideal compensation. Rate-compensating devices have the additional advantage of only requiring a small volume and they are relatively independent of the environmental conditions in which they are operating such as temperature, humidity, and altitude. This is advantageous for control systems operating in space vehicles, airplanes, and ships. For example, if a phase-lead network were used in a hostile environment, the possible use of very large resistors and capacitors could result in very large time-constant variations causing a stable control system to behave as a conditionally stable control system.

In trading off the advantages and disadvantages of minor-loop feedback compensation, it should be emphasized that although they result in systems having the highest performance, this approach is usually also the most expensive. However, in some cases, it may be the only and "best" way to go.

In the practical world of control-system applications, each case is not so straightforward that the control-system engineer can merely go to Table 2.2 and choose an approach clearly. Conflicting system requirements will usually direct the control-system engineer to more than one possible solution, and neither one of them may be the "best" method. In the final design, the choice of the compensating approach will usually be a compromise based on the tradeoffs considered.

The control-system engineer will usually select the approach based on his or her experience, subjective personal preferences, and the availability of components. As was discussed in the case of choosing the "best" method for stability analysis in Section 6.22[‡], all of the compensation methods should be considered before one is selected. Then the "best" or most effective technique can be selected by the control-system engineer for the final design.

2.11. ILLUSTRATIVE PROBLEMS AND SOLUTIONS

This section provides a set of illustrative problems and their solutions to supplement the material presented in Chapter 2.

I2.1. A closed-loop unity–feedback control system used to position a load has a forward transfer function given by

$$G(s) = \frac{10}{2s+1}.$$

The time constant of the open-loop process is 2 sec. Determine the time constant of the closed-loop control system.

SOLUTION:

$$\frac{C(s)}{R(s)} = \frac{G(s)}{1+G(s)H(s)} = \frac{\dfrac{10}{2s+1}}{1+\dfrac{10}{2s+1}} = \frac{10}{2s+11}.$$

Therefore,

$$\frac{C(s)}{R(s)} = \frac{10}{11\left(\dfrac{2}{11}s+1\right)} = \frac{10}{11(0.1818s+1)}.$$

So, the time constant of the closed-loop control system is 0.1818 sec.

I2.2. We wish to analyze the performance of a unity-feedback second-order control system whose forward transfer function represents a process, $G_p(s)$, given by:

$$G_p(s) = \frac{500}{s(s+10)}.$$

(a) Determine the gain crossover frequency, ω_c, analytically without using the graphical Bode plot.

(b) Determine the phase margin and gain margin of this control system analytically without using the graphical Bode plot.

(c) A phase-lead network compensation, $G_c(s)$, given by

$$G_c(s) = \frac{(1+aTs)}{(1+Ts)}$$

is to be added in series with the process' transfer function, $G_p(s)$. Determine the values of a and T in order that the zero factor of $G_c(s)$ cancels the pole of $G_p(s)$ at $s = -10$, and the damping ratio of the control system is unity.

(d) Determine the gain crossover frequency, ω_c, analytically without using the graphical Bode plot for the compensated control system.

(e) Determine the phase margin and gain margin of this control system analytically without using the graphical Bode plot.

SOLUTION: (a)
$$\left|\frac{500}{j\omega(j\omega+10)}\right| = 1,$$
$$\left|\frac{500}{-\omega^2+10j\omega}\right| = 1.$$

Solving, we find that the gain crossover frequency, ω_c, equals 21.27 radians per second.

(b) $\phi(\omega) = -90° - \tan^{-1}\frac{\omega_c}{10} = -90° - \tan^{-1} 2.127 = -155.7°.$

Therefore, the phase margin, γ, equals $180° + \phi(\omega) = 24.3°$.

The gain margin equals infinity because the phase of $\phi(\omega)$ is always less than $-180°$.

(c) $G(s) = G_c(s)G_p(s) = \dfrac{500(1+aTs)}{s(s+10)(1+Ts)} = \dfrac{500aT(s+1/aT)}{Ts(s+10)(s+1/T)}.$

Therefore, we make
$$\frac{1}{aT} = 10.$$

The characteristic equation of the resulting control system is determined as follows:

$$\frac{C(s)}{R(s)} = \frac{\dfrac{500a}{s(s+\frac{1}{T})}}{1+\dfrac{500a}{s(s+\frac{1}{T})}} = \frac{500a}{s^2+\dfrac{1}{T}s+500a}.$$

Therefore, comparing the denominator of this equation with Eq. (B.3):
$$\omega_n^2 = 500a$$
$$2\zeta\omega_n = \frac{1}{T}$$

where $\zeta = 1$. Solving these two equations simultaneously and the previous result that $1/aT = 10$, we find that $a = 20$ and $T = 0.005$. Therefore,
$$G_c(s) = \frac{1+0.1s}{1+0.005s},$$
$$G(s) = G_c(s)G_0(s) = \frac{10,000}{s(s+200)}.$$

(d)

$$\frac{10,000}{j\omega(j\omega + 200)} = 1,$$

$$\frac{10,000}{-\omega^2 + 200j\omega} = 1.$$

Solving, we find that the gain crossover frequency of the compensated control system, ω_c, equals 48.6 rad/sec.

(e)

$$\phi(\omega) = -90° - \tan^{-1}\frac{\omega_c}{200} = -90° - \tan^{-1} 0.243 = -90° - 13.66° = -103.66°.$$

Therefore, the phase margin, γ, equals $180° + \phi(\omega) = 76.34°$. The gain margin equals infinity because the phase of $\phi(\omega)$ is always less than 180°.

I2.3. Due to the inherent instability of a control system containing only a double integration, we wish to add a phase-lead network to the following control system:

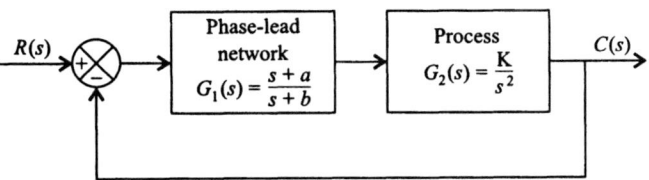

Figure I2.3i

Due to accuracy considerations, we desire the gain of the control system, K, to be 229.6. Due to component considerations, we wish the pole of the phase-lead network, b, to be located at 27. Using the root-locus method, determine the value of the zero of the phase-lead network, a, if the dominant poles of this control system are located at $-5.288 + 5.288j$ and $-5.288 - 5.288j$.

SOLUTION: The root locus for this problem is given by the plot in Figure I2.3ii.

We know that the absolute value of the open-loop transfer function, $G_1(s)G_2(s)$ must equal one. Therefore,

$$\left|\frac{K(s+a)}{s^2(s+27)}\right| = 1.$$

This equation can be written as follows for $K = 229.6$ and at $s = -5.288 + 5.288j$:

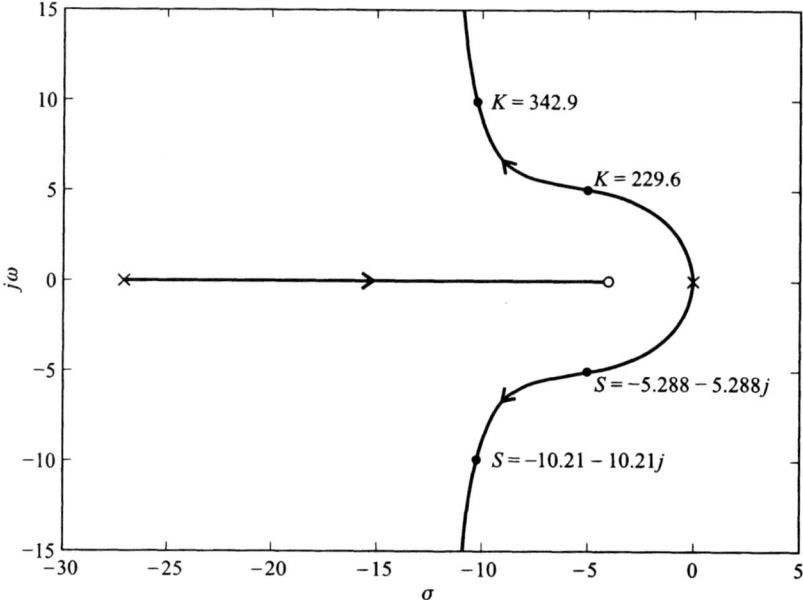

Figure I.23ii

$$\frac{229.6\sqrt{5.288^2 + (a - 5.288)^2}}{\sqrt{(5.288^2 + 5.288^2)^2}\sqrt{5.288^2 + (27 - 5.288)^2}} = 1$$

Therefore, $a = -4$.

PROBLEMS

2.1. Determine the circuit structure, the values of resistance and capitance, the gains of any amplifiers required, and the complex-plane plot for first-order networks having the following characteristics:

(a) Phase lead of 60° at $\omega = 4$ rad/sec, a minimum input impedance of 50,000 Ω, and an attentuation of 10 dB at dc.

(b) Phase lag of 60° at $\omega = 4$ rad/sec, a minimum input impedance of 50,000 Ω, and a high-frequency attenuation of 10 dB.

(c) A phase-lag-lead network having an attenuation of 10 dB for a frequency range of $\omega = 1$ to $\omega = 10$ rad/sec and an input impedance of 50,000 Ω.

In all cases, limit the maximum values of resistance to 1 MΩ and capitance to 10 μF. Furthermore, assume that the loads on the networks have essentially infinite impedance.

2.2. The system illustrated in Figure P2.2 consists of a unity-feedback loop containing a minor-rate-feedback loop.

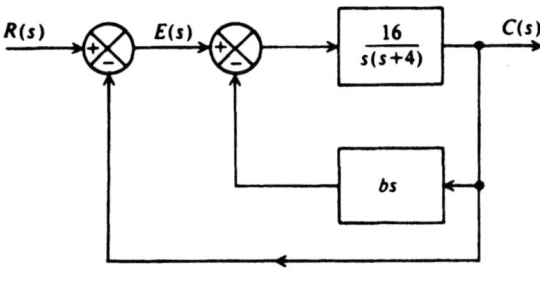

Figure P2.2

(a) Without any rate feedback ($b = 0$), determine the damping ratio, undamped natural frequency, peak overshoot of the system to a unit step input, and the steady-state error resulting from a unit ramp input.

(b) Determine the rate-feedback constant b which will increase the equivalent damping ratio of the system to 0.8.

(c) With rate feedback and a damping ratio of 0.8, determine the maximum percent overshoot of the system to a unit step input and the steady-state error resulting from a unit ramp input.

(d) Illustrate how the resulting steady-state error of the system with rate feedback to ramp inputs can be reduced to the same level if rate feedback were not used, and still maintain a damping factor of 0.8.

2.3. Repeat Problem 2.2 for the forward transfer function of the system given by $20/s(1 + s)$.

2.4. Figure P2.4 illustrates the block diagram of a roll control system used to limit the roll rate excursions of a missile by providing sufficient dynamic reaction to disturbing moments [12]. The disturbance moments result from changes in bank angle and steering control deflections. The basic limitation which determines the effectiveness of the roll-control system is the response of the aileron servo.

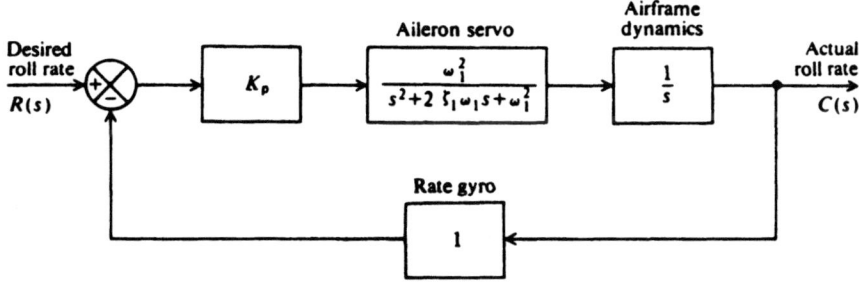

Figure P2.4

(a) Determine the transfer function, $C(s)/R(s)$, of the system illustrated in Figure P2.4.

(b) Because the transient response is governed by a pair of dominant complex-conjugate poles, specify the requirements of the aileron servo parameters in order that the equivalent damping ratio of the system is approximately 0.5, and the equivalent undamped natural frequency of the system is approximately 4 rad/sec.

2.5. A unity-feedback system has a forward transfer function given by

$$G(s) = \frac{28(1+0.05s)}{s(1+s)}.$$

It is desired to compensate this system so that the resulting damping ratio is unity (critically damped).

(a) Using the classical approach, determine the time constant of a cascaded phase-lead network, containing a zero factor only, that can achieve this.

(b) Using the classical approach, determine the rate-feedback constant of a minor rate-feedback loop which can achieve critical damping.

2.6. Repeat Problem 2.5 for the forward transfer function of the system given by

$$\frac{100(1+0.1s)}{s(1+10s)}.$$

2.7. A phase-lag–lead network, whose general transfer function is given by Eq. (2.9), and whose gain and phase characteristics are illustrated in Figure 2.13, provides a phase lag at low frequencies and a phase lead at high frequencies. For the following phase-lag–lead network,

$$G(s) = K \frac{(1+T_1 S)(1+T_2 s)}{(\alpha + T_1 s)\left(\frac{1}{\alpha} + T_2 s\right)}.$$

Find the frequency ω_0 where the phase angle of $G(j\omega)$ becomes zero. For frequencies less than ω_0, this network acts as a phase-lag network; for frequencies greater than ω_0, this network acts as a phase-lead network.

2.8. It is desired that a unity feedback control system whose forward transfer function is $G(s) = 20/[s(1+0.2s)(1+0.5s)]$ have a minimum phase margin of 45° and a minimum gain margin of 20 dB.

(a) Specify the time constant of a phase-lead network (or networks) that can achieve this.

(b) Repeat part (a) for a phase-lag network.

2.9. It is desired that a unity feedback control system whose forward transfer function is $G(s) = 20/[s(1+0.5s)(1+00.001s)]$ have a phase margin of 60°

and a gain margin of 60 dB. Determine the time constants of a phase-lag network (or networks) that can achieve this.

2.10. The temperature-control loop of a nuclear power plant is illustrated in Figure P2.10. The transfer function of the nuclear reactor can be adequately represented by

$$G_R(s) = \frac{e^{-0.2s}}{0.4s + 1}.$$

A time delay (or transportation lag) is included in this transfer function to account for the time required to transport the fluid from the reactor to the measurement point. A proportional-plus-integral (PI) controller is used for compensation.

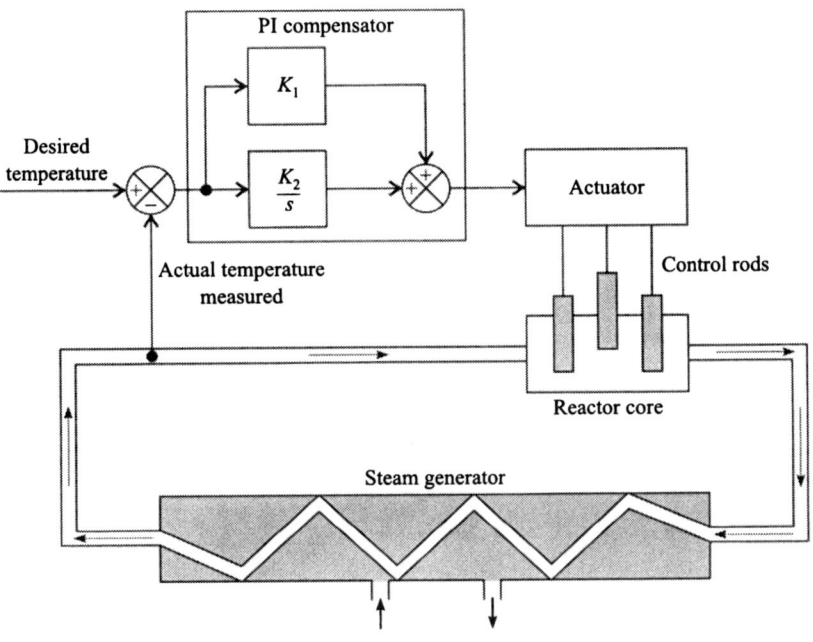

Figure P2.10

Using the Bode diagram, determine the values of K_1 and K_2 in order to achieve a phase margin of 30° and a gain margin of 4 dB.

2.11. It is desired that a unity feedback control system whose forward transfer function is $G(s) = 1.5/[s^2(0.1s + 1)]$ have a phase margin of 45° at the crossover frequency. Determine the stabilizing element required to achieve this.

2.12. It is desired that a unity feedback control system whose forward transfer function is $G(s) = 0.15/[s^2(0.1s + 1)]$ have a phase margin of 45° at the crossover frequency. Determine the stabilizing element required to achieve this.

104 LINEAR CONTROL-SYSTEM COMPENSATION AND DESIGN

2.13. The H.S. Denison, shown in Figure P2.13a, the first large hydrofoil seacraft built and operated in the United States [13] was designed and built by the Grumman Aerospace Corporation for the Maritime Administration of the U.S. Department of Commerce. The 80-ton hydrofoil is capable of operating at speeds of 60 knots in seas containing waves nine feet high. A simplified schematic of the automatic control system of the H.S. Denison is illustrated in Figure P2.13b. It consists of transducers for sensing craft motions and a

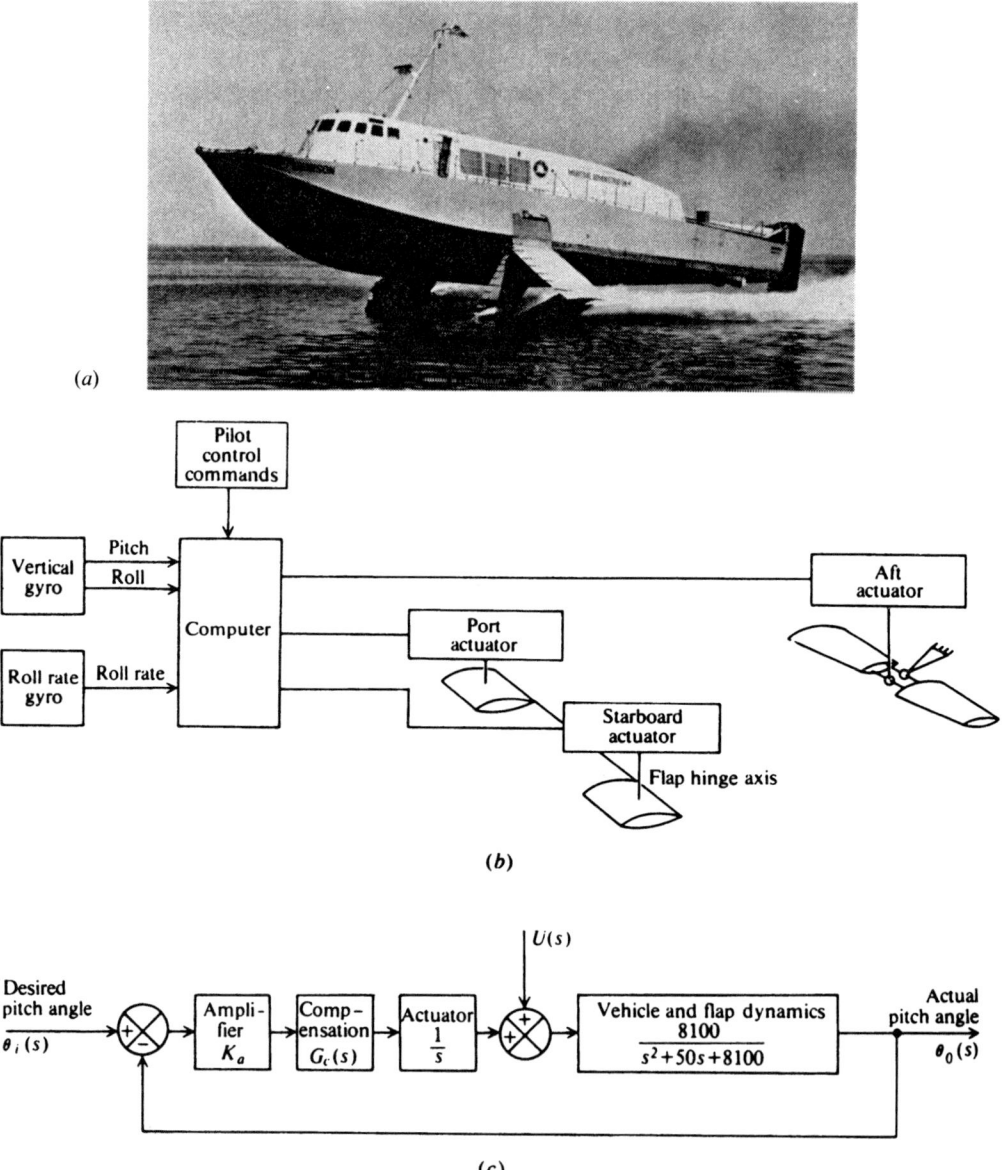

Figure P2.13 (a) Photograph of H. S. Denison. (Courtesy of Grumman Aerospace Corporation) (b) The automatic control system. (c) Block diagram of the pitch-control system.

computer for transmitting commands to the electrohydraulic actuators [13]. Heave rate is fed symmetrically to the forward flaps; roll and roll rate are fed differentially to the forward flaps; pitch rate is fed to the stern foil. The stabilization control system maintains level flight by means of two main surface piercing foils located ahead of the center of gravity and an all-movable submerged foil aft. An equivalent block diagram of the pitch-control system is illustrated in Figure 2.13c. It is desired that the craft maintain a constant level of travel despite a wave disturbance $U(s)$ whose energy is concentrated at 1 rad/sec. Assume that the specifications require that the pitch loop maintain a gain of 40 dB at 1 rad/sec in order to minimize the wave disturbance, and a crossover of 10 rad/sec for adequate response time. In addition, it is desired to have a phase margin of at least 50° at the gain crossover frequency of 10 rad/sec and a gain margin of 12 dB. Select the amplifier gain K_a and compensation network $G_c(s)$ in order to achieve these requirements.

2.14. It is desired that a unity feedback control system whose forward transfer function is $G(s) = 99/[(1 + 0.1s)(1 + 0.2s)(1 + 1.5s)]$ have a phase margin of 45° at the gain crossover frequency. Determine the compensation required and the resulting gain crossover frequency to meet this specification.

2.15. A unity-feedback control system has a forward transfer function given by

$$G_0(s) = \frac{9}{s^2}.$$

(a) Determine the gain crossover frequency and the phase margin of this control system.

(b) A phase-lead network, as defined in Eq. (2.6), is to be added in cascade with $G_0(s)$ so that the phase margin is 70°. Determine the phase-lead network which will achieve this phase margin.

2.16. A remotely piloted aircraft (RPA) for reconnaissance purposes over heavily defended terrain is to be controlled from a ground station. Use of the RPA will eliminate loss of human life if the aircraft is destroyed due to enemy action. The conceptual diagram of the RPA system is illustrated in Figure P2.16a.

An equivalent block diagram of the pitch attitude axis of the RPA system is illustrated in Figure P2.16b.

The transportation lag T_1 represents the delay caused by the man-in-the-loop at the ground station and the time it takes to transmit the signal from the ground station to the RPA. The transportation lag T_2 represents the time it takes for the return signal to be received by the ground station from the RPA. Assume that $T_1 = 0.3$ sec and that $T_2 = 0.05$ sec.

(a) Determine analytically the gain crossover frequency needed to achieve a phase margin of 50°.

106 LINEAR CONTROL-SYSTEM COMPENSATION AND DESIGN

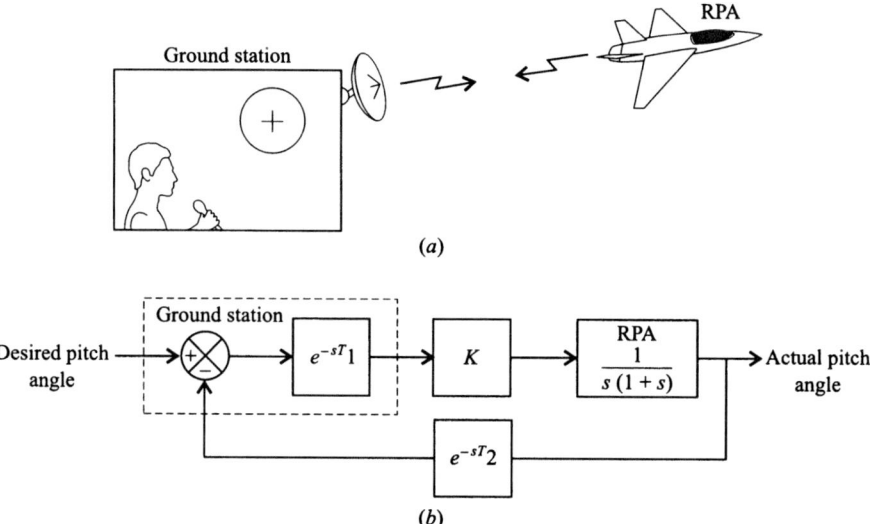

Figure P2.16

(b) Without drawing a Bode diagram, determine the value of K needed to obtain the gain crossover frequency obtained in part (a).

2.17. Space vehicles, such as the space shuttle, using wings to maneuver while reentering the Earth's atmosphere present an interesting control problem.

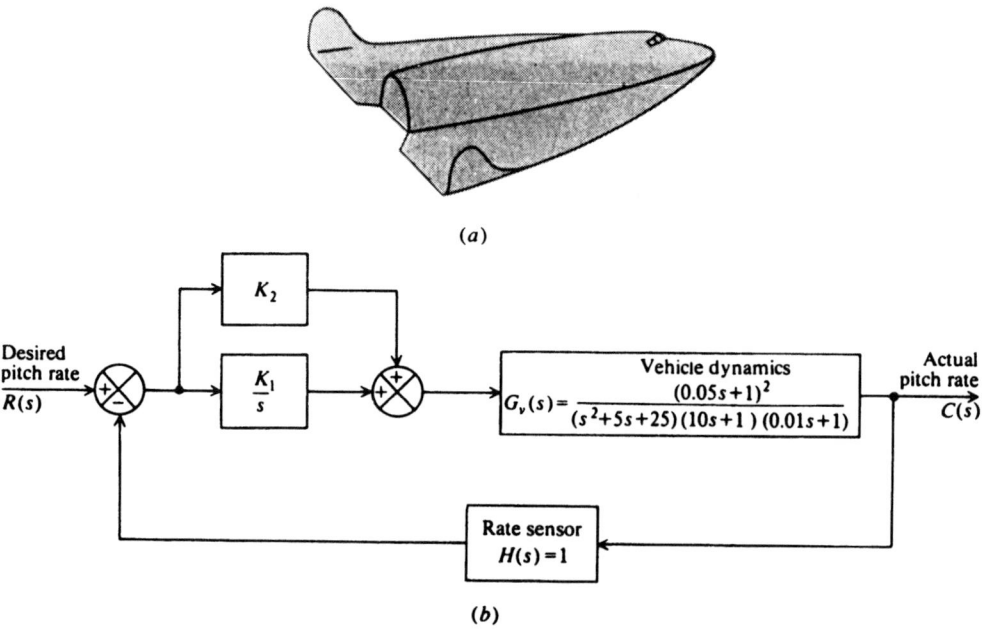

Figure P2.17

Figure P2.17a illustrates a conceptual design of such a system and Figure P2.17b indicates the block diagram of its pitch-rate control system [14].

(a) Draw the Bode diagram of this system with $K_1 = 1$ and $K_2 = 0$. What are the resulting gain and phase margins?

(b) Select the values of K_1 and K_2 which will result in a gain crossover frequency of 1 rad/sec, a phase margin of at least 30° and a gain margin of at least 16 dB.

2.18. It is desired that a unity feedback control system whose forward transfer function is $G(s) = 20/[s(1+s)(1+0.1s)]$ have a phase margin of 55° at the gain crossover frequency, and a gain margin of 25 dB. Determine a phase-lend network (s) and the resulting gain crossover frequency to meet this specification.

2.19. The design of the Lunar Excursion Module (LEM) shown in Figure P2.19a, was an extremely interesting problem [15]. The control, guidance, and navigation for the LEM are provided by an all-digital system from the sensors to the gas-jet propulsion units. For purposes of this analysis, the vehicle dynamics can be approximated by a double integration, as indicated in Figure P2.19b, which illustrates one axis of the attitude-control system. In addition, the torque $T(s)$ is assumed to be proportional to the control signal $U(s)$. Assume that $J = 0.25$ and

$$T(s) = 2U(s).$$

Utilizing the Bode diagram for solution, determine a lead-compensation network $G_c(s)$ which will result in a crossover frequency of 5.1 rad/sec and a phase margin of 60°.

2.20. It is desired to add cascade compensation to a unity feedback control system whose forward transfer function is $G(s) = 60(1+0.5s)/[s(1+5s)]$ in order that the peak overshoot to a step input be approximately 12%.

(a) Using the Nichols chart, design a phase-lead network which can achieve this.

(b) With the compensation network chosen in part (a), determine the closed-loop amplitude and phase-frequency response.

2.21. It is desired to add cascade compensation to a unity feedback control system whose forward transfer function is $G(s) = 19.1/[s(1+s)(1+0.1s)]$ in order that $M_p = 0.75$ while the same steady-state error is maintained.

(a) Design a phase-lead network to achieve this.

(b) With the compensation network chosen in part (a), determine the closed-loop amplitude and phase frequency response.

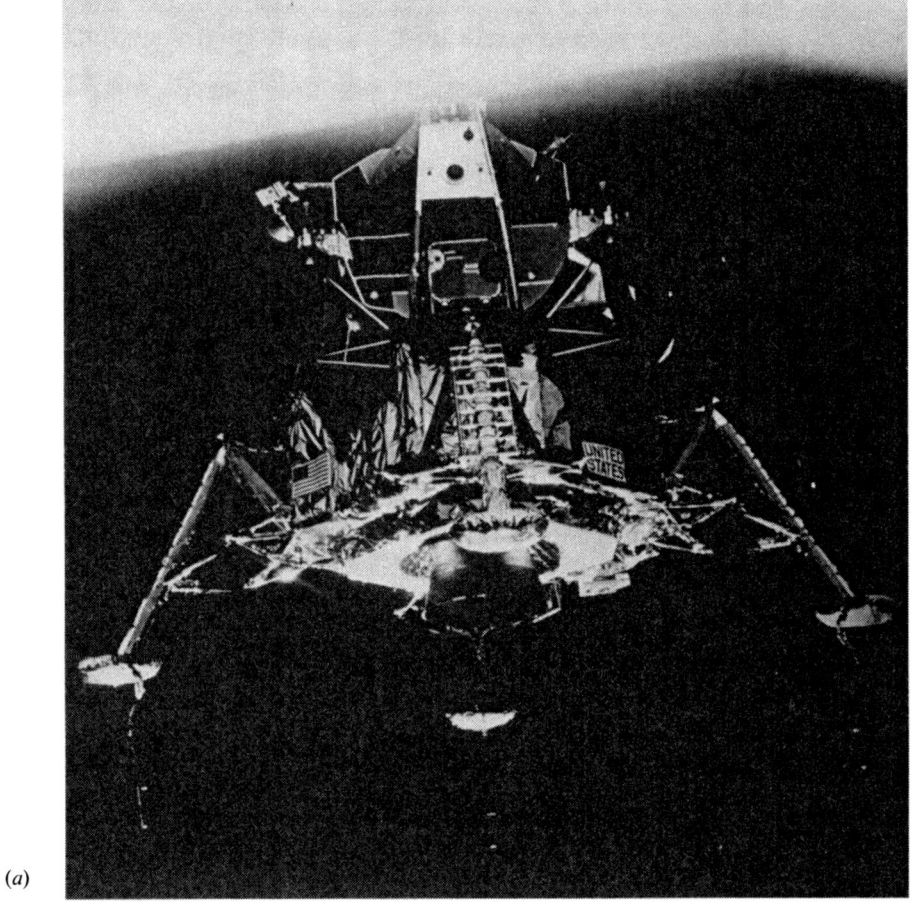

(a)

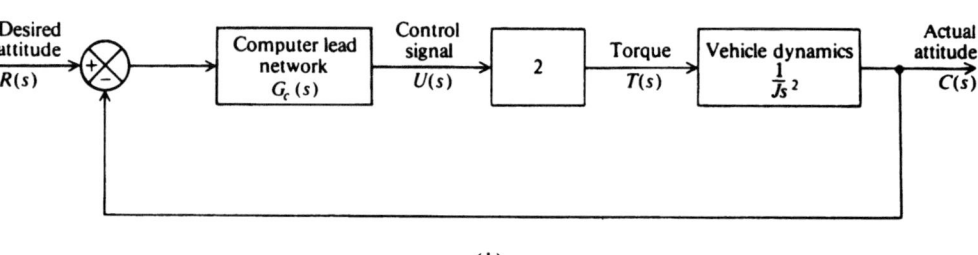

(b)

Figure P2.19

2.22. It is desired that the system considered in Problem 2.21 have a peak overshoot of approximately 15% to a step input.

(a) Utilizing a minor rate-feedback loop, specify the tachometer constant which can achieve this.

(b) What will be the resulting system steady-state error to a unit ramp input with the minor rate-feedback loop added?

(c) Utilizing a simple high-pass *RC* filter in cascade with the tachometer, determine the time constant of the network and the tachometer constant which will result in a 15% overshoot to a step input.

(d) What will be the steady-state error to a unit ramp when the high-pass filter is cascaded with the tachometer?

2.23. It is desired that a unity feedback control system whose forward transfer function is $G(s) = K/[s(1 + 0.1s)(1 + s)]$ have a damping ratio of 0.75 for the dominant complex roots. Using the root-locus method, determine the increase in gain and phase-lag network $(s + n\alpha)/(s + \alpha)$ which can achieve this. Assume $K_v = 15$.

2.24. A unity-feedback control system has a forward transfer function given by:

$$G(s) = \frac{100}{s(1 + 0.1s)(1 + 0.01s)}$$

$$H(s) = 1.$$

(a) Draw the Bode diagram of this control system and determine the resulting phase margin and gain margin.

(b) To achieve an acceptable transient response for this system, the phase margin should be approximately 30° and the gain margin 20 dB. In addition, a sinusoidal disturbance is present at 0.1 rad/sec, and a gain of 60 dB is required at this frequency to nullify its effect. Design a passive compensation network which will achieve the desired transient response and accuracy, and also minimize the noise susceptibility of the system.

2.25. The block diagram of one axis of a robotic positioning system that uses rate feedback for compensation is illustrated in Figure P2.25.

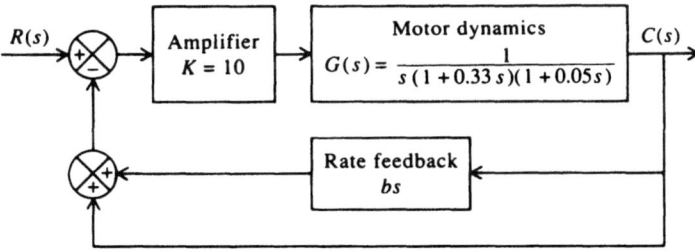

Figure P2.25

(a) Without any rate feedback ($b = 0$), determine the gain crossover frequency, phase margin and gain margin of the uncompensated system using the Bode-diagram method.

(b) For proper operation of the robot, a minimum phase margin of 65° and a minimum gain margin of 90 dB are desired. Using the Bode diagram, determine the amount of rate-feedback constant b which will achieve these requirements.

2.26. The system shown in Figure P2.26 contains a proportional plus integral (PI) controller. The PI controller contains a zero term at $s = -K$ and a pole at $s = 0$. Therefore, the PI controller has infinite gain at zero frequency, and it behaves as a phase-lag network. This feature improves the steady-state characteristics.

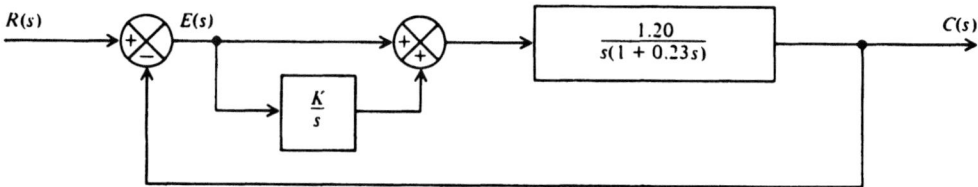

Figure P2.26

(a) Determine the steady-state error of this system to a unit step input.

(b) Determine the steady-state error of this system to a unit ramp input.

(c) Determine the range of gain K for which this system is stable.

2.27. A negative-feedback system containing unity feedback has a forward transfer function given by

$$G(s) = \frac{K(s+A)}{s(s+2)(s+4)}.$$

The zero factor $(s + A)$ in the numerator of this transfer is to be used to compensate the system. Utilizing the root locus, analyze the effects on system stability of the following values of A:

(a) $A = 1$, (b) $A = 2$, (c) $A = 3$,
(d) $A = 4$, (e) $A = 6$.

What conclusions can you draw from your results on the best value of A for compensating this system? What happens if A is greater than 6?

2.28. A unity-feedback control system has a forward transfer function given by:

$$G_0(s) = \frac{1}{(s+1)(s+8)}.$$

We wish to use a proportional plus integral (PI) controller in cascade with $G_0(s)$ to compensate this control system. The transfer function of this compensator is given by:

$$G_c(s) = 1 + \frac{K}{s}.$$

Select K so that the pair of dominant closed-loop poles for this control system is located at $s = -0.5 \pm 0.5j$.

2.29. A unity-feedback control system has the following forward transfer function:

$$G(s) = \frac{K(s+\alpha)}{s^2(s+2)}.$$

Determine the values of α so that the root locus will have zero breakaway points, not including the one at s equal zero.

2.30. The forward transfer function of a unity-feedback control system is given by the following:

$$G(s) = \frac{K}{s^2(s+2)(s+4)}.$$

(a) Construct the root-locus diagram for $0 \leq K \leq \infty$, and show all pertinent values on the root locus. What conclusions can you reach from this root-locus diagram?

(b) The system will be stabilized by means of a rate-feedback element (e.g., tachometer) added in parallel to the unity feedback so that the total feedback transfer function becomes:

$$H(s) = 1 + s.$$

Construct the root locus of this compensated control system for $0 \leq K \leq \infty$, and show all pertinent values on the root locus.

2.31. The transfer functions of a negative feedback system are given by the following:

$$G(s) = \frac{K}{s(s^2 + 6s + 10)},$$
$$H(s) = 1.$$

(a) Sketch the root locus.

(b) Determine $C(s)/R(s)$, with the denominator in factored form, if a damping ratio of 0.5 is required for the dominant roots.

2.32. Determine the increase in gain and phase-lag network compensation required to stabilize a unity-feedback system whose forward transfer function is given by

$$G(s) = \frac{K}{s(s+3)(s+4)}.$$

The requirement for the system damping ratio is 0.707, and the velocity constant is 100/sec.

2.33. A unity-feedback control system has a forward transfer function given by

$$G(s) = \frac{K(s+1)}{s(s+3)(s+6)}.$$

It is desired that the system have a velocity constant of 15 and a damping ratio of 0.75. Using the root-locus method, determine the increase in gain and phase-lag network $(s+n\alpha)/(s+\alpha)$ which can achieve this, assuming that the transient response is governed by a pair of dominant complex-conjugate poles.

2.34. A unity-feedback control system has a forward transfer function given by

$$G(s) = \frac{K(s+1)}{s(s^2+6s+9)}.$$

It is desired that the system have a velocity constant of 150 and a damping ratio of 0.75. Using the root-locus method, determine the increase in gain and phase-lag network, represented by $(s+na)/(s+a)$, which can achieve this. Assume that the transient response is governed by a pair of dominant complex-conjugate poles. Show all pertinent points on the root locus before and after compensation.

2.35. The block diagram of a positioning system is shown in Figure P2.35.

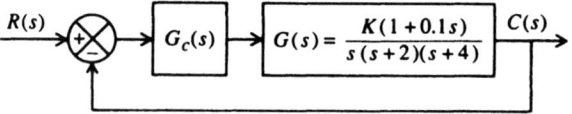

Figure P2.35

(a) Without any compensation, $G_c(s) = 1$, draw the root locus of the uncompensated system. On this diagram determine and show the following clearly:
- Point (s) of breakaway from the real axis
- Crossing (s) of the imaginary axis
- K_{max}
- All asymptotes.

(b) It is desired that the system have a velocity constant of 4 and a damping ratio of 0.707. Using the root-locus method, determine the compensation $G_c(s)$ which can achieve this. Assume that the transient response is governed by a pair of dominant complex-conjugate poles. Show all pertinent changes and points on the root locus after compensation. (An exact recalculation of the point (s) of breakaway, the crossing of the imaginary axis, and K_{max} are not necessary.)

2.36. In order to obtain a control system which has an infinite velocity constant, a control-system engineer designs the control system of Figure P2.36, which he or she recognizes will need some compensation network, $G_c(s)$.

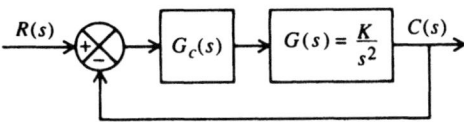

Figure P2.36

(a) Without any compensation, $G_c(s) = 1$, draw the root locus of this control system. From a practical viewpoint, is it stable?

(b) The control-system engineer desires to use a passive compensation network for $G_c(s)$, either a phase-lag or a phase-lead network. Assume that the resistors and capacitors that are available can provide a zero at -1 and a pole at -2, *or* a pole at -1 and a zero at -2. As the control-system engineer, which would you select and why? Draw the root locus for your stabilized control system (with either a phase-lag or a phase-lead network). Show all pertinent points on your root locus.

2.37. The signal-flow graph of the temperature-control loop for a xylene chemical process [16] is shown in Figure P2.37. The temperature of the process, $C(s)$, is related to the heat supplied to the process by the quadratic transfer function $G_p(s)$. Temperature is measured by a sensor having a pole at $s = -0.3$, and the output of the sensor in the form of air pressure is compared with the desired value of temperature as indicated by the reference pressure $R(s)$. The pressure difference (a measure of temperature error) actuates a pneumatic controller which provides as its output a pneumatic actuating signal applied to a steam valve. The valve, in turn, controls the flow of heat to the xylene column in order to minimize the error.

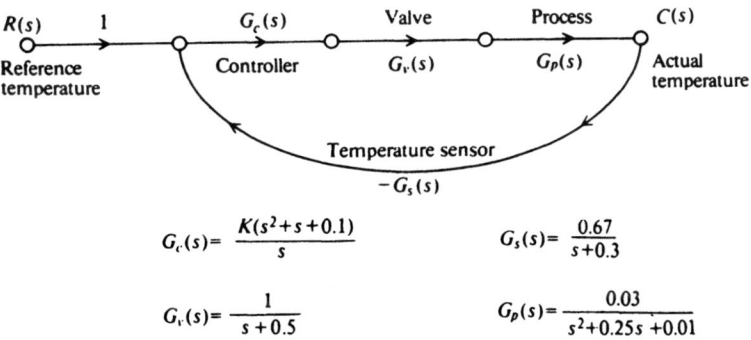

Figure P2.37

(a) Draw the root locus for this system.

(b) Determine the required gain K for a damping ratio of 0.5.

2.38. It is desired that a unity feedback control system whose forward transfer function is $G(s) = K(1+s)/[s^2(1+0.1s)]$ have a damping ratio of 0.75 for the dominant complex roots. Using the root-locus method, determine the value of K which can achieve this.

2.39. A turbine speed-control system is illustrated in Figure P2.39. Assume that the transfer function of the control valve, turbine, and speed converter are:

$$G_1(s) = \frac{1}{s+0.1},$$

$$G_2(s) = \frac{0.5}{s^2+3s+2},$$

$$H(s) = 1.$$

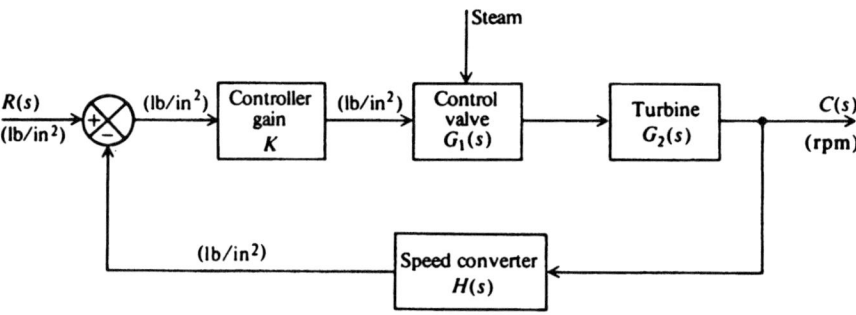

Figure P2.39

Assuming that the transient response is governed by a pair of dominant complex-conjugate poles, determine the value of the controller gain K in order that the system has a damping ratio of 0.5.

2.40. It is desired that a unity feedback control system whose forward transfer function is $G(s) = K(s+1)/[s(s^2+8s+16)]$ have a damping ratio of 0.75 for the dominant complex roots. Using the root-locus method:

(a) Determine the value of K which can achieve this.

(b) Determine the values of ω_p and M_p for the value of K found in part (a) using the Nichols chart.

2.41. Determine the gain needed for a unity feedback control system whose forward transfer function is $G(s) = K(s+0.1)^2/[s^2(s^2+9s+20)]$ which can achieve a damping ratio of 0.75.

2.42. Unlike fixed-wing aircraft, which possess a moderate degree of inherent stability, the helicopter is very unstable and requires the use of feedback loops for stabilization. A typical control system involves the use of an inner automatic stabilization loop and an outer loop which is controlled by the pilot, who inserts commands into it based on attitude errors displayed to the pilot.

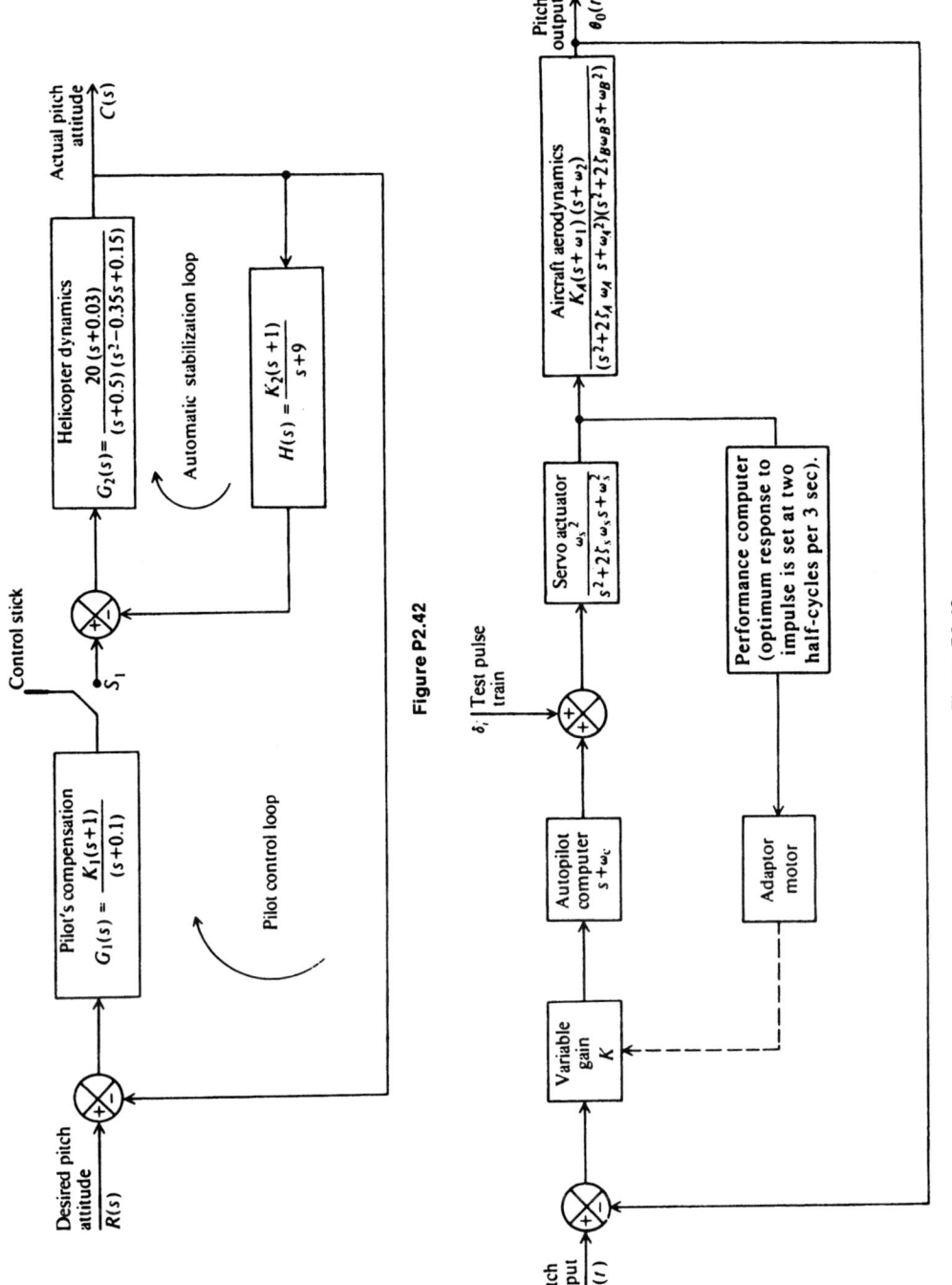

Figure P2.42

Figure P2.43

Figure P2.42 illustrates the pitch-control system used on the S-55 helicopter [17]. When the pilot is not utilizing the control stick, the switch S_1 is open, which disengages the pilot control loop. The model of the pilot's transfer function, $G_1(s)$, includes a gain factor, an anticipation time constant of 1 sec, and an error-smoothing time constant of 10 sec [17].

(a) With the pilot control loop open, plot the root locus for the automatic stabilization loop and determine the gain K_2 which results in a damping ratio of 0.5 for the dominant complex roots.

(b) Draw the root locus of the pilot control loop with K_2 set at the value determined in part (a). Determine the value of the pilot's gain compensation factor K_1 in order that the pilot control loop have a damping ratio of 0.5.

2.43. Many modern control systems are designed to be adaptive, in order that they can achieve a desired response in the presence of extreme changes in the system parameters and major external disturbances. Adaptive control systems are usually characterized by devices that automatically measure the dynamics of the controlled system and by other devices that automatically adjust the characteristics of the controlled elements based on a comparison of the measurements with some optimum figure of merit. Figure P2.43 illustrates an adaptive pitch flight control system [18]. It attempts to measure the exact location of a pair of dominant, variable, servo actuator poles that move in the complex plane, as a function of the flight conditions. The adaptive feature overcomes this problem by adjusting the gain in order to keep the location of these sensitive poles fixed in the complex plane. A test impulse train is injected into the system when the error is small. The performance computer determines the transient response of the system and compares it with an optimum desired response that is set at 2 half cycles of a transient response over a 3-sec interval of time. The performance computer is designed so that a count less than 2 over a 3-sec period will cause the adaptor motor to increase K, and a count greater than 2 over a 3-sec period will cause the adaptor motor to decrease K. Assume that the poles and zeros of the system are located in the complex plane as follows:

Table P2.43. Helicopter Control System Parameters

	Poles	Zeros	Gain
Aircraft aerodynamics	$P_a = -4 \pm j2$ $P_B = -1 \pm j1$	$\omega_1 = -3$ $\omega_2 = -2$	$K_A = 0.02$
Servo actuator	$P_C = -3 \pm j6$	—	—
Autopilot computer	—	$\omega_c = -6$	—

(a) Draw the root locus of this system.

(b) Determine the variable gain K that will result in a damping ratio of 0.3, assuming that the transient response is governed by a pair of dominant complex-conjugate poles.

2.44. Repeat Problem 2.33 for

$$G(s) = \frac{K(s+1)}{s(s^2 + 6s + 9)}.$$

REFERENCES

1. W. R. Ahrendt, *Servomechanism Practice*. McGraw-Hill, New York, 1954.
2. J. G. Truxal, *Automatic Feedback Control System Synthesis*. McGraw-Hill, New York, 1955.
3. R. E. Kalman, "On the general theory of control systems." In *Proceedings of the First International Congress of Automatic Control*, Moscow, 1960.
4. R. E. Kalman, Y. C. Ho, and K. S. Navendra, "Controllability of linear dynamical systems," *Contrib. Diff. Equat.*, **1**, 189–213 (1961).
5. H. Lauer, R. N., Lesnick, and L. E. Matson, *Servomechanism Fundamentals*. McGraw-Hill, New York, 1960.
6. J. G. Truxal, ed. *Control Engineer's Handbook*. McGraw-Hill, New York, 1955.
7. H. W. Bode, *Network Analysis and Feedback Amplifier Design*. Van Nostrand, New York, 1945.
8. L. A. Gould, *Chemical Process Control: Theory and Applications*. Addison-Wesley, Reading, MA, 1969.
9. H. Chestnut and R. W. Mayer, *Servomechanisms and Regulating System Design*, 2nd ed., Vol. 1, Wiley, New York, 1959.
10. J. L. Bower and P. M. Schultheiss, *Introduction to the Design of Servomechanisms*. Wiley, New York, 1958.
11. S. M. Shinners, "How to approach the stability analysis and compensation of control systems." *Control Eng.* **25**(5), 62–67 (1978).
12. W. K. Waymeyer and R. W. Sporing, "Closed loop adaptation applied to missile control," In *Proceedings of the 1962 Joint Automatic Control Conference*. New York, p. 18-33.
13. R. M. Rose, *The Rough Water Performance of the H. S. Denison*, Paper No. 64–197. Am. Inst. Aeronaut. Astronaut., Washington, DC, 1984.
14. R. P. Kotfile and S. S. Oseder, "Stabilization and control of maneuvering reentry vehicle." *Sperry Eng. Rev.* **18**, 2–10 (1965).
15. F. Doennebrink and J. Russel, "LEM stabilization and control system." *AIAA/ION Guidance and Control Conference, 1965*, pp. 430–441.
16. W. A. Lynch and J. G. Truxal, *Principles of Electronic Instrumentation*. McGraw-Hill, New York, 1962.
17. L. Kaufman, "Helicopter control stick steering," *Sperry Eng. Rev.* **11**, 41–48 (1958).
18. F. C. Gregory, ed. *Proceedings of the Self-Adaptive Flight Control Systems Symposium*, WADC Techn. Rep. No. 59-49, ASTIA Doc. No. AD 209389, 1959.
19. S. M. Shinners, *Techniques of System Engineering*. McGraw-Hill, New York, 1967.

3
MODERN CONTROL-SYSTEM DESIGN USING STATE-SPACE, POLE PLACEMENT, ACKERMANN'S FORMULA, ESTIMATION, ROBUST CONTROL, AND H^∞ TECHNIQUES

3.1. INTRODUCTION

State-space analysis was introduced in Chapter 1, and has been used in parallel with the classical frequency-domain analyses techniques presented in Chapter 2. The state-space approach is applicable to a wider class of problems such as multiple-input/multiple-output (MIMO) control systems. Chapter 2 applied the frequency-domain approaches such as the Bode diagram, and the root locus to linear control-system design.

In the design of a control system, the question arises as to where to place the closed-loop roots. In Section 2.9 which presented the root-locus method, we could specify where to place the dominant-pair of complex-conjugate roots in order to obtain a desired transient response. However, we could not do so with great certainty because we were never sure what effect the higher-order poles would have on the second-order approximation.

The control-system design engineer desires to have design methods available which would enable the design to proceed by specifying all of the closed-loop poles of higher-order control systems. Unfortunately, the frequency-domain design methods presented in Chapter 2 do not permit the control-system engineer to specify all poles in control systems which are higher than two because they do not provide a sufficient number of unknowns for solving uniquely for the specified closed-loop poles. This problem is overcome using state-space methods which provide additional adjustable parameters, and methods for determining these parameters.

This chapter presents a modern control-system design method using state-space techniques known as *pole placement* or *pole assignment*. This design technique is similar to what we did in Section 2.9 where we placed two dominant complex-conjugate poles of the closed-loop transfer function in desired locations in order to obtain desirable transient responses. However, in this chapter, we will show how pole placement allows the control-system engineer to place all of the poles of the closed-loop transfer function in desirable locations. Ackermann's formula is also presented for designs using pole placement for application in those control systems that require feedback from state variables which are not phase variables (where each subsequent state variable is defined as the derivative of the previous state variable). A practical problem arises with the pole placement method involving cost and the availability of determining (measuring) all of the system variables needed for obtaining a solution. In many practical control systems, all of the system state variables may not be available due to cost considerations, environmental considerations (e.g., nuclear power plant control systems), and the availability of transducers to measure certain states. For these cases, it is necessary for the control-system engineer to estimate the state variables that cannot be measured from the state variables that can be measured. Therefore, in addition to pole placement, this chapter also presents the very important subjects of controllability, observability, and estimation.

This chapter on modern control-system design also presents the design of robust control systems. Robust control systems are concerned with determining a stabilizing controller that achieves feedback performance in terms of stability and accuracy requirements, but the controller must achieve performance that is robust (insensitive) to plant uncertainty, parameter variation, and external disturbances. The design of two-degrees-of-freedom compensation control systems exhibiting desirable robustness to plant uncertainty, parameter variation, and external disturbances is presented.

This chapter concludes with an introduction to H^∞ control concepts which is a new technique that emerged in the 1980s that combines both the frequency- and time-domain approaches to provide a unified design approach. The H^∞ approach has dominated the trend of control-system development in the 1980s and 1990s. The H^∞ control-system design approach expands on the concept of robustness presented in this chapter, sensitivity (presented in Chapter 1), together with the frequency and state-variable domain techniques presented in this book. The H^∞ approach is applied to determine the optimum sensitivity for control systems.

3.2. POLE-PLACEMENT DESIGN USING LINEAR-STATE-VARIABLE FEEDBACK

Having presented methods for designing linear control systems using classical techniques, let us now look at the problem of specifying pole placement from the viewpoint of state-variable feedback [1]. In order to do this, let us first look at the basic feedback problem illustrated in Figure 3.1. This figure illustrates the concept of feeding back the states of the process in addition to that of the output. Because a linear process can be characterized by the phase-variable canonical equations

$$\dot{\mathbf{x}}(t) = \mathbf{P}\mathbf{x}(t) + \mathbf{b}u(t), \tag{3.1}$$

$$c(t) = \mathbf{L}\mathbf{x}(t), \tag{3.2}$$

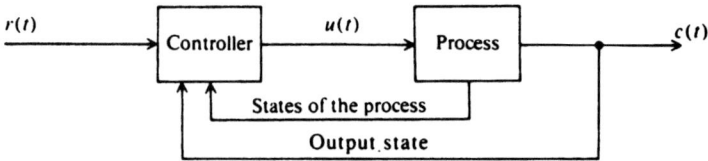

Figure 3.1 General feedback system problem illustrating feedback of the output state and the states of the process.

let us consider the configuration of Figure 3.2. It is important to observe from this figure that the control signal is generated from a knowledge of the reference input $r(t)$ and the state variables $\mathbf{x}(t)$. Note that $r(t)$, $u(t)$, and $c(t)$ represent scalars.

In general, the control input u can be represented as

$$u(t) = f(\mathbf{x}(t), r(t)).$$

Rather than considering the controller in such a broad sense, let us consider the specific condition of linear state-variable feedback where the controller weights the sum of the state variables in a linear manner. In addition, it is assumed that the controller provides a linear gain K which multiplies the difference between the reference input and the linear weighted sum of state variables fed back. Therefore, $u(t)$ can be represented as

$$u(t) = K[r(t) - (h_1 x_1(t) + h_2 x_2(t) + h_3 x_3(t) \cdots + h_n x_n(t))], \tag{3.3}$$

where h_i is defined as the ith feedback coefficient. In matrix form, $u(t)$ can be represented as

$$u(t) = K[r(t) - \mathbf{h}\mathbf{x}(t)], \tag{3.4}$$

where

$$\mathbf{h} = [h_1 \; h_2 \; h_3 \cdots h_n], \tag{3.5}$$

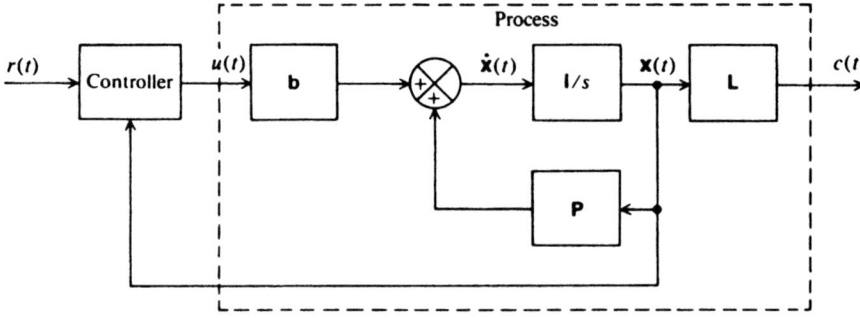

Figure 3.2 General feedback system with state-variable feedback.

3.2. POLE-PLACEMENT DESIGN USING LINEAR-STATE-VARIABLE FEEDBACK

$$\mathbf{x}(t) = \begin{bmatrix} x_1(t) \\ x_2(t) \\ x_3(t) \\ \vdots \\ x_n(t) \end{bmatrix}. \tag{3.6}$$

Figure 3.3 presents a matrix representation of the concept of linear-state-variable feedback, and Figure 3.4 is an example of a physical representation of a typical system as implied by Figure 3.3. In the following discussion, it is assumed that all state variables are directly available for measurement and control. In practice, this is not always possible, and techniques for modifying and extending the design procedure presented, to the case where all the state variables are not available, are also discussed in Sections 3.6 and 3.7.

How does linear feedback of the state variables affect the behavior of the process given by Eqs. (3.1) and (3.2)? This can easily be determined by substituting Eq. (3.4) into Eq. (3.1):

$$\dot{\mathbf{x}}(t) = \mathbf{P}\mathbf{x}(t) + \mathbf{b}[K(r(t) - \mathbf{h}\mathbf{x}(t))]. \tag{3.7}$$

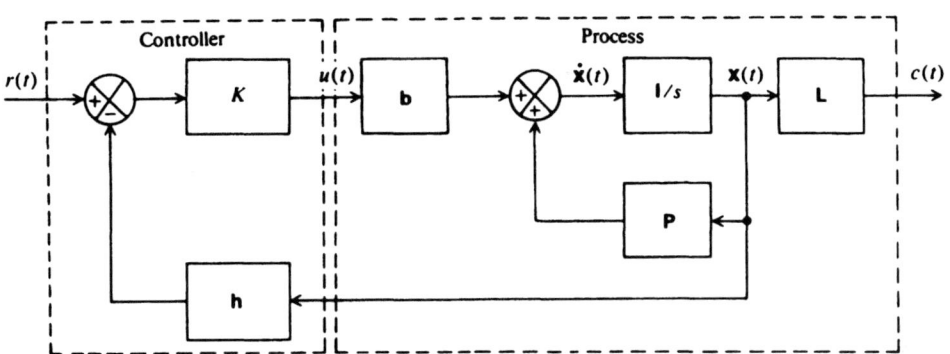

Figure 3.3 Linear-state-variable feedback representation.

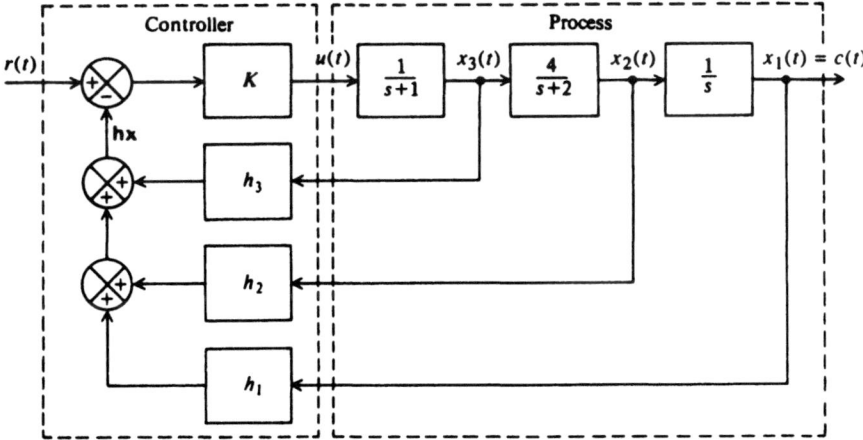

Figure 3.4 Example of pole placement design using linear-state-variable feedback.

Simplifying Eq. (3.7), and incorporating Eq. (3.2), we obtain the closed-loop equation

$$\dot{x}(t) = P_h x(t) + K b r(t), \qquad (3.8)$$

$$c(t) = Lx, \qquad (3.9)$$

where

$$P_h = P - K b h \qquad (3.10)$$

is the closed-loop-system matrix. Comparing Eq. (3.1) and (3.2) with (3.8) and (3.9), we observe that they are identical except that the P matrix has been replaced by P_h and $u(t)$ becomes $Kr(t)$.

How can we relate the closed-loop-system matrix P_h to the closed-loop transfer function, $C(s)/R(s)$? This can be accomplished by taking the Laplace transform of Eqs. (3.8) and (3.9). Because the results will be used to find a transfer function, all initial conditions are assumed to be zero:

$$sX(s) = P_h X(s) + K b R(s), \qquad (3.11)$$
$$C(s) = L X(s). \qquad (3.12)$$

Solving for $X(s)$ from Eq. (3.11), we get

$$X(s) = K[sI - P_h]^{-1} b R(s). \qquad (3.13)$$

The inverse matrix $[sI - P_h]^{-1}$ is defined as the closed-loop resolvent matrix, $\Phi_h(s)$, where

$$\Phi_h(s) = [sI - P_h]^{-1}. \qquad (3.14)$$

Therefore, Eq. (3.13) may be rewritten as

$$X(s) = K \Phi_h(s) b R(s). \qquad (3.15)$$

Substituting Eq. (3.15) into Eq. (3.12), we obtain a relation between $C(s)$ and $R(s)$:

$$C(s) = K L \Phi_h(s) b R(s). \qquad (3.16)$$

Therefore, the closed-loop transfer function in terms of the closed-loop resolvent matrix is given by

$$\frac{C(s)}{R(s)} = K L \Phi_h(s) b \qquad (3.17)$$

In addition, the characteristic equation in terms of the closed-loop-system matrix can also easily be determined, simply by substituting the numerator and denominator portions of the inverse matrix, $\Phi_h(s)$:

$$\frac{C(s)}{R(s)} = \frac{KL[\text{adj}(sI - P_h)]b}{\det(sI - P_h)}. \qquad (3.18)$$

3.2. POLE-PLACEMENT DESIGN USING LINEAR-STATE-VARIABLE FEEDBACK

The corresponding characteristic equation of the closed-loop system in terms of the closed-loop system matrix is given by

$$\det(s\mathbf{I} - \mathbf{P}_h) = 0. \quad (3.19)$$

Because we are concerned with synthesizing control systems in terms of pole placement design using linear-state-variable feedback concepts, we would like to force the system illustrated in Figure 3.3 into the generalized form illustrated in Figure 3.5, and study its properties. Let us first consider the derivation of $H(s)$. From Figure 3.5 we observe that

$$H(s) = \frac{\mathbf{h}X(s)}{C(s)}. \quad (3.20)$$

Substituting Eq. (3.12) into Eq. (3.20), we obtain

$$H(s) = \frac{\mathbf{h}X(s)}{\mathbf{L}X(s)}. \quad (3.21)$$

After substitution of Eq. (3.15) for $X(s)$, Eq. (3.21) becomes

$$H(s) = \frac{\mathbf{h}\boldsymbol{\Phi}_h(s)\mathbf{b}}{\mathbf{L}\boldsymbol{\Phi}_h(s)\mathbf{b}}. \quad (3.22)$$

The term $G(s)$ can also be derived in terms of $\boldsymbol{\Phi}_h(s)$. The closed-loop transfer function of the system illustrated in Figure 3.5 is given by

$$\frac{C(s)}{R(s)} = \frac{KG(s)}{1 + KG(s)H(s)}. \quad (3.23)$$

Substituting Eqs. (3.17) and (3.22) into Eq. (3.23), we obtain the expression

$$G(s) = \frac{\mathbf{L}\boldsymbol{\Phi}_h(s)\mathbf{b}}{1 - K\boldsymbol{\Phi}_h(s)\mathbf{b}}. \quad (3.24)$$

Combining Eqs. (3.22) and (3.24), the open-loop transfer function is found to be given by

$$KG(s)H(s) = \frac{K\mathbf{h}\boldsymbol{\Phi}_h(s)\mathbf{b}}{1 - K\mathbf{H}\boldsymbol{\Phi}_h(s)\mathbf{b}}. \quad (3.25)$$

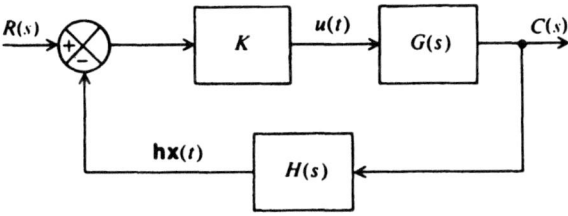

Figure 3.5 An equivalent model of Figure 3.3.

Let us compare Eqs. (3.17), (3.22)–(3.24), and (3.25) in order to draw conclusions regarding $G(s)$, $H(s)$, the open-loop transfer function $KG(s)H(s)$, and the closed-loop transfer function $C(s)/R(s)$. These characteristics will be important for designing systems using pole placement techniques with linear-state-variable feedback techniques in the following section. Based on these five equations, we can state the following properties:

1. The poles of $KG(s)H(s)$ are the poles of $G(s)$.
2. The zeros of $C(s)/R(s)$ are the zeros of $G(s)$.
3. The pole-zero excess of $C(s)/R(s)$ must be equal to the pole-zero excess of $G(s)$.

With these properties as a basis, we consider in the following section the design of control systems from the viewpoint of linear-state-variable feedback.

3.3. CONTROLLER DESIGN USING POLE PLACEMENT AND LINEAR-STATE-VARIABLE FEEDBACK TECHNIQUES

The preceding section has indicated several important relationships between open-loop and closed-loop transfer functions. This is very important in the design of control systems for the case where the closed-loop transfer function is specified and it is desired to determine the open-loop transfer function. A typical problem might specify the desired velocity constant; then use is made of Eq. (C.29) in Appendix C which gave the velocity constant in terms of the closed-loop poles and zeros. The problem is to determine the resulting linear-state-variable feedback system.

Let us illustrate the procedure by considering the following problem. It is desired that the closed-loop characteristics of a unity-feedback control system be given by the following parameters:

$$\omega_n = 50 \text{ rad/sec}, \quad K_v = 35/\text{sec}, \quad \zeta = 0.707.$$

What form of closed-loop transfer function will satisfy these requirements? Let us first try a simple quadratic control system having a pair of complex-conjugate poles. From Eq. (C.31) of Appendix C, such a system has a velocity constant given by

$$K_v = \frac{\omega_n}{2\zeta} = \frac{50}{2(0.707)} = 35.7/\text{sec}. \tag{3.26}$$

Therefore, a simple quadratic control system having a pair of complex-conjugate poles will satisfy these specifications. From Eq. (B.18) of Appendix B,

$$\cos \alpha = \zeta.$$

For a damping ratio of 0.707, $\alpha = 45°$ and the relations among the complex-conjugate poles, ω_n and ζ are illustrated in Figure 3.6. Therefore, the closed-loop control system is given by

$$\frac{C(s)}{R(s)} = \frac{\omega_n^2}{s^2 + 2\zeta\omega_n s + \omega_n^2}. \tag{3.27}$$

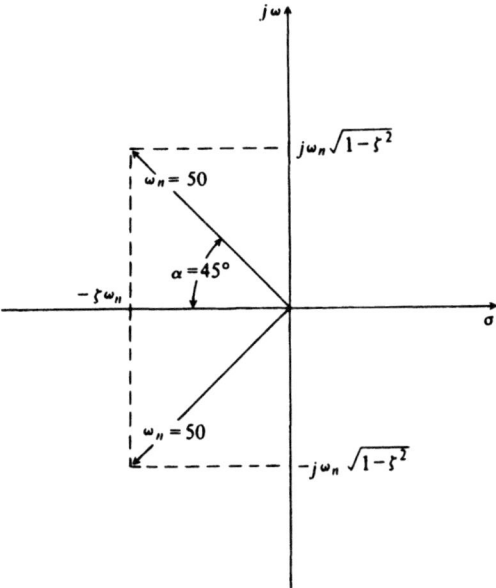

Figure 3.6 Closed-loop poles.

By substituting $\zeta = 0.707$ and $\omega_n = 50$ into Eq. (3.27), we obtain the following desired closed-loop transfer function:

$$\frac{C(s)}{R(s)} = \frac{2500}{s^2 + 70.7s + 2500}. \quad (3.28)$$

Let us assume that the open-loop process that is being controlled is illustrated in Figure 3.7. The corresponding state-variable representation is readily found to be

$$\dot{\mathbf{x}}(t) = \begin{bmatrix} 0 & 1 \\ 0 & -70 \end{bmatrix} \mathbf{x}(t) + \begin{bmatrix} 0 \\ 1 \end{bmatrix} u(t), \quad (3.29)$$

$$c(t) = \begin{bmatrix} 1 & 0 \end{bmatrix} \mathbf{x}(t) \quad (3.30)$$

where

$$\mathbf{x}(t) = \begin{bmatrix} x_1(t) \\ x_2(t) \end{bmatrix} = \begin{bmatrix} c(t) \\ \dot{c}(t) \end{bmatrix}.$$

The resulting linear-state-variable feedback representation is illustrated in Figure 3.8. This feedback represenation can be simplified by the configuration illustrated

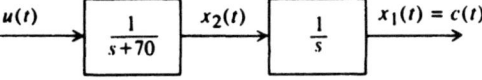

Figure 3.7 Open-loop process to be controlled.

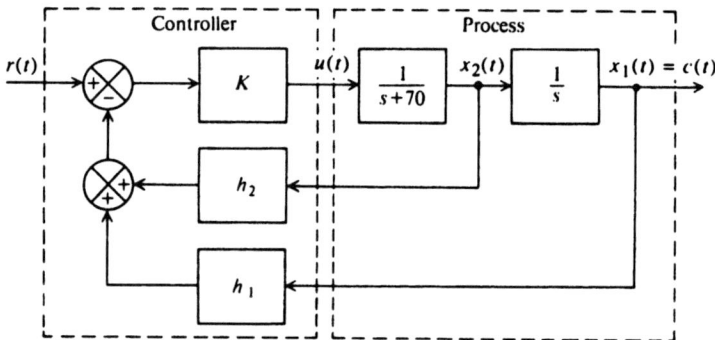

Figure 3.8 State-variable feedback representation of system.

in Figure 3.9. The resulting closed-loop transfer function is given by

$$\frac{C(s)}{R(s)} = \frac{K/[s(s+70)]}{1 + (h_1 + h_2 s)K/[s(s+70)]}, \tag{3.31}$$

which can be reduced to the following expression:

$$\frac{C(s)}{R(s)} = \frac{K}{s^2 + (70 + h_2 K)s + K h_1}. \tag{3.32}$$

The values of K, h_1 and h_2 can be found from Eqs. (3.28) and (3.32). The following set of simultaneous equations result:

$$K = 2500, \tag{3.33}$$

$$70 + h_2 K = 70.7, \tag{3.34}$$

$$K h_1 = 2500. \tag{3.35}$$

We have three equations and three unknowns. Solving, we find that $h_1 = 1$, $K = 2500$, and $h_2 = 2.8 \times 10^{-4}$. The final step is to draw the root locus and examine the relative stability, and the sensitivity as a function of slight gain variations. For this simple system, the final step is not necessary.

Although this simple example has been solved using block diagrams and transfer functions, it could also have been solved using the matrix-algebra approach. To illustrate this, let us pick up this problem from Eq. (3.28) which is the desirable closed-loop transfer function. We want to determine the closed-loop transfer function

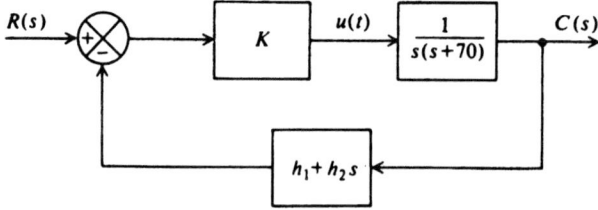

Figure 3.9 Equivalent configuration of Figure 3.8.

for the linear-state-variable-feedback control system using Eq. (3.18) and knowledge of the **P** and **B** matrices from Eq. (3.29), and the matrix **L** from Eq. (3.30) as follows:

$$\frac{C(s)}{R(s)} = \frac{K\mathbf{L}[\text{adj}(s\mathbf{I} - \mathbf{P} + K\mathbf{b}\mathbf{h})]\mathbf{b}}{\det(s\mathbf{I} + \mathbf{P} + K\mathbf{b}\mathbf{h})} = \frac{K[1\ 0]\,\text{adj}\begin{bmatrix} s & -1 \\ Kh_1 & s+70+Kh_2 \end{bmatrix}\begin{bmatrix} 0 \\ 1 \end{bmatrix}}{s^2 + (70 + Kh_2)s + Kh_1}. \quad (3.36)$$

Simplifying Eq. (3.36), we obtain the following:

$$\frac{C(s)}{R(s)} = \frac{K[1\ 0]\begin{bmatrix} s+70+Kh_2 & 1 \\ -Kh_1 & s \end{bmatrix}\begin{bmatrix} 0 \\ 1 \end{bmatrix}}{s^2 + (70 + Kh_2)s + Kh_1}. \quad (3.37)$$

Simplifying Eq. (3.37) results in the following expression for the closed-loop transfer function of the system:

$$\frac{C(s)}{R(s)} = \frac{K}{s^2 + (70 + Kh_2)s + Kh_1}. \quad (3.38)$$

Equation (3.38) is identical to the closed-loop transfer function we obtained in Eq. (3.32) which was obtained from the block diagram shown in Figure 3.8. Therefore, we repeat the process of setting like terms equal to each other from Eq. (3.38) and Eq. (3.28). The resulting three simultaneous equations of (3.33) through (3.35) will be identical, and the resulting parameters of $K = 1$, $h_1 = 1$, and $h_2 = 2.8 \times 10^{-4}$ will be the same as found before.

With this fundamental example as a basis, the general pole placement design procedure can be formulated as follows:

1. Determine the desired closed-loop transfer function based on the discussion of Appendix C.
2. Determine the representation of the process to be controlled.
3. Represent the closed-loop system in terms of an equivalent linear-state-variable-feedback configuration.
4. Determine the closed-loop transfer function $C(s)/R(s)$ from the equivalent model in terms of K and **h**.
5. Equate the $C(s)/R(s)$ expressions from Steps 1 and 4 and determine K and **h**.*
6. Plot the resulting root locus of $KG(s)H(s)$ and evaluate the relative stability and sensitivity as a function of gain variations.

Let us apply this pole placement procedure next to the following more complex example. The problem concerns the control of a process in a unity-feedback closed-loop system whose transfer function is given by

$$G(s) = \frac{1}{s(s+1)(s+10)}. \quad (3.39)$$

---

*This assumes that all of the states are measurable. Sections 3.7 and 3.8 discuss the approach to be taken when all the states are not measurable.

It is assumed that the transient response of the system is governed by a pair of dominant complex-conjugate poles, and that the following parameters are desired:

$$K_v = 0.93,$$
$$\zeta = 0.707,$$
$$\omega_n = 1 \text{ rad/sec}.$$

What should the closed-loop transfer function be? From Eq. (3.26), a pair of complex-conjugate poles in the denominator would only have a velocity constant given by

$$K_v = \frac{\omega_n}{2\zeta} = 0.707. \tag{3.40}$$

Therefore, a simple pair of complex-conjugate poles is inadequate to meet the velocity constant requirement of 0.93. By examining Eq. (C.29), we conclude that a zero Z must be added to the closed-loop transfer function. How many poles should the closed-loop system have? Because

$$(N_{P_c} - N_{Z_c})_{C/R} = (N_{P_0} - N_{Z_0})_G, \tag{3.41}$$

where

N_{P_c} = number of closed-loop poles = ?
N_{Z_c} = number of closed-loop zeros = 1,
N_{P_0} = number of open-loop poles = 3,
N_{Z_0} = number of open-loop zeros = 0

therefore,

$$(N_{P_c} - 1) = (3 - 0), \tag{3.42}$$

and

$$N_{P_c} = 4. \tag{3.43}$$

Since the resulting unity-feedback, closed-loop transfer function has to have one zero and four poles, it has the following general form:

$$\frac{C(s)}{R(s)} = \frac{\omega_n^2 P_3 P_4}{Z} \frac{(s+Z)}{(s^2 + 2\zeta\omega_n s + \omega_n^2)(s+P_3)(s+P_4)}. \tag{3.44}$$

The value of the zero Z can be found from Eq. (5.35) as follows:

$$\frac{1}{K_v} = \frac{2\zeta}{\omega_n} + \frac{1}{P_3} + \frac{1}{P_4} - \frac{1}{Z}. \tag{3.45}$$

Due to external overall system factors in which this feedback system is to operate, it is assumed that the poles at P_3 and P_4 are specified to occur at 9 and 16, respectively. Therefore,

$$\frac{1}{0.93} = \frac{2(0.707)}{1} + \frac{1}{9} + \frac{1}{16} - \frac{1}{Z}, \tag{3.46}$$

so that $Z = 2$, and the desired closed-loop transfer function is given by

$$\frac{C(s)}{R(s)} = \frac{72(s+2)}{(s^2 + 1.414s + 1)(s+9)(s+16)} \tag{3.47}$$

or

$$\frac{C(s)}{R(s)} = \frac{72(s+2)}{s^4 + 26.4s^3 + 180.4s^2 + 299s + 144}. \tag{3.48}$$

Because the zeros of $G(s)$ must be the same as that of $C(s)/R(s)$, we must also add the factor $(s+2)$ to the numerator of $G(s)$. Then, to satisfy Eq. (3.41), we must add a pole factor $(s+\alpha)$ to the denominator of $G(s)$. The resulting compensating network to be added to $G(s)$ is given by

$$G_c(s) = \frac{s+2}{s+\alpha}, \tag{3.49}$$

where α is a pole of the open-loop transfer function which is to be determined. The resulting linear-state-variable feedback system is illustrated in Figure 3.10. The problem remaining is to select the values of K, α, and $\mathbf{h}$.

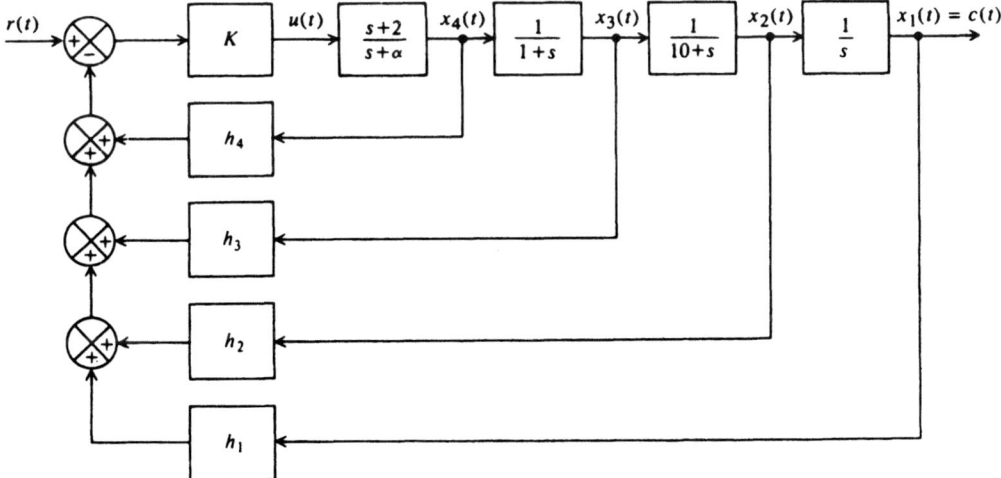

Figure 3.10 State-variable feedback representation of system.

An equivalent block diagram of this system is illustrated in Figure 3.11. The resulting closed-loop transfer function from this equivalent model is given by

$$\frac{C(s)}{R(s)} = \frac{K(s+2)}{\{(Kh_4+1)s^4 + [K(h_3+13h_4)+911+\alpha)]s^3 \\ +[K(h_2+12h_3+32h_4)+(10+11\alpha)]s^2 \\ +[K(h_1+2h_2+20h_3+20h_4)+10\alpha]s + 2Kh_1\}} \quad (3.50)$$

Equating the two forms of $C(s)/R(s)$ given by Eqs. (3.48) and (3.50), the following set of equations is obtained:

$$K = 72,$$
$$Kh_4 + 1 = 1,$$
$$K(h_3 + 13h_4) + (11 + \alpha) = 26.4,$$
$$K(h_2 + 12h_3 + 32h_4) + (10 + 11\alpha) = 180.4,$$
$$K(h_1 + 2h_2 + 20h_3 + 20h_4) + 10\alpha = 229,$$
$$2Kh_1 = 144.$$

Notice that we have six simultaneous equations with six unknowns (K, h_1, h_2, h_3, h_4, and α). Solving these equations, we obtain the following expressions:

$$\left.\begin{array}{lll} K = 72, & h_1 = 1, & h_2 = 0.0121 \\ h_3 = 0.0017, & h_4 = 0, & \alpha = 15.28 \end{array}\right\} \quad (3.51)$$

From Eq. (3.49), the resulting compensation network, $G_c(s)$, is given by

$$G_c(s) = \frac{s+2}{s+15.28} \quad (3.52)$$

which is a phase-lead network.

It is important to emphasize that α could have turned out to be negative for a different set of specifications This would be undesirable, because it would result in a zero in the right-half plane; this system would be unstable. In other cases, the system might be conditionally stable.

Our results of this pole placement example can be evaluated most conveniently on a root-locus plot. To obtain the root locus of the compensated system, the open-loop

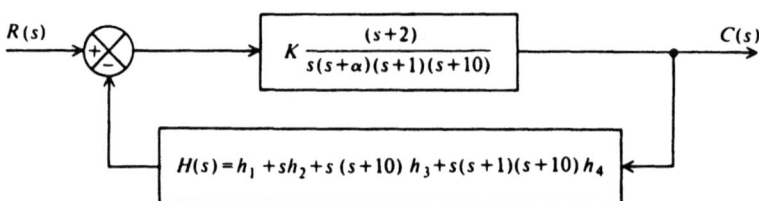

Figure 3.11 Equivalent block diagram for system illustrated in Figure 3.10.

transfer function will be obtained. For the values of the parameters found in Eq. (3.51), $H(s)$ results in the following:

$$H(s) = 0.0017[s + (8.5 + j22.4)][s + (8.5 - j22.4)]. \tag{3.53}$$

Combining Eqs. (3.39), (3.52), and (3.53) we obtain the following transfer function for the open-loop system:

$$KG_c(s)G(s)H(s) = 0.0017 \frac{K(s+2)[s + (8.5 + j22.4)][s + (8.5 - j22.4)]}{s(s + 15.28)(s + 10)(s + 1)} \tag{3.54}$$

The resulting root locus is plotted in Figure 3.12. Observe that the resulting root locus is stable for all values of K from zero to infinity. The locations of the dominant complex-conjugate roots for $K = 72$ are indicated.

It is important to emphasize again that the discussion of linear-state-variable feedback in this and the preceding section has assumed that all of the state variables are accessible. This is not always the case. This is analyzed further in Sections 3.6 and 3.7.

3.4. CONTROLLABILITY

The presentation of linear-state-variable feedback in Sections 3.2 and 3.3 assumed that all of the states are observable and measurable, and available to accept control signals (controllable). The concepts of controllability [2–4], and observability also play a very important role in optimal control theory (presented in the accompanying volume). Before designing a control system, we must determine whether it is

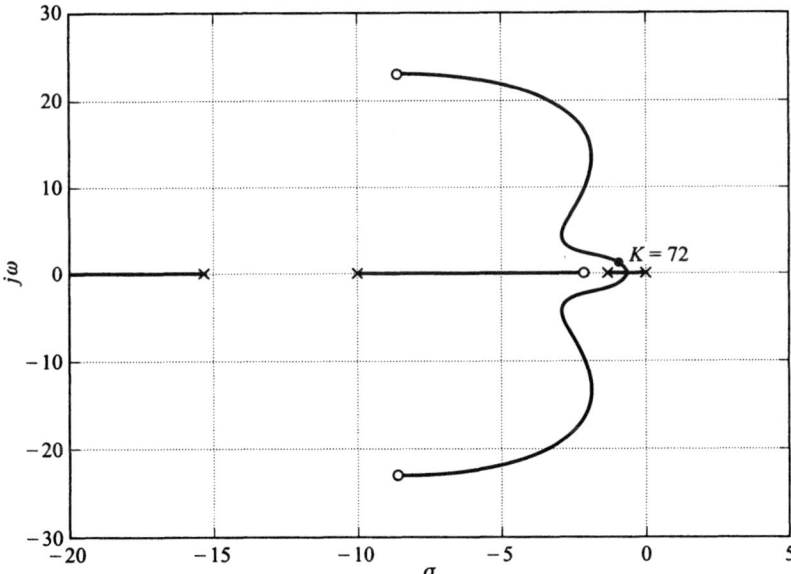

Figure 3.12 Root locus for the system of Figure 3.11 with the parameters of Eq. (3.51).

132 MODERN CONTROL-SYSTEM DESIGN USING VARIOUS TECHNIQUES

controllable and its states are observable, since the conditions on controllability and observability often govern the existence of a solution to an optimal control system. Kalman [2,3] first introduced the concepts of controllability and observability in 1960. These concepts are basic in modern optimal control theory. This section develops mathematical tests to determine controllability, and the following section presents mathematical tests for determining observability.

In order to introduce the concept of controllability, let us consider the simple open-loop system illustrated in Figure 3.13. A system is completely controllable if there exists a control which transfers every initial state at $t = t_0$ to any final state at $t = T$ for all t_0 and T. Qualitatively, this means that the system $G(s)$ is controllable if every state variable of G can be affected by the input signal $\mathbf{u}(t)$. However, if one (or several) of the state variables is (or are) not affected by $\mathbf{u}(t)$, then this (or these) state variable(s) cannot be controlled in a finite amount of time by $\mathbf{u}(t)$ and the system is not completely controllable.

A. Controllability by Inspection

As an example of a system which is not completely controllable, let us consider the signal-flow diagram illustrated in Figure 3.14. This system contains four states, only two of which are affected by $u(t)$. This input only affects the states $x_1(t)$ and $x_2(t)$. It has no effect on $x_3(t)$ and $x_4(t)$. Therefore, $x_3(t)$ and $x_4(t)$ are uncontrollable. This means that it is impossible for $u(t)$ to change $x_3(t)$ from initial state $x_3(0)$ to final state $x_3(T)$ in a finite time interval T and the system is not completely controllable.

B. The Controllability Matrix

Let us now consider this problem more precisely and establish a mathematical criterion for determining whether a system is controllable. We limit our discussion to linear constant systems. Assume that the system is described by

$$\dot{\mathbf{x}}(t) = \mathbf{P}\mathbf{x}(t) + \mathbf{B}\mathbf{u}(t), \tag{3.55}$$

$$\mathbf{c}(t) = \mathbf{L}\mathbf{x}(t). \tag{3.56}$$

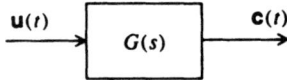

Figure 3.13 Open-loop system containing several inputs and outputs.

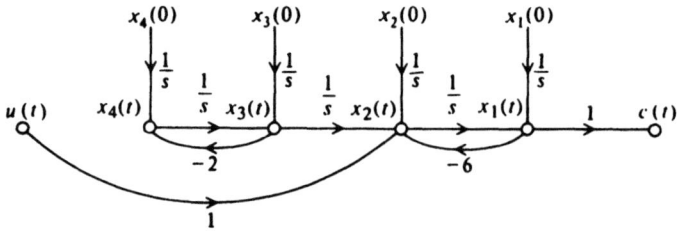

Figure 8.14 Signal-flow graph of a system that is not completely controllable.

3.4. CONTROLLABILITY

The solution of Eq. (3.55) can be expressed as Eq. (2.265)‡:

$$\mathbf{x}(t) = \mathbf{\Phi}(t - t_0)\mathbf{x}(t_0) + \int_{t_0}^{t} \mathbf{\Phi}(t - \tau)\mathbf{B}\mathbf{u}(\tau)d\tau, \quad \text{where } t \geqslant t_0. \tag{3.57}$$

Let us assume that the desired final state of our system at $t = t_f$ is zero:

$$\mathbf{x}(t_f) = \mathbf{0}. \tag{3.58}$$

Using Eqs. (2.259)‡ and (2.261)‡, we can write Eq. (3.57) as

$$\mathbf{x}(t_0) = -\int_{t_0}^{t_f} \mathbf{\Phi}(t_0 - \tau)\mathbf{B}\mathbf{u}(\tau)d\tau. \tag{3.59}$$

The state-transition matrix can be expressed from Eq. (2.322)‡ as

$$\mathbf{\Phi}(t) = e^{\mathbf{P}t} = \sum_{n=0}^{m-1} \alpha_n(t)\mathbf{P}^n. \tag{3.60}$$

Here, $\mathbf{x}$ is an $m \times 1$ vector, $\mathbf{P}$ is an $m \times m$ matrix, $\mathbf{u}$ is an $r \times 1$ vector, $\mathbf{B}$ is an $m \times r$ matrix, and $\alpha_n(t)$ is a scalar function of t. (This form results from application of the Cayley–Hamilton theorem). Substituting Eq (3.60) into Eq. (3.59), we obtain the following expression for $\mathbf{x}(t_0)$:

$$\mathbf{x}(t_0) = -\int_{t_0}^{t_f} \sum_{n=0}^{m-1} \alpha_n(t_0 - \tau)\mathbf{P}^n \mathbf{B}\mathbf{u}(\tau)d\tau. \tag{3.61}$$

Because the matrices $\mathbf{P}$ and $\mathbf{B}$ are not functions of τ, we can rewrite Eq. (3.61) as

$$\mathbf{x}(t_0) = -\sum_{n=0}^{m-1} \mathbf{P}^n \mathbf{B} \int_{t_0}^{t_f} \alpha_n(t_0 - \tau)\mathbf{u}(\tau)d\tau. \tag{3.62}$$

Equation (3.62) can be rewritten as

$$\mathbf{x}(t_0) = -[\mathbf{B} \quad \mathbf{PB} \quad \mathbf{P}^2\mathbf{B} \quad \mathbf{P}^3\mathbf{B} \cdots \mathbf{P}^{m-1}\mathbf{B}] \begin{bmatrix} \mathbf{A}_0 \\ \mathbf{A}_1 \\ \mathbf{A}_2 \\ \vdots \\ \mathbf{A}_{m-1} \end{bmatrix}, \tag{3.63}$$

where

$$\mathbf{A}_n = \int_{t_0}^{t_f} \alpha_n(t_0 - \tau)\mathbf{u}(\tau)d\tau. \tag{3.64}$$

If we define

$$\mathbf{D} = [\mathbf{B} \quad \mathbf{PB} \quad \mathbf{P}^2\mathbf{B} \quad \mathbf{P}^3\mathbf{B} \cdots \mathbf{P}^{m-1}\mathbf{B}] \tag{3.65}$$

$$\mathbf{A} = [\mathbf{A}_0 \quad \mathbf{A}_1 \quad \mathbf{A}_2 \cdots \mathbf{A}_{m-1}]^T, \tag{3.66}$$

where **D**, the controllability matrix, is an $m \times mr$ matrix and **A** is an $mr \times 1$ vector, then Eq. (3.63) becomes

$$\mathbf{x}(t_0) = -\mathbf{DA}. \qquad (3.67)$$

For a given initial state $\mathbf{x}(t_0)$, the input **u** can be found to drive the state to $\mathbf{x}(t_f) = \mathbf{0}$ for a finite time interval $t_f - t_0$ if Eq. (3.67) has a solution. A unique solution occurs only if there is a set of m linearly independent column vectors in the matrix **D**. If **u** is a scalar, then **D** is an $m \times m$ square matrix, and Eq. (3.67) represents a set of m linearly independent equations which have a solution if **D** is nonsingular or the determinant of **D** is not zero. The controllability criterion thus states that the system of Eqs. (3.55) and (3.56) is completely controllable if **D** [see Eq. (3.65)] contains m linearly independent column vectors or, if **u** is a scalar, **D** is nonsingular.

In order to illustrate this mathematical controllability concept, consider a second-order system where

$$\mathbf{P} = \begin{bmatrix} -4 & 1 \\ 0 & -2 \end{bmatrix}$$

and

$$\mathbf{B} = \begin{bmatrix} 1 \\ 0 \end{bmatrix}.$$

Then, from Eq. (3.65),

$$\mathbf{D} = [\mathbf{B} \quad \mathbf{PB}] = \begin{bmatrix} 1 & -4 \\ 0 & 0 \end{bmatrix}. \qquad (3.68)$$

The resulting matrix **D** is singular (its determinant is zero) and the system is therefore not completely controllable.

As a second example, consider a second-order system where

$$\mathbf{P} = \begin{bmatrix} -3 & 2 \\ 4 & 1 \end{bmatrix}$$

and

$$\mathbf{B} = \begin{bmatrix} 0 \\ 1 \end{bmatrix}.$$

Then, from Eq. (3.65)

$$\mathbf{D} = [\mathbf{B} \quad \mathbf{PB}] = \begin{bmatrix} 0 & 2 \\ 1 & 1 \end{bmatrix}.$$

The resulting matrix **D** is nonsingular and the system, therefore, is completely controllable.

3.5. OBSERVABILITY

In order to introduce the concept of observability [2–4], let us again consider the simple open-loop system illustrated in Figure 3.13. A system is completely observable if, given the control and the output over the interval $t_0 \leqslant t \leqslant T$, one can determine the initial state $\mathbf{x}(t_0)$. Qualitatively, the system G is observable if every state variable of G affects some of the outputs in $\mathbf{c}$. It is very often desirable to determine information regarding the system states based on measurements of $\mathbf{c}$. However, if we cannot observe one or more of the states from the measurements of $\mathbf{c}$, then the system is not completely observable. We had assumed the systems were observable in our discussion of pole placement using linear-state variable feedback in Sections 3.2 and 3.3.

A. Observability by Inspection

As an example of a system which is not completely observable, let us consider the signal-flow diagram illustrated in Figure 3.15. This system contains four states, only two of which are observable. The states $x_3(t)$ and $x_4(t)$ are not connected to the output $c(t)$ in any manner. Therefore, $x_3(t)$ and $x_4(t)$ are not observable and the system is not completely observable.

B. The Observability Matrix

Let us now consider this problem more precisely and establish a mathematical criterion for determining whether a system is completely observable. Again, we limit our discussion to linear constant systems of the form

$$\dot{\mathbf{x}}(t) = \mathbf{P}\mathbf{x}(t) + \mathbf{B}\mathbf{u}(t), \tag{3.69}$$

$$\mathbf{c}(t) = \mathbf{L}\mathbf{x}(t) \tag{3.70}$$

where $\mathbf{x}$ is an $m \times 1$ vector, $\mathbf{P}$ is an $m \times m$ matrix, $\mathbf{u}$ is an $r \times 1$ vector, $\mathbf{B}$ is an $m \times r$ matrix, $\mathbf{c}$ is a $p \times 1$ vector, and $\mathbf{L}$ is a $p \times m$ matrix. The solution of Eq. (3.69) is given by [see Eq. (3.57)]

$$\mathbf{x}(t) = \mathbf{\Phi}(t - t_0)\mathbf{x}(t_0) + \int_{t_0}^{t} \mathbf{\Phi}(t - \tau)\mathbf{B}\mathbf{u}(\tau)d\tau. \tag{3.71}$$

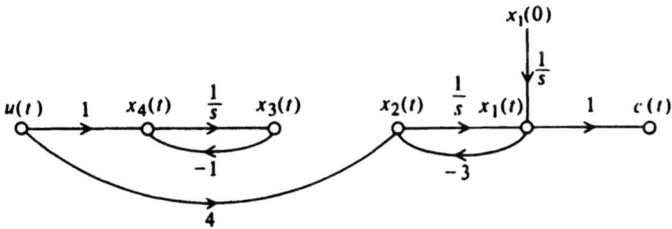

Figure 3.15 Signal-flow graph of a system that is not completely observable.

It will now be shown that observability depends on the matrices **P** and **L**. Substituting Eq. (3.71) into Eq. (3.70), we obtain

$$\mathbf{c}(t) = \mathbf{L}\mathbf{\Phi}(t - t_0)\mathbf{x}(t_0) + \mathbf{L} \int_{t_0}^{t} \mathbf{\Phi}(t - \tau)\mathbf{B}\mathbf{u}(\tau)d\tau, \text{ where } t \geq t_0. \quad (3.72)$$

From the definition of observability we can see that the observability of $\mathbf{x}(t_0)$ depends on the term $\mathbf{L}\mathbf{\Phi}(t - t_0)\mathbf{x}(t_0)$. Therefore, the output $\mathbf{c}(t)$ when $\mathbf{u} = \mathbf{0}$ is given by

$$\mathbf{c}(t) = \mathbf{L}\mathbf{\Phi}(t - t_0)\mathbf{x}(t_0). \quad (3.73)$$

Substituting Eq. (3.60) into Eq. (3.73), we obtain the following expression for the output $\mathbf{c}(t)$:

$$\mathbf{c}(t) = \sum_{n=0}^{m-1} \alpha_n(t - t_0)\mathbf{L}\mathbf{P}^n\mathbf{x}(t_0). \quad (3.74)$$

Equation (3.74) indicates that if the output $\mathbf{c}(t)$ is known over the time interval $t_0 \leq t \leq T$, then $\mathbf{x}(t_0)$ is uniquely determined from this equation if $\mathbf{x}(t_0)$ is a linear combination of $(\mathbf{L}_j\mathbf{P}^n)^T$ for $n = 0, 1, 2, \ldots, m - 1$, and $j = 1, 2, 3, \ldots, r$. The matrix $\mathbf{L}_j$ is the $1 \times m$ matrix formed by the elements of the jth row of **L**. Because $(\mathbf{L}_j\mathbf{P}^n)^T = (\mathbf{P}^T)^n \mathbf{L}_j^T$, we let **U** be the $m \times mr$ matrix defined by

$$\mathbf{U} = [\mathbf{L}^T \quad \mathbf{P}^T\mathbf{L}^T \quad (\mathbf{P}^T)^2\mathbf{L}^T \quad (\mathbf{P}^T)^3\mathbf{L}^T \cdots (\mathbf{P}^T)^{m-1}\mathbf{L}^T]. \quad (3.75)$$

The observability criterion states that the system is completely observable if there is a set of m linearly independent column vectors in the observability matrix **U**.

In order to illustrate the observability concept mathematically, consider a second-order system where

$$\mathbf{P} = \begin{bmatrix} -4 & 0 \\ 0 & -2 \end{bmatrix}$$

and

$$\mathbf{L} = [1 \quad 0].$$

Therefore,

$$\mathbf{L}^T = \begin{bmatrix} 1 \\ 0 \end{bmatrix}$$

and

$$\mathbf{P}^T\mathbf{L}^T = \begin{bmatrix} -4 & 0 \\ 0 & -2 \end{bmatrix}\begin{bmatrix} 1 \\ 0 \end{bmatrix} = \begin{bmatrix} -4 \\ 0 \end{bmatrix}.$$

Substituting these values into Eq. (3.75), we obtain

$$U = [L^T \quad P^T L^T] = \begin{bmatrix} 1 & -4 \\ 0 & 0 \end{bmatrix}.$$

Because U is singular, the system is not completely observable.

As a second example, consider the second-order system where

$$P = \begin{bmatrix} 4 & -2 \\ 4 & -2 \end{bmatrix}$$

and

$$L = [1 \quad 1].$$

Therefore,

$$L^T = \begin{bmatrix} 1 \\ 1 \end{bmatrix}$$

and

$$P^T L^T = \begin{bmatrix} 4 & 4 \\ -2 & -2 \end{bmatrix} \begin{bmatrix} 1 \\ 1 \end{bmatrix} = \begin{bmatrix} 8 \\ -4 \end{bmatrix}.$$

Substituting these values into Eq. (3.75), we obtain

$$U = [L^T \quad P^T L^T] = \begin{bmatrix} 1 & 8 \\ 1 & -4 \end{bmatrix}.$$

Because U has two independent columns, the system is completely observable.

3.6. ACKERMANN'S FORMULA FOR DESIGN USING POLE PLACEMENT [5–7]

In addition to the method of matching the coefficients of the desired characteristic equation with the coefficients of det $(sI - P_h)$ as given by Eq. (3.19), Ackermann has developed a competing method. The pole placement method using the matching of coefficients of the desired characteristic equation with the coefficients of Eq. (3.19) is very useful for control systems which are represented in phase-variable form, where *phase variable* refers to systems where each subsequent state variable is defined as the derivative of the previous state variable. Some control systems require feedback from state variables which are not phase variables. Such high-order control systems can lead to very complex calculations for the feedback gains. Ackermann's method simplifies this problem by transforming the control system to phase variables, determining the feedback gains, and transforming the designed control system back to its original state-variable representation.

Let us represent a control system which is not represented in phase-variable form by the following:

$$\dot{\mathbf{y}}(t) = \mathbf{P}\mathbf{y}(t) + \mathbf{B}u(t), \quad (3.76)$$

$$c(t) = \mathbf{L}\mathbf{y}(t). \quad (3.77)$$

We will assume that the controllability matrix [see Eq. (3.65)] can be represented by

$$\mathbf{D}_y = [\mathbf{B} \ldots \mathbf{PB} \ldots \mathbf{P}^2\mathbf{B} \ldots \mathbf{P}^{m-1}\mathbf{B}]. \quad (3.78)$$

Subscript y is used to designate the original, non-phase-variable, controllability matrix. We will next assume that the control system can be transformed into the phase-variable representation using the following transformation:

$$\mathbf{y}(t) = \mathbf{A}\mathbf{x}(t). \quad (3.79)$$

Substituting Eq. (3.79) into Eq. (3.76) and Eq. (3.77), we obtain:

$$\dot{\mathbf{x}}(t) = \mathbf{A}^{-1}\mathbf{P}\mathbf{A}\mathbf{x}(t) + \mathbf{A}^{-1}\mathbf{B}u(t), \quad (3.80)$$

$$c(t) = \mathbf{L}\mathbf{A}\mathbf{x}(t). \quad (3.81)$$

Using the transformation of Eq. (3.79), the controllability matrix for the transformed system defined by Eqs. (3.80) and (3.81) is:

$$\mathbf{D}_x = [\mathbf{A}^{-1}\mathbf{B} \ldots (\mathbf{A}^{-1}\mathbf{P}\mathbf{A})(\mathbf{A}^{-1}\mathbf{B}) \ldots (\mathbf{A}^{-1}\mathbf{P}\mathbf{A})^2(\mathbf{A}^{-1}\mathbf{B}) \ldots (\mathbf{A}^{-1}\mathbf{P}\mathbf{A})^{m-1}(\mathbf{A}^{-1}\mathbf{B})]. \quad (3.82)$$

The subscript x is used to designate the phase-variable form of the transformed control system. Equation (3.82) can be simplified to

$$\mathbf{D}_x = \mathbf{A}^{-1}[\mathbf{B} \ldots \mathbf{PB} \ldots \mathbf{P}^2\mathbf{B} \ldots \mathbf{P}^{m-1}\mathbf{B}]. \quad (3.83)$$

Therefore, substituting Eq. (3.78) into Eq. (3.83), we find that

$$\mathbf{D}_x = \mathbf{A}^{-1}\mathbf{D}_y. \quad (3.84)$$

The transformation matrix $\mathbf{A}$ can be found from

$$\mathbf{A} = \mathbf{D}_y \mathbf{D}_x^{-1}. \quad (3.85)$$

Therefore, the transformation matrix $\mathbf{A}$ can be determined from the controllability matrices defined by Eqs. (3.65) and (3.83). Once the control system is transformed to the phase-variable form, the feedback gains $\mathbf{h}$ can be determined as described in Section 3.2.

Returning to Eq. (3.4) to represent $u(t)$ for the control system in phase-variable form,

$$u(t) = K[r(t) - \mathbf{h}\mathbf{x}(t)] \tag{3.86}$$

the following is obtained by substituting Eq. (3.86) into Eq. (3.80):

$$\dot{\mathbf{x}}(t) = \mathbf{A}^{-1}\mathbf{P}\mathbf{A}x(t) + \mathbf{A}^{-1}\mathbf{B}(K[r(t) - \mathbf{h}\mathbf{x}(t)]). \tag{3.87}$$

Equation (3.87) can be simplified to the following phase-variable state equation:

$$\dot{\mathbf{x}}(t) = (\mathbf{A}^{-1}\mathbf{P}\mathbf{A} - K\mathbf{A}^{-1}\mathbf{B}\mathbf{h})\mathbf{x}(t) + \mathbf{A}^{-1}\mathbf{B}Kr(t). \tag{3.88}$$

The output equation remains as shown in Eq. (3.81):

$$c(t) = \mathbf{L}\mathbf{A}\mathbf{x}(t). \tag{3.89}$$

Because Eqs. (3.88) and (3.89) are in phase-variable form, the rules for pole placement developed in Section 3.2 for phase-variable systems are valid for this representation. We now have to transform Eqs. (3.88) and (3.89) back to the original state and output equation representation by using the transformation provided by Eq. (3.79):

$$\mathbf{x}(t) = \mathbf{A}^{-1}\mathbf{y}(t). \tag{3.90}$$

Substituting Eq. (3.90) into Eq. (3.88), we obtain the following:

$$\mathbf{A}^{-1}\dot{\mathbf{y}}(t) = (\mathbf{A}^{-1}\mathbf{P}\mathbf{A} - K\mathbf{A}^{-1}\mathbf{B}\mathbf{h})\mathbf{A}^{-1}\mathbf{y}(t) + \mathbf{A}^{-1}\mathbf{B}Kr(t). \tag{3.91}$$

Therefore, Eq. (3.91) reduces to the following state equation:

$$\dot{\mathbf{y}}(t) = \mathbf{P}\mathbf{y}(t) - K\mathbf{B}\mathbf{h}\mathbf{A}^{-1}\mathbf{y}(t) + \mathbf{B}Kr(t). \tag{3.92}$$

The output equation is found by substituting Eq. (3.79) in Eq. (3.89):

$$c(t) = \mathbf{L}\mathbf{A}\mathbf{A}^{-1}\mathbf{y}(t) = \mathbf{L}\mathbf{y}(t). \tag{3.93}$$

By comparing Eq. (3.92) with Eqs. (3.8) and (3.10), we find that the state-variable feedback constants for the original system are

$$\mathbf{h}_y = \mathbf{h}_x \mathbf{A}^{-1}. \tag{3.94}$$

A. Example Applying Ackermann's Formula for Design using Pole Placement

We wish to apply the pole placement concepts using Ackermann's design formula to a system that uses linear-state-variable feedback. The system specifications require an overshoot of 4.33% and a settling time of 6 sec. The process transfer function is

140 MODERN CONTROL-SYSTEM DESIGN USING VARIOUS TECHNIQUES

$$\frac{C(s)}{U(s)} = \frac{(s+3)}{(s+2)(s+1)(s+4)} = \frac{(s+3)}{s^3 + 7s^2 + 14s + 8} \quad (3.95)$$

and the control system's signal-flow graph is given in Figure 3.16.

The state equation for the process illustrated in Figure 3.16 is:

$$\dot{\mathbf{y}}(t) = \mathbf{P}_y \mathbf{y}(t) + \mathbf{B}_y u(t) = \begin{bmatrix} \dot{y}_1(t) \\ \dot{y}_2(t) \\ \dot{y}_3(t) \end{bmatrix} = \begin{bmatrix} -4 & 1 & 0 \\ 0 & -1 & 1 \\ 0 & 0 & -2 \end{bmatrix} \begin{bmatrix} y_1(t) \\ y_2(t) \\ y_3(t) \end{bmatrix} + \begin{bmatrix} 0 \\ 0 \\ 1 \end{bmatrix} u(t). \quad (3.96)$$

Since we are going to need the controllability matrix $\mathbf{D}_y$ to convert this original system to phase-variable form [see Eq. (3.85)], let us compute this controllability matrix for this original system:

$$\mathbf{D}_y = [\mathbf{B}_y \quad \mathbf{P}_y \mathbf{B}_y \quad \mathbf{P}_y^2 \mathbf{B}_y] = \begin{bmatrix} 0 & 0 & 1 \\ 0 & 1 & -3 \\ 1 & -2 & 4 \end{bmatrix}. \quad (3.97)$$

The value of the determinant of $\mathbf{D}_y$ is -1, it is nonsingular, and the original control system is controllable as expected from inspection of Figure 3.16.

The next step is to transform the original system to the phase-variable form. This can easily be obtained from the transfer function $C(s)/U(s)$, which can be determined from Figure 3.16:

$$\frac{C(s)}{U(s)} = \frac{\frac{3}{s^3} + \frac{1}{s^2}}{1 + \left[\frac{2}{s} + \frac{1}{s} + \frac{4}{s}\right] + \left[\left(\frac{2}{s}\right)\left(\frac{1}{s}\right) + \left(\frac{2}{s}\right)\left(\frac{4}{s}\right) + \left(\frac{1}{s}\right)\left(\frac{4}{s}\right)\right] + \left[\left(\frac{2}{s}\right)\left(\frac{1}{s}\right)\left(\frac{4}{s}\right)\right]}. \quad (3.98)$$

Eq. (3.98) can be reduced to the following [as stated in Eq. (3.95)]:

$$\frac{C(s)}{U(s)} = \frac{(s+3)}{s^3 + 7s^2 + 14s + 8}. \quad (3.99)$$

We can determine the phase-variable state equations which represent the transfer function given by Eq. (3.99). Defining $x_1(t) = c(t)$, $x_2(t) = \dot{c}(t)$, and $x_3(t) = \ddot{c}(t)$, we obtain the following:

$$\dot{x}_1(t) = x_2(t) \quad (3.100)$$

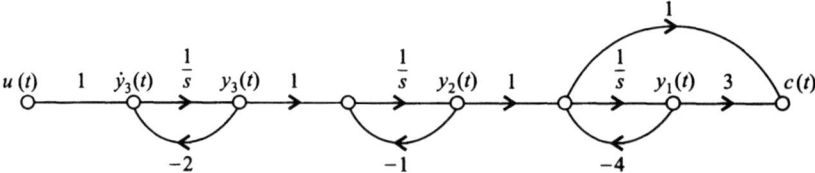

Figure 3.16 Signal-flow graph representation of process whose transfer function is defined in Eq. (3.95).

3.6. ACKERMANN'S FORMULA FOR DESIGN USING POLE PLACEMENT

$$\dot{x}_2(t) = x_3(t) \tag{3.101}$$

$$\dot{x}_3(t) = -8x_1(t) - 14x_2(t) - 7x_3(t) + 3u(t) + \dot{u}(t) \tag{3.102}$$

Therefore, the state and output equations in matrix vector form of the phase-variable control system are given by:

$$\dot{\mathbf{x}}(t) = \mathbf{P}_x \mathbf{x}(t) + \mathbf{B}_x u(t) = \begin{bmatrix} 0 & 1 & 0 \\ 0 & 0 & 1 \\ -8 & -14 & -7 \end{bmatrix} \begin{bmatrix} x_1(t) \\ x_2(t) \\ x_3(t) \end{bmatrix} + \begin{bmatrix} 0 \\ 0 \\ 1 \end{bmatrix} (3u(t) + \dot{u}(t)), \tag{3.103}$$

$$c(t) = \begin{bmatrix} 3 & 1 & 0 \end{bmatrix} \begin{bmatrix} x_1(t) \\ x_2(t) \\ x_3(t) \end{bmatrix} \tag{3.104}$$

The resulting controllability matrix for the phase-variable form can be obtained from Eqs. (3.65) and (3.103) and is given by:

$$\mathbf{D}_x = [\mathbf{B}_x \; \mathbf{P}_x\mathbf{B}_x \; \mathbf{P}_x^2\mathbf{B}_x] = \begin{bmatrix} 0 & 0 & 1 \\ 0 & 1 & -7 \\ 1 & -7 & 35 \end{bmatrix}. \tag{3.105}$$

Note that the determinant of $\mathbf{D}_x$ is -1 indicating that the determinant is nonsingular, and the phase-variable form is also controllable, as expected. The transformation matrix, defined in Eq. (3.85), can be determined from Eqs. (3.97) and (3.105) as follows:

$$\mathbf{A} = \mathbf{D}_y \mathbf{D}_x^{-1} = \begin{bmatrix} 0 & 0 & 1 \\ 0 & 1 & -3 \\ 1 & -2 & 4 \end{bmatrix} \begin{bmatrix} 14 & 7 & 1 \\ 7 & 1 & 0 \\ 1 & 0 & 0 \end{bmatrix} = \begin{bmatrix} 1 & 0 & 0 \\ 4 & 1 & 0 \\ 4 & 5 & 1 \end{bmatrix} \tag{3.106}$$

The next step in the procedure is to design the controller (see Figure 3.3) using the phase-variable representation, after which we will transform the design back to the original system by using Eq. (3.106). The 4.33 percent overshoot specified can be obtained from a second-order control system having a damping ratio of 0.707 [see Eq. (B.33)]. The settling time of 6 sec can be obtained from a second-order control system having a damping ratio of 0.707 and a $\omega_n = 0.943$ [see Eq. (5.41)†]. Therefore, the following second-order control system can meet the design specifications:

$$\frac{C(s)}{R(s)} = \frac{\omega_n^2}{s^2 + 2\zeta\omega_n s + \omega_n^2} = \frac{0.889}{s^2 + 1.333s + 0.889}. \tag{3.107}$$

Since it was shown in Section 3.2 that zeros of closed-loop systems are zeros of open-loop systems, then we can select the third pole at $s = -3$ which will also cancel the zero at $s = -3$. Therefore, the characteristic equation of the desired closed-loop system is

$$(s^2 + 1.333s + 0.889)(s + 3) = s^3 + 4.333s^2 + 4.888s + 2.667 = 0. \tag{3.108}$$

The state and output equations for the phase-variable form with linear-state-variable-feedback can be obtained from Eqs. (3.8), (3.9), and (3.10) as follows:

$$\dot{x}(t) = [\mathbf{P}_x - K\mathbf{b}_x\mathbf{h}_x]\mathbf{x}(t) + Kr(t)\mathbf{b}_x =$$

$$\left[\begin{bmatrix} 0 & 1 & 0 \\ 0 & 0 & 1 \\ -8 & -14 & -7 \end{bmatrix} - \begin{bmatrix} 0 & 0 & 0 \\ 0 & 0 & 0 \\ Kh_1 & Kh_2 & Kh_3 \end{bmatrix}\right] \begin{bmatrix} x_1(t) \\ x_2(t) \\ x_3(t) \end{bmatrix} + Ku(t)\begin{bmatrix} 0 \\ 0 \\ 1 \end{bmatrix}, \quad (3.109)$$

$$c(t) = [3 \quad 1 \quad 0]\mathbf{x}(t) \quad (3.110)$$

Eq. (3.109) reduces to the following:

$$\dot{\mathbf{x}}(t) = \begin{bmatrix} 0 & 1 & 0 \\ 0 & 0 & 1 \\ -8 - Kh_1 & -14 - Kh_2 & -7 - Kh_3 \end{bmatrix}\mathbf{x}(t) + \begin{bmatrix} 0 \\ 0 \\ Ku(t) \end{bmatrix}. \quad (3.111)$$

The resulting characteristic equation can be obtained from Eqs. (3.10) and (3.19):

$$\det(s\mathbf{I} - \mathbf{P}_h) = \det(s\mathbf{I} - \mathbf{P} + K\mathbf{b}\mathbf{h}) = 0. \quad (3.112)$$

Substituting $-(\mathbf{P}_x - K\mathbf{b}_x\mathbf{h}_x)$ from Eq. (3.111) into Eq. (3.112), we obtain the following:

$$\begin{vmatrix} s & -1 & 0 \\ 0 & s & -1 \\ 8 + Kh_1 & 14 + Kh_2 & s + 7 + Kh_3 \end{vmatrix} = 0. \quad (3.113)$$

The resulting characteristic equation is given by:

$$s^3 + (7 + Kh_3)s^2 + (14 + Kh_2)s + (8 + Kh_1) = 0. \quad (3.114)$$

Comparing Eq. (3.114) and Eq. (3.108), we obtain the following three equations:

$$7 + Kh_3 = 4.333; \quad Kh_3 = -2.667, \quad (3.115)$$
$$14 + KH_2 = 4.888; \quad Kh_2 = -9.112, \quad (3.116)$$
$$8 + Kh_1 = 2.667; \quad Kh_1 = -5.333. \quad (3.117)$$

Therefore,

$$K\mathbf{h}_x = [-5.333 \quad -9.112 \quad -2.667] \quad (3.118)$$

We will now transform $K\mathbf{h}_x$ as shown in Eq. (3.118) back to the original system using Eq. (3.94) and Eq. (3.106) as follows:

$$\mathbf{h}_y = \mathbf{h}_x \mathbf{A}^{-1} =$$

$$[-5.333 \quad -9.112 \quad -2.667]\begin{bmatrix} 1 & 0 & 0 \\ 4 & 1 & 0 \\ 4 & 5 & 1 \end{bmatrix}^{-1} = [-11.557 \quad 4.223 \quad -2.667]. \quad (3.119)$$

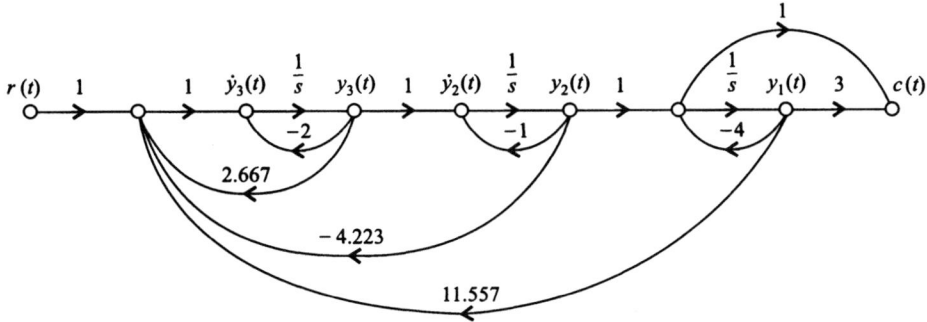

Figure 3.17 Resulting closed-loop control system with linear-state-variable-feedback designed using Ackermann's Formula.

The resulting closed-loop control system with linear-state-variable-feedback is shown in Figure 3.17.

3.7. ESTIMATOR DESIGN IN CONJUNCTION WITH THE POLE PLACEMENT APPROACH USING LINEAR-STATE-VARIABLE FEEDBACK

In the discussion of Sections 3.2 and 3.3 on linear-state-variable feedback, it was assumed that all of the states are observable and measurable, and available to accept control signals (controllable). As Sections 3.4 and 3.5 on controllability and observability have shown, some states of a feedback control system may not always be controllable and/or observable. In some systems, the system may be observable, but all of the states may not be measured due to physical limitations (e.g., chemical process control systems), or it may be due to cost restrictions that limit the use of costly sensors needed to measure all of the states. It is assumed in this section that the system is observable (no part of the system is disconnected physically from the output), but measurements are being made only on some of the states, and we wish to estimate all of the states.

Let us focus attention on the process portion of the system illustrated in Figure 3.3. We wish to use the closed-loop estimator system shown in Figure 3.18 for determining an estimate of the state vector $\mathbf{x}(t)$ and output $c(t)$ [8]. This estimator feeds back the difference between the measured output $c(t)$ and the estimated output $\hat{c}(t)$ that is obtained from a model of the process. Therefore,

$$\dot{\hat{\mathbf{x}}}(t) = \mathbf{P}\hat{\mathbf{x}}(t) + \mathbf{b}u(t) + \mathbf{M}(c(t) - \mathbf{L}\hat{\mathbf{x}}(t)), \qquad (3.120)$$

where $\mathbf{M}$ defines the gain factors m_i, which are selected to obtain desirable error characteristics of the state vector $\mathbf{x}$, and $\hat{\mathbf{x}}(t)$ represents the estimate of the state $\mathbf{x}(t)$:

$$\mathbf{M} = [m_1, m_2, \ldots, m_n]^T. \qquad (3.121)$$

The error in the state estimate, $\tilde{\mathbf{x}}(t)$, can be derived from

$$\tilde{\mathbf{x}}(t) = \mathbf{x}(t) - \hat{\mathbf{x}}(t). \qquad (3.122)$$

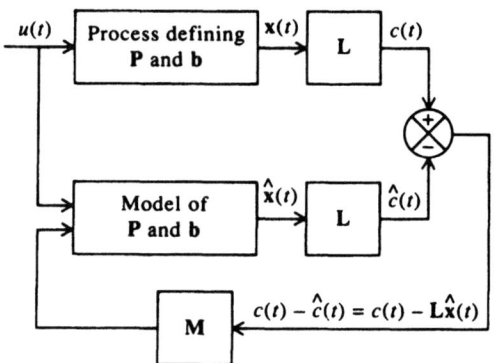

Figure 3.18 An estimator system.

The derivative of the $\tilde{\mathbf{x}}(t)$ can be obtained by subtracting $\dot{\hat{\mathbf{x}}}(t)$ [given by Eq. (3.120)] from $\dot{\mathbf{x}}(t)$ given by the system dynamics:

$$\dot{\mathbf{x}}(t) = \mathbf{P}\mathbf{x}(t) + \mathbf{b}u(t) \tag{3.123}$$

Subtracting Eq. (3.120) from Eq. (3.123), we obtain the following:

$$\dot{\mathbf{x}}(t) = \mathbf{P}\mathbf{x}(t) + \mathbf{b}u(t), \tag{3.124}$$

$$\dot{\hat{\mathbf{x}}}(t) = \mathbf{P}\hat{\mathbf{x}}(t) + \mathbf{b}u(t) + \mathbf{M}(c(t) - \mathbf{L}\hat{\mathbf{x}}(t)) \tag{3.125}$$

$$\dot{\mathbf{x}}(t) - \dot{\hat{\mathbf{x}}}(t) = \mathbf{P}(\mathbf{x}(t) - \hat{\mathbf{x}}(t)) - \mathbf{M}(c(t) - \mathbf{L}\hat{\mathbf{x}}(t)). \tag{3.126}$$

Substituting Eq. (3.2),

$$c(t) = \mathbf{L}\mathbf{x}(t) \tag{3.127}$$

into Eq. (3.126), we obtain

$$\dot{\mathbf{x}}(t) - \dot{\hat{\mathbf{x}}}(t) = \mathbf{P}(\mathbf{x}(t) - \hat{\mathbf{x}}(t)) - \mathbf{ML}(\mathbf{x}(t) - \hat{\mathbf{x}}(t)). \tag{3.128}$$

Using the definition of error in the state estimate as given by (3.122), Eq. (3.128) reduces to

$$\dot{\tilde{\mathbf{x}}}(t) = \mathbf{P}\tilde{\mathbf{x}}(t) - \mathbf{ML}\tilde{\mathbf{x}}(t)$$

or

$$\dot{\tilde{\mathbf{x}}}(t) = (\mathbf{P} - \mathbf{ML})\tilde{\mathbf{x}}(t). \tag{3.129}$$

It is shown in Section 6.2[‡] on the state-variable determination of the characteristic equation that a state equation, as given by Eq. (6.15)[‡],

$$\dot{\mathbf{x}}(t) = \mathbf{A}\mathbf{x}(t) \tag{3.130}$$

3.7. ESTIMATOR DESIGN USING LINEAR-STATE-VARIABLE FEEDBACK

has a characteristic equation given by Eq. (6.22)‡. Similarly, the characteristic equation of Eq. (3.129) is given by

$$|s\mathbf{I} - (\mathbf{P} - \mathbf{ML})| = 0. \tag{3.131}$$

The objective of the control-system engineer is to select $\mathbf{P} - \mathbf{ML}$ so that it has stable roots in order for $\tilde{\mathbf{x}}(t)$ to decay to zero. It is also desirable to have the root location produce a fast transient response so that the estimation error decays very fast to zero. Notice that the estimation error $\tilde{\mathbf{x}}(t)$ will converge to zero independent of the forcing function $u(t)$. Therefore, stability is determined from the homogeneous solution to the system (with $u(t) = 0$), as opposed to its particular solution (with $u(t)$ finite).

The design procedure for determining $\mathbf{M}$ is to specify the desired location of the estimator roots (e.g., $\alpha_1, \alpha_2, \ldots, \alpha_n$) from which the desired estimator characteristic equation can be specified:

$$(s + \alpha_1)(s + \alpha_2) \cdots (s + \alpha_n) = 0. \tag{3.132}$$

We can then solve for $\mathbf{M}$ by comparing the coefficients in Eqs (3.131) and (3.132).

Let us consider the design of $\mathbf{M}$ for a simple second-order system whose differential equation is given by

$$\ddot{c}(t) + \dot{c}(t) + c(t) = u(t).$$

Defining its two states as

$$x_1(t) = c(t)$$
$$x_2(t) = \dot{c}(t),$$

we obtain its state equations as

$$\dot{x}_1(t) = x_2(t)$$
$$\dot{x}_2(t) = -x_1(t) - x_2(t) + u(t),$$

and its output equation as

$$c(t) = [1 \quad 0]\begin{bmatrix} x_1(t) \\ x_2(t) \end{bmatrix}. \tag{3.133}$$

Therefore, for this system, we obtain $\mathbf{P}, \mathbf{b}$, and $\mathbf{L}$ to be as follows:

$$\mathbf{P} = \begin{bmatrix} 0 & 1 \\ -1 & -1 \end{bmatrix}, \quad \mathbf{b} = \begin{bmatrix} 0 \\ 1 \end{bmatrix},$$
$$\mathbf{L} = [1 \quad 0]. \tag{8.134}$$

Substituting Eq. (3.134) into Eq. (3.131), we obtain the following:

$$\left| \begin{bmatrix} s & 0 \\ 0 & s \end{bmatrix} - \begin{bmatrix} 0 & 1 \\ -1 & -1 \end{bmatrix} + \begin{bmatrix} m_1 \\ m_2 \end{bmatrix} [1 \quad 0] \right| = 0, \tag{3.135}$$

146 MODERN CONTROL-SYSTEM DESIGN USING VARIOUS TECHNIQUES

$$\left| \begin{bmatrix} s & 0 \\ 0 & s \end{bmatrix} - \begin{bmatrix} 0 & 1 \\ -1 & -1 \end{bmatrix} + \begin{bmatrix} m_1 & 0 \\ m_2 & 0 \end{bmatrix} \right| = 0, \qquad (3.136)$$

$$\begin{vmatrix} s+m_1 & -1 \\ 1+m_2 & s+1 \end{vmatrix} = 0. \qquad (3.137)$$

The resulting characteristic equation is given by

$$s^2 + (m_1 + 1)s + m_1 + m_2 + 1 = 0. \qquad (3.138)$$

Where do we desire to place the two second-order roots? The primary goal is to design the estimator to be very fast compared to that of the controller. Therefore, let us assume that the controller is critically damped and the controller's second-order characteristic equation has a pair of real roots located at $s = \beta_1 = \beta_2 = 2$. Let us assume that we wish the estimator to be critically damped and have the two second-order roots of the estimator located at $s = \alpha_1 = \alpha_2 = 20$ in Eq. (3.132), which will ensure a very fast response compared to that of the controller. Therefore, Eq. (3.132) for this example becomes

$$(s + 20)^2 = 0,$$
$$s^2 + 40s + 400 = 0. \qquad (8.139)$$

Comparing the coefficients of Eqs. (3.138) and (3.139), we obtain the following two sumultaneous equations to solve:

$$m_1 + 1 = 40, \qquad (3.140)$$
$$m_1 + m_2 + 1 = 400. \qquad (3.141)$$

Solving Eqs. (3.140) and (3.141), we obtain:

$$\mathbf{M} = \begin{bmatrix} m_1 \\ m_2 \end{bmatrix} = \begin{bmatrix} 39 \\ 360 \end{bmatrix}.$$

Designing a combined compensator of a controller and estimator is illustrated in Section 3.8.

3.8. COMBINED COMPENSATOR DESIGN INCLUDING A CONTROLLER AND AN ESTIMATOR FOR A REGULATOR SYSTEM

This section considers the combined compensator design of a controller and estimator for a regulator in which the reference input equals zero, and for the case where the reference input is finite. The block diagram of this regulator is shown in Figure 3.19, which combines the concepts illustrated in Figure 3.3 for the controller and Figure 3.18 for the estimator. We wish to determine in this system the effect of using the estimated state vector $\hat{x}(t)$, instead of $x(t)$ on the system dynamics [8].

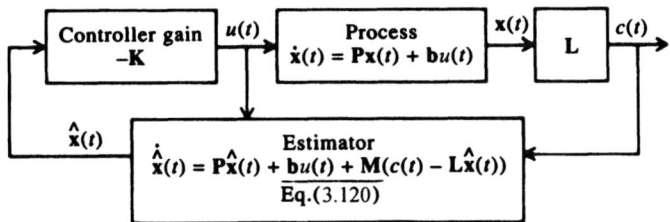

Figure 3.19 Regulator system (where the reference input $r = 0$) containing combined controller and estimator.

Let us consider the effect of driving the controller with $\hat{x}(t)$ instead of $x(t)$. From the process dynamics shown in Figure 3.3,

$$\dot{x}(t) = Px(t) + bu(t) \tag{3.142}$$

From Figure 3.19, we also know that

$$u(t) = -K\hat{x}(t). \tag{3.143}$$

Therefore, substituting Eq. (3.143) into Eq. (3.142), we obtain

$$\dot{x}(t) = Px(t) - bK\hat{x}(t). \tag{3.144}$$

In terms of the state estimation error $\tilde{x}(t)$ defined in Eq. (3.122), Eq. (3.144) becomes

$$\dot{x}(t) = Px(t) - bK(x(t) - \tilde{x}(t)). \tag{3.145}$$

Combining Eqs. (3.145) and (3.129), we obtain an overall equation for the state vector x and its error $\tilde{x}$ as follows:

$$\begin{bmatrix} \dot{x}(t) \\ \dot{\tilde{x}}(t) \end{bmatrix} = \begin{bmatrix} P - bK & bK \\ 0 & P - ML \end{bmatrix} \begin{bmatrix} x(t) \\ \tilde{x}(t) \end{bmatrix}. \tag{3.146}$$

The characteristic equation of this closed-loop combined controller and estimator system can be obtained in a manner similar to that obtained for the estimator alone [see Eq. (3.131)]:

$$\begin{vmatrix} sI - P + bK & bK \\ 0 & sI - P + ML \end{vmatrix} = 0. \tag{3.147}$$

We can write this equation as

$$\det[sI - P + bK]\det[sI - P + ML] = 0 \tag{3.148}$$

The first determinant specifies the characteristic equation of the controller, and the second determinant specified the characteristic equation of the estimator [which is identical to Eq. (3.131)]. As we did for the case of the estimator alone in Section 3.7, Eq. (3.132), we can now define the combined desirable location of estimator roots

$$(s + \alpha_1)(s + \alpha_2) \cdots (s + \alpha_n)$$

and controller roots

$$(s + \beta_1)(s + \beta_2) \cdots (s + \beta_n)$$

and specify the combined estimator and controller's characteristic equation as

$$[(s + \beta_1)(s + \beta_2) \cdots (s + \beta_n)][(s + \alpha_1)(s + \alpha_2) \cdots (s + \alpha_n)] = 0. \quad (3.149)$$

We can now simultaneously determine the controller gain **K** and estimator coefficients **M** by setting Eqs. (3.148) and (3.149) equal to each other:

$$\det[s\mathbf{I} - \mathbf{P} + \mathbf{bK}]\det[s\mathbf{I} - \mathbf{P} + \mathbf{ML}] = [(s + \beta_1) \cdots (s + \beta_n)]$$
$$\times [(s + \alpha_1) \cdots (s + \alpha_n)]. \quad (3.150)$$

Therefore, we can observe from Eq. (3.150) that the roots of the combined controller and estimator is the sum of the controller and estimator roots found independently [8]. The primary concept to recognize in the combined compensator design for the controller and estimator is to make the estimator respond much faster than the controller, as we do not want the system's transient response limited by that of the estimator (which we can control by proper design).

It is useful to compare the modern pole placement method using the state-variable feedback method with the conventional transfer-function method for the design of the compensator as we have done in Section 3.2, when we found the open-loop transfer function $KG(s)H(s)$ in terms of **h**, $\Phi_h(s)$ and **b** in Eq. (3.25). We wish to find the transfer function of the combined controller and estimator, $U(s)/C(s)$. Let us reconsider the estimator equation, Eq. (3.120), and incorporate it in the control-law equation, Eq. (3.143), because the controller is part of the compensator:

$$\dot{\hat{\mathbf{x}}}(t) = \mathbf{P}\hat{\mathbf{x}}(t) - \mathbf{bK}\hat{\mathbf{x}}(t) + \mathbf{M}(c(t) - \mathbf{L}\hat{\mathbf{x}}(t)). \quad (3.151)$$

Simplifying, we obtain

$$\dot{\hat{\mathbf{x}}}(t) = (\mathbf{P} - \mathbf{bK} - \mathbf{ML})\hat{\mathbf{x}}(t) + \mathbf{M}c(t). \quad (3.152)$$

Let us compare Eq. (3.152) with the state equation of the process:

$$\dot{\mathbf{x}}(t) = \mathbf{P}\mathbf{x}(t) + \mathbf{b}u(t), \quad (3.153)$$

whose characteristic equation we know is given by

$$|s\mathbf{I} - \mathbf{P}| = 0, \quad (3.154)$$

using the same reasoning we used in obtaining Eqs. (6.22)[‡], (3.19), and (3.131). By comparing Eqs. (3.152) and (3.153), we obtain the characteristic equation of the compensator as follows:

$$|s\mathbf{I} - (\mathbf{P} - \mathbf{bK} - \mathbf{ML})| = 0. \qquad (3.155)$$

The resulting compensator may not result in a stable system because the roots of Eq. (3.155) have not been specified in advance. This is similar to our results in Section 3.3, where we found that design using linear-state-variable feeback did not ensure a stable system.

Before finding the transfer function representing the compensator, $U(s)/C(s)$, we first find the transfer function of the process from Eq. (3.153):

$$s\mathbf{X}(s) - \mathbf{x}(0) = \mathbf{P}\mathbf{X}(s) + \mathbf{b}U(s). \qquad (3.156)$$

Simplifying,

$$(s\mathbf{I} - \mathbf{P})\mathbf{X}(s) = \mathbf{b}U(s) + \mathbf{x}(0). \qquad (3.157)$$

Solving for $\mathbf{X}(s)$, we obtain

$$\mathbf{X}(s) = [s\mathbf{I} - \mathbf{P}]^{-1}\mathbf{b}U(s) + [s\mathbf{I} - \mathbf{P}]^{-1}\mathbf{x}(0). \qquad (3.158)$$

Since

$$c(t) = \mathbf{L}\mathbf{x}(t)$$

and its Laplace transform is

$$C(s) = \mathbf{L}\mathbf{X}(s) \qquad (3.159)$$

we can combine Eqs. (3.158) and (3.159) to relate the output $C(s)$ and input $U(s)$ of the process.

$$C(s) = \mathbf{L}[s\mathbf{I} - \mathbf{P}]^{-1}\mathbf{b}U(s) + \mathbf{L}[s\mathbf{I} - \mathbf{P}]^{-1}\mathbf{x}(0). \qquad (3.160)$$

In order to find the transfer function of the process $C(s)/U(s)$, we assume that the initial condition, $\mathbf{x}(0)$, equals zero. Therefore,

$$\frac{C(s)}{U(s)} = \mathbf{L}[s\mathbf{I} - \mathbf{P}]^{-1}\mathbf{b}. \qquad (3.161)$$

By analogy, we compare Eqs. (3.153) [and its resulting transfer function given by Eq. (3.161)] with Eq. (3.152), and conclude that the transfer function of the compensator defined by Eq. (3.152) is given by

$$\frac{U(s)}{C(s)} = -\mathbf{K}[s\mathbf{I} - \mathbf{P} + \mathbf{bK} + \mathbf{ML}]^{-1}\mathbf{M}. \qquad (3.162)$$

When we determine the transfer function $U(s)/C(s)$ using this procedure, we will find that the resulting transfer function will result in a phase-lead network, a phase-lag network, or a phase-lag-lead network.

150 MODERN CONTROL-SYSTEM DESIGN USING VARIOUS TECHNIQUES

To illustrate this approach for obtaining the transfer function of the compensator, consider a process whose transfer function is given by

$$C(s)/U(s) = G(s) = 1/s^2. \tag{3.163}$$

Therefore,

$$\ddot{c}(t) = u(t).$$

Defining the state variables of this second-order system as

$$x_1(t) = c(t),$$
$$x_2(t) = \dot{c}(t),$$

we obtain the state equation to be

$$\begin{bmatrix} \dot{x}_1(t) \\ \dot{x}_2(t) \end{bmatrix} = \underbrace{\begin{bmatrix} 0 & 1 \\ 0 & 0 \end{bmatrix}}_{\mathbf{P}} \begin{bmatrix} x_1(t) \\ x_2(t) \end{bmatrix} + \underbrace{\begin{bmatrix} 0 \\ 1 \end{bmatrix}}_{\mathbf{b}} u(t), \tag{3.164}$$

and the output equation is

$$c(t) = \underbrace{\begin{bmatrix} 1 & 0 \end{bmatrix}}_{\mathbf{L}} \begin{bmatrix} x_1(t) \\ x_2(t) \end{bmatrix}. \tag{3.165}$$

Let us assume that the design specification for the controller is

$$\omega_n = \sqrt{3},$$
$$\zeta = 0.58.$$

Therefore, the complex-conjugate roots of the controller are located at $-1 \pm j1.414$ and $\alpha_c(s)$ for the controller is given by

$$\alpha_c(s) = (s+1-j1.414)(s+1+j1.414) = s^2 + 2s + 3 \tag{3.166}$$

We can determine the controller gain $\mathbf{K}$ by equating like powers of s from Eq. (3.166) and that part of Eq. (3.148) concerned with the controller:

$$|s\mathbf{I} - \mathbf{P} + \mathbf{bK}| = 0. \tag{3.167}$$

Substituting for $\mathbf{P}$ and $\mathbf{b}$ from Eq. (3.164) into Eq. (3.167), we obtain the following:

$$\left| \begin{bmatrix} s & 0 \\ 0 & s \end{bmatrix} - \begin{bmatrix} 0 & 1 \\ 0 & 0 \end{bmatrix} + \begin{bmatrix} 0 \\ 1 \end{bmatrix} [K_1 \quad K_2] \right| = 0. \tag{3.168}$$

This simplifies to

$$\begin{vmatrix} s & -1 \\ K_1 & s+K_2 \end{vmatrix} = 0, \tag{3.169}$$

3.8. COMBINED COMPENSATOR DESIGN

from which we obtain the characteristic equation of the controller:

$$s^2 + K_2 s + K_1 = 0. \qquad (3.170)$$

Comparing like coefficients in Eqs. (3.166) and (3.170), we find the controller gains to be $K_1 = 3$ and $K_2 = 2$:

$$\mathbf{K} = [3 \quad 2]. \qquad (3.171)$$

As discussed previously, we desire the estimator to have a much faster response than the controller. Therefore, let us assume the design specification of the estimator to be

$$\omega_n = 17,$$
$$\zeta = 0.5.$$

Therefore, the complex-conjugate roots of the estimator are located at $-8.5 \pm j14.7$, and $\alpha_e(s)$ for the estimator is given by

$$\alpha_e(s) = (s + 8.5 + j14.7)(s + 8.5 - j14.7) = s^2 + 17s + 288.3. \qquad (3.172)$$

The resulting estimator feedback gain matrix is found from Eq. (3.131) [for the estimator portion of Eq. (3.148)] as follows:

$$|s\mathbf{I} - (\mathbf{P} - \mathbf{ML})| = 0. \qquad (3.173)$$

Substituting the matrix values into Eq. (3.173), we obtain

$$\left| \begin{bmatrix} s & 0 \\ 0 & s \end{bmatrix} - \begin{bmatrix} 0 & 1 \\ 0 & 0 \end{bmatrix} + \begin{bmatrix} m_1 \\ m_2 \end{bmatrix} [1 \quad 0] \right| = 0, \qquad (3.174)$$

which reduces to:

$$\begin{vmatrix} s + m_1 & -1 \\ m_2 & s \end{vmatrix} = 0.$$

The resulting characteristic equation in terms of m_1 and m_2 is given by

$$s^2 + m_1 s + m_2 = 0. \qquad (3.175)$$

Setting like coefficients in Eqs. (3.172) and (3.175) equal to each other, we obtain $m_1 = 17$ and $m_2 = 288.3$:

$$\mathbf{M} = \begin{bmatrix} 17 \\ 288.3 \end{bmatrix}. \qquad (3.176)$$

The resulting compensator transfer function is obtained by substituting parameters obtained from Eqs. (3.164), (3.165), (3.171), and (3.176) into Eq. (3.162) as follows:

$$\frac{U(s)}{C(s)} = -[3 \quad 2] \left[\begin{bmatrix} s & 0 \\ 0 & s \end{bmatrix} - \begin{bmatrix} 0 & 1 \\ 0 & 0 \end{bmatrix} + \begin{bmatrix} 0 \\ 1 \end{bmatrix} [3 \quad 2] + \begin{bmatrix} 17 \\ 288.3 \end{bmatrix} [1 \quad 0] \right]^{-1} \begin{bmatrix} 17 \\ 288.3 \end{bmatrix}. \qquad (3.177)$$

This equation reduces to

$$\frac{U(s)}{C(s)} = -[3 \quad 2] \frac{\begin{bmatrix} s+2 & 1 \\ -291.3 & s+17 \end{bmatrix} \begin{bmatrix} 17 \\ 288.3 \end{bmatrix}}{s^2 + 19s + 325.3}.$$

Upon further simplification, we obtain the following:

$$\frac{U(s)}{C(s)} = -\frac{[3s - 576.6 \quad 2s + 37] \begin{bmatrix} 17 \\ 288.3 \end{bmatrix}}{(s + 9.5 + j15.3)(s + 9.5 - j15.3)}. \tag{3.178}$$

The resulting transfer function of the compensator is given by the following equation:

$$G_{\text{comp}} = \frac{U(s)}{C(s)} = -\frac{627.6(s + 1.38)}{(s + 9.5 + j15.3)(s + 9.5 - j15.3)}. \tag{3.179}$$

Analysis of Eq. (3.179) shows that the compensator has the form of a phase-lead network with a zero at -1.38 and with two complex-conjugate poles as opposed to a simple pole as defined by the conventional phase-lead network of Eq. (2.6).

We can analyze the resulting system using the conventional root-locus and Bode-diagram methods. To obtain the open-loop transfer function for analyses, we combine Eqs. (3.163) and (3.179):

$$G(s)G_{\text{comp}}(s) = -\frac{627.6(s + 1.38)}{s^2(s + 9.5 + j15.3)(s + 9.5 - j15.3)}. \tag{3.180}$$

Replacing the specific gain of -627.6 with the variable gain K, the root locus can be evaluated from

$$G(s)G_{\text{comp}}(s) = \frac{K(s + 1.38)}{s^2(s + 9.5 + j15.3)(s + 9.5 - j15.3)}, \tag{3.181}$$

which is shown in Figure 3.20. Observe that the root locus goes through the roots chosen in Eqs. (3.166) and (3.172) when $K = 627.6$. These roots are shown in Figure 3.20 by solid dots. This figure was obtained using **MATLAB**, and is contained in the M-file that is part of my AMCSTD Toolbox which can be retrieved from The MathWorks anonymous FTP server at ftp://ftp.mathworks.com/pub/books/advshinners.

To draw the Bode diagram, we consider the modified form of Eq. (3.180):

$$G(s)G_{\text{comp}}(s) = \frac{866.1(0.725s + 1)}{s^2(s^2 + 19s + 325.3)}. \tag{3.182}$$

The resulting Bode diagram is drawn from the following simplification to Eq. (3.182):

$$G(s)G_{\text{comp}}(s) = \frac{2.66(0.725s + 1)}{s^2} \cdot \frac{325.3}{s^2 + 19s + 325.3}. \tag{3.183}$$

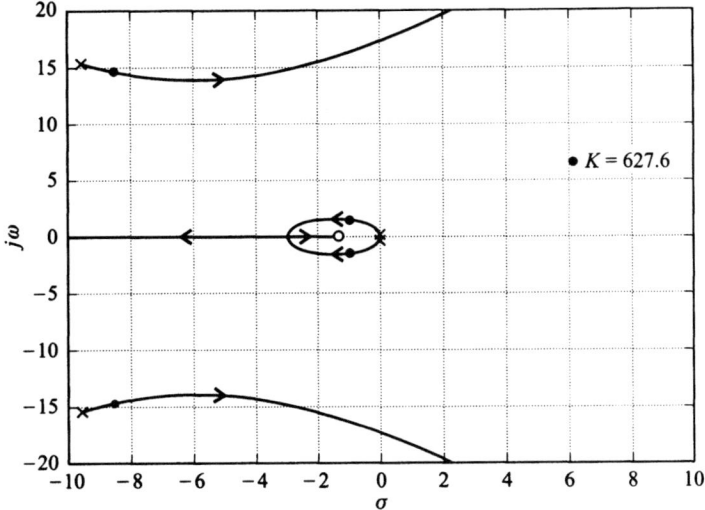

Figure 3.20 Root locus for compensated system of Figure 3.19 where $G(s)G_{comp}(s) = (K(s + 1.38))/(s^2(s + 9.5 + j15.3)(s + 9.5 - j15.3))$.

The quadratic poles in the denominator have an undamped natural frequency ω_n and a damping ratio ζ given by

$$\omega_n = \sqrt{325.3} = 18.04,$$
$$\zeta = \frac{19}{2(18.04)} = 0.527.$$

The resulting Bode diagram is shown in Figure 3.21. This figure was also obtained using MATLAB and is contained in the M-file that is part of my AMCSTD Toolbox.

We conclude that the uncompensated transfer function

$$G(s) = 1/s^2$$

has its phase margin increased from 0 to 51.07 degrees at its gain crossover frequency of 2.28 rad/sec., and its gain margin is increased from minus infinity to 19.13 dB (phase crossover frequency occurs at $\omega = 17.32$ rad/sec) when we use the compensator of Eq. (3.179). Notice that the crossover frequency of 2.28 rad/sec is approximately consistent with the controller's closed-loop roors of $\omega_n = \sqrt{3} = 1.732$ rad/sec and $\zeta = 0.85$. This is a reasonable result, as the slower roots of the controller are more dominant than the faster estimator roots, on the system response.

A complete case study for the design of a combined controller and estimator for a regulator, using the techniques presented in Section 3.7 and 3.8, is presented in Chapter 7.

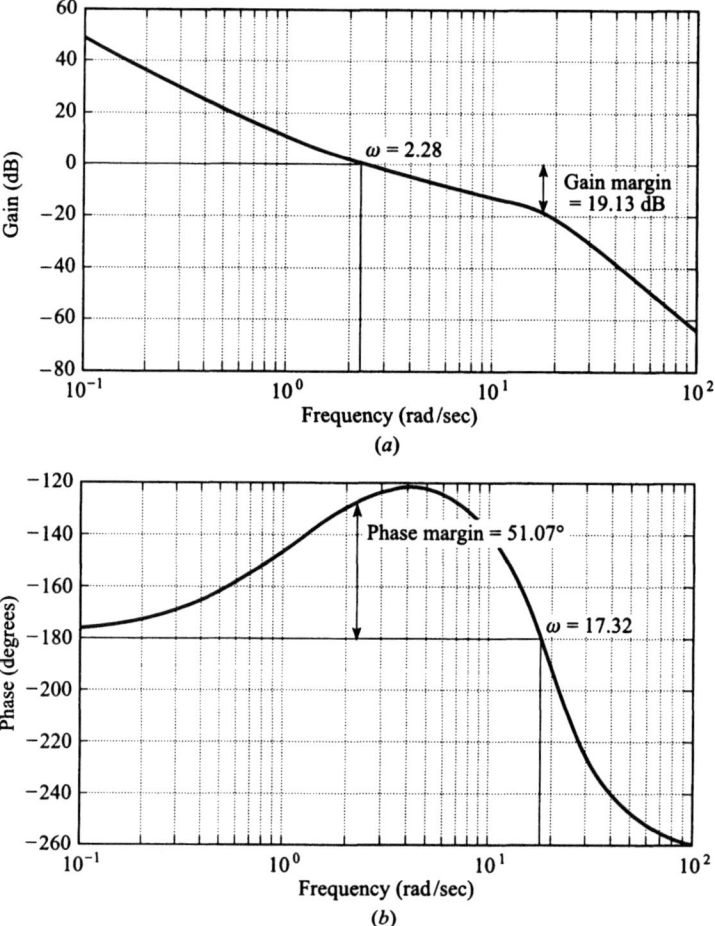

Figure 3.21 Bode diagram for compensated system of Figure 3.19 where $G(s)G_{comp}(s) = ((2.66(0.725s + 1))/s^2)(325.3/(s^2 + 19s + 325.3))$.

3.9. EXTENSION OF COMBINED COMPENSATOR DESIGN INCLUDING A CONTROLLER AND AN ESTIMATOR FOR SYSTEMS CONTAINING A REFERENCE INPUT

How can we extend the concepts developed in the previous section for regulator design, shown in Figure 3.19 (where the reference input $r(t) = 0$), to the more general problem where the reference input exists? Several methods exist which can be used to design such a system [8,9]. This section will consider the configuration illustrated in Figure 3.22. The design goal of the approach to be presented is to have the state-estimation error $\tilde{x}(t)$ be independent of the reference input $r(t)$ (e.g., $\tilde{x}$ should be uncontrollable from $r(t)$). This is a very important consideration, as we do not want the state-estimation error to be dependent on the type and level of the input.

Let us reconsider the controller equation (3.143),

$$u(t) = -\mathbf{K}\hat{\mathbf{x}}(t) \tag{3.184}$$

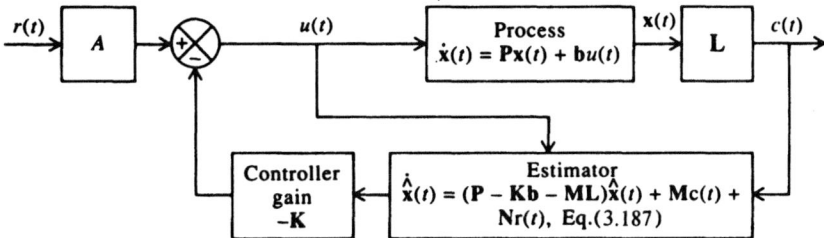

Figure 3.22 Addition of the reference input to the system shown in Figure 3.19 containing a combined controller and estimator.

and the estimator equation (3.152)

$$\dot{\hat{x}}(t) = (P - Kb - ML)\hat{x}(t) + Mc(t). \quad (3.185)$$

The reference input $r(t)$ will be introduced to these equations by adding a term $Ar(t)$ to the controller equation (3.184), and a term $Nr(t)$ to the estimator equation (3.185) (where N is a vector). Therefore, Eqs. (3.184) and (3.185) become

$$u(t) = -K\hat{x}(t) + Ar(t), \quad (3.186)$$

$$\dot{\hat{x}}(t) = (P - Kb - ML)\hat{x}(t) + Mc(t) + Nr(t). \quad (3.187)$$

What kind of system do Eqs. (3.186) and (3.187) infer? Does it result in the configuration of Figure 3.22? To answer this question, let us first substitute

$$c(t) = Lx(t) \quad (3.188)$$

into Eq. (3.187) and, thereby, eliminate the output $c(t)$ from Eq. (3.187):

$$\dot{\hat{x}}(t) = (P - Kb - ML)\hat{x}(t) + MLx(t) + Nr(t). \quad (3.189)$$

To find the estimation error (which we want to be independent of $r(t)$), let us difference $\dot{\hat{x}}$ [from Eq. (3.189)] and the state equation

$$\dot{x}(t) = Px(t) + bu(t). \quad (3.190)$$

We will first substitute Eq. (3.186) into Eq. (3.190):

$$\dot{x}(t) = Px(t) + b(-K\hat{x}(t) + Ar(t)). \quad (3.191)$$

Subtracting Eq. (3.189) from Eq. (3.191), we obtain the following for the derivative of the estimation error:

$$\dot{\tilde{x}}(t) = \dot{x}(t) - \dot{\hat{x}}(t) = Px(t) + b(-K\hat{x}(t) + Ar(t))$$
$$- (P - Kb - ML)\hat{x}(t) - MLx(t) - Nr(t).$$

Simplifying, we obtain the following equation:

$$\dot{\tilde{x}}(t) = (P - ML)\tilde{x}(t) + (bA - N)r(t). \tag{3.192}$$

In order to eliminate $r(t)$ from Eq. (3.192), it is necessary that

$$bA = N. \tag{3.193}$$

Therefore, the design criterion of the control-system engineer is to invoke Eq. (3.193). Substituting Eqs. (3.186) and (3.193) into Eq. (3.187), we obtain the following:

$$\dot{\hat{x}}(t) = P\hat{x}(t) + b(u(t) - Ar(t)) - ML\hat{x}(t) + Mc(t) + bAr(t). \tag{3.194}$$

Simplifying, we find that

$$\dot{\hat{x}}(t) = (P - ML)\hat{x}(t) + Mc(t) + bu(t) \tag{3.195}$$

or

$$\dot{\hat{x}}(t) = P\hat{x}(t) + bu(t) + M(c(t) - L\hat{x}(t)). \tag{3.196}$$

Notice that Eq. (3.196) is the same estimator equation defined in Eq. (3.120). It is important to emphasize that this occurs only when $bA = N$ as defined in Eq. (3.193). Therefore we conclude that the introduction of the reference input by adding the term $Ar(t)$ in the controller equation (3.186) and a term $Nr(t)$ to Eq. (3.187) results in the configuration shown in Figure 3.22.

Complete design examples for the design of the controller, estimator, and compensator, with their associated root-locus and Bode-diagram analyses of the resulting design are found in Chapter 7. In Section 7.5, the state-variable design for the controller and full-order estimator for a space vehicle is presented. In Problem 7.6, the state-variable design for the controller and full-order estimator of a chemical process control system is analyzed.

3.10. ROBUST CONTROL SYSTEMS [10–14]

Robust control is concerned with determining a stabilizing controller that achieves feedback performance in terms of stability and accuracy requirements, but the control must achieve the performance that is robust (insensitive) to plant uncertainty, parameter variation, and external disturbances. We know from the previous discussion in this book that feedback reduces the effects of external disturbances (Section 1.7) and parameter variations (Section 1.7). However, this is only achieved with relatively high loop gain which limits stability. Robust control is basically the same problem that was addressed in the 1930s by Black, Bode, and Nyquist. Modern robust control revolves around the feedback configurations illustrated in Figures 2.5, 2.6, and 2.7 which illustrate two-degrees-of-freedom compensation systems.

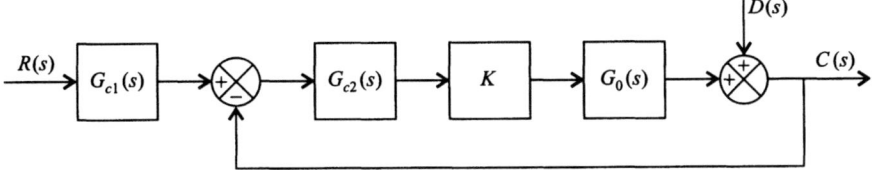

Figure 3.23 Control system with a two-degrees-of-freedom series controller $G_{c1}(s)$ and a forward-loop-controller $G_{c2}(s)$.

Let us consider the control system illustrated in Figure 3.23 which contains a disturbance $D(s)$, and it contains a two-degrees-of-freedom series controller $G_{c1}(s)$ and a forward-loop controller $G_{c2}(s)$. In this control system's operation, the amplifier gain, K, can vary.

For this control system, the overall transfer function, $C(s)/R(s)$, is given by

$$H(s) = \frac{C(s)}{R(s)} = \frac{G_{c1}(s)G_{c2}(s)KG_0(s)}{1 + KG_{c2}(s)G_0(s)}. \qquad (3.197)$$

The transfer function relating the disturbance $D(s)$ to the output $C(s)$ is given by

$$N(s) = \frac{C(s)}{D(s)} = \frac{1}{1 + KG_{c2}(s)G(s)}. \qquad (3.198)$$

The design approach in robust control systems is to choose the controller $G_{c1}(s)$ so that the desired closed-loop transfer function $H(s)$ is obtained, and to choose the controller $G_{c2}(s)$ so that the output, $C(s)$, is insensitive to the disturbance $D(s)$ over the frequency range in which $D(s)$ is dominant.

The sensitivity of $H(s)$ due to variations of K is given by

$$S_K^{H(s)}(s) = \frac{\frac{dH(s)}{H(s)}}{\frac{dK}{K}} = \frac{K}{H(s)} \frac{dH(s)}{dK}. \qquad (3.199)$$

For the control system of Figure 3.23,

$$\frac{dH}{dK} = \frac{G_{c1}(s)G_{c2}(s)G_0(s)}{[1 + KG_{c2}(s)G_0(s)]^2}. \qquad (3.200)$$

Substituting Eqs. (3.197) and (3.200) into Eq. (3.199), we obtain the following:

$$S_K^{H(s)}(s) = \frac{1}{1 + KG_{c2}(s)G_0(s)}. \qquad (3.201)$$

It is important to recognize that for this control system, both $C(s)/D(s)$ given by Eq. (3.198) and the sensitivity of $H(s)$ with respect to K given by Eq. (3.201) are identical. This is a very important result which implies that we can use the same control-system techniques to suppress the affect of the disturbance $D(s)$ and robustness (insensitivity) with respect to variations of K.

A. Design Example Illustrating Robustness and Disturbance Rejection

We will next analyze how the two-degrees-of-freedom control system illustrated in Figure 3.23 can achieve a high-gain which will satisfy the robustness and performance requirements while, at the same time, minimizing the affects of the disturbance. We will analyze the system of Figure 3.23 with the following transfer function which represents a fourth-order process $KG_0(s)$:

$$KG_0(s) = \frac{142(7+s)}{s(1+s)(5+s)(20+s)}. \tag{3.202}$$

We will assume at this point that $G_{c1}(s) = G_{c2}(s) = 1$. Our problem is to investigate the affect of the variation of K. The transfer function of $KG_{c2}(s)G_0(s)$ is given by:

$$KG_{c2}(s)G_0(s) = \frac{142(7+s)}{s(1+s)(5+s)(20+s)}. \tag{3.203}$$

We want to consider the variation of the gain from 142 to double that amount (i.e., 248) and to one-half that amount (i.e., 71).

Because this control system acts as a low-pass filter, the sensitivity of $H(s)$ with respect to K is poor. The bandwidth of this control system with $K = 142$ is only 3.2 rad/sec, while the sensitivity of $H(s)$ with respect to K is expected to be greater than one at frequencies greater than 3.2 rad/sec. Figure 3.24 illustrates the unit step response of the system when $K = 142$ (the nominal value), $K = 248$, and $K = 71$. Table 3.1 lists the characteristics of the unit step transient responses and the characteristic equation roots of this control system which were obtained using

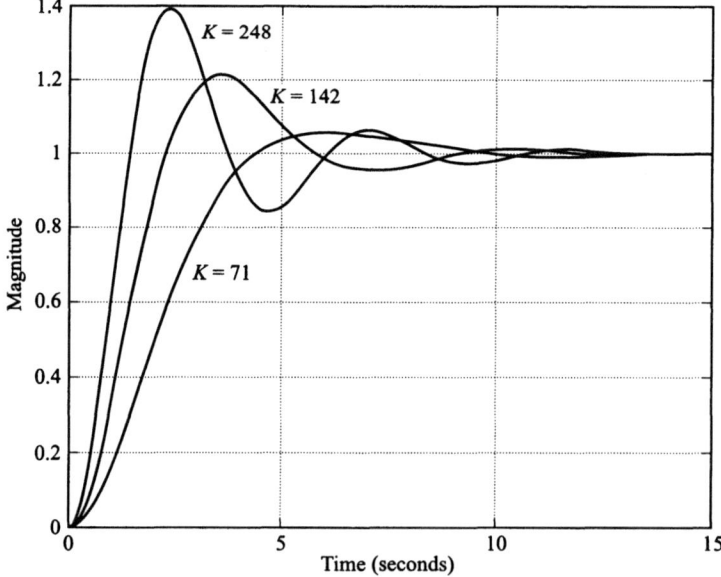

Figure 3.24 Unit step response for system of Figure 3.23 with $KG_{c2}(s)G_0(s)$ given by Eq. (3.203) and $G_{c1}(s) = 1$.

Table 3.1. Characteristics of the Control System Illustrated in Figure 3.23 where $KG_{c2}(s)G_0(s)$ is given by Eq. (3.203), and $G_{c1}(s) = 1$

K	Damping ratio	Roots of characteristic equation
248	0.277	$-5.1640, -20.0649, -0.3856 \pm j1.3348$
142	0.444	$-5.0878, -20.0325, -0.4398 \pm j0.8875$
71	0.666	$-5.0456, -20.01629, -0.4691 \pm j0.5245$

MATLAB. Observe that variations of K from its nominal value of 142 result in considerable variation in the damping ratio and the transient responses of this control system. Figure 3.25 illustrates the root loci and the location of the closed-loop, complex-conjugate, roots for the three cases being analyzed.

The design approach for this robust controller, $G_{c2}(s)$, is to place two zeros at (or near) the desired complex, conjugate-loop, poles at $-0.4398 \pm j0.8875$ for the nominal gain case of $K = 142$. Therefore,

$$G_{c2}(s) = \frac{(s + 0.4398 + j0.8875)(s + 0.4398 - j0.8875)}{0.98} \tag{3.204}$$

or

$$G_{c2}(s) = \frac{s^2 + 0.88s + 0.98}{0.98}. \tag{3.205}$$

We will approximate $G_{c2}(s)$ as follows:

$$G_{c2}(s) = s^2 + 0.88s + 1. \tag{3.206}$$

Therefore, the forward-path transfer function of this control system with $KG_0(s)$ given by Eq. (3.202) and $G_{c2}(s)$ given by Eq. (3.206) is:

$$KG_{c2}(s)G_0(s) = \frac{K(7+s)(s^2 + 0.88s + 1)}{s(1+s)(5+s)(20+s)}. \tag{3.207}$$

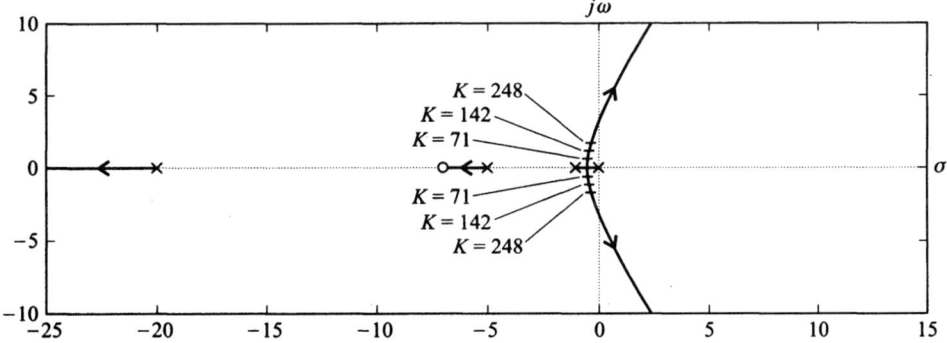

Figure 3.25 Root-locus plot for system of Figure 3.23 with $KG_{c2}(s)G_0(s)$ given by Eq. (3.203) and $G_{c1}(s) = 1$.

Table 3.2 lists the damping ratio and the characteristic equation roots of this control system, obtained using MATLAB, with the forward-loop transfer function given by Eq. (3.207). Observe that the range of the damping ratios are much closer (0.548 to 0.808) than they were before the addition of the robust controller $G_{c2}(s)$ and shown in Table 3.1 (where the damping previously varied from 0.277 to 0.666).

Figure 3.26 illustrates the root loci with the robust controller $G_{c2}(s)$ added, and the location of the closed-loop, complex-conjugate, roots for the three cases. Observe from this root locus that by locating the two zeros of the forward-loop controller $G_{c2}(s)$ near the desired characteristic equation complex-conjugate roots for $K = 142$, the sensitivity of this control system becomes much better.

It was shown in Section 3.2 during the discussion on the concept of liner-state-variable feedback that for the system illustrated in Figure 3.3, the zeros of $C(s)/R(s)$ are the zeros of $G(s)$ based on comparing Eqs. (3.17) and (3.24). Therefore, in the system we are currently analyzing in Figure 3.23, the zeros of the forward-path transfer function $KG_{c2}(s)G_0(s)$ are identical to the zeros of the closed-loop transfer function. Therefore, the closed-loop zeros of $G_{c2}(s)$ in Eq. (3.206) come close to canceling the effect of the complex-conjugate, closed-loop poles. Therefore, it is necessary to also add the series controller $G_{c1}(s)$, as illustrated in Figure 3.23, so that $G_{c1}(s)$ contains poles to cancel the zeros of $s^2 + 0.88 + 1$ of the closed-loop transfer function. Therefore, the transfer function of the forward-loop controller, $G_{c1}(s)$, is given by:

$$G_{c1}(s) = \frac{1}{s^2 + 0.88s + 1}. \tag{3.208}$$

Table 3.2. Characteristics of the Control System Illustrated in Figure 3.23 with the Forward-Loop Controller $G_{c2}(s)$ added and where $KG_{c2}(s)G_0(s)$ is given by Eq. (3.207)

K	Damping ratio	Roots of characteristic equation
248	0.548	$-6.3715, -47.3367, -0.4459 \pm j0.6814$
142	0.644	$-6.0358, -33.3566, -0.4538 \pm j0.5392$
71	0.808	$-5.6914, -26.5282, -0.4652 \pm j0.3387$

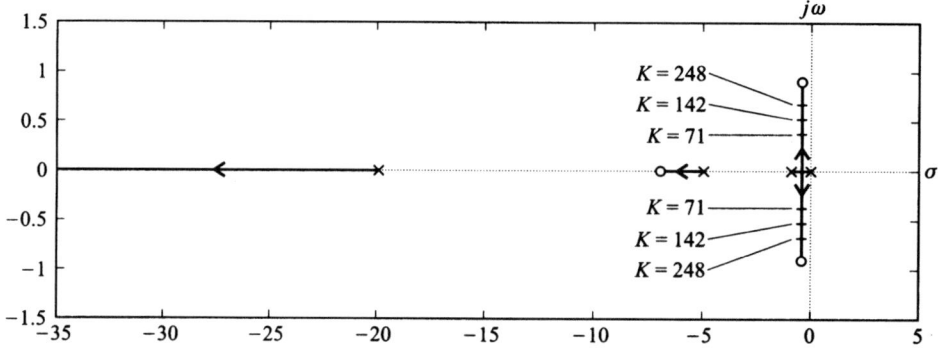

Figure 3.26 Root-locus plot for system shown in Figure 3.23 with $KG_{c2}(s)G_0(s)$ given by Eq. (3.207).

The unit step response of this control system with the forward-path transfer function of the control system given by Eq. (3.207), with $K = 71$, 142, and 248, and with the forward-loop controller transfer function given by Eq. (3.208) is illustrated in Figure 3.27. Comparing these unit step responses with those in Figure 3.24, we conclude that this control system has been made to be much less sensitive to variations in K. For example, the maximum percent overshoot of the transient responses for the original system illustrated in Figure 3.24 ranged from 5.6% (for $K = 71$) to 39% (for $K = 248$). However, the control system designed to be robust has a transient response as illustrated in Figure 3.27 which shows that the maximum percent overshoot varies from 1.4% (for $K = 71$) to only 13% (for $K = 248$). In addition, as we pointed out before in comparing Eqs. (3.198) and (3.201), which are identical, the robustness with respect to variations in K will also provide disturbance suppression using the same control-system techniques.

Since the disturbance suppression attributes are a function of frequency, let us examine the frequency characteristics of the $C(s)/D(s)$ transfer function. Substituting Eq. (3.203) into Eq. (3.198), we obtain the following transfer function for $C(s)/D(s)$ for the case where the forward-loop controller $G_{c2}(s)$ has a gain equal to one:

$$\frac{C(s)}{D(s)} = \frac{s^4 + 26s^3 + 125s^2 + 100s}{s^4 + 26s^3 + 125s^2 + 242s + 994}. \qquad (3.209)$$

Substituting Eq. (3.207) into Eq. (3.198), we obtain the following transfer function for $C(s)/D(s)$ for the case where the forward-loop controller $G_{c2}(s)$ has the transfer function given by Eq. (3.206)

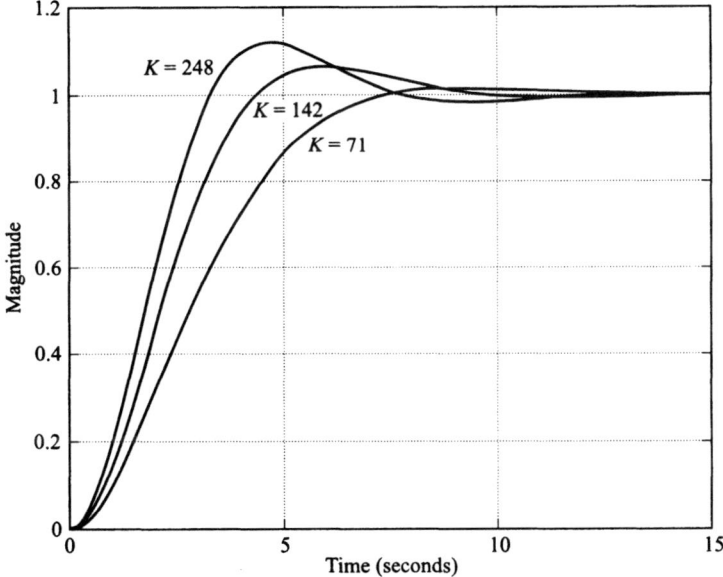

Figure 3.27 Unit step response for system of Figure 3.23 with $KG_{c2}(s)G_0(s)$ given by Eq. (3.207) and $G_{c1}(s)$ given by Eq. (3.208).

$$\frac{C(s)}{D(s)} = \frac{s^4 + 26s^3 + 125s^2 + 100s}{s^4 + 168s^3 + 1244s^2 + 1117s + 994}. \qquad (3.210)$$

Figure 3.28 is a plot of the frequency response of the disturbance suppression transfer fucntion $C(s)/D(s)$ as defined in Eqs. (3.209) and (3.210). It shows that the disturbance suppression of the control system at low frequencies is approximately the same with and without $G_{c2}(s)$ in the control system. However, the addition of $G_{c2}(s)$, as defined by Eq. (3.206), greatly improves the disturbance suppression attributes of the control system at high frequencies. This is consistent with our expectations.

Robust control-system design principles are being applied to many, modern, practical control systems. For example, the reader is referred to Reference 15 which presents the design for a robust control system for preventing car skidding. A complete case study for the design of a robust control system for controlling the flaps of a hydrofoil is presented in Section 7.7.

3.11. AN INTRODUCTION TO H^∞ CONTROL CONCEPTS [16,17]

In the presentation of this book, the classical frequency-domain approach and the modern state-variable time-domain approach have been presented in parallel. It has

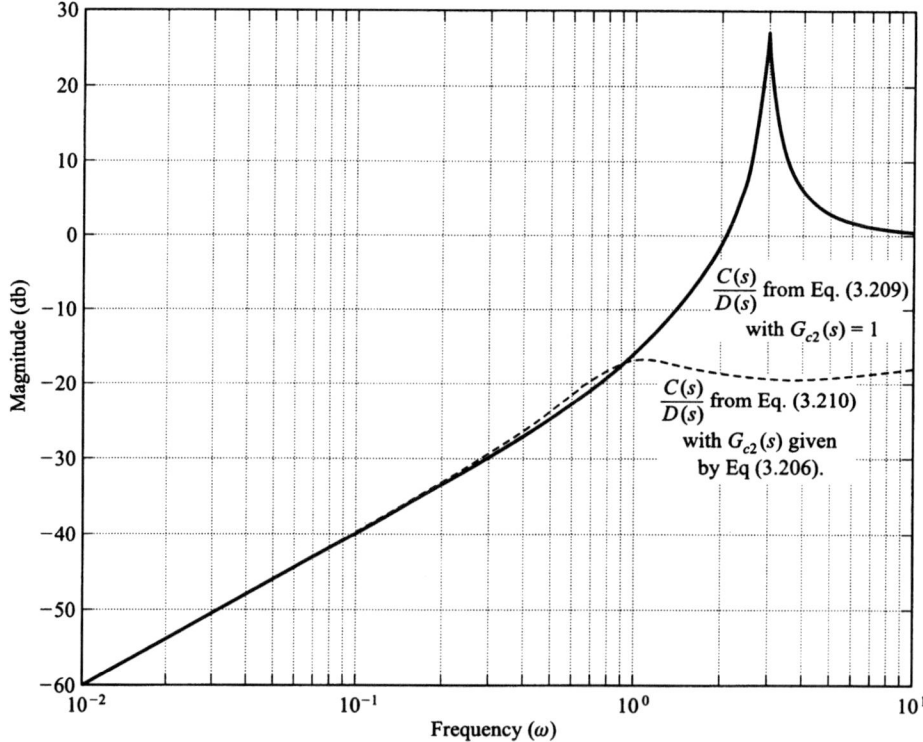

Figure 3.28 Frequency response of the disturbance suppression transfer functions defined in Eqs. (3.209) and (3.210).

been shown that they complement each other. Prior to the 1960s, the frequency-domain approach predominated. With the advent of the space race, the availability of practical digital computers, modern optimal control theory, and the state-variable approach in the early 1960s, the pendulum swung to the time-domain approach. The 1960s and 1970s saw an abundant amount of work performed on applying modern optimal control theory, which is presented in Chapter 6 in this book. In the early 1980s, a new technique has emerged known as H^∞ control theory which combines both the fequency- and time-domain approaches to provide a unified answer. Zames is given credit for its introduction with his paper in the *IEEE Transactions on Automatic Control* [16]. The H^∞ approach has dominated the trend of control-system development in the 1980s and 1990s. A complete treatment of the subject of H^∞ is complex, and beyond the scope of this book. However, we can expand the concepts of robustness (introduced in Section 3.10) and sensitivity (introduced in Chapter 1), together with the frequency and state-variable domain techniques presented in this book to introduce the basic concepts of H^∞ control theory and apply it to some simple problems. This is the objective of this and the following sections.

A. Sensitivity of Control Systems Containing Disturbances and Measurement Noise

Let us extend our understanding of sensitivity developed in Chapter 1 to the control system illustrated in Figure 3.29 which contains a disturbance $U(s)$ and feedback sensor measurement noise $N(s)$. Although noise is a stochastic process, we will assume in this analysis that it can be represented as a deterministic process. This initial analysis will focus on the single-input single-output (SISO) system shown in Figure 3.29 which contains a disturbance $U(s)$ and feedback sensor measurement noise $N(s)$. We will then discuss the extension of these results to multiple-input multiple-output (MIMO) systems.

Using Mason's theorem, we can write the following relationship for the output $C(s)$ by inspection of Figure 3.29:

$$C(s) = \frac{G(s)}{1+G(s)}[R(s) - N(s)] + \frac{1}{1+G(s)} U(s). \qquad (3.211)$$

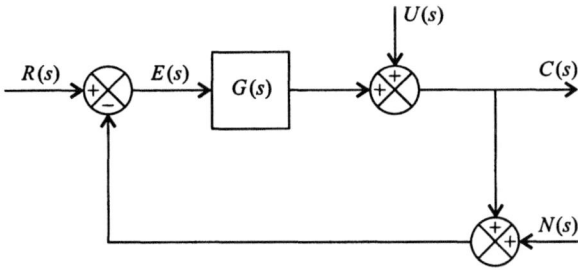

Figure 3.29 Control system containing a disturbance $U(s)$ and sensor measurement observation noise $N(s)$.

We can also use Mason's theorem to write the following relationship for the control-system error $E(s)$ by inspection of Figure 3.29:

$$E(s) = \frac{1}{1 + G(s)}[R(s) - U(s) - N(s)]. \tag{3.212}$$

From the analysis in Section 1.7 on sensitivity, we know that the *sensitivity* of the transfer function $C(s)/R(s)$ to changes in $G(s)$ for the control system shown in Figure 3.29 is given by the following expression:

$$S_{G(s)}^{C(s)/R(s)} = \frac{1}{1 + G(s)}. \tag{3.213}$$

It is interesting to observe that this sensitivity is also the transfer function from $U(s)$ to $-E(s)$:

$$-\frac{E(s)}{U(s)} = S_{G(s)}^{C(s)/R(s)} = \frac{1}{1 + G(s)}. \tag{3.214}$$

In the MIMO case, the sensitivity function in Eq. (3.213) is modified so that $G(s)$ becomes the matrix of transfer functions between the inputs and the outputs, and the value of 1 is replaced with the identity matrix.

The transfer function $C(s)/R(s)$ of the control system shown in Figure 3.29 is given by the following:

$$\frac{C(s)}{R(s)} = \frac{G(s)}{1 + G(s)}. \tag{3.215}$$

This expression is also known as the *complementary sensitivity function* which is defined as follows:

$$T(s) = \frac{G(s)}{1 + G(s)}. \tag{3.216}$$

It is very important to recognize that the sum of the sensitivity function given by Eq. (3.213), and the complementary sensitivity function given by Eq. (3.216) equals one:

$$S(s) + T(s) = 1. \tag{3.217}$$

We can express the error function $E(s)$ in Eq. (3.212) in terms of the sensitivity function. Let us assume that the feedback sensor measurement noise is zero. Therefore, substituting Eq. (3.213) into Eq. (3.212), we obtain the following:

$$E(s) = S(s)[R(s) - U(s)]. \tag{3.218}$$

Equation (3.218) is a very important relationship because it states that we should make the sensitivity function S small to make the control system error $E(s)$ small. In addition, we know that since Eq. (3.213) represents the sensitivity function as well as

3.11. AN INTRODUCTION TO H^∞ CONTROL CONCEPTS

the transfer function between $U(s)$ and $-E(s)$ [see Eq. (3.214)], we want the sensitivity function $S(s)$ to be as small as possible for both disturbance rejection and for making the control-system error small.

The conclusion of this analysis is that we want to make $S(s)$ as small as possible. However, is this possible over the entire range of frequencies? Let us analyze Eq. (3.213) to determine if this is feasible. Since $G(s)$ approaches zero as s approaches infinity in practical systems, then

$$\lim_{s \to \infty} S(s) = 1. \qquad (3.219)$$

The result of Eq. (3.213) is that we can only make the sensitivity function $S(s)$ small over low and mid-range frequencies, but not at high frequencies.

Ideally, what would we like to make the complementary sensitivity function? Analyzing Eq. (3.217), we would like to make the complementary sensitivity function $T(s)$ equal to one because that would then result in the sensitivity function $S(s)$ being equal to zero. However, we know from Eq. (3.216) that $T(s)$ approaches zero as $G(s)$ approaches infinity. Therefore, we can design the complementary sensitivity function $T(s)$ to approximate one at low and mid-range frequencies, but not at high frequencies. Figure 3.30 illustrates the representative frequency responses of the sensitivity function $S(s)$ and the complementary sensitivity function $T(s)$.

The transfer function relating the sensor measurement noise $N(s)$ to the output $C(s)$ can be obtained from Eq. (3.211) by setting $R(s) = U(s) = 0$:

$$\frac{C(s)}{N(s)} = -\frac{G(s)}{1 + G(s)}. \qquad (3.220)$$

Observe that Eq. (3.220) is the negative of the complementary sensitivity function $T(s)$ defined in Eq. (3.216). Therefore,

$$T(s) = -\frac{C(s)}{N(s)} = \frac{G(s)}{1 + G(s)}. \qquad (3.221)$$

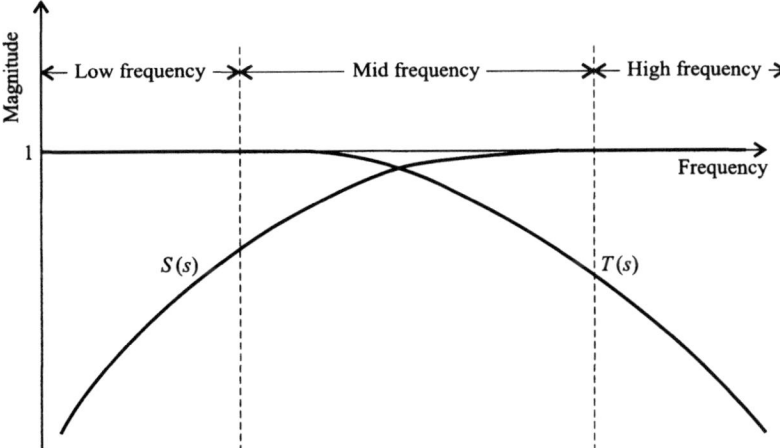

Figure 3.30 Representative sensitivity and complementary sensitivity frequency responses.

This results in a problem to the control-system engineer because we want the transfer function $C(s)/N(s)$ to be as small as possible, but the desirable $T(s)$ is large at low and mid-range frequencies as illustrated in Figure 3.30. Equation (3.217) showed that we would want to make $T(s) = 1$ ideally in order to drive the sensitivity function $S(s) = 0$. Therefore, the control-system engineer must make a tradeoff here between allowable feedback sensor measurement noise and the complementary sensitivity function (which also affects the sensitivity function [see Eq. (3.217)]. The basic trade-off is to determine allowable noise and permissible sensitivity.

B. Desirable Control-System Transfer-Function Characteristics

From Figure 3.29 and Eq. (3.211), we can state the relationship between the reference input $R(s)$, the disturbance $U(s)$, and the feedback sensor measurement noise $N(s)$ with the output $C(s)$ as follows:

$$C(s) = H_r(s)R(s) + H_u(s)U(s) + H_n(s)N(s) \qquad (3.222)$$

where

$$H_r(s) = \frac{C(s)}{R(s)} = \frac{G(s)}{1 + G(s)}, \qquad (3.223)$$

$$H_u(s) = \frac{C(s)}{U(s)} = \frac{1}{1 + G(s)}, \qquad (3.224)$$

$$H_n(s) = \frac{C(s)}{N(s)} = -\frac{G(s)}{1 + G(s)}. \qquad (3.225)$$

In terms of the error function $E(s)$, as defined in Eq. (3.212), we define $E(s)$ as follows:

$$E(s) = [1 - H_r(s)]R(s) - H_u(s)U(s) + H_n(s)N(s). \qquad (3.226)$$

It is interesting to examine the effect of the feedback sensor noise on the control system. To do this, let us assume that all the other external inputs are zero:

$$R(s) = U(s) = 0. \qquad (3.227)$$

Let us assume that $n(t)$ has a Fourier transform although, in practice, noise usually does not have a Fourier transform. Using Parseval's theorem, we can determine the integral-squared error (ISE):

$$\int_0^\infty e^2(t)\,dt = \frac{1}{2\pi}\int_{-\infty}^\infty |H_n(j\omega)|^2 |N(j\omega)|^2\,d\omega. \qquad (3.228)$$

By focusing on the square of the error, the ISE penalizes both positive and negative values of the error. The function $|N(j\omega)|^2$ is defined as the energy density spectrum of ω.

Let us try to define the desirable frequency characteristics from a practical viewpoint for $H_r(j\omega)$, $H_u(j\omega)$, and $H_n(j\omega)$. The reference input $R(s)$ and the disturbance $U(s)$ are usually low-frequency signals. Therefore, it is desired to have $H_r(j\omega) \approx 1$ and $H_u(j\omega) \approx 0$ at low frequencies. In practice, $H_r(j\omega) \approx 0$ and $H_u(j\omega)$ and $H_n(j\omega) \approx 1$ at high frequencies. In practice, we try to make $H_r(j\omega) \approx 1$, and $H_u(j\omega)$ and $H_n(j\omega) \approx 0$ over the frequency range from 0 to the gain crossover frequency ω_c. These are ideals and, in practice, we can only approximate these. For example, Figure I1.7-3 shows the closed-loop frequency response of Figure I1.7-1. This closed-loop frequency response corresponds to $H_r(s)$ for the system shown in Figure I1.7-1.

In practice, $G(s)$ is much greater than one at low frequencies. Therefore, Eq. (3.223) reduces to

$$H_r(s) = \frac{C(s)}{R(s)} = \frac{G(s)}{1 + G(s)} \approx 1 \tag{3.229}$$

and Eqs. (3.224) and (3.225) reduce to

$$H_u(s) = \frac{C(s)}{U(s)} = \frac{1}{1 + G(s)} \approx \frac{1}{G(s)}, \tag{3.230a}$$

$$H_n(s) = \frac{C(s)}{N(s)} = -\frac{G(s)}{1 + G(s)} \approx -1 \tag{3.230b}$$

at low frequencies. Therefore, a high loop gain is very desirable for frequencies in the passband (defined as frequencies between 0 and ω_c) because it then approximates $H_r(j\omega) \approx 1$ and $H_u(j\omega)$ is very small. However, the sensor measurement observation noise remains a problem because $H_n(s)$ is approximately -1. The control engineer has two choices regarding $N(s)$: (a) Design the sensor so that its measurement noise is very small; (b) tradeoff how high the loop gain is designed so that it minimizes the effect of the disturbance input $U(s)$, but the loop gain should not be too high so that the effect of the noise is tolerable.

C. Extension of Sensitivity and Complementary Sensitivity Concepts to Multivariable Control Systems

Many modern control systems contain several inputs and outputs. Therefore, it is important to extend our understanding of sensitivity and complementary sensitivity to multiple-input multiple-output (MIMO) control systems.

Figure 3.31 illustrates the block diagram of a MIMO control system where $\mathbf{r}(t)$ and $\mathbf{c}(t)$ are vectors, and $\mathbf{D}(s)$ and $\mathbf{G}(s)$ are matrices. The dimensions of $\mathbf{D}(s)$ are $n \times i$, and the dimensions of $\mathbf{G}(s)$ are $i \times n$, where $n = \dim[\mathbf{C}(s)]$ and $i = \dim[\mathbf{m}(s)]$. By analogy to Eq. (3.213) for the SISO control system, the sensitivity function for the MIMO control system is given by:

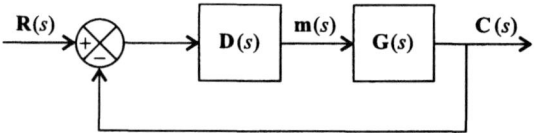

Figure 3.31 Block diagram of a MIMO control system.

$$S_{G(s)D(s)}^{C(s)/R(s)} = [I + D(s)G(s)]^{-1}. \qquad (3.231)$$

By analogy to Eq. (3.216) for the SISO control system, the complementary sensitivity function for the MIMO control system is given by

$$T(s) = [I + D(s)G(s)]^{-1} D(s)G(s). \qquad (3.232)$$

The sensitivity and complementary sensitivity functions are both $n \times n$ matrices. By analogy to Eq. (3.217) for the SISO case, the sum of the sensitivity and complementary functions for the MIMO case is given by

$$S(s) + T(s) = I \qquad (3.233)$$

where **I** represents the identity matrix.

D. Example for Finding S(s) and T(s) for a MIMO Control System

The open-loop transfer function matrix for a two-input, two-output control system to be analyzed is given by

$$D(s)G(s) = \begin{bmatrix} \dfrac{1}{s} & \dfrac{1}{s+1} \\ 1 & \dfrac{2}{s(s+2)} \end{bmatrix}. \qquad (3.234)$$

Let us determine the sensitivity and the complementary sensitivity functions.

The sensitivity function for this MIMO control system is obtained from Eq. (3.231) as follows:

$$S(s) = [I + D(s)G(s)]^{-1} = \begin{bmatrix} \dfrac{1}{s}+1 & \dfrac{1}{s+1} \\ 1 & \dfrac{2}{s(s+2)}+1 \end{bmatrix}^{-1}. \qquad (3.235)$$

Therefore,

$$S(s) = \dfrac{1}{-s^3 - 2s^2 + 2s + 2} \begin{bmatrix} 2s(s+1) & -s^2(s+2) \\ -s^2(s+1)(s+2) & s(s+1)(s+2) \end{bmatrix}. \qquad (3.236)$$

The complementary sensitivity function $\mathbf{T}(s)$ can be found from

$$\mathbf{T}(s) = \mathbf{I} - \mathbf{S}(s) = \begin{bmatrix} 1 & 0 \\ 0 & 1 \end{bmatrix} - \frac{1}{-s^3 - 2s^2 + 2s + 2} \begin{bmatrix} 2s(s+1) & -s^2(s+2) \\ -s^2(s+1)(s+2) & s(s+1)(s+2) \end{bmatrix}. \quad (3.237)$$

Simplifying this expression, we obtain

$$\mathbf{T}(s) = \frac{1}{-s^3 - 2s^2 + 2s + 2} \begin{bmatrix} -s^3 - 4s^2 + 2 & s^2(s+2) \\ s^2(s+1)(s+2) & -2s^3 - 5s^2 + 2 \end{bmatrix}. \quad (3.238)$$

3.12. FOUNDATIONS OF H^∞ CONTROL THEORY

Based on the preceding results, the concepts of modern H^∞ control theory will be presented. The very basic problem that H^∞ as presented by Zames in Reference 16 focuses on is sensitivity reduction of feedback control systems as an optimization problem, and it is separated from the problem of stabilization. The technique is concerned with the effects of feedback on uncertainty, where the uncertainty may be in the form of an additive disturbance $U(s)$ as illustrated in Figure 3.29. H^∞ control theory approaches the problem from the point of view of classical sensitivity theory, which has been presented in Chapter 1 and subsections 3.11A–3.11D, with the difference that feedback will not only reduce but also optimize sensitivity in an appropriate sense.

H^∞ control theory is a complex subject. The purpose of presenting it in this book, in a clear and cohesive manner, is to introduce it and motivate the reader with an interest in this field to review fome of the recent papers which have been written on this subject[18–21]. In its basic form H^∞ control theory attempts to minimize the supremum function over the entire frequency range

$$u = \sup_\omega |S(j\omega)W(j\omega)| \quad (3.239)$$

where $S(j\omega)$ is the sensitivity function and $W(j\omega)$ is a weighting function. We can view the magnitude of the product $S(j\omega)W(j\omega)$ as the magnitude of the weighted sensitivity. The weighting function emphasizes that low sensitivity is more important at low frequencies than higher frequencies. Therefore, by emphasizing the minimization of the magnitude of the product, the result is that the weighting function is greatest at those frequencies where the sensitivity is the smallest—namely at low frequencies.

The general solution to Eq. (3.239), obtained from function analysis, has the following general form [21]:

$$S(s)W(s) = kB(s) \quad (3.240)$$

where k is a constant and $B(s)$ is known as the Blaschke product. For the problem being considered in this book where the process $G(s)$ is stable (e.g., all of its poles are in the left-half of the s-plane and are of minimum phase), then $B(s) = 1$ and the constant k is any desirable, small, real number. Therefore, Eq. (3.240) states that we want to make the weighting function large for those low frequencies where we want to make the sensitivity small. Conversely, we want to make the weighting function small for those high frequencies where the sensitivity is large.

Let us examine in detail the result when $G(s)$ has all its poles in the left-half of the s-plane, and $B(s) = 1$. Therefore,

$$|S(j\omega)| = \frac{|k|}{|W(j\omega)|} \quad (3.241)$$

which implies that the shape of the magnitude of the sensitivity curve is the inverse of the weighting function. Therefore, the shape of the optimum $|S(j\omega)|$ is independent of the process $G(s)$, and only depends on the magnitudes of the constant k and the weighting function $W(j\omega)$.

A. Application of the Theory to an Example

Let us try to pull all the concepts presented in this section together by giving an example. We will consider a simple SISO control system, and determine its sensitivity function, S, complementary sensitivity function, T, and the weighting function, W. We will also assume that the constant k in Eq. (3.240) is equal to one.

For the illustrative example, let us consider a unity-feedback control system whose forward transfer function $G(s)$ is given by

$$G(s) = \frac{2}{s(s+1)}. \quad (3.242)$$

Substituting Eq. (3.242) into Eqs. (3.213), (3.216), and (3.241), we can calculate the sensitivity S, the complementary sensitivity T, and the weighting function W, respectively. The result is shown in Figure 3.32. The result agrees with the theoretical results expected from the presentation of this section. The sensitivity S is very small at low frequencies, and approaches one at very high frequencies. The weighting function W is the inverse of the sensitivity function, and is very large at low frequencies and also approaches one at very high frequencies. The complementary sensitivity function equals one minus the sensitivity function (see Eq. (3.217)), and equals one at very low frequencies and is very small at very high frequencies.

B. The H^∞ Process for the General SISO Case

Let us assume that in the general SISO case, the process $G(s)$ has poles and zeros in the right half-plane. For this general case, the H^∞ process would proceed in the following step-by-step manner:

1. The weighting function $W(s)$ would be selected which would have the following characteristics:

   - It must have no zeros in the right half-plane.

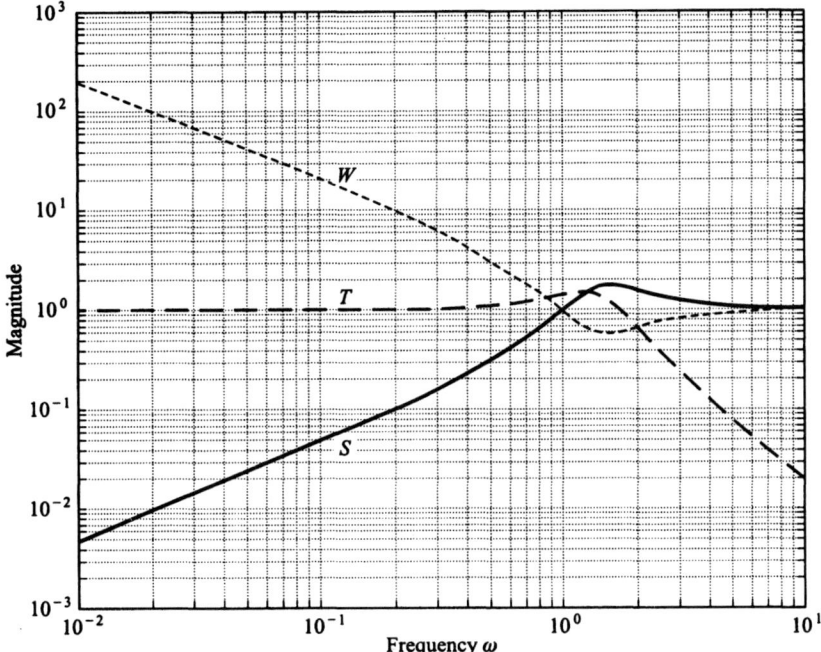

Figure 3.32 S, T and W for a unity-feedback system with $G(s) = 2/s(s+1)$.

- It must contain any poles that $G(s)$ has on the $j\omega$ axis.
- It must not contain $j\omega$ poles at zeros of $G(s)$.

2. Determine the Blaschke product $B(s)$ from knowledge of the poles of $G(s)$. If $G(s)$ has no poles in the right half-plane as illustrated in the preceding example in subsection 3.11A, $B(s) = 1$. Let us next consider the general case where $G(s)$ has poles in the right half-plane. Therefore, from Eq. (3.240),

$$S(p_n) = k\frac{B(p_n)}{W(p_n)} = 0, \quad n = 1, 2, 3, \ldots, N \tag{3.243}$$

where p_n are poles of $G(s)$ in the right half-plane.

3. $S(p_n)$ has to satisfy the following interpolation conditions for simple poles and zeros of $G(s)$ in the right half-plane:

- S equals zero at right-half poles, and one at right-half zeros.
- T is one at right-half poles and zero at right-half zeros.

This can be shown as follows. Let us consider the control system in Figure 3.33. The sensitivity of $C(s)/R(s)$ to $D(s)G(s)$ is given by

$$S^{C(s)/R(s)}_{D(s)G(s)} = \frac{1}{1 + D(s)G(s)}. \tag{3.244}$$

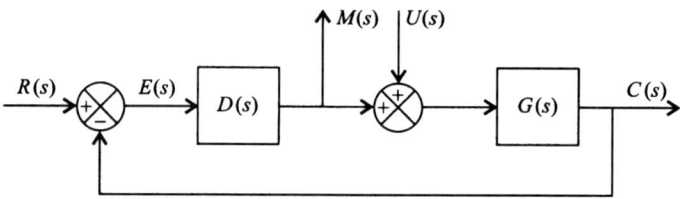

Figure 3.33 General SISO control system containing a reference input, $R(s)$, and a disturbance signal, $U(s)$.

The complementary sensitivity function is given by

$$T(s) = 1 - S(s) = \frac{D(s)G(s)}{1 + D(s)G(s)}. \tag{3.245}$$

The transfer functions between the inputs $R(s)$ and $U(s)$, and the outputs $M(s)$ and $C(s)$ are given as follows:

$$M(s) = \frac{D(s)}{1 + D(s)G(s)} R(s) - \frac{D(s)G(s)}{1 + D(s)G(s)} U(s). \tag{3.246}$$

In terms of $T(s)$ given by Eq. (3.245), Eq. (3.246) can be written as

$$M(s) = \frac{T(s)}{G(s)} R(s) - T(s)U(s). \tag{3.247}$$

Therefore,

$$C(s) = G(s)[M(s) + U(s)] = G(s)\left[\frac{D(s)}{1 + D(s)G(s)} R(s) - \frac{D(s)G(s)}{1 + D(s)G(s)} U(s) + U(s)\right]. \tag{3.248}$$

Equation (3.248) can be reduced to the following:

$$C(s) = \frac{D(s)G(s)}{1 + D(s)G(s)} R(s) + \frac{G(s)}{1 + D(s)G(s)} U(s) \tag{3.249}$$

In terms of $T(s)$ and $S(s)$, Eq. (3.249) can be rewritten as follows:

$$C(s) = T(s)R(s) + G(s)S(s)U(s) \tag{3.250}$$

Analysis of Eq. (3.247) reveals the following:

- If $T(s)/G(s)$ is to be stable, then $T(s)$ must cancel the right-half zero(s) of $G(s)$.

Analysis of Eq. (3.250) reveals the following:

- If $G(s)S(s)$ is to be stable, then $S(s)$ must cancel the pole(s) in the right-half plane of $G(s)$.

These two very important *interpolation conditions* imply that for simple zeros and poles which are in the right half-plane, $S(s)$ equals one at the right-half-plane zeros, and $S(s)$ equals zero at the right-half plane poles. Similarly, $T(s)$ equals zero at the right-half plane zeros, and $T(s)$ equals one at the right-half-plane poles in accordance with Eq. (3.217).

4. Form the Blaschke product. Returning to Eq. (3.243), and recognizing that $S(p_n)$ must satisfy these interpolation conditions, what should $B(p_n)$ and $W(p_n)$ be? Let us assume that $G(s)$ has poles in the right half of the s-plane, and let p_n, $n = 1, 2, \ldots, N$, be poles of $G(s)$. Therefore, Eq. (3.243) must be satisfied for the interpolation conditions. Because the weighting function $W(s)$ has no poles in the right half-plane, it cannot equal infinity. Therefore, it is necessary to have the Blaschke product $B(s)$ contain the zeros of $S(p_n)$: $B(p_n) = 0$, $n = 1, 2, 3 \ldots, N$. This means that the Blaschke product accounts for these zeros as follows:

$$B(s) = B_p(s)B'(s) \tag{3.251}$$

where

$$B_p(s) = \left[\frac{(-s+p_1)}{(s+p_1)}\right] \cdots \left[\frac{(-s+p_N)}{(s+p_N)}\right]. \tag{3.252}$$

We can find $B'(s)$ from the interpolation conditions previously shown. For example, let us assume that these zeros in the right half-plane which force $S(s)$ to equal one at these zeros are $z_1, z_2, z_3, \ldots, z_M$. From Eqs. (3.243) and (3.251),

$$S(z_m)W(z_m) = W(z_m) = kB'(z_m)B_p(z_m), \quad m = 1, 2, 3, \ldots, M. \tag{3.253}$$

This equation can be simplified to

$$kB'(z_m) = \frac{W(z_m)}{B_p(z_m)}, \quad m = 1, 2, 3, \ldots, M. \tag{3.254}$$

Equation (3.254) represents M equations which must be solved. It can be shown that if $M = 1$, then $B'(s) = 1$. If M is greater than one, then

$$B'(s) = \left(\frac{-s+c_1}{s+c_1}\right)\left(\frac{-s+c_2}{s+c_2}\right) \cdots \left(\frac{-s+c_{M-1}}{s+c_{M-1}}\right) \tag{3.255}$$

where c_i are complex values with the positive real parts chosen, together with the value of k, to satisfy Eq. (3.254).

5. Solve for $S(s)$. Having solved $B'(s)$, the optimal sensitivity $S(s)$ is then calculated from the following equation which is obtained by substituting Eq. (3.251) into Eq. (3.240):

$$S(s) = \frac{kB'(s)B_p(s)}{W(s)}. \tag{3.256}$$

174 MODERN CONTROL-SYSTEM DESIGN USING VARIOUS TECHNIQUES

The resulting $S(s)$ is stable because $B'(s)$ and $B_p(s)$ are stable, and since $W(s)$ has no poles in the right half-plane. Because $|B(j\omega)| = 1$

$$|S(j\omega)| = \frac{|k|}{|W(j\omega)|}. \tag{3.257}$$

Therefore, this equation shows that the magnitude of the sensitivity curve is the inverse of the magnitude of the weighting function curve. It is also interesting to observe that the optimal $|S(j\omega)|$ is independent explicitly of $G(s)$. However, $G(s)$ does effect the value of k.

C. An Example of the SISO Case Containing a Pole in the Right Half-Plane

Consider the control system illustrated in Figure 3.33 where

$$D(s)G(s) = \frac{(-s+2)}{(-s+3)(s+4)} \tag{3.258}$$

and the weighting function $W(s)$, shown in Figure 3.34, is given by

$$W(s) = \frac{(0.2Ts+1)}{(2Ts+1)}. \tag{3.259}$$

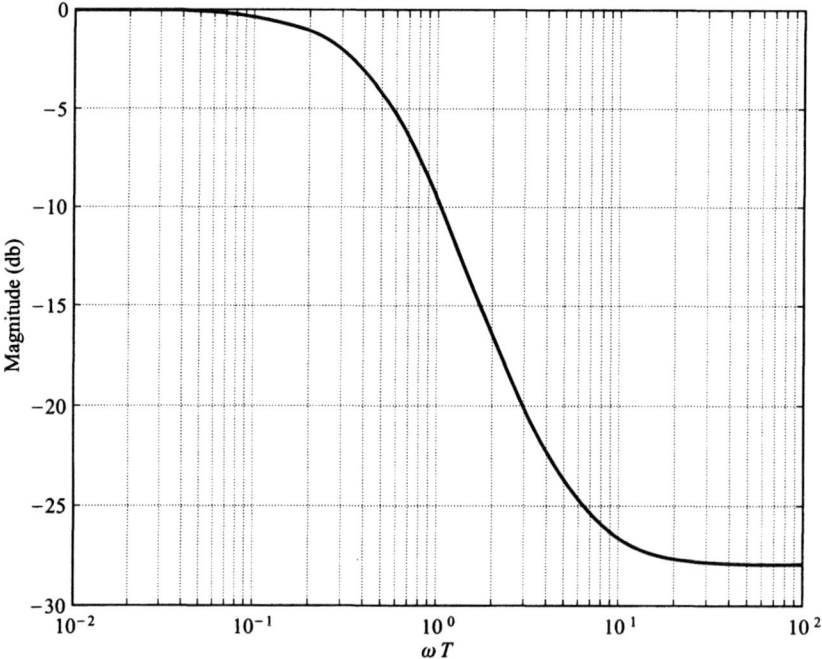

Figure 3.34 Weighting function $W(s)$.

This weighting function gives more weight to low frequencies (up to the bandwidth 1/2T) than high frequencies. From Eq. (3.252), the right-half pole of $G(s)$ at $s = 3$ results in $B_p(s)$ given by

$$B_p(s) = \frac{-s+3}{s+3}. \qquad (3.260)$$

Because $D(s)G(s)$ has only one zero at $s = 2$ in the right half-plane of the s-plane, then $\mathbf{B}'(s) = 1$. Therefore, from Eq. (3.254), we can find k as follows:

$$k = \frac{W(2)}{B_p(2)} = \frac{(0.4T+1)\,(5)}{(4T+1)\,(1)} = 5\frac{(0.4T+1)}{(4T+1)}. \qquad (3.261)$$

Substituting k from Eq. (3.261), $B_p(s)$ from Eq. (3.260), and $W(s)$ from Eq. (3.259) into Eq. (3.256) we can solve for the optimal sensitivity $S(s)$:

$$S(s) = 5\frac{(0.4T+1)}{(4T+1)}\frac{(-s+3)}{(s+3)}\frac{(2Ts+1)}{(0.2Ts+1)}. \qquad (3.262)$$

We can obtain the absolute value of the optimal sensitivity function from Eq. (3.257) as follows:

$$|S(j\omega)| = 5\frac{(0.4T+1)}{(4T+1)}\sqrt{\frac{(4\omega^2 T^2+1)}{(0.04\omega^2 T^2+1)}}. \qquad (3.263)$$

Figure 3.35 illustrates the absolute value of the sensitivity function obtained from Eq. (3.263) for values of $T = 0.2, 2,$ and 20. It is concluded from this curve that the sensitivity is smaller (better) at low frequencies if the bandwidth is not too high (or T is not too small).

3.13. LINEAR ALGEBRAIC ASPECTS OF CONTROL-SYSTEM DESIGN COMPUTATIONS [22–25]

The fundamental problems of linear control-system theory were reconsidered at the end of the 1960s and the beginning of the 1970s. The algebraic characteristics of the control-system problems were reestablished and have resulted in a better understanding of the foundations of linear control-system theory.

This section focuses on the interplay between some recent results and methodologies in numerical linear algebra and their application to problems arising in control systems. Let us reconsider the phase-variable canonical form representation of a control system:

$$\dot{\mathbf{x}}(t) = \mathbf{P}\mathbf{x}(t) + \mathbf{B}\mathbf{u}(t), \qquad (3.264)$$

$$\mathbf{c}(t) = \mathbf{L}\mathbf{x}(t). \qquad (3.265)$$

176 MODERN CONTROL-SYSTEM DESIGN USING VARIOUS TECHNIQUES

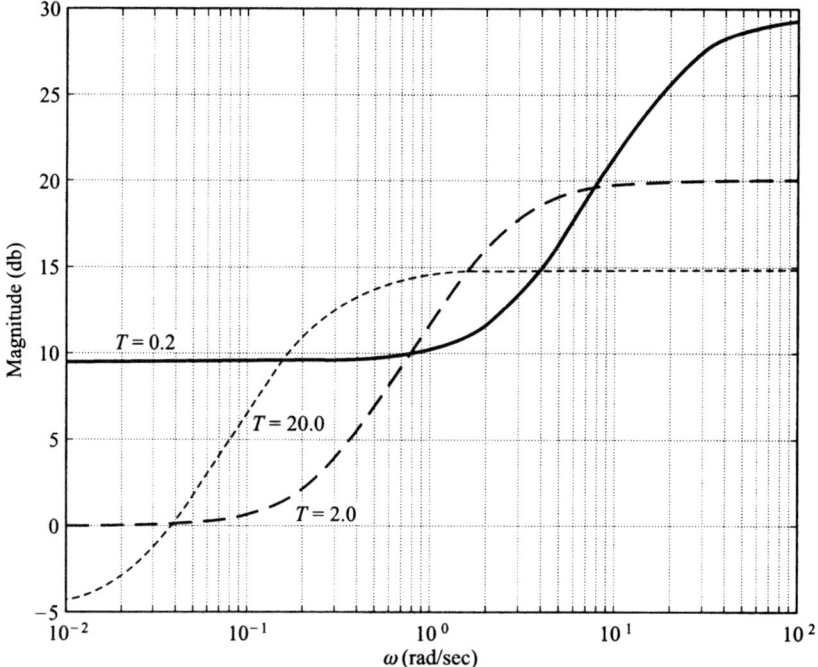

Figure 3.35 Optimal sensitivity frequency characteristic as a function of T.

We can represent this canonical form representation in terms of its frequency response by using Eqs. (3.10) and (3.18):

$$\frac{C(s)}{R(s)} = \frac{K\mathbf{L}[(s\mathbf{I} - \mathbf{P} + K\mathbf{bh})]\mathbf{b}}{\det(s\mathbf{I} - \mathbf{P} + K\mathbf{bh})}. \qquad (3.266)$$

The problem of computing $C(s)/R(s)$, or $C(j\omega)/R(j\omega)$, from knowledge of K, $\mathbf{P}$, $\mathbf{b}$, $\mathbf{L}$, and $\mathbf{h}$, efficiently is a practical problem of great interest. Reference 23 presents an efficient and generally applicable algorithm to solve this problem. The approach recommended performs an initial reduction of $\mathbf{P}$ to upper Hessenberg form $\mathbf{H}$, rather than solving the linear equation (with dense, unstructured $\mathbf{P}$) $(j\omega\mathbf{I} - \mathbf{P})\mathbf{X} = \mathbf{b}$, which would require $O(n^3)$ operations for each successive value of ω. The orthogonal matrices used to effect the Hessenberg form of $\mathbf{P}$ are incorporated into $\mathbf{b}$ and $\mathbf{L}$ giving $\tilde{\mathbf{b}}$ and $\tilde{\mathbf{L}}$. Therefore, as ω varies, the coefficient matrix in the linear equation $(j\omega\mathbf{I} - \mathbf{H})\mathbf{X} = \tilde{\mathbf{b}}$ remains in upper Hessenberg form. The advantage is that $\mathbf{X}$ can be determined in $O(n^2)$ operations rather than $O(n^3)$ as before, which is a very significant saving. In addition, this methodology is numerically very stable and has the advantage of being independent of the eigenstructure (possibly ill-conditioned) of $\mathbf{P}$.

This methodology can be extended to state-variable models in implicit form. For example, let us replace Eq. (3.264) by

$$\mathbf{E}\dot{\mathbf{x}}(t) = \mathbf{P}\mathbf{x}(t) + \mathbf{b}u(t) \qquad (3.267)$$

Therefore, Eq. (3.266) can be replaced with the following:

$$\frac{C(s)}{R(s)} = \frac{K\mathbf{L}[adj(s\mathbf{E} - \mathbf{P} + K\mathbf{bh})]\mathbf{b}}{\det(s\mathbf{I} - \mathbf{P} + K\mathbf{bh})}. \quad (3.268)$$

We can use the initial triangular/Hessenberg reduction to again reduce the problem to one of updating the diagonal of a Hessenberg matrix and, therefore, an $O(n^2)$ linear equation problem. [24].

3.14. ILLUSTRATIVE PROBLEMS AND SOLUTIONS

This section provides a set of illustrative problems and their solutions to supplement the material presented in Chapter 3.

I3.1. The state and output equations of a second-order control system are given by the following:

$$\frac{dx_1(t)}{dt} = -3x_1 + 4u(t),$$
$$\frac{dx_2(t)}{dt} = -x_2(t) + u(t),$$
$$c(t) = x_1(t)$$

where $x_1(t)$ and $x_2(t)$ represent the system states, $c(t)$ is the system's output, and $u(t)$ represents its input.

(a) Determine whether the system is controllable.

(b) Determine whether the system is observable.

SOLUTION: (a) From Eq. (3.65) controllability can be determined for this second-order system from:

$$\mathbf{D} = [\mathbf{B} \quad \mathbf{PB}].$$

The phase variable canonical form of the state and output equations can be written as:

$$\begin{bmatrix} \dot{x}_1(t) \\ \dot{x}_2(t) \end{bmatrix} = \begin{bmatrix} -3 & 0 \\ 0 & -1 \end{bmatrix} \begin{bmatrix} x_1(t) \\ x_2(t) \end{bmatrix} + \begin{bmatrix} 4 \\ 1 \end{bmatrix} u(t),$$

$$c(t) = \begin{bmatrix} 1 & 0 \end{bmatrix} \begin{bmatrix} x_1(t) \\ x_2(t) \end{bmatrix}$$

Therefore the companion matrix, **P** is given by

$$\mathbf{P} = \begin{bmatrix} -3 & 0 \\ 0 & -1 \end{bmatrix}$$

and the input vector, **B**, is given by

$$\mathbf{B} = \begin{bmatrix} 4 \\ 1 \end{bmatrix}$$

and the output matrix is given by:

$$\mathbf{L} = \begin{bmatrix} 1 & 0 \end{bmatrix}$$

Therefore, the matrix **D** is given by:

$$\mathbf{D} = \begin{bmatrix} 4 & -12 \\ 1 & -1 \end{bmatrix}$$

and the system is controllable.

(b) Observability can be determined for this second-order control system from Eq. (3.75) where

$$\mathbf{U} = [\mathbf{L}^T \quad \mathbf{P}^T \mathbf{L}^T]$$

Therefore, the matrix **U** is given by:

$$\mathbf{U} = \begin{bmatrix} 1 & -3 \\ 0 & 0 \end{bmatrix}.$$

So, the system is unobservable because **U** is singular.

I3.2. Synthesize a system using linear-state-variable feedback that has a closed-loop transfer function given by

$$\frac{C(s)}{R(s)} = \frac{10(s+6)}{s^3 + 4s^2 + 10s + 20}.$$

Assume that the proces to be controlled has a transfer function given by

$$G(s) = \frac{2}{s(0.5s+1)}.$$

In your solution, show the following:

(a) Synthesis of the linear-state-variable-feedback system.

(b) Identification of any compensation network needed to satisfy the synthesis. What kind of network is it?

(c) Check of the stability of the resulting system synthesized using the root-locus method.

3.14. ILLUSTRATIVE PROBLEMS AND SOLUTIONS

SOLUTION: (a)

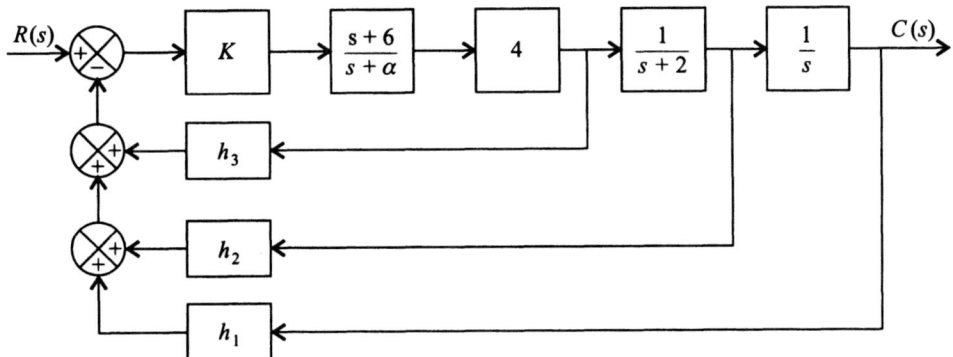

Figure I3.2(i)

$$\frac{C(s)}{R(s)} = \frac{\dfrac{4K(s+6)}{s(s+2)(s+\alpha)}}{1 + \dfrac{4K(s+6)h_3}{(s+\alpha)} + \dfrac{4K(s+6)h_2}{(s+\alpha)(s+2)} + \dfrac{4K(s+6)h_1}{s(s+\alpha)(s+2)}}.$$

This equation can be reduced to the following:

$$\frac{C(s)}{R(s)} = \frac{4K(s+6)}{(1+4kh_3)s^3 + (\alpha + 2 + 32Kh_3 + 4Kh_2)s^2 + (2\alpha + 48Kh_3 + 24Kh_2 + 4Kh_1)s + 24Kh_1}.$$

Comparing the coefficients of this equation for the synthesized linear-state-variable-feedback system and that of the desired closed-loop transfer function given by

$$\frac{C(s)}{R(s)} = \frac{10(s+60)}{s^3 + 4s^2 + 10s + 20}$$

we obtain the following five simultaneous equations to be solved:

$$10 = 4K,$$
$$1 + 4Kh_3 = 1,$$
$$\alpha + 2 + 32Kh_3 + 4KH_2 = 4,$$
$$2\alpha + 48Kh_3 + 24Kh_2 + 4Kh_1 = 10,$$
$$24Kh_1 = 20.$$

Therefore, we obtain the following results:

$$K = 2.5; \quad h_1 = 0.33; \quad h_2 = 0.0675; \quad h_3 = 0; \quad \alpha = 1.325.$$

(b) The compensation network is given by

$$\frac{(s+6)}{(s+1.325)}$$

which is a phase-lag network.

(c) The root locus is drawn from the following transfer function:

$$G(s)H(s) = \frac{4(s+6)}{s(s+2)(s+\alpha)}[h_1 + sh_2] = \frac{K(s+6)(s+4.9)}{s(s+1.325)(s+2)}$$

The following root-locus diagram shows that this control system is stable for $0 < K < \infty$:

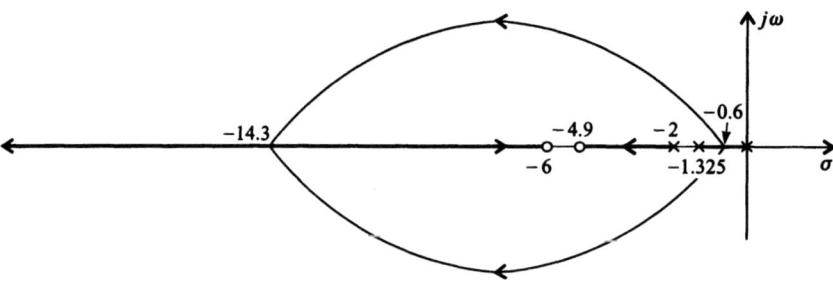

Figure I3.2(ii)

I3.3. Design a third-order controller whose three roots are located at the following locations in the s-plane: -1; -1; -12. The transfer function of the system to be controlled is given by

$$\frac{C(s)}{U(s)} = G(s) = \frac{1}{s(s+1)(s+2)}.$$

Determine the controller gain matrix, **K**.

SOLUTION: The desired location of the controller roots are located at:

$$\alpha_c(s) = (s+1)^2(s+12) = s^3 + 14s^2 + 25s + 12 \qquad (13.3\text{-}1)$$

We need to determine the companion matrix, **P**, and the input gain vector, **b**, so that we can solve for the controller gain matrix, **K** from Eq. (3.167). The states of this third-order control system are defined as follows:

Let $x_1 = c(t)$; $x_2 = \dot{c}(t)$; $x_3 = \ddot{c}(t)$. Therefore, the state equations are given by:

$$\dot{x}_1(t) = x_2(t),$$
$$\dot{x}_2(t) = x_3(t),$$
$$\dot{x}_3(t) = -2x_2(t) - 3x_3(t) + u(t).$$

3.14. ILLUSTRATIVE PROBLEMS AND SOLUTIONS

The state equations in their vector and matrix format are given by:

$$\begin{bmatrix} \dot{x}_1(t) \\ \dot{x}_2(t) \\ \dot{x}_3(t) \end{bmatrix} = \begin{bmatrix} 0 & 1 & 0 \\ 0 & 0 & 1 \\ 0 & -2 & -3 \end{bmatrix} \begin{bmatrix} x_1(t) \\ x_2(t) \\ x_3(t) \end{bmatrix} + \begin{bmatrix} 0 \\ 0 \\ 1 \end{bmatrix} u(t).$$

Therefore, the companion matrix, **P**, is given by:

$$\mathbf{P} = \begin{bmatrix} 0 & 1 & 0 \\ 0 & 0 & 1 \\ 0 & -2 & -3 \end{bmatrix}.$$

The controller gain **K** can be determined from Eq. (3.167) as follows:

$$|s\mathbf{I} - \mathbf{P} + \mathbf{bK}| = 0.$$

Substituting into Eq. (3.167), we obtain the following:

$$\left| \begin{bmatrix} s & 0 & 0 \\ 0 & s & 0 \\ 0 & 0 & s \end{bmatrix} - \begin{bmatrix} 0 & 1 & 0 \\ 0 & 0 & 1 \\ 0 & -2 & -3 \end{bmatrix} + \begin{bmatrix} 0 \\ 0 \\ 1 \end{bmatrix} [K_1 \quad K_2 \quad K_3] \right| = 0.$$

This can be simpified to

$$\begin{vmatrix} s & -1 & 0 \\ 0 & s & -1 \\ K_1 & 2+K_2 & s+3+K_3 \end{vmatrix} = 0$$

which can be reduced to the following:

$$s^3 + (3 + K_3)s^2 + (2 + K_2)s + K_1 = 0. \tag{13.3-2}$$

Setting coefficients of Eqs. (I3.3-1) and (I3.3-2) equal to each other, we obtain the following:

$$14 = 3 + K_3,$$
$$25 = 2 + K_2,$$
$$12 = K_1.$$

Therefore, we solve these three equations and find that:

$$K_1 = 12,$$
$$K_2 = 23,$$
$$K_3 = 11.$$

Therefore, the controller gain matrix, **K**, is given by:

$$\mathbf{K} = [12 \quad 23 \quad 11].$$

13.4. Design a third-order estimator whose three roots are located at the following locations in the s-plane: -3.5; -3.5; -15. The transfer function of the process is given by

$$G(s)H(s) = \frac{1}{s(s+3)(s+4)}$$

Determine the estimator gain vector **M**.

SOLUTION: The desired location of the estimator roots is at:

$$\alpha_e(s) = (s+3.5)^2(s+15) = s^3 + 22s^2 + 117.25s + 183.75. \quad (13.4\text{-}1)$$

We need to determine the companion matrix, **P**, the input gain vector, **b**, and the output matrix, **L**, so that we can determine the estimator gain factors in the **M** matrix from Eq. (3.131). The states of this third-order control system are defined as follows:

Let $x_1(t) = c(t)$; $x_2(t) = \dot{c}(t)$; $x_3(t) = \ddot{c}(t)$. Therefore, the state equations are given by:

$$\dot{x}_1(t) = x_2(t),$$
$$\dot{x}_2(t) = x_3(t),$$
$$\dot{x}_3(t) = -12x_2(t) - 7x_3(t) + r(t).$$

The state equations in their vector and matrix form are given by:

$$\begin{bmatrix} \dot{x}_1(t) \\ \dot{x}_2(t) \\ \dot{x}_3(t) \end{bmatrix} = \begin{bmatrix} 0 & 1 & 0 \\ 0 & 0 & 1 \\ 0 & -12 & -7 \end{bmatrix} \begin{bmatrix} x_1(t) \\ x_2(t) \\ x_3(t) \end{bmatrix} + \begin{bmatrix} 0 \\ 0 \\ 1 \end{bmatrix} r(t).$$

Therefore, the companion matrix, **P** and the input vector, **b**, are

$$\mathbf{P} = \begin{bmatrix} 0 & 1 & 0 \\ 0 & 0 & 1 \\ 0 & -12 & -7 \end{bmatrix}; \quad \mathbf{b} = \begin{bmatrix} 0 \\ 0 \\ 1 \end{bmatrix}.$$

The output equation is given by:

$$c(t) = \begin{bmatrix} 1 & 0 & 0 \end{bmatrix} \begin{bmatrix} x_1(t) \\ x_2(t) \\ x_3(t) \end{bmatrix}.$$

Therefore, the output matrix is given by

$$\mathbf{L} = \begin{bmatrix} 1 & 0 & 0 \end{bmatrix}.$$

The **M** matrix can be determined from Eq. (3.131) as follows:

$$|s\mathbf{I} - (\mathbf{P} - \mathbf{ML})| = 0$$

Substituting into Eq. (3.131), we obtain the following:

$$\left| \begin{bmatrix} s & 0 & 0 \\ 0 & s & 0 \\ 0 & 0 & s \end{bmatrix} - \begin{bmatrix} 0 & 0 & 1 \\ 0 & 0 & 1 \\ 0 & -12 & -7 \end{bmatrix} + \begin{bmatrix} m_1 \\ m_2 \\ m_3 \end{bmatrix} \begin{bmatrix} 1 & 0 & 0 \end{bmatrix} \right| = 0.$$

This can be simplified to

$$\begin{vmatrix} s + m_1 & -1 & 0 \\ m_2 & s & -1 \\ m_3 & 12 & s+7 \end{vmatrix} = 0$$

which can be reduced to the following:

$$s^3 + (7 + m_1)s^2 + (12 + 7m_1 + m_2)s + (12m_1 + 7m_2 + m_3) = 0. \qquad (I3.4\text{-}2)$$

Setting like coefficients of Eqs. (I3.4-1) and (I3.4-2), equal to each other, we obtain the following:

$$22 = 7 + m_1,$$
$$117.25 = 12 + 7m_1 + m_2,$$
$$183.75 = 12m_1 + 7m_2 + m_3.$$

Therefore, we solve these equations and find that:

$$m_1 = 15,$$
$$m_2 = 0.25,$$
$$m_3 = 2.$$

Therefore, the **M** vector is given by:

$$\mathbf{M} = \begin{bmatrix} 15 \\ 0.25 \\ 2 \end{bmatrix}.$$

I3.5. Repeat the problem solved in Section 3.6 using Ackermann's Formula for a process transfer function given by

$$\frac{C(s)}{U(s)} = \frac{(s+3)}{(s+4)(s+2)(s+8)}$$

and the control system's signal flow graph is given by:

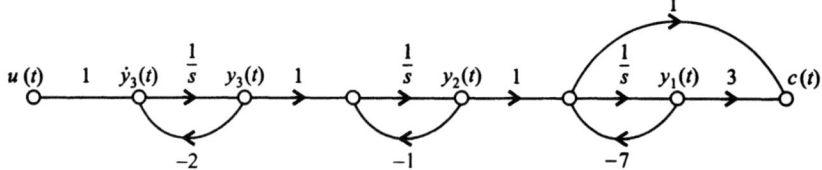

Figure 13.5(l)

(a) Determine the state equations for the process.

(b) Determine the controllability matrix for this original system.

(c) Transform the original system to the phase variable form, and determine the state and output equations.

(d) Determine the controllability matrix for the phase variable form.

(e) Determine the transformation matrix **A**.

(f) Design a contoller assuming that the dominant pair of complex-conjugage roots has a damping ratio of 0.707, and it results in a settling time of 4 sec. Select the third pole at the location of the zero of the process to be controlled. What is the resulting characteristic equation of the desired closed-loop control system?

(g) Determine the state and ouput equations for the phase-variable form with linear-state-variable-feedback.

(h) What is the resulting characteristic equation for the set of equations found in part (g)?

(i) Determine the linear-state-variable-feedback constants for the control system in phase-variable form.

(j) Transform the linear-state-variable-feedback constants back to the original system using the transformation matrix **A**.

(k) Draw the resulting closed-loop control system with linear-state-variable-feedback.

SOLUTION: (a)

$$\dot{\mathbf{y}}(t) = \mathbf{P}_y \mathbf{y}(t) + \mathbf{B}_y u(t) = \begin{bmatrix} \dot{y}_1(t) \\ \dot{y}_2(t) \\ \dot{y}_3(t) \end{bmatrix} = \begin{bmatrix} -7 & 1 & 0 \\ 0 & -1 & 1 \\ 0 & 0 & -2 \end{bmatrix} + \begin{bmatrix} y_1(t) \\ y_2(t) \\ y_3(t) \end{bmatrix} + \begin{bmatrix} 0 \\ 0 \\ 1 \end{bmatrix} u(t)$$

(b)
$$\mathbf{D}_y = [\mathbf{B}_y \quad \mathbf{P}_y \mathbf{B}_y \quad \mathbf{P}_y^2 \mathbf{B}_y] = \begin{bmatrix} 0 & 0 & 1 \\ 0 & 1 & -3 \\ 1 & -2 & 4 \end{bmatrix}$$

The value of the determinant of $\mathbf{D}_y$ is -1, it's nonsingular, and the original control system is controllable as expected from inspection of the signal-flow graph.

(c) From the given signal-flow graph:

$$\frac{C(s)}{U(s)} = \frac{\frac{3}{s^3} + \frac{1}{s^2}}{1 + \left[\frac{2}{s} + \frac{1}{s} + \frac{7}{s}\right] + \left[\left(\frac{2}{s}\right)\left(\frac{1}{s}\right) + \left(\frac{2}{s}\right)\left(\frac{7}{s}\right) + \left(\frac{1}{s}\right)\left(\frac{7}{s}\right)\right] + \left[\left(\frac{2}{s}\right)\left(\frac{1}{s}\right)\left(\frac{7}{s}\right)\right]}$$

This transfer function can be reduced to the following:

$$\frac{C(s)}{U(s)} = \frac{(s+3)}{s^3 + 10s^2 + 23s + 14}$$

Defining $x_1(t) = c(t)$, $x_2(t) = \dot{c}(t)$, and $x_3(t) = \ddot{c}(t)$, we obtain

$$\dot{x}_1(t) = x_2(t)$$
$$\dot{x}_2(t) = x_3(t)$$
$$\dot{x}_3(t) = -14x_1(t) - 23x_2(t) - 10x_3(t) + 3u(t) + \dot{u}(t)$$

Therefore, the state and output equations in matrix vector form of the phase variable control system are given by:

$$\dot{\mathbf{x}}(t) = \mathbf{P}_x \mathbf{x}(t) + \mathbf{B}_x u(t) = \begin{bmatrix} 0 & 1 & 0 \\ 0 & 0 & 1 \\ -14 & -23 & -10 \end{bmatrix} \begin{bmatrix} x_1(t) \\ x_2(t) \\ x_3(t) \end{bmatrix} + \begin{bmatrix} 0 \\ 0 \\ 1 \end{bmatrix} (3u(t) + \dot{u}(t))$$

$$c(t) = \begin{bmatrix} 3 & 1 & 0 \end{bmatrix} \begin{bmatrix} x_1(t) \\ x_2(t) \\ x_3(t) \end{bmatrix}$$

(d) The resulting controllability matrix for the phase variable form can be obtained from

$$\mathbf{D}_x = [\mathbf{B}_x \quad \mathbf{P}_x \mathbf{B}_x \quad \mathbf{P}_x^2 \mathbf{B}_x] = \begin{bmatrix} 0 & 0 & 1 \\ 0 & 1 & -10 \\ 1 & -10 & 77 \end{bmatrix}$$

Note that the determinant of $\mathbf{D}_x$ is -1 indicating that the determinant is nonsingular, and the phase variable form is also controllable, as expected.

(e) The transformation matrix, defined in Eq. (3.85) is

$$\mathbf{A} = \mathbf{D}_y \mathbf{D}_x^{-1} = \begin{bmatrix} 0 & 0 & 1 \\ 0 & 1 & -3 \\ 1 & -2 & 4 \end{bmatrix} \begin{bmatrix} 23 & 10 & 1 \\ 10 & 1 & 0 \\ 1 & 0 & 0 \end{bmatrix} = \begin{bmatrix} 1 & 0 & 0 \\ 7 & 1 & 0 \\ 7 & 8 & 1 \end{bmatrix}$$

(f) A second-order control system having a damping ratio of 0.707 with a settling time of 4 seconds results in a $\omega_n = 1.414$ [see Eq. (5.41)‡]. Therefore, the following second-order control system can meet the design specifications:

$$\frac{C(s)}{R(s)} = \frac{\omega_n^2}{s^2 + 2\zeta\omega_n s + \omega_n^2} = \frac{2}{s^2 + 2s + 2}$$

With third pole selected at $s = 3$, which also cancels the zero at $s = -3$, the characteristic equation of the desired closed-loop system is given by:

$$(s^2 + 2s + 2)(s + 3) = s^3 + 5s^2 + 8s + 6 = 0$$

(g) The state and output equations for the phase-variable form with linear-state-variable-feedback can be obtained from Eqs. (3.8) and (3.10) as follows:

$$\dot{\mathbf{x}}(t) = [\mathbf{P}_x - K\mathbf{b}_x\mathbf{h}_x]\mathbf{x}(t) + Kr(t)\mathbf{b}_x = \begin{bmatrix} 0 & 1 & 0 \\ 0 & 0 & 1 \\ -14 & -23 & -10 \end{bmatrix} - \begin{bmatrix} 0 & 0 & 0 \\ 0 & 0 & 0 \\ Kh_1 & Kh_2 & Kh_3 \end{bmatrix}$$

$$\times \begin{bmatrix} x_1(t) \\ x_2(t) \\ x_3(t) \end{bmatrix} + Kr(t) \begin{bmatrix} 0 \\ 0 \\ 1 \end{bmatrix}$$

$$c(t) = [3 \quad 1 \quad 0]\mathbf{x}(t)$$

The state equation reduces to the following:

$$\dot{\mathbf{x}}(t) = \begin{bmatrix} 0 & 1 & 0 \\ 0 & 0 & 1 \\ -14 - Kh_1 & -23 - Kh_2 & -10 - Kh_3 \end{bmatrix} \mathbf{x}(t) + \begin{bmatrix} 0 \\ 0 \\ Kr(t) \end{bmatrix}$$

(h) The resulting characteristic equation can be obtained from Eqs. (3.10) and (3.19):

$$\det(s\mathbf{I} - \mathbf{P}_h) = (s\mathbf{I} - \mathbf{P} + K\mathbf{b}\mathbf{h}) = 0$$

Therefore,

$$\begin{vmatrix} s & -1 & 0 \\ 0 & s & -1 \\ 14 + Kh_1 & 23 + Kh_2 & s + 10 + Kh_3 \end{vmatrix} = 0$$

The resulting characteristic equation is given by:

$$s^3 + (10 + Kh_3)s^2 + (23 + Kh_2)s + (14 + Kh_1) = 0$$

(i) Comparing this characteristic equation (see part h) with the characteristic equation of the desired closed-loop system (see part f), results in the following:

$$10 + Kh_3 = 5; \quad Kh_3 = -5$$
$$23 + Kh_2 = 8; \quad Kh_2 = -15$$
$$14 + Kh_1 = 6; \quad Kh_1 = -8$$

Therefore,

$$K\mathbf{h}_x = [-8 \quad -15 \quad -5]$$

(j) We will now transform $K\mathbf{h}_x$ back to the original system using Eq. (3.94) as follows:

$$\mathbf{h}_y = \mathbf{h}_x \mathbf{A}^{-1} = [-8 \quad -15 \quad -5] \begin{bmatrix} 1 & 0 & 0 \\ -7 & 1 & 0 \\ 49 & -8 & 1 \end{bmatrix}^{-1} = [-148 \quad 25 \quad -5]$$

(k) The resulting closed-loop control system with linear-state-variable-feedback is the following:

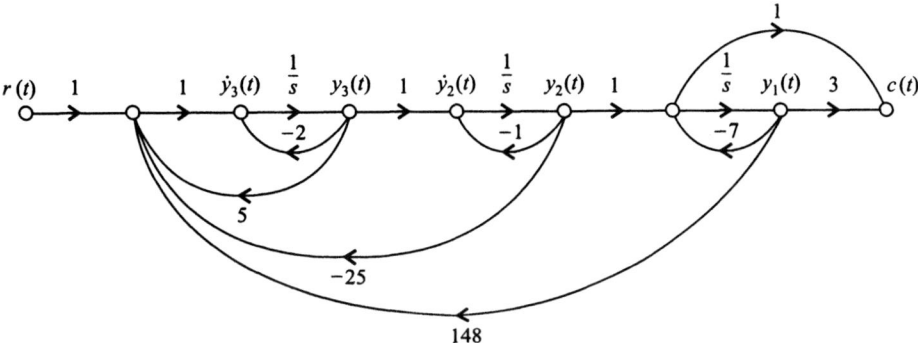

Figure I3.5(ii) Resulting closed-loop control system wth linear-state-variable-feedback designed using Ackermann's formula.

I3.6. Design a combined compensator of a regulator containing a controller and estimator for the system illustrated in Figure I3.6(i).

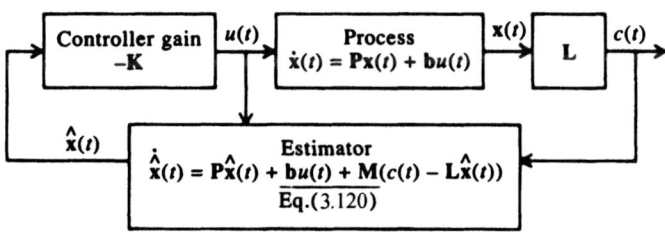

Figure I3.6(i)

MODERN CONTROL-SYSTEM DESIGN USING VARIOUS TECHNIQUES

Assume that the transfer function of the process is given by

$$\frac{C(s)}{U(s)} = G(s) = \frac{4}{(s+2)^2}.$$

Assume that the design specification of the controller is that it is critically damped with $\omega_n = 3$ rad/sec, and that the estimator is also critically damped with $\omega_n = 30$ rad/sec.

(a) Find the controller's gain coefficients' matrix.

(b) Find the estimator's coefficients' vector.

(c) Determine the transfer function of the compensator for the combined controller and estimator, $U(s)/C(s)$.

(d) *Sketch* the root locus of the compensated system. From your sketch, is the system unstable, stable, or conditionally stable?

SOLUTION: **(a)** The desired location of the controller roots is at:

$$\alpha_c(s) = (s+3)^2 = s^2 + 6s + 9 \qquad (13.6\text{-}1)$$

We need to determine the companion matrix, **P**, and the input gain vector, **b**, so that we can solve for the controller gain matrix, **K**, from Eq. (3.167). The states of this second-order control system are defined as follows:

Let $x_1(t) = c(t)$; $x_2(t) = \dot{c}(t)$. Therefore, the state equations are given by:

$$\dot{x}_1(t) = x_2(t)$$
$$\dot{x}_2(t) = -4x_1(t) - 4x_2(t) + 4u(t).$$

The state and output equations in their vector and matrix format are given by:

$$\begin{bmatrix} \dot{x}_1(t) \\ \dot{x}_2(t) \end{bmatrix} = \begin{bmatrix} 0 & 1 \\ -4 & -4 \end{bmatrix} \begin{bmatrix} x_1(t) \\ x_2(t) \end{bmatrix} + \begin{bmatrix} 0 \\ 4 \end{bmatrix} u(t)$$

$$c(t) = \begin{bmatrix} 1 & 0 \end{bmatrix} \begin{bmatrix} x_1(t) \\ x_2(t) \end{bmatrix}.$$

Therefore, the companion matrix, **P**, the input vector, **b**, and the output matrix, **L** are given by:

$$\mathbf{P} = \begin{bmatrix} 0 & 1 \\ -4 & -4 \end{bmatrix}; \quad \mathbf{b} = \begin{bmatrix} 0 \\ 4 \end{bmatrix}; \quad \mathbf{L} = \begin{bmatrix} 1 & 0 \end{bmatrix}.$$

3.14. ILLUSTRATIVE PROBLEMS AND SOLUTIONS

The controller gain **K** can be determined from Eq. (3.167) as follows:

$$|s\mathbf{I} - \mathbf{P} + \mathbf{bK}| = 0.$$

Substituting into Eq. (3.167), we obtain the following:

$$\left| \begin{bmatrix} s & 0 \\ 0 & s \end{bmatrix} - \begin{bmatrix} 0 & 1 \\ -4 & -4 \end{bmatrix} + \begin{bmatrix} 0 \\ 4 \end{bmatrix} [K_1 \ \ K_2] \right| = 0.$$

This equation can be simplified to

$$\begin{vmatrix} s & -1 \\ 4 + 4K_1 & s + 4 + 4K_2 \end{vmatrix} = 0.$$

which can be reduced to the following:

$$s^2 + 4(1 + K_2)s + 4(1 + K_1) = 0. \qquad (13.6\text{-}2)$$

Setting like coefficients of Eqs. (I3.6-1) and (I3.6-2) equal to each other, we obtain the following:

$$6 = 4(1 + K_2),$$
$$9 = 4(1 + K_1).$$

Therefore, we solve these two equations and find that:

$$K_1 = 1.25$$
$$K_2 = 0.5$$

Therefore, the controller gain matrix, **K**, is given by:

$$\mathbf{K} = [1.25 \ \ 0.5]$$

(b) The desired location of the estimator roots is at:

$$\alpha_e(s) = (s + 30)^2 = s^2 + 60s + 900. \qquad (13.6\text{-}3)$$

The **M** matrix can be determined from Eq. (3.131) as follows:

$$|s\mathbf{I} - (\mathbf{P} - \mathbf{ML})| = 0$$

Substituting into Eq. (3.131), we obtain the following:

$$\left| \begin{bmatrix} s & 0 \\ 0 & s \end{bmatrix} - \begin{bmatrix} 0 & 1 \\ -4 & -4 \end{bmatrix} + \begin{bmatrix} m_1 \\ m_2 \end{bmatrix} [1 \ \ 0] \right| = 0.$$

This can be simplified to

$$\begin{vmatrix} s + m_1 & -1 \\ 4 + m_2 & s + 4 \end{vmatrix} = 0$$

which can be reduced to the following:

$$s^2 + (m_1 + 4)s + 4m_1 + 4 + m_2 = 0 \qquad (13.6\text{-}4)$$

Setting like coefficients of Eqs. (13.6-3) and (13.6-4) equal to each other, we obtain the following:

$$60 = m_1 + 4; \quad m_1 = 56$$
$$900 = 4m_1 + 4 + 4m_2; \quad m_2 = 672.$$

Therefore, the **M** vector is given by:

$$\mathbf{M} = \begin{bmatrix} 56 \\ 672 \end{bmatrix}$$

(c) The transfer function of the compensator, $U(s)/C(s)$ can be determined from Eq. (3.162):

$$\frac{U(s)}{C(s)} = -\mathbf{K}[s\mathbf{I} - \mathbf{P} + \mathbf{bK} + \mathbf{ML}]^{-1}\mathbf{M}$$

Substituting values for **K**, **P**, **B**, and **L** found in part (a), and **M** found in part (b), we obtain the following:

$$\frac{U(s)}{C(s)} = -[1.25 \ 0.5]\left[\begin{bmatrix} s & 0 \\ 0 & s \end{bmatrix} - \begin{bmatrix} 0 & 1 \\ -4 & -4 \end{bmatrix} + \begin{bmatrix} 0 \\ 4 \end{bmatrix}[1.25 \ 0.5]\right.$$

$$\left. + \begin{bmatrix} 56 \\ 672 \end{bmatrix}[1 \ 0]\right]^{-1} \begin{bmatrix} 56 \\ 672 \end{bmatrix}$$

which reduces to

$$\frac{U(s)}{C(s)} = -[1.25 \ 0.5]\begin{bmatrix} s+56 & -1 \\ 681 & s+6 \end{bmatrix}^{-1} \begin{bmatrix} 56 \\ 672 \end{bmatrix}.$$

Further reduction reduces this equation to the following:

$$\frac{U(s)}{C(s)} = \frac{[1.25 \ 0.5]\begin{bmatrix} s+6 & 1 \\ -681 & s+56 \end{bmatrix}\begin{bmatrix} 56 \\ 672 \end{bmatrix}}{s^2 + 62s + 1017}.$$

This equation reduces the following expression for the transfer function of the compensator for the combined controller and estimator, $U(s)/C(s)$:

$$\frac{U(s)}{C(s)} = -\frac{406(s + 2.48)}{s^2 + 62s + 1017}.$$

This is a phase-lead network having the zero at $s = -2.48$.

(d) The root locus for the compensated system is plotted from the following expression:

$$G(s)G_{\text{comp}}(s) = G(s)\frac{U(s)}{C(s)} = -\frac{1}{(s+2)^2}\frac{406(s+2.48)}{(s+31+j7.48)(s+31-j7.48)}.$$

The root locus shown in Figure I3.6(ii) shows that this is a conditionally stable system that is stable for $< K < 61,102$. Observe that the root locus goes through the roots chosen for the controller roots ($s = -3, -3$) and estimator roots ($s = -30, -30$).

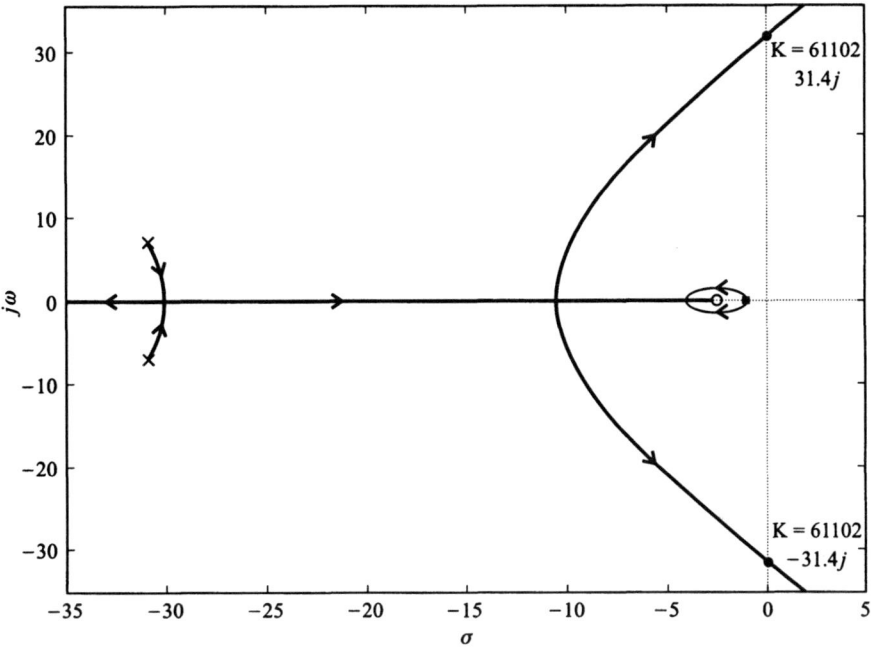

Figure I3.6(ii)

13.7. The system of Problem 13.6 is modified to respond to a reference input, $r(t)$, as shown in the following block diagram:

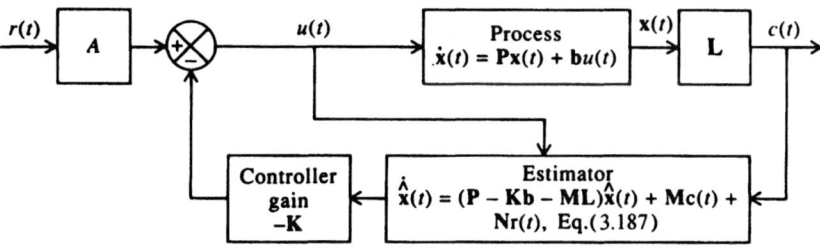

Figure I3.7

Assume that the process, combined compensator containing a controller and an estimator, and the specifications for the controller and estimator are exactly the same as in Problem I3.6. Assuming that $A = 4$, determine the requirement on **M** so that the state-estimation error is independent of the reference input, $r(t)$.

SOLUTION: We can determine the requirement on **N** so that the state-estimation error is independent of the reference input, $r(t)$, from Eq. (3.193):

$$\mathbf{b}A = \mathbf{N}.$$

From Problem I3.6, we know that the input vector, **b**, is given by

$$\mathbf{b} = \begin{bmatrix} 0 \\ 4 \end{bmatrix}.$$

Substituting **b** and $A = 4$ into Eq. (3.193), we obtain the following:

$$\begin{bmatrix} 0 \\ 4 \end{bmatrix} 4 = \begin{bmatrix} n_1 \\ n_2 \end{bmatrix}$$

Therefore, **N** is given by

$$\mathbf{N} = \begin{bmatrix} 0 \\ 16 \end{bmatrix}.$$

I3.8. We wish to design a robust control system containing a two-degrees-of-freedom series controller $G_{cl}(s)$ and a forward-loop-controller $G_{c2}(s)$ which is illustrated as follows:

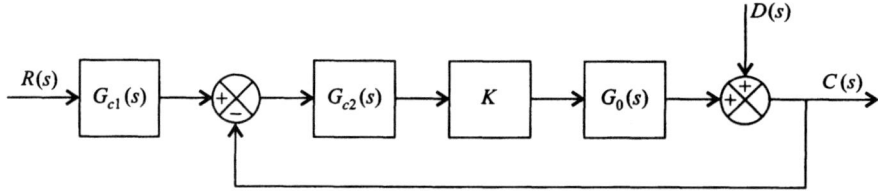

Figure I3.8(i)

The transfer function $KG_0(s)$ is

$$KG_0(s) = \frac{150(s + 12)}{s(s + 2)(s + 8)(s + 25)}. \tag{I3.8-1}$$

The gain of this transfer function can vary significantly, and the value of 150 is only the nominal amount. During its operation, the gain has been known to

3.14. ILLUSTRATIVE PROBLEMS AND SOLUTIONS

go as low as 75 and as high as 300. The objective of the control-system engineer is to design the robust control system shown so that the effect of this wide gain variation is minimized on the control system's transient response.

(a) Determine the control system's transient response to a unit step input for $K = 75$, 150, and 300, assuming that $G_{c1}(s) = G_{c2}(s) = 1$.

(b) Draw the root locus for the conditions defined in part (a).

(c) From the root locus drawn in part (b), determine the location of the roots for the closed loop system for gains of 75, 150, and 300, and the damping and overshoot, for the dominant set of complex-conjugate roots, for the three sets of gains being considered.

(d) Design the robust controller $G_{c2}(s)$.

(e) Determine the transfer function $KG_{c2}(s)G_0(s)$.

(f) Plot the root locus for the transfer function determined in part (e), and determine the damping, overshoot, and the roots of the characteristic equations for gains of 75, 150, and 300.

(g) Determine the design of the forward-loop controller $G_{c1}(s)$ in this two-degrees-of-freedom control system.

(h) With $G_{c1}(s)$, as determined in part (g) added to this control system, determine the transient response of this control system to a unit step input for $K = 75$, 150, and 300.

(i) What conclusions can you reach from the resulting transient responses found in parts (a) and (h)?

SOLUTION: (a) Figure I3.8(ii) illustrates the unit step responses for this control system for $K = 75$, 150, and 300 which was obtained using MATLAB. Observe from this figure that the peak overshoot varies from 27.3% for $K = 75$, to 69.3% for $K = 150$, and 76.3% for $K = 300$.

(b) The root locus obtained from MATLAB is shown in Figure I3.8(iii).

(c) Table I3.8(i) lists the damping, overshoot, and the characteristic equation roots for the dominant set of complex-conjugate roots of this control system.

Observe from this table the considerable variation in the damping ratio and the maximum percent overshoot (based on the dominant complex-conjugate roots) for the three cases being analyzed.

(d) The robust controller has the following pair of complex-conjugate zeros to cancel the poles of the system where $K = 150$:

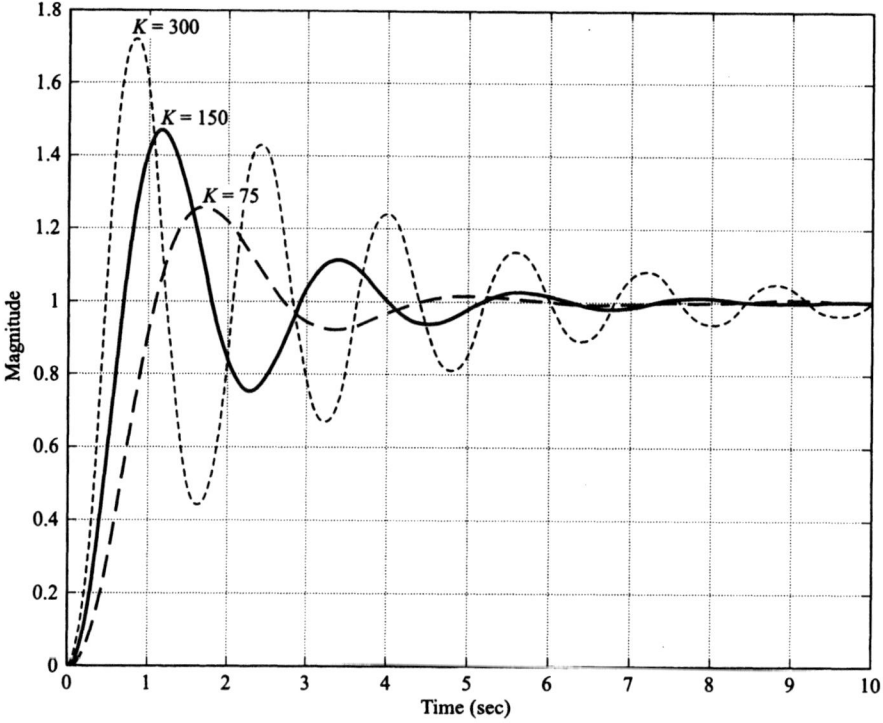

Figure I3.8(ii) Unit step response of $KG(s) = K(s + 12)/s(s + 2)(s + 8)(s + 25)$.

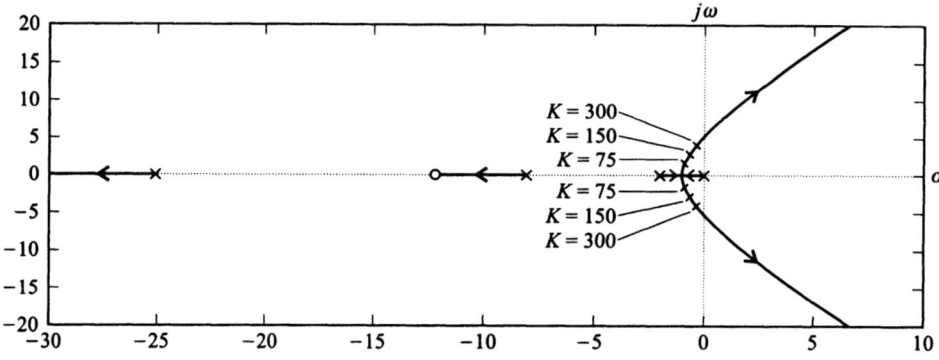

Figure I3.8(iii) Root-locus plot for system shown in Figure I3.8(i) for $KG_0(s)$ as defined in Eq. (I3.8-1) and $G_{c1}(s) = G_{c2}(s) = 1$.

Table I3.8(i)

K	Damping ratio	Max % overshoot	Roots of characteristic equation
300	0.08577	76.3	$-8.9274, -25.3890, -0.3418 \pm j3.9707$
150	0.21528	69.3	$-8.5592, -25.1969, -0.6219 \pm j2.8213$
75	0.38176	27.3	$-8.3154, -25.0991, -0.7928 \pm j1.9193$

3.14. ILLUSTRATIVE PROBLEMS AND SOLUTIONS

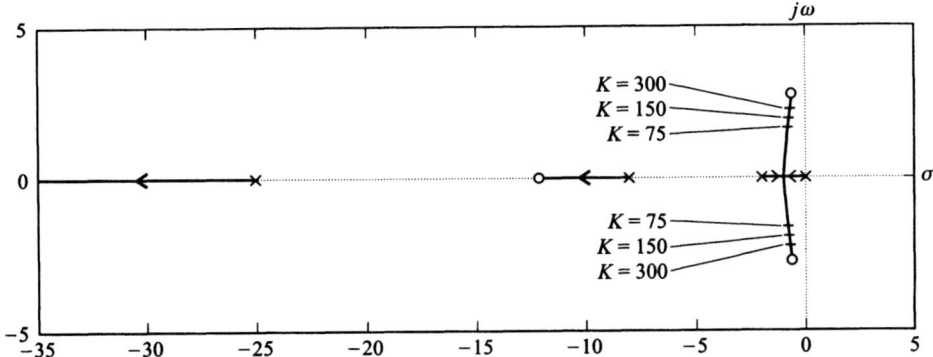

Figure I3.8(iv) Root-locus plot for system shown in Figure I3.8(i) with $KG_{c2}(s)G_0(s)$ given by Eq. (I3.8-3).

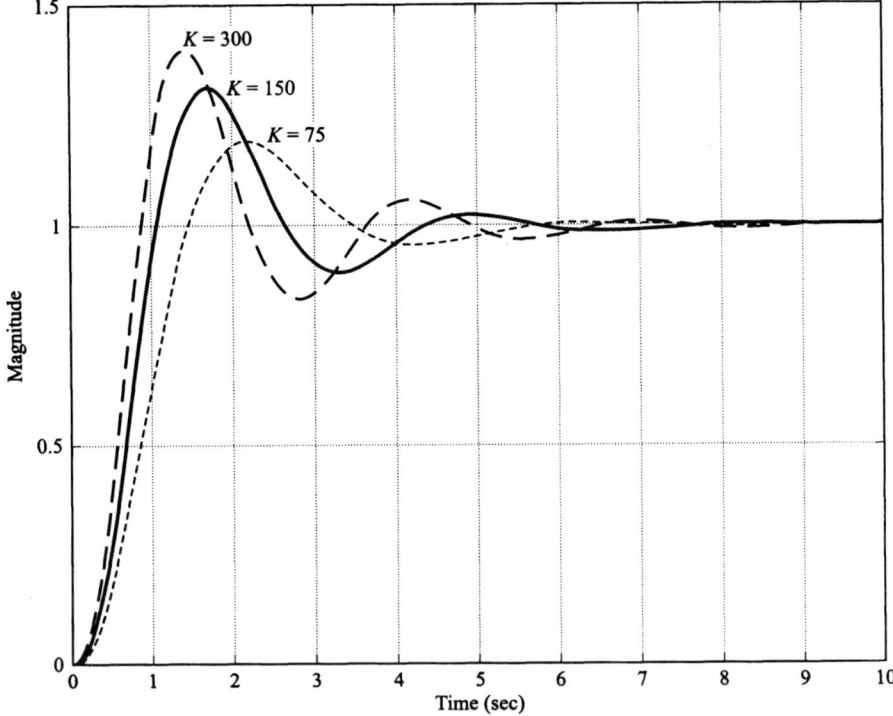

Figure I3.8(v) Unit step response of $KG(s) = K(s+12)/s(s+2)(s+8)(s+25)$ with robust control.

Table I3.8(ii)

K	Damping ratio	Max. % overshoot	Roots of characteristic equation
300	0.27822	40.2%	$-10.9999, -58.6295, -0.6573 \pm j2.2694$
150	0.34267	31.7%	$-10.3725, -41.1929, -0.7033 \pm j1.9282$
75	0.4575	19.9%	$-9.6604, -32.7831, -0.7713 \pm j1.4990$

$$G_{c2}(s) = \frac{s^2 + 1.2439s + 8.3463}{8.3463} = \frac{(s + 0.6219 + 2.8263j)(s + 0.6219 - 2.8263j)}{8.3463} \quad (13.8\text{--}2)$$

(e)

$$KG_{c2}(s)G_0(s) = K\frac{(s^2 + 1.2439s + 8.3463)}{8.3463} \cdot \frac{(s+12)}{s(s+2)(s+8)(s+25)}. \quad (13.8\text{--}3)$$

(f) The root locus, which was obtained using MATLAB, is illustrated in Figure I3.8(iv). From this root locus, the damping ratio, maximum percent overshoot, and roots of the characteristic equation are shown in Table I3.8(ii):

(g) The design of the forward-loop controller $G_{c1}(s)$ is the reciprocal of the robust controller $G_{c2}(s)$:

$$G_{c1}(s) = \frac{8.3463}{s^2 + 1.2439s + 8.3463}. \quad (13.8\text{--}4)$$

(h) The resulting transient response of this control system to a unit step input for $K = 75, 150$, and 300 is illustrated in Figure I3.8(v).

(i) The resulting transient response illustrated in Figure I3.8(v) shows that the maximum percent overshoots with the robust control design have been greatly reduced from those illustrated in Figure I3.8(ii). The new robust control-system design truly exhibits admirable robust control features. Table I3.8(iii) compares the transient response with and without robust control:

Table I3.8(iii)

K	Maximum percent overshoot without robust control	Maximum percent overshoot with robust control
300	76.3%	40.2%
150	69.3%	31.7%
75	27.3%	19.9%

In addition to the reduction in the maximum percent overshoots, observe that the range of the maximum percent overshoots is reduced from 76.3%/27.3% = 2.79 without robust control to 40.2/19.9% = 2.02. That is a very significant reduction which robust control makes possible. Therefore, the robust system is much less sensitive to variations in K. In addition, the robustness with respect to variations in K will also provide disturbance suppression to $D(s)$.

13.9. The optimal sensitivity function for the control system illustrated in Figure 18.9(i) is to be determined in the H^∞ sense.

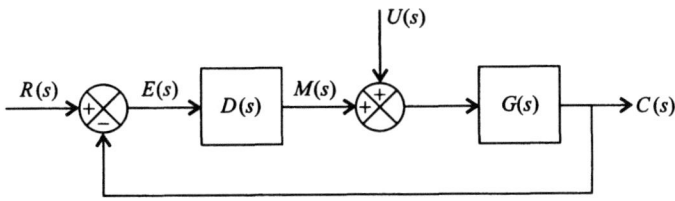

Figure 13.9(i)

The transfer function of the process $G(s)$ is given by

$$G(s) = \frac{(-s+4)}{(-s+5)(s+6)}.$$

The weighting function $W(s)$ is given by

$$W(s) = \frac{(0.3Ts+1)}{(3Ts+1)}.$$

(a) Determine the value of $B_p(s)$.

(b) Determine the value of the constant k.

(c) Determine the value of the optimal sensitivity function $S(s)$.

(d) Determine the absolute value of the optimal sensitivity function and plot the results for $T = 0.2, 2,$ and 20 sec.

(e) What conclusions can you reach from these curves?

SOLUTION: **(a)** From Eq. (3.252), the right-half pole of $G(s)$ at $s = 5$ results in

$$B_p(s) = \frac{-s+5}{s+5}.$$

(b) Because $G(s)$ has only one zero in the right half-plane of the s-plane, then $B'(s) = 1$. Therefore, from Eq. (3.254), we can find k as follows:

$$k = \frac{W(4)}{B_p(4)} = \frac{[0.3(4)T+1]}{[3(4)T+1]}\frac{9}{1} = 9\frac{(1.2T+1)}{(12T+1)}.$$

(c) Substituting into Eq. (3.256) for k, $B'(s)$, $B_p(s)$, and $W(s)$, we obtain the following:

$$S(s) = 9\frac{(1.2T+1)}{(12T+1)}\frac{(-s+5)}{(s+5)}\frac{(3Ts+1)}{(0.3Ts+1)}.$$

(d) The absolute value of the optimal sensitivity function can be obtained from Eq. (3.257) as follows:

$$|S(j\omega)| = 9\frac{(1.2T+1)}{(12T+1)}\sqrt{\frac{(9\omega^2 T^2 + 1)}{(0.09\omega^2 T^2 + 1)}}.$$

The absolute value of the sensitivity function is plotted in Figure I3.9(ii) for the following values of T: 0.2, 2, and 20.

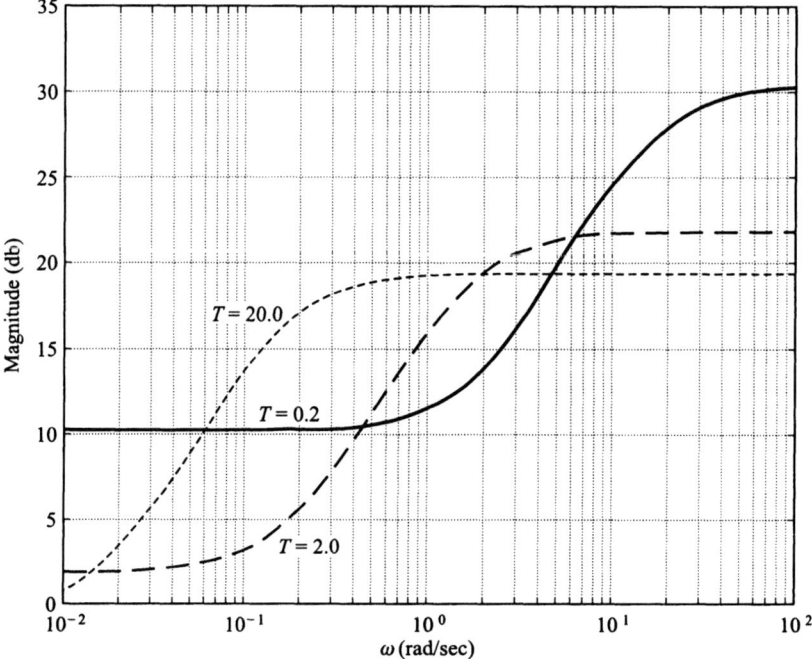

Figure I3.9(ii) Optimal sensitivity frequency characteristic as a function of T.

(e) It is concluded from these curves that the sensitivity is smaller (better) at low frequencies if the bandwidth is not too high (or T is not too small).

PROBLEMS

3.1. The control system illustrated in Figure P3.1(i) contains linear-state-variable-feedback elements h_1 and h_2.

(a) Determine the gain K and the linear-state-variable-feedback constants h_1 and h_2 so that the resulting control system represents a zero steady-state

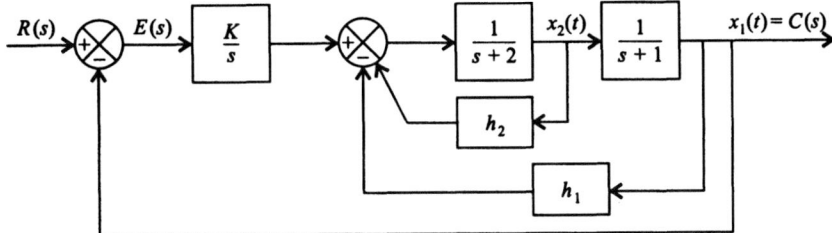

Figure P3.1(i)

step error system and its characteristic equation contains roots at $-2+j, -2-j$, and -8.

(b) The same performance can be obtained as in part (a) if we implement a series controller, $G_c(s)$, instead of using linear-state-variable-feedback as illustrated in the configuration shown in Figure P3.1(ii).

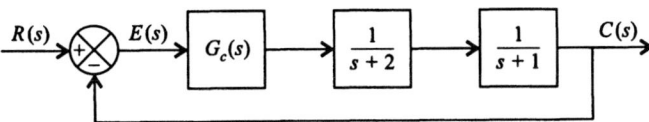

Figure P3.1(ii)

Determine the transfer function of $G_c(s)$ in terms of K, h_1, and h_2 obtained in part (a) and the other system parameters provided.

3.2. A control system containing a controller and a process are illustrated in the block diagram in Figure P3.2(i).

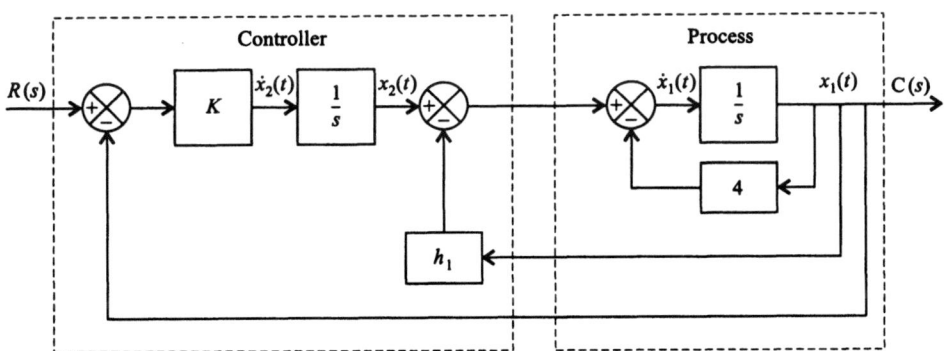

Figure P3.2(i)

(a) Determine the state equations of this control system.

(b) Determine the characteristic equation from knowledge of **P**.

(c) Determine the constant h_1 and the gain K if the roots of the characteristic equation are at -6 and -8.

(d) Instead of using the controller configuration, the control-system engineer wishes to design a "proportional plus integral controller" whose transfer function, $G_c(s)$, is given by:

$$G_c(s) = K_p + \frac{K_I}{s}$$

and is shown in figure P3.2(ii).

Determine the values of K_p and K_I so that the roots of the characteristic equation are also at -6 and -8.

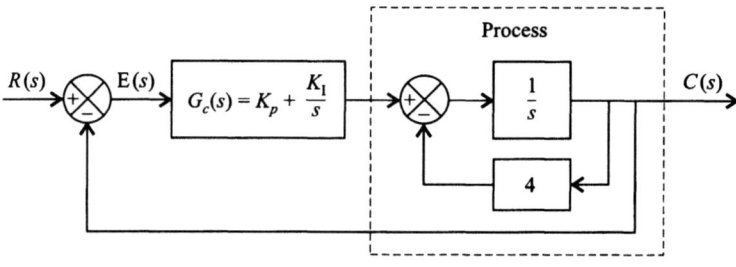

Figure P3.2(ii)

(e) Which design would you select, the controller configuration of part (a) or the proportional plus integral controller of part (d).

3.3. The controllability and observability of the control system shown in Figure P3.3 is to be determined.

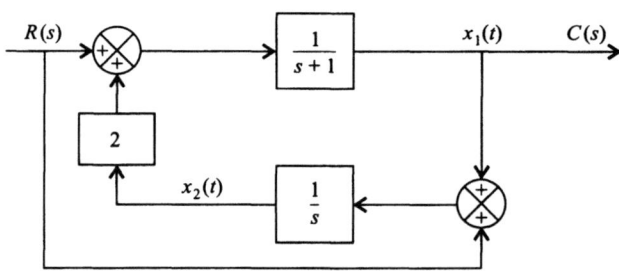

Figure P3.3

(a) Find the state and output equations of this control system.
(b) Determine the **D** matrix. Is the system controllable?
(c) Determine the **U** matrix. Is the system observable?

3.4. Synthesize a system utilizing linear-state-variable feedback that has closed-loop poles existing at $-9, 0$ and $-16, 0$ and can satisfy the following specifications:

$$K_v = 1, \quad \zeta = 0.707, \quad \omega_n = 1.$$

It is assumed that the process to be controlled has a transfer function given by $G(s) = 20/[s(s+1)(s+10)]$, and the transient response is governed by a pair of dominant complex-conjugate poles.

3.5. Repeat Problem 3.4 for the following specifications:

$$\frac{C(s)}{R(s)} = \frac{72(s+2)}{(s^2 + 1.414s + 1)(s+9)(s+16)},$$

$$G(s) = \frac{10}{s(1+5s)(1+0.5s)}.$$

3.6. Repeat Problem 3.4 for the following specifications:

$$\frac{C(s)}{R(s)} = \frac{8}{s^3 + 6s^2 + 10s + 8},$$

$$G(s) = \frac{1}{s(s+1)}.$$

3.7. It is desired to synthesize a system using linear-state-variable feedback which has a closed-loop pole at $-4, 0$ and can satisfy the following specifications:

$$K_v = 1, \quad \zeta = 0.707, \quad \omega_n = 1.$$

It is assumed that the process to be controlled has a transfer function given by

$$G(s) = \frac{20}{s(s+1)},$$

and the transient response is governed by a pair of dominant complex-conjugate poles.

(a) Can the system specifications of velocity constant be met with a simple second-order system?

(b) What must be added to the closed-loop system transfer function?

(c) What must be added to the open-loop system transfer function?

(d) Synthesize the block-diagram configuration of the linear-state-variable feedback configuration for the resulting system.

(e) Determine all unknown network constants and linear-state-variable feedback coefficients.

(f) Draw the root locus, and determine whether the resulting synthesized system is stable. If it shows that the system is unstable, what might be done to stabilize the system?

3.8. It is desired to synthesize a system using linear-state-variable feedback which can satisfy the following specifications:

$$K_v = 100/\text{sec}, \quad \omega_n = 100, \quad \zeta = 0.5.$$

The process to be controlled has a transfer function given by

$$G(s) = \frac{1}{s(2s+1)}$$

and assume that the transient response is governed by a pair of dominant complex-conjugate poles.

(a) Can the system specifications be met with a simple second-order system?

(b) Synthesize the block-diagram configuration of the linear-state-variable feedback for the resulting system.

(c) Determine all unknown constants (if any) and all linear-state-variable feedback coefficients.

(d) Determine whether the resulting synthesized system is stable using the root-locus method for solution.

3.9. Synthesize a system using linear-state-variable feedback that has a closed-loop transfer function given by

$$\frac{C(s)}{R(s)} = \frac{10(s+4)}{s^3 + 6s^2 + 22s + 40}.$$

It is assumed that the process to be controlled has a transfer function given by

$$G(s) = \frac{4}{s(0.5s+1)}$$

In your solution, show the following:

(a) Synthesis of the linear-state-variable feedback system.

(b) Identification of any compensation network needed to satisfy the synthesis.

(c) Check of the stability of the resulting system synthesized using the root-locus method.

3.10. A control system is defined by the following:

$$\mathbf{P} = \begin{bmatrix} -1 & 0 & 0 \\ 1 & -1 & 0 \\ 0 & 0 & 1 \end{bmatrix}, \quad \mathbf{L} = \begin{bmatrix} 0 & 1 & 1 \end{bmatrix}, \quad \mathbf{B} = \begin{bmatrix} 0 \\ 1 \\ 0 \end{bmatrix}$$

(a) Determine whether the system is controllable.

(b) Determine whether the system is observable.

3.11. The state and output equations of a second-order system are given by the following:

$$\dot{x}_1(t) = -8x_1(t) + 4u(t),$$
$$\dot{x}_2(t) = -2x_2(t) + u(t),$$
$$c(t) = x_1(t)$$

where $x_1(t)$ and $x_2(t)$ represent the system state, $c(t)$ is its output, and $u(t)$ is its input.

(a) Determine whether the system is controllable.

(b) Determine whether the system is observable.

3.12. A control system is defined by the following:

$$\mathbf{P} = \begin{bmatrix} -1 & 0 & 0 \\ 0 & -2 & 1 \\ 0 & 0 & -2 \end{bmatrix}, \quad \mathbf{L} = \begin{bmatrix} 1 & 0 & 0 \end{bmatrix}, \quad \mathbf{B} = \begin{bmatrix} 1 \\ 0 \\ 1 \end{bmatrix}.$$

(a) Determine whether the system is controllable.

(b) Determine whether the system is observable.

3.13. Design a second-order controller which is to be critically damped, and whose two second-order roots are located at -7 in the s-plane. The transfer function of the system to be controlled is given by

$$\frac{C(s)}{U(s)} = G(s) = \frac{1}{(s+1)^2}.$$

Find the vector **K**.

3.14. Repeat Problem 3.13 if the controller is critically damped, but its two second-order roots are now located at -4 instead of -7.

3.15. Design a second-order estimator for a system which is to have a damping ratio of 0.5, and whose estimator has two second-order roots located at -30 in the s-plane. Find the vector **M**. Assume that $\omega_n = 1$ rad/sec.

3.16. Repeat Problem 3.15 if the two estimator roots are located at -15 instead of -30.

3.17. Design a second-order estimator which satisfies the ITAE criterion for a zero steady-state step error system. Assume that $\omega_n = 25$ rad/sec.

3.18. We wish to design a regulator system (where the reference input equals zero) containing a combined controller and estimator as illustrated in Figure 3.19. The process' transfer function is given by the following expression:

$$\frac{C(s)}{U(s)} = G(s) = \frac{1}{s(s+1)}.$$

(a) Determine the state and output equations of this process.

(b) We wish to design the controller whose design specifications are for $\omega_n = 2$ rad/sec and for $\zeta = 1$. Determine the controller gain matrix, **K**.

(c) We wish to design the estimator to have a much faster response than the controller. Therefore, we will design the estimator to have $\omega_n = 20$ rad/sec and $\zeta = 1$. Determine the estimator gain matrix **M**.

(d) Determine the resulting compensator's transfer function, $U(s)/C(s)$.

(e) We wish to extend the design of this sytem to the case of a system containing a reference input, $r(t)$, as shown in Figure 3.22. Determine the vector **N** so that the estimator error is independent of $r(t)$. Assume that $A = 10$.

3.19. Repeat the problem solved in Section 3.6 using Ackermann's Formula for a process transfer function given by

$$\frac{C(s)}{U(s)} = \frac{(s+2)}{(s+1)(s+4)(s+8)}$$

and the control system's signal-flow graph is shown in Figure P3.19.

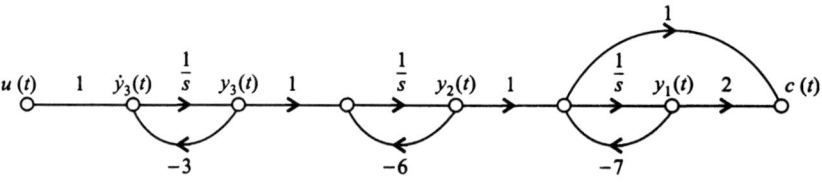

Figure P3.19

(a) Determine the state equations for the process.

(b) Determine the controllablity matrix for this original system. Is the system controllable?

(c) Transform the original system to the phase-variable form, and determine the state and output equations.

(d) Determine the controllability matrix for the phase variable form.

(e) Determine the transformation matrix **A**.

(f) Design a controller assuming that the dominant pair of complex-conjugate roots has a damping ratio of 0.707, and it results in a settling time of 4 sec. Select the third pole at the location of the zero of the process to be controlled. What is the resulting characteristic equation of the desired closed-loop control system?

(g) Determine the state and output equations for the phase-variable form with linear-state-variable feedback.

(h) What is the resulting characteristic equation for the set of equations found in part (g)?

(i) Determine the linear-state-variable feedback constants for the control system in phase-variable form.

(j) Transform the linear-state-variable feedback constants back to the original system using the transformation matrix **A**.

(k) Draw the resulting closed-loop control system with linear-state-variable feedback.

3.20. Determine the closed-loop transfer function of the resulting closed-loop control system developed using Ackermann's Formula and shown in Figure 3.17, and compare it with the desired closed-loop transfer function given by Eq. (3.107). Do they agree?

3.21. The design of the positioning system of a tracking radar is a very interesting control-system problem. Figure P3.21(a) illustrates a tracking radar which has wind torque disturbances acting on it, and the equivalent block diagram for one axis of the tracking radar's positioning loop is illustrated in Figure P3.21(b). The block diagram illustrated represents a two-degrees-of-freedom robust control system which has the dual capability of minimizing the effects of the wind torque disturbance $D(s)$ while also being robust (insensitive) to variations in the gain K. The transfer function for the forward-loop transfer function for the tracking radar $G_0(s)$ is given by

$$G_0(s) = \frac{(1+0.4s)}{s(1+s)(1+0.15s)}.$$

(a) Determine the transient response of this control system to a unit step input at $R(s)$ assuming that $G_{c1}(s) = G_{c2}(s) = 1$ and K has a nominal value of 40, but can also vary as low as 10 and as high as 160.

(b) Draw the root locus for the conditions in part (a). Determine the location of the closed roots for $K = 10, 40$, and 160, and determine the damping ratio for the dominant complex-conjugate roots for these three cases.

(c) Design $G_{c2}(s)$ so that it contains the zeros which cancel the complex-conjugate roots for the case of nominal gain 40.

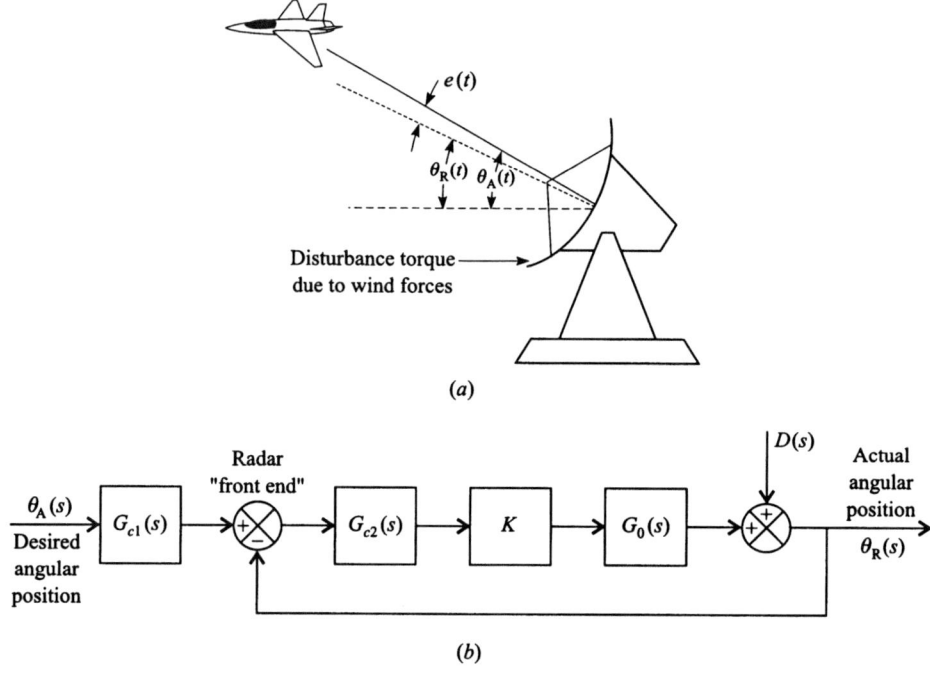

Figure P3.21 A tracking radar conceptually illustrating disturbance torques due to wind forces (a), and the equivalent block diagram of the tracking radar's positioning loop (b).

(d) Draw the root locus for the conditions of part (c). Determine the location of the closed-loop roots for $K = 10$, 40, and 160, and determine the damping ratio for the dominant complex-conjugate roots for these three cases.

(e) Design the robust controller $G_{c1}(s)$, and plot the resulting transient response of this control system containing $G_{c1}(s)$ and $G_{c2}(s)$, which was designed in part (c).

(f) How do the transient responses of part (e) compare to the transient responses of part (a)? Discuss the robustness of your resulting design to variations in the gain K.

3.22. The optimum sensitivity for the control system illustrated in Figure P3.22 is to be determined in the H^∞ sense. The proces $G(s)$ has a transfer function given by

$$G(s) = \frac{(-s+3)}{(-s+4)(s+5)}.$$

The weighting function, $W(s)$, is given by

$$W(s) = \frac{(0.5Ts+1)}{(5Ts+1)}$$

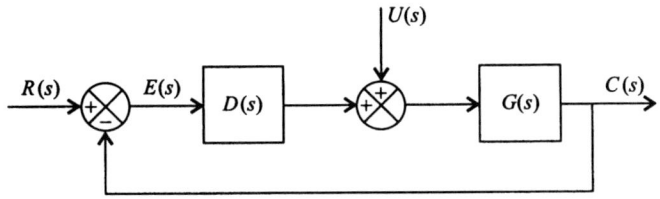

Figure P3.22

(a) Determine $B_p(s)$.

(b) Determine $B'(s)$.

(c) Determine k.

(d) Determine $S(s)$.

(e) Determine the magnitude of the optimum value of the sensitivity function, $|S(j\omega)|$, as a function of T.

(f) From your results for $|S(j\omega)|$ in part (e), plot the magnitude of the optimum sensitivity function as a function of ω for $T = 0.2$, 2 and 20.

REFERENCES

1. J. L. Melsa and D. G. Schultz, *Linear Control Systems*. McGraw-Hill, New York, 1969.
2. R. E. Kalman, "On the general theory of control systems." In *Proceedings of the First International Congress of Automatic Control*, Moscow, 1960.
3. R. E. Kalman, Y. C. Ho, and K. S. Navendra, "Controllability of linear dynamical systems." *Contrib. Differ. Equat.*, 189–213 (1961).
4. E. G. Gilbert, "Controllability and observability in multivariable control systems." *J. Control*, Ser. (A) 1, Soc. Ind. Appl. Math., 128–151 (1963).
5. J. E. Ackermann, "Der Entwurf linearer regelungs Systems in Zustandstraum." *Regelungstech, Prozess-Datenverarb*, 7, 297–300 (1972).
6. K. Ogatta, *Modern Control Engineering*, 3rd ed., Prentice-Hall, Englewood Cliffs, NJ, 1997.
7. N. S. Nise, *Conrrol Systems Engineering*, 2nd ed., Addison–Wesley, Reading, MA, 1995.
8. G. F. Franklin, J. D. Powell, and A. Emami-Naeini, *Feedback Control of Dynamic Systems*, 2nd ed. Addison–Wesley, Reading, MA, 1991.
9. A. Emami-Naeini and G. F. Franklin, "Zero assignment in the multivariable robust servomechanism," *Proc IEEE Conf. Dec. Control.*, 1982, pp. 891–893.
10. Modern CACSD Using the Robust Control Toolbox by R. Y. Chiang and M. G. Safonou, *Proc. Conf. on Aerospace and Computational Control*, Oxnard, CA, August 28–30, 1989.
11. P. Dorato, *Robust Control*, IEEE Press, New York, 1987.
12. P. Dorato, "Case Studies in Robust Control Design." *IEEE Proceedings of the Decision and Control Conference*, December 1990, pp. 2030–3031.
13. B. C. Kuo, *Automatic Control Systems*. 7th ed., Prentice-Hall, Englewood Cliffs, NJ, 1995.

14. P. A. Ioannou and J. Sun, *Robust Adaptive Control*. Prentice-Hall Professional Technical Reference, Des Moines, IA, 1995.
15. J. Ackermann, "Robust Control Prevents Car Skidding." *IEEE Control Systems*, **17**, pp. 23–31 (1997).
16. G. Zames, "Feedback and Optimal Sensitivity; Model Reference Transformations, Multiplicative Seminorms, and Approximate Inverses," *IEEE Transactions on Automatic Control*, **AC-26**, 301–320 (1981).
17. P. R. Belanger, *Control Engineering*. Saunders College Publishing, Harcourt Brace and Company, Fort Worth, 1995.
18. Z. Gajic and M.Lelic, *Modern Control System Engineering*. Prentice-Hall, 1996.
19. B. A. Francis, J. W. Helton, and G. Zames, "H^∞-Optimal Feedback Controllers for Linear Multivariable Systems. *IEEE Trans. Automatic Control*, **AC-29**, 888–900 (1984).
20. J. C. Doyle, K. Glover, P. O. Khargonekhav, and B. A. Francis,"State Space Solutions to Standard H^2 and H^∞ Control Problems." *IEEE Trans. Automatic Control*, **34**, 831–847 (1989).
21. A. F. A. Serrarens, M. J. G. van de Molengraft, J. J. Kok, and L. van der Steen, "H-Infinity Control for Suppressing Stick-Slip in Oil Well Drillstrings." *IEEE Control Systems*, **18-2**, 19–30 (1998).
22. A. J. Laub, "Numerical Linear Algebra Aspects of Control Design Computations." *IEEE Trans. Automatic Control*, **AC-30**, 97–108 (1985).
23. K. J. Åström and B. Wittenmark, *Computer-Controlled Systems, Theory and Design*. Prentice-Hall, Englewood Cliffs, NJ, 1990.
24. A. J. Laub, "Efficient multivariable frequency response calculations." *IEEE Trans. Automatic Control*, **AC-26**, 407–408 (1981).
25. C. B. Moler and G. W. Stewart, "An algorithm for generalized matrix eigenvalue problems." *SIAM J. Numer. Anal.*, **10**, 241–256 (1973).

4

DIGITAL CONTROL-SYSTEM ANALYSIS AND DESIGN

4.1. INTRODUCTION

This chapter is concerned with a class of feedback control systems in which the signal at one or more points appears as a train of pulses rather than as a continuous function. Such a system is known as a digital control system, in contrast to a continuous control system, where the signal is continuous everywhere. The controlling information, determined by the amplitude of the pulses, is present only at discrete instants of time, which in this chapter, are assumed to be equally spaced. Because of the inherent time delay introduced by the sampling operation, stabilization of the digital control system becomes a more complex problem than with continuous systems requiring special techniques for analysis and design.

Digital control systems are used in a wide variety of applications. The impetus for the use of this class of control systems has been the ready availability of digital computers, and the improvement in cost and reliability of digital computers. Digital control systems are very common today and are used in controlling robots [1], navigation systems for aircraft and ships, aircraft autopilots, mass transit vehicles, chemical process control systems, and automation systems for various applications.

There are several excellent books available on digital control systems that are dedicated to this subject only [2–5]. The purpose of this chapter is to present digital control systems to the student and the practicing engineer to a level permitting the analysis and design of digital control systems. We will extend the concepts developed in this book for the analysis and design of continuous control systems to digital control systems. We will focus attention in this chapter on that class of digital control systems where the digital computer is connected to the control elements and controlled system by means of analog-to-digital (A/D) and digital-to-analog (D/A) converters as illustrated in Figure 4.1. Therefore, this system contains both discrete signals [$r(kT)$, $e(kT)$, $b(kT)$], and continuous signals [$u(t)$, $m(t)$, $c(t)$]. A system having both discrete and continuous signals is defined as a sampled-data system [2].

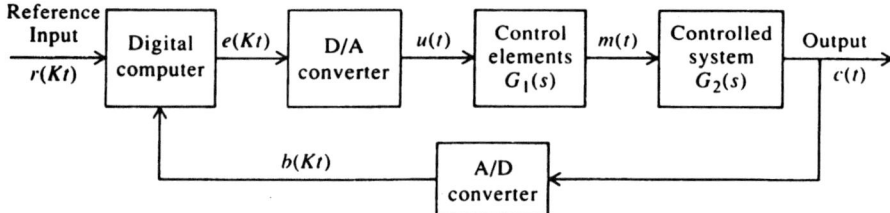

Figure 4.1 A sampled-data system.

The sampling operation can best be described by considering a variable of interest, $e(t)$, which is shown in Figure 4.2a. Let us assume that this function is sampled at equal intervals of time T, so that a sampled function can be described by the sequence of numbers given by

$$e(0), e(T), e(2T), e(3T), e(4T), \ldots, e(nT).$$

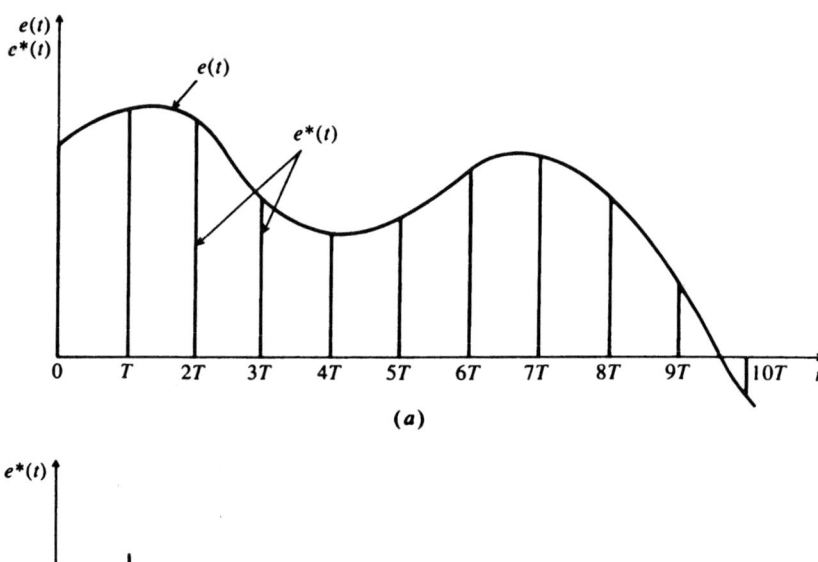

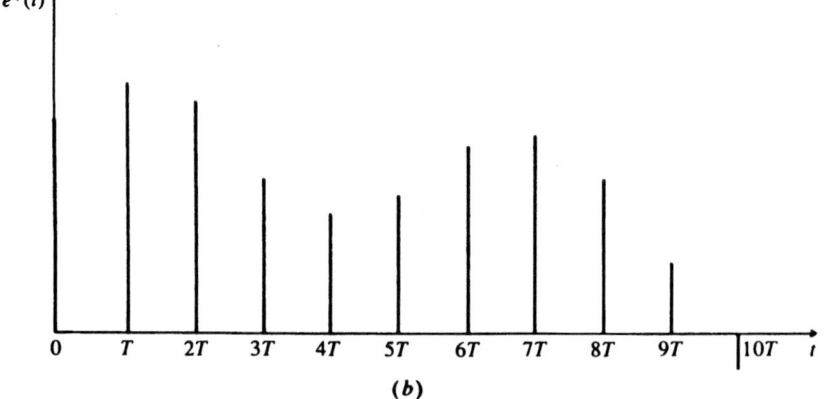

Figure 4.2 (a) Sampling a variable, $e(t)$. (b) Sampled function, $e^*(t) = e(kT)$.

This resultant series of numbers, $e(kT)$ gives a limited description of the function $e(t)$. Specifically, the value of $e(t)$ at times $0, T, 2T, 3T, 4T$, and nT seconds are known. Its value at other times can only be approximated by means of interpolation techniques.

In practical control systems, it is necessary that the discrete data be reconstructed during the intervals between samples, because the controlled elements and controlled system components are continuous, or analog, type of devices. Data extrapolators, or predictors, are usually quite simple devices which predict values during sampling intervals based upon either the last one, two, or three pieces of data. Very rarely are more than three pieces of information used. Figure 4.2b illustrates the sampled information which will be designated as $e^*(t)$ or $e(kT)$. It is common practice in sampled-data theory to denote the sampled version of time with an asterisk. Figure 4.3 illustrates the information after it is reconstructed by means of a simple holding device, which simply remembers the last piece of information until a new sample is obtained. This type of holding circuit is known as a zero-order hold. The reconstructed function is designated as $e_R(t)$.

This chapter discusses the basic concepts and characteristics of sampled-data systems and illustrates the techniques for analyzing and designing such systems. An analysis of the sampling process, data extrapolators, stability, and compensation techniques will be presented. In addition, the relations between variables of the control system will be obtained by use of a new transformation which is denoted as the z transformation. The use of the Laplace transformation in sampled-data control systems is quite cumbersome, because we are working with difference equations [2], rather than differential equations as in continuous systems. The transformation is very useful in the analysis and design of sampled-data control systems. Several practical design problems are presented for illustrative purposes wherever appropriate.

4.2. CHARACTERISTICS OF SAMPLING

We will analyze in this section the operation of an ideal sampler which produces a discrete-time signal from a continuous-time signal. For purposes of this analysis, let

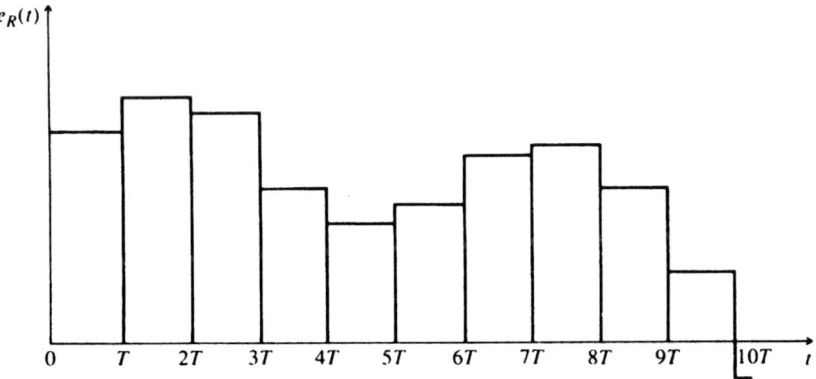

Figure 4.3 Reconstruction of $e^*(t)$ by means of a zero-order hold.

us consider the following sampler, which samples the continuous signal $e(t)$ and produces the discrete-time signal $e^*(t)$:

$$e(t) \underset{}{\overset{T}{\diagup}} e^*(t)$$

We shall further assume that the closure of the switch is of very short duration as compared to the time between closures. Therefore, the value of the function at the output of the switch is the instantaneous values of the function $e(t)$ when the switch is closed.

In order to obtain a clearer mathematical picture of the sampling process, we can think of the sampler as a device which multiplies some input intelligence signal $e(t)$ by a sampling function $s(t)$. This is mathematically equivalent to a modulation process where the sampling function represents the carrier that is being modulated by the input intelligence signal. This can be expressed as

$$e^*(t) = e(t)s(t). \qquad (4.1)$$

From the viewpoint of mathematical simplicity, it is very desirable to define the sampling function as a series of unit impulses that are characterized by having an infinitely narrow width and an infinite amplitude. Its total area is defined as unity. However, practical samplers remain closed for a finite length of time and may result in an actual sampling pulse whose area may be a finite number A. For this case, the only modification that is necessary to the unit-impulse approximation is to replace the unit area impulse with one whose area is A. In practice, the only real requirement for using the unit-impulse approximation is that the sampling duration be small compared to the time constant of the system. This is usually the case, because feedback control systems generally represent low-pass filters. In the succeeding discussion, we shall assume that the sampling function is a series of unit impulses that can be expressed as

$$s(t) = \sum_{n=-\infty}^{n=+\infty} \delta(t - nT), \qquad (4.2)$$

where $\delta(t - nT)$ represents an impulse having unit area at time $t = nT$.

The Laplace transform of the unit-impulse modulated function $e^*(t)$ can be easily determined. Assuming that the function $e(t)$ is zero for negative time, then $e^*(t)$ can be expressed as

$$e^*(t) = \sum_{n=0}^{\infty} e(nT) \, \delta(t - nT), \qquad (4.3)$$

where $e(nT)$ is the value of $e(t)$ at the sampling instant nT, and $\delta(t - nT)$ represents a unit impulse occurring at the instant nT. The Laplace transform of Eq. (4.3) can be written as

$$E^*(s) = \sum_{n=0}^{\infty} e(nT) e^{-nTs}. \qquad (4.4)$$

4.2. CHARACTERISTICS OF SAMPLING

The Laplace transform for a sampler having equal sampling intervals and whose switching can be represented by a series of unit impulses is an infinite summation which can be expressed in closed form. This is easily demonstrated by considering the case where the function $e(t)$ is a simple unit step, $U(t)$. For this case, all values of $e(nT)$ are equal to unity for positive time and zero for negative times. The Laplace transform of $e^*(t)$, given by Eq. (4.3), can be expressed as

$$E^*(s) = \sum_{n=0}^{\infty} e^{-nTs} = 1 + e^{-Ts} + e^{-2Ts} + e^{-3Ts} + \cdots + e^{-nTs}. \tag{4.5}$$

Equation (4.5) is recognizable as a geometric progression in e^{-Ts}, which can be expressed as

$$E^*(s) = \frac{1}{1 - e^{-Ts}} \quad \text{for} \quad |e^{-Ts}| < 1. \tag{4.6}$$

The Laplace transform for the unit-impulse sampling device can be obtained in another form, which makes use of the fact that the sampling function $s(t)$ is periodic and can be expressed as a Fourier series. (Fourier series were reviewed in Section 2.3‡, and will also be used in Section 5.8 when we derive describing functions.) Expanding $s(t)$ into a Fourier series, we get

$$s(t) = \sum_{n=-\infty}^{\infty} C_n e^{j(2\pi n/T)t}, \tag{4.7}$$

where C_n represents the Fourier coefficients of the exponential series. These coefficients can be easily determined for the unit-impulse series from the integral

$$C_n = \frac{1}{T} \int_{-T/2}^{T/2} s(t) e^{-j(2\pi n/T)t} dt. \tag{4.8}$$

Because the area of any impulse equals unity, the integral also equals unity, and all of the Fourier coefficients can be expressed by

$$C_n = \frac{1}{T} \tag{4.9}$$

Substituting Eq. (4.9) into (4.7), we obtain the expression for the Fourier series of the sampling function:

$$s(t) = \frac{1}{T} \sum_{n=-\infty}^{\infty} e^{j(2\pi n/T)t}. \tag{4.10}$$

The expression for the Fourier series of the sampled function, $e^*(t)$, can easily be obtained by substituting Eq. (4.10) into Eq. (4.1). The result is

$$e^*(t) = \frac{1}{T} \sum_{n=-\infty}^{\infty} e(t) e^{j(2\pi/T)nt}. \tag{4.11}$$

Notice that the term $2\pi/T$ in the exponent of Eq. (4.11) represents the radian frequency of sampling and will be defined as ω_s. Therefore $e^*(t)$ can be expressed as

$$e^*(t) = \frac{1}{T} \sum_{n=-\infty}^{\infty} e(t)e^{jn\omega_s t}. \qquad (4.12)$$

The Laplace transform of Eq. (4.12) is given by [use was made of the frequency shifting theorem given by Eq. (2.72)‡]:

$$E^*(s) = \frac{1}{T} \sum_{n=-\infty}^{\infty} E(s - jn\omega_s). \qquad (4.13)$$

Let us now utilize Eq. (4.13) in order to determine $E^*(s)$ for the case where $e(t)$ is a unit step. The result will then be compared with that obtained previously, given by Eq. (4.6). Because the Laplace transform of a unit step is $1/s$, the Laplace transform of the sampled function is given by

$$E^*(s) = \frac{1}{T} \sum_{n=-\infty}^{\infty} \frac{1}{s - jn\omega_s}. \qquad (4.14)$$

Notice that Eq. (4.14) is an infinite summation and cannot be expressed in closed form as was Eq. (4.6). However, Eq. (4.14) does show very clearly the periodic properties of the Laplace transform of the sampled function. From Eq. (4.14), the frequency spectrum can be expressed as

$$E^*(j\omega) = \frac{1}{T} \sum_{n=-\infty}^{\infty} \frac{1}{j(\omega - n\omega_s)}. \qquad (4.15)$$

It appears that the spectrum is not band limited and exact reproduction of the sampled signal is not possible.

In general, the frequency spectrum of a sampled function can be expressed as

$$E^*(j\omega) = \frac{1}{T} \sum_{n=-\infty}^{\infty} E[j(\omega - n\omega_s)]. \qquad (4.16)$$

Notice that the frequency spectrum is periodic in ω_s. If the input intelligence signal $e(t)$ has a spectral distribution as shown in Figure 4.4, then the output of the sampler will have a frequency spectrum as shown in Figure 4.5. The bandwidth of the input signal is designated ω_c.

At this point in our development of sampled data theory, the relation between ω_c and ω_s will be considered. Referring to Figure 4.5, it can be seen that the function $e(t)$ can be recovered if the sampled function, $e^*(t)$, is passed through an ideal low-pass filter as shown by the dashed line. Shannon [6] has shown that if the highest input frequency is ω_c, then the sampling frequency must be at least $2\omega_c$ in order that the input function $e(t)$ can be recovered undistorted. If the sampling frequency is less than $2\omega_c$, then the spectral distribution from 0 to ω_c would overlap partly the region

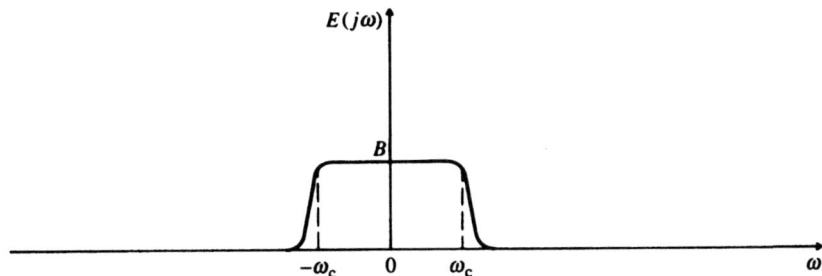

Figure 4.4 Input function frequency spectrum.

from ω_s to $\omega_s - \omega_c$. This would result in the filter recovering the input signal distorted by the overlapped portion of the first sideband (alias frequency), and aliasing occurs. Practical considerations often dictate a sampling frequency much higher than twice the highest input frequency. In addition, because signals generally do not have finite frequency spectrums and filters do not have ideal frequency response characteristics as shown by the dashed line in Figure 4.5, the recovered signal always has a certain amount of distortion. This distortion is referred to as ripple. It occurs at the sampling frequency and its harmonics should be treated as noise of the sampled-data control system.

4.3. DATA EXTRAPOLATORS

A data extrapolator is a device that reconstructs a sampled function into a continuous signal based upon a knowledge of past samples. This device, which is also referred to as a data hold, follows the sampler in practical feedback control systems. Figure 4.6 illustrates a simple system having a sampler, data hold, and a process which is to be controlled. The reconstructed signal, $e_R(t)$, is piecewise continuous and is applied to the process.

Data extrapolators are classified according to the number of prior samples that are required for predicting the sampled function during waiting intervals. As was illustrated in Figure 4.3, an extrapolator which only depends on the value of the sampled function at the beginning of a sampling interval is known as a zero-order hold. It is similar in operation to an electronic clamping circuit that maintains its output level equal to the magnitude of an input pulse and then resets itself when a new pulse is applied. An extrapolator that depends on two prior samples is known as a first-order hold. Figure 4.7 illustrates a reconstructed time function $e_R(t)$ produced by such an extrapolation from a signal $e(t)$. Practical systems usually do not use data extrapolators that are higher than first order, because they introduce an excessive phase lag into the feedback control system. In addition, they usually increase the effects of noise and are more complex and costly.

The transfer function of a zero-order hold can be easily determined from the impulsive response. This can be obtained from the superposition of a positive unit step at zero time and a negative unit step T seconds later, where T is the sampling period. This is illustrated in Figure 4.8. The impulsive response can be expressed in the time domain as

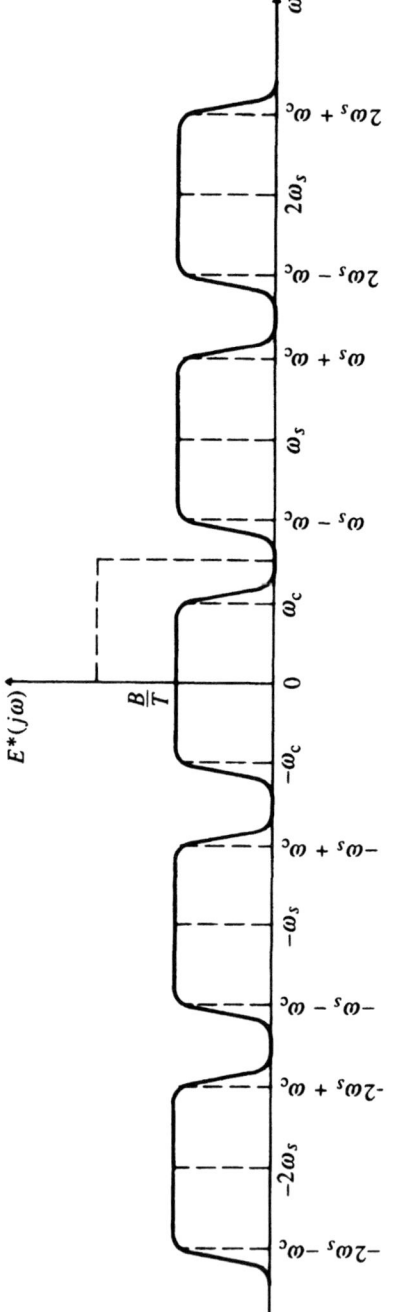

Figure 4.5 Sampled function frequency spectrum.

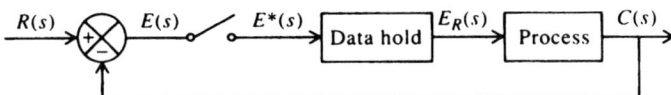

Figure 4.6 Simple sampled-data feedback control system.

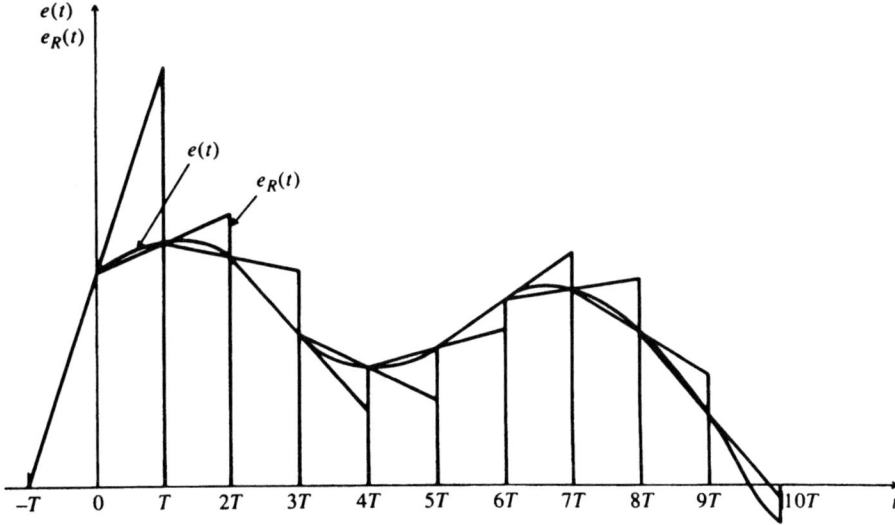

Figure 4.7 Reconstruction of a signal by means of a first-order hold.

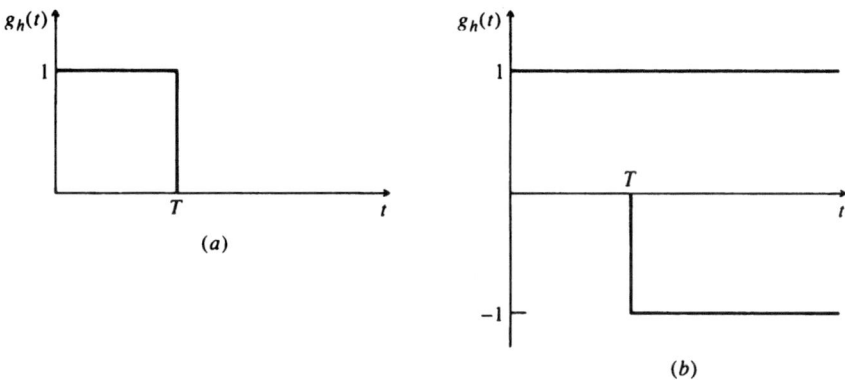

Figure 4.8 (a) Zero-order hold impulsive response. (b) Equivalent step-function components.

$$g_h(t) = U(t) - U(t - T). \tag{4.17}$$

The Laplace transform of Eq. (4.17) is

$$G_h(s) = \frac{1}{s} - \frac{1}{s}e^{-Ts}. \tag{4.18}$$

This can be simplified to

$$G_h(s) = \frac{1 - e^{-Ts}}{s}. \tag{4.19}$$

Equation (4.19) is quite useful for studying the effects of a zero-order hold in a feedback control system.

The frequency response of the zero-order hold can be expressed as

$$G_h(j\omega) = \frac{1 - e^{-j\omega T}}{j\omega}. \tag{4.20}$$

In order to obtain an expression for the amplitude and phase characteristics of a zero-order hold, Eq. (4.20) is rearranged successively as follows:

$$G_h(j\omega) = \frac{e^{-j(\omega T/2)}[e^{j(\omega T/2)} - e^{-j(\omega T/2)}]}{j\omega}. \tag{4.21}$$

$$G_h(j\omega) = \frac{2}{\omega} \frac{e^{-j(\omega T/2)}[e^{j(\omega T/2)} - e^{-j(\omega T/2)}]}{2j}. \tag{4.22}$$

Observing that $[e^{j(\omega T/2)} - e^{-j(\omega T/2)}]/2j$ is equal to $\sin(\omega T/2)$, Eq. (4.22) can be simplified to

$$G_h(j\omega) = \frac{2}{\omega}(e^{-j(\omega T/2)})\sin\frac{\omega T}{2}. \tag{4.23}$$

Multiplying numerator and denominator by T, a familiar $\sin x/x$ term results:

$$G_h(j\omega) = T\frac{\sin(\omega T/2)}{\omega T/2} \bigg/\!\!\underline{-\frac{\omega T}{2}}. \tag{4.24}$$

The amplitude and phase of $G_h(j\omega)$ are sketched in Figure 4.9. The frequency characteristics of this extrapolating device are similar to that of a low-pass filter where full cutoff occurs at $2\pi n/T$ rad/sec, where n is an integer. Of primary concern to the control engineer is the phase shift introduced by the zero-order hold. This appreciable phase lag may cause a feedback system, which might ordinarily be stable in its continuous form, to become unstable.

Let us next determine the characteristics of a first-order holding device. Figure 4.7 shows that the extrapolated function of a first-order hold is linear and has a slope determined by the last two samples. The transfer function of such an extrapolating device can be obtained in a manner similar to that used for the zero-order hold. The impulsive response of a first-order hold is shown in Figure 4.10a. Its equivalent step and ramp function components are sketched in Figure 4.10b. By superposition the Laplace transform of the impulsive response can be expressed as

$$G_h(s) = \frac{1}{s} + \frac{1}{Ts^2} - \frac{2}{s}e^{-Ts} - \frac{2}{Ts^2}e^{-Ts} + \frac{1}{s}e^{-2Ts} + \frac{1}{Ts^2}e^{-2Ts}. \tag{4.25}$$

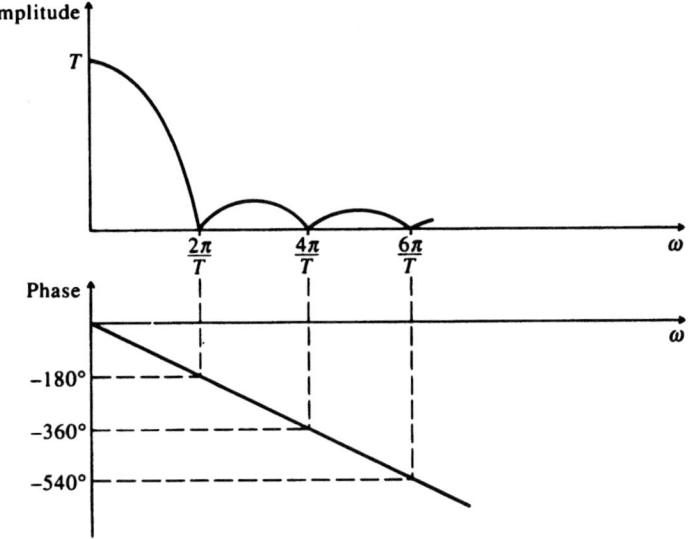

Figure 4.9 Amplitude and phase response of a zero-order hold.

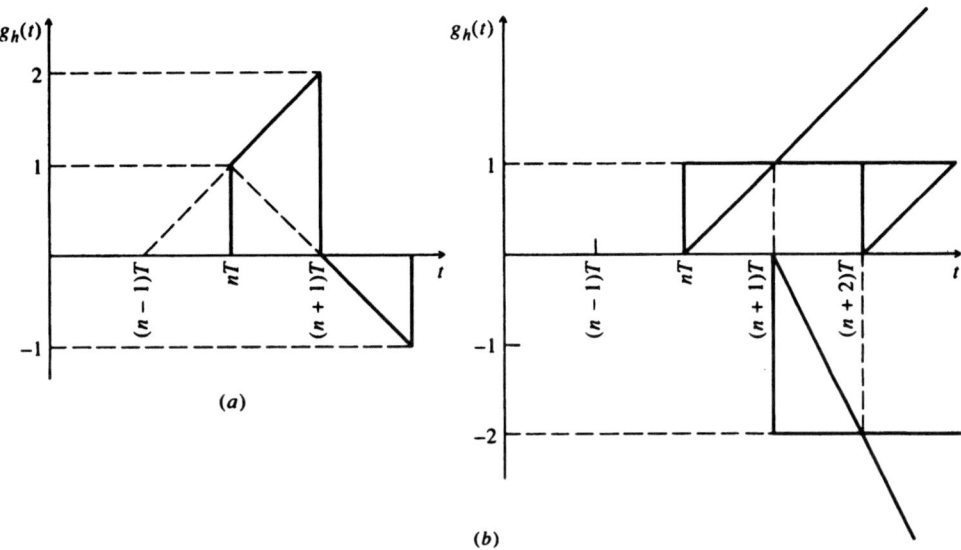

Figure 4.10 (a) First-order hold impulsive response. (b) Equivalent step-and-ramp-function components.

This can be simplified to

$$G_h(s) = T(1 + Ts)\left[\frac{1 - e^{-Ts}}{Ts}\right]^2 \qquad (4.26)$$

Equation (4.26) is quite useful for studying the effects of a first-order hold in a feedback control system.

The frequency characteristics of the first-order hold can be expressed as

$$G_h(j\omega) = T(1 + j\omega T)\left[\frac{1 - e^{-j\omega T}}{j\omega T}\right]^2. \quad (4.27)$$

In order to obtain the amplitude and phase characteristics, Eq. (4.27) is rearranged as follows:

$$G_h(j\omega) = T\sqrt{1 + \omega^2 T^2}(e^{-j\omega T})\frac{4}{\omega^2 T^2}\left[\frac{e^{j(\omega T/2)} - e^{-j(\omega T/2)}}{2j}\right]^2 \underline{/\tan^{-1}\omega T} \quad (4.28)$$

Observing that $[(e^{j(\omega T/2)} - e^{-j(\omega T/2)})/2j]^2$ equals $\sin^2(\omega T/2)$, Eq. (4.28) can be simplified to

$$G_h(j\omega) = T\sqrt{1 + \omega^2 T^2}\left[\frac{\sin(\omega T/2)}{\omega T/2}\right]^2 \underline{/\tan^{-1}\omega T - \omega T}. \quad (4.29)$$

The amplitude and phase of $G_h(j\omega)$ are sketched in Figure 4.11.

Observe that the amplitude response of this extrapolating device is similar to that of the zero-order hold. However, the phase shift introduced by the first-order hold is almost twice that of the zero-order hold. For example, at $\omega = 2\pi/T$, the first-order hold introduces a phase shift of $-2\pi + \tan^{-1} 2\pi(-279.04°)$, and the zero-order hold introduces a phase shift of $-180°$. This greatly affects the stability and is a serious disadvantage of the more complex extrapolator. In addition, the higher-order extrapolator transmits a greater amount of the higher-frequency components with a resultant higher noise level in the system. For these reasons, higher-order holding devices are not usually included in practical feedback control systems. Their only advantage is that they are capable of perfectly reproducing functions (in the steady

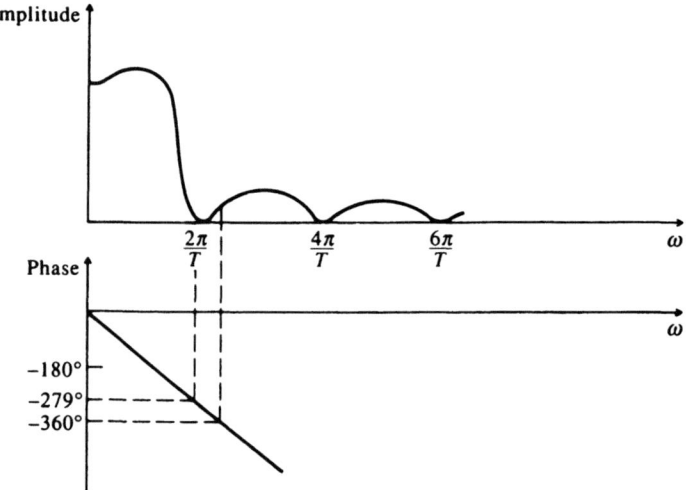

Figure 4.11 Amplitude and phase response of a first-order hold.

state) which are higher-order derivatives of time. For example, a zero-order hold is only capable of reproducing a step function perfectly, while a first-order hold can reproduce a ramp function perfectly. However, the zero-order hold is adequate for most practical applications.

4.4. z-TRANSFORM THEORY

The analysis of continuous feedback control systems is considerably simplified by the use of the Laplace transform. The transfer function of simple networks and complicated processes can be represented by an algebraic function that is a ratio of two polynomials in s. A sampler, however, produces an equivalent transfer function which is the ratio of two polynomials in e^{Ts}. In Section 4.2, Eq. (4.4) showed this effect for the Laplace transform of $e(t)$. This result stems from the fact that we are working with difference equations in sampled-data control systems instead of differential equations as in continuous systems. It is possible to obtain an algebraic relationship, similar in form to the Laplace transform, if e^{Ts} is replaced by some arbitrary symbol such as z. This is the basis of the z-transform theory [2–5]. The analysis of sampled-data control systems is very similar to that of continuous systems when the z-transform is employed.

The z transform of a sampler can be simply obtained by substituting $z = e^{Ts}$ into Eq. (4.4). This results in a power series of $1/z$ as follows:

$$E(z) = \sum_{n=0}^{\infty} e(nT) z^{-n}. \tag{4.30}$$

This definition of the z transform is denoted as the one-sided z transform, which assumes that $e(t) = 0$ for $t < 0$, or $e(nT) = 0$ for $n < 0$. Equation (4.30) results in terms that are negative powers of z^{-1}, and it uniquely defines the z transform. For functions of $e(t)$ that have values for $-\infty < t < \infty$, and for $e(nT)$ that have values for $n = 0, \pm 1, \pm 2, \pm 3, \ldots$, we use the two-sided z transform:

$$E(z) = \sum_{n=-\infty}^{\infty} e(nT) z^{-n}. \tag{4.31}$$

Equation (4.31) results in terms that contain both positive and negative powers of z. To uniquely define the z transform in this case, we must also define bounds on the magnitude of z for which Eq. (4.31) converges. Otherwise, it is possible for more than one sequence to have the same z transform. For example, let us consider the following two sequences:

$$x_n = \begin{cases} -a^n, & n < 0 \\ 0, & n \geq 0, \end{cases}$$

$$y_n = \begin{cases} 0, & n < 0 \\ a^n, & n \geq 0. \end{cases}$$

Both of these sequences have the same z transform:

$$X(z) = -\sum_{n=1}^{\infty}(az^{-1})^{-n} = \frac{1}{1-az^{-1}},$$

$$Y(z) = \sum_{n=0}^{\infty} a^n z^{-n} = \frac{1}{1-az^{-1}}.$$

By also defining the region of convergence for $X(z)$ to be

$$\left|\frac{1}{az^{-1}}\right| < 1 \quad \text{or} \quad |z| < |a|,$$

and the region of convergence for $Y(z)$ to be

$$|az^{-1}| < 1 \quad \text{or} \quad |z| > |a|,$$

then the z transforms of $X(z)$ and $Y(z)$ converge and are unique definitions. As we are not interested in the practical world in functions which exist for negative times, we will only consider the one-sided z transform in this chapter.

Notice that the z transform describes the values of the time function at the sampling instants only. It offers no information on the behavior of the function between the sampled times.

In order to analyze or synthesize a feedback control system containing a sampler, it is also necessary to obtain the z transforms of all other elements in the closed loop. Stability can then be determined using the complex z-plane just as the complex s-plane (Nyquist diagram) was used in the case of continuous systems. For purposes of clarity, a brief comparison of the conditions for stability will next be made for these two planes.

We had concluded from Eqs. (4.14)–(4.16) that the frequency spectrum of a sampled function is periodic in s with period $j\omega_s$. Therefore, it is important to recognize that if a function $E(s)$ has a pole (zero) at $s = s_a$, then $E^*(s)$ has poles (zeros) at $s = s_a + jn\omega_s$, where $n = 0, \pm 1, \pm 2, \ldots, \pm\infty$. Therefore, for every pole (or zero) at $s = s_a$ in the s-plane, the sampled function $E^*(s)$ has the same value at all periodic frequency points $s_a + jn\omega_s$, where n is an integer. Figure 4.12 illustrates a periodic function $E^*(s)$, where $E(s)$ has a pole (designated as X) at $s = s_a = -\sigma_a + j\omega_a$ and the sampling operation generates a pole in $E^*(s)$ at $-\sigma_a + j(\omega_a + n\omega_s)$ where n is an integer. In practice, because the low-pass filter characteristics of the process and the zero-order hold (or first-order hold) greatly attenuate the response of the poles and zeros in the complementary strips, we will only consider the poles and zeros in the major strip for analysis and design in this chapter.

Let us focus attention on the primary strip of the s-plane shown in Figure 4.13a. We wish to map the contour $S - T - U - V - W$ of the primary strip in the s-plane into the z-plane. Using the definition of the z transform, we can obtain the necessary relationship as follows:

$$z = e^{Ts} = e^{(\sigma \pm j\omega)T} = e^{\sigma T} e^{\pm j\omega T} = e^{\sigma T} \underline{/\pm \omega T}. \qquad (4.32)$$

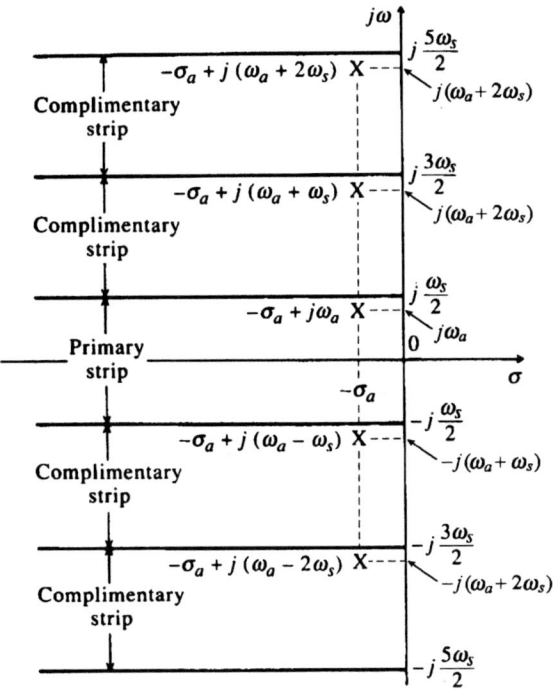

Figure 4.12 Location of a complex pole $E(s)$ at $-\sigma_a + j\omega_a$ (labeled X) in the primary strip, and the location of additional poles of $E^*(s)$, caused by sampling, in the complementary strips.

The results are shown in Figure 4.13b, and the corresponding sequence of points S–T–U–V–W–S in the s-plane and z-planes are illustrated. It is important to recognize that because the area of any complementary strip in the s-plane is mapped into the same unit circle in the z-plane, the correspondence between the s- and z-planes is not unique. Therefore, a single point on the z-plane corresponds to an infinite number of points in the s-plane. However, a single point in the s-plane corresponds to a single point in the z-plane.

We can now reach a very important conclusion regarding stability in the z-plane. Because we observe from Figure 4.13 that the left half of the s-plane is mapped into the unit circle of the z-plane, and the $j\omega$ axis in the s-plane maps onto the unit circle in the z-plane, we conclude that all the closed-loop poles or roots of the characteristic equation of the sampled-data system must lie within the unit circle in the z-plane for stability. A pole located at $z = 1$ in the z-plane corresponds to a pole at $s = 0$ in the s-plane and is permissible for stability. Closed-loop zeros do not affect absolute stability and can be located anywhere in the z-plane. However, as the zeros approach $z = 1$, they tend to increase the percent overshoot and rise time [2].

In order to use the Bode diagram and the simplicity of the logarithmic plots, which are based on the property that the stability boundary is the imaginary axis in the s-plane, we have to transform the z transform into the w-plane, where

$$w = \frac{2}{T} \cdot \frac{z-1}{z+1}.$$

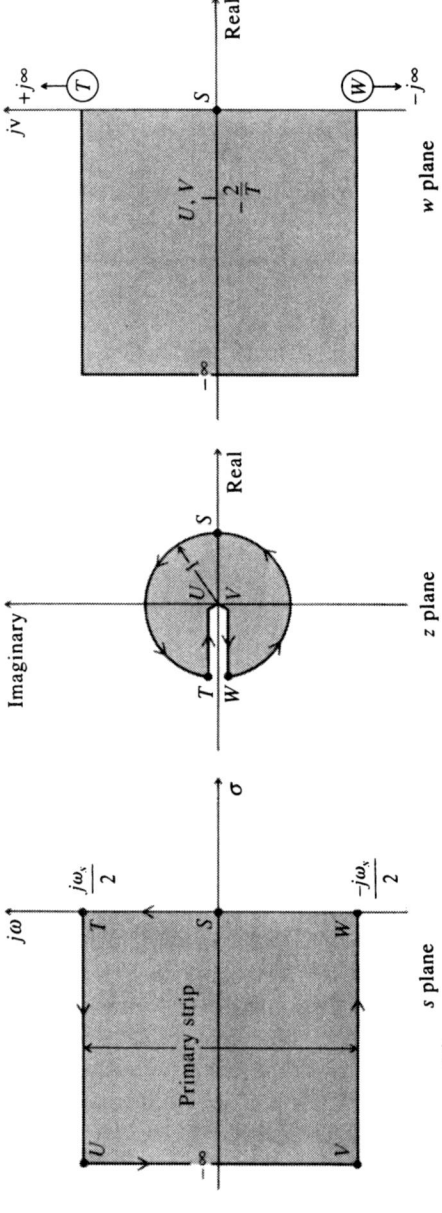

Figure 4.13 Mapping of the primary strip in the left half of the s-plane into the z-plane and from the z-plane to the w-plane. (a) Primary strip in the left half of the s-plane. (b) Mapping of the primary strip of the s-plane into the z-plane. (c) Mapping of the unit circle in the z-plane into the w-plane.

Using this transformation, the unit circle of the z-plane is transformed into the imaginary axis of the w-plane. This is illustrated in Figure 4.13c. Therefore, as ω varies from 0 to $j\omega_s/2$ along the $j\omega$ axis in the s-plane, w varies from 0 to infinity along the positive jv axis in the w-plane, where the fictitious frequency in the w-plane is defined to be v where $w = jv$. The Bode diagram method in the w-plane is presented in detail in Section 4.9.

A. Techniques of z Transformation

The z transformation for various types of time functions will now be considered. These can be found through a direct application of Eq. (4.4) or (4.30). Attention will be focused on those types of functions which are common to feedback control systems. Several examples will be used to illustrate the techniques prior to tabulating several useful transform pairs:

1. The z Transform of a Unit Step

$$r(t) = U(t) = 1 \quad [r(t) = 0 \quad \text{for} \quad t < 0],$$

$$R^*(s) = \sum_{n=0}^{\infty} r(nT)e^{-nTs},$$

where $r(nT) = 1$, for $n = 0, 1, 2, \ldots$,

$$\therefore R^*(s) = \sum_{n=0}^{\infty} e^{-nTs} = 1 + e^{-Ts} + e^{-2Ts} + e^{-3Ts} + \cdots.$$

$$R(z) = 1 + z^{-1} + z^{-2} + z^{-3} + \cdots,$$

This series is convergent for $|z| > 1$, and yields

$$R(z) = \frac{1}{1 - z^{-1}}, \quad |z^{-1}| < 1,$$
$$R(z) = \frac{z}{z - 1}, \quad |z| > 1. \quad (4.33)$$

2. The z Transform of a Unit Ramp

$$r(t) = t, \quad [r(t) = 0 \quad \text{for} \quad t < 0],$$

$$R^*(s) = \sum_{n=0}^{\infty} r(nT)e^{-nTs},$$

where $r(nT) = 0, T, 2T, 3T, 4T$, etc., for $n = 0, 1, 2, \ldots$,

$$\therefore R^*(s) = Te^{-Ts} + 2Te^{-2Ts} + 3Te^{-3Ts} + 4Te^{-4Ts} + \cdots;$$
$$R(z) = Tz^{-1} + 2Tz^{-2} + 3Tz^{-3} + 4Tz^{-4} + \cdots.$$

226 DIGITAL CONTROL-SYSTEM ANALYSIS AND DESIGN

This series is convergent for $|z| > T$, and yields

$$R(z) = \frac{Tz}{(z-1)^2}. \tag{4.34}$$

3. The z Transform of an Exponential Decay

$$r(t) = e^{-at} \quad [r(t) = 0 \quad \text{for} \quad t < 0],$$

$$R^*(s) = \sum_{n=0}^{\infty} r(nT)e^{-nTs},$$

where $r(nT) = e^{-anT}$, for $n = 0, 1, 2, \ldots,$

$$\therefore R^*(s) = \sum_{n=0}^{\infty} e^{-nT(a+s)} = 1 + e^{-aT}e^{-Ts} + e^{-2aT}e^{-2Ts} + e^{-3aT}e^{-3Ts} + \cdots;$$

$$R(z) = 1 + e^{-aT}z^{-1} + e^{-2aT}z^{-2} + e^{-3aT}z^{-3} + \cdots.$$

This series is convergent for $|z| > e^{-aT}$, and yields

$$R(z) = \frac{1}{1 - e^{-aT}z^{-1}},$$
$$R(z) = \frac{z}{z - e^{-aT}}. \tag{4.35}$$

Once the z transform for any function is obtained and tabulated, it is not necessary that it be derived again. Tabulations of the foregoing results, and other transform pairs useful to the control engineer, appear in Table 4.1.

Because it is necessary for the control engineer to obtain the z transform for all of the elements of a feedback control system when analyzing a sampled-data system, it is a useful exercise to obtain the z transform for a typical transfer function that the process may have. Attention will be focused on the technique of application for the z transform of a third-order process whose transfer function is given by

$$G(s) = \frac{1}{s^2(s+1)}. \tag{4.36}$$

By means of partial fraction expansion, this transfer function can be expressed as

$$G(s) = \frac{1}{s^2} - \frac{1}{s} + \frac{1}{s+1}. \tag{4.37}$$

The z transform can be obtained, by inspection, from Table 4.1:

$$G(z) = \frac{Tz}{(z-1)^2} - \frac{z}{z-1} + \frac{z}{z-e^{-T}}. \tag{4.38}$$

Table 4.1. Table of z Transforms

No.	Time function	Laplace transform	z transform
1	$\delta(t)$	1	1
2	$\delta(t - nT)$	e^{-nTs}	$\dfrac{1}{z^n}$
3	$U(t)$	$\dfrac{1}{s}$	$\dfrac{z}{z-1}$
4	t	$\dfrac{1}{s^2}$	$\dfrac{Tz}{(z-1)^2}$
5	$\dfrac{t^2}{2}$	$\dfrac{1}{s^3}$	$\dfrac{T^2 z(z+1)}{2(z-1)^3}$
6	$\dfrac{t^3}{6}$	$\dfrac{1}{s^4}$	$\dfrac{T^3 z(z^2 + 4z + 1)}{6(z-1)^4}$
7	e^{-at}	$\dfrac{1}{s+a}$	$\dfrac{z}{z - e^{-aT}}$
8	te^{-at}	$\dfrac{1}{(s+a)^2}$	$\dfrac{Tze^{-aT}}{(z - e^{-aT})^2}$
9	$\dfrac{1}{2} t^2 e^{-at}$	$\dfrac{1}{(s+a)^3}$	$\dfrac{T^2 z e^{-aT}(z + e^{-aT})}{(z - e^{-aT})^3}$
10	$1 - e^{-at}$	$\dfrac{a}{s(s+a)}$	$\dfrac{z(1 - e^{-aT})}{(z - 1)(z - e^{-aT})}$
11	$\sin at$	$\dfrac{a}{s^2 + a^2}$	$\dfrac{z \sin aT}{z^2 - 2z \cos aT + 1}$
12	$\cos at$	$\dfrac{s}{s^2 + a^2}$	$\dfrac{z^2 - z \cos aT}{z^2 - 2z \cos aT + 1}$
13	$e^{-at} f(t)$	$F(s + a)$	$F(e^{aT} z)$
14	$a^k, k \geq 0$	—	$\dfrac{z}{z - a}$

B. Obtaining the Inverse z Transformation

In general, there are four methods for obtaining the inverse z transform. These are by means of partial fraction expansion, power-series expansion, the inversion integral, and the digital computer computational method. Each of these techniques is outlined as follows:

1. Partial Fraction Expansion. The z transform is factored into its components so that each term of the expansion can be obtained, by inspection, from Table 4.1 as was illustrated in the preceding example. The technique for partial fraction expansion is also applicable to the z transform. For example, consider the following transfer function:

$$E(z) = \frac{z}{(z-2)(z-3)}. \tag{4.39}$$

The partial fraction expansion is obtained as follows:

$$\frac{E(z)}{z} = \frac{A}{z-2} + \frac{B}{z-3} \tag{4.40}$$

where

$$A = \left.\frac{1}{z-3}\right]_{z=2} = -1 \tag{4.41}$$

$$B = \left.\frac{1}{z-2}\right]_{z=3} = 1 \tag{4.42}$$

Therefore,

$$E(z) = -\frac{z}{z-2} + \frac{z}{z-3}. \tag{4.43}$$

From Table 4.1, the inverse z transformation of $E(z)$ is given by:

$$e(nT) = -(2)^{nT} + (3)^{nT}, \quad n = 0, 1, 2, \cdots, \tag{4.44}$$

This can also be written in the following form:

$$e(k) = -(2)^k + (3)^k, \quad k \geq 0 \tag{4.45}$$

2. Power-Series Expansion. The z transform can be expanded into an inverse power series in z by expanding Eq. (4.30). The result is

$$E(z) = \sum_{n=0}^{\infty} e(nT)z^{-n} = \frac{e(0)}{z^0} + \frac{e(T)}{z^1} + \frac{e(2T)}{z^2} + \frac{e(3T)}{z^3} + \cdots. \tag{4.46}$$

The coefficient of z^{-n}, $e(nT)$, is the value of the variable at the nth sensing instant, and the coefficients can be used directly to find the time function at the sampling instants. For example, let us reconsider Eq. (4.39) and obtain its inverse z transformation using the power series expansion as follows:

$$E(z) = \frac{z}{(z-2)(z-3)} = \frac{z}{z^2 - 5z + 6}. \tag{4.47}$$

Dividing denominator into numerator, we obtain the following:

$$z^2 - 5z + 6 \overline{\smash{\big)}\begin{array}{l} z^{-1} + 5z^{-2} + 19z^{-3} + 65z^{-4} + 211z^{-5} + \cdots \\ z \\ \underline{z - 5 + 6z^{-1}} \\ 5 - 6z^{-1} \\ \underline{5 - 25z^{-1} + 30z^{-2}} \\ 19z^{-1} - 30z^{-2} \\ \underline{19z^{-1} - 95z^{-2} + 114z^{-3}} \\ \phantom{z - 5 - 19z^{-1} - } 65z^{-2} - 114z^{-3} \\ \phantom{z - 5 - 19z^{-1} - } \underline{65z^{-2} - 325z^{-3} + 510z^{-4}} \\ \phantom{z - 5 - 19z^{-1} - 65z^{-2} - } 211z^{-3} - 510z^{-4}. \end{array}}$$

Therefore,

$$E(z) = z^{-1} + 5z^{-2} + 19z^{-3} + 65z^{-4} + 211z^{-5} + \cdots . \qquad (4.48)$$

From Table 4.1, the inverse z transform of $E(z)$ is given by:

$$\begin{aligned} e(nT) = &\delta(t - T) + 5\delta(t - 2T) + 19\delta(t - 3T) \\ &+ 65\delta(t - 4T) + 211\delta(t - 5T) + \cdots, \quad n = 0, 1, 2, \cdots \end{aligned} \qquad (4.49)$$

This can also be written in the following form:

$$e(k) = \delta(t - 1) + 5\delta(t - 2) + 19\delta(t - 3) + 65\delta(t - 4) + 211\delta(t - 5) + \cdots, \quad k \geq 0. \qquad (4.50)$$

Let us check our results from Eq. (4.50) with that of Eq. (4.45). For example, let us check the value of Eq. (4.45) at $k = 3$:

$$e(3) = -(2)^3 + (3)^3 = 19.$$

This result agrees with the value at $k = 3$ in Eq. (4.50). Let us also check the value of Eq. (4.45) at $k = 5$:

$$e(5) = -(2)^5 + (3)^5 = 211.$$

This result also agrees with the value at $k = 5$ in Eq. (4.50). We, therefore, conclude that Eq. (4.50) is the series form of the closed-form solution obtained in Eq. (4.45).

3. Computation using a Digital Computer. A digital computer can be a very powerful tool for obtaining the inverse z transform. The approach involves converting the z transform into a difference equation that can then be solved using a digital computer. For example, let us reconsider the transfer function which was evaluated in Eqs. (4.39) and (4.47):

$$E(z) = \frac{z}{(z - 2)(z - 3)}. \qquad (4.51)$$

We wish to obtain the impulsive response of such a system. From Table 4.1, we know that the z transform of a unit impulse is 1. Therefore, with $R(z)$ representing a unit impulse whose z transform is 1, we obtain

$$E(z) = \frac{z}{(z-2)(z-3)} R(z) = \frac{z}{z^2 - 5z + 6} R(z) \tag{4.52}$$

from which

$$E(z)[z^2 - 5z + 6] = zR(z). \tag{4.53}$$

Using Table 4.1, the difference equation corresponding to this equation is

$$e(n+2) - 5e(n+1) + 6e(n) = r(n+1) \tag{4.54}$$

where the unit impulse implies that $r(0) = 1$ and $r(n) = 0$ for all other n. The solution to this difference equation depends on the initial conditions $e(0)$ and $e(1)$. The value of $e(0)$ can be obtained by substituting $n = -2$ into the difference equation for $e(n+2)$; the result is $e(0) = 0$. Similarly, the value of $e(1)$ can be obtained by substituting $n = -1$ into the difference equation; the result is $e(1) = 1$.

The problem of solving for $e(n)$ reduces to solving the difference equation subject to the following initial conditions: $e(0) = 0$; $e(1) = 1$; $r(0) = 1$; $r(n) = 0$ for all other n. A program, written in Basic, for solving the difference equation subject to the initial conditions is shown in Table 4.2. The output obtained from the digital computer for this problem is shown in Table 4.3. The results were then checked with the solutions for this problem that were obtained using the previous inversion integral method. The results were identical. The program shown in Table 4.2 is a very useful program for solving difference equations, and can be modified for other problems by changing the initial conditions on line 100 (which is read in on line 10) and the elements of the difference equation on line 30.

C. Example of Simple, Open-Loop Sampled-Data System

Let us now extend our development of z-transform theory to the case of a simple, open-loop, sampled-data system illustrated in Figure 4.14a. The input function $r(t)$ is

Table 4.2. Computer Program (in Basic) for Solving the Difference Equation $e(n+2) - 5e(n+1) + 6e(n) = r(n+1)$ having the Following Initial Conditions: $e(0) = 0$; $e(1) = 1$; $r(0) = 1$; $r(n) = 0$ for all other n

10	READ E2,E1,R1,R0
20	For N = 0 TO 14
30	E0 = R1 + 5*E1 − 6*E2
40	PRINT N,E0
50	E2 = E1
60	E1 = E0
70	R1 = R0
80	R0 = 0
90	NEXT N
100	DATA 0,0,0,1

Table 4.3. Output of Program in Table 4.2

n	$e(n)$
0	0
1	1
2	5
3	19
4	65
5	211
6	665
7	2059
8	6305
9	19171
10	58025
11	175099
12	527345
13	1586131
14	4766585

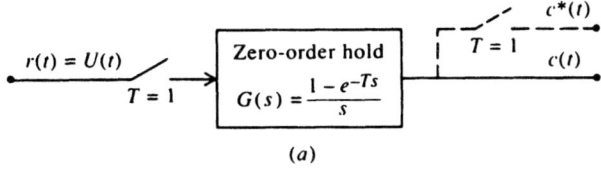

(a)

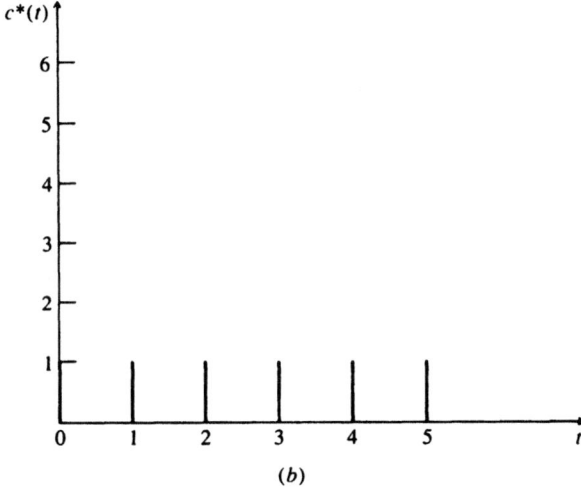

(b)

Figure 4.14 Analysis of a simple, open-loop, sampled-data system.

a unit step. The sampling rate T is 1 sec. The output $c^*(t)$ is to be assumed sampled by an imaginary sampler synchronized with the real one. Let us apply the techniques of z-transform theory in order to obtain $R(z)$, $C(z)$, and $c^*(t)$.

1. Computation of R(z)

$$r(t) = U(t).$$

$R(z)$ can be obtained from Table 4.1 (and our previous derivation):

$$R(z) = \frac{z}{z-1}. \tag{4.55}$$

2. Computation of C(z).
The z transform of the output of a network is equal to the z transform of the input multiplied by the z-transfer function of the network. Therefore, for Figure 4.14a,

$$C(z) = G(z)R(z). \tag{4.56}$$

The z transform of the zero-order hold, shown in Figure 4.14a, is given by

$$G(z) = Z\left\{\frac{1 - e^{-s}}{s}\right\} = (1 - z^{-1})\left(\frac{z}{z-1}\right) = 1. \tag{4.57}$$

where Z indicates the z transform corresponding to the bracketed Laplace-transform term. Substituting Eqs. (4.55) and (4.57) into Eq. (4.56) we obtain the expression

$$C(z) = (1)\left(\frac{z}{z-1}\right) = \frac{z}{z-1}. \tag{4.58}$$

Dividing denominator into numerator, we obtain the series given by

$$C(z) = 1 + z^{-1} + z^{-2} + z^{-3} + z^{-4} + z^{-5} + \cdots. \tag{4.59}$$

3. Computation of c*(t).
The value of $c^*(t)$ can be obtained from the power-series expansion given by Eq. (4.5). Utilizing Table 4.1, we obtain

$$c^*(t) = \delta(t) + \delta(t-1) + \delta(t-2) + \delta(t-3) + \delta(t-4) + \delta(t-5) + \cdots.$$

The output $c^*(t)$ is illustrated in Figure 4.14b.

D. Other Properties

There are certain other properties of the z transform that are worthy of mention. The initial- and final-value theorems relate the initial and final values of a time function to its associated z transform. They are given by

$$c(0) = \lim_{z \to \infty} C(z) \tag{4.60}$$

and

$$c(\infty) = \lim_{z \to 1}(1 - z^{-1})C(z). \quad (4.61)$$

Applying the initial- and final-value theorems as given by Eqs. (4.60) and (4.61), respectively, to the problem illustrated in Figure 4.14, we obtain the following. (a) Initial value:

$$c(0) = \lim_{z \to \infty} C(z), \quad (4.62)$$

where $C(z)$ is given by Eq. (4.59).

$$\therefore c(0) = 1. \quad (4.63)$$

This result agrees with the sketch of Figure 4.14b. (b) Final value:

$$c(\infty) = \lim_{z \to 1}(1 - z^{-1})C(z), \quad (4.64)$$

where $C(z)$ is given by Eq. (4.58).

$$\therefore c(\infty) = 1. \quad (4.65)$$

This result also agrees with the sketch of Figure 4.14b.

E. Advanced z-Transform Method

A very powerful tool that can be used to evaluate the response between sampling instants is the advanced z-transform method [7]. This technique essentially consists of modifying the ordinary z transform by inserting a fictitious time delay into the sampled-data control system. The resultant transform describes pulse sequences that are delayed from the time functions by nonintegral multiples of the sampling frequency. By varying the amount of delay, the information contained in the continuous signal between sampling instants can be studied.

The advanced z transform can best be understood by referring to the time function in Figure 4.15, which is delayed by αT sec:

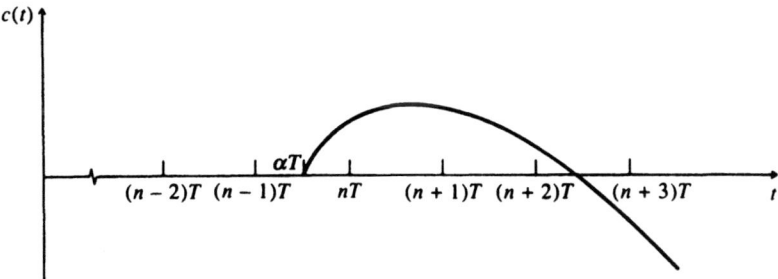

Figure 4.15 Time function delayed by αT sec.

234 DIGITAL CONTROL-SYSTEM ANALYSIS AND DESIGN

If α is an integer, the result is redundant, because the z transform of the resulting function is merely $z^{-\alpha}C(z)$. The value of n is chosen as the next highest integer after α, making it an advance. The difference of nT and αT is defined as Δ where

$$\Delta T = nT - \alpha T \tag{4.66}$$

or

$$\Delta = n - \alpha. \tag{4.67}$$

Δ is assumed to be a positive number between zero and unity.

Assuming that $C(s)$ is the Laplace transform of $c(t)$, then the Laplace transform of the delayed function, $C(s, \alpha)$, is given by

$$\mathscr{L}[c(t - \alpha T)] = C(s, \alpha) = C(s)e^{-\alpha Ts}. \tag{4.68}$$

Substituting Eq. (4.67) into Eq. (4.68), the following expression can be obtained:

$$C(s, \Delta) = e^{-nTs}C(s)e^{\Delta Ts}. \tag{4.69}$$

The advanced z transform corresponding to Eq. (4.69) is given by

$$C(z, \Delta) = z^{-n}Z[C(s)e^{\Delta Ts}]. \tag{4.70}$$

where Z indicates the z transform corresponding to the bracketed Laplace-transform term.

From the definition of the z transform, the advanced z transform can be expressed in the following general form:

$$C(z, \Delta) = \sum_{n=0}^{\infty} c(n + \Delta)Tz^{-n}. \tag{4.71}$$

As an example of the advanced z-transform derivation for a time function, the case of an exponential decay, $e^{-\alpha T}$, that is advanced by $0.2T$ sec will be considered. For this problem,

$$c(t + 0.2T) = e^{-a(t+0.2T)}. \tag{4.72}$$

Utilizing Eq. (4.72), the advanced z transform corresponding to this expression is given by

$$C(z, 0.2) = \sum_{n=0}^{\infty} e^{-a(n+0.2)T}z^{-n}. \tag{4.73}$$

This can be simplified to

$$C(z, 0.2) = e^{-0.2aT}\sum_{n=0}^{\infty} e^{-anT}z^{-n}. \tag{4.74}$$

The infinite geometric progression corresponding to Eq. (4.74) can be expressed in closed form by

$$C(z, 0.2) = \frac{e^{-0.2aT}}{1 - e^{-aT}z^{-1}}. \qquad (4.75)$$

Utilizing this approach, a table of advanced z transforms for several functions can be derived. Table 9.4 lists the advanced z transforms for some commonly found functions.

The inverse of the advanced z transform enables the control-system engineer to evaluate the system's response between samples directly. For example, let us consider a system where $G(s) = 1/(s+1)$, and let us further assume that the input is a unit step. From Table 4.4, the corresponding advanced z transform corresponding to the input $R(s)$ is given by

$$R(z, \Delta) = \frac{1}{1 - z^{-1}}, \qquad (4.76)$$

and that for the control element $G(s)$ is given by

$$G(z, \Delta) = \frac{e^{-\Delta T}}{1 - e^{-T}z^{-1}}, \qquad (4.77)$$

Table 4.4. Table of Advanced z Transforms

	Time function	Laplace transform	Advanced z transform
1	$\delta(t)$	1	0
2	$\delta(t - nT)$	e^{-nTs}	$z^{-n-1+\Delta}$
3	$U(t)$	$\frac{1}{s}$	$\frac{1}{1 - z^{-1}}$
4	t	$\frac{1}{s^2}$	$\frac{\Delta Tz}{z - 1} + \frac{Tz}{(z - 1)^2}$
5	e^{-at}	$\frac{1}{s + a}$	$\frac{ze^{-a\Delta T}}{z - e^{-aT}}$
6	te^{-at}	$\frac{1}{(s + a)^2}$	$\frac{Tze^{-a\Delta T}[e^{-aT} + \Delta(z - e^{-aT})]}{(z - e^{-aT})^2}$
7	$\sin at$	$\frac{a}{s^2 + a^2}$	$\frac{z^2 \sin a\Delta T + z \sin(1 - \Delta)aT}{z^2 - 2z \cos aT + 1}$
8	$\cos at$	$\frac{s}{s^2 + a^2}$	$\frac{z^2 \sin a\Delta T - z \cos(1 - \Delta)aT}{z^2 - 2z \cos aT + 1}$

The advanced z transform of the output, $C(z, \Delta)$, is given by

$$C(z, \Delta) = R(z, \Delta)G(z, \Delta) \tag{4.78}$$

or

$$C(z, \Delta) = \frac{e^{-\Delta T}}{(1 - z^{-1})(1 - e^{-T}z^{-1})}. \tag{4.79}$$

By keeping Δ as a constant parameter, the inversion of Eq. (4.79) results in the following expression:

$$c(nT, \Delta) = \left(\frac{1 - e^{-T}e^{-nT}}{1 - e^{-T}}\right)e^{-\Delta T}. \tag{4.80}$$

Therefore, if the response following the nth sampling interval is desired, the value of n is substituted into this expression and Δ is allowed to vary from zero to unity. As would be expected, the continuous function between any sampling interval is an exponential decaying term whose initial value is that in the parentheses.

4.5. z-TRANSFORM BLOCK-DIAGRAM ALGEBRA

The procedure for determining the z-transfer function of sampled-data systems is complicated by the absence of samplers between elements, and it is impossible to obtain a direct analogy with the rules governing the reduction of continuous systems.

Let us initially consider two elements in cascade separated by a sampler as is illustrated in Figure 4.16. The z transform of the output from the second sampler, $C_1(z)$, is

$$C_1(z) = R(z)G_1(z). \tag{4.81}$$

The z transform of the output from the last sampler, in terms of $C_1(z)$, is

$$C_2(z) = C_1(z)G_2(z). \tag{4.82}$$

Substituting Eq. (4.82) into Eq. (4.81), we obtain

$$C_2(z) = R(z)G_1(z)G_2(z) \tag{4.83}$$

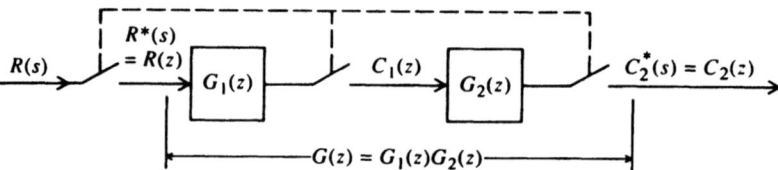

Figure 4.16 Cascaded elements separated by a synchronous sampler.

4.5. z-TRANSFORM BLOCK-DIAGRAM ALGEBRA

and conclude that the z transforms of cascaded elements separated by samplers equal the product of the z transforms of the individual cascaded elements.

Let us next consider the configuration where a sampler does not separate cascaded elements, as is illustrated in Figure 4.17. Notice that the second element is driven by the value $c_1(t)$ at the sampling instants, and also between the sampling instants. For this case, the z transform of the output is given by

$$C_2(z) = R(z)G_1G_2(z), \tag{4.84}$$

where $G_1G_2(z)$ represents the z transform of the cascaded combination corresponding to the transfer function $G_1(s)G_2(s)$. It is very important to emphasize here that $G_1G_2(z)$ is very different from $G_1(z)G_2(z)$. A very simple example will illustrate the difference implied by Eqs. (4.83) and (4.84). Consider that

$$G_1(s) = \frac{1}{s+a}$$

and

$$G_2(s) = \frac{1}{s+b}. \tag{4.85}$$

For the system illustrated in Figure 4.16,

$$G(z) = G_1(z)G_2(z).$$

From Table 4.1, the individual z transforms are

$$G_1(z) = \frac{z}{z - e^{-aT}}$$

and

$$G_2(z) = \frac{z}{z - e^{-bT}}.$$

Substituting these z transforms into Eq. (4.83), we obtain

$$G(z) = \frac{z^2}{(z - e^{-aT})(z - e^{-bT})}. \tag{4.86}$$

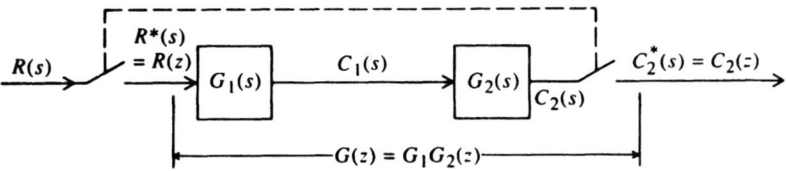

Figure 4.17 Cascaded continuous elements not separated by a sampler.

For the connection illustrated in Figure 4.17,

$$G(z) = G_1G_2(z). \tag{4.87}$$

The term $G_1G_2(z)$ corresponds to the z transform of $G_1(s)G_2(s)$, where

$$G_1(s)G_2(s) = \frac{1}{(s+a)(s+b)}. \tag{4.88}$$

By means of partial fraction expansion, Eq. (4.88) can be written as

$$G_1(s)G_2(s) = \frac{1}{b-a}\frac{1}{s+a} + \frac{1}{a-b}\frac{1}{s+b}. \tag{4.89}$$

The z transform of Eq. (4.89), $G_1G_2(z)$, can be obtained by inspection from Table 4.1 as follows:

$$G_1G_2(z) = \frac{1}{b-a}\frac{z}{z-e^{-aT}} + \frac{1}{a-b}\frac{z}{z-e^{-bT}}. \tag{4.90}$$

Observe the great difference between Eqs. (4.90) and (4.86).

Let us next determine the z transform of the output for several versions of sampled-data feedback control systems. We shall specifically derive $C(z)$ for the systems illustrated in Figures 4.18a, b, c, and d.

A. Sampler in Error Channel (Figure 4.18a)

The closed-loop transfer function can be derived in the following manner:

$$E(s) = R(s) - G(s)H(s)E^*(s). \tag{4.91}$$

Because the sampler is a linear device and superposition is valid, we can write

$$E^*(s) = R^*(s) - GH^*(s)E^*(s). \tag{4.92}$$

Solving Eq. (4.92) for $E^*(s)$, we obtain

$$E^*(s) = \frac{R^*(s)}{1 + GH^*(s)}. \tag{4.93}$$

Because

$$C(s) = E^*(s)G(s), \tag{4.94}$$

we obtain the value of $C(s)$ as

$$C(s) = \frac{R^*(s)G(s)}{1 + GH^*(s)}. \tag{4.95}$$

Taking the z transform of Eq. (4.95), the following results:

$$C(z) = \frac{R(z)G(z)}{1 + GH(z)}. \qquad (4.96)$$

Equation (4.96) gives the value of the output at the sampling instants for the configuration of Figure 4.18a.

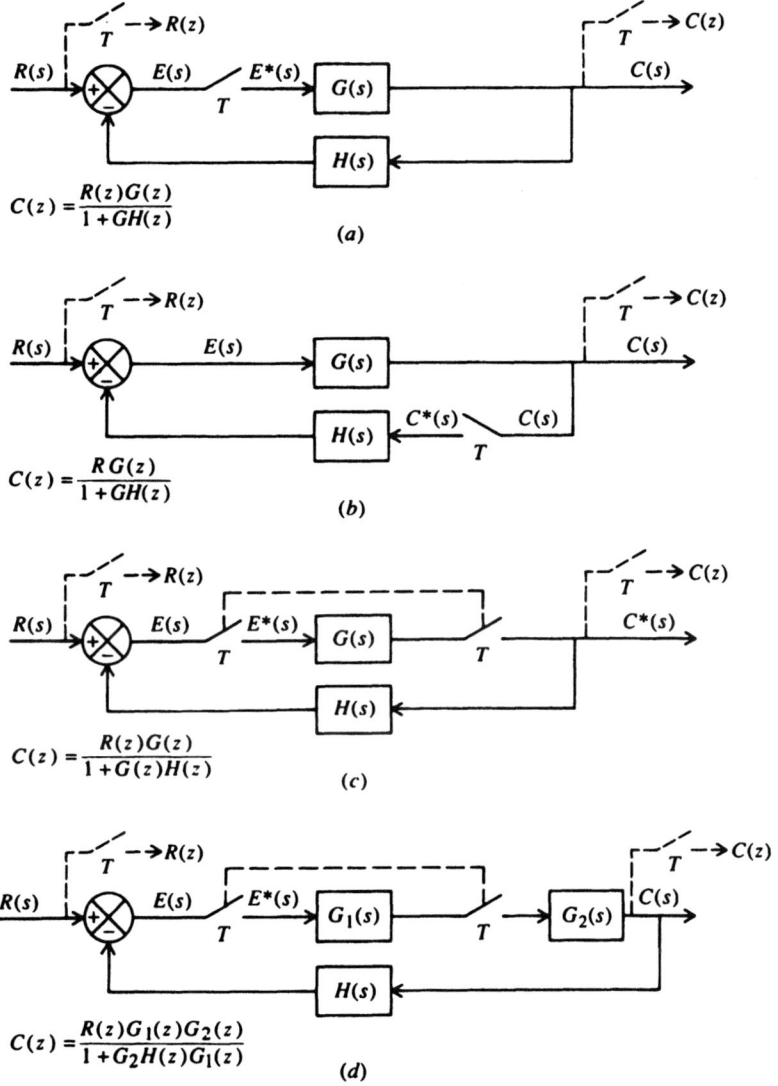

Figure 4.18 Various versions of sampled-data control systems. (a) Sampler in error channel. (b) Sampler in feedback loop. (c) Synchronous samplers in forward loop. (d) Synchronous samplers and cascaded elements in forward loop.

B. Sampler in Feedback Loop (Figure 4.18b)

The z transfer function of the output can be derived in the following manner:

$$E(s) = R(s) - H(s)C^*(s) \qquad (4.97)$$

and

$$C(s) = G(s)E(s) \qquad (4.98)$$

Substituting Eq. (4.97) into Eq. (4.98), we obtain

$$C(s) = R(s)G(s) - G(s)H(s)C^*(s). \qquad (4.99)$$

Taking the z transform of Eq. (4.99) the following results:

$$C(z) = RG(z) - GH(z)C(z). \qquad (4.100)$$

Simplifying, we obtain

$$C(z) = \frac{RG(z)}{1 + GH(z)}. \qquad (4.101)$$

Equation (4.101) gives the value of the output at the sampling instants for the configuration of Figure 4.18b.

C. Synchronous Samplers in Forward Loop (Figure 4.18c)

The z transfer function of the output can be derived in the following manner:

$$E(s) = R(s) - G^*(s)H(s)E^*(s). \qquad (4.102)$$

Because the sampler is a linear device, and superposition is valid, we can write

$$E^*(s) = R^*(s) - G^*(s)H^*(s)E^*(s). \qquad (4.103)$$

Solving Eq. (4.103) for $E^*(s)$, we obtain

$$E^*(s) = \frac{R^*(s)}{1 + G^*(s)H^*(s)}. \qquad (4.104)$$

In addition, because

$$C^*(s) = E^*(s)G^*(s), \qquad (4.105)$$

we obtain the relation between $C^*(s)$ and $R^*(s)$ as

$$C^*(s) = \frac{R^*(s)G^*(s)}{1 + G^*(s)H^*(s)}. \qquad (4.106)$$

Taking the z transform of Eq. (4.106), the following results:

$$C(z) = \frac{R(z)G(z)}{1 + G(z)H(z)}. \quad (4.107)$$

Equation (4.107) gives the value of the output at the sampling instants for the configuration of Figure 4.18c.

D. Synchronous Samplers and Cascaded Elements in Forward Loop (Figure 4.18d)

The z-transfer function of the output can be derived in the following manner:

$$E(s) = R(s) - G_2(s)H(s)G_1^*(s)E^*(s). \quad (4.108)$$

Because the sampler is a linear device and superposition is valid, we can write

$$E^*(s) = R^*(s) - G_2H^*(s)G_1^*(s)E^*(s). \quad (4.109)$$

Solving Eq. (4.109) for $E^*(s)$, we obtain

$$E^*(s) = \frac{R^*(s)}{1 + G_2H^*(s)G_1^*(s)}. \quad (4.110)$$

In addition, because

$$C(s) = E^*(s)G_1^*(s)G_2(s), \quad (4.111)$$

we obtain the value of $C(s)$ as

$$C(s) = \frac{R^*(s)G_1^*(s)G_2(s)}{1 + G_2H^*(s)G_1^*(s)}. \quad (4.112)$$

Taking the z transform of Eq. (4.112) the following results:

$$C(z) = \frac{R(z)G_1(z)G_2(z)}{1 + G_2H(z)G_1(z)}. \quad (4.113)$$

Equation (4.113) gives the value of the output at the sampling instants for the configuration of Figure 4.18d.

It is important to recognize that the concepts of signal-flow graphs developed in Sections 2.14–2.16[‡] for continuous systems can be extended to discrete systems with some modifications. In particular, Mason's theorem stated in Eq. (2.135)[‡] can be used for discrete systems with some modifications, based on the results shown in this section. The following rules should be used in adapting Mason's theorem for discrete systems:

1. If the input $R(s)$ is not separated from the first element in the forward part of the loop [e.g., $G(s)$] by a sampler, then the z transform of the input and the

242 DIGITAL CONTROL-SYSTEM ANALYSIS AND DESIGN

first element cannot be separated, and a term $RG(z)$ will result. Therefore, a transfer function of output/input cannot be obtained (e.g., Figure 4.18*b*, Eq. [4.101]).

2. For an element in the forward or feedback part of the loop to occur as a stand-alone transfer function in terms of z [e.g., $G(z)$], then that element must be separated from all other elements by a sampler at its input and output. For example, $G(s)$ in Figure 4.18*c* and $G_1(s)$ in Figure 4.18*d* satisfied this requirement.

3. If an element in the forward or feedback part of the loop is not separated from an adjacent element (or the input) by a sampler, then it is necessary to take the z transform of the combined transfer function of the two elements (or the element and the input). We saw examples of this in Figures 4.18*a* and *b* [$G(s)$ and $H(s)$], $G_2(s)$ and $H(s)$ in Figure 4.18*d*, and $R(s)$ and $G(s)$ in Figure 4.18*b* of this section.

It is left as an exercise to the reader to check the results of the four examples illustrated in this section with Mason's theorem using the three modifications stated.

4.6. CHARACTERISTIC RESPONSE OF A SAMPLER AND ZERO-ORDER HOLD COMBINATION

The tools developed so far have made it possible for us to determine the characteristic response of the sampler and zero-order hold combination illustrated in Figure 4.19 to various types of inputs. We shall apply the z-transform theory in order to determine its response to inputs of a unit step and ramp. We shall assume that the sampling time T is 1 sec and that the transfer function $G_2(s)$ is given by

$$G_2(s) = \frac{1}{s+1}. \qquad (4.114)$$

The z-transfer function of this open-loop system, corresponding to the transfer function $G_1(s)G_2(s)$, is denoted by $G_{12}(z)$. The zero-order hold transfer function $G_1(s)$ was derived previously, and is given by Eq. (4.19) as

$$G_1(s) = \frac{1 - e^{-Ts}}{s}. \qquad (4.115)$$

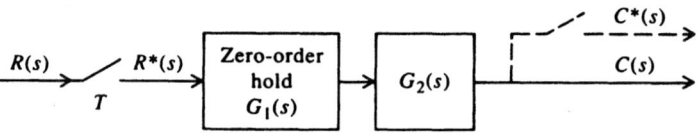

Figure 4.19 Sampler, zero-order hold and process $G_2(s)$ in cascade.

4.6. CHARACTERISTIC RESPONSE OF A SAMPLER AND ZERO-ORDER HOLD COMBINATION

Because $T = 1$, Eq. (4.115) reduces to

$$G_1(s) = \frac{1 - e^{-s}}{s}. \tag{4.116}$$

Therefore,

$$G_1(s)G_2(s) = \left(\frac{1 - e^{-s}}{s}\right)\left(\frac{1}{s+1}\right). \tag{4.117}$$

By means of a partial fraction expansion, Eq. (4.117) can be written as

$$G_1(s)G_2(s) = (1 - e^{-s})\left(\frac{1}{s} - \frac{1}{s+1}\right). \tag{4.118}$$

The z transform of Eq. (4.118) can be obtained from Table 4.1 by inspection. Its value is

$$G_{12}(z) = (1 - z^{-1})\left(\frac{z}{z-1} - \frac{z}{z-e^{-1}}\right). \tag{4.119}$$

This relation can be simplified to the following:

$$G_{12}(z) = \frac{0.63}{z - 0.37}. \tag{4.120}$$

For a given input, whose z transform is $R(z)$, the z transform of the output can be expressed as

$$C(z) = R(z)G_{12}(z). \tag{4.121}$$

The characteristic response to inputs of a unit step and ramp are computed as follows:

A. Unit Step Input

$$r(t) = U(t)$$

From Table 4.1,

$$R(z) = \frac{z}{z - 1}. \tag{4.122}$$

Substituting Eqs. (4.120) and (4.122) into Eq. (4.121), we obtain

$$C(z) = \frac{0.63z}{(z - 1)(z - 0.37)}. \tag{4.123}$$

244 DIGITAL CONTROL-SYSTEM ANALYSIS AND DESIGN

This can also be expressed as

$$C(z) = \frac{0.63z}{z^2 - 1.37z + 0.37}. \tag{4.124}$$

The inverse z transform can be obtained by expanding Eq. (4.124) into a power-series expansion in z and then taking the inverse z transform by utilizing Table 4.1. The expression for a power-series expansion in z is given by

$$C(z) = 0.63z^{-1} + 0.86z^{-2} + 0.95z^{-3} + 0.98z^{-4} + z^{-5} + z^{-6} + z^{-7} + \cdots. \tag{4.125}$$

By utilizing Table 4.1 the inverse z transform can be expressed as

$$\begin{aligned} c^*(t) = {} & 0.63\delta(t-1) + 0.86\delta(t-2) + 0.95\delta(t-3) + 0.98\delta(t-4) \\ & + \delta(t-5) + \delta(t-6) + \delta(t-7) + \cdots. \end{aligned} \tag{4.126}$$

The result is sketched in Figure 4.20.

B. Unit Ramp Input

$$r(t) = tU(t).$$

From Table 4.1.

$$R(z) = \frac{Tz}{(z-1)^2}. \tag{4.127}$$

Because $T = 1$, Eq. (4.127) reduces to

$$R(z) = \frac{z}{(z-1)^2}. \tag{4.128}$$

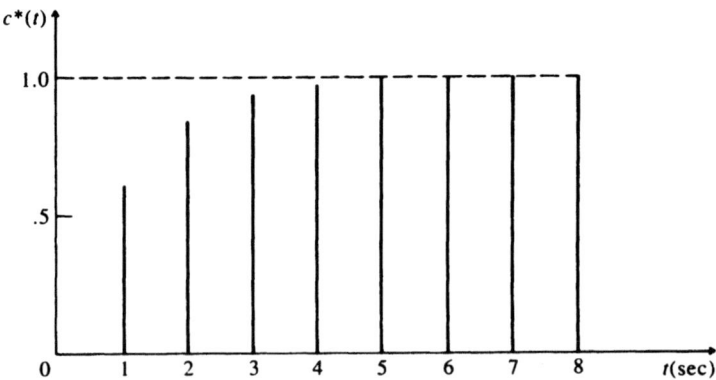

Figure 4.20 Output response of the circuit shown in Figure 4.19 to a unit step input.

Substituting Eqs. (4.120) and (4.128) into Eq. (4.121), we obtain

$$C(z) = \frac{0.63z}{(z-1)^2(z-0.37)}. \quad (4.129)$$

This can also be expressed as

$$C(z) = \frac{0.63z}{z^3 - 2.37z^2 + 1.74z - 0.37}. \quad (4.130)$$

The inverse z transform can be obtained by expanding Eq. (4.130) into a power-series expansion in z and then taking the inverse z transform by utilizing Table 4.1. The expression for a power-series expansion in z is given by

$$C(z) = 0.63z^{-2} + 1.5z^{-3} + 2.45z^{-4} + 3.43z^{-5} + 4.4z^{-6} + \cdots. \quad (4.131)$$

By utilizing Table 4.1 the inverse z transform can be expressed as

$$c^*(t) = 0.63\delta(t-2) + 1.5\delta(t-3) + 2.45\delta(t-4) + 3.43\delta(t-5) + 4.4\delta(t-6) + \cdots. \quad (4.132)$$

The result is sketched in Figure 4.21.

From the results illustrated in Figures 4.20 and 4.21, we conclude that the zero-order hold can respond with zero steady-state error to a unit step input but not to higher-order inputs such as a ramp. This is to be expected from our previous discussion of data extrapolators in Section 4.3.

4.7. STABILITY ANALYSIS USING THE NYQUIST DIAGRAM

The Nyquist stability criterion for continuous systems was described in Section 1.7. This section will show how to extend the Nyquist diagram for continuous systems to discrete systems. For the discrete case, we will show that the ideas are identical to the continuous case with the exception that the contours enclosing the unstable region of the z-plane are outside the unit circle as shown in Figure 4.22. This stability concept

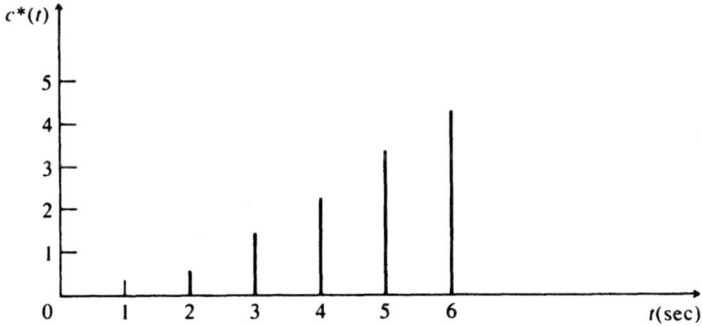

Figure 4.21 Output response of the circuit shown in Figure 4.19 to a unit-ramp input.

246 DIGITAL CONTROL-SYSTEM ANALYSIS AND DESIGN

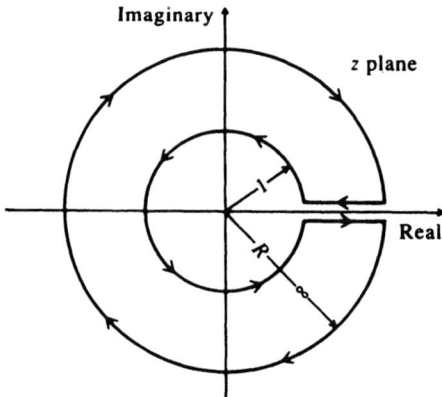

Figure 4.22 Contour used to enclose poles in the z-plane outside the unit circle.

was introduced in Section 4.4 (see Figures 4.12 and 4.13). The corresponding Nyquist stability criterion of Eq. (6.63)[‡] for continuous systems is modified for discrete systems to

$$N = Z - P \qquad (4.133)$$

where N is the number of clockwise encirclements of the $-1 + j0$ point for the $GH(z)$ or $G(z)H(z)$ contour for z taking values shown on Figure 4.22, P is the number of unstable poles of $GH(z)$ or $G(z)H(z)$, and Z is the number of unstable roots of the characteristic equation. Phase and gain margins defined for continuous systems in Eqs. (6.67)[‡] and (6.69)[‡] and Section 1.7 for the $G(s)H(s)$ plane, respectively, are exactly the same for discrete systems in the $GH(z)$ or the $G(z)H(z)$ plane.

It is interesting to observe from Figure 4.18 that the system transfer functions for the various versions of sampled-data control systems generally can be written as

$$H(z) = \frac{C(z)}{R(z)} = \frac{M(z)}{1 + N(z)}, \qquad (4.134)$$

where $N(z) = z$ transform of the open loop.

For the configuration of Figure 4.18a, where the sampler is in the error channel, $M(z) = G(z)$ and $N(z)$ corresponds to the z transform of the continuous Laplace transfer function $G(s)H(s)$. Therefore,

$$H(z) = \frac{C(z)}{R(z)} = \frac{G(z)}{1 + GH(z)}. \qquad (4.135)$$

For purposes of analyzing stability using the Nyquist diagram, we shall study the form given in Eq. (4.135). However, it should be emphasized that the technique used is applicable to any configuration of sampled-data control system.

The characteristic equation of a sampled-data control system, given by the denominator of Eq. (4.134) is

$$1 + N(z) = 0. \qquad (4.136)$$

For the case of the sampled-data control system having a sampler in the error channel, the characteristic equation is given by

$$1 + GH(z) = 0. \tag{4.137}$$

In accordance with the discussion of stability in Section 4.4, a closed contour having a radius of unity is examined in the z-plane. The area enclosed by the contour covers the entire region outside the unit circle. This is illustrated in Figure 4.22. The map of this contour, in the $1 + GH(z)$ plane, indicates by its enclosure of the origin the difference between the zeros and poles of this function in accordance with Cauchy's theorem. Specifically, the number of times the map of this contour encloses the origin equals the difference between the number of poles and zeros. This is the same as for the Nyquist diagrams applied to continuous systems which were discussed in Section 1.7.

It is important to realize that if $GH(z)$ is a stable function, then $1 + GH(z)$ will also be stable. In sampled-data systems, it is convenient to shift the imaginary axis to the point $(-1, 0)$ just as was done in the case of continuous systems. When the axis is shifted in this manner, it is necessary to consider only $GH(z)$ and study the enclosure of the point $(-1, 0)$ rather than the origin.

Practical sampled-data control systems usually contain one or more integrations among the continuous elements. From Table 4.1, it can be seen that the z transform of an integration results in a pole in the z-plane at the point $(1, 0)$. This pole would cause a discontinuity if the contour being mapped were exactly the one shown in Figure 4.22. Therefore, it is necessary to generate a small semicircular detour around the pole at $z = 1$ in order to establish a connection between the segments of the contour. This is illustrated in Figure 4.23. The detour is conventionally oriented in order to include the pole inside the unit circle. Due to the fact that practical values of $GH(z)$ vanish for infinite values of z, and because the segment of the contour along the real axis cancels itself in the limit, only the unit circle and its detours are usually mapped. This is illustrated in Figure 4.24.

We next apply the Nyquist stability criterion to the configuration illustrated in Figure 4.18a for two different values of the process, $G_2(s)$.

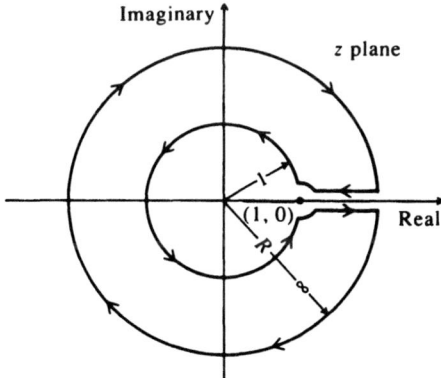

Figure 4.23 Contour used to enclose poles in the z-plane outside the unit circle when integrations are present.

248 DIGITAL CONTROL-SYSTEM ANALYSIS AND DESIGN

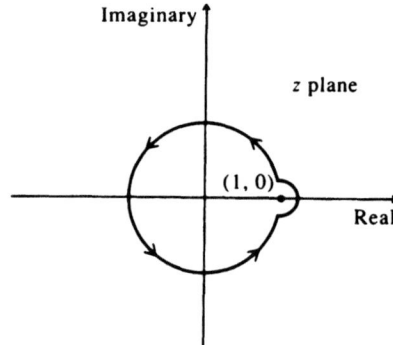

Figure 4.24 Reduced contour for mapping practical values of GH(z).

A. System of Figure 4.25

The sampled-data control system illustrated in Figure 4.25 contains a sampler in the error channel followed by a zero-order circuit and a pure integrator. We wish to determine whether the system is stable when $K = 1$, the maximum value of K before the system becomes unstable, and the system's response to a unit-step input. The sampling time will be assumed to be 1 sec.

1. System Stability (K = 1). The system stability can be determined by examining the locus of $GH(z)$ in the $GH(z)$ plane. The Laplace transfer function corresponding to $G_1(s)G_2(s)H(s)$ is

$$G_1(s)G_2(s)H(s) = \left(\frac{1-e^{-Ts}}{s}\right)\left(\frac{1}{s}\right)(1) = \frac{1-e^{-s}}{s^2}. \quad (4.138)$$

From Table 4.1, the z-transfer function $GH(z)$ corresponding to Eq. (4.138) is

$$GH(z) = \frac{1}{z-1}. \quad (4.139)$$

The locus of $GH(z)$ is shown in Figure 4.26. Because the locus does not enclose the $-1 + j0$ point, the Nyquist criterion is satisfied and the system is stable.

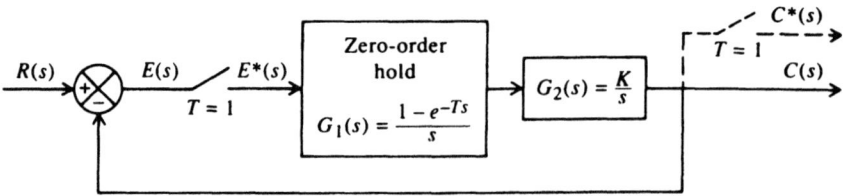

Figure 4.25 Sampled-data control system containing a zero-order hold and an integrator.

4.7. STABILITY ANALYSIS USING THE NYQUIST DIAGRAM

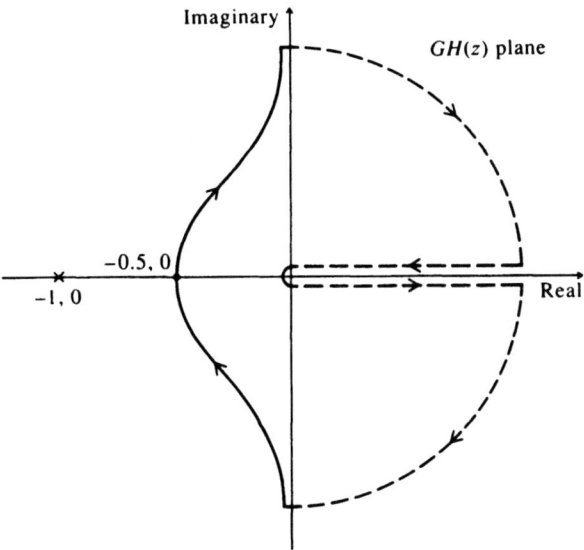

Figure 4.26 Locus in GH(z) plane of system shown in Figure 4.25.

2. Maximum Value of Gain. The maximum value of gain K that is theoretically possible before the system becomes unstable can also be found from the locus of $GH(z)$ in the $GH(z)$ plane. This is illustrated in Figure 4.26. It can easily be seen that if the gain were doubled, the locus would pass through the $-1, 0$ point. Therefore, the maximum value of K, before the system becomes unstable, is 2. Note that the gain margin is 6 dB.

3. System Response to a Unit Step. From Figure 4.18a, the z transform of the system output equals

$$C(z) = \frac{R(z)G(z)}{1 + GH(z)}. \quad (4.140)$$

From Table 4.1, the z transform of a unit step equals

$$R(z) = \frac{z}{z - 1}. \quad (4.141)$$

Substituting Eqs. (4.139) and (4.141) into Eq. (4.140), we obtain

$$C(z) = \frac{1}{z - 1}. \quad (4.142)$$

By expanding Eq. (4.142) into a power series in z, and then taking the inverse z transform, the output response can be plotted. The result is

$$C(z) = z^{-1} + z^{-2} + z^{-3} + z^{-4} + \cdots \quad (4.143)$$

or

$$c^*(t) = \delta(t-1) + \delta(t-2) + \delta(t-3) + \delta(t-4) + \cdots, \quad (4.144)$$

and is plotted in Figure 4.27. Therefore, the system behaves as a critically damped system in its response to a unit step input.

B. System of Figure 4.28

Let us next consider the sampled-data control system illustrated in Figure 4.28. It contains a sampler in the error channel followed by a zero-order hold circuit and a second-order process. For this configuration, we wish to determine whether the system is stable when $K = 1$, the maximum value of K before the system becomes unstable, and the system's response to a unit step input. The sampling rate will be assumed to be 1 sec.

1. System Stability ($K = 1$). The solution to this problem is similar to the previous problem. The Laplace transformation function $G_1(s)G_2(s)H(s)$ equals

$$G_1(s)G_2(s)H(s) = \left(\frac{1-e^{-Ts}}{s}\right)\left(\frac{1}{s(s+2)}\right)(1) = \frac{1-e^{-s}}{s^2(s+2)}. \quad (4.145)$$

By utilizing a partial fraction expansion of the denominator and Table 4.1, the z-transfer function $GH(z)$ corresponding to Eq. (4.145) can be obtained. Its value is

$$GH(z) = \frac{0.284(z+0.524)}{(z-1)(z-0.136)}. \quad (4.146)$$

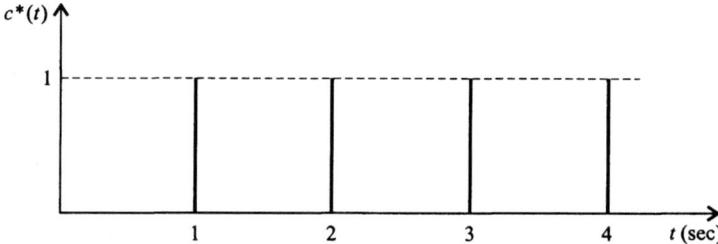

Figure 4.27 Output response of the system shown in Figure 4.25 to a unit step input.

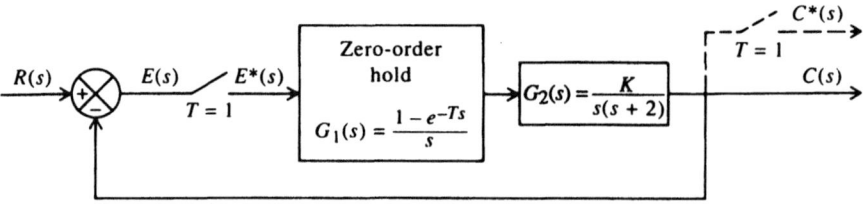

Figure 4.28 Sampled-data control system containing a zero-order hold and a second-order process.

The locus of $GH(z)$ is shown in Figure 4.29. Because the locus does not enclose the $-1 + j0$ point, the Nyquist criterion is satisfied and the system is stable.

2. Maximum Value of Gain. The maximum value of gain K that is theoretically possible before the system becomes unstable can be found from the locus of $GH(z)$ in the $GH(z)$ plane. This is illustrated in Figure 4.29. It can easily be seen that if the gain were increased by a factor of $1/0.170 = 5.88$, the locus would pass through the $-1, 0$ point. Therefore, the maximum value of K, before the system becomes unstable, is 5.88. Note that the gain margin is 15.39 dB.

3. System Response to a Unit Step. The system response to a unit step can easily be determined by substituting Eqs. (4.141) and (4.146) into Eq. (4.140). The result is

$$C(z) = \frac{0.284z^2 + 0.148z}{z^3 - 1.85z^2 + 1.13z - 0.284}. \tag{4.147}$$

By expanding Eq. (4.147) into a power series in z and then taking the inverse z transform, the output response can be plotted. The result is

$$\begin{aligned} C(z) = {} & 0.284z^{-1} + 0.674z^{-2} + 0.92z^{-3} + 1.02z^{-4} \\ & + 1.03z^{-5} + 1.01z^{-6} + z^{-7} + z^{-8} + z^{-9} + z^{-10} + \cdots \end{aligned} \tag{4.148}$$

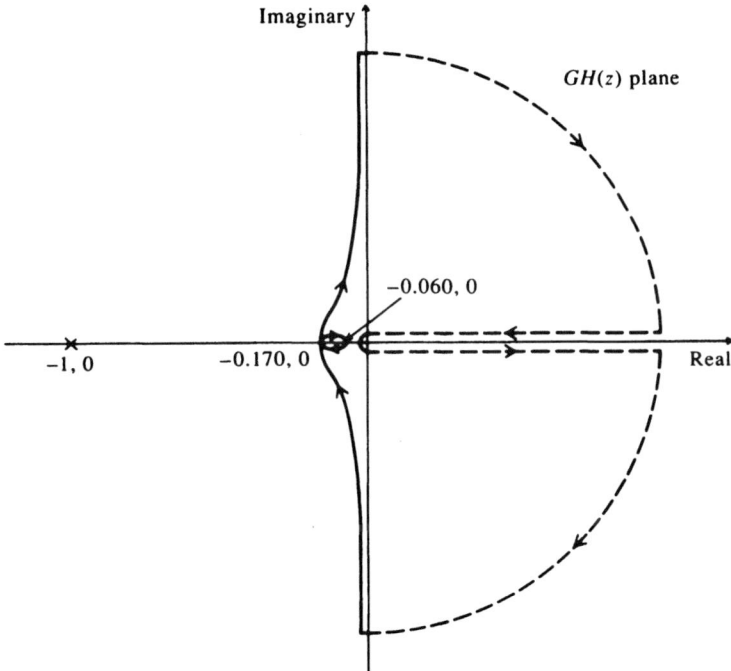

Figure 4.29 Locus in the $GH(z)$ plane.

or

$$c^*(t) = 0.284\delta(t-1) + 0.674\delta(t-2) + 0.92\delta(t-3)$$
$$+ 1.02\delta(t-4) + 1.03\delta(t-5) + 1.01\delta(t-6) \quad (4.149)$$
$$+ \delta(t-7) + \delta(t-8) + \delta(t-9) + \delta(t-10) + \cdots,$$

and is plotted in Figure 4.30. Therefore, the system has an overshoot of only 3% in its response to the unit step input, and has zero steady-state error.

4.8. STABILITY DETERMINATION USING MATHEMATICAL TESTS

Stability of sampled-data control systems can also be determined through the application of several mathematical tests. This section will present the modified Routh–Hurwitz criterion, the Schur–Cohn stability criterion, and Jury's stability criterion.

A. The Modified Routh–Hurwitz Criterion

Stability of sampled-data control systems cannot be determined by means of a direct application of the Routh–Hurwitz criterion. However, it is possible to apply a transformation to the characteristic equation z that will transform the region outside the unit circle in the z-plane to the right half of an auxiliary plane, and the region inside the unit circle to the left half of this plane. This transformation, known as the bilinear transformation, has been applied to discrete control systems [3–5, 8]. The auxiliary plane, denoted as the w-plane (see Figure 4.13), is defined by the following relationship:

$$z = \frac{w+1}{w-1} \quad (4.150)$$

or

$$w = \frac{z+1}{z-1}. \quad (4.151)$$

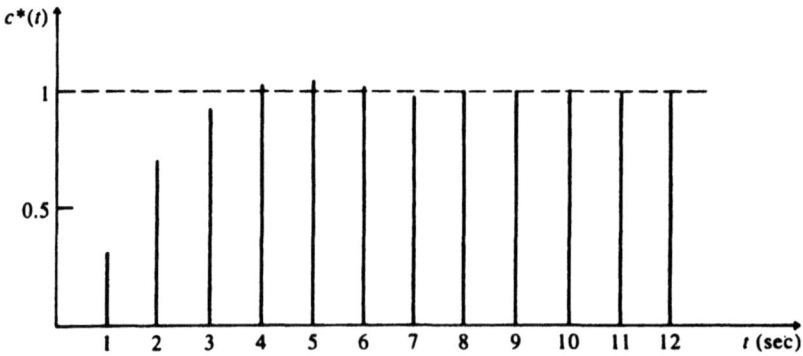

Figure 4.30 Output response of the system shown in Figure 4.28 to a unit step input.

If the Routh–Hurwitz critereon is applied to the characteristic equation, that is, subjected to the transformation of Eq. (4.150), conditions can be found that cause the roots of the transformed equation to lie in the left half of the w-plane. Therefore, it is possible to determine if the roots of the characteristic equation lie within the unit circle of the z-plane using this transformation. The procedure is illustrated next for the system having the form shown in Figure 4.18a.

As an example, let us consider the following closed-loop system transfer function:

$$H(z) = \frac{C(z)}{R(z)} = \frac{0.092(z + 0.718)}{z^2 - 1.2z + 0.32}, \quad (4.152)$$

whose characteristic equation is given by

$$z^2 - 1.2z + 0.32 = 0. \quad (4.153)$$

Using the bilinear transformation of Eq. (4.150) on Eq. (4.153), we obtain

$$\left(\frac{w+1}{w-1}\right)^2 - 1.2\left(\frac{w+1}{w-1}\right) + 0.32 = 0. \quad (4.154)$$

Simplifying this expression, we obtain

$$w^2 + 11.3w + 21 = 0. \quad (4.155)$$

Application of the Routh–Hurwitz criterion to Eq. (4.155) yields the following array (see Section 6.3‡):

$$\begin{array}{c|cc} w^2 & 1 & 21 \\ w^1 & 11.3 & 0. \\ w^0 & 21 & \end{array} \quad (4.156)$$

Because there are no sign inversions in the left column of the resulting array, this system has no poles in the right half of the w-plane, or outside the unit circle of the z-plane, and is stable. As a check, we can factor the second-order characteristic equation of Eq. (4.153) and determine that the closed-loop roots in the z-plane are located within the unit circle at $z = 0.4$ and 0.8.

B. The Schur–Cohn Stability Criterion

Although the modified Routh–Hurwitz criterion is simple to apply in principle, it is quite laborious for higher-order systems. For these cases, the Schur–Cohn [9, 10] stability criterion and Jury's [2, 10, 11] stability criterion are quite useful.

The Schur–Cohn stability criterion states that a sampled data control system is stable if the sequence of Schur–Cohn determinants, $1, \Delta_1, \Delta_2, \ldots, \Delta_j, \ldots, \Delta_n$, has n variations in sign. Another way of stating this particular stability criterion is that

$$\Delta_j < 0 \quad \text{for } j \text{ odd} \quad (4.157)$$

and

$$\Delta_j > 0 \quad \text{for } j \text{ even} \tag{4.158}$$

in order for the system to be stable.

Assuming that the characteristic equation of the linear sampled data control system is given by

$$a_n z^n + a_{n-1} z^{n-1} + a_{n-2} z^{n-2} + a_{n-3} z^{n-3} + \cdots + a_3 z^3 + a_2 z^2 + a_1 z + a_0 = 0. \tag{4.159}$$

Then the jth Schur–Cohn determinant is defined as

$$\Delta_j = \begin{vmatrix} a_0 & 0 & 0 & 0 & \cdots & 0 & a_n & a_{n-1} & a_{n-2} & a_{n-3} & \cdots & a_{n-j+1} \\ a_1 & a_0 & 0 & 0 & \cdots & 0 & 0 & a_n & a_{n-1} & a_{n-2} & \cdots & a_{n-j+2} \\ a_2 & a_1 & a_0 & 0 & \cdots & 0 & 0 & 0 & a_n & a_{n-1} & \cdots & a_{n-j+3} \\ a_3 & a_2 & a_1 & a_0 & \cdots & 0 & 0 & 0 & 0 & a_n & \cdots & a_{n-j+4} \\ \vdots & \vdots & \vdots & \vdots & & \vdots & \vdots & \vdots & \vdots & \vdots & & \vdots \\ a_{j-1} & a_{j-2} & a_{j-3} & a_{j-4} & \cdots & a_0 & 0 & 0 & 0 & 0 & & a_n \\ \hline \bar{a}_n & 0 & 0 & 0 & \cdots & 0 & \bar{a}_0 & \bar{a}_1 & \bar{a}_2 & \bar{a}_3 & \cdots & \bar{a}_{j-1} \\ \bar{a}_{n-1} & \bar{a}_n & 0 & 0 & \cdots & 0 & 0 & \bar{a}_0 & \bar{a}_1 & \bar{a}_2 & \cdots & \bar{a}_{j-2} \\ \bar{a}_{n-2} & \bar{a}_{n-1} & \bar{a}_n & 0 & \cdots & 0 & 0 & 0 & \bar{a}_0 & \bar{a}_1 & \cdots & \bar{a}_{j-3} \\ \bar{a}_{n-3} & \bar{a}_{n-2} & \bar{a}_{n-1} & \bar{a}_n & \cdots & 0 & 0 & 0 & 0 & \bar{a}_0 & \cdots & \bar{a}_{j-4} \\ \vdots & \vdots & \vdots & \vdots & & \vdots & \vdots & \vdots & \vdots & \vdots & & \vdots \\ \bar{a}_{n-j+1} & \bar{a}_{n-j+2} & \bar{a}_{n-j+3} & \bar{a}_{n-j+4} & \cdots & \bar{a}_n & 0 & 0 & 0 & 0 & \cdots & \bar{a}_0 \end{vmatrix}, \tag{4.160}$$

where $j = 1, 2, 3, 4, \ldots, n$ and $\bar{a}_j$ is the conjugate of a_j. Assuming that all of the coefficients of the characteristic equation are real, then the determinant Δ_j is symmetrical with respect to its principal diagonal.

It is interesting to observe that if the conditions of the Schur–Cohn stability criterion are not satisfied, then we merely know that the characteristic equation has at least one root which lies outside the unit circle in the z-plane. Unlike the Routh–Hurwitz criterion, this criterion does not indicate the number of roots which lie on or outside the unit circle.

The Schur–Cohn stability criterion will next be applied to a simple system of the form shown in Figure 4.18a, where

$$H(z) = \frac{C(z)}{R(z)} = \frac{2(z+1)}{z^2 - 2.5z + 1}, \tag{4.161}$$

whose characteristic equation is given by

$$z^2 - 2.5z + 1 = 0. \tag{4.162}$$

The Schur–Cohn determinant Δ_1 of this equation is given by

$$\Delta_1 = \begin{vmatrix} a_0 & a_n \\ \bar{a}_n & \bar{a}_0 \end{vmatrix} = \begin{vmatrix} 1 & 1 \\ 1 & 1 \end{vmatrix} = 0. \tag{4.163}$$

Therefore, the sequence of Schur–Cohn determinants for this simple case is 1, 0.

Because Δ_1 is not less than zero, this system is unstable based on the Schur–Cohn stability criterion. As a check, we can factor the second-order characteristic equation of Eq. (4.162) and determine that the closed-loop roots in the z-plane are located at $z = 0.5$ and 2. The root at 2 is outside the unit circle, and verifies that this system is unstable.

C. Jury's Stability Criterion

The stability criterion devised by Jury [2, 10, 11] is much simpler to apply than the Schur–Cohn stability criterion. The first step in Jury's stability criterion is the formation of the following table from the characteristic equation:

$$a_0 z^n + a_1 z^{n-1} + \cdots + a_n = 0$$

Row	z^0	z^1	z^2	$\cdots$	z^{n-j}	$\cdots$	z^{n-1}	z^n
1	a_0	a_1	a_2	$\cdots$	a_{n-j}	$\cdots$	a_{n-1}	a_n
2	a_n	a_{n-1}	a_{n-2}	$\cdots$	a_j	$\cdots$	a_1	a_0
3	b_0	b_1	b_2	$\cdots$	b_{n-j}	$\cdots$	b_{n-1}	.
4	b_{n-1}	b_{n-2}	b_{n-3}	$\cdots$	b_j	$\cdots$	b_0	.
5	c_0	c_1	c_2	$\cdots$	c_{n-2}			
6	c_{n-2}	c_{n-3}	c_{n-4}	$\cdots$	c_0			

Note that the elements of even rows consist of the coefficients of odd rows written in reverse order. The values in the third row are formed from the second-order determinants using the first column of the first two rows with each of the other columns from these rows, and starting from the right and dividing by the coefficient a_0. Therefore, the terms b_0, b_1, and b_j, are given by

$$b_0 = a_0 - \frac{a_n}{a_0} a_n,$$

$$b_1 = a_1 - \frac{a_n}{a_0} a_{n-1},$$

$$b_j = a_j - \frac{a_n}{a_0} a_{n-j}.$$

The fourth row is obtained by reversing the elements in the third row, and the process is repeated. The elements in the fifth row are given by

$$c_j = b_j - \frac{b_{n-1}}{b_0} b_{n-1-j}.$$

Jury's stability criterion requires that all the terms of the odd rows in the left column be positive for the system to be stable (has all roots inside the unit circle of the z plane). This is a sufficient and necessary condition for all the roots of the characteristic equation to lie inside the unit circle in the z-plane.

As an example for applying Jury's stability criterion, let us consider the following characteristic equation:

$$z^3 - 1.1z^2 + 0.01z + 0.4 = 0. \tag{4.164}$$

The resulting Jury array is given by

Row

1	1	-1.1	0.01	0.4
2	0.4	0.01	-1.1	1
3	$b_0 = \dfrac{1}{1}\begin{vmatrix} 1 & 0.4 \\ 0.4 & 1 \end{vmatrix} = 0.84$	$b_1 = \dfrac{1}{1}\begin{vmatrix} 1 & 0.01 \\ 0.4 & -1.1 \end{vmatrix} = -1.104$	$b_2 = \dfrac{1}{1}\begin{vmatrix} 1 & -1.1 \\ 0.4 & 0.01 \end{vmatrix} = 0.45$	
4	0.45	-1.104	0.84	
5	$c_0 = \dfrac{1}{0.84}\begin{vmatrix} 0.84 & 0.45 \\ 0.45 & 0.84 \end{vmatrix} = 0.599$	$c_1 = \dfrac{1}{0.84}\begin{vmatrix} 0.84 & -1.104 \\ 0.45 & -1.104 \end{vmatrix} = -0.513$		
6	-0.513	0.599		
7	$d_0 = \dfrac{1}{0.599}\begin{vmatrix} 0.599 & -0.513 \\ -0.513 & 0.599 \end{vmatrix} = 0.1595$			

Because the terms in the left column of the odd rows are positive, the system having the characteristic equation of Eq. (4.164) is stable. As a check, we can determine that the roots of this third-order characteristic equation are at -0.4973 and $0.7987 \pm 0.408j$. Because all of the roots are inside the unit circle in the z-plane, the system is stable.

The Routh–Hurwitz criterion mathematical test does not show how to improve design. It merely tells the control-system engineer whether the system is stable or unstable. The same is true for the three mathematical tests presented in this section for discrete systems. The modified Routh–Hurwitz criterion, the Schur–Cohn stability criterion, and Jury's stability criterion tell the control-system engineer whether a discrete control system is stable or not, but give no information on the location of the roots in the z-plane. For example, if any of these three mathematical tests indicate that the system is stable, we do not know how close the roots are to the unit circle in the z-plane. If the mathematical tests indicate that the system is unstable, we do not know how distant the roots are from the unit circle in the z-plane. The main attributes of these mathematical tests are to serve as a check on other design methods and to give a quick answer to the question of stability.

4.9. STABILITY ANALYSIS AND DESIGN USING THE BODE DIAGRAM

The usefulness of the Bode diagram, as presented in Chapter 1 for continuous systems, is lost for discrete systems in the z-plane due to the relationship between z and s:

$$z = e^{Ts}. \tag{4.165}$$

However, we can extend the usefulness of the Bode diagram for continuous systems to discrete systems by means of the bilinear transformation, which was presented in Section 4.8, Eq. (4.151),

$$w = \frac{z+1}{z-1}, \qquad (4.166)$$

and used for the modified Routh–Hurwitz criterion. Before we can use this transformation constructively for the Bode diagram, a modification must be made to Eq. (4.166), because this definition of the bilinear transformation lacks the desirable property that as the sampling time T approaches zero, we want w to approach s. This limitation can be overcome by defining a modified bilinear transformation [2–5], where

$$w = \frac{2}{T}\frac{z-1}{z+1}. \qquad (4.167)$$

For example, with the definition of Eq. (4.167) the relationship between w and s, as T approaches zero, is given by

$$w]_{T \to 0} = \lim_{T \to 0} \left[\frac{2}{T}\frac{z-1}{z+1} \right] = \frac{2}{T} \lim_{T \to 0} \left[\frac{e^{sT}-1}{e^{sT}+1} \right]$$

$$= \frac{2}{T} \lim_{T \to 0} \left[\frac{sT + (sT)^2/2! + \cdots}{2 + sT + (sT)^2/2! + \cdots} \right]$$

$$w]_{T \to 0} \to s. \qquad (9.168)$$

Therefore, the definition of Eq. (4.167) has the desirable property that w approaches s as T approaches zero. In this section on the extension of the Bode diagram to discrete systems, we will use the modified bilinear transformation defined by Eq. (4.167).

By means of the z transformation and the w transformation, the primary strip of the left half of the s-plane shown in Figure 4.13 is mapped inside of the z-plane's unit circle and then mapped into the left half of the w-plane. Therefore, as s varies from 0 to $j\omega_s/2$ along the $j\omega$ axis in the s-plane, w varies from zero to infinity along the positive jv axis in the w-plane. We designate the fictitious frequency on the w-plane to be v, where

$$w = jv. \qquad (4.169)$$

After we make the transformation of $G(z)$ to $G(w)$, we can then treat $G(w)$ as a conventional transfer function in w. By replacing w by jv, we can then use the conventional Bode diagram for analyzing the transfer function in terms of w. Therefore, all of the straight-line approximation magnitude plots we used to draw the Bode diagram for continuous systems in Chapter 1 and Chapter 2 are adaptable to discrete systems in the w-plane.

We will illustrate the approach through an example. Let us consider the system shown in Figure 4.31, which contains a zero-order hold and a process which is

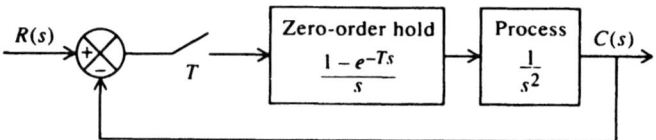

Figure 4.31 A sampled-data control system containing a zero-order hold and a process which is a double integration.

represented by a double integration. The resulting open-loop transfer function is given by

$$G(s) = \frac{1-e^{-Ts}}{s}\frac{1}{s^2}. \quad (4.170)$$

Its z transform is given by

$$G(z) = (1-z^{-1})Z(1/s^3). \quad (4.171)$$

From Table 4.1, we find that the z transform of $1/s^3$ is given by

$$Z(1/s^3) = \frac{T^2}{2}\frac{z(z+1)}{(z-1)^3}. \quad (4.172)$$

Substituting Eq. (4.172) into Eq. (4.171), we obtain the following z transform:

$$G(z) = \left(\frac{(z-1)}{z}\right)\left(\frac{T^2}{2}\frac{z(z+1)}{(z-1)^3}\right). \quad (4.173)$$

Simplification of Eq. (4.173) results in the following:

$$G(z) = \frac{T^2}{2}\frac{(z+1)}{(z-1)^2}. \quad (4.174)$$

To find the w transform of Eq. (4.174) we solve Eq. (4.167) for z in terms of w

$$z = \frac{1+(T/2)w}{1-(T/2)w} \quad (4.175)$$

and substitute it into Eq. (4.174)

$$G(w) = (T^2/2)\left[\frac{[1+(T/2)w]/[1-(T/2)w]+1}{([1+(T/2)w]/[1-(T/2)w]-1)^2}\right] \quad (4.176)$$

Simplifying Eq. (4.176), we obtain

$$G(w) = \frac{(1-w/(2/T))}{w^2}. \quad (4.177)$$

4.9. STABILITY ANALYSIS AND DESIGN USING THE BODE DIAGRAM

For this example, let us assume that the sampling time T is 0.26 sec. Therefore, Eq. (4.177) reduces to

$$G(w) = \frac{(1 - w/7.69)}{w^2}. \tag{4.178}$$

In terms of the frequency v, where $w = jv$, Eq. (4.178) can be written as

$$G(jv) = \frac{(1 - jv/7.69)}{(jv)^2}. \tag{4.179}$$

There are several interesting things to observe in comparing Eq. (4.178) [or (4.179)] with the process to be controlled:

$$G(s) = 1/s^2. \tag{4.180}$$

The gains of both transfer functions are the same and they will be in all cases. In addition, the denominators are identical in this case, although this may not be true in all cases. However, as T approaches zero, the denominators will be identical in all cases. The zero term in the numerator in the right half-plane is due to the sample and zero-order hold, and is a function of the sampling rate T. It is important to recognize that although this zero term provides a negative phase contribution (similar to a pole's effect) because it is in the right half-plane, it does contribute a + 20 dB/decade slope (at frequencies greater than its break frequency) to the magnitude plot.

The resulting Bode diagram for Eq. (4.179) is shown in Figures 4.32a and b. It has a gain crossover frequency of 1.017 rad/sec and is unstable with a phase margin of −7.536 degrees. To stabilize this system, we add a phase-lead network $G_D(w)$, as shown in Figure 4.33. The design specifications for this system are a gain crossover frequency of 7 rad/sec, a phase margin of approximately 35 degrees, and a gain margin of approximately 3 dB. The design procedure is to add a zero at 1 rad/sec prior to the crossover frequency of 7 rad/sec, because the initial slope of the Bode diagram in Figure 4.31a is −40 dB decade. (Recall Bode's first theorem in Section 2.6). The pole due to the phase-lead network is placed at a frequency much greater than the crossover of 7 rad/sec. Analysis of Figure 4.32a, indicates that the gain is −31.18 dB at the desired crossover frequency of 7 rad/sec (see Figure 4.32a). Therefore, the gain of the system must be increased by 31.18 dB to achieve a crossover frequency of the compensated system at 7 rad/sec. Through trial and error, it was found that by placing the pole of the phase-lead network at 100 rad/sec, the resulting phase margin is 35.33 degrees, and the gain margin is 3.399 dB (phase crossover frequency occurs at 25.72 rad/sec). These are very close to the design specifications and are acceptable. Of the 31.18 dB gain increase, the phase-lead network addition provides + 17 dB at 7 rad/sec. Therefore, the gain has to be increased by 14.18 dB, or 5.14. The resulting attenuation of 0.01 due to the phase-lead network can be made up for by increasing the system's amplitude gain by 100.

The resulting transfer function of the phase-lead network is given by

$$G_D(jv) = 5.14 \frac{(1 + jv)}{(1 + 0.01jv)}. \tag{4.181}$$

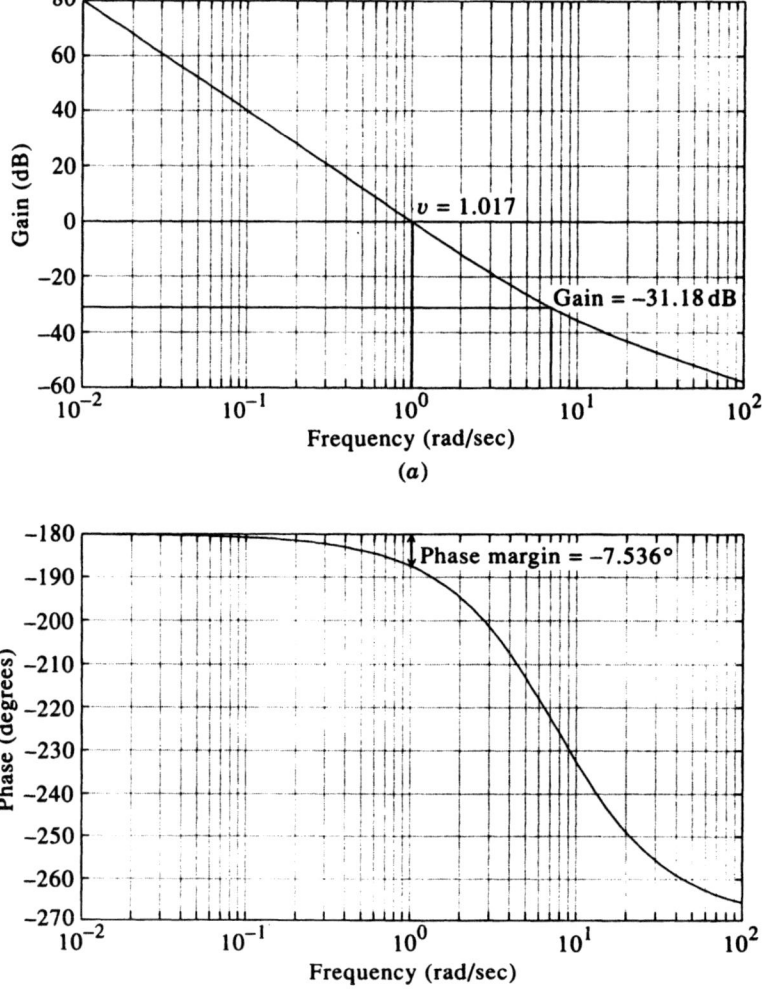

Figure 4.32 Bode diagram for system shown in Figure 4.31.

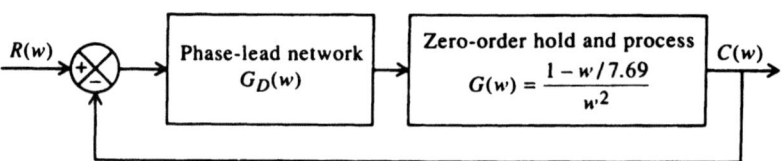

Figure 4.33 A phase-lead network added to stabilize the system of Figure 4.31.

The resulting transfer function of the compensated system is given by

$$G_D(jv)G(jv) = 5.14 \frac{(1+jv)}{(1+0.01jv)} \frac{(1-jv/7.69)}{(jv)^2} \qquad (4.182)$$

and the Bode diagram is shown in Figures 4.34a and b. Figures 4.32 and 4.34 were obtained using MATLAB and are contained in the M-files that are part of my *Advanced Modern Control System Theory and Design* (AMCSTD) Toolbox which can be retrieved free from The MathWorks, Inc. anonymous FTP server at ftp://ftp.mathworks.com/pub/books/advshinners.

In conclusion, we see from this section that the w transform maps the inside of the unit circle in the z-plane into the left-half of the w-plane. The result is that the s-plane and the w-plane are similar over the regions of interest. Therefore, the conventional straight-line approximations to the magnitude curve, and the concepts of phase and gain margins are valid for the Bode diagram for discrete systems when we apply the w transform.

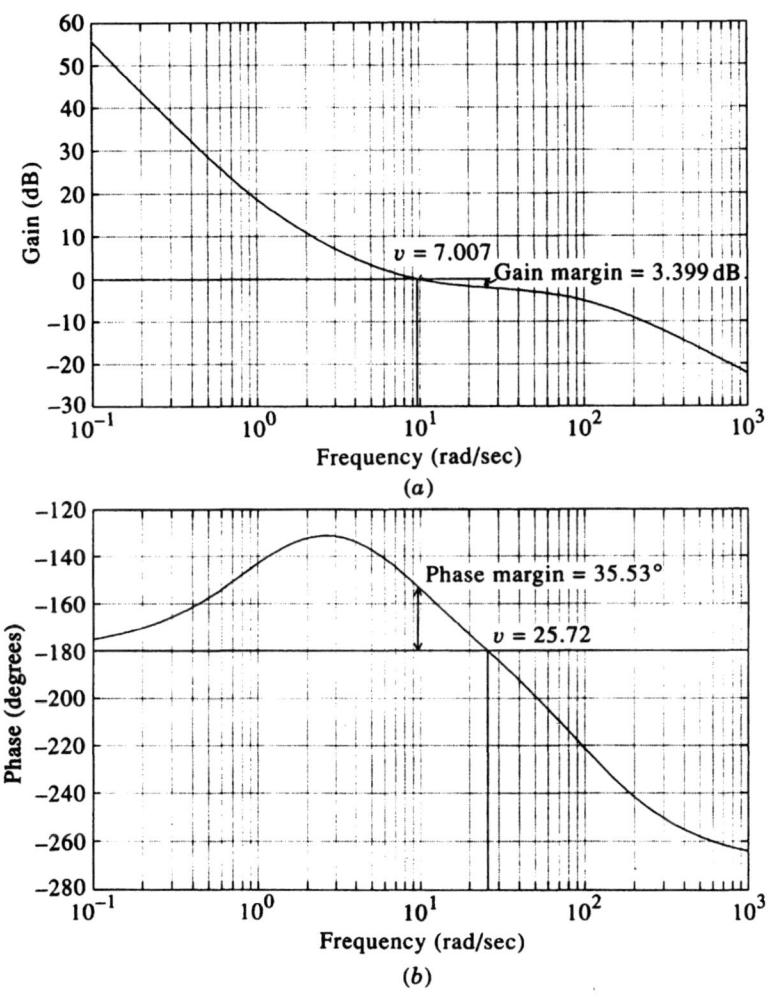

Figure 4.34 Bode diagram of compensated system of Figure 4.33.

262 DIGITAL CONTROL-SYSTEM ANALYSIS AND DESIGN

4.10. STABILITY ANALYSIS AND DESIGN USING THE ROOT-LOCUS DIAGRAM

We will extend the root-locus concept, developed for continuous systems in Section 1.7 and Section 2.9, to sampled-data systems in this section. Let us consider the sampled-data system of Figure 4.35. From the discussion of z-transform block-diagram algebra in Section 4.5, we recognize that its transfer function is given by the following:

$$\frac{C(z)}{R(z)} = \frac{KG(z)}{1 + KGH(z)}. \tag{4.183}$$

The characteristic equation of this system is given by

$$1 + KGH(z) = 0. \tag{4.184}$$

Equation (4.184) is analogous to the characteristic equation found in the s-plane root locus. The root locus is a graphical technique for determining the roots of the closed-loop characteristic equation of a system in the s-plane as a function of the static gain. We can extend that definition to sampled-data systems by replacing the words "s-plane" with "z-plane." Because the characteristic equation of sampled-data systems in the z-plane [Eq. (4.184)] has the same form as characteristic equations of continuous systems in the s-plane, the rules for drawing the root locus shown in Section 6.14‡ for the s-plane are exactly the same for the z-plane. For example, rule 4 for determining portions of the real axis that are part of the root locus, rules 5 and 6 for asymptotic construction, and rule 7 for the points of breakaway and break-in are all the same as those developed for the s-plane.

Although the rules of construction of the root locus in the s- and z-planes are the same, differences lie in interpreting the results. For example, the stable region in the z-plane is in the interior of the unit circle, while the stable region in the s-plane is in the left half-plane. Therefore, the pole locations have different interpretations in the s- and z-planes.

We will illustrate the approach of using the root locus for the analysis and design of a sampled-data system through an example. Let us assume that the sampling period T is held constant and the gain K varies. The root locus may also be plotted as a function of the sampling period T with K held constant. This case is much more

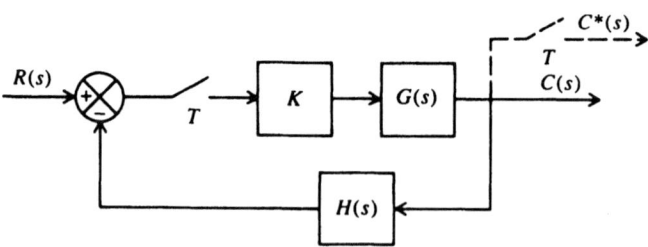

Figure 4.35 A sampled-data control system with a sampler in the error path.

4.10. STABILITY ANALYSIS AND DESIGN USING THE ROOT-LOCUS DIAGRAM

complex and beyond the scope of this chapter. This subject is treated in books dedicated to digital control systems such as References 2, 3, and 12.

Let us analyze the negative-feedback sampled-data system illustrated in Figure 4.36. It contains an amplifier having a gain K, a zero-order hold and a process whose transfer function is $1/[s(s+0.05)]$. The z transform of this zero-order hold and process is given by:

$$G(z) = Z\left\{\left(\frac{1-e^{-Ts}}{s^2}\right)\left(\frac{1}{s+0.05}\right)\right\}. \quad (4.185)$$

We will assume in this problem that the sampling time T is 1.43 seconds. Therefore, Eq. (4.185) reduces to

$$G(z) = \frac{(z+0.98)}{(z-1)(z-0.93)}, \quad T = 1.43. \quad (4.186)$$

We wish to plot the root locus of the open-loop transfer function given by

$$KG(z) = \frac{K(z+0.98)}{(z-1)(z-0.93)}. \quad (4.187)$$

The rules of construction from Section 6.14‡ that are needed to plot the root locus are as follows

Rule 1. This is a second-order system and, therefore, there will be two loci.

Rule 2. The loci start ($K=0$) at the two poles located at $z=1$ and $z=0.93$. One locus terminates ($K=\infty$) at the finite zero at $z=-0.98$, and the second locus terminates at a zero at infinity.

Rule 3. Complex portions occur as complex-conjugate pairs.

Rule 4. Portions of the real axis are part of the root locus if an odd number of poles and zeros exist to the right. Therefore, the real axis from -0.98 to $-\infty$ is part of the root locus. In adition, the real axis from 0.93 to 1 has to be part of the root locus.

Rules 5 and 6. The asymptotic rules are not needed in this problem.

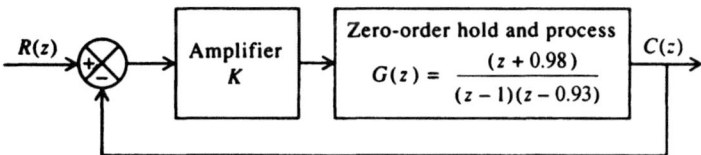

Figure 4.36 A sampled-data control system containing an amplifier, zero-order hold and a process whose transfer function is $1/[s(s+0.05)]$.

Rule 7. The points of breakaway and break-in are determined from the characteristic equation as follows:

$$1 + KG(z) = 1 + \frac{K(z+0.98)}{(z-1)(z-0.93)} = 0. \tag{4.188}$$

We let the real part of z equal σ' and solve for $K(\sigma')$:

$$K(\sigma') = -\frac{(\sigma'-1)(\sigma'-0.93)}{(\sigma'+0.98)}. \tag{4.189}$$

Taking the derivative of $K(\sigma')$ with respect to σ' and setting it equal to zero,

$$\frac{dK(\sigma')}{d\sigma'} = 0, \tag{4.190}$$

we obtain the following roots:

$$\begin{aligned}\sigma_1' &= -2.92,\\ \sigma_2' &= 0.965.\end{aligned} \tag{4.191}$$

Therefore, there is a breakaway point at $0.965, 0$ and a break-in point at $-2.92, 0$.

Rule 8. The crossing of the imaginary axis and the value of K when it crosses the imaginary axis is obtained using this rule. The characteristic equation is obtained from Eq. (4.188) to be

$$(z-1)(z-0.93) + K(z+0.98) = 0 \tag{4.192}$$

or

$$z^2 + z(K - 1.93) + (0.93 + 0.98K) = 0. \tag{4.193}$$

Using the conventional Routh–Hurwitz criterion as presented in Section 6.3[‡], we obtain the following array:

$$\begin{array}{l|ll} z^2 & 1 & 0.93 + 0.98K \\ z^1 & K - 1.93 & \\ z^0 & 0.93 + 0.98K & \end{array}$$

Therefore, from row 2 we find that $K = 1.93$ when the root locus crosses the imaginary axis. It is important to recognize that this is not K_{max} as defined in Section 1.7 for the root locus of continuous systems. It is permissible to use the conventional Routh–Hurwitz criterion here, and not have to use the bilinear transformation presented in Section 4.8, because we are not trying to use rule 8 in this problem to find K_{max}. We only want to find K when the root locus crosses the imaginary axis and the value of the imaginary axis at crossing. This point can be found from the real terms of Eq. (4.193) and knowing that $K = 1.93$ when the root locus crosses the imaginary axis as follows:

$$z^2 + (0.93 + 0.98K) = 0. \tag{4.194}$$

Because $K = 1.93$, we obtain

$$z^2 + (0.93 + 1.89) = 0,$$
$$z = \pm j1.679.$$

Therefore, the root locus crosses the imaginary axis at $\pm j1.679$.

Rules 9–12. These rules are not needed for this problem.

The resulting root locus, which is shown in Figure 4.37, was obtained using MATLAB (see Section 1.7) and is contained in the M-file that is part of my AMCSTD Toolbox which can be retrieved free from The MathWorks, Inc. anonymous FTP server at ftp://ftp.mathworks.com/pub/books/advshinners. The root-locus diagram illustrated shows that the system is conditionally stable for only very low gains from 0 to 0.07143.

Let us next consider the compensation of this system. How would we proceed to design a compensation network which would make this system stable for a wider range of gain? It is apparent that we must cancel one of the poles at $z = 0.93$ with a zero placed there. The pole of this compensating network can be placed along the real axis between ± 1. A pole at 0.2 was selected, and the resulting network is a phase-lead network. The resulting transfer function of this phase-lead network is given by

$$G_D(z) = \frac{z - 0.93}{z - 0.2}.$$

The block diagram of the compensated system is shown in Figure 4.38. Therefore, the open-loop transfer function of the compensated system is given by

$$KG_D(z)G(z) = \frac{K(z - 0.93)(z + 0.98)}{(z - 0.2)(z - 1)(z - 0.93)}, \tag{4.195}$$

which can be reduced to the following:

$$KG_D(z)G(z) = \frac{K(z + 0.98)}{(z - 0.2)(z - 1)}. \tag{4.196}$$

The root locus of the compensated system, shown in Figure 4.39, was obtained using MATLAB (see Section 1.7) and is also contained in the M-file that is part of the AMCSTD Toolbox. The point of breakaway (σ'_2) and break-in (σ'_1) can be obtained using rule 7:

$$1 + \frac{K(z + 0.98)}{(z - 0.2)(z - 1)} = 0. \tag{4.197}$$

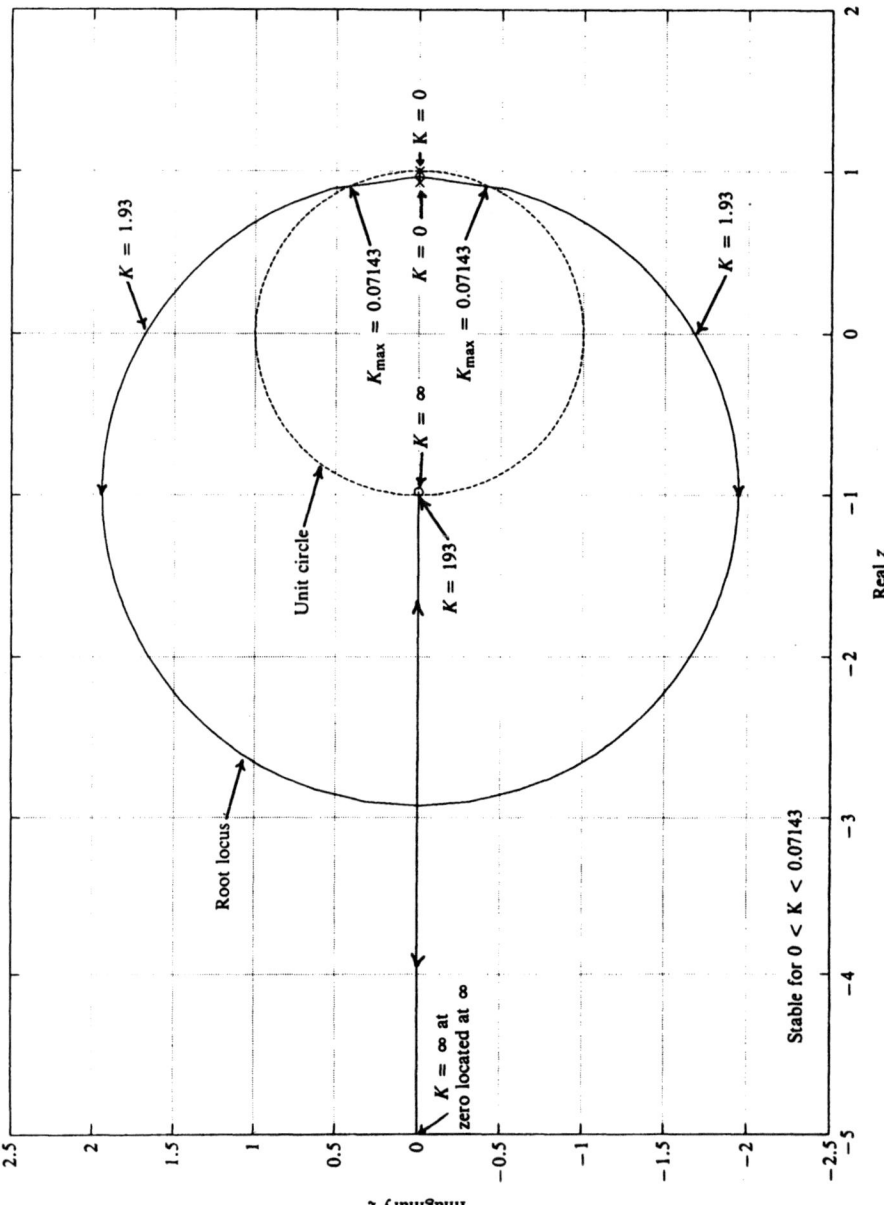

Figure 4.37 Root locus for the system of Figure 4.36.

4.10. STABILITY ANALYSIS AND DESIGN USING THE ROOT-LOCUS DIAGRAM

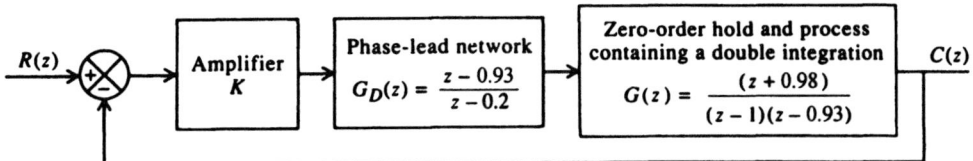

Figure 4.38 System of Figure 4.36 compensated with a phase-lead network, $G_D(z)$.

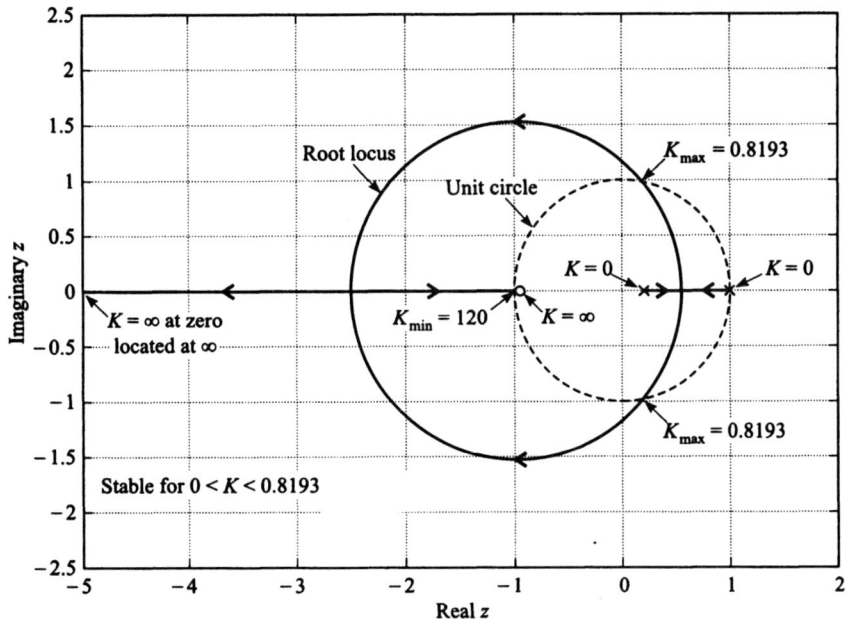

Figure 4.39 Root locus for the system of Figure 4.38.

We let the real part of z equal σ' and solve for $K(\sigma')$:

$$K(\sigma') = \frac{-(\sigma' - 0.2)(\sigma' - 1)}{(\sigma' + 0.98)}. \tag{4.198}$$

Taking the derivative of $K(\sigma')$ with respect to σ' and setting it equal to zero,

$$\frac{dK(\sigma')}{d\sigma'} = 0, \tag{4.199}$$

we obtain the roots 0.55 (breakaway) and -2.50 (break-in). Therefore, the compensated system is conditionally stable as shown in Figure 4.39. The maximum value of K, where the root locus intersects the unit circle, will be obtained using the modified Routh–Hurwitz criterion and the bilinear transformation presented in Section 4.8:

$$z = \frac{w+1}{w-1}. \tag{4.200}$$

Substituting Eq. (4.200) into the characteristic Eq. (4.197), we obtain the following Routh–Hurwitz array in the w-plane:

$$
\begin{array}{l|ll}
w^2 & 1.976K & 2.4 - 0.0236. \\
w^1 & 1.6 - 1.9528K & \\
w^0 & 2.4 - 0.0236K &
\end{array}
$$

Setting the second and third rows of the array equal to zero, we get

$$K_{\max} = 0.9193.$$

Therefore, the system is stable for $0 < K < 0.8193$

What value of K should we select for a desired damping ratio for the compensated system? To answer this question, we note an important difference between finding the damping ratio for a continuous system in the s-plane and a digital control system in the z-plane. We saw in Figure B.3 and Eq. (B.18) that the damping ratio for a second-order system was given in the s-plane by

$$\cos \alpha = \zeta. \tag{4.201}$$

In the s-plane, the loci for a constant damping ratio ζ, as ω_n is varied from zero to infinity, is a straight line, as illustrated in Figure B.3. In the z-plane, the loci for a constant damping ratio ζ, as ω_n is varied from zero to infinity, is a logarithmic spiral as illustrated in Figure 4.40. This figure was obtained using MATLAB (see Section 1.7) and is contained in the M-file that is part of the AMCSTD Toolbox which can be retrieved by the reader free from The MathWorks, Inc. anonymous FTP server at ftp://ftp.mathworks.com/pub/books/advshinners.

The reason that the loci for a constant damping ratio ζ in the z-plane is a logarithmic spiral will now be proven. From Figure B.3, the location of the complex-conjugate root in the upper half of the s-plane is given by

$$s = -\zeta\omega_n + j\omega_n\sqrt{1 - \zeta^2}. \tag{4.202}$$

Let us transform Eq. (4.202) into the z-plane using $z = e^{Ts}$ as follows:

$$z = e^{Ts} = e^{(-\zeta\omega_n + j\omega_n\sqrt{1-\zeta^2})T}. \tag{4.203}$$

We can rewrite Eq. (4.203) as

$$z = e^{-\zeta\omega_n T} e^{j\omega_n T \sqrt{1-\zeta^2}}, \tag{4.204}$$

and simplify it to

$$z = e^{-\zeta\omega_n T} \underline{/\omega_n T \sqrt{1 - \zeta^2}}. \tag{4.205}$$

Equation (4.205) is in polar form. Because we are assuming that the damping ratio and the sampling time T are constant, then as ω_n increases, the amplitude of

4.10. STABILITY ANALYSIS AND DESIGN USING THE ROOT-LOCUS DIAGRAM 269

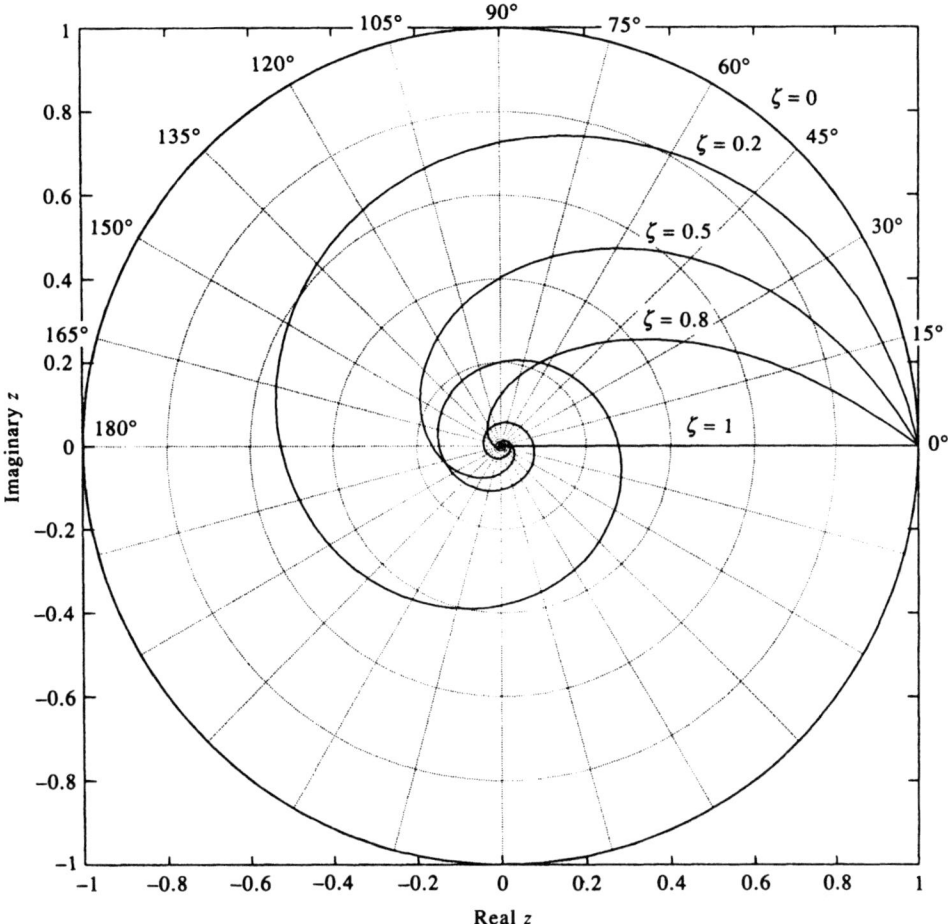

Figure 4.40 z plane with constant damping ratio loci.

z decreases exponentially as the phase angle rotates. Therefore, Eq. (4.205) is the equation of a logarithmic spiral. For example, when $\omega_n = 0$, $z = 1 \underline{/0°}$, and when $\omega_n \to \infty$, $z \to 0$ as illustrated in Figure 4.40.

Let us return to our design problem, and assume that we desire a damping ratio of 0.45. The root locus in Figure 4.39 is redrawn in Figure 4.41 with the constant damping ratio spiral of 0.45 superimposed on the root locus. From the intersection of the root locus and the constant damping ratio loci of 0.45, we see that the root location needed to achieve this damping is at

$$z = 0.472 + j0.477. \tag{4.206}$$

(The other root is located at $z = 0.472 - j0.477$.) We can now proceed to find the value of K needed from the characteristic equation with these root locations as follows:

$$1 + KG_D(z)G(z) = 1 + \frac{K(z + 0.98)}{(z - 0.2)(z - 1)} = 0. \tag{4.207}$$

270 DIGITAL CONTROL-SYSTEM ANALYSIS AND DESIGN

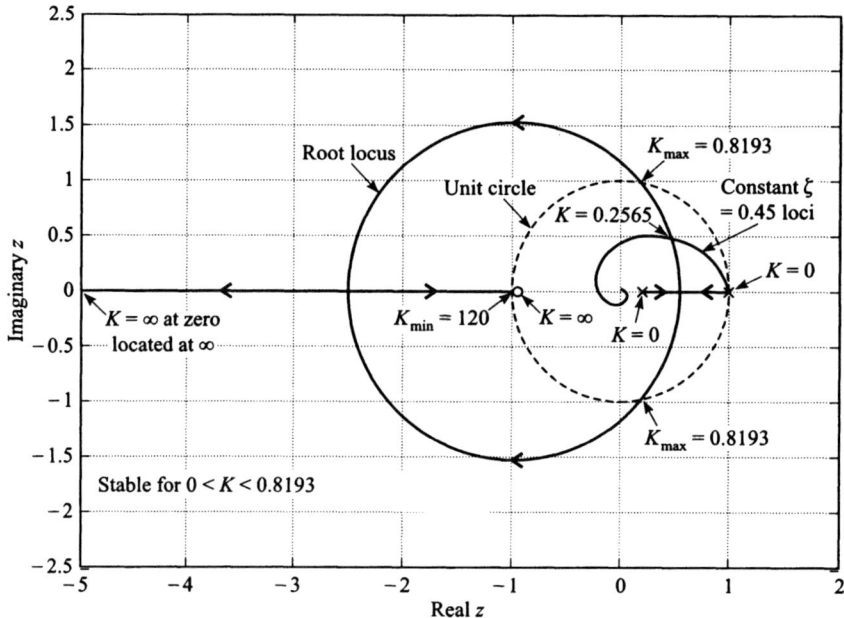

Figure 4.41 Root locus for the system of Figure 4.38 with a loci for a constant damping ratio of 0.45 superimposed.

Solving for K,

$$K = -\frac{(z - 0.2)(z - 1)}{(z + 0.98)}. \tag{4.208}$$

Substituting Eq. (4.206) into (4.208), we obtain the following:

$$K = -\frac{(0.472 + j0.477 - 0.2)(0.472 + j0.477 - 1)}{(0.472 + j0.477 + 1)}. \tag{4.209}$$

Solving for K, we obtain

$$K = 0.2565. \tag{4.210}$$

We wish to check that this value of K will result in a damping ratio of 0.45. The percent overshoot of the system can be determined by solving for $C(z)$ from the following equation:

$$\frac{C(z)}{R(z)} = \frac{KG_D(z)G(z)}{1 + KG_D(z)G(z)}. \tag{4.211}$$

The inverse z transform will be obtained to find $c^*(t)$, from which the maximum percent overshoot (and consequent damping) of the system will be determined. Substituting Eqs. (4.196) and (4.210) into Eq. (4.211), we obtain

$$\frac{C(z)}{R(z)} = \frac{0.2565z + 0.25}{z^2 - 0.9435z + 0.45}. \tag{4.212}$$

To find the percent overshoot, we will solve Eq. (4.212) for $C(z)$ for a unit step response, where

$$R(z) = \frac{z}{z-1}. \tag{4.213}$$

Substituting Eq. (4.213) into Eq. (4.212), we obtain

$$C(z) = \frac{0.2565z^2 + 0.25z}{z^3 - 1.9435z^2 + 1.394z - 0.45}. \tag{4.214}$$

Dividing denominator into numerator, we obtain the z transform of the output:

$$\begin{aligned}C(z) = &\ 0.2565z^{-1} + 0.749z^{-2} + 1.098z^{-3} \\ &+ 1.2056z^{-4} + 1.149z^{-5} + 1.049z^{-6} + 0.9875z^{-7} + \cdots.\end{aligned} \tag{4.215}$$

Remember that our sampling time, T, in this problem is 1.43, as shown in Eq. (4.186), and z^{-1} represents one cycle of delay. Taking the inverse z transform of Eq. (4.215), we obtain the output at sampling times, $c^*(t)$:

$$\begin{aligned}c^*(t) = &\ 0.2565\delta(t-T) + 0.749\delta(t-2T) + 1.098\delta(t-3T) + 1.2056\delta(t-4T) \\ &+ 1.149\delta(t-5T) + 1.049\delta(t-6T) + 0.9875\delta(t-7T) + \cdots,\end{aligned} \tag{4.216}$$

where T is the sampling time of 1.43 seconds. This result is plotted in Figure 4.42. We conclude from this problem that the resulting design has a 20.56% overshoot, which corresponds to a damping ratio of 0.45 [see Eq. (B.33) and Figure B.4]. Therefore, we have achieved our design objective.

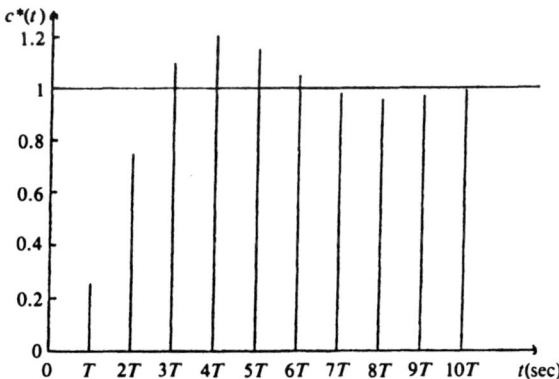

Figure 4.42 Response of the systems shown in Figure 4.38 to a unit-step input, where $T = 1.43$.

4.11. BODE AND ROOT-LOCUS DIAGRAMS FOR DISCRETE TIME SYSTEMS USING MATLAB [13]

The control-system engineer has much flexibility in converting between continuous time to discrete time, and obtaining the Bode and root-locus diagrams for discrete time systems using MATLAB, the professional Control System Toolbox, and the AMCSTD Toolbox.

A. Conversion between Continuous and Discrete Time

The AMCSTD Toolbox has a very useful utility for converting between continuous and discrete time, and accomplishing the bilinear transformation. This utility is denoted as "polysbst." Its application in going from $G(z)$, as given by Eq. (4.174), to $G(w)$, as given by Eq. (4.178), will next be illustrated:

POLYSBST : Substitute the variable of a polynomial with another polynomial

function [newnum] = polysubt (num, subnum);
function [newnum, newden] = polysubt (num, den, subnum, subden);
Substitute the variable in the original polynomial (num or num/den) with a polynomial expression (subnum or subnum/subden). The result is returned as a polynomial (newnum or newnum/newden).

Using $G(z) = (T^{*2}/2)(z + 1)/(z - 1)^2$ and $T = 0.26$ from Eq. (4.174), given $G(w) = (1 - w)/7.69)/w^2$ as shown in Eq. (4.178):

$t = 0.26$; num = $(t^{*2}/2) * [1 \ 1]$; den = conv([1 -1], [1 -1]);
Using Bi-Linear Transformation: $z = (1 + (t/2)w)/(1 - (t/2)w)$
[numw, denw] = polysbst (num, den, [t/2 1], [−t/2 1])
numw =
    −0.0088   0.0676
denw =
    0.0676   0   0

which simplifies to $G(w) = (1 - w/7.69)/w^2$ as shown in Eq. (4.178).

The Control Toolbox accommodates many types of conversions:

C2DM (or C2D) allows conversion from continuous to discrete domain.
D2CM (or D2C) allows conversion from discrete to continuous domain.

With the Control System Toolbox, this would be accomplished using D2CM:

[numw,denw] = d2cm (num,den,t,'tustin')

The resulting output from the Control System Toolbox would be:

numw =
    0   −0.1300   1.0000

denw =
    1.0000   0   −0.0000

This simplifies to $G(w) = (1 - w/7.69)/w^2$ as shown in Eq. (4.178).

B. Bode Diagram

The application and use of the Bode diagram for discrete control systems is the same as for continuous systems, except that we have to convert from the z-domain to the w-domain, and the frequency in the discrete time case is the fictitious frequency "v" where $w = jv$. Otherwise, the Bode diagram is drawn the same, and phase and gain margins have the same interpretation. Therefore, everything stated in Section 1.7 for obtaining the Bode diagram of continuous systems using MATLAB is valid here too for discrete-time systems.

C. Root-Locus Diagram

The construction of the root-locus diagram for discrete time systems is the same as for continuous systems. Therefore, everything stated in Section 1.7 for generating the root-locus diagram for continuous sytems is valid here too for discrete-time systems. The primary difference is in the interpretation of the root locus for discrete-time systems. We have shown in Section 4.10 that the unit circle in the z-plane defines the region of stability, as compared to the imaginary axis of the s-plane for continuous systems. The AMCSTD Toolbox has the following special functions which are great aids in constructing and analyzing the root locus for discrete-time systems:

1. *rlaxis* determines portions of the real axis that are part of the root locus.
2. *rlpoba* determines the points of breakaway and break-in, and the values of gain at these points.
3. *rootmag* is especially useful in the use of the root locus for discrete-time systems. It determines gain and values of the roots at a particular distance from the origin of the z-plane. This is particularly useful in finding the gain and root values when the root locus crosses the unit circle which is the stability boundary for discrete systems.
4. *rootangl* determines the gain and values of the roots at a particular angle from the origin.

4.12. RAGAZZINI'S METHOD

A very useful method for the design of sampled-data control systems is Ragazzini's method [2, 7]. Unlike other approaches for analysis and design, such as the Bode diagram (in the w-plane) and the root locus which was extended from continuous systems to discrete systems, Ragazzini's method was developed specifically for the design of sampled-data control systems.

Let us next consider stabilizing a sampled-data control system by means of a digital controller, as is illustrated in Figure 4.43. The object of this device is to compute a sequence of output numbers $e_2^*(t)$, which are linearly related to the input number sequence $e_1^*(t)$, in order to obtain a desired system response. The digital controller may contain active elements or simple passive networks. When this device contains active elements, the control engineer is completely free in choosing any stabilizing function in order that the desired system response may be obtained. The procedure for finding $D(z)$ is known as the direct design method of Ragazzini [2, 7].

We shall assume that the digital controller is a linear device. Therefore, the output and input number sequences are related linearly. The linear relationship which exists between the input and the output sequence, $e_1^*(t)$ and $e_2^*(t)$, respectively, can be expressed as

$$X_0 e_1(nT) + X_1 e_1[(n-1)T] + X_2 e_1[(n-2)T] + X_3 e_1[(n-3)T] + \cdots + X_l e_1[(n-l)T]$$
$$= e_2(nT) + Y_1 e_2[(n-1)T] + Y_2 e_2[(n-2)T] + \cdots + Y_m e_2[(n-m)T] \quad (4.217)$$

where X_l and Y_m represent constant terms. The z transform of this equation can be written as

$$E_1(z)[X_0 + X_1 z^{-1} + X_2 z^{-2} + X_3 z^{-3} + \cdots + X_l z^{-l}]$$
$$= E_2(z)[1 + Y_1 z^{-1} + Y_2 z^{-2} + \cdots + Y_m z^{-m}]. \quad (4.218)$$

The z transfer function of the digital controller, $D(z)$, is defined as

$$D(z) = \frac{E_2(z)}{E_1(z)} \quad (4.219)$$

From Eq. (4.218), $D(z)$ can be expressed as

$$D(z) = \frac{X_0 + X_1 z^{-1} + X_2 z^{-2} + \cdots + X_l z^{-l}}{1 + Y_1 z^{-1} + Y_2 z^{-2} + \cdots + Y_m z^{-m}} \quad (4.220)$$

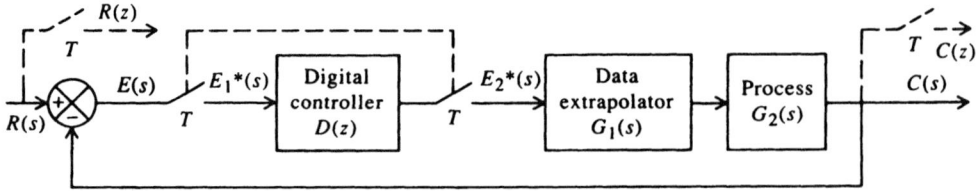

Figure 4.43 Sampled-data control system utilizing digital controller compensation.

In order that the digital controller be physically realizable, the denominator of Eq. (4.220) must contain the term 1.

For the system in Figure 4.43, the system transfer function $H(z)$ can be expressed as

$$H(z) = \frac{C(z)}{R(z)} = \frac{D(z)G(z)}{1 + D(z)G(z)}. \qquad (4.221)$$

The term $G(z)$ is the z transform corresponding to the Laplace transfer function $G_1(s)G_2(s)$. In order that a specified system transfer function $H(z)$ may be realized, the various constants of $D(z)$ must be appropriately chosen. From Eq. (4.221) the value of $D(z)$ corresponding to a specified value of $H(z)$ is given by

$$D(z) = \frac{1}{G(z)} \frac{H(z)}{1 - H(z)}. \qquad (4.222)$$

It is important to realize that the primary purpose of the digital controller is to cancel any undesirable poles and zeros of the uncompensated system and replace them with poles and zeros which will result in a desired system reponse. It is obvious from Eq. (4.221) that this can be accomplished by designing $D(z)$ in order that its zeros correspond to the poles of $G(z)$ and its poles correspond to the zeros of $G(z)$, which lie on or outside the unit circle of the z-plane. However, it has been shown that this method of compensation is not practical because any slight change in the parameters of $D(z)$ or $G(z)$ may result in having the poles and zeros of $G(z)$, lying on or outside the unit circle of the z-plane, being uncanceled [7].

A better method for compensation may be obtained by considering Eq. (4.222). Poles and zeros of $G(z)$ that lie on or outside the unit circle in the z-plane, may be canceled by specifying $1 - H(z)$ and $H(z)$ in order that they cancel the poles and zeros, respectively. The following four rules must be pursued when specifying a system transfer function $H(z)$, which will result in a stable system:

1. The specified system transfer function $H(z)$ must contain as its zeros all the zeros of $G(z)$ that lie on or outside the unit circle in the z-plane.
2. $1 - H(z)$ must contain as its zeros all of the poles of $G(z)$ that lie on or outside the unit circle in the z-plane.
3. In order that $D(z)$ be physically realizable (causal), it should not have a pole at infinity as $z \to \infty$. Therefore, from Eq. (4.222) if $G(z)$ has a zero at infinity, it is necessary that $H(z)$ have a zero there, too, in order to prevent $D(z)$ from having a pole at infinity.
4. $H(z)$ must be specified in order that the steady-state error, resulting from the application of an input having the form

$$R(z) = \frac{A(z)}{(1 - z^{-1})^m}, \qquad (4.223)$$

is zero. From Table 4.1, this type of input can represent a step, ramp, acceleration, etc., depending upon the value of m. $A(z)$ represents a polynomial in

z^{-1} which does not contain factors of the form $1 - z^{-1}$. From the relationships given by

$$E_1(z) = R(z) - C(z) \qquad (4.224)$$

and

$$H(z) = \frac{C(z)}{R(z)}, \qquad (4.225)$$

the z transform of the system error can be expressed as

$$E_1(z) = R(z)[1 - H(z)]. \qquad (4.226)$$

Applying the final-value theorem (see Eq. (4.61)), to Eq. (4.226), we obtain

$$e_1(\infty) = \lim_{z \to 1} [(1 - z^{-1})R(z)(1 - H(z))]. \qquad (4.227)$$

Substituting Eq. (4.223) into Eq. (4.227) the following expression results:

$$e_1(\infty) = \lim_{z \to 1} \left[(1 - z^{-1}) \frac{A(z)}{(1 - z^{-1})^m} (1 - H(z)) \right]. \qquad (4.228)$$

It is easily seen from Eq. (4.228) that the steady-state error, for inputs of the form given by Eq. (4.223) will be zero if $1 - H(z)$ satisfies the relationship

$$1 - H(z) = (1 - z^{-1})^m F(z), \qquad (4.229)$$

where $F(z)$ is an unspecified ratio of polynomials in z^{-1} and m is the order of the denominator of the input, $R(z)$. When $F(z)$ is unity, a "minimal prototype" response function results and the order of $H(z)$ in z^{-1} is a minimum. However, a minimal prototype response function can only be utilized when $H(z)$ does not contain any zeros on or outside the unit circle of the z-plane. A little further thought reveals that a system in which $m = 1$ will respond to a unit step with zero steady-state error, and when $m = 2$ it will respond to a unit ramp with zero-state error, etc. In addition, a system utilizing a minimal prototype will respond with zero steady-state error to lower-order input functions.

We shall next apply the theory developed to several systems of the types shown in Figure 4.43, where the data extrapolator is a zero-order hold. The sampling rate will be assumed equal to 1 sec. The techniques used to choose the value of $D(z)$ and the system's response to various inputs will be illustrated.

A. Example of Ragazzini's Method when the Process $G_2(s)$ is an Integrator

Consider a system which has a process that is an integrator:

$$G_2(s) = \frac{1}{s}. \qquad (4.230)$$

4.12. RAGAZZINI'S METHOD

1. Design of D(z). The first step of the procedure is to determine the z transform $G(z)$, corresponding to the Laplace transfer function given by

$$G(s) = G_1(s)G_2(s) = \left(\frac{1-e^{-s}}{s}\right)\left(\frac{1}{s}\right) = \frac{1-e^{-s}}{s^2}. \tag{4.231}$$

From Table 4.1,

$$G(z) = \frac{z^{-1}}{1-z^{-1}}. \tag{4.232}$$

Examination of $G(z)$ shows that it contains a permissible pole at $z = 1$. Applying rule numbers 1 and 3, we can specify the system transfer function as

$$H(z) = K_1 z^{-1}. \tag{4.233}$$

Assuming that zero steady-state error to a unit step input is desired, the following relationship is obtained from rule 4, Eq. (4.229):

$$1 - H(z) = (1 - z^{-1})F(z). \tag{4.234}$$

From rule number 4, $F(z)$ is set equal to 1 to obtain a minimal prototype system. Therefore, Eq. (4.234) can be expressed as

$$1 - H(z) = 1 - z^{-1}. \tag{4.235}$$

Substituting Eq. (4.233) into Eq. (4.235) we obtain

$$1 - K_1 z^{-1} = 1 - z^{-1}. \tag{4.236}$$

From Eq. (4.236) the solution for K_1 is

$$K_1 = 1. \tag{4.237}$$

Substituting Eq. (4.237) into Eq. (4.233), we obtain

$$H(z) = z^{-1}. \tag{4.238}$$

It is now possible to compute the value of $D(z)$. Substituting Eq. (4.232) and (4.238) into Eq. (4.222), we obtain

$$D(z) = \left(\frac{1-z^{-1}}{z^{-1}}\right)\left(\frac{z^{-1}}{1-z^{-1}}\right). \tag{4.239}$$

This expression reduces to

$$D(z) = 1. \tag{4.240}$$

2. System Response to a Unit Step. The z-transfer function of the system is given by

$$H(z) = \frac{C(z)}{R(z)}. \tag{4.241}$$

From Table 4.1, the z transform of a unit step input

$$R(z) = \frac{1}{1 - z^{-1}}. \tag{4.242}$$

By substituting Eqs. (4.238) and (4.242) into Eq. (4.241) the z transform of the output response is obtained. The resulting expression is

$$C(z) = \frac{z^{-1}}{1 - z^{-1}}. \tag{4.243}$$

By expanding Eq. (4.243) into a power series in z, and then taking the inverse z transform, the output response can be sketched. The result is

$$C(z) = z^{-1} + z^{-2} + z^{-3} + \cdots \tag{4.244}$$

or

$$c^*(t) = \delta(t - 1) + \delta(t - 2) + \delta(t - 3) + \cdots \tag{4.245}$$

and is plotted in Figure 4.44. Observe that this system follows a unit step input with zero steady-state error based on our previous choice of $1 - H(z)$ in accordance with rule 4.

B. Example of Ragazzini's Method when the Process $G_2(s)$ is a Double Integration

The next system we will consider contains a process which is a double integrator and may be represented by the transfer function

$$G_2(s) = \frac{1}{s^2}. \tag{4.246}$$

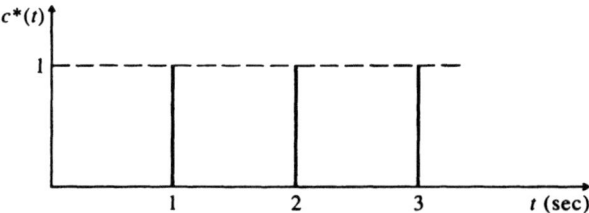

Figure 4.44 Response of the system shown in Figure 4.43 containing a zero-order hold and an integrating process, designed in accordance with Ragazzini's method, to a unit step input.

1. Design of D(z). Utilizing the same procedure that was followed in the previous problem, we will first obtain $G(z)$. The expression for $G(s)$ is

$$G(s) = \frac{1-e^{-s}}{s^3}. \tag{4.247}$$

Utilizing Table 4.1, we obtain the z transform as

$$G(z) = \frac{(1+z^{-1})z^{-1}}{2(1-z^{-1})^2}. \tag{4.248}$$

Examination of $G(z)$ shows that it contains a zero at $z = -1$ and two poles at $z = 1$. Applying rule numbers 1 and 3, we can specify the system transfer function as

$$H(z) = (1+z^{-1})(K_1 z^{-1} + K_2 z^{-2}). \tag{4.249}$$

Assuming that zero steady-state error to a unit ramp is desired and we wish to cancel the two poles of $G(z)$ at $z = 1$, the following relationship is obtained from rules 2 and 4:

$$1 - H(z) = (1-z^{-1})^2 F(z). \tag{4.250}$$

Observe from rule 4 that this expression automatically insures that the system will respond to a unit ramp with zero steady-state error. However, a minimal prototype response function cannot be specified, because $G(z)$ contains a zero on the unit circle in the z-plane. Instead, a larger minimum finite settling response time will be obtained. Therefore Eq. (4.250) can be rewritten as

$$1 - H(z) = (1-z^{-1})^2(1 + \gamma_1 z^{-1} + \gamma_2 z^{-2} + \gamma_3 z^{-3} + \cdots). \tag{4.251}$$

Substituting Eq. (4.249) into Eq. (4.251), we obtain the expression

$$1 - (1+z^{-1})(K_1 z^{-1} + K_2 z^{-2}) = (1-z^{-1})^2(1 + \gamma_1 z^{-1} + \gamma_2 z^{-2} + \cdots). \tag{4.252}$$

Here γ_2, γ_3, etc., will be considered to be zero in order to obtain a solution for the constants of the equation. Equating coefficients of terms having the same power of z together, we then obtain three simultaneous equations, relating the constants of Eq. (4.252), If γ_2, γ_3, etc., were considered, then we would have more unknowns than equations and we would obtain a non-minimum finite settling time. The results are as follows:

$$K_1 = 1.25, \qquad K_2 = -0.75, \qquad \gamma_1 = 0.75. \tag{4.253}$$

Substituting Eq. (4.253) into Eqs. (4.249) and (4.251), we obtain

$$1 - H(z) = (1-z^{-1})^2(1 + 0.75z^{-1}), \tag{4.254}$$
$$H(z) = (1+z^{-1})(1.25z^{-1} - 0.75z^{-2}). \tag{4.255}$$

It is now possible to compute the value of $D(z)$. Substituting Eqs. (4.248), (4.254), and (4.255), into Eq. (4.222) we obtain the expression

$$D(z) = \left[\frac{2(1-z^{-1})^2}{(1+z^{-1})(z^{-1})}\right]\left[\frac{(1+z^{-1})(1.25z^{-1}-0.75z^{-2})}{(1-z^{-1})^2(1+0.75z^{-1})}\right]. \tag{4.256}$$

This expression can be reduced to

$$D(z) = \frac{2.5 - 1.5z^{-1}}{1 + 0.75z^{-1}}. \tag{4.257}$$

2. System Response to a Unit Step. By substituting Eqs. (4.242) and (4.255) into Eq. (4.241) the z transform of the output response can be obtained. The resulting expression can be reduced to

$$C(z) = \frac{1.25z^{-1} + 0.5z^{-2} - 0.75z^{-3}}{1 - z^{-1}}. \tag{4.258}$$

By expanding Eq. (4.258) into a power series in z and then taking the inverse z transform, the output response can be obtained. The result is

$$C(z) = 1.25z^{-1} + 1.75z^{-2} + z^{-3} + z^{-4} + \cdots \tag{4.259}$$

or

$$c^*(t) = 1.25\delta(t-1) + 1.75\delta(t-2) + \delta(t-3) + \delta(t-4) + \cdots \tag{4.260}$$

and is shown in Figure 4.45. Notice that the steady-state error of this system to a unit step input is zero. This is what we should expect based on our previous choice of $1 - H(z)$ in accordance with rule 4.

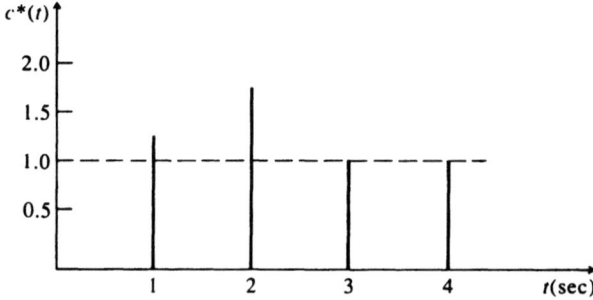

Figure 4.45 Response of the system shown in Figure 4.43 containing a zero-order hold and a double integrating process, designed in accordance with Ragazzini's method, to a unit step input.

3. System Response to a Unit Ramp. From Table 4.1, the z transform of a unit ramp is given by the expression

$$R(z) = \frac{Tz^{-1}}{(1 - z^{-1})^2}. \quad (4.261)$$

Becuase we are assuming a sampling time of 1 sec, Eq. (4.261) can be reduced to

$$R(z) = \frac{z^{-1}}{(1 - z^{-1})^2}. \quad (4.262)$$

By substituting Eqs. (4.255) and (4.262) into Eq. (4.241), the z transform of the output response is obtained. The resulting expression can be reduced to

$$C(z) = \frac{2.5z^{-2} + z^{-3} - 1.5z^{-4}}{2 - 4z^{-1} + 2z^{-2}}. \quad (4.263)$$

By expanding Eq. (4.263) into a power series in z and then taking the inverse z transform, the output response can be obtained. The result is

$$C(z) = 1.25z^{-2} + 3z^{-3} + 4z^{-4} + 5z^{-5} + 6z^{-6} + \cdots \quad (4.264)$$

or

$$c^*(t) = 1.25\delta(t - 2) + 3\delta(t - 3) + 4\delta(t - 4) + 5\delta(t - 5) + \cdots \quad (4.265)$$

and is plotted in Figure 4.46. Notice that the steady-state error of this system to a unit ramp input is zero because of our choice of $1 - H(z)$ based on rule 4.

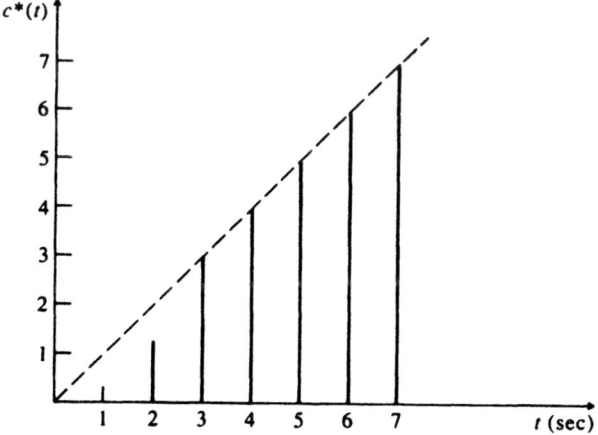

Figure 4.46 Response of the system shown in Figure 4.43 containing a zero-order hold and a double-integrating process, designed in accordance with Ragazzini's method, to a unit ramp input.

282 DIGITAL CONTROL-SYSTEM ANALYSIS AND DESIGN

A comparison of Figures 4.45 and 4.46 indicates that the system responds relatively smoothly to the type of input it was designed for, specifically a ramp, and poorly for lower-order inputs such as a step. For example, it exhibits an overshoot of 75% when responding to a step input before finally settling to its steady-state value. It is interesting to note that in the previous example, which consisted of a sampled-data system containing a single integrator, the response did not exhibit an overshoot to a step input because we originally designed it to respond specifically to this type of input.

C. Damping Effect of the Staleness Factor

A method of overcoming the severe overshoots of sampled-data control systems to inputs that are of a lower order than they are specifically designed for is to introduce a term to the system transfer function known as the "staleness factor." Implementation of this factor results in softening the system's response to a wide variety of inputs. The staleness factor term is added to the original system transfer function in the following manner:

$$H_s(z) = \frac{H(z)}{(1 - Cz^{-1})^N}, \tag{4.266}$$

where $H(z)$ = original z-transfer function of the system, $H_s(z)$ = z-transfer function of system with the staleness factor term added, C = staleness factor constant whose value ranges from $+1$ to -1 for stable systems, and N = exponent which can be any positive value.

Because Bertram [14] has shown that not very much is gained by designing N greater than unity, we shall assume that $N = 1$. The value for the staleness factor, C, can be chosen by analytical optimizing procedures or by laboratory trial-and-error techniques. Generally, it can be stated that as C approaches unity, maximum damping is produced, and as C approaches zero, the damping is decreased. We shall illustrate the damping effect of the staleness factor term upon the response of the sampled-data control system considered previously (control system of Figure 4.43 where $G_2(s)$ is a double integration) by introducing a staleness factor of 0.3 and comparing the system's response with that illustrated in Figures 4.45 and 4.46.

1. Design of D(z). Utilizing the same procedure that was followed previously, we shall first compute the value of $D(z)$. The introduction of the staleness factor term,

$$1 - 0.3z^{-1},$$

modifies Eq. (4.249) and (4.251) as follows:

$$H_s(z) = \frac{(1 + z^{-1})(K_1 z^{-1} + K_2 z^{-2})}{1 - 0.3z^{-1}}, \tag{4.267}$$

$$1 - H_s(z) = \frac{(1 - z^{-1})^2 (1 + \gamma_1 z^{-1} + \gamma_2 z^{-2} + \gamma_3 z^{-3} + \cdots)}{1 - 0.3z^{-1}}. \tag{4.268}$$

Substituting Eq. (4.267) into (4.268), we obtain the expression

$$1 - \frac{(1+z^{-1})(K_1 z^{-1} + K_2 z^{-2})}{1 - 0.3z^{-1}} = \frac{(1-z^{-1})^2(1 + \gamma_1 z^{-1} + \gamma_2 z^{-2} + \cdots)}{1 - 0.3z^{-1}}. \quad (4.269)$$

As in the previous example, the constants γ_2, γ_3, etc., need not be considered. Equating coefficients of terms having the same power of z together, we obtain three simultaneous equations relating the constants of Eq. (4.269). The results are as follows:

$$K_1 = 1.025,$$
$$K_2 = -0.675, \quad (4.270)$$
$$\gamma_1 = 0.675.$$

Substituting Eq. (4.270) into Eq. (4.267) and (4.268), we obtain

$$H_s(z) = \frac{(1+z^{-1})(1.025z^{-1} - 0.675z^{-2})}{1 - 0.3z^{-1}}, \quad (4.271)$$

$$1 - H_s(z) = \frac{(1-z^{-1})^2(1 + 0.675z^{-1})}{1 - 0.3z^{-1}}. \quad (4.272)$$

The value of $D(z)$ can now be computed by substituting Eqs. (4.248), (4.271) and (4.272) into Eq. (4.222). The resultant expression is

$$D(z) = \left[\frac{2(1-z^{-1})^2}{(1+z^{-1})(z^{-1})}\right]\left[\frac{(1+z^{-1})(1.025z^{-1} - 0.675z^{-2})}{(1-z^{-1})^2(1 + 0.675z^{-1})}\right]. \quad (4.273)$$

This can be reduced to

$$D(z) = \frac{2.05 - 1.35z^{-1}}{1 + 0.675z^{-1}}. \quad (4.274)$$

A comparison of Eqs. (4.257) and (4.274) shows that the values of the various coefficients are different but that the complexity of the digital controller is the same. Therefore, the introduction of the staleness factor term may be obtained by merely readjusting the programming of the digital controller and does not require a more complex device.

2. System Response to a Unit Step. By substituting Eqs. (4.242) and (4.271) into Eq. (4.241) the z transform of the output response is obtained. The resulting expression can be reduced to

$$C(z) = \frac{1.025z^{-1} + 0.35z^{-2} - 0.675z^{-3}}{1 - 1.3z^{-1} + 0.3z^{-2}}. \quad (4.275)$$

By expanding Eq. (4.275) into a power series in z and then taking the inverse z transform, the output response can be obtained. The result is

$$C(z) = 1.025z^{-1} + 1.68z^{-2} + 1.2z^{-3} + 1.07z^{-4} + 1.02z^{-5}$$
$$+ 1.01z^{-6} + z^{-7} + z^{-8} + \cdots \quad (4.276)$$

or

$$c^*(t) = 1.025\delta(t-1) + 1.68\delta(t-2) + 1.2\delta(t-3) + 1.07\delta(t-4)$$
$$+ 1.02\delta(t-5) + 1.01\delta(t-6) + \delta(t-7) + \delta(t-8) + \cdots \quad (4.277)$$

and is sketched in Figure 4.47. Comparing Figures 4.45 and 4.47, we can clearly see the damping or "softening" effect of the staleness factor upon the system's response. The maximum percent overshoot is reduced from 75% to 68%. By varying C between ± 1, an optimum value can be found.

3. System Response to a Unit Ramp. The z transform of the output response to a unit ramp input can be obtained by substituting Eqs. (4.262) and (4.271) into Eq. (4.241). The resulting expression can be reduced to

$$C(z) = \frac{1.025z^{-2} + 0.35z^{-3} - 0.675z^{-4}}{1 - 2.3z^{-1} + 1.6z^{-2} - 0.3z^{-3}}. \quad (4.278)$$

By expanding Eq. (4.278) into a power series in z and then taking the inverse z transform, the output response can be obtained. The result is

$$C(z) = 1.025z^{-2} + 2.71z^{-3} + 3.93z^{-4} + 5z^{-5} + 6z^{-6} + 7z^{-7} + \cdots \quad (4.279)$$

or

$$c^*(t) = 1.025\delta(t-2) + 2.71\delta(t-3) + 3.93\delta(t-4)$$
$$+ 5\delta(t-5) + 6\delta(t-6) + 7\delta(t-7) + \cdots \quad (4.280)$$

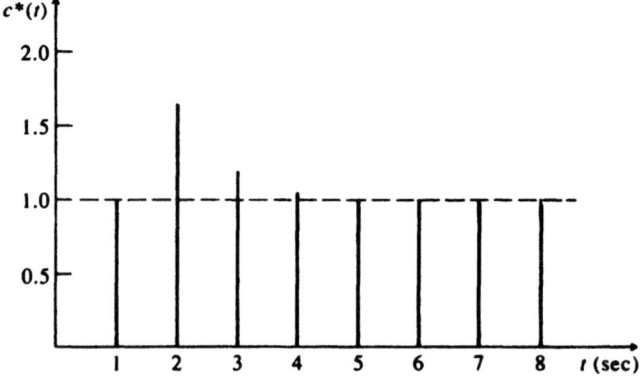

Figure 4.47 Use of the staleness-factor term for softening the response of a system containing a zero-order hold and a double-integrating process to a unit step input.

and is plotted in Figure 4.48. Comparing Figures 4.46 and 4.48, we can clearly see the damping or softening effect of the staleness factor upon the system's response. Although it is true that the implementation of the staleness factor term results in the system taking a little longer to respond to a ramp input, it must be remembered that this solution is a compromise which enables the system to respond less sharply to a step input.

4.13. THE DIGITIZATION PROCESS AND THE DESIGN OF DIGITAL FILTERS [2]

Up until this chapter on digital control-system analysis and design, the analysis and design has focused on continuous or analog control systems. When compensating networks were determined in Chapters 2 and 3 for linear continuous control systems, the resulting compensating network's form was either continuous or analog. In practice, however, we know that the control system will be implemented digitally. Therefore, how do we digitize the resulting continuous (or analog) compensating network? That is the subject of this section.

Let us consider the control system illustrated in Figure 4.49. We wish to implement the continuous (or analog) form of $D(s)$ digitally by obtaining $D(z)$ so that $D(z)$ duplicates the phase and amplitude characteristics of $D(s)$ as closely as possible. However, we must recognize that $D(z)$ cannot exactly duplicate the phase and ampli-

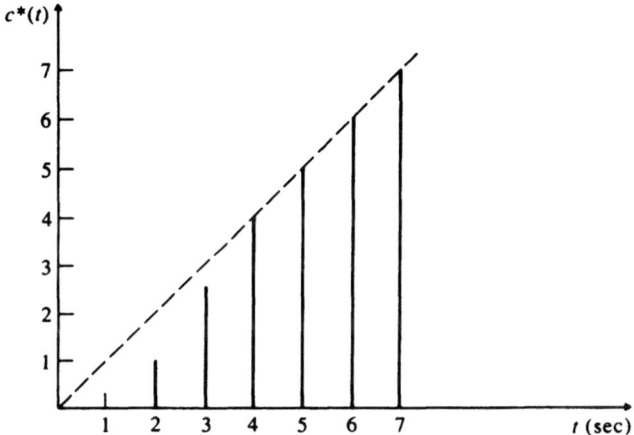

Figure 4.48 Use of the staleness-factor term for softening the response of a system containing a zero-order hold and a double-integrating process to a unit ramp input.

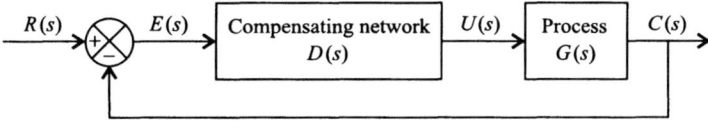

Figure 4.49 A continuous (or analog) control system.

tude characteristics of $D(s)$ because $D(s)$ acts on the complete time history of $E(s)$ while $D(z)$ only acts upon the samples of $E(s)$.

There are many digital approximations to this problem. For example, one approach is that of numerical integration. This is illustrated in Problem 4.41 which illustrates numerical integration by summing rectangles over a continuous function. Problem 4.42 illustrates numerical integration by summing trapezoidal areas under a continuous function. Both methods result in a z-transform transfer function relating $E(z)$ and $U(z)$, $D(z)$, whose phase and amplitude characteristics are very close to that of $D(s)$.

We will focus attention in this section on the practical *matched pole-zero method* which is based on the relation between the s- and z-planes given by Eq. (4.32).

$$z = e^{Ts} \tag{4.281}$$

The procedure is to obtain the z transform of the sampled function of the continuous transfer function $D(s)$ and relate the poles of $D(z)$ to the poles of $D(s)$ according to Eq. (4.281). The same reasoning is applied to the zeros of $D(z)$ and the zeros of $D(s)$. The rules of the matched pole-zero method are as follows

Rule 1. Match all poles and zeros of $D(s)$ and $D(z)$ according to Eq. (4.281). For example, if $D(s)$ has a pole at $s = -2$, then $D(z)$ has a pole according to $z = e^{-2T}$.

Rule 2. Match the gain of $D(s)$ and $D(z)$ at a critical frequency. Usually, the critical frequency selected is dc. Therefore, for this case we match the gains of

$$D(s)]_{s=0} = D(z)]_{z=1}. \tag{4.282}$$

Rule 3. Add $(1 + z^{-1})^n$ as appropriate only if the numerator is of lower order that of the denominator. This has the effect of mapping zeros of $D(s)$ at s equals infinity to the point z equals minus one for $D(z)$.

Let us illustrate the matched pole-zero with an example. It is assumed that the compensating network, $D(s)$, in Figure 4.49 represents a phase-lead network designed to add 36.88° phase lead at a critical frequency of $\omega_1 = 1$ rad/sec:

$$D(s) = 0.25 \frac{(2s + 1)}{(0.5s + 1)}. \tag{4.283}$$

It will also be assumed that the sampling time that the digitized network will operate at is 0.5 sec. Following the three rules presented, $D(z)$ is obtained as follows

Rule 1. Pole at $s = -2$ transforms into the z-plane as

$$z = e^{Ts} = e^{-(0.5)(2)} = e^{-1} = 0.37. \tag{4.284}$$

Therefore, the corresponding pole in the z-plane is $(z - 0.37)$.

Zero at $s = -0.5$ transforms into the z-plane as

$$z = e^{Ts} = e^{-(0.5)(0.5)} = e^{-0.25} = 0.78. \qquad (4.285)$$

Therefore, the corresponding zero in the z-plane is $(z - 0.78)$.

Rule 2. The gain of $D(z)$ is made to match the gain of $D(s)$ at dc:

$$D(s)]_{s=0} = D(z)]_{z=1} \qquad (4.286)$$

The gain of $D(s)$ at dc is:

$$D(s)]_{s=0} = 0.25 \frac{(2s+1)}{(0.5s+1)} \bigg]_{s=0} = 0.25. \qquad (4.287)$$

Therefore,

$$D(z)]_{z=1} = 0.25 = K \frac{(z-0.78)}{(z-0.37)} \bigg]_{z=1} \qquad (4.288)$$

Therefore,

$$K = 0.72. \qquad (4.289)$$

Combining the results of Eqs. (4.284), (4.285), and (4.289), $D(z)$ is given by

$$D(z) = 0.72 \frac{(z-0.78)}{(z-0.37)}. \qquad (4.290)$$

The next question to answer is whether the resulting digitized network, $D(z)$, is stable from the resulting pole and zero plot shown in Figure 4.50. The location of the pole illustrated in this figure shows that $D(z)$ is stable.

The last step of the design is to determine the phase shift and amplitude characteristics resulting from the digitized network $D(z)$ and compare it with that of the continuous network $D(s)$ at the critical frequency of $\omega_1 = 1$ rad/sec. To determine

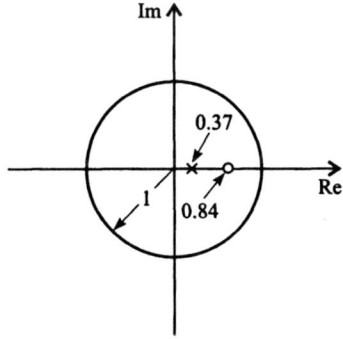

Figure 4.50 Location of pole and zero of $D(z) = 0.72\frac{(z-0.78)}{(z-0.37)}$

this for the digitized network $D(z)$ of Eq. (4.290), we must first determine z corresponding to the critical frequency $\omega_1 = 1$ rad/sec and the sampling time $T = 0.5$ sec.

$$z_1 = e^{Ts_1} = e^{jT\omega_1} = e^{j(0.5)(1)} = e^{0.5j}. \tag{4.291}$$

Since

$$e^{jx} = \cos(x) + j\sin(x) \tag{4.292}$$

then

$$z_1 = \cos(0.5) + j\sin(0.5) = 0.88 + j0.48. \tag{4.293}$$

Substituting Eq. (4.293) into Eq. (4.290), we obtain the following:

$$D(z) = \frac{0.72(z - 0.78)}{(z - 0.37)} = \frac{0.72(0.88 + j0.48 - 0.78)}{(0.88 + j0.48 - 0.37)} = \frac{0.72(0.1 + j0.48)}{(0.51 + j0.48)}. \tag{4.294}$$

Equation (4.294) can be simplified to the following:

$$D(z) = 0.5 \underline{/35.1°}. \tag{4.295}$$

Therefore, this method gives a very close fit between the phase of $D(s)$ and $D(z)$ at 1 rad/sec, and an exact amplitude fit at dc.

The resulting difference equation, which must be programmed into the digital computer, is obtained from Eq. (4.290) as follows:

$$D(z) = \frac{U(z)}{E(z)} = \frac{0.72(z - 0.78)}{(z - 0.37)} = \frac{(0.72z - 0.5616)}{(z - 0.37)} = \frac{(0.72 - 0.5616z^{-1})}{(1 - 0.37z^{-1})}. \tag{4.296}$$

Therefore,

$$u(n) - 0.37u(n - 1) = 0.72e(n) - 0.561e(n - 1) \tag{4.297}$$

or,

$$u(n) = 0.37u(n - 1) + 0.72e(n) - 0.561e(n - 1). \tag{4.298}$$

The resulting digital control-system version, corresponding to the continuous (or analog) version of Figure 4.49 is illustrated in Figure 4.51.

There are several other methods that can be used to digitize a continuous (or analog) network. We have mentioned the numerical integration methods which are presented in Problems 4.41 and 4.42. Another more sophisticated method is the hold equivalence method which is discussed in Reference 13. Based on my experience, I prefer to use the matched pole-zero method. It is very simple to use and usually provides very close matches between the continuous (or analog) network and the digital case.

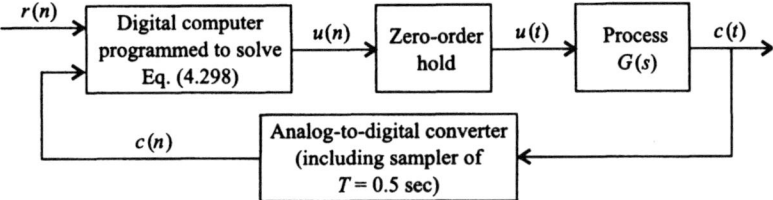

Figure 4.51 Digital control system corresponding to the continuous (or analog) version of Figure 4.49.

4.14. SUMMARY

This chapter has extended the concept of linear continuous system analysis and design, presented in the first eight chapters of this book, to digital control systems. Sampling characteristics were analyzed in the time and frequency domains. The use of zero-order and first-order holds for digital-to-analog conversion in sampled-data systems were presented and analyzed. The use of z-transform theory was presented for simplifying the analysis of discrete systems. Its use permitted the analysis of digital control systems using techniques similar to those used for continuous systems. For example, Mason's signal-flow graph and theorem were equally applicable to the analysis of sampled-data block diagrams.

The concept of stability in the z-plane was presented, where the unit circle in the z-plane was shown to be analogous to the imaginary axis in the s-plane for stability analysis. Three mathematical tests were analyzed: modified Routh–Hurwitz, Schur–Cohn, and Jury's stability criteria. The modified Routh–Hurwitz criterion in the w-plane was very similar to the conventional Routh–Hurwitz criterion used for continuous systems in the s-plane. The Schur–Cohn and Jury tests were unique to discrete systems. Both the Bode diagram and the root-locus concepts for continuous system analysis and design, presented in Chapter 1 and Chapter 2, were extended to discrete systems. By the use of the w transform for the Bode diagram, we were able to use analysis techniques similar to those used for continuous systems in the s-plane. We found that all the rules of root-locus construction for continuous systems in the s-plane, presented in Section 6.14[‡], were also valid for discrete systems in the z-plane. The design methods used in this chapter for discrete systems using the Bode and root-locus methods relied on the techniques presented in Chapter 2 for continuous systems. We concluded the design methodology for discrete systems with Ragazzini's direct design method, which is a technique unique to discrete systems.

It is important for the control-system engineer to be able to think and be adept from the dual continuous and discrete-system viewpoints. This capability was emphasized in this chapter, as we related the techniques presented in this chapter on digital control systems to those of the continuous world presented previously in this book. As digital computer control systems are becoming increasingly prevalent in industrial and military applications, the control-system engineer should find the techniques presented in this chapter very useful in practice.

A complete design case study of a sampled-data control systems for controlling the temperature of a liquid in a tank is presented in Section 7.6. The system uses a microcomputer to control the position of a solenoid valve, which then controls the

290 DIGITAL CONTROL-SYSTEM ANALYSIS AND DESIGN

quantity of steam into the tank coil. In this way, the microcomputer controls the temperature of the liquid contained in the tank. This is an interesting problem that integrates the various procedures illustrated in this chapter into a practical design example.

4.15. ILLUSTRATIVE PROBLEMS AND SOLUTIONS

This section provides a set of illustrative problems and their solutions to supplement the material presented in Chapter 4.

I4.1. A sinusoidal function

$$r(t) = 2 \sin 4t$$

is sampled every $T = 0.2$ sec. Determine the z transform of the resultant sampled sequence.

SOLUTION: From item 11 of Table 4.1:

$$R(z) = \frac{z \sin aT}{z^2 - 2z \cos aT + 1}$$

For this problem, $T = 0.2$, $a = 4$, and we must multiply this expression by 2. Therefore:

$$R(z) = 2 \frac{z \sin(0.2)(4)}{z^2 - 2z \cos(0.2)(4) + 1}$$

Therefore:

$$R(z) = 2 \frac{z(0.717)}{z^2 - 2z(0.697) + 1}.$$

This expression can be simplified to the following:

$$R(z) = \frac{1.435z}{z^2 - 1.393z + 1}.$$

I4.2. Determinte the z transform for the following:

$$F(s) = \frac{10}{s(s+4)^2}.$$

4.15. ILLUSTRATIVE PROBLEMS AND SOLUTIONS

SOLUTION: Using partial fraction expansion of $F(s)$, we obtain the following:

$$F(s) = \frac{A}{s} + \frac{B}{(s+4)} + \frac{C}{(s+4)^2}$$

where the constants are given by the following:

$$A = \frac{10}{(s+4)^2}\bigg]_{s=0} = 0.625$$

$$C = \frac{10}{s}\bigg]_{s=-4} = -2.5$$

$$B = \frac{d}{ds}\frac{10}{s}\bigg]_{s=-4} = -10\frac{1}{s^2}\bigg]_{s=-4} = -0.625$$

Therefore,

$$F(s) = \frac{0.625}{s} - \frac{0.625}{(s+4)} - \frac{2.5}{(s+4)^2}.$$

From Table 4.1, we find the z transform of this expression to be:

$$F(z) = 0.625\frac{z}{z-1} - 0.625\frac{z}{z-e^{-4T}} - 2.5\frac{Tze^{-4T}}{(z-e^{-4T})^2}.$$

I4.3. Determine the z transform of the function $r(t)$ drawn in the following figure. Assume that the sampling period is one second.

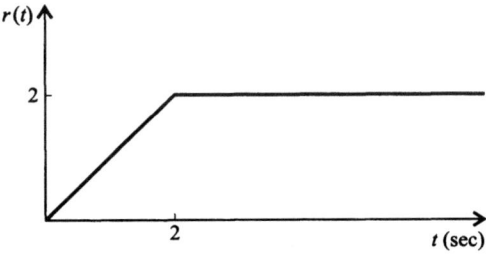

Figure I4.3

SOLUTION: $y(t) = t - (t-2)U(t-2), \quad t \geq 0.$

Therefore,

$$Y(z) = Z[t] - Z[(t-2)U(t-2)]$$

From Table 4.1, we have:

$$Y(z) = \frac{z^{-1}}{(1-z^{-1})^2} - \frac{z^{-1}}{(1-z^{-1})^2} z^{-2}.$$

Therefore, the z transform of $y(t)$ is given by:

$$Y(z) = \frac{z^{-1}(1-z^{-2})}{(1-z^{-1})^2}.$$

14.4. The z transform of a sampled-function, $r(t)$, is given by:

$$R(z) = \frac{z^3}{z^3 - 4.5z^2 + 2.5z - 6}.$$

Determine the z transform of $r(t-3T)U(t-3T)$.

SOLUTION: $Z[r(t-3T)U(t-3T)] = z^{-3}R(z).$

Therefore, the z transform of $r(t-3T)U(t-3T)$ is given by:

$$z^{-3}R(z) = \frac{1}{z^3 - 4.5z^2 + 2.5z - 6}.$$

14.5. Determine the inverse z transform of

$$E(z) = \frac{z(z-0.8)}{(z-1)(z-0.5)}$$

using:

(a) Partial fraction expansion.
(b) Power series expansion.

How do your results compare?

SOLUTION: (a)

$$\frac{E(z)}{z} = \frac{A}{z-1} + \frac{B}{z-0.5}$$

where

$$A = \frac{z-0.8}{z-0.5}\bigg]_{z=1} = 0.4$$

$$B = \frac{z-0.8}{z-1}\bigg]_{z=0.5} = 0.6$$

Therefore,

$$E(z) = 0.4\frac{z}{z-1} + 0.6\frac{z}{z-0.5}.$$

From Table 4.1, the inverse z transform of $E(z)$ is given by:

$$e(k) = 0.4 + 0.6(0.5)^k, \quad k \geq 0$$

(b) The z transform of $E(z)$ can be written as:

$$E(z) = \frac{z^2 - 0.8z}{z^2 - 1.5z + 0.5}.$$

Dividing denominator into numerator, we obtain the following:

$$z^2 - 1.5z + 0.5 \overline{\smash{\big)}\begin{array}{l} 1 + 0.7z^{-1} + 0.55z^{-2} + 0.475z^{-3} + 0.4375z^{-4} + \cdots \\ z^2 - 0.8z \\ \underline{z^2 - 1.5z + 0.5} \\ 0.7z - 0.5 \\ \underline{0.7z - 1.05 + 0.35z^{-1}} \\ 0.55 - 0.35z^{-1} \\ \underline{0.55 - 0.825z^{-1} + 0.275z^{-2}} \\ 0.475z^{-1} - 0.275z^{-2} \\ \underline{0.475z^{-1} - 0.712z^{-2} + \cdots} \\ 0.4375z^{-2} - \cdots \end{array}}$$

This is the series form corresponding to the closed-loop form obtained in part (a). The results in parts (a) and (b) are in agreement.

I4.6. The difference equation representing a digital process is given by the following expression:

$$c(k+2) - c(k+1) + (2/9)c(k) = r(k)$$

where

$$r(k) = 1, k = 0, 1, 2, 3, \ldots$$
$$c(0) = c(1) = 0$$

(a) Determine $C(z)$.

(b) Determine $c(k)$ from finding the inverse z transformation of $C(z)$.

(c) Find $c(100,000)$.

SOLUTION: **(a)** the z transform of the given difference equation is given by:

$$z^2 C(z) - zC(z) + \frac{2}{9}C(z) = \frac{z}{z-1}.$$

Solving for $C(z)$, we obtain the following:

$$C(z) = \frac{z}{(z-1)\left(z-\frac{1}{3}\right)\left(z-\frac{2}{3}\right)}.$$

(b) Using partial fraction expansion, we obtain the following:

$$C(z) = A\frac{z}{z-1} + B\frac{z}{z-\frac{1}{3}} + C\frac{z}{z-\frac{2}{3}}$$

where

$$A = \left[\frac{1}{\left(z-\frac{1}{3}\right)\left(z-\frac{2}{3}\right)}\right]_{z=1} = \frac{9}{2},$$

$$B = \left[\frac{1}{(z-1)\left(z-\frac{2}{3}\right)}\right]_{z=1/3} = \frac{9}{2},$$

$$C = \left[\frac{1}{(z-1)\left(z-\frac{1}{3}\right)}\right]_{z=2/3} = -9.$$

Therefore, $C(z)$ is given by:

$$C(z) = \frac{9}{2}\frac{z}{z-1} + \frac{9}{2}\frac{z}{\left(z-\frac{1}{3}\right)} - 9\frac{z}{\left(z-\frac{2}{3}\right)}.$$

The inverse z transform of $C(z)$ can be obtained from Table 4.1 as follows:

$$c(k) = \frac{9}{2} + \frac{9}{2}\left(\frac{1}{3}\right)^k - 9\left(\frac{2}{3}\right)^k, \quad k \geq 0.$$

(c)

$$c(100,000) \approx \frac{9}{2}.$$

I4.7. The difference equation representing a digital process is given by the following expression:

$$c(k+2) - 5c(k+1) + 6c(k) = r(k)$$

where

$$r(k) = 1, \quad k = 0, 1, 2, 3, \ldots$$

$$c(0) = c(1) = 0$$

Determine $c(k)$.

SOLUTION:
$$z^2 C(z) - 5z C(z) + 6C(z) = \frac{z}{z-1}.$$

Solving for $C(z)$, we obtain:

$$C(z) = \frac{z}{(z-1)(z-2)(z-3)}.$$

$C(z)$ can be expanded into partial fractions as follows:

$$\frac{C(z)}{z} = A \frac{1}{z-1} + B \frac{1}{z-2} + C \frac{1}{z-3}$$

where

$$A = \frac{1}{(z-2)(z-3)}\bigg]_{z=1} = 0.5,$$

$$B = \frac{1}{(z-1)(z-3)}\bigg]_{z=2} = -1,$$

$$C = \frac{1}{(z-1)(z-2)}\bigg]_{z=3} = 0.5.$$

Therefore,

$$C(z) = 0.5 \frac{z}{z-1} - \frac{z}{z-2} + 0.5 \frac{z}{z-3}.$$

Using Table 4.1, the inverse z transform of this equation is given by:

$$c(k) = 0.5 - (2)^k + 0.5(3)^k \ldots, \quad k \geq 0.$$

14.8. A system can be represented by the following,

$$c(k) - 1.2c(k-1) + 0.5c(k-2) = K[r(k-1) + 0.8r(k-2)]$$

where K represents the system gain. Determine the z-transform transfer function, $C(z)/R(z)$, of this difference equation.

SOLUTION:
$$C(z) - 1.2z^{-1}C(z) + 0.5z^{-2}C(z) = K[z^{-1}R(z) + 0.8z^{-2}R(z)]$$

$$C(z)[1 - 1.2z^{-1} + 0.5z^{-2}] = KR(z)[z^{-1} + 0.8z^{-2}].$$

Therefore, the transfer function, $C(z)/R(z)$ is given by:

$$\frac{C(z)}{R(z)} = \frac{K[z^{-1} + 0.8z^{-2}]}{1 - 1.2z^{-1} + 0.5z^{-2}}.$$

14.9. The transfer function of a linear, time-invariant discrete-data system is unknown. To determine its transfer function,

$$G(z) = \frac{C(z)}{R(z)}$$

it is subjected to an input sequence given by

$$r(nT) = 1 \quad \text{for} \quad n \geq 0.$$

The resulting output is modeled by the following sequence:

$$c(nT) = 1 - e^{-4nT} \quad \text{for} \quad n \geq 0.$$

Determine the transfer function $G(z)$.

SOLUTION: From Eq. (4.30):

$$C(z) = \sum_{n=0}^{\infty} c(nT)z^{-n}.$$

Therefore,

$$C(z) = \sum_{n=0}^{\infty} (1 - e^{-4nT})z^{-n} = \frac{z}{z-1} - \frac{z}{z - e^{-4T}}.$$

This can be simplified to:

$$C(z) = \frac{z(1 - e^{-4T})}{(z-1)(z - e^{-4T})}.$$

Since

$$R(z) = \frac{z}{z-1}$$

then, the transfer function, $G(z)$, is given by:

$$G(z) = \frac{C(z)}{R(z)} = \frac{\dfrac{z(1-e^{-4T})}{(z-1)(z-e^{-4T})}}{\dfrac{z}{z-1}}.$$

Therefore,

$$G(z) = \frac{1-e^{-4T}}{z-e^{-4T}}.$$

I4.10. A sinusoidal signal of 6 Hz is applied to an ideal sampler and zero-order hold combination where the sampling rate is 10 Hz. Determine all frequencies present in the output which are less than 50 Hz.

SOLUTION:

Output components	Frequencies at output (Hz)
f_1	6
$f_s - f_1$	4
$f_s + f_1$	16
$2f_s - f_1$	14
$2f_s + f_1$	26
$3f_s - f_1$	24
$3f_s + f_1$	36
$4f_s - f_1$	34
$4f_s + f_1$	46
$5f_s - f_1$	44

I4.11. The responses of two open-loop control systems to an exponential input, $r(t) = e^{-t}U(t)$, are to be compared.

(a) Determine $c(nT)$ for the system shown in Figure I4.11a. Carry your answer out to the first four sampling instants.

(b) Repeat part (a) for the continuous control system shown in Figure I4.11b, and determine $c(t)$ at these same first five sampling instants.

(c) From your results in parts (a) and (b), compare the outputs of these two systems. Explain any differences.

298 DIGITAL CONTROL-SYSTEM ANALYSIS AND DESIGN

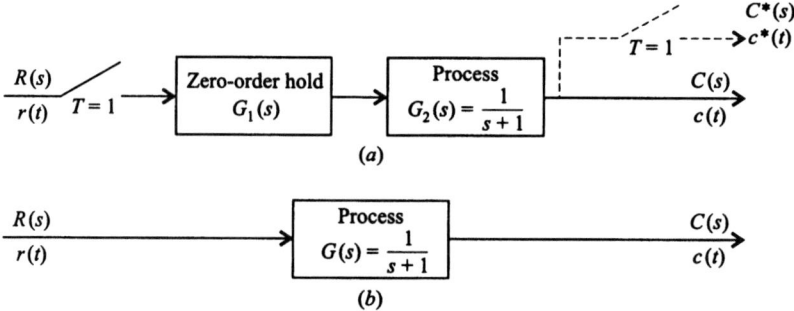

Figure I4.11

SOLUTION: (a)
$$G_1(s)G_2(s) = \frac{1-e^{-s}}{s} \cdot \frac{1}{s+1}$$

By means of partial-fraction expansion, this can be written as

$$G_1(s)G_2(s) = (1-e^{-s})\left(\frac{1}{s} - \frac{1}{s+1}\right).$$

(This agrees with Eq. (4.118) in Section 4.6 of this book.) The z transform of this expression can be obtained from Table 4.1 by inspection as follows:

$$G_{12}(z) = (1-z^{-1})\left(\frac{z}{z-1} - \frac{z}{z-e^{-1}}\right) = \frac{0.63}{z-0.37}.$$

Therefore,

$$C(z) = R(z)G(z) = \frac{z}{z-e^{-1}} \cdot \frac{0.63}{z-0.37} = \frac{0.63z}{(z-0.37)^2} = \frac{0.63z}{z^2 - 0.74z + 0.137}.$$

Dividing the denominator into the numerator, we obtain the following series:

$$\begin{array}{r}
0.63z^{-1} + 0.466z^{-2} + 0.259z^{-3} + 0.128z^{-4} + \cdots \\
z^2 - 0.74z + 0.137 \overline{\smash{)}0.63z} \\
\underline{0.63z - 0.466 + 0.86z^{-1}} \\
0.466 - 0.86z^{-1} \\
\underline{0.466 - 0.345z^{-1} + 0.064z^{-2}} \\
0.259z^{-1} - 0.064z^{-2} \\
\underline{0.259z^{-1} - 0.192z^{-2} + 0.035z^{-3}} \\
0.128z^{-2} - 0.035z^{-3} \\
0.128z^{-2} - \cdots
\end{array}$$

Therefore,

$$C(z) = 0.63z^{-1} + 0.466z^{-2} + 0.259z^{-3} + 0.128z^{-4} + \ldots$$

By utilizing Table 4.1, the inverse z transform can be expressed as

$$c(nT) = 0.63\delta(t-1) + 0.466\delta(t-2) + 0.259\delta(t-3) + 0.128\delta(t-4) + \ldots$$

(b) $$C(s) = R(s)G(s) = \left(\frac{1}{s+1}\right)\left(\frac{1}{s+1}\right) = \frac{1}{(s+1)^2}.$$

From inspection of Appendix A, eighth item (for $n = 2$), the inverse Laplace transform is given by:

$$c(t) = \frac{1}{(2-1)!} t^{2-1} e^{-t} = te^{-t}.$$

Therefore, in terms of this function sampled every second, we can write this as

$$c(nT) = (nT)e^{-(nT)} = 0.37\delta(t-1) + 0.271\delta(t-2) + 0.15\delta(t-3) + 0.073\delta(t-4) + \ldots$$

(c) The values of $c(nT)$ in parts (a) and (b) differ because the input energy to

$$G(s) = \frac{1}{s+1}$$

is different in parts (a) and (b) of this problem.

14.12. An open-loop digital control system containing a sampler whose sampling time is 1 Hz, a digital filter, a zero-order hold, and a process is illustrated:

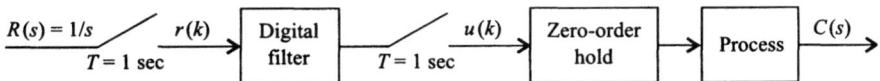

Figure I4.12

Assume that the digital filter solves the following difference equation:

$$u(k+1) = r(k+1) - 0.5r(k) + 0.8u(k)$$

The transfer function of the process is given by

$$G(s) = \frac{1}{(s+1)(s+4)}.$$

Determine the transfer function, $C(z)/R(z)$.

SOLUTION: The z transform of the difference equation is given by

$$zU(z) = zR(z) - 0.5R(z) + 0.8U(z).$$

Therefore, the z-transform transfer function of the digital filter is given by

$$\frac{U(z)}{R(z)} = \frac{z - 0.5}{z - 0.8}$$

The z transform of the zero-order hold and the process is given by

$$Z\left[\frac{1 - e^{-Ts}}{s} \cdot \frac{1}{(s+1)(s+4)}\right] = (1 - z^{-1})Z\left[\frac{1}{s(s+1)(s+4)}\right].$$

Using partial fraction expansion, we obtain the following z transform for the zero-order hold and the process:

$$(1 - z^{-1})Z\left[\frac{0.25}{s} - \frac{0.33}{s+1} + \frac{0.083}{s+4}\right] = \left(\frac{z-1}{z}\right)\left[\frac{0.25z}{z-1} - \frac{0.33z}{z-0.368} + \frac{0.0832z}{z-0.018}\right].$$

Therefore,

$$\frac{C(z)}{R(z)} = \left(\frac{z-0.5}{z-0.8}\right)(z-1)\left[0.25\frac{1}{z-1} - 0.33\frac{1}{z-0.368} + 0.83\frac{1}{z-0.018}\right].$$

$$\frac{C(z)}{R(z)} = \left(\frac{z-0.5}{z-0.8}\right)\left[\frac{(z^2 - 1.368z - 0.0928)}{(4z^2 - 1.544z + 0.0264)}\right].$$

14.13. Determine the transfer function $C(z)/R(z)$ for the discrete-data control system illustrated. Assume that the sampling period is one second.

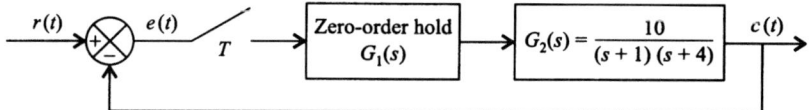

Figure 14.13

SOLUTION: From Problem 14.12, the z transform of $G_1(s)$ and $G_2(s)$ is given by

$$G_{12}(z) = 10\left[\frac{z^2 - 1.368z - 0.0928}{(4z^2 - 1.544z + 0.0264)}\right].$$

Therefore,

$$\frac{C(z)}{R(z)} = \frac{G_{12}(z)}{1 + G_{12}(z)} = \frac{(10z^2 - 13.68z - 0.928)}{(14z^2 - 15.224z - 0.902)}.$$

14.14. Determine the form of the transient response of discrete-time systems whose characteristic equations are given by:

(a) $z^2 - 1.5z + 0.56 = 0$.

(b) $z^2 - 0.49 = 0$.

(c) $z^2 - 1 = 0$.

(d) $z^2 - 1.3z + 0.36 = 0$.

SOLUTION: (a) $\quad z^2 - 1.5z + 0.56 = (z - 0.8)(z - 0.7) = 0$.

Using Table 4.1, we find:
$$c(k) = A(0.8)^k + B(0.7)^k.$$

(b) $\quad z^2 - 0.49 = (z + 0.7)(z - 0.7) = 0$.

Using Table 4.1, we find:
$$c(k) = A(0.7)^k + B(-0.7)^k.$$

(c) $\quad z^2 - 1 = (z + 1)(z - 1) = 0$.

Using Table 4.1, we find:
$$c(k) = A(1)^k + B(-1)^k.$$

(d) $\quad z^2 - 1.3z + 0.36 = (z - 0.9)(z - 0.4) = 0$.

Using Table 4.1, we find:
$$c(k) = A(0.9)^k + B(0.4)^k.$$

14.15. A digital control system is described by the difference (or state) equation given by

$$x(n + 1) = (0.4 - 0.8K)x(n) + Kr(n)$$

where $r(n)$ represents the input and $x(n)$ represents the state variable. Determine the value of K for the system to be stable.

SOLUTION: Taking the z transform of this difference (or state) equation, we obtain:

$$zX(z) = (0.4 - 0.8K)X(z) + KR(z).$$

Therefore, the transfer function, $X(z)/R(z)$ is given by:

$$\frac{X(z)}{R(z)} = \frac{K}{z - 0.4 + 0.8K}.$$

The resulting characteristic equation is given by:

$$z - 0.4 + 0.8K = 0.$$

The root of this characteristic equation is given by

$$z = 0.4 - 0.8K.$$

For stability, it is necessary that

$$|\ 0.4 - 0.8K\ | < 1$$

or

$$0.4 - 0.8K > -1 \quad \text{and} \quad 0.4 - 0.8K < 1$$
$$\therefore K < 1.75 \quad \text{and} \quad \therefore K > -0.75.$$

Therefore, for stability, it is necessary that

$$-0.75 < K < 1.75.$$

14.16. A unity-feedback digital control system has a forward transfer function given by

$$G(z) = \frac{K(z + 0.5)(z + 0.7)}{(z - 0.4)(z - 1.5)}.$$

The Nyquist diagram in the $GH(z)$ plane for a specific value of gain K is illustrated.

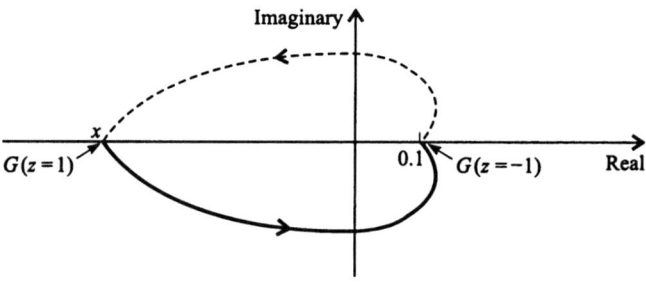

Figure 14.16

(a) Determine the value of gain K that this Nyquist diagram is drawn for.

(b) Determine the location of point "x."

(c) Is the system stable?

SOLUTION: (a) At $z = -1$:

$$G(-1) = 0.1 = \frac{K(-0.5)(-0.3)}{(-1.4)(-2.5)}.$$

Solving for K:

$$K = \frac{0.1(1.4)(2.5)}{(0.5)(0.3)} = 2.33.$$

(b) At $z = 1$:

$$G(1) = \frac{2.33(1.5)(1.7)}{(0.6)(-0.5)} = -19.83 = x.$$

(c) $$N = Z - P$$

where $N = -1$ because there is one counterclockwise encirclement of the $-1, 0$ point, and $P = 1$ because there is one pole of $GH(z)$ outside the unit circle at $z = 1.5$. Therefore,

$$-1 = Z - 1$$

Therefore, $Z = 0$ and the system is stable.

14.17. We wish to analyze the performance of a unity-feedback sampled-data control system whose forward transfer function, $G_p(s)$, is given by

$$G_p(s) = \frac{500}{s(s+10)}$$

and it contains a zero-order hold and a sampler whose sampling frequency will be designed to be one-hundredth of the gain crossover frequency.

(a) Determine the gain crossover frequency, ω_c, analytically without using the graphical Bode plot.

(b) Determine the phase margin and gain margin of this control system analytically without using the graphical Bode plot.

(c) A phase-lead network compensation, $G_c(s)$, given by

$$G_c(s) = \frac{(1+aTs)}{(1+Ts)}$$

is to be added in series with the process' transfer function, $G_p(s)$. Determine the values of a and T in order that the zero factor of $G_c(s)$ cancels the pole of $G_p(s)$ at $s = -10$, and the damping ratio is unity.

(d) Determine the gain crossover frequency, ω_c, analytically without using the graphical Bode plot.

(e) Determine the phase margin and gain margin of this control system analytically without using the graphical Bode plot.

SOLUTION: **(a)** This system can be approximated and analyzed as a continuous system because the sampling frequency is so fast compared to the gain crossover frequency (see Figure 4.9). Therefore,

$$\left| \frac{50}{j\omega_c(0.1j\omega_c + 1)} \right| = 1.$$

Therefore,

$$\omega_c = 21.27 \text{ rad/sec}.$$

(b)
$$\phi = -90° - \tan^{-1}\frac{0.1\omega_c}{1} = -155.7°.$$

Therefore, the phase margin, γ, is given by

$$\gamma = 180° - 155.7° = 24.3°.$$

The gain margin is infinity.

(c)
$$G_c(s)G_p(s) = \frac{500(1 + aTs)}{s(s + 10)(1 + Ts)} = \frac{500(aT)\left(s + \dfrac{1}{aT}\right)}{Ts(s + 10)\left(s + \dfrac{1}{T}\right)}.$$

Therefore, it is necessary that

$$\frac{1}{aT} = 10$$

so that the zero factor of $G_c(s)$ cancels the pole of $G_p(s)$. Therefore,

$$G_c(s)G_p(s) = \frac{500a}{s\left(s + \dfrac{1}{T}\right)}.$$

The transfer function of the closed-loop system is given by:

$$\frac{C(s)}{R(s)} = \frac{\dfrac{500a}{s\left(s + \dfrac{1}{T}\right)}}{1 + \dfrac{500a}{s\left(s + \dfrac{1}{T}\right)}} = \frac{500a}{s^2 + \dfrac{1}{T}s + 500a}.$$

Comparing this equation with that of Eq. (B.3), we obtain the following:

$$\omega_n^2 = 500a$$

$$2\zeta\omega_n = \frac{1}{T}$$

where the damping ratio $\zeta = 1$ (given). Solving these two equations simultaneously, we obtain the following:

$$a = 20 \text{ and } T = 0.005$$

Therefore, the phase-lead compensation network is given by:

$$G_c(s) = \frac{1 + 0.1s}{1 + 0.005s}$$

and

$$G_c(s)G_p(s) = \frac{(500)(20)}{s(s+200)} = \frac{10{,}000}{s(s+200)}.$$

(d)
$$\left| \frac{10{,}000}{j\omega_c(j\omega_c + 200)} \right| = 1.$$

Therefore,

$$\left| \frac{10{,}000}{-\omega_c^2 + 200j\omega_c} \right| = 1.$$

Therefore,

$$\omega_c = 48.6 \text{ rad/sec}.$$

(e)
$$\phi = -90° - \tan^{-1}\frac{0.005\omega_c}{1} = -90° - \tan^{-1}(0.005)(48.6)$$
$$= -90° - 13.66° = -103.66°.$$

Therefore, the phase margin, γ, is given by:

$$\gamma = 180° - 103.66° = 76.34°.$$

The gain margin equals infinity.

I4.18. A sampled-data control system is shown in Figure I4.18(i). It contains a first-order hold $G_1(s)$, and a process $G_2(s)$ where

$$G_2(s) = \frac{2}{(1+s)^2}$$

and the sampling time, T, is 1 sec.

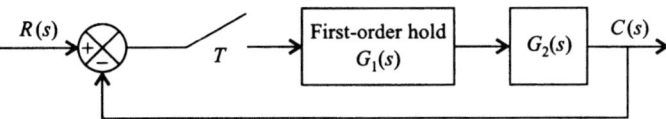

Figure I4.18(i)

306   DIGITAL CONTROL-SYSTEM ANALYSIS AND DESIGN

(a) Draw the Bode diagram of this sampled-data control system, and determine the gain crossover, phase margin, gain margin, and the phase crossover frequency.

(b) A minimum phase margin of 65°, and a minimum gain margin of 6.5 dB are desired. Determine the gain reduction required to achieve these design values.

SOLUTION: (a) $G_1(s)G_2(s) = \left(\frac{(1+s)(1-e^{-s})^2}{s^2}\right)\left(\frac{2}{(1+s)^2}\right) = \frac{2(1-e^{-s})^2}{s^2(1+s)}.$

The z transform of this expression is given by

$$Z[G_1(s)G_2(s)] = Z\left[\frac{2(1-e^{-s})^2}{s^2(1+s)}\right] = 2(1-z^{-1})^2 Z\left[\frac{1}{s^2(1+s)}\right].$$

The partial-fraction expansion of the last bracketed term is given by

$$\frac{1}{s^2(1+s)} = \frac{1}{s^2} - \frac{1}{s} + \frac{1}{s+1}.$$

Therefore, the z transform of $G_1(s)G_2(s)$ is given by

$$G_{12}(z) = 2(1-z^{-1})^2 Z\left[\frac{1}{s^2} - \frac{1}{s} + \frac{1}{s+1}\right]$$

$$= 2\left(\frac{z-1}{z}\right)^2 \left[\frac{z}{(z-1)^2} - \frac{z}{z-1} + \frac{z}{z-e^{-1}}\right]$$

$$G_{12}(z) = \frac{0.7358(z+0.7181)}{z(z-0.3679)}. \qquad (14.18\text{-}1)$$

Using Eq. (4.175) with $T = 1$ sec, we obtain the following relationship between the z- and the w-planes:

$$z = \frac{1+\frac{w}{2}}{1-\frac{w}{2}} \qquad (14.18\text{-}2)$$

Substituting (14.18-2) into Eq. (14.18-1), we obtain the following:

$$G_{12}(w) = \frac{2.02(1+0.082w)(1-0.5w)}{(1+1.0821w)(1+0.5w)}.$$

The resulting Bode diagram is shown in Figure I4.18(ii). It shows that the gain crossover frequency = 1.6197 rad/sec, the phase margin =

49.2689°, the gain margin = 5.5387 dB, and the phase crossover frequency = 3.5211 rad/sec.

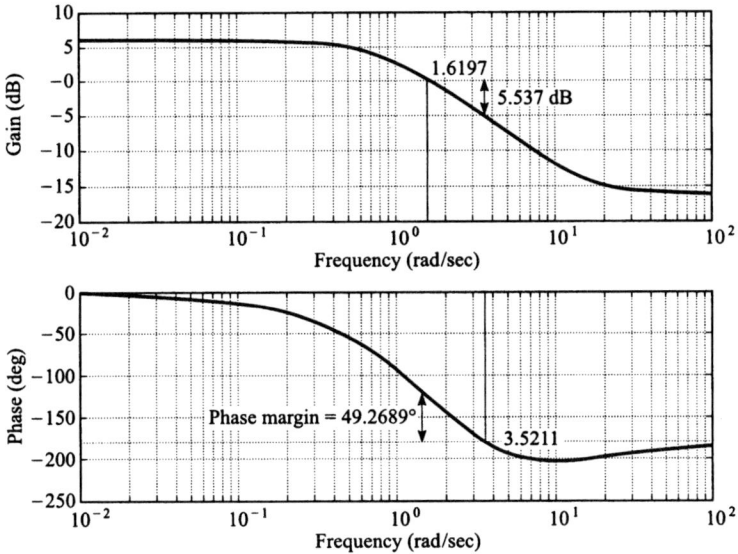

Figure I4.18(ii) Bode Diagram with $K = 2$.

(b) When the gain is reduced from 2.02 to 1.78, we achieve the following: gain crossover frequency = 1.3079 rad/sec; phase margin = 65°; gain margin = 6.83 dB; phase crossover frequency = 3.5211 rad/sec (see Bode diagram in Figure I4.18(iii)).

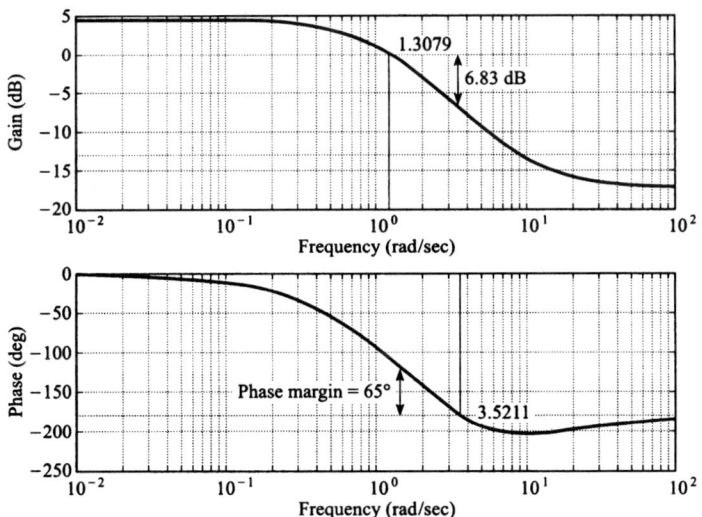

Figure I4.18(iii) Bode diagram with $K = 1.78$.

14.19. The performance of the following digital control system is to be determined from knowledge of the root locus:

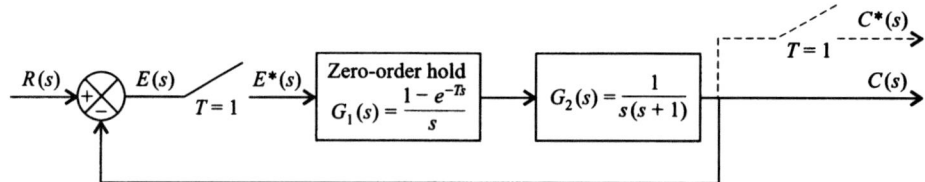

Figure I4.19

(a) Determine the z transform of the open-loop system.

(b) Determine the closed-loop transfer function, $C(z)/R(z)$.

(c) Determine and locate in the z-plane the closed-loop poles.

(d) Determine the damping ratio of this closed-loop control system from knowledge of the exact location of the roots and by using Figure 4.40.

(e) Determine the damping ratio of a comparable analog control system in which the sampler and zero-order hold are removed.

(f) What conclusions can you reach by comparing your results in parts (d) and (e).

SOLUTION: (a)
$$G_{12}(z) = Z[G_1(s)G_2(s)] = Z\left[\left(\frac{1-e^{-s}}{s}\right)\left(\frac{1}{s(s+1)}\right)\right],$$

$$G_{12}(z) = (1-z^{-1})Z\left(\frac{1}{s^2(s+1)}\right),$$

$$G_{12}(z) = \left(\frac{z-1}{z}\right)Z\left(\frac{1}{s^2} - \frac{1}{s} + \frac{1}{s+1}\right)$$

$$= \left(\frac{z-1}{z}\right)\left[\frac{z}{(z-1)^2} - \frac{z}{z-1} + \frac{z}{z-e^{-1}}\right],$$

$$G_{12}(z) = \frac{0.367(z+0.72)}{(z-1)(z-0.367)}.$$

(b)
$$\frac{C(z)}{R(z)} = \frac{G_{12}(z)}{1+G_{12}(z)} = \frac{\frac{0.367(z+0.72)}{(z-1)(z-0.367)}}{1+\frac{0.367(z+0.72)}{(z-1)(z-0.367)}},$$

$$\frac{C(z)}{R(z)} = \frac{0.367z+0.264}{z^2-z+0.632}.$$

(c) The characteristic equation is given by

$$z^2 - z + 0.632 = 0.$$

Therefore, the roots are located in the z-plane at

$$z_1 = 0.5 + 0.618j \text{ and } z_2 = 0.5 - 0.618j.$$

(d) From Figure 4.40, the damping ratio equals 0.25.

(e)
$$\frac{C(s)}{R(s)} = \frac{G_2(s)}{1 + G_2(s)} = \frac{1}{s^2 + s + 1} = \frac{\omega_n^2}{s^2 + 2\delta\omega_n s + \omega_n^2}.$$

Setting like coefficients toether, we obtain the following two simultaneous equations to solve:

$$\omega_n^2 = 1,$$

$$2\delta\omega_n = 1.$$

Solving these two equations simultaneously, we obtain

$$\omega_n = 1 \quad \text{and} \quad \delta = 0.5.$$

(f) The analog version of this control system, which does not contain the sampler and zero-order hold, has a damping ratio of 0.5. The digital version has a damping ratio of 0.25. The sampler and zero-order hold have the effect of increasing the phase lag in the control system, decreasing its relative stability, and decreasing its damping ratio.

I4.20. A sampled-data control system is shown in Figure I4.20(i). It conains a zero-order hold $G_1(s)$, and a process $G_2(s)$ where

$$G_2(s) = \frac{K}{(s+1)(s+2)}.$$

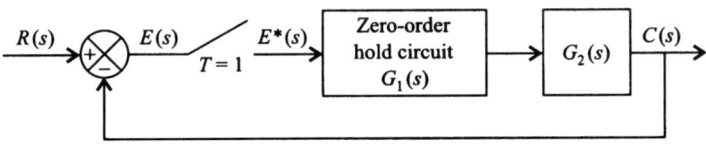

Figure I4.20(i)

(a) Draw the root locus of this sampled-data control system and find K_{max} if the sampling time, T, is 1 sec.

(b) The value of K_{max} can be increased if the sampling time is made faster. Draw the root locus of this sampled-data control system if the sampling time, T, is halved to 0.5 sec and find K_{max} for this system.

310   DIGITAL CONTROL-SYSTEM ANALYSIS AND DESIGN

SOLUTION: (a) $G_{12}(z) = Z[G_1(s)G_2(s)] = Z\left[\left(\frac{1-e^{-Ts}}{s}\right)\left(\frac{K}{(s+1)(s+2)}\right)\right]$, (I4.20-1)

$$G_{12}(z) = K(1-z^{-1})Z\left[\frac{0.5}{s} - \frac{1}{s+1} + \frac{0.5}{s+2}\right]$$
$$= K\left(\frac{z-1}{z}\right)\left[\frac{0.5z}{z-1} - \frac{z}{z-e^{-1}} + \frac{0.5z}{z-e^{-2}}\right] \quad (I4.20\text{-}2)$$

$$G_{12}(z) = \frac{0.2K(z+0.375)}{(z-0.368)(z-0.135)}. \quad (I4.20\text{-}3)$$

The root-locus plot for this part is shown in Figure I4.20(ii). It was obtained using MATLAB, and shows that $K_{\max}$ equals 12.44 at a break-in point located at $s = -0.9914, 0$.

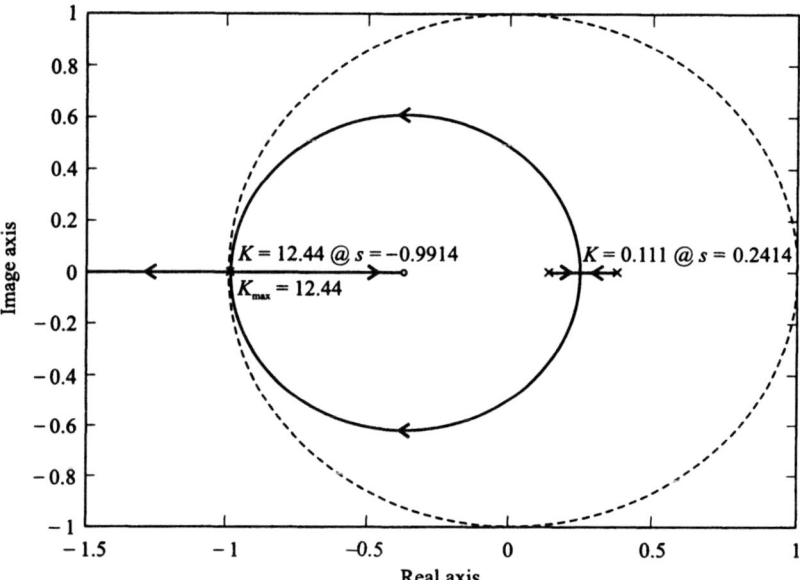

Figure I4.20(ii)   Root locus for the case where the sampling time = 1 sec.

(b) With a sampling rate of 0.5 sec, Eq. (I4.20-2) is modified to the following:

$$G_{12}(z) = K\left(\frac{z-1}{z}\right)\left[\frac{0.5z}{z-1} - \frac{z}{z-e^{-0.5}} + \frac{0.5z}{z-e^{-1}}\right]. \quad (I4.20\text{-}4)$$

Therefore,

$$G_{12}(z) = \frac{0.85K(z+0.47)}{(z-0.6)(z-0.368)} \quad (I4.20\text{-}5)$$

The root-locus plot for this part is shown in Figure I4.20(iii). It was obtained using MATLAB, and shows that the system is stable for the following conditions:

$$0 < K < 19.5.$$

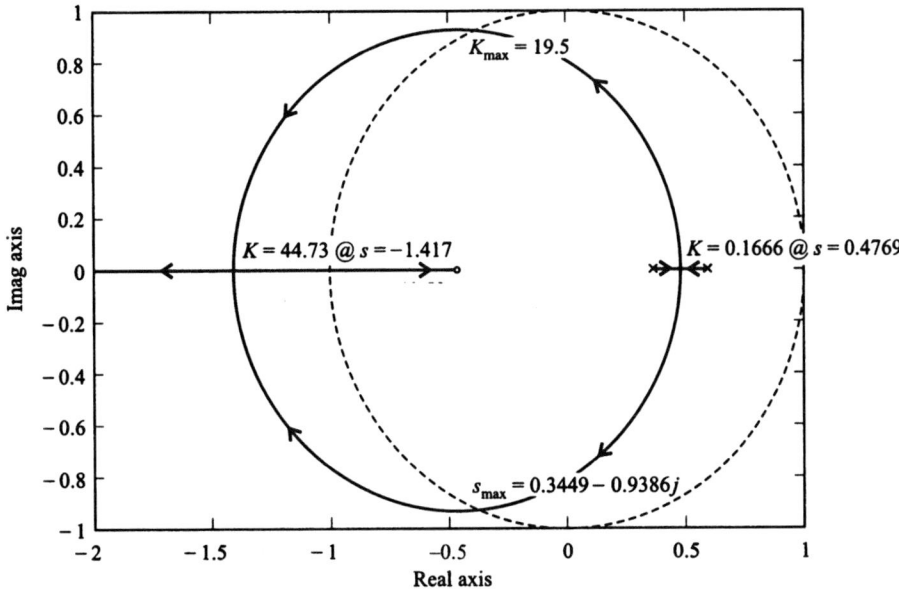

Figure I4.20(iii) Root locus for the case where the sampling time = 0.5 sec.

PROBLEMS

4.1. Prove item 8 of Table 4.1.

4.2. Prove item 12 of Table 4.1.

4.3. Find the z transform of

$$f(t) = \frac{\sin t}{t}.$$

4.4. Determine the z transform of

$$X(s) = \frac{4s}{(s+1)^2(s+2)}$$

by using the partial fraction method.

4.5. Determine the z transform in closed form for the following function:

$$r(t) = e^{-a(t-3T)}u(t-3T).$$

4.6. Determine the inverse z transform of

$$E(z) = \frac{z(z-0.4)}{(z-1)(z-0.2)}$$

using:

(a) Partial fraction expansion.

(b) Power series expansion.

How do your results compare?

4.7. Determine the inverse z transform of

$$E(z) = \frac{z}{(z-1)(z-0.5)}$$

using:

(a) Partial fraction expansion.

(b) Power series expansion.

4.8. A triangular input, $r(t)$ is applied to the low-pass filter shown in Figure P4.8.

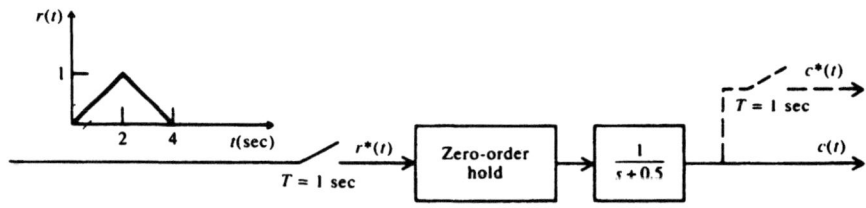

Figure P4.8

(a) What is the z transform of $r^*(t)$?

(b) What is the z transform of $c^*(t)$?

(c) Find $c^*(t)$.

(d) Apply the initial- and final-value theorems to obtain $c(0)$ and $c(\infty)$, respectively.

4.9. Repeat Problem 4.8 if the input $r(t)$ is given by the waveform shown in Figure P4.9.

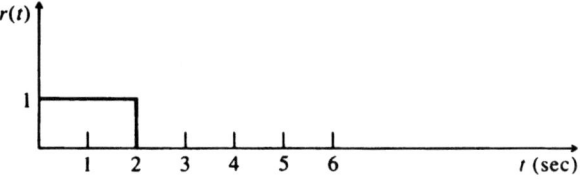

Figure P4.9

4.10. Determine the z transforms corresponding to the following transfer functions:

(a) $G(s) = \dfrac{1}{s(s+1)^2}$,

(b) $G(s) = \dfrac{1}{(s+1)(s+2)}$,

(c) $G(s) = \dfrac{1}{s^2 + 10s + 25}$.

4.11. Determine the value of $c(k)$ for the following difference equation by using the z-transform method:

$$c(k) + 4c(k-1) + 3c(k-2) = u(k)$$

where

$$u(k) = \begin{cases} 1, & k = 0, 1 \\ 0, & k \geq 2 \end{cases}$$
$$c(-1) = c(-2) = 0.$$

4.12. The difference equation representing a digital process is given by the following expression:

$$c(k+2) - 3c(k-1) + 2c(k) = r(k)$$

where

$$r(k) = 1, \quad k = 0, 1, 2, 3, \ldots$$
$$c(0) = c(1) = 0.$$

Determine $c(k)$.

4.13. Derive the z transform of the output, $C(z)$, for the sampled-data feedback control systems shown in Figure P4.13.

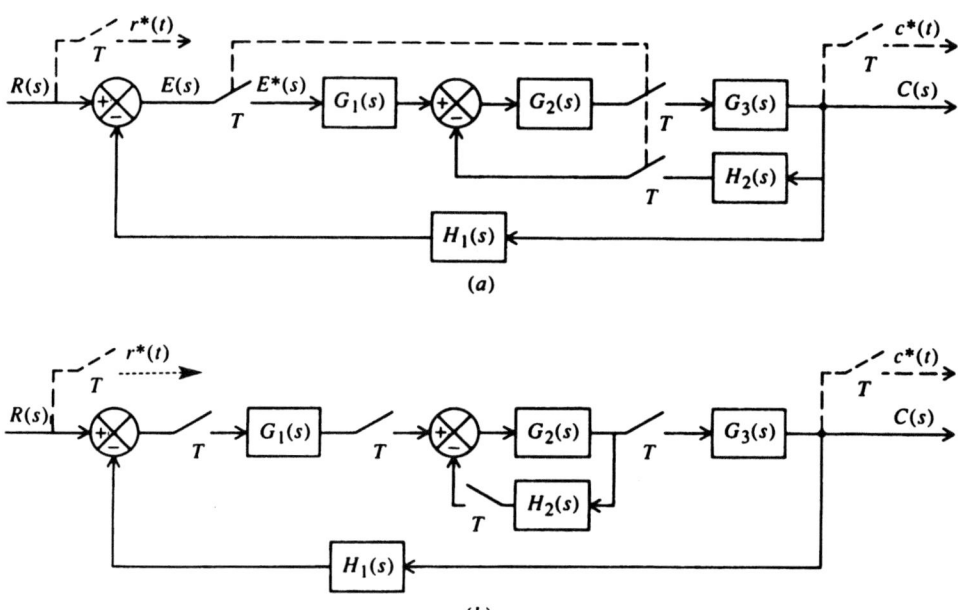

Figure P4.13

4.14. Determine the transfer function $C(z)/R(z)$ for the system shown in Figure P4.14.

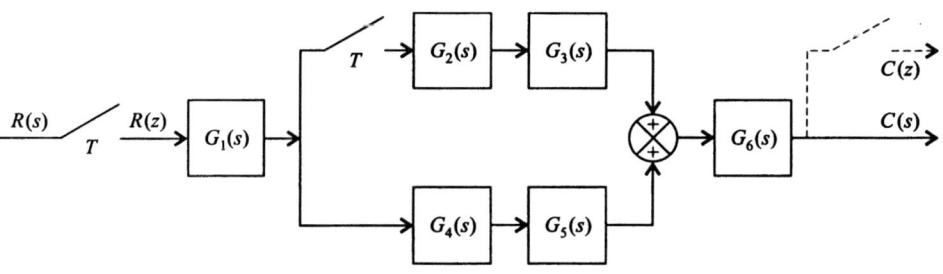

Figure P4.14

4.15. A second-order digital filter having the configuration shown in Figure P4.15 is to be designed:

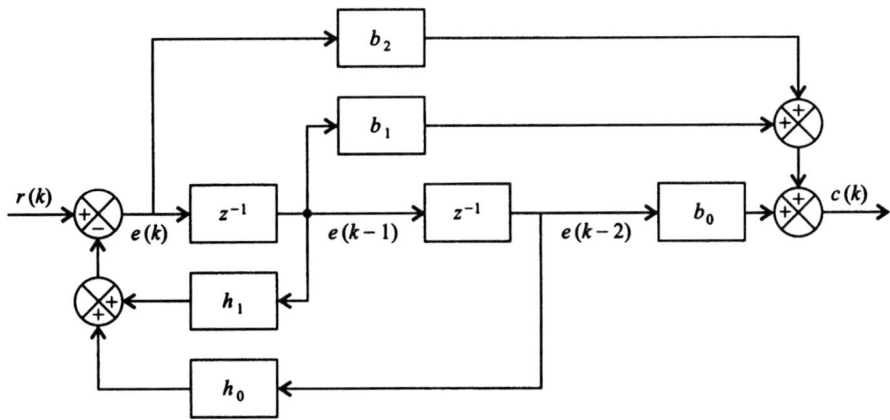

Figure P4.15

It is desired that the z-transform transfer function of this filter should be:

$$\frac{C(z)}{R(z)} = \frac{z^2 - 4z + 2}{z^2 - 3z + 2}.$$

(a) Determine the values of the coefficients b_0, b_1, b_2, h_0, and h_1 to satisfy the desired filter's transfer function.

(b) Determine the two difference equations required to describe this second-order digital filter. [Hint: Write one difference equation at $e(k)$ in terms of $r(k)$, $e(k-1)$, and $e(k-2)$; write the second difference equation at $c(k)$ in terms of $e(k)$, $e(k-1)$, and $e(k-2)$].

4.16. The linear system shown in Figure P4.16 is subjected to an input which is sampled and held by a zero-order holding circuit. Assume that the impulse sampling approximation is applicable, and that the sampling interval T, is 1 sec.

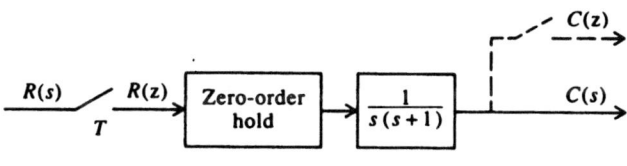

Figure P4.16

(a) Determine the z transfer function relating the sampled input $R(z)$ and the sampled output $C(z)$.

(b) For a step function at the input $R(s)$, determine the z transform of the output $C(z)$. Sketch the impulse sequence, $c^*(t)$.

(c) For a unit-ramp input, determine the z transform of the output $C(z)$. Sketch the impulse sequence $c^*(t)$.

4.17. Repeat Problem 4.16 if the sampling interval is changed to 0.1 and 10 sec.

4.18. The system in Figure P4.18 is to be analyzed.

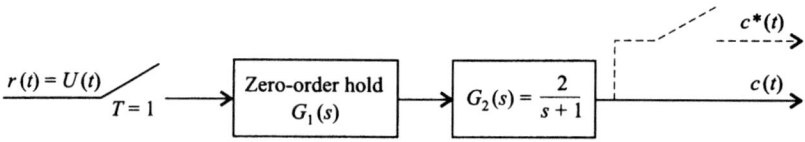

Figure P4.18

(a) Determine the value of $G(z) = G_{12}(z)$.

(b) Determine the value of $C(z)$ for a unit step input at $R(z)$.

(c) Determine the value of $c^*(t)$.

(d) Determine the value of $c(0)$ using the initial-value theorem. Check your answer from your result in part (c).

(e) Determine the value of $c(\infty)$ using the final-value theorem. Check your answer from your result in part (c).

4.19. Consider the open-loop control system in Figure P4.19 where the sampling time is 1 sec.

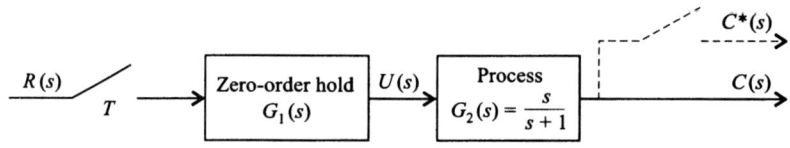

Figure P4.19

(a) Determine the output response at the sampling instant when the input, $R(s)$, is a unit step. Sketch $c^*(t)$.

(b) Check your result of part (a) by considering the input to the process, $U(s)$, and then determining $c(t)$ from continuous-data methods.

4.20. Repeat Problem 4.16 if the extrapolating circuit is a first-order hold.

4.21. Repeat Problem 4.17 for the condition of Problem 4.20.

4.22. The sampled-data control system illustrated in Figure P4.22 contains a sampler in the error channel followed by a zero-order hold circuit and a linear continuous plant whose transfer function $G_2(s)$ is given by

$$G_2(s) = \frac{K}{s^2(s+4)}.$$

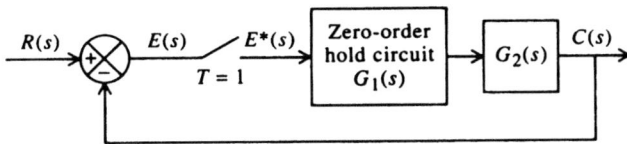

Figure P4.22

(a) Using the Nyquist diagram, determine if the system is stable when $K = 1$.

(b) Check your answer to part (a) using the bilinear transformation.

4.23. Repeat Problem 4.22 if

$$G_2(s) = \frac{K(s+2)}{s(s+4)(s+8)}.$$

In addition, find the maximum value of K before the system becomes unstable, and the system's response to a unit step input and a unit ramp input.

4.24. A feedback control system has a forward transfer function given by

$$G(s) = \frac{K}{s^2(s+1)}.$$

The error channel and feedback channel are sampled at a sampling interval of 1 sec as shown in the block diagram of Figure P4.24. The error and output impulse trains are extrapolated by means of zero-order holding circuits.

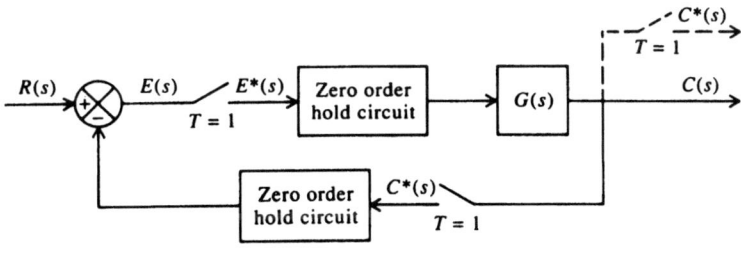

Figure P4.24

(a) Determine the open-loop z-transfer functions $GH(z)$.

(b) Using the Nyquist diagram, determine whether the system is stable when $K = 1$.

4.25. Find $f(0)$ and $f(\infty)$ in each case:

(a) $F(z) = \dfrac{(z+1)(z+2)}{(z-0.33)(z^2-0.5z+0.5)}$,

(b) $F(z) = \dfrac{(z-2)^2}{(z+0.33)(z+0.5)(z-0.67)}$,

(c) $F(z) = \dfrac{(z-1)}{z^2-3z+1}$.

4.26. Consider the open-loop, sampled-data control system illustrated in Figure P4.26. Assume that the sampling rate T is 1 sec and that the input $r(t)$ is a unit pulse lasting for 5 sec. Using z-transform theory, determine the following:

(a) The z transform of the input, $R(z)$.

(b) The z transform of the output, $C(z)$.

(c) The value of $c^*(t)$.

(d) Compute and sketch the value of $c(t)$ using the advanced z transform. Assume that $\Delta = 0.25, 0.5,$ and 0.75.

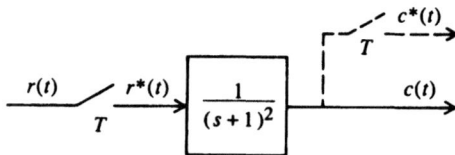

Figure P4.26

4.27. Using the bilinear transformation, determine the stability of sampled-data control systems whose characteristic equations are given by

(a) $z^2 + 1.5z - 1 = 0$.

(b) $z^2 - z + 0.25 = 0$.

(c) $z^3 - 3z^2 + 2.25z - 0.5 = 0$.

4.28. Repeat Problem 4.27 using the Schur–Cohn stability criterion.

4.29. The z transform of a system's response is given by

$$C(z) = \dfrac{z}{(z-1)(z-0.5)^2}.$$

(a) Determine the resulting response given by the final-value theorem.

(b) Find the value of $c(k)$ by taking the inverse transform of $C(z)$ using partial-fraction expansion and a table of z transforms.

(c) Do your answers to parts (a) and (b) agree?

4.30. Determine the maximum value of gain K that can be designed into the sampled-data control systems illustrated in Figure P4.30 before the system becomes unstable. Assume that the sampling period is 1 sec.

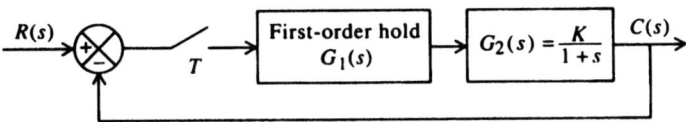

Figure P4.30

4.31. Consider the digital control system in Figure P4.31.

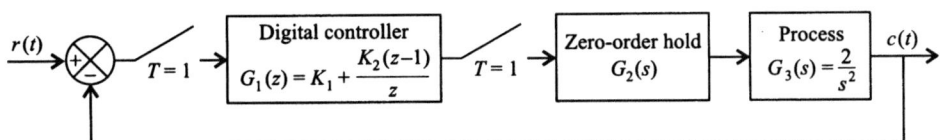

Figure P4.31

(a) Determine the characteristic equation for this system.

(b) It is desired to have two of the roots of the characteristic equation located at $z = 0.4$ and $z = 0.4$. Determine the values of K_1 and K_2 to achieve this.

(c) Determine the location of the third root.

4.32. It is desired to design a discrete compensation network for the sampled-data control system illustrated in Figure P4.32 using the Bode-diagram method. The process to be controlled has the following transfer function

$$G(s) = \frac{1}{s(1+s)}$$

and contains a zero-order hold. Assume that the sampling rate of the control system is 0.5 sec.

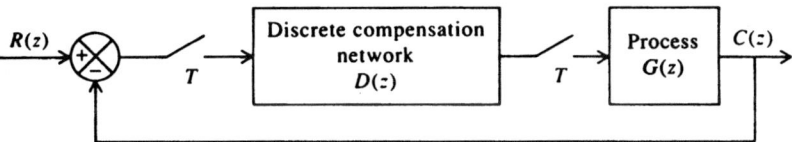

Figure P4.32

(a) Draw the Bode diagram in the w-plane for the uncompensated system. Determine the resulting bandwidth and phase margin.

(b) Design the discrete compensation network so that the resulting bandwidth is approximately the same, but the phase margin is increased to approximately 55°.

4.33. Repeat the sampled-data problem analyzed in Figure 4.31 of Section 4.9, with the sampling rate increased from 0.26 sec to 0.056 sec. What effect does the faster sampling rate have on the resulting system? Design the compensated system for a gain crossover frequency of 1 rad/sec and a minimum phase margin of 25°.

4.34. Consider the digital control system in Figure P4.34.

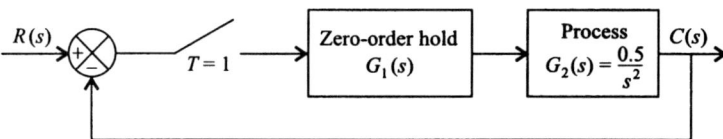

Figure P4.34

(a) Determine the open-loop transfer function, $G_1G_2(z)$.

(b) Determine the closed-loop transfer function, $C(z)/R(z)$.

(c) Determine the characteristic equation of this system.

(d) Determine the location of the roots of the characteristic equation.

(e) Is this system stable?

4.35. Repeat the analysis of the digital control system illustrated in Figure 4.31 with the sampling time decreased to 0.1 sec from 0.26 sec. Determine the following:

(a) $GH(z)$

(b) $GH(v)$

(c) From the Bode diagram, determine the gain crossover frequency, the phase margin, and the gain margin. Is the system stable or unstable?

4.36. The block diagram of a sampled-data control system is illustrated in Figure P4.36. Assume that the sampling rate is 1 sec.

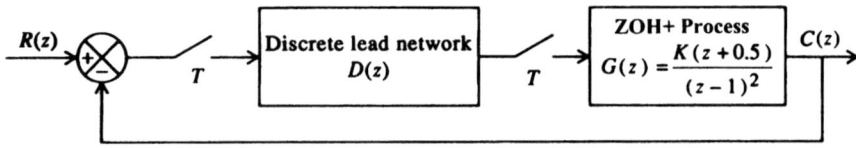

Figure P4.36

(a) Assuming that $D(z) = 1$, sketch the root locus of this system as a function of K. Label all pertinent points. Is the system stable or unstable?

(b) Design a discrete lead network $D(z)$, so that the dominant poles are at $\zeta = 0.7$ and $\theta = \pm 36°$ (the angle made by the radial lines shown emanating from the origin) in Figure 4.40.

(c) Plot the closed-loop response to a unit step input for the compensated system. How does the resulting overshoot compare to that predicted for a $\zeta = 0.7$?

4.37. It is desired to compensate a sampled-data control system using Ragazzini's method. Assume that it is desired to respond perfectly to a ramp, and that a minimal prototype system is desired. For the system shown in Figure P4.37, assume that the data extrapolator is a zero-order hold, the sampling rate is 1 sec, and $G_2(s)$ is given by

$$G_2(s) = \frac{2}{s(s+1)}.$$

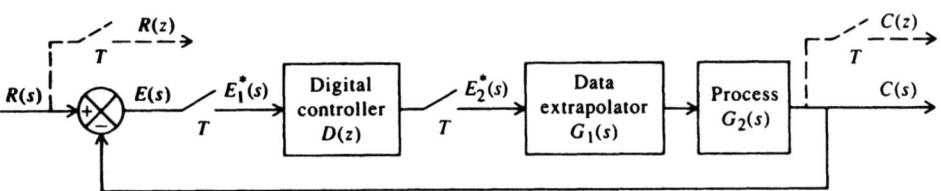

Figure P4.37

(a) Applying Ragazzini's method, design $D(z)$.

(b) Determine and sketch the response to a unit step input.

(c) Determine and sketch the response to a unit ramp input.

4.38. Illustrate how the response of the problem analyzed in Problem 4.37 can be softened by the addition of a staleness factor term where the staleness factor C is given by

(a) $C = 0.2$.

(b) $C = 0.8$.

4.39. Repeat Problem 4.36 if the transfer function of the process is given by

$$G(z) = \frac{K(z+0.2)}{(z-1)^2}.$$

4.40. Repeat Problem 4.32 if the transfer function of the process is given by

$$G(s) = \frac{3}{s(0.25s+1)}.$$

4.41. Numerical integration can be approximated by summing rectangular areas under a continuous function, $y(t)$, as shown in Figure P4.41.

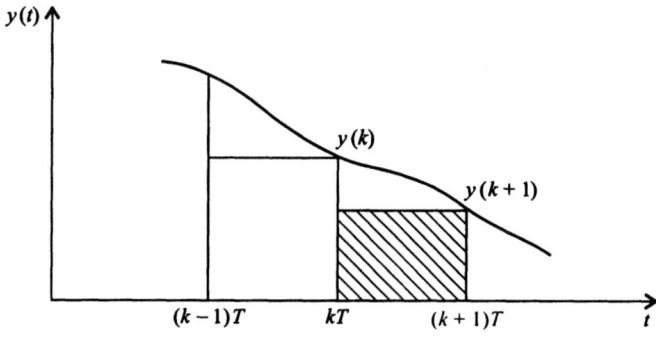

Figure P4.41

(a) Defining the integral of $y(t)$ as $x(t)$, write the difference equation relating $x[(k+1)]$, $x(k)$, and $y[(k+1)]$ for this method.

(b) Determine the transfer function for this rule of numerical integration, $X(z)/Y(z)$.

4.42. Numerical integration can be approximated by summing trapezoidal areas under a continuous function $y(t)$ as shown in Figure P4.42.

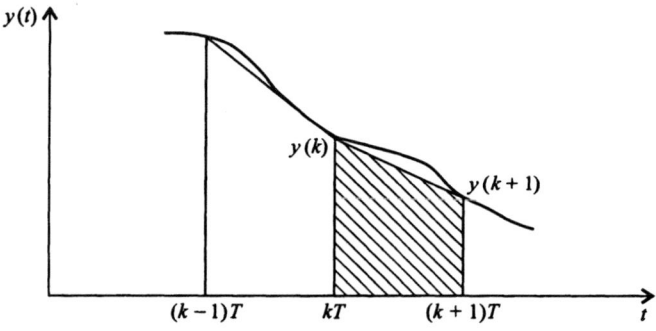

Figure P4.42

(a) Defining the integral of $y(t)$ as $x(t)$, write the difference equation relating $x(k+1)$, $x(k)$, and $y(k+1)$ for this method.

(b) Determine the transfer function for this rule of numerical integration, $X(z)/Y(z)$.

4.43. A continuous-type phase-lead network, used to compensate a continuous control system, is given by the following transfer function:

$$D(s) = 0.50 \frac{(4s+1)}{(2s+1)}.$$

It produces a phase-lead of 19.33° at a critical frequency of 0.4 rad/sec.

(a) Determine the corresponding digitized phase-lead network assuming that the sampling time is 1 sec, and we wish to match the gains at dc.

(b) Is the digitized phase-lead network stable?

(c) Compare the phase lead of the digitized phase-lead network with that of the continuous-type phase-lead network at the critical frequency of 0.4 rad/sec.

4.44. The digital control system shown in Figure P4.44 contains a process containing a double integration. Assume that the sampling time T is $\sqrt{2}$ sec.

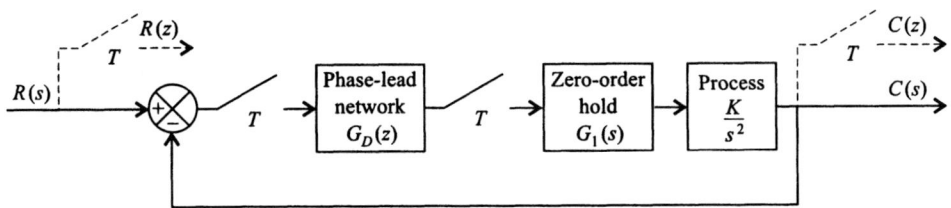

Figure P4.44

(a) Determine the z transform of the zero-order hold and the double integration. Assume that the compensating phase-lead network has a gain of one in this part.

(b) Plot the root locus for this control system. Determine all pertinent points of the root locus including points of breakaway and break-in along the real axis. What does the root locus indicate regarding the stability of this control system?

(c) Compensate this control system by the addition of a phase lead network, $G_D(s)$, so that the control system is conditionally stable.

(d) Draw the root locus for the compensated control system and indicate all pertinent points of the root locus. Determine the maximum value of K where the root locus intersects the unit circle using the modified Routh–Hurwitz criterion.

(e) For a desired damping ratio of 0.45, determine the value of K which will achieve this damping ratio.

(f) Determine the response of the compensated control system with the value of K found in part (e) to a unit step input. What is the maximum percent overshoot to the unit step input? What damping ratio does this maximum percent overshoot correspond to? Does this damping ratio agree with the desired damping ratio of 0.45?

(g) What would happen if noise entered this digital control system at the input to the process?

REFERENCES

1. J. J. Craig, *Introduction to Robotics.* Addison-Wesley, Reading, MA, 1986.
2. G. F. Franklin, J. D. Powell, and M. L. Workman, *Digital Control of Dynamic Systems,* 3rd ed. Addison-Wesley, Reading, MA, 1994.
3. K. Ogata, *Discrete-Time Control Systems.* Prentice-Hall, Englewood Cliffs, NJ, 1987.
4. C. H. Houpis and G. B. Lamont, *Digital Control Systems Theory, Hardware, Software.* McGraw-Hill, New York, 1985.
5. C. L. Phillips and H. T. Nagle, Jr., *Digital Control System Analysis and Design.* Prentice-Hall, Englewood Cliffs, NJ, 1984.
6. C. Shannon, "The philosophy of pulse code modulation." *Proc. IRE* **36**(11), 1324–1331 (1948).
7. J. R. Ragazzini and G. F. Franklin, *Sampled-Data Control Systems.* McGraw-Hill, New York, 1958.
8. R. L. Oldenbourg and R. Sartorious, *The Dynamics of Automatic Control,* Chapter 5. American Society of Mechanical Engineers, New York, 1948.
9. E. I. Jury, "A simplified stability criterion for linear discrete systems." *Proc. IRE* **50**, 1493 (1962).
10. E. I. Jury and J. Blanchard, "A stability test for linear discrete systems in tabular form. *Proc. IRE* **49**, 1947 (1961).
11. E. I. Jury, *Theory and Application of the z-Transform Method.* Wiley, New York, 1964.
12. B. C. Kuo, *Analysis and Synthesis of Sampled-Data Control Systems.* Prentice-Hall, Englewood Cliffs, NJ, 1963.
13. MATLAB™ for MS-DOS Personal Computers *User's Guide*, Control System Toolbox. MathWorks, Inc., Natick, MA 1997.
14. J. E. Bertram, "Factors in the design of digital controllers for sampled-data feedback control systems." *Trans. Am. Inst. Electr. Eng.*, Pap. 56-209 (1956).

5

NONLINEAR CONTROL-SYSTEM DESIGN

5.1. INTRODUCTION

The feedback control-system design methods presented in previous chapters were restricted to linear constant systems, that is, systems that can be represented by linear differential equations with constant coefficients. In practice, linear systems possess the property of linearity only over a certain range of operation; all physical systems are nonlinear to some degree. Therefore it is important that one acquire a facility for analyzing control systems with varying degrees of nonlinearity.

Any attempt to restrict attention strictly to linear systems can only result in severe complications in system design. To operate linearly over a wide range of variation of signal amplitude and frequency would require components of an extremely high quality; such a system would probably be impractical from the viewpoints of cost, space, and weight. In addition, the restriction of linearity severely limits the system characteristics that can be realized.

In practice, linear operation is required only for small deviations about a quiescent operating point. The saturation of amplifying devices having large deviations about the quiescent operating point is usually acceptable. The presence of nonlinearities in the form of dead zones for small deviations about the quiescent operating point is also usually acceptable. In both cases one attempts to limit the effects of nonlinearities to acceptable tolerances, as it is impractical to eliminate the problem entirely.

It is worth noting that nonlinearities may be intentionally introduced into a system in order to compensate for the effects of other undesirable nonlinearities, or to obtain better performance than could be achieved using linear elements only. A simple example of an intentional nonlinearity is the use of nonlinear damping (see Section 4.5‡) to optimize response as a function of the error [1]. The on-off contactor (relay) servo, where full torque is applied as soon as the error exceeds a specified value, is another case of an intentionally nonlinear system.

The purpose of this chapter is to examine the broad aspects of nonlinear systems. We first study the characteristics of nonlinearities and then present several methods for analysis and design of nonlinear control systems. We follow in Chapter 6 with several illustrations of synthesis of nonlinear systems having intentional nonlinearities utilizing optimal control theory. A case study of a positioning system containing nonlinearities is presented in Chapter 7.

We should emphasize here that methods of analyzing nonlinear systems have not progressed as rapidly as have techniques for anlayzing linear systems. Comparatively speaking, at the present time we are still in the development stage. However, the various methods presented in this chapter will enable one to analyze and synthesize nonlinear control systems quantitatively.

5.2. NONLINEAR DIFFERENTIAL EQUATIONS

A linear differential equation of the nth order, with constant coefficients, is written

$$A_n \frac{d^n y(t)}{dt^n} + A_{n-1} \frac{d^{n-1} y(t)}{dt^{n-1}} + \cdots + A_0 y(t) = x(t), \qquad (5.1)$$

where $x(t)$ represents the input to the system, t represents time and is the independent variable, $y(t)$ represents the dependent variable, or the output of the system, and $A_n, A_{n-1}, \ldots, A_0$ are constants.

This equation is of the form derived for several representative mechanical and electrical systems in Chapter 3[‡]. For example, Eq. (3.19)[‡], which is repeated below, gave the differential equation of motion for a mechanical system which consists of a force $f(t)$ applied to a mass, damper, and spring:

$$M \frac{d^2 y(t)}{dt^2} + B \frac{dy(t)}{dt} + Ky(t) = f(t).$$

The mass of the system is represented by the constant M, the damping factor by the constant B, and the spring constant by K.

Detailed solutions for the class of differential equations having the form shown in Eq. (5.1) are available. They have been studied extensively, and several powerful techniques, such as the Laplace transformation, exist for their solution. All the analytical methods discussed in Chapter 1 and Chapter 2 are based on systems that can be represented by simple differential equations having this general form.

If any of the coefficients $A_n, A_{n-1}, \ldots, A_0$ are functions of the independent variable time, then the linear differential equation is said to have variable coefficients. In this case, the differential equation takes the following form:

$$A_n(t) \frac{d^n y(t)}{dt^n} + A_{n-1}(t) \frac{d^{n-1} y(t)}{dt^{n-1}} + \cdots + A_0(t)y(t) = x(t), \qquad (5.2)$$

where $A_n, A_{n-1}, \ldots, A_0$ are all functions of time. Except in special cases (such as when the coefficients are polynomials), the solution of linear time-variable equations is quite complex.

If the coefficients of the differential equation are functions of the dependent variable $y(t)$, then a nonlinear differential equation results. Its general form is

$$A_n \frac{d^n y(t)}{dt^n} + A_{n-1} \frac{d^{n-1} y(t)}{dt^{n-1}} + \cdots + A_1 \frac{dy(t)}{dt} + A_0 y(t)$$
$$+ \epsilon f\left(y(t), \frac{dy(t)}{dt}, \ldots, \frac{d^{n-1} y(t)}{dt^{n-1}}\right) = x(t), \qquad (5.3)$$

where $x(t)$ represents the input to the system, t represents time and is the independent variable, $y(t)$ represents the dependent variable and the output of the system, $A_n, A_{n-1}, \ldots, A_0$ are constants, ϵ is a constant indicating the degree of nonlinearity present, and $f(y(t), dy(t)/dt, \ldots, d^{n-1} y(t)/dt^{n-1})$ is a nonlinear function.

Notice that if $\epsilon = 0$, Eq. (5.3) reduces to Eq. (5.1), which represents a linear differential equation having constant coefficients. This leads us to the qualitative rule that a small amount of nonlinearity in a system means that ϵ is small in comparison with the coefficients $A_n, A_{n-1}, \ldots, A_0$. In addition, a large amount of nonlinearity means that ϵ is large compared with $A_n, A_{n-1}, \ldots, A_0$.

5.3. PROPERTIES OF LINEAR SYSTEMS THAT ARE NOT VALID FOR NONLINEAR SYSTEMS

Several inherent properties of linear systems, which greatly simplify the solution for this class of systems, are not valid for nonlinear systems. The fact that nonlinear systems do not have these properties further complicates their analysis.

Superposition is a fundamental property of linear systems. As a matter of fact, this property is the basis of the definition of a linear system. The principle of superposition states that if $c_1(t)$ is the response of a system to $r_1(t)$ and $c_2(t)$ is its response to $r_2(t)$, then the system's response to $a_1 r_1(t) + a_2 r_2(t)$ is $a_1 c_1(t) + a_2 c_2(t)$. Unfortunately, the superposition principle does not apply to nonlinear systems. Therefore, several mathematical procedures used in the design of linear systems cannot be used for nonlinear systems.

Stability of linear systems has been shown (in previous chapters) to depend only on the system's parameters. The stability of nonlinear systems, however, depends on the intial conditions and the nature of the input signal as well as the system parameters. One cannot expect a nonlinear system that exhibits a stable response to one type of input to have a stable response to other types of input. We shall shortly illustrate nonlinear systems that are stable for very small or very large signals, but not for both.

We normally expect the output of a linear system, excited by a sinusoidal signal, to have the same frequency as the input, although its amplitude and phase may differ. However, the output of nonlinear systems usually contains additional frequency components and may, in fact, not contain the input frequency.

For linear systems, interchanging two elements in cascade does not affect behavior. This is not true if one of the elements is nonlinear.

The question of stability is clearly defined for linear constant systems: A system is either stable or unstable. An unstable linear constant system has an output that grows without bound, either exponentially or in an oscillatory mode with the envelope of the oscillations increasing exponentially. In the nonlinear systems, system instability may means a constant-amplitude output having an arbitrary waveform. It is important to emphasize that an oscillator is stable according to Liapunov [2,3]. The exponentially decaying system, which we have referred to in this book as being stable, is described by Liapunov as being asymptotically stable.*

5.4. UNIQUE CHARACTERISTICS OF NONLINEAR SYSTEMS

This section describes in detail some of the unusual characteristics that are peculiar to nonlinear systems. These phenomena, which do not occur in linear systems, may be desirable or undesirable depending on the application. We discuss specifically the following behavior: limit cycle, soft and hard self-excitation, hysteresis, jump resonance, and subharmonic generation.

Limit cycles are oscillations of fixed amplitude and period that occur in nonlinear systems. Depending on whether the oscillation converges or diverges from the conditions represented, limit cycles can be either stable or unstable. It is possible that conditionally stable systems may contain both a stable and an unstable limit cycle. The occurrence of limit cycles in nonlinear systems makes it necessary to define instability† in terms of acceptable magnitudes of oscillation, because a very small nonlinear oscillation may not be detrimental to the performance of a system.

Self-excited oscillations occurring in systems that are unstable in the presence of very small signals are called *soft self-excitation*. Self-excited oscillations occurring in systems that are unstable in the presence of very large signals are called *hard-self-excitation*. Because soft and hard types of oscillation can occur, the control engineer must specify the dynamic range of operation completely when designing a nonlinear system. A feedback control system containing an element having saturation characteristics, such as illustrated in Figure 5.1a, could exhibit soft self-excitation. A feedback control system containing an element having dead-zone characteristics, such as illustrated in Figure 5.1b, could exhibit hard self-excitation.

Hysteresis is a nonlinear phenomenon that is most usually associated with magnetization curves or backlash of gear trains. A conventional magnetization curve whose path depends on whether the magnetizing force **H** is increasing or decreasing is shown in Figure 5.1c.

Jump resonance [4], another form of hysteresis, is of considerable interest. It exhibits itself in the closed-loop frequency response of certain nonlinear systems, as illustrated in Figure 5.1d. As the frequency ω is increased and the input amplitude R is held constant, the response follows the curve *AFB*. At point *B*, a small change in frequency results in a discontinuous jump to point *C*. The response then follows the

---

*Asymptotic stability and nonlinear system stability classified in terms of a regional basis are discussed in Section 5.20, where Liapunov's stability criterion is presented.
†As far as control-system design is concerned, a steady oscillation is treated as being unstable.

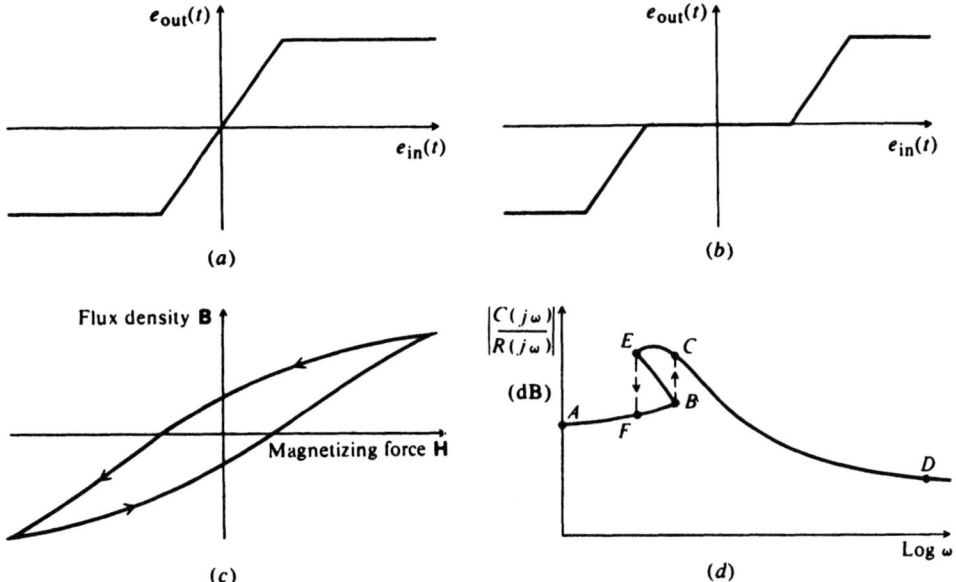

Figure 5.1 (a) Saturation characteristics. (b) Dead-zone characteristics. (c) Conventional hysteresis loop. (d) Closed-loop response of a system with jump resonance.

curve to point D upon further increase in frequency. As the frequency is decreased from point D, the response follows the curve to points C and E. At point E, a small change in frequency results in a discontinuous jump to point F. The response follows the curve to point A for further decreases in frequency. Observe from this description that the response never actually follows the segment BE. This portion of the curve represents a condition of unstable equilibrium. The system must be of second order or higher for the phenomenon of jump resonance to occur.

Subharmonic generation [5] refers to nonlinear systems whose output contains subharmonics of the input's sinusoidal excitation frequency. The transition from normal harmonic operation to subharmonic operation is usually quite sudden. Once the subharmonic operation is established, however, it is usually quite stable. In general, if sinuosoidal signals f_1 and f_2 are added and their sum is applied to a nonlinear device, the output contains frequency components $af_1 \pm bf_2$, where a and b assume all possible integers including zero.

5.5. METHODS AVAILABLE FOR ANALYZING NONLINEAR SYSTEMS

Several tools are available for the analysis of nonlinear systems. All these techniques depend on the severity of the nonlinearity and/or the order of the system under consideration. We consider all of the useful and popular techiques in this chapter and illustrate their practical application. The chapter concludes with the presentation of guidelines for selecting the "best" nonlinear control-system method for the analysis and design of a particular problem in Section 5.23.

The analysis of nonlinear systems is concerned with the existence and effects of limit cycles, soft and hard self-excitation, hysteresis, jump resonance, and subharmonic generation. In addition, the response to specific input functions must be determined. The major difficulty of analyzing nonlinear systems is that no single technique is generally applicable to all problems.

Quasilinear systems, where the deviation from linearity is not too large, permit the use of certain *linearizing approximations* [6]. The *describing-function* approach, which is applicable to nonlinear systems of any order and is concerned with discovering limit cycles, simplifies the problem by assuming that the input to the nonlinear system is sinusoidal and the only significant frequency component of the output is that component having the same frequency as the input [7–10].

Nonlinear systems can often be approximated by several linear regions. The *piecewise-linear* approach permits the segmented linearization of any nonlinearity for any order of system. The *phase-plane method* is a very useful technique for analyzing the response of a second-order nonlinear system [10–13]. *Liapunov's stability methods* are very powerful techniques for determining the steady-state stability of nonlinear systems based on generalizations of energy notions [2]. *Popov's* method is very useful for determining the stability of time-invariant, nonlinear systems. *The generalized circle criterion* is applicable to time-variable, nonlinear systems whose linear portion is not necessarily open-loop stable [16,17].

Systems of very high order having several nonlinearities have hardly been dealt with in general analytical terms. This problem usually requires the use of *numerical methods* utilizing *digital computers* for a solution. It is worth emphasizing at this point that any nonlinear differential equation can be solved by these techiques provided many small increments are used [21–24]. However, the resulting solution is valid only for the specific problem being considered. It is very difficult to extend the result and obtain a general solution which can be used for other problems.

As a final check on the stability of nonlinear control systems, I always recommend that the system be simulated [25,26]. It will aid in overcoming such factors as possible uncertainty regarding the validity of assumptions, and to analytic difficulties caused by system complexity. That is further emphasized in Section 5.23.

5.6. LINEARIZING APPROXIMATIONS

In quasilinear systems, where the deviation from linearity is not too great, linear approximations may permit the extension of ordinary linear concepts. This approach acknowledges that certain system characteristics change from operating point to operating point, but it assumes linearity in the neighborhood of a specific operating point. The technique of linearizing approximations is universally used by the engineer and may be more familiar to the reader under the names *small-signal theory* and/or *theory of small perturbations*.

Linearizing approximations were utilized when we discussed the two-phase ac servomotor in Section 3.4‡. For this device, Figure 3.16‡ illustrated the quasilinear characteristics relating developed torque and speed. However, by approximating the torque speed curves with straight lines, the linear differential equation (3.98)‡ was formulated. We then obtained the transfer function of the two-phase ac servomotor, assuming that it was a linear device. It is left as an exercise to the reader in Problem 5.1 to determine the effect of various linearizing approximations.

The effects of a small amount of nonlinearity can be studied analytically by considering small perturbations or changes in the variables about some average value of the variables. This can be represented analytically by [see Eq. (5.3)]

$$A_n \frac{d^n y(t)}{dt^n} + A_{n-1} \frac{d^{n-1} y(t)}{dt^{n-1}} + \cdots + A_1 \frac{dy(t)}{dt} + A_0 y(t)$$
$$+ \epsilon f\left(y(t), \frac{dy(t)}{dt}, \ldots, \frac{d^{n-1} y(t)}{dt^{n-1}}\right) = x(t). \quad (5.4)$$

An expansion of the solution to this differential equation, for small nonlinearities, can be written as a power series in ϵ as

$$y(t) = y_{(0)}(t) + \epsilon y_{(1)}(t) + \epsilon^2 y_{(2)}(t) + \epsilon^3 y_{(3)}(t) + \cdots. \quad (5.5)$$

From this equation, $y(t)$ may be interpreted as being composed of a linear component $y_{(0)}(t)$ and several deviation factors: $\epsilon y_{(1)}(t) + \epsilon^2 y_{(2)}(t) + \cdots$. Assuming that ϵ is very small, the nonlinear components will not seriously affect the system's behavior if a linear approximation is assumed. Therefore, within the realm of reasonable engineering approximations, the control engineer may be able to extend linear theory for certain feedback control systems which exhibit a small amount of nonlinearity. It is very interesting that this is just the reason linear theory has had such good results, even though practical systems are never purely linear.

Linearization techniques can also be applied to those problems where it is desired to linearize nonlinear equations by limiting attention to small perturbations about a reference state [6]. This technique is often used in the design of space navigation and control systems where it is desired to maintain a space vehicle along a specified reference trajectory. It will now be shown how the corrective control forces required to keep the vehicle on the desired flight trajectory can be synthesized from a set of linear differential equations, although the basic differential equations describing the reference flight trajectory are nonlinear.

To illustrate this, let us assume that the equation of the system is given by

$$\dot{\mathbf{x}}(t) = \mathbf{f}(\mathbf{x}(t)\mathbf{u}(t)), \quad (5.6)$$

where the function $\mathbf{f}$ is nonlinear. Figure 5.2 illustrates the reference trajectory of the space vehicle (solid line) which satisfies the equation

$$\dot{\mathbf{x}}^0(t) = \mathbf{f}(\mathbf{x}^0(t), \mathbf{u}^0(t)), \quad (5.7)$$

where the superscript zero refers to parameters occurring along the reference trajectory. These reference parameters are related to the parameters of the actual trajectory (dashed line) as follows:

$$\mathbf{x}(t) = \mathbf{x}^0(t) + \delta \mathbf{x}(t), \quad (5.8)$$
$$\mathbf{u}(t) = \mathbf{u}^0(t) + \delta \mathbf{u}(t). \quad (5.9)$$

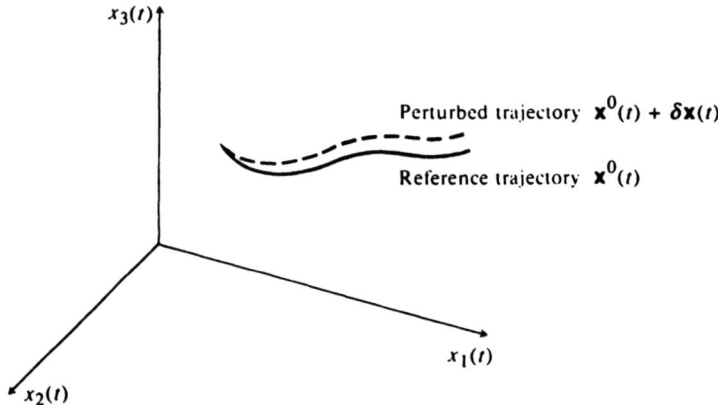

Figure 5.2 Reference and perturbed trajectories of a space vehicle [6].

Figure 5.2 illustrates the reference and actual trajectories, where the actual state $\mathbf{x}(t)$ is perturbed from the reference state $\mathbf{x}^0(t)$ by $\delta\mathbf{x}(t)$. Physically, this means that the actual trajectory of the space vehicle is perturbed, or slightly different, from that of the desired reference trajectory. The vector $\delta\mathbf{u}(t)$ represents the deviation of the control input from the desired reference input $\mathbf{u}^0(t)$ which would result in the desired system response $\mathbf{x}^0(t)$.

What kind of relationship can we derive between $\mathbf{x}^0(t)$, $\delta\mathbf{x}(t)$, $\mathbf{u}^0(t)$, and $\delta\mathbf{u}(t)$? The basic nonlinear equation of the system,

$$\dot{\mathbf{x}}(t) = \mathbf{f}(\mathbf{x}(t), \mathbf{u}(t))$$

can be expressed as

$$\frac{d}{dt}(\mathbf{x}^0(t) + \delta\mathbf{x}(t)) = \dot{\mathbf{x}}^0(t) + \delta\dot{\mathbf{x}}(t) = f(\mathbf{x}^0(t) + \delta\mathbf{x}(t), \mathbf{u}^0(t) + \delta\mathbf{u}(t)).$$

Because we are assuming that the actual perturbations of the system are small, we can expand the jth component of this equation in a Taylor series about the reference trajectory as follows:

$$\dot{x}_j^0(t) + \delta\dot{x}_j(t) \approx f_j(\mathbf{x}^0(t), \mathbf{u}^0(t)) + \frac{\partial f_j}{\partial x_1}\delta x_1(t) + \cdots + \frac{\partial f_j}{\partial x_m}\delta x_m(t)$$
$$+ \frac{\partial f_j}{\partial u_1}\delta u_1(t) + \cdots + \frac{\partial f_j}{\partial u_n}\delta u_n(t). \quad (5.10)$$

Using Eq. (5.7), we can rewrite Eq. (5.10) as follows:

$$\delta\dot{x}_j(t) \approx \left(\frac{\partial f_j}{\partial x_1}\right)^0 \delta x_1(t) + \cdots + \left(\frac{\partial f_j}{\partial x_m}\right)^0 \delta x_m(t) + \left(\frac{\partial f_j}{\partial u_1}\right)^0 \delta u_1(t)$$
$$+ \cdots + \left(\frac{\partial f_j}{\partial u_n}\right)^0 \delta u_n(t), \quad (5.11)$$

where $j = 1, 2, 3, \ldots, m$.

Equation (5.11) can be simplified by utilizing the Jacobian matrices* that are defined as follows:

$$\mathbf{A} = \begin{bmatrix} \frac{\partial f_1}{\partial x_1} & \frac{\partial f_1}{\partial x_2} & \cdots & \frac{\partial f_1}{\partial x_m} \\ \frac{\partial f_2}{\partial x_1} & \frac{\partial f_2}{\partial x_2} & \cdots & \frac{\partial f_2}{\partial x_m} \\ \vdots & \vdots & & \vdots \\ \frac{\partial f_n}{\partial x_1} & \frac{\partial f_n}{\partial x_2} & \cdots & \frac{\partial f_n}{\partial x_m} \end{bmatrix}_{\substack{x=x^0 \\ u=u^0}}, \quad \mathbf{B} = \begin{bmatrix} \frac{\partial f_1}{\partial u_1} & \frac{\partial f_1}{\partial u_2} & \cdots & \frac{\partial f_1}{\partial u_n} \\ \frac{\partial f_2}{\partial u_1} & \frac{\partial f_2}{\partial u_2} & \cdots & \frac{\partial f_2}{\partial u_n} \\ \vdots & \vdots & & \vdots \\ \frac{\partial f_n}{\partial u_1} & \frac{\partial f_n}{\partial u_2} & \cdots & \frac{\partial f_n}{\partial u_n} \end{bmatrix}_{\substack{x=x^0 \\ u=u^0}} \quad (5.12)$$

It is important to emphasize that all of the partial derivatives in the Jacobian matrices are evaluated along the reference trajectory of the space vehicle. Based on the Jacobian matrices, Eq. (5.11) can be rewritten in the following simplified form:

$$\delta \dot{\mathbf{x}}(t) \approx \mathbf{A}\, \delta \mathbf{x}(t) + \mathbf{B}\, \delta \mathbf{u}(t). \quad (5.13)$$

This resulting equation is very important. It states that the differential equation describing the perturbations about the reference trajectory are approximately linear, although the basic system differential equations describing the reference flight trajectory are nonlinear. Therefore, we have succeeded in linearizing the problem.

We can also linearize a system if we can adapt it to behave like a linear system. In order to demonstrate this, let us consider a two-position relay that controls the rotation of a motor in either direction. It is assumed that the control voltage applied by the relay to the motor, $e_c(t)$, is given by

$$e_c(t) = E \sin \omega t \quad (5.14)$$

and the resulting motor torque developed, $T(t)$, would be a square wave due to the switching action. Both $e_c(t)$ and $T(t)$ are illustrated in Figure 5.3. Observe from this figure that the average, or mean, value of both functions is zero.

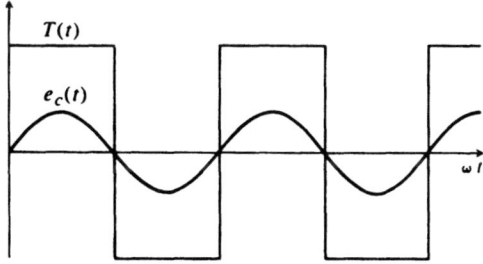

Figure 5.3 Relay-controlled motor characteristics—Case 1.

---

*Note that when linearizing along a trajectory, the matrices **A** and **B** will be functions of time.

Let us next assume that the control voltage has a finite mean value E_0, where

$$e_c(t) = E_0 + E \sin \omega t. \quad (5.15)$$

For this case, the torque is a periodic function whose mean value is some nonzero value T_0, because time intervals where $e_c(t)$ is positive or negative are not equal, as indicated in Figure 5.4. Note that E_0 is also a function of time, but it is assumed that it is very slowly varying compared with ω. Assuming, in addition, that $E_0 \ll E$, it can easily be shown that the mean value of T_0 is given by the linear relationship

$$T_0 = \frac{2T}{\pi E} E_0. \quad (5.16)$$

Therefore, the mean value of the torque T_0 is proportional to the mean value of the controlling voltage.

This is a very important result. It shows that by means of a nonlinear element like a relay, a linear relationship can be obtained between the mean value of the controlling voltage and the mean value of the motor torque developed. The basic linearization technique utilized has taken the mean value of the applied voltage to the relay as the input, and superimposed on it a sinuosidal function of time whose amplitude and frequency are very high relative to that of the input.

In the following section, we extend our linearization concepts and attempt to apply them to nonlinear systems. Although the notion of a transfer function is inapplicable for nonlinear systems, an equivalent approximate transfer characteristic for a nonlinear device can be derived which can be manipulated as a transfer function under certain circumstances. We define this approximate transfer characteristic as the *describing function*. It is a very useful notion and is frequently employed in practice.

5.7. DESCRIBING-FUNCTION CONCEPT

The use of describing functions is an attempt to extend the very powerful transfer-function approach of linear systems to nonlinear systems [7,8,10]. A describing

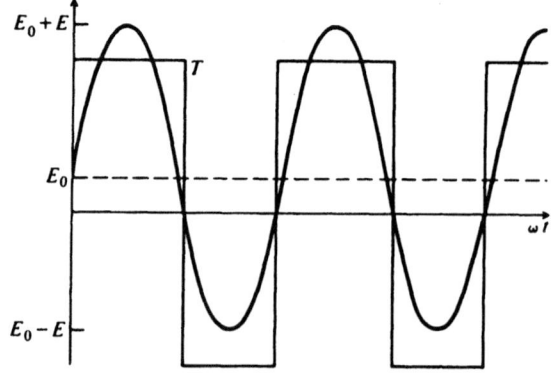

Figure 5.4 Relay-controlled motor characteristics—Case 2.

function is defined as the ratio of the fundamental component of the output of a nonlinear device to the amplitude of a sinusoidal input signal. In general, the describing function depends on the input signal's amplitude and frequency and is complex because phase shift may occur between the input and the fundamental component of the output. We study the describing function method of analysis and compare it with the transfer-function concept for linear systems.

If the input to a nonlinear element is a sinusoidal signal, the describing-function analysis assumes that the output is a periodic signal having the same fundamental period as that of the input signal. Therefore, the analysis is concerned only with the fundamental component of the output waveform. All harmonics, subharmonics, and any dc component are neglected. This assumption is reasonable, because the harmonic terms are often small when compared witht he fundamental term. In addition, a feedback control system usually provides additional attenuation of the harmonic terms because of its inherent filtering action of high frequencies in the region of the harmonic terms. Many nonlinear elements do not generate a dc term because they are symmetrical, nor do they generate any subharmonic terms [5]. Therefore, in many (but not all) situations the fundamental term is the only significant component of the output of the nonlinear element. In addition, it is assumed that there is only one nonlinear element in the feedback control system and that it is not varying. If a system contains more than one nonlinearity, we must lump all the nonlinearities together and obtain an overall describing function.

An examination of these limitations indicates that the describing function is based on a restricted mathematical foundation. The technique does give, however, reasonable results and does have an advantage that it can be used for systems of any order and is fairly simply to apply. It is recommended that the results always be verified with another technique or a computer simulation [25,26].

Given its limitations, the describing-function technique is still a very useful tool for analyzing and designing nonlinear systems. The describing function should be thought of as a generalized transfer function for nonlinear systems.

In order to derive a mathematical expression for the describing function, let us consider the general nonlinear system illustrated in Figure 5.5. In accordance with the definition of the describing function, let us assume that the input to the nonlinear element $N(M, \omega)$ is given by

$$m(\omega t) = M \sin \omega t. \tag{5.17}$$

In general, the steady-state output of the nonlinear device can be represented by the series

$$n(\omega t) = N_1 \sin(\omega t + \phi_1) + N_2 \sin(2\omega t + \phi_2) + N_3 \sin(3\omega t + \phi_3) + \cdots. \tag{5.18}$$

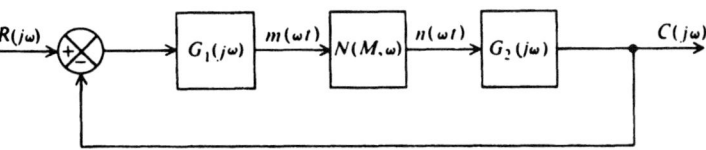

Figure 5.5 General nonlinear system.

By definition, the describing function is

$$N(M, \omega) = \frac{N_1}{M} e^{j\phi_1}. \tag{5.19}$$

Notice that the describing function depends on the amplitude and frequency of the input signal. The nonlinear element is thus considered to have a gain and phase shift varying with the amplitude and frequency of the input signal.

5.8. DERIVATION OF DESCRIBING FUNCTIONS FOR COMMON NONLINEARITIES

The describing functions for several common nonlinearities are derived in this section [8–10,27]. The procedure most commonly used is to determine the Fourier series of the output waveshape from the nonlinear device and consider only the fundamental component. Let us consider the nonlinear element $N(M, \omega)$ in an overall feedback control system, as shown in Figure 5.5. Assuming that the input $m(\omega t)$ is given by a sinusoidal signal where

$$m(\omega t) = M \sin \omega t, \tag{5.20}$$

we can represent the output waveshape by a Fourier series given by the expression

$$n(\omega t) = \frac{A_0}{2} + \sum_{K=1}^{K=\infty} A_K \cos K\omega t + \sum_{K=1}^{K=\infty} B_K \sin K\omega t, \tag{5.21}$$

where

$$A_K = \frac{2}{T} \int_{-T/2}^{T/2} n(\omega t) \cos K\omega t \, d(\omega t), \quad K = 0, 1, 2, \ldots, \tag{5.22}$$

$$B_K = \frac{2}{T} \int_{-T/2}^{T/2} n(\omega t) \sin K\omega t \, d(\omega t), \quad K = 1, 2, 3, \ldots. \tag{5.23}$$

In general, if $n(\omega t) = -n(-\omega t)$, then the function is odd and $A_K = 0$. In addition, if $n(\omega t) = n(-\omega t)$, then the function is even and $B_K = 0$.

Because we are only concerned with the fundamental component of the output, it is necessary to determine only A_1 and B_1. For $m(\omega t) = M \sin \omega t$, the describing function can then be obtained from the expression

$$N(M, \omega) = \frac{B_1}{M} + j\frac{A_1}{M} = \left[\left(\frac{B_1}{M}\right)^2 + \left(\frac{A_1}{M}\right)^2\right]^{1/2} \bigg/ \tan^{-1} \frac{A_1}{B_1}. \tag{5.24}$$

The control engineer is usually concerned with the nonlinearities due to dead zones, saturation, backlash, on-off relay-control systems, Coulomb friction, and stiction. We specifically derive and catalog their describing functions so that a handy reference for some common describing functions will be available. In addition, the procedure illustrated should enable one to develop the facility for calculating the describing function of any nonlinearity encountered.

A. Describing Function of a Dead Zone

Figure 5.6 illustrates the dead-zone characteristics. The relationships between input and output of this nonlinearity can be expressed by the equations

$$n(\omega t) = 0 \qquad \text{for } -D < m < D, \quad (5.25)$$
$$n(\omega t) = K_1 M(\sin \omega t - \sin \omega t_1) \qquad \text{for } m > D, \quad (5.26)$$
$$n(\omega t) = K_1 M(\sin \omega t + \sin \omega t_1) \qquad \text{for } m < -D. \quad (5.27)$$

Figure 5.7 illustrates typical input and output waveshapes. Notice that the output is an odd function, and therefore $A_K = 0$. The symmetry over the four quarters of the period allows us to evaluate the expression for the Fourier coefficient, B_1, by taking four times the integral over one quarter of a cycle, as follows:

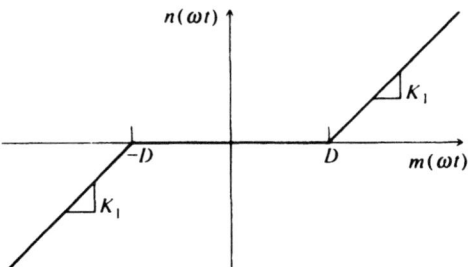

Figure 5.6 Nonlinear characteristics of a dead zone.

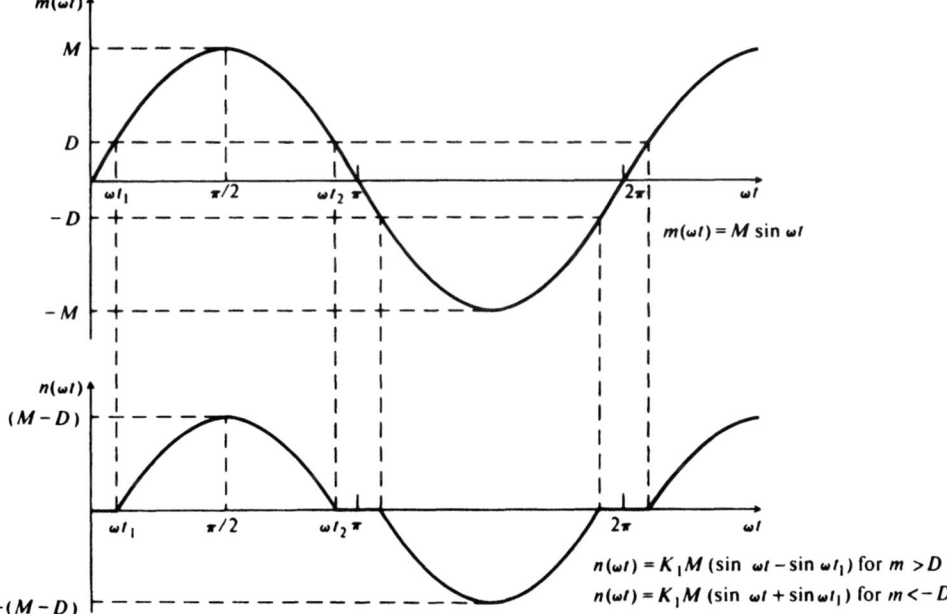

Figure 5.7 Input and output waveshape from a nonlinear device having the dead-zone characteristics shown in Figure 5.6.

$$B_1 = \frac{4}{\pi} \int_0^{\pi/2} n(\omega t) \sin \omega t \, d(\omega t). \tag{5.28}$$

Substituting Eqs. (5.25) and (5.26) into Eq. (5.28), we obtain

$$B_1 = \frac{4}{\pi} \left[\int_0^{\omega t_1} (0) \sin \omega t \, d(\omega t) + \int_{\omega t_1}^{\pi/2} K_1 M (\sin \omega t - \sin \omega t_1) \sin \omega t \, d(\omega t) \right], \tag{5.29}$$

where

$$\omega t_1 = \sin^{-1}(D/M).$$

Evaluation of Eq. (5.29) results in the expression

$$B_1 = \frac{2K_1 M}{\pi} \left(\frac{\pi}{2} - \frac{D}{M} \cos \sin^{-1} \frac{D}{M} - \sin^{-1} \frac{D}{M} \right). \tag{5.30}$$

From Eq. (5.30), because the describing function is the ratio of the amplitude of the fundamental component of the output B_1 to M, it can be expressed as

$$N_{dz}(M) = \frac{B_1}{M} = \frac{2K_1}{\pi} \left(\frac{\pi}{2} - \frac{D}{M} \cos \sin^{-1} \frac{D}{M} - \sin^{-1} \frac{D}{M} \right). \tag{5.31}$$

Notice that the describing function for a dead zone is only a function of the amplitude of the input and not of frequency. Figure 5.8, obtained from Eq. (5.31), is a sketch of the normalized value of the describing function $N(M)/K_1$ as a function of the ratio D/M. For very small values of D/M the normalized describing function approaches unity. For values of $D/M \geq 1$ it equals zero, which implies that the input must be greater than the dead-zone magnitude in order to obtain an output. Notice that the describing function for a dead zone does not have any phase shift.

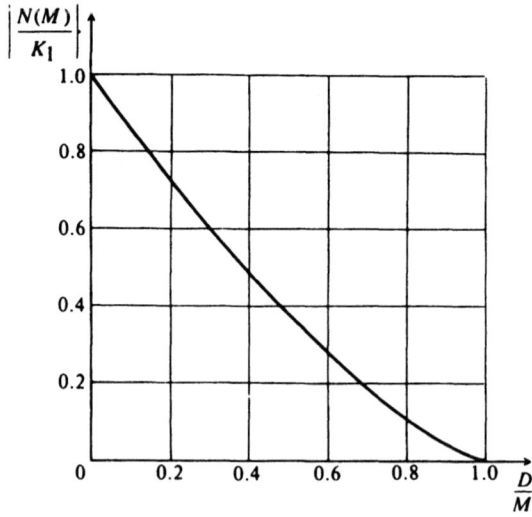

Figure 5.8 Normalized describing function for a dead zone.

B. Describing Function of Saturation

Figure 5.9 illustrates the nonlinear saturation characteristic. The relationship between input and output of this nonlinearity can be expressed by the following equations

$$n(\omega t) = K_1 M \sin \omega t \quad \text{for } -S < m < S, \tag{5.32}$$
$$n(\omega t) = K_1 S \quad \text{for } m(\omega t) > S,$$
$$n(\omega t) = -K_1 S \quad \text{for } m(\omega t) < -S. \tag{5.33}$$

Figure 5.10 illustrates typical input and output waveshapes.

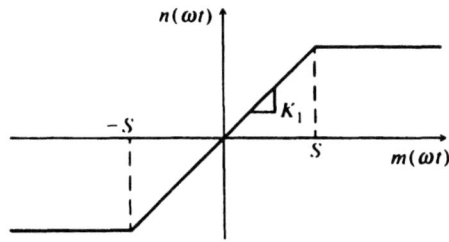

Figure 5.9 Nonlinear characteristics of saturation.

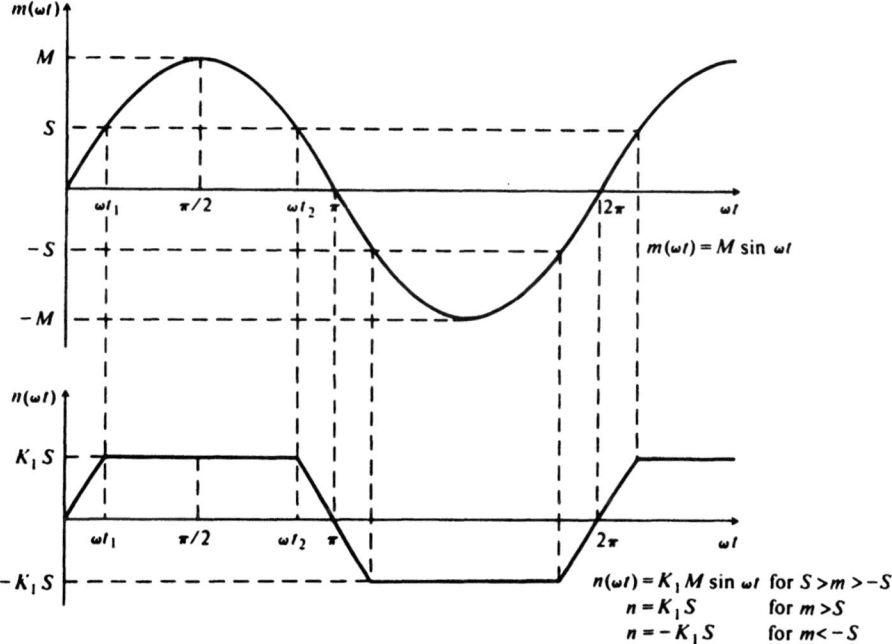

Figure 5.10 Input and output waveshape from a nonlinear device having saturation characteristics shown in Figure 5.9.

Notice that the output waveshape is an odd function for the case of saturation just as it was for the case of a dead zone. Therefore, the expression for the Fourier coefficient B_1 of the output waveshape is

$$B_1 = \frac{2}{T} \int_{-T/2}^{T/2} n(\omega t) \sin \omega t \, d(\omega t). \qquad (5.34)$$

As was true for a dead zone, the expression for the Fourier coefficient B_1 can be obtained by taking four times the integral over one quarter of a cycle because of the symmetry over the four quarters of the period. This results in the expression

$$B_1 = \frac{4}{\pi} \int_0^{\pi/2} n(\omega t) \sin \omega t \, d(\omega t). \qquad (5.35)$$

Substituting Eqs. (5.32) and (5.33) into Eq. (5.35), we obtain

$$B_1 = \frac{4}{\pi} \left[\int_0^{\omega t_1} K_1 M \sin \omega t \sin \omega t \, d(\omega t) + \int_{\omega t_1}^{\pi/2} K_1 M \sin \omega t_1 \sin \omega t \, d(\omega t) \right], \qquad (5.36)$$

where

$$\omega t_1 = \sin^{-1} \frac{S}{M}. \qquad (5.36)$$

Evaluation of Eq. (5.36) results in the expression

$$B_1 = \frac{2K_1 M}{\pi} \left(\frac{S}{M} \cos \sin^{-1} \frac{S}{M} + \sin^{-1} \frac{S}{M} \right). \qquad (5.37)$$

Because the describing function is defined as the ratio of the amplitude of the fundamental component of the output, B_1 to M, the describing function can be expressed as

$$N_{\text{sat}}(M) = \frac{B_1}{M} = \frac{2K_1}{\pi} \left(\frac{S}{M} \cos \sin^{-1} \frac{S}{M} + \sin^{-1} \frac{S}{M} \right). \qquad (5.38)$$

Notice that the describing function for saturation is only a function of the amplitude of the input and not of frequency. Figure 5.11, obtained from Eq. (5.38) is a sketch of the normalized value of the describing function $N(M)/K_1$ as a function of the ratio S/M. For very small values of S/M, the normalized describing function approaches zero. For values of $S/M \geqslant 1$ it equals unity, which implies that the output is unaffected by the saturation level if $S \geqslant M$. Notice that the describing function for saturation does not introduce any phase shift.

It is important to emphasize that the describing functions for dead zone and saturation could have been obtained from one nonlinear characteristic containing both types of nonlinearities. Then the resulting describing function could be reduced to the describing function of dead zone [Eq. (5.31)] by letting the satura-

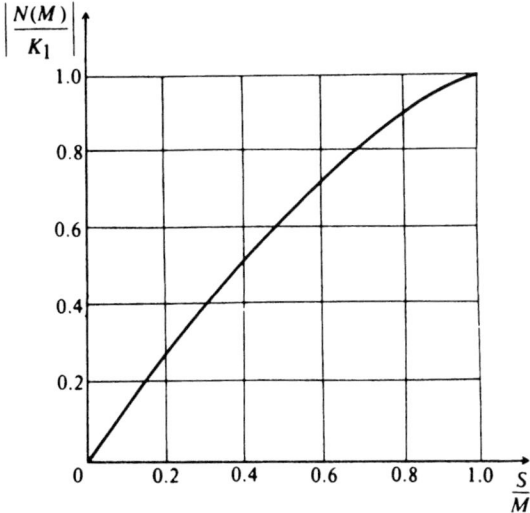

Figure 5.11 Normalized describing function for saturation.

tion level approach infinity, or it could be reduced to the describing function of saturation [Eq. (5.38)] by letting the dead-zone width approach zero. It is recommended that the reader derive Eqs. (5.31) and (5.38) using this approach (see Problem 5.2).

C. Describing Function of Backlash

Backlash or mechanical hysteresis is due to the difference in motion between an increasing and a decreasing output. Figure 5.12 illustrates a model of backlash, and Figure 5.13 illustrates its characteristics. The source of backlash that usually receives the most attention is the "looseness" inherent in mechanical gearing.

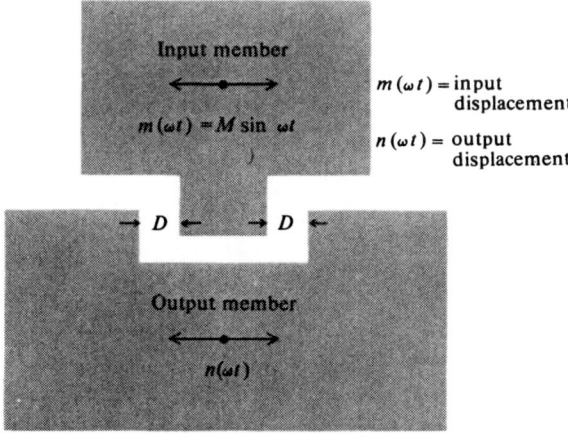

Figure 5.12 Physical model of backlash.

342 NONLINEAR CONTROL-SYSTEM DESIGN

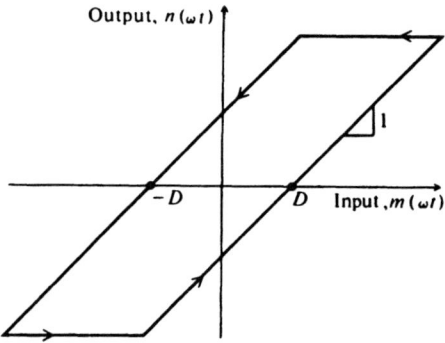

Figure 5.13 Characteristics of backlash.

From Figure 5.14, the relationship between inut and output for this nonlinearity can be expressed by the following equations:

$$n(\omega t) = -(M - D) \qquad \text{for } 0 \leq \omega t < \omega t_1 \tag{5.39}$$

$$n(\omega t) = M \sin \omega t - D \qquad \text{for } \omega t_1 \leq \omega t < \frac{\pi}{2}, \tag{5.40}$$

$$n(\omega t) = M - D \qquad \text{for } \frac{\pi}{2} \leq \omega t < \omega t_2, \tag{5.41}$$

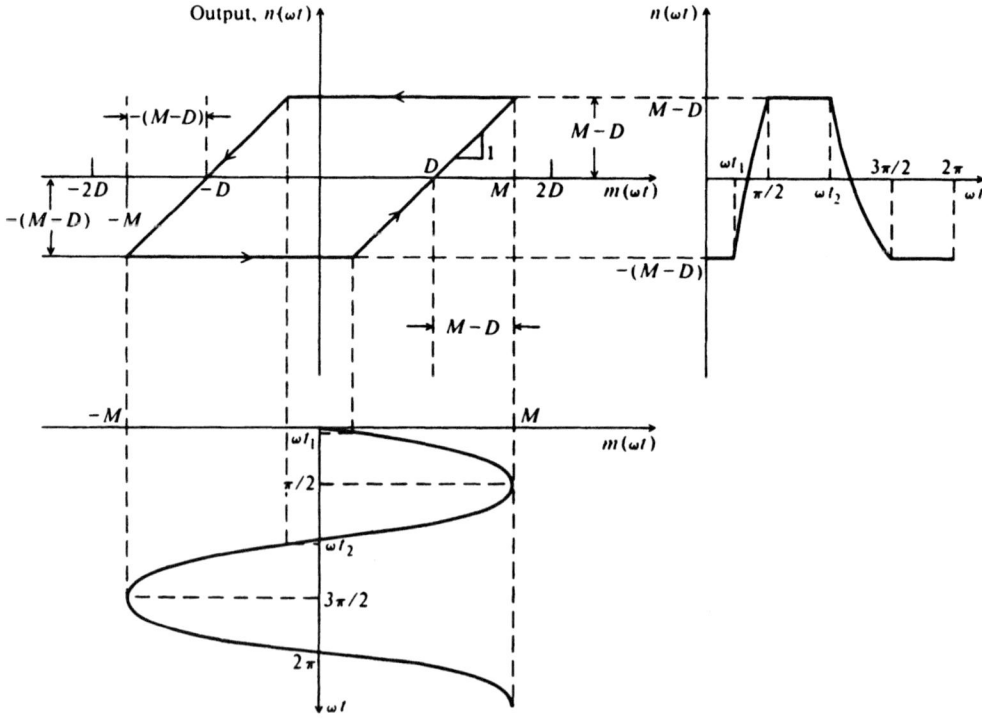

Figure 5.14 Backlash characteristics where $D < M < 2D$.

$$n(\omega t) = M \sin \omega t + D \quad \text{for } \omega t_2 \leq \omega t < \frac{3\pi}{2}, \tag{5.42}$$

$$n(\omega t) = -(M - D) \quad \text{for } \frac{3\pi}{2} \leq \omega t \leq 2\pi. \tag{5.43}$$

Notice that the output waveshape is neither an odd nor an even function. This means that the Fourier series of the output waveshape has A_K and B_K components. Because the describing function is concerned only with the fundamental component of the output, however, we are interested only in A_1 and B_1. From Eqs. (5.39) through (5.43), we can express A_1 as

$$A_1 = \frac{2}{2\pi} \left[\int_0^{\omega t_1} -(M-D) \cos \omega t \, d(\omega t) + \int_{\omega t_1}^{\pi/2} (M \sin \omega t - D) \cos \omega t \, d(\omega t) \right.$$
$$+ \int_{\pi/2}^{\omega t_2} (M-D) \cos \omega t \, d(\omega t) + \int_{\omega t_2}^{3\pi/2} (M \sin \omega t + D) \cos \omega t \, d(\omega t)$$
$$\left. + \int_{3\pi/2}^{2\pi} -(M-D) \cos \omega t \, d(\omega t) \right], \tag{5.44}$$

where

$$\omega t_1 = \sin^{-1}\left(\frac{2D}{M} - 1\right) \tag{5.45}$$

and

$$\omega t_2 = \omega t_1 + \pi, \tag{5.46}$$

and B_1 can be expressed as

$$B_1 = \frac{2}{2\pi} \left[\int_0^{\omega t_1} -(M-D) \sin \omega t \, d(\omega t) + \int_{\omega t_1}^{\pi/2} (M \sin \omega t - D) \sin \omega t \, d(\omega t) \right.$$
$$+ \int_{\pi/2}^{\omega t_2} (M-D) \sin \omega t \, d(\omega t) + \int_{\omega t_2}^{3\pi/2} (M \sin \omega t + D) \sin \omega t \, d(\omega t)$$
$$\left. + \int_{3\pi/2}^{2\pi} -(M-D) \sin \omega t \, d(\omega t) \right], \tag{5.47}$$

Integrating Eqs. (5.44) and (5.47), we obtain the following results:

$$A_1 = \frac{2D}{\pi M}\left(\frac{2D}{M} - 2\right) M, \tag{5.48}$$

$$B_1 = \frac{1}{\pi}\left[\frac{\pi}{2} - \sin^{-1}\left(\frac{2D}{M} - 1\right) - \left(\frac{2D}{M} - 1\right)\cos \sin^{-1}\left(\frac{2D}{M} - 1\right)\right] M. \tag{5.49}$$

Therefore, the describing function for backlash is given by

$$N_{\text{backlash}}(M) = \frac{1}{M}\sqrt{A_1^2 + B_1^2} \; \underline{/\tan^{-1}(A_1/B_1)}. \tag{5.50}$$

344 NONLINEAR CONTROL-SYSTEM DESIGN

This expression is valid only when the positive slope of the backlash characteristic as shown in Figure 5.14 is unity. If it is any other value, such as K_1, then Eq. (5.50) is modified, because the right-hand sides of the defining equations (5.40) and (5.42) would have to be modified.

Notice the describing function for backlash is only a function of the amplitude of the input and not of the frequency. Figures 5.15 and 5.16, which have been obtained from Eqs. (5.48) through (5.50), are sketches of the amplitude and phase characteristics, respectively, of the describing function as a function of the ratio D/M. Notice that a phase lag occurs at low input amplitudes. This phase lag may introduce problems of stability in a feedback system.

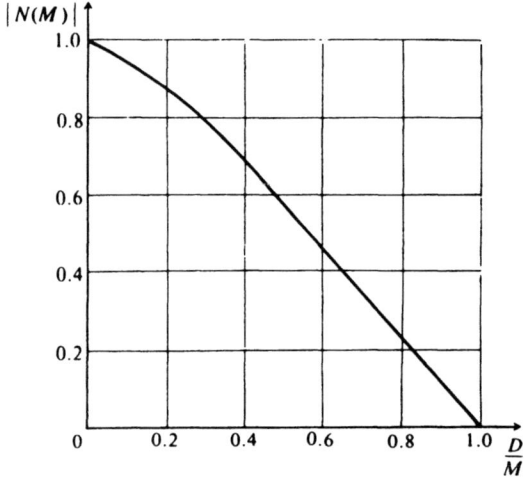

Figure 5.15 Amplitude characteristics for the describing function of backlash.

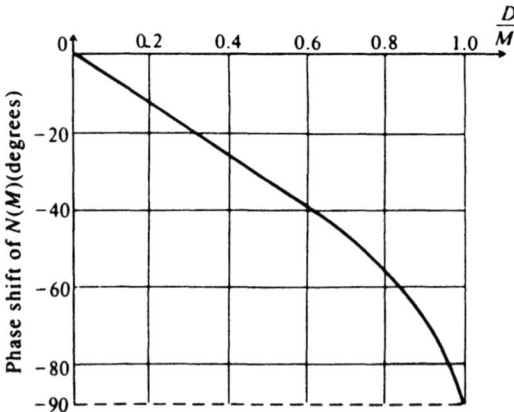

Figure 5.16 Phase characteristics for the describing function of backlash.

D. Describing Function of an On-Off Element Having Hysteresis

A class of systems of great practical importance is that of on-off control systems. In these systems, as soon as the error signal exceeds a certain level, a relay switches on full corrective torque having proper polarity. When the error falls below a certain level, all the corrective torque is removed. These simple and relatively inexpensive devices find many practical uses in thermostatic control of heat, in automobile voltage regulators, in aircraft and space-vehicle control applications where space and weight limitations are very critical, and so on.

The heart of an on-off control system, the relay or contactor, has a variety of characteristics. For the purpose of deriving a describing function, we consider a three-position contactor exhibiting hysteresis characteristics; this includes the two-position contactor as a limiting case. Figure 5.17 illustrates the input–output characteristic and the waveshapes for such a device. The hysteresis effect occurs because of the different values of control signal required for corrective torque application and its removal. Torque is applied when the control signal reaches $\pm(D+h)$, but it is not removed until the control signal equals $\pm D$. The relationship between input and output for this nonlinearity can be expressed by the following equations:

$$n(\omega t) = 0 \quad \text{for} \quad 0 \leqslant \omega t < \omega t_1, \tag{5.51}$$

$$n(\omega t) = K_1 \quad \text{for} \quad \omega t_1 \leqslant \omega t < \omega t_2, \tag{5.52}$$

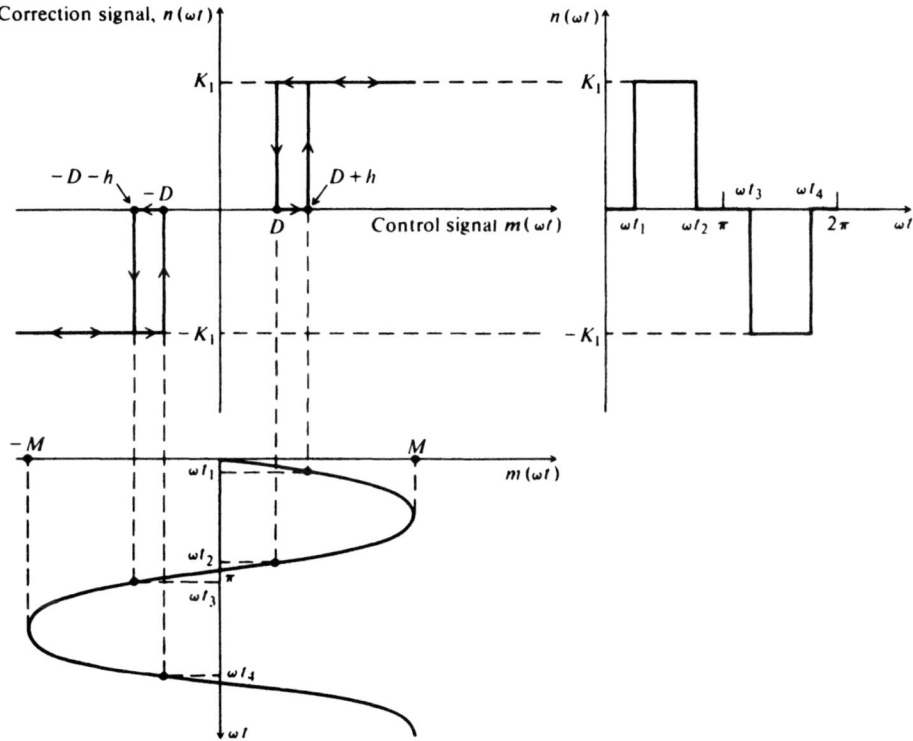

Figure 5.17 On-off characteristics for a three-position contactor having hysteresis.

346 NONLINEAR CONTROL-SYSTEM DESIGN

$$n(\omega t) = 0 \qquad \text{for } \omega t_2 \leqslant \omega t \leqslant \omega t_3, \qquad (5.53)$$

$$n(\omega t) = -K_1 \qquad \text{for } \omega t_3 \leqslant \omega t < \omega t_4, \qquad (5.54)$$

Notice that the output waveshape is neither an odd nor an even function. The Fourier series of the output waveshape, therefore, contains A_K and B_K components. From Eqs. (5.51)–(5.54), we can express A_1 as

$$A_1 = \frac{2}{2\pi}\left[\int_0^{\omega t_1}(0)\cos\omega t\, d(\omega t) + \int_{\omega t_1}^{\omega t_2}(K_1)\cos\omega t\, d(\omega t)\right.$$
$$\left.+ \int_{\omega t_2}^{\omega t_3}(0)\cos\omega t\, d(\omega t) + \int_{\omega t_3}^{\omega t_4}(-K_1)\cos\omega t\, d(\omega t)\right], \qquad (5.55)$$

where

$$\omega t_1 = \sin^{-1}\frac{D+h}{M}, \qquad (5.56)$$

$$\omega t_2 = \pi - \sin^{-1}\frac{D}{M}, \qquad (5.57)$$

$$\omega t_3 = \omega t_1 + \pi, \qquad (5.58)$$

$$\omega t_4 = \omega t_2 + \pi, \qquad (5.59)$$

and B_1 as

$$B_1 = \frac{2}{2\pi}\left[\int_0^{\omega t_1}(0)\sin\omega t\, d(\omega t) + \int_{\omega t_1}^{\omega t_2}(K_1)\sin\omega t\, d(\omega t)\right.$$
$$\left.+ \int_{\omega t_2}^{\omega t_3}(0)\sin\omega t\, d(\omega t) + \int_{\omega t_3}^{\omega t_4}(-K_1)\sin\omega t\, d(\omega t)\right]. \qquad (5.60)$$

Integrating Eqs. (5.55) and (5.60), we obtain the expressions

$$A_1 = -\frac{2K_1}{\pi}\left(\frac{h}{M}\right), \qquad (5.61)$$

$$B_1 = \frac{2K_1}{\pi}\left[\cos\sin^{-1}\frac{D+h}{M} - \cos\left(\pi - \sin^{-1}\frac{D}{M}\right)\right]. \qquad (5.62)$$

The describing function is given by the expression

$$DN_{\text{on-off}}(M) = \frac{D}{M}\sqrt{A_1^2 + B_1^2}\;\bigg/\;\tan^{-1}\frac{A_1}{B_1}. \qquad (5.63)$$

Notice that the describing function for this device is a function only of the amplitude of the input and not of frequency. Figures 5.18 and 5.19, which have been obtained from Eqs. (5.61) and (5.62), are sketches of the normalized amplitude and phase characteristics of the describing function as a function of the ratio M/D, respectively. Notice that the phase lag is zero when hysteresis is not present and grows progressively worse as the hysteresis increases.

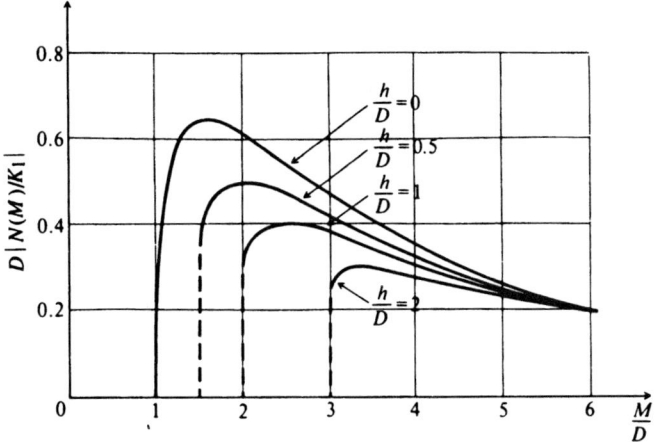

Figure 5.18 Amplitude characteristics for the describing function of an on-off device having hysteresis.

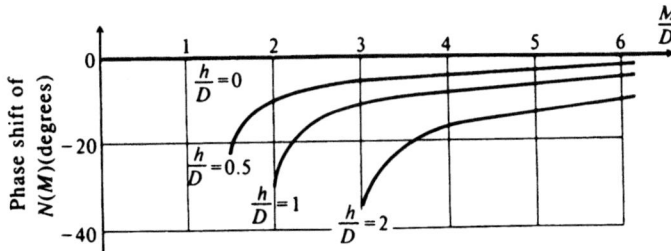

Figure 5.19 Phase characteristics for the describing function of an on-off device having hysteresis.

E. Describing Function of Coulomb Friction and Stiction

In Chapter 3‡ we considered the effect of damping in linear systems. Damping, a form of friction, is known as *viscous friction*. Its characteristic is that its magnitude is always proportional to velocity, as illustrated in Figure 5.20a. The damping factor B is the slope of this characteristic. Another type of frictional force commonly found in control systems is known as *Coulomb friction*. Unlike viscous friction, it is not proportional to velocity but is a constant force that always opposes the velocity. This nonlinear phenomenon is illustrated in Figure 5.20b, where the Coulomb friction force is denoted by $\pm F_c$. Another nonlinear form of frictional force, known as static friction or *stiction*, is the value of the frictional force at zero velocity. It is usually denoted by $\pm F_s$. Figure 5.20c illustrates the composite frictional-force characteristics generally encountered when controlling some load.

To determine the describing function of Coulomb friction, we can express the relationship between the input and output as

$$n(\omega t) = m(\omega t) \pm F_c, \tag{5.64}$$

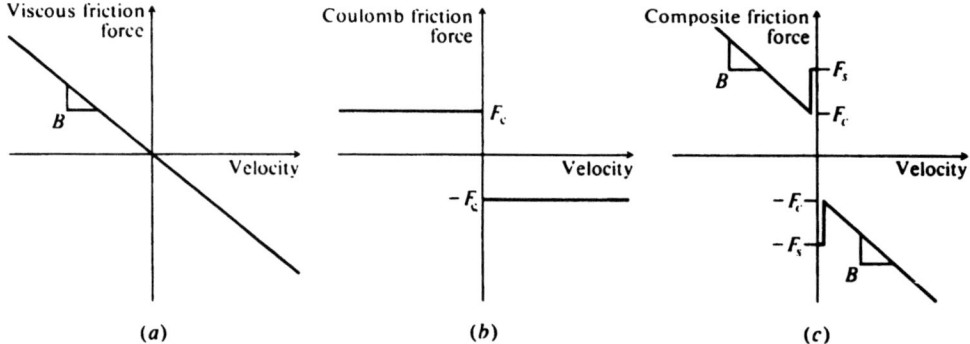

Figure 5.20 (a) Viscous friction characteristics. (b) Coulomb friction characteristics. (c) Composite friction characteristics illustrating viscous, Coulomb, and static friction (stiction).

where

$m(\omega t)$ = applied force,
F_c = force necessary to overcome Coulomb friction,
$n(\omega t)$ = output force.

The corresponding steady-state waveforms are given by Figure 5.21. It should be noted that the discontinuities of the output waveform correspond to zero velocity, because the force required to overcome Coulomb friction changes sign at these instants. The relationship between the input and output forces is

$$n(\omega t) = M \sin \omega t + F_c \quad \text{for } -\omega t_1 < \omega t < \omega t_2, \tag{5.65}$$

$$n(\omega t) = M \sin \omega t - F_c \quad \text{for } \omega t_2 < \omega t < \pi, \tag{5.66}$$

where

$$\omega t_1 = \sin^{-1} \gamma, \quad \omega t_2 = \cos^{-1}(\pi \gamma / 2), \quad \gamma = F_c / M.$$

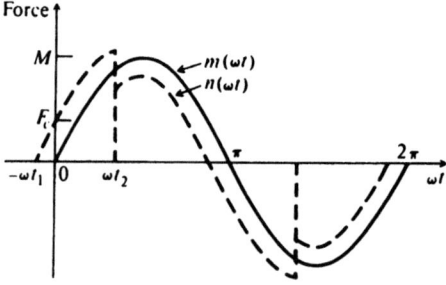

Figure 5.21 Coulomb friction characteristics.

5.8. DERIVATION OF DESCRIBING FUNCTIONS FOR COMMON NONLINEARITIES

The fundamental components of the output A_1 and B_1 are

$$A_1 = \frac{2}{\pi} \int_{-\omega t_1}^{\omega t_2} M(\sin \omega t + \gamma) \cos \omega t\, d(\omega t)$$
$$+ \frac{2}{\pi} \int_{\omega t_2}^{\pi - \omega t_1} M(\sin \omega t - \gamma) \cos \omega t\, d(\omega t), \tag{5.67}$$

$$B_1 = \frac{2}{\pi} \int_{-\omega t_1}^{\omega t_2} M(\sin \omega t + \gamma) \sin \omega t\, d(\omega t)$$
$$+ \frac{2}{\pi} \int_{\omega t_2}^{\pi - \omega t_1} M(\sin \omega t - \gamma) \sin \omega t\, d(\omega t). \tag{5.68}$$

Integrating Eqs. (5.67) and (5.68), we obtain the following expressions:

$$A_1 = 2M\gamma \left(\frac{4}{\pi^2} - \gamma^2\right)^{1/2}, \tag{5.69}$$

$$B_1 = M[1 - 2\gamma^2]. \tag{5.70}$$

The resulting expression for Coulomb friction is

$$N_c(\gamma) = \left[1 - 4\left(1 - \frac{4}{\pi^2}\right)\gamma^2\right]^{1/2} \bigg/ \tan^{-1} \frac{2\gamma(4/\pi^2 - \gamma^2)^{1/2}}{1 - 2\gamma^2},$$
$$\text{for } \gamma \leqslant 0.536, \tag{5.71}$$

$$N_c(\gamma) = \frac{1}{\pi} \{[\pi - (\omega t_1 - \omega t_2) - \sin \omega t_1(\cos \omega t_1 + \cos \omega t_2)$$
$$- \cos \omega t_2(\sin \omega t_1 + \sin \omega t_2)]^2 + [\sin \omega t_1 + \sin \omega t_2]\}^{1/2}$$
$$\times \bigg/ \tan^{-1} \frac{\sin \omega t_1 + \sin \omega t_2}{\{\pi - (\omega t_1 - \omega t_2) - \sin \omega t_1(\cos \omega t_1 + \cos \omega t_2)},$$
$$- \cos \omega t_2(\sin \omega t_1 + \sin \omega t_2)\}$$
$$\text{for } \gamma > 0.536. \tag{5.72}$$

Observe that the describing function for Coulomb friction depends only on the amplitude of the input and not on its frequency. The gain-phase relationship of the describing function for Coulomb friction is illustrated in Figure 5.22.

The describing function for the simultaneous occurrence of both Coulomb friction and stiction is considered next. Waveform relationships between the applied force $m(\omega t)$, the output force $n(\omega t)$, and the forces necessary to overcome Coulomb friction and stiction F_c and F_s are given in Figure 5.23. The expressions for the fundamental components of the Fourier coefficients are

$$A_1 = \frac{2}{\pi} \int_{\omega t_2}^{\pi + \omega t_3} M(\sin \omega t - \gamma) \cos \omega t\, d(\omega t), \tag{5.73}$$

$$B_1 = \frac{2}{\pi} \int_{\omega t_2}^{\pi + \omega t_3} M(\sin \omega t - \gamma) \sin \omega t\, d(\omega t), \tag{5.74}$$

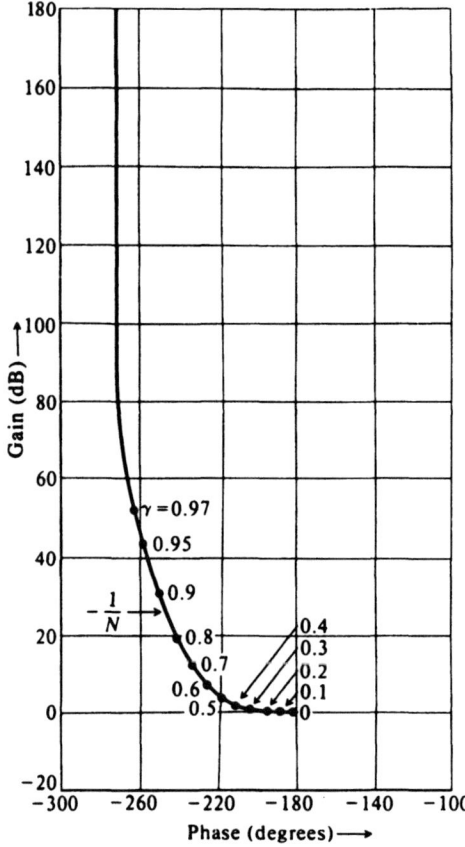

Figure 5.22 Gain-phase characteristics for the describing function of Coulomb friction.

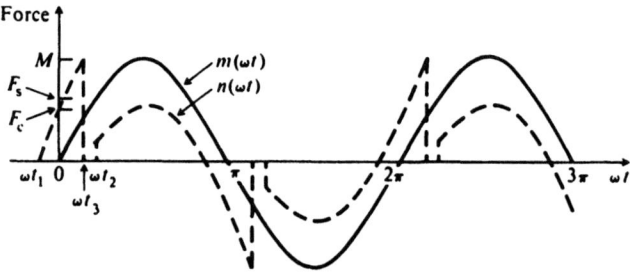

Figure 5.23 Couloumb friction and stiction characteristics.

where

$$\omega t_1 = \sin^{-1} \gamma,$$
$$\omega t_2 = \sin^{-1}(F_s/M),$$
$$\omega t_3 = \cos^{-1}(\pi\gamma/2),$$
$$\gamma = F_c/M.$$

The describing function for the combined case of Coulomb friction and stiction is obtained by integrating Eqs. (5.73) and (5.74). The resultant expression is

$$N(\gamma) = \frac{1}{\pi} \{[\pi - (\omega t_2 - \omega t_3) - \cos \omega t_2 (2 \sin \omega t_1 - \sin \omega t_2)$$
$$- \cos \omega t_3 (2 \sin \omega t_1 + \sin \omega t_3)]^2$$
$$+ [\sin \omega t_3 (2 \sin \omega t_1 + \sin \omega t_3) + \sin \omega t_2 (2 \sin \omega t_1 - \sin \omega t_2)]^2\}^{1/2}$$
$$\times \tan^{-1} \frac{\sin \omega t_3 (2 \sin \omega t_1 + \sin \omega t_3) + \sin \omega t_2 (2 \sin \omega t_1 - \sin \omega t_2)}{\pi - (\omega t_2 - \omega t_3) - \cos \omega t_2 (2 \sin \omega t_1 - \sin \omega t_2)}$$
$$- \cos \omega t_3 (2 \sin \omega t_1 + \sin \omega t_3)$$

(5.75)

Observe from this expression that the describing function for the combined case of Coulomb friction and stiction is a function of the amplitude of the input and the relative magnitudes of friction but not of frequency. Figure 5.24 illustrates the gain-phase relationship of the describing function for the combined case of Coulomb friction and stiction.

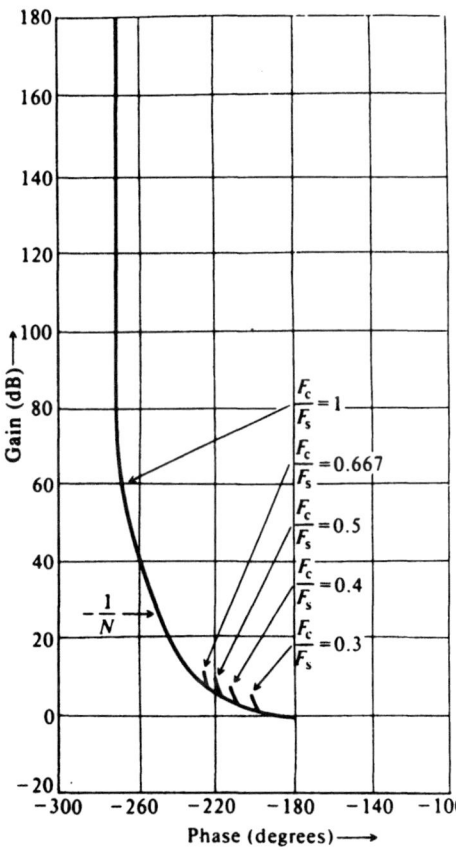

Figure 5.24 Gain-phase characteristics for the describing function of Coulomb friction and stiction.

5.9. USE OF THE DESCRIBING FUNCTION TO PREDICT OSCILLATIONS

The describing function of a nonlinear element can be utilized to determine the existence of a limit cycle in a nonlinear control system in an approximating manner [28]. Let us reconsider the general nonlinear system illustrated in Figure 5.5. If we assume that the describing function of the nonlinearity is $N(M, \omega)$ and $R(j\omega) = 0$, let us determine the conditions under which an assumed oscillation $m(\omega t)$ can be sustained, where

$$m(\omega t) = M \cos \omega t. \qquad (5.76)$$

The fundamental component of $n(\omega t)$ is given by

$$n_1(\omega t) = |N(M, \omega)| M \cos[\omega t + \phi_1(M, \omega)]. \qquad (5.77)$$

Expressing Eqs. (5.76) and (5.77) as the real part of a complex exponential, we obtain the following expressions:

$$m(\omega t) = \text{Re}[M e^{j\omega t}],$$
$$n_1(\omega t) = \text{Re}[|N(M, \omega)| M e^{j(\omega t + \phi_1)}].$$

The output of the linear elements is given by

$$\text{Re}[|N(M, \omega)| M |G(j\omega)| e^{j(\omega t + \phi_1 + \alpha)}], \qquad (5.78)$$

where

$$G(j\omega) = G_1(j\omega) G_2(j\omega),$$
$$\alpha = \text{phase shift due to } G(j\omega).$$

Equation (5.78) must equal $-m(\omega t)$ if the initial assumption is to hold. Thus, dropping the real-part notation, we obtain the following expression:

$$-M e^{j\omega t} = |N(M, \omega)| M |G(j\omega)| e^{j(t + \phi_1 + \alpha)}.$$

This can be rewritten as

$$[|N(M, \omega)||G(j\omega)| e^{j(\phi_1 + \alpha)} + 1] M e^{j\omega t} = 0. \qquad (5.79)$$

For a sustained oscillation, the bracketed term in Eq. (5.79) must vanish, since $M \neq 0$. Therefore,

$$1 + N(M, \omega) G(j\omega) = 0 \qquad (5.80)$$

and any combination of values of input amplitude M and frequency ω which satisfy the equation

$$G(j\omega) = -\frac{1}{N(M, \omega)} \qquad (5.81)$$

are capable of providing a limit cycle. If a combination of amplitude and frequency can be found which satisfies Eq. (5.81), the feedback control system can have a sustained oscillation.

From experience, I have found the gain-phase plot to be the most revealing technique for stability analysis when utilizing the describing-function method. Two separate sets of loci, corresponding to $G(j\omega)$ and $-1/N(M, \omega)$ of Eq. (5.81), are sketched on the gain-phase plot. Intersections of the two loci indicate possible solutions to Eq. (5.81) and yield information as to the magnitude and frequency ω of sustained oscillation. If no intersections result, an oscillation is unlikely. (Remember, the describing function is an approximation.) The distance to a possible intersection can be used as a criterion of closeness to oscillation. This is next illustrated on the gain-phase diagram for several representative systems.

A. Example 5.9-1

Figure 5.25a illustrates an analysis via the gain-phase diagram for a nonlinear system containing a dead zone where $K_1 = 1$. The figure illustrates a stable system where a dead zone is present and

$$G(j\omega)H(j\omega) = \frac{2}{j\omega(1 + 0.5j\omega)(1 + 0.1j\omega)}. \quad (5.82)$$

Figure 5.25b illustrates a system where a dead zone is present and

$$G(j\omega)H(j\omega) = \frac{17}{j\omega(1 + 0.5j\omega)(1 + 0.1j\omega)}. \quad (5.83)$$

An intersection occurs at a frequency of approximately 4.4 rad/sec and a value of D/M of 0.09. This is to be interpreted as the frequency and amplitude which satisfy Eq. (5.81) and which result in a limit cycle. Notice that the system is unstable from a linear viewpoint, because the $-1, 0$ point is enclosed.

The interpretation of this situation is quite illuminating. Because the normalized describing function of a dead zone is less than one, its effect is to reduce the overall system gain. When multiplied by the transfer function given by Eq. (5.82), which would produce a stable feedback system by itself, its effect is to make the system even more stable. Therefore, in order to illustrate a limit cycle in this nonlinear system containing a dead zone, it is necessary to illustrate a system that produces an unstable control system from a linear viewpoint as well. The frequency function of Eq. (5.83) satisfies this requirement, as indicated in Figure 5.25b.

It is interesting to observe that this nonlinear control system can only be unstable if we also make it unstable from a linear viewpoint. The next question to answer is how will it behave in an unstable mode, from the linear or nonlinear aspect? A linear system that is unstable has an unbounded response to a bounded input. (Review the BIBO concept presented in Section 6.1.[‡]) A nonlinear system that is unstable can have a stable or unstable limit cycle. The limit cycle in this problem is a stable one, based on the rule presented in the following problem (see Figure 5.27). Therefore, nonlinear theory states that the system would oscillate at a fixed frequency ω and amplitude M defined by the intersection of the two curves in Figure 5.25b. However,

354 NONLINEAR CONTROL-SYSTEM DESIGN

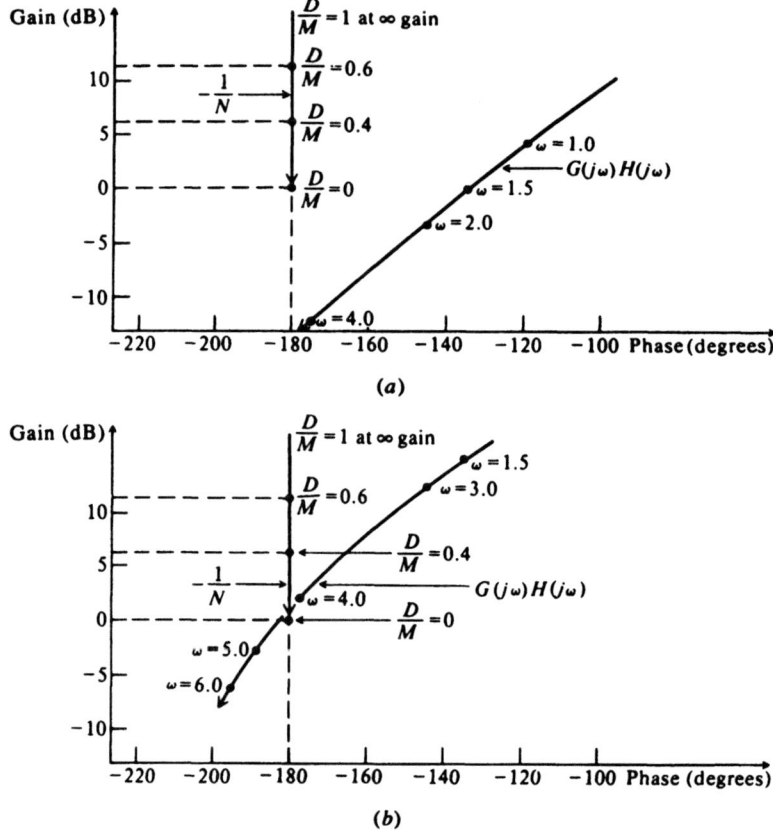

Figure 5.25 (a) Gain-phase diagram stability analysis of nonlinear system containing a dead zone, where

$$G(j\omega)H(j\omega) = \frac{2}{j\omega(1+0.5j\omega)(1+0.1j\omega)}.$$

(b) Gain-phase diagram stability analysis of nonlinear system containing a dead zone, where

$$G(j\omega)H(j\omega) = \frac{17}{j\omega(1+0.5j\omega)(1+0.1j\omega)}.$$

linear theory states that the system would have an unbounded output to any bounded input (including noise), and the linear theory viewpoint would be dominant in this problem.

B. Example 5.9-2

Figure 5.26 illustrates the gain-phase diagram analysis for a nonlinear system containing backlash, where

$$G(j\omega)H(j\omega) = \frac{1.5}{j\omega(1+j\omega)^2}. \tag{5.84}$$

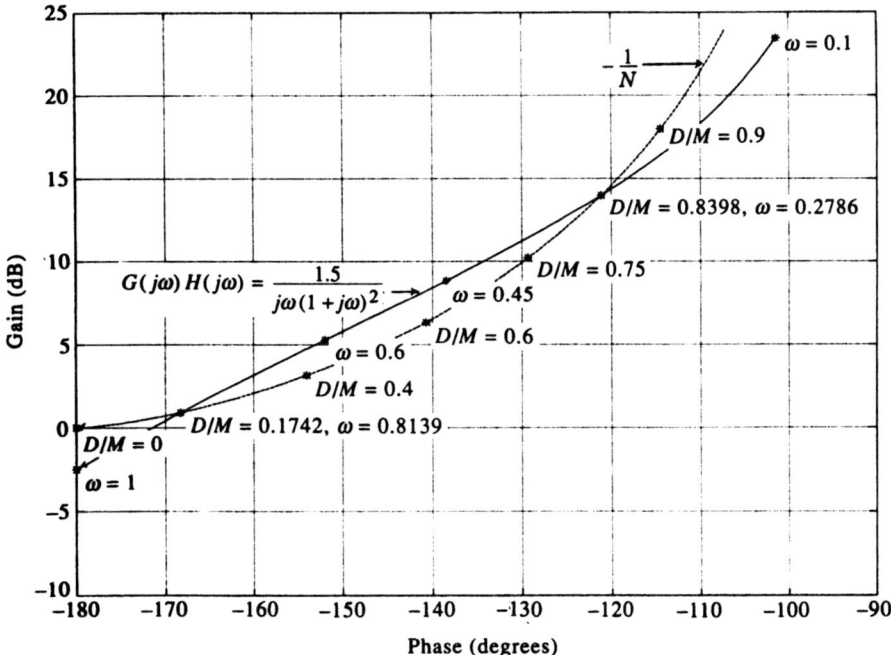

Figure 5.26 Gain-phase diagram analysis of nonlinear system containing backlash.

This gain-phase diagram was obtained using MATLAB, [29–31] and is contained in the M-file that is part of my AMCSTD Toolbox disk which can be retrieved free from The MathWorks, Inc. anonymous FTP server at ftp://ftp.mathworks.com/pub/books/advshinners. Notice that the system has two points of intersection corresponding to a pair of limit cycles. They occur at $\omega = 0.8139$, $D/M = 0.1742$ and $\omega = 0.2786$, $D/M = 0.8398$. These two limit cycles must now be examined to determine whether they are stable or unstable.

A generalized rule can be established for determining whether a limit cycle is stable or unstable [32]. If the two loci are assigned a sense of direction so that the linear locus $G(j\omega)$ is pointing in the direction of increasing frequency and the nonlinear locus $-1/N$ is pointing in the direction of increasing amplitude M (decreasing D/M in our example), then a stable limit cycle occurs when the nonlinear locus appears to an observer, stationed on the linear locus and facing in the direction of increasing frequency, to cross from left to right in the direction of increasing amplitude M. If the opposite occurs, then the limit cycle is unstable and the state of the system is divergent. As an example of applying this rule, consider the gain-phase plot of Figure 5.27. Here, we see that there are two unstable limit cycles (divergent states) and three stable limit cycles (convergent states). The unstable limit cycles cannot maintain themselves in the presence of minute disturbances; the stable limit cycles will always maintain themselves in the presence of disturbances.

Applying this generalized rule for determining whether the limit cycles in Figure 5.26 are stable or unstable, we find that $\omega = 0.8139$, $D/M = 0.1742$ corresponds to a stable limit cycle and the other intersection corresponds to an unstable limit cycle. The stable limit cycle corresponds to a convergent point because disturbances at

356 NONLINEAR CONTROL-SYSTEM DESIGN

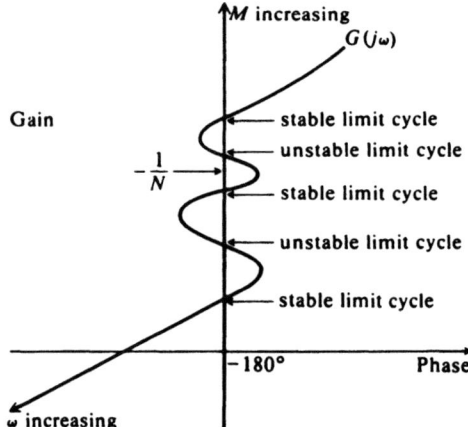

Figure 5.27 Illustration of stable and unstable limit cycles.

either side tend to converge to these conditions. This is contrasted with the unstable limit cycle at $\omega = 0.2786$, $D/M = 0.8398$, which is a divergent point, because disturbances that are not large enough to reach this intersection will decay, and disturbances that are larger will result in oscillations that tend to increase in amplitude until the stable limit cycle is reached.

The describing function analysis presented in this book uses the following special functions, which are part of my AMCSTD Toolbox: relays;back_lsh.m;dead_zn.m. They determine the following:

1. back_lsh.m determines the value of $N(M)$ in Eq. (5.50) for the describing function of backlash, and is then used for the gain-phase plot. Values of D/M and ω at the various limit-cycle intersections, if any, are then provided as an output after the values for the linear transfer function, $G(j\omega)$, and an initial estimate of where D/M is at the intersection are provided as inputs to BACK_LSH.M.

2. dead_zn.m determines the value of $N(M)$ in Eq. (5.31) for the describing function of dead zone, and is then used for the gain-phase plot. Values of D/M and ω at the various limit-cycle intersections, if any, are then provided as an output after the values for the linear transfer function, $G(j\omega)$, and an initial estimate of where D/M is at the intersection are provided as inputs to dead_zm.m.

3. relays determine the value of $N(M)$ in Eq. (5.63) for the describing function of an on-off element having hysteresis, which can be used for the gain-phase plot. Values at the various limit cycle intersections, if any, can be determined after the values for the linear transfer function, $G(j\omega)$, are added to the gain-phase plot.

The calculations performed in back_lsh.m and dead_zn.m are based on the describing function definitions, and are analytical and very accurate as opposed to those results obtained using graphical techniques. The determination of $N(M)$ by relays for a nonlinear on-off element having hysteresis is based on this nonlinear device's describing-function definition, but the solution for limit cycles using relays is based on graphical techniques. These three functions could be used as illustrations by the student to implment functions for the other nonlinear characteristics presented (e.g., saturation). My AMCSTD Toolbox can be retrieved free from The MathWorks, Inc. anonymous FTP server at ftp://ftp.mathworks.com/pub/books/advshinners.

5.10. COMPENSATION AND DESIGN OF NONLINEAR CONTROL SYSTEMS USING DESCRIBING FUNCTIONS

The purpose of this section is to illustrate the procedure for compensating for undesirable nonlinearities in a system. As an example, we consider the nonlinearity to be backlash. The gain-phase plot will be used for analysis, although we could use the Nyquist diagram just as well.

Oscillating input signals, having frequencies very much greater in magnitude than the system bandwidth, can be used to maintain the output of a system containing backlash at its correct average value. This techique is known as *dither*. It is effective in reducing the influence of very small amplitude nonlinearities. However, the resultant increased wear on the system is a serious disadvantage of this simple approach to the problem. The effect of dither on the describing function is analyzed in References [33–37] where the "dual input describing function" is discussed.* For systems which are to operate continuously for long periods of time, it is necessary to utilize other approaches that will eliminate backlash. Reducing the system gain, adding phase-lead networks, and introducing rate feedback in parallel with position feedback are all relatively simple methods that can be used to minimize the effects of backlash. We next demonstrate the theoretical effects of these techniques on backlash [1,28].

Let us reconsider the nonlinear system analysis of Figure 5.26. This sketch illustrates that a nonlinear system having the form shown in Figure 5.5, where

$$G(j\omega) = G_1(j\omega)G_2(j\omega) = \frac{1.5}{j\omega(1+j\omega)^2}, \quad (5.85)$$

and having a backlash element, was indeed oscillatory. We demonstrate how this system can be stabilized by each of the following electrical methods:

(a) Reducing the system gain.
(b) Adding a phase-lead network.
(c) Introducing rate feedback in parallel with position feedback.

---

*The dual-input describing function is a modified describing function which is dependent on two frequency components: the intelligence signal and the dither signal.

A. Reducing the System Gain

In order to consider the effects of gain changes, let us rewrite $G(j\omega)$ of Eq. (5.80) as

$$G(j\omega) = KG'(j\omega), \qquad (5.86)$$

where K represents the system gain and $G'(j\omega)$ represents only the poles and zeros of the linear part of the system. Therefore, Eq. (5.81) can be rewritten as

$$G'(j\omega) = -\frac{1}{KN(M,\omega)}. \qquad (5.87)$$

By reducing the gain K, the limit cycle is eliminated because the curve of $-1/KN$ is moved upward. Figure 5.28 illustrates how the oscillation can be eliminated by reducing the system gain from 1.5 to unity. At a gain setting of approximately 1.3, the curves of $G'(j\omega)$ and $-1/KN$ just clear each other. A gain setting of unity was chosen in order to maintain some margin of safety.

B. Addition of a Phase-Lead Network

A passive phase-lead network can also be used to eliminate the oscillation. The transfer function of this network is given by

$$G(j\omega)_{\text{lead}} = \frac{1+\alpha T(j\omega)}{1+T(j\omega)}, \qquad (5.88)$$

where $\alpha T > T$. (The attenuation $1/\alpha$ is nullified by increasing the system gain by α.) The compensated value of $G(j\omega)$, $G(j\omega)_{\text{comp}}$, is given by

$$G(j\omega)_{\text{comp}} = \frac{1.5}{j\omega(1+j\omega)^2} \times \frac{1+\alpha T(j\omega)}{1+T(j\omega)}. \qquad (5.89)$$

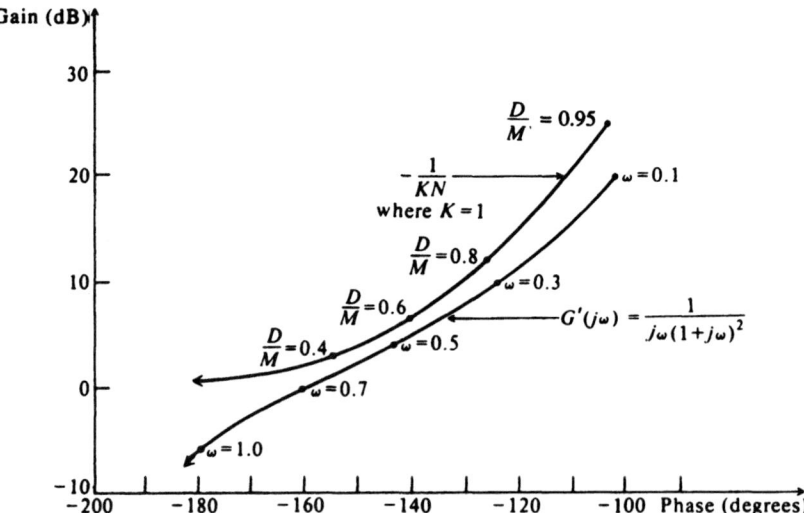

Figure 5.28 Compensation of a nonlinear system containing backlash, by reduction of the system gain.

5.10. COMPENSATION AND DESIGN OF NONLINEAR CONTROL SYSTEMS USING DESCRIBING FUNCTIONS

The best approach for designing the phase-lead network is to use Eqs. (2.7) and (2.8), which are repeated here for solving for ω_{max} and ϕ_{max}, respectively:

$$\omega_{max} = \frac{1}{T\sqrt{\alpha}}, \quad (5.90)$$

$$\phi_{max} = \sin^{-1}\frac{\alpha - 1}{\alpha + 1}. \quad (5.91)$$

To choose ω_{max} and ϕ_{max}, we analyze Figures 2.11 and 5.26. We will assume that the $1/\alpha$ attenuation of the phase-lead network, shown in Figure 2.11, is compensated for by the control system's amplifier. In selecting ω_{max} and ϕ_{max}, we must take into account from Figure 2.11 that the gain characteristics of the phase-lead network increased with frequency, while the low-frequency portion (below $1/\alpha T$) will be unaffected (because the $1/\alpha$ attenuation is compensated for). Therefore, it is advantageous to place ω_{max} at a relatively high frequency such as near the stable limit cycle at $\omega = 0.8139$ and $D/M = 0.1742$, rather than at the frequency where the two curves have maximum phase overlap (at approximately $\omega = 0.7$). Accordingly, we will add an additional phase of 29° at $\omega_{max} = 0.9$. The result is $\alpha T = 1.9$ and $T = 0.65$. For this system, a value of $\alpha T = 1.9$ and $T = 0.65$ will eliminate the limit cycles, as illustrated in Figure 5.29. It is important to recognize, however, that this is not a unique solution. It is only one of many possible solutions, as can be noted from studying the gain-phase diagram.

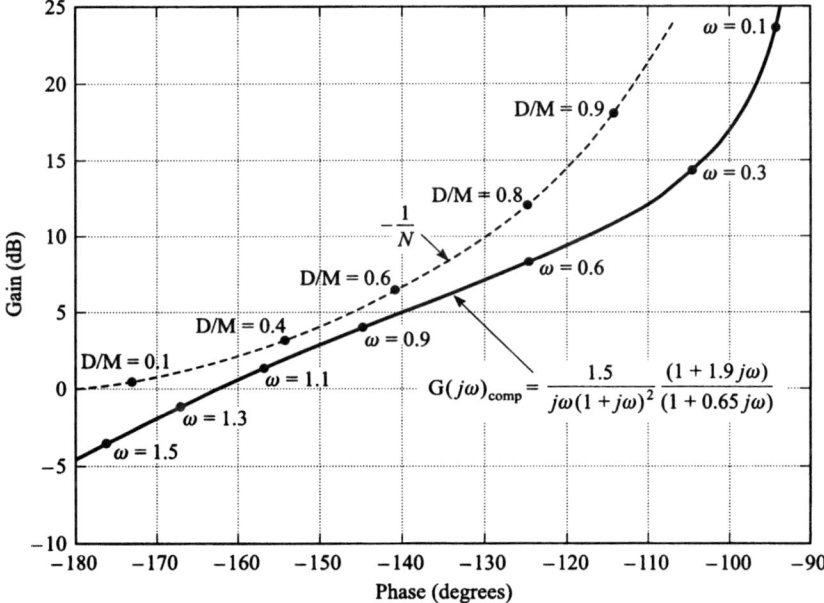

Figure 5.29 Compensation of a nonlinear system containing backlash, by addition of a phase-lead network.

C. Introduction of Rate Feedback

Addition of rate feedback in parallel with position feedback can also be a very effective method for eliminating the oscillation. For this configuration, which is illustrated in Figure 5.30, the value of the system feedback element $H(j\omega)$ is $1 + bj\omega$ which represents a pure zero. Therefore, the oscillation criterion for this configuration must be modified from that given by Eq. (5.81) to the following expression:

$$G(j\omega)H(j\omega) = -\frac{1}{N(M,\omega)}. \tag{5.92}$$

With rate feedback, the value of $G(j\omega)H(j\omega)$ for the system being considered is given by

$$G(j\omega)H(j\omega) = \frac{1.5}{j\omega(1+j\omega)^2}(1+bj\omega). \tag{5.93}$$

The best approach for selecting b is to recognize from Figure 5.30 that the rate feedback in parallel with position feedback results in a pure zero term $(1 + bj\omega)$ as shown in Eq. (5.93). Therefore, it contributes a phase lead of $\tan^{-1} b\omega$. Analysis of Figure 5.26 indicates that the maximum phase overlap is approximately 9° at $\omega = 0.7$. We will design for an additional 7° for safety, and we can solve for b from

$$(9° + 7°) = \tan^{-1} b(0.7). \tag{5.94}$$

Therefore,

$$b = 0.4, \tag{5.95}$$

A value of $b = 0.4$ will eliminate the limit cycle, as is illustrated in Figure 5.31, and a relatively safe margin is achieved. At a value of b approximately equal to 0.25, the curves of $G(j\omega)H(j\omega)$ and $-1/N$ just clear each other.

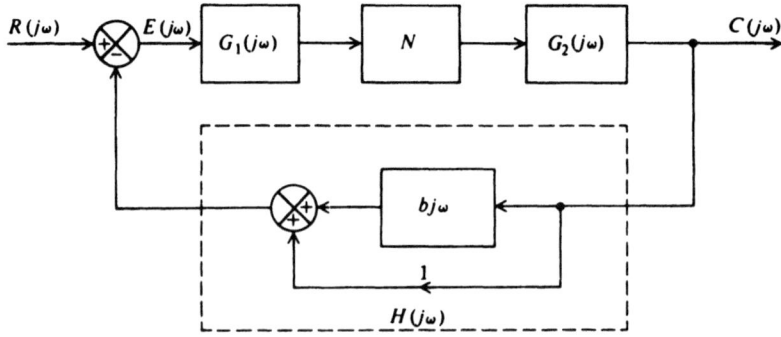

Figure 5.30 Nonlinear system containing rate and position feedback.

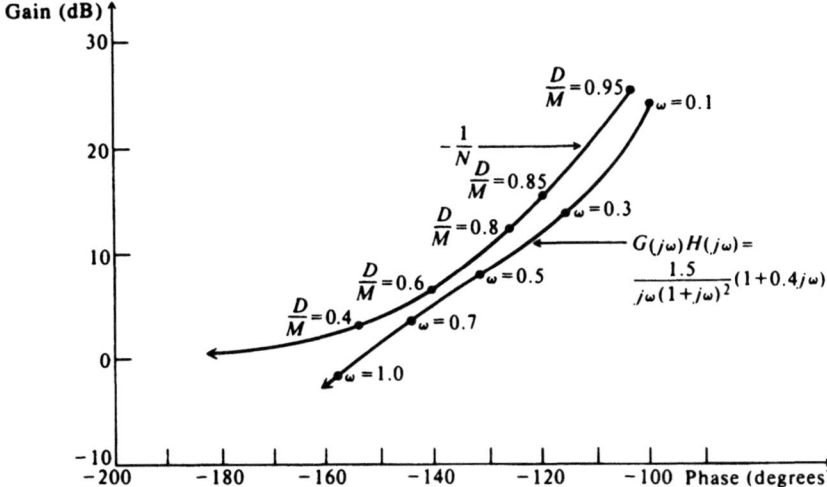

Figure 5.31 Compensation of a nonlinear system containing backlash, by the introduction of rate feedback.

A complete case study for the design of the positioning system of a tracking radar that uses conventional linear techniques and the describing function is presented jointly in Section 7.3. In this problem we show the separate linear and nonlinear considerations for the design of an actual control system, and their joint relationships.

5.11. DESCRIBING-FUNCTION ANALYSIS AND DESIGN USING MATLAB [29]

The use of MATLAB to analyze linear control sytems has been abundantly illustrated in previous chapters of this book. In this section, the application of MATLAB for analyzing the application of the describing functions for analysis and design is illustrated.

The AMCSTD Toolbox provides the following describing function utilities to supplement MATLAB and the Control System Toolbox:

- "deadzone" provides the normalized gain (N) for a given D/M and it also allows the refinement of a solution to the intersection of the dead-zone curve ($-1/N$) and a given transfer function for the linear portion of the control system. The AMCSTD Toolbox's command for this utility is

$$\text{dead\_zn}$$

- "backlash" provides the gain (N) for any given D/M, and it also allows the refinement of a solution to the intersection of the backlash curve ($-1/N$) and a given transfer function for the linear portion of the control systems. The AMCSTD Toolbox's command for this utility is

$$\text{back\_lsh}$$

- "relays" is a hysteresis utility with relay being a subset of its actual capability. It provides the gain (*N*) for any given set(s) of *M*, *D*, and *H* or *H/D*. It also allows the refinement of a solution to the intersection of the hysteresis curve ($-1/N$) and a given transfer equation for the lienar portion of the control system. The AMCSTD Toolbox's command for this utility is

relays

The MathWorks has available a Nonlinear Control Design Toolbox for nonlinear control system analysis, which is illustrated in the AMCSTD Toolbox. The DEMO section of the AMCSTD Toolbox has many illustrations of the analysis of nonlinear systems using MATLAB.

A. Example 5.11-1

Let us now demonstrate the MATLAB program which provides the describing-function analysis illustrated in Figure 5.26. This is also illustrated in the DEMO example 5.2 in the AMCSTD Toolbox.

We will reconsider the nonlinear control system whose linear portion is represented by Eq. (5.84), which is repeated here,

$$G(s)H(s) = \frac{1.5}{s(1+s)^2} \tag{5.96}$$

and whose nonlinear portion is backlash. The MATLAB program to illustrate the backlash curves corresponding to Figures 5.15 and 5.16 is illustrated in the MATLAB program shown in Table 5.1, and is repeated in Figure 5.32. The MATLAB program to plot the describing function analysis shown in Figure 5.26 is given by the MATLAB program shown in Table 5.2, and is repeated in Figure 5.33.

Table 5.1. MATLAB Program for Obtaining the Describing Function of Backlash

```
dm = linspace(.999,0);
n = back_lsh(dm);
subplot(211);
plot(dm,abs(n));
grid;
title('Backlash Amplitude');
xlabel('D/M')
ylabel('lN(M)l');
subplot(212);
plot(dm, angle(n)*180/pi);
grid;
title('Backlash Phase');
xlabel('D/M');
ylabel('Phase(Deg)')
```

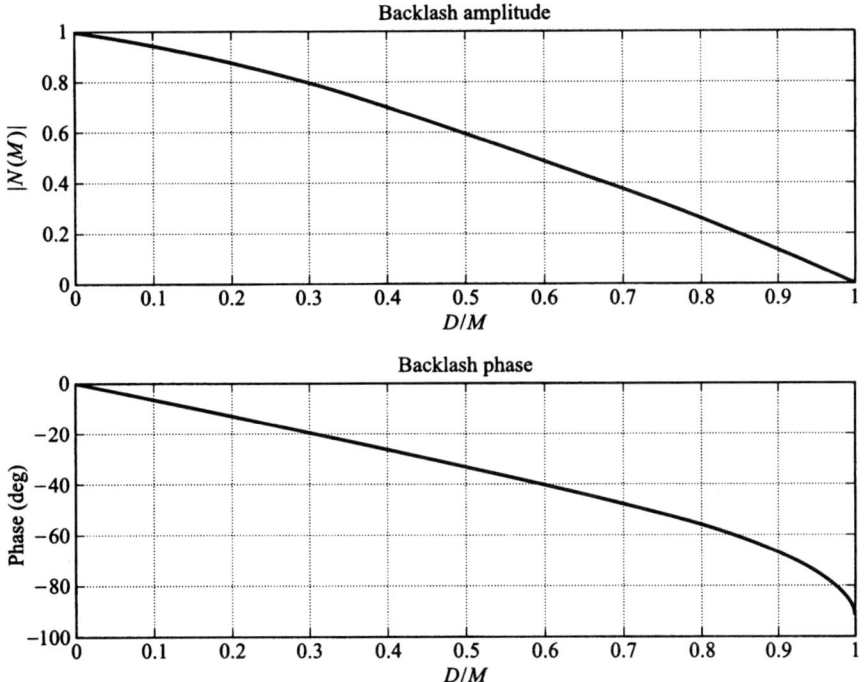

Figure 5.32 Describing-function for backlash.

Table 5.2. MATLAB Program for Obtaining the Describing Function Analysis for a Control System Containing Backlash and whose Linear Portion is Defined by Eq. 5.96

```
num = [0 1.5];
den = [1 1];
w = logspace(-1,0);
[mag,ph] = bode(num,den,w);
magdb=20*log10(mag);
dm=linspace(.95,0);
n = back_lsh(dm);
ninv = (0*n-1)./n;
xlabel('Phase(degrees)')
ylabel('Gain(dB)');
w = [.1 .45 .6 1];
dm = [.9 .75 .6 .4 0];
grid;
title('Nonlinear Analysis for System Containing Backlash')
```

5.12. DIGITAL COMPUTER PROGRAMS FOR PERFORMING THE DESCRIBING-FUNCTION ANALYSIS

For those who do not have access to MATLAB, which has now become an industry standard (more or less), this book also provides another digital computer program for performing the describing function analysis. This approach for providing both

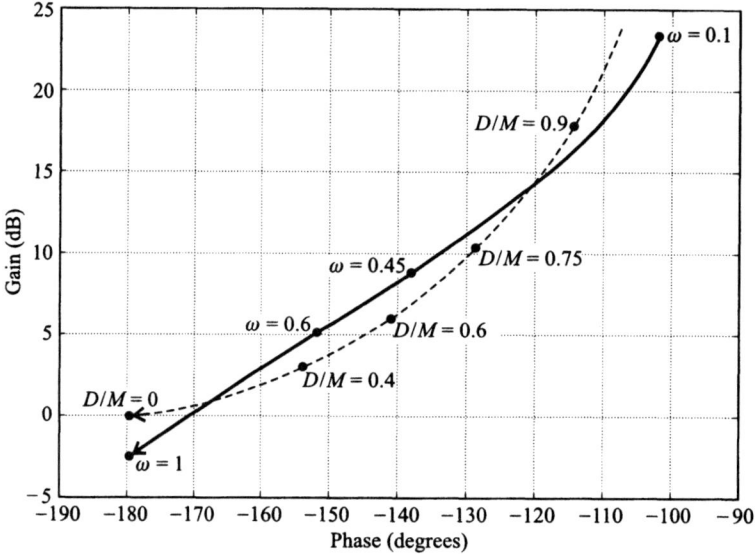

Figure 5.33 Describing-function analysis for a nonlinear control system containing backlash, whose linear portion is described by Eq. (5.84).

MATLAB and other digital computer programs which are not dependent on COTS (commercial-off-the-shelf software) is used throughout this book.

As shown in Sections 5.10 and 5.11 with the use of MATLAB, the digital computer is very useful for the construction of the describing function [25,38]. The computer can compute $-1/N$ and, as indicated in Chapter 1, $G(j\omega)$. This section illustrates the procedure used to analyze and compensate a practical nonlinear system containing backlash with the aid of a working digital computer program. It is based on the method illustrated in Sections 5.7 through 5.10. The Basic language will be used for this program [39,40].

Let us consider the system illustrated in Figure 5.5, where

$$G(j\omega) = \frac{4(1+3j\omega)}{j\omega(1+2j\omega)^2} \tag{5.97}$$

and N corresponds to backlash. We know from Eq. (5.50) that

$$N_{\text{backlash}}(M) = \frac{1}{M}\sqrt{A^2+B^2} \;\; \underline{/\tan^{-1}(A/B)}, \tag{5.98}$$

where

$$A = \frac{2D}{\pi M}\left(\frac{2D}{M}-2\right)M,$$

$$B = \frac{1}{\pi}\left[\frac{\pi}{2} - \sin^{-1}\left(\frac{2D}{M}-1\right) - \left(\frac{2D}{M}-1\right)\cos\sin^{-1}\left(\frac{2D}{M}-1\right)\right]M.$$

Note that the subscripts have been dropped from the A and B terms for simplicity. The coding symbols used are as follows:

$$X = D/M, \quad S = \sin^{-1} C = \tan^{-1}\left(\frac{C}{\sqrt{1-C^2}}\right), \quad P = -\pi - \tan^{-1}(A/B),$$

$$C = 2X - 1, \quad G = 20 \log_{10} \frac{1}{N}.$$

Figure 5.34a illustrates the logic flow diagram for developing the program for computing $-1/N$. Table 5.3 illustrates the actual program for computing $-1/N$. The reader should compare these two in order fully to understand the digital computer's program. Table 5.4 illustrates the computer run for calculating $-1/N$. Notice that is was necessary to compute additional values of $0.9 < D/M < 0.99$ at the end of the computer run since it was found that a limit cycle existed in this region. Figure 5.34b illustrates the logic flow diagram for determining $G(j\omega)$. In the coding G represents the gain of $G(j\omega)$, P represents its phase and W represents ω. Table 5.5 illustrates the actual program for computing $G(j\omega)$ and Table 5.6 illustrates the computer run for calculating $G(j\omega)$.

The values for $-1/N$ and $G(j\omega)$ are illustrated on the gain-phase diagram in Figure 5.35. It indicates limit cycles at $\omega = 1.36$, $D/M = 0.41$ (stable) and $\omega = 0.27$, $D/M = 0.945$ (unstable). As illustrated previously in Section 5.10, this nonlinear system can be compensated by lowering the gain, by the addition of a phase-lead network or rate feedback. A similar analysis indicates that at $K = 1.65$, a phase-lead network given by

$$G(j\omega) = \frac{1 + 1.2j\omega}{1 + 0.4j\omega} \tag{5.99}$$

Table 5.3. Computer Program for Computing $-1/N$ (Basic Program)

```
1     REM DESCRIBING FUNCTION FOR BACKLASH
10    PRINT ''D/M'', ''GAIN(DB)'', ''PHASE(DEGREES)'', ''N''
20    READ X1,X2,D
30    FOR X=X1 TO X2 STEP D
40    LET C=2*X-1
50    LET S=ATN(C/SQR(1-C↑2))
60    LET A=1.27324*X*(X-1)
70    LET B=0.31831*(1.570796-S-C*COS(S))
80    LET N=SQR(A↑2+B↑2)
90    LET G=20*0.43429448*LOG(1/N)
100   LET P=-180-57.29578*ATN(A/B)
110   PRINT X,G,P,N
120   NEXT X
130   DATA 0.05, 0.95, 0.05
140   END
OK
```

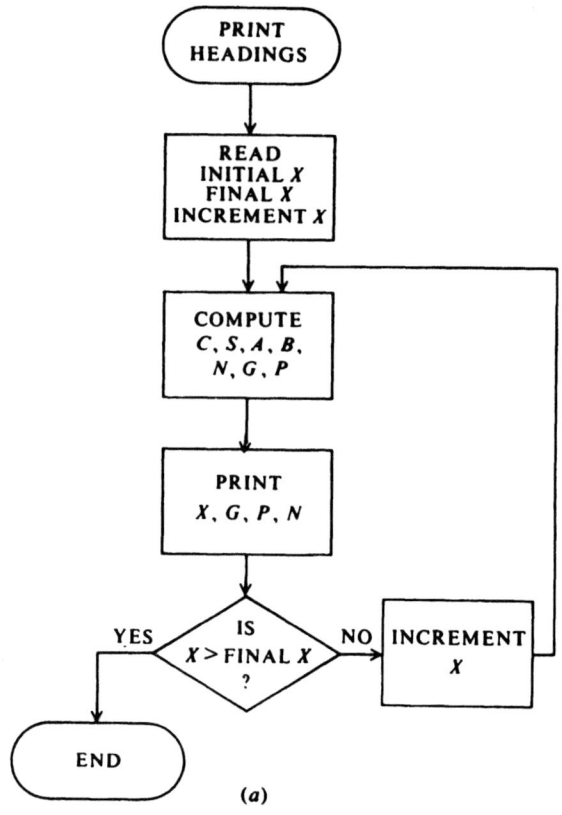

Figure 5.34 (a) Logic flow diagram for determination of $-1/N$.

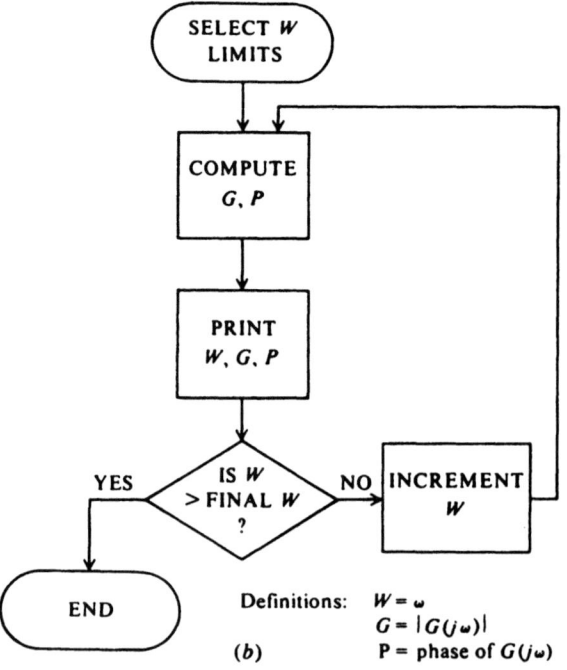

Definitions: $W = \omega$
$G = |G(j\omega)|$
P = phase of $G(j\omega)$

(b)

Figure 5.34 (b) Logic flow diagram for determination of $G(j\omega)$.

Table 5.4. Results of Computer Analysis for −1/N
RUN

Balash D/M	Gain (dB)(−1/N)	Phase (degrees)	N
0.05	0.147435	−176.473	0.983169
0.1	0.401232	−173.107	0.954857
0.15	0.720751	−169.841	0.92037
0.2	1.0957	−166.638	0.881485
0.25	1.52297	−163.472	0.839173
0.3	2.00298	−160.322	0.794056
0.35	2.53864	−157.17	0.746565
0.4	3.13505	−153.998	0.697024
0.45	3.79968	−150.787	0.645678
0.5	4.54295	−147.518	0.592724
0.55	5.37919	−144.17	0.53832
0.6	6.32826	−140.714	0.4826
0.65	7.41848	−137.119	0.425673
0.7	8.69166	−133.34	0.367635
0.75	10.213	−129.314	0.308567
0.8	12.092	−124.95	0.248541
0.85	14.5345	−120.088	0.187619
0.9	18.0025	−114.426	0.125856
0.95	23.9717	−107.175	6.33018E-2

READY
130 DATA 0.9, 0.99, 0.01*

RUN

Balash D/M	Gain (dB)(−1/N)	Phase (degrees)	N
0.9	18.0025	−114.426	0.125856
0.91	18.9072	−113.146	0.113407
0.92	19.9199	−111.798	0.100927
0.93	21.0694	−110.367	8.84157E-2
0.94	22.3982	−108.836	7.58739E-2
0.95	23.9717	−107.175	6.33018E-2
0.96	25.8999	−105.345	5.06999E-2
0.97	28.3887	−103.275	3.80685E-2
0.98	31.9007	−100.827	2.54078E-2
0.99	37.9115	−97.6472	1.27182E-2

*Modification to Address 130 of Computer Program shown in Table 5.3.

and rate feedback having a constant of 0.47 result in a 3-dB margin of safety. The system compensated with a phase-lead network and rate feedback is illustrated in Figure 5.35.

368 NONLINEAR CONTROL-SYSTEM DESIGN

Table 5.5. Computer Program for Computing $G(j\omega)$ (Basic Program)

```
10      PRINT ''OMEGA'', ''GAIN(DB)'', ''PHASE(DEG)''
20      READ K
30      READ W1,W2,D
40      FOR W=W1 TO W2 STEP D
50      LET G=4.3429448*LOG(K*K*(1+9*W*W)/(W*W*(1+4*W*W)↑2))
60      LET P=-90+57.29578*(ATN(3*W)-2*ATN(2*W))
70      PRINT W,G,P
80      NEXT W
90      DATA 4,0.1,2.0,0.05
200     END
```

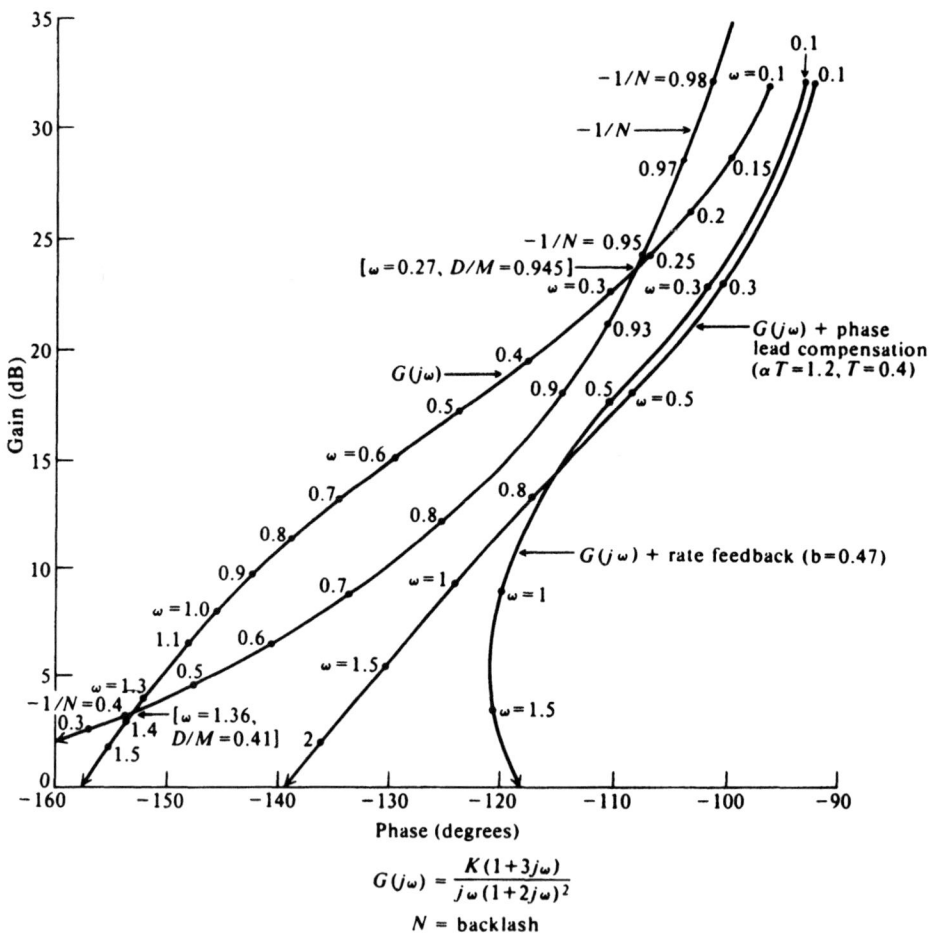

$$G(j\omega) = \frac{K(1+3j\omega)}{j\omega(1+2j\omega)^2}$$

N = backlash

Figure 5.35 Describing-function analysis based on computer program results.

Table 5.6. Results of Computer Analysis for $G(j\omega)$

RUN

Omega	Gain (dB)	Phase (deg)
0.1	32.0748	−95.9206
0.15	28.5717	−99.1707
0.2	26.0668	−102.639
0.25	24.0824	−106.26
0.3	22.4048	−109.94
0.35	20.9235	−113.587
0.4	19.577	−117.125
0.45	18.3297	−120.503
0.5	17.16	−123.69
0.55	16.0544	−126.671
0.6	15.004	−129.443
0.65	14.0026	−132.012
0.7	13.0454	−134.388
0.75	12.1288	−136.582
0.8	11.2499	−138.609
0.85	10.4059	−140.482
0.9	9.59458	−142.214
0.95	8.81385	−143.818
1.0	8.0618	−145.305
1.05	7.33669	−146.686
1.1	6.63691	−147.97
1.15	5.96097	−149.167
1.2	5.3075	−150.284
1.25	4.67521	−151.329
1.3	4.06293	−152.306
1.35	3.46955	−153.223
1.4	2.89405	−154.085
1.45	2.33549	−154.895
1.5	1.79296	−155.659
1.55	1.26566	−156.379
1.6	0.752819	−157.06
1.65	0.253712	−157.704
1.7	−0.232325	−158.315
1.75	−0.705916	−158.894
1.8	−1.16765	−159.443
1.85	−1.61806	−159.966
1.9	−2.05767	−160.464
1.95	−2.48695	−160.938
2.0	−2.90636	−161.39

5.13. PIECEWISE-LINEAR APPROXIMATIONS

Approximating any nonlinearity by means of piecewise-linear segmentation is a very useful tool for analysis. Each segment leads to a relatively simple linear differential equation which can be solved by conventional linear techniques. This method, which is not limited to quasilinear systems, has the advantage of yielding an exact solution

370 NONLINEAR CONTROL-SYSTEM DESIGN

for nonlinearities of any order, if the nonlinearity is itself piecewise linear or can be approximated by piecewise-linear segments. We illustrate its application by means of an example.

Let us consider saturation. Figure 5.36 illustrates a simple feedback control system containing an integrator and an amplifier which saturates. The amplifier gain is 5 over an input voltage range of ±1 V. for input voltages greater than this, the amplifier saturates. It is quite evident that two distinct linear operating regions for the amplifier exist. Each of these linear regions can be considered separately in a piecewise-linear manner in order to obtain the composite response of the system.

For the unsaturated region, the relationships depicting the system operation are

$$e(t) = r(t) - c(t), \tag{5.100}$$

$$f(t) = 5e(t), \tag{5.101}$$

$$c(t) = \int f(t)\,dt. \tag{5.102}$$

During saturation, Eqs. (5.100) and (5.102) are still valid. However, (5.101) changes to

$$f(t) = 5 \quad \text{for } e(t) > 1, \tag{5.103}$$

$$f(t) = -5 \quad \text{for } e(t) < -1. \tag{5.104}$$

Assuming zero initial conditions and a step input of 10 V, the expression for the output during the saturated region of operation $c_{\text{sat}}(t)$ is given by

$$c_{\text{sat}}(t) = \int_0^t 5\,dt = 5t. \tag{5.105}$$

The expression for the output during the unsaturated region is given by

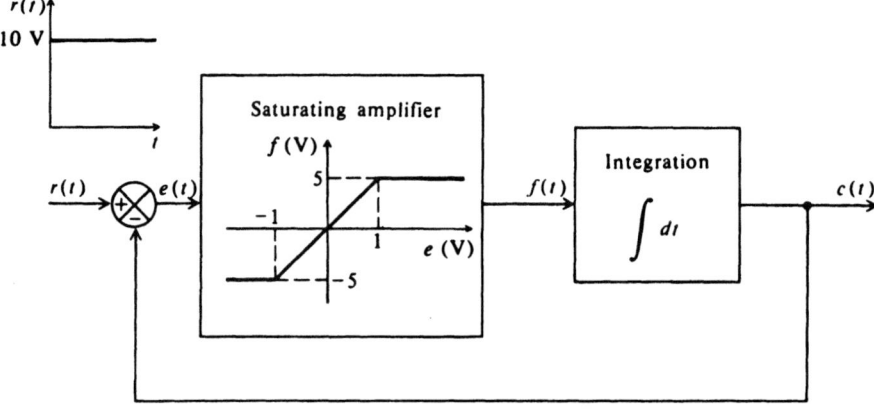

Figure 5.36 Feedback control system containing saturation.

$$c_{us}(t) = 5\int_{t_1}^{t}(10-c(t))\,dt, \quad \text{or} \quad \frac{dc_{us}(t)}{dt} + 5c(t) = 50. \tag{5.106}$$

The time t_1 is the time at which the amplifier becomes unsaturated. When $c = 9$, $e = 1$, and t_1 is 1.8 sec. Using conventional techniques, the solution to Eq. (5.106) is

$$c_{us}(t) = 10 - e^{-5(t-1.8)}. \tag{5.107}$$

The initial value for this region, $c_{us}(0)$, is the same as the final value of the saturated region, $c_{sat}(1.8) = 9$; this continuity of the output is imposed by the integrator. Therefore, the composite solution for this problem, obtained by a piecewise-linear analysis, is

$$c_{sat}(t) = 5t \quad \text{for } 0 \leqslant t \leqslant 1.8, \tag{5.108}$$
$$c_{us}(t) = 10 - e^{-5(t-1.8)} \quad \text{for } t > 1.8. \tag{5.109}$$

The response of this system to a step input of 10 V is sketched in Figure 5.37.

The piecewise-linear approach illustrated in the preceding problem can be extended to very complex nonlinearities. It is important to emphasize that the boundary conditions between linear regions are continuous whenever the transfer function following the nonlinearity is a proper rational function. The resulting differential equation for each segmented region is linear and can be easily solved by conventional linear techniques.

5.14. STATE-VARIABLE ANALYSIS: THE PHASE PLANE

A useful technique of applying the state-space approach to nonlinear systems is the phase-plane method [10,11,41]. It is a technique for analyzing the transient response of a nonlinear control system to an external input or for solving an initial-condition problem. The phase-plane method is limited to second-order systems. The variation of the displacement is plotted against velocity on a graph known as the *phase-plane*. A curve for a specific step input is known as a *trajectory*. A set of curves of displacment versus velocity of a specific system, which are repeated for several input values

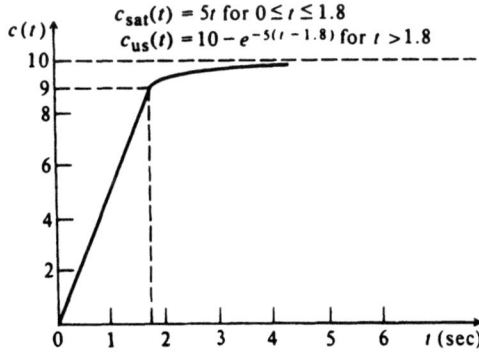

Figure 5.37 Step response of saturating system obtained from piecewise-linear analysis.

(or initial conditions) and are plotted on the same phase plane, is called a *phase portrait*.

The starting point of a trajectory is the initial displacement, $x(0)$, and velocity, $(dx(0)/dt)$. The future path of the trajectory after it leaves its initial starting point represents the behavior of the control system for an input excitation. If the trajectory approaches infinity on the phase plane, the system is unstable. If the phase trajectory approaches the vicinity of the origin, however, the system is stable. If the trajectory circles the origin continuously in a closed curve after excitation, a sustained oscillation known as a *limit cycle exists*.

The phase-plane method is specifically concerned with the solution of second-order, nonlinear differential equations having the following general form:

$$\frac{d^2 x(t)}{dt^2} + A_1\left(x, \frac{dx}{dt}\right)\frac{dx(t)}{dt} + A_2\left(x, \frac{dx}{dt}\right)x(t) = u(t). \qquad (5.110)$$

For our initial analysis, we will assume that the forcing function, $u(t)$, equals zero and that the initial conditions, $x(0)$ and $(dx(0)/dt)$, represent the input to the system. Equation (5.110) immediately emphasizes some serious limitations of the phase-plane method. It is only useful for analyzing second-order systems. It is interesting to compare the limitations of the phase-plane method with those of the describing-function analysis, where the response to sinusoidal inputs of systems having any order could be determined. The phase-plane method could be extended to higher-order systems, but it is too impractical.*

The following section discusses the techniques which can be used for constructing the phase portrait from the differential equation of a system. Then we examine the properties and interpretation of the phase portrait. The procedure to be followed for applying the phase-plane approach to design problems, and a computer program for obtaining the phase plane are then presented.

5.15. CONSTRUCTION OF THE PHASE PORTRAIT

Four procedures that can be used to construct the phase portrait of a system are:

1. direct solution of the differential equation;
2. transformation of the second-order differential equation to a first-order equation;
3. the method of isoclines;
4. digital computer program.

The first two methods are analytical techniques; the third is graphical; the last is presented in Section 5.19. We shall describe these methods next, together with some illustrative examples.

---

*A system described by an nth-order differential equation requires an n-dimensional phase space with a knowledge of n initial conditions. It is, indeed, ar arduous task to visualize this for third- and higher-order systems, and is rarely used.

A. Direct Solution of the Differential Equation

This is the most straightforward method for obtaining the phase portrait. It is usually the most useless method from a practical viewpoint, because we do not have to resort to the phase-plane representation for a solution if the differential equation is integrable. However, this method does provide an understanding of the situation.

The procedure is to solve the differential equation for the dependent variable $x(t)$. The solution for $x(t)$ is then differentiated in order to obtain the derivative of the dependent variable, $dx(t)/dt$. The dependent variable, time, is then eliminated between the two resulting equations. A single equation that relates $x(t)$ and $dx(t)/dt$ results, and can be used to plot the phase portrait directly. However, the approach has the disadvantage that it requires the solution of a nonlinear, second-order, differential equation.

Let us illustrate the procedure in detail by first considering a simple linear system where a torque $T(t)$ was applied to a body having a moment of inertia J, a twisting shaft having a stiffness factor K, and a damper having a damping factor B. Applying Newton's second law of motion to this system resulted in the following relationship:

$$J\frac{d^2\theta(t)}{dt^2} + B\frac{d\theta(t)}{dt} + K\theta(t) = T(t), \tag{5.111}$$

where $\theta(t)$ is the displacement. The differential equation is second order, and the phase-plane method is certainly applicable. Assuming that the system is unexcited, the resulting differential equation is given by

$$\frac{d^2\theta(t)}{dt^2} + \frac{B}{J}\frac{d\theta(t)}{dt} + K\frac{\theta(t)}{J} = 0. \tag{5.112}$$

Equation (5.112) can be written in terms of damping ratio ζ and undamped natural frequency ω_n, as follows.

$$\frac{d^2\theta(t)}{dt^2} + 2\zeta\omega_n\frac{d\theta(t)}{dt} + \omega_n^2\theta(t) = 0, \tag{5.113}$$

where

$$B/J = 2\zeta\omega_n,$$
$$K/J = \omega_n^2.$$

Before sketching the phase portrait, let us consider the simpler case of an undamped system where $\zeta = 0$. For this situation, Eq. (5.113) reduces to

$$\frac{d^2\theta(t)}{dt^2} + \omega_n^2\theta(t) = 0. \tag{5.114}$$

From elementary calculus, the solution to Eq. (5.114) is that of simple harmonic motion:

$$\theta(t) = R\sin(\omega_n t + \phi). \tag{5.115}$$

In order to obtain a relationship between $\theta(t)$ and $d\theta(t)/dt$, we differentiate Eq. (5.115) and then eliminate time between the resulting equations, as follows:

$$\frac{d\theta(t)}{dt} = R\omega_n \cos(\omega_n t + \phi). \tag{5.116}$$

Eliminating time between Eqs. (5.115) and (5.116), we obtain the expression

$$\left[\frac{1}{\omega_n}\frac{d\theta(t)}{dt}\right]^2 + \theta^2(t) = R^2. \tag{5.117}$$

The phase portrait for this system can be drawn directly from Eq. (5.117). Observe that Eq. (5.117) describes a family of concentric circles in the $(1/\omega_n)(d\theta(t)/dt)$ versus $\theta(t)$ plane having a radius of R. Therefore, the phase portrait for this system (shown in Figure 5.38) is a family of concentric circles if a normalized ordinate axis is used. If the ordinate axis is not normalized, a family of ellipses results. Any set of initial conditions, such as points R_1, R_2, R_3, and R_4, specifies a particular circle. For $t > 0$, the motion is in the indicated direction. The origin is defined as a center, and is discussed further in Section 5.16.

Let us next sketch the phase portrait for the case where the damping ratio of this system is finite. The general solution for Eq. (5.113), when it is excited by a step input and has a set of initial conditions, was derived in Chapter 4[†] [see Eq. (4.45)[‡]]. The form of the solution for this system when it is unexcited by a step input and only has a set of initial conditions, $\theta(t_0)$ and $\dot{\theta}(t_0)$, present is given by

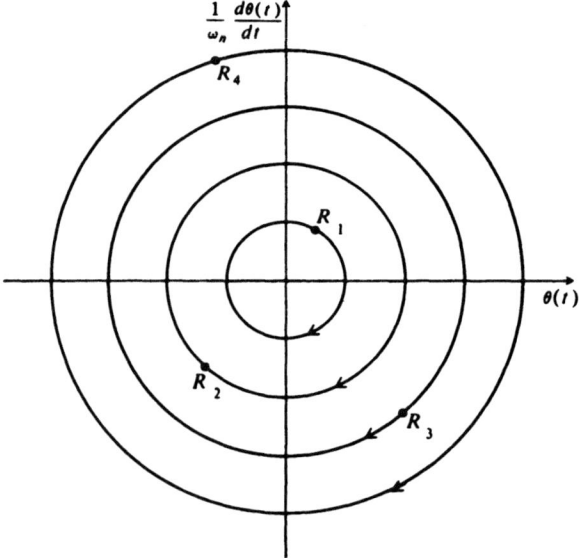

Figure 5.38 Phase portrait for a second-order linear system having no damping. The origin is defined as a center.

5.15. CONSTRUCTION OF THE PHASE PORTRAIT

$$\theta(t) = \frac{\omega_n^2}{\sqrt{1-\zeta^2}} e^{-\zeta\omega_n(t-t_0)} \sin[\omega_n\sqrt{1-\zeta^2}(t-t_0) + \phi_1]\theta(t_0)$$

$$+ \frac{\omega_n}{\sqrt{1-\zeta^2}} e^{-\zeta\omega_n(t-t_0)} \sin[\omega_n\sqrt{1-\zeta^2}(t-t_0)]\dot{\theta}(t_0). \quad (5.118)$$

Proceeding as in the previous example, the derivative of Eq. (5.118) is obtained:

$$\dot{\theta}(t) = \omega_n^3 e^{-\zeta\omega_n(t-t_0)} \cos[\omega_n\sqrt{1-\zeta^2}(t-t_0) + \phi_1]\theta(t_0)$$

$$- \frac{\zeta\omega_n^3}{\sqrt{1-\zeta^2}} e^{-\zeta\omega_n(t-t_0)} \sin[\omega_n\sqrt{1-\zeta^2}(t-t_0) + \phi_1]\theta(t_0)$$

$$+ \omega_n^2 e^{-\zeta\omega_n(t-t_0)} \cos[\omega_n\sqrt{1-\zeta^2}(t-t_0)]\dot{\theta}(t_0)$$

$$- \frac{\zeta\omega_n^2}{\sqrt{1-\zeta^2}} e^{-\zeta\omega_n(t-t_0)} \sin[\omega_n\sqrt{1-\zeta^2}(t-t_0)]\dot{\theta}(t_0). \quad (5.119)$$

The complexity of Eqs. (5.118) and (5.119) makes it quite difficult to eliminate time between them. Therefore we use an alternative approach. Equations (5.118) and (5.119) will be evaluated for several values of time to obtain the corresponding coordinates in a normalized phase plane. The result is plotted in Figure 5.39 for a damping ratio of 0.7. Notice that the phase portrait is a collection of noncrossing paths describing system behavior for all possible initial conditions. Because all the trajectories approach the origin of this system where $u(t) = 0$, the system is stable, as we would expect it to be. The origin is defined as a stable node, and is discussed further in Section 5.16.

The projection of a specific trajectory onto the abscissa gives the variation of $\theta(t)$ with time, and its projection onto the ordinate gives the variation of $d\theta(t)/dt$ with

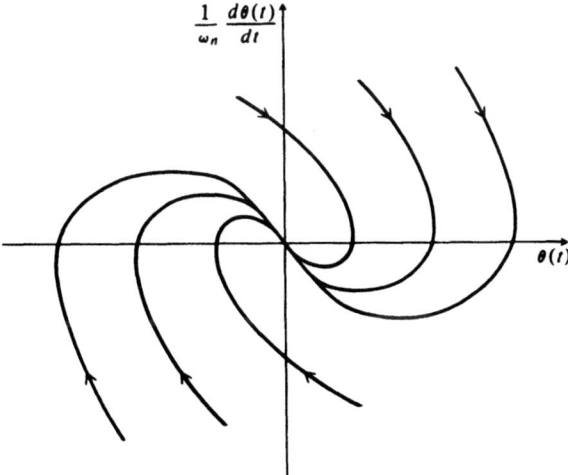

Figure 5.39 Phase portrait of a second-order linear system having damping. The origin is defined as a stable node.

376 NONLINEAR CONTROL-SYSTEM DESIGN

time. Time appears in the portrait implicitly as a parameter along all phase trajectories. We discuss its computation later in Section 5.16.

Let us next illustrate the application of this method to a relatively simple nonlinear differential equation. We consider the problem of nonlinear friction which was discussed previously in Section 5.8. We first determine the phase portrait for Coulomb friction and then extend it to the case of stiction. Consider a system containing Coulomb friction $\pm F_c$, the moment of inertia of the system being J and the spring constant K. The differential equation of the motion of the system is given by

$$J \frac{d^2 c(t)}{dt^2} + Kc(t) \pm F_c = 0. \quad (5.120)$$

Defining

$$\omega = (K/J)^{1/2}, \quad (5.121)$$

$$\gamma = \frac{F_c}{J\omega^2}, \quad (5.122)$$

$$\alpha(t) = \begin{cases} c(t) - \gamma & \text{when } \dot{c}(t) < 0, \\ c(t) + \gamma & \text{when } \dot{c}(t) > 0 \end{cases} \quad (5.123)$$

the normalized equation of motion is

$$\ddot{\alpha}(t) + \omega^2 \alpha(t) = 0. \quad (5.124)$$

The phase portrait for Eq. (5.124) in the $\alpha(t)$ versus $\dot{\alpha}(t)$ plane is the same as that for the linear, second-order system considered previously (see Figure 5.38). However, in the $\dot{c}(t)/\omega$ versus $c(t)$ plane, the phase portrait for Eq. (5.124) is quite different, and can be obtained by considering Eq. (5.124) in two parts: one for the upper half-plane where $\dot{c}(t)/\omega > 0$, and the other for the lower half-plane, where $\dot{c}(t)/\omega < 0$. A little thought shows that the phase portrait in the upper half-plane is a family of semicircles, centered about $c(t) = -\gamma$, since $c(t)$ is related to $\alpha(t)$ merely by a simple translation. In a similar manner, the phase portrait for the lower half-plane is a family of semicircles centered about $c(t) = \gamma$. Figure 5.40 illustrates the phase portrait.

It is interesting to observe from Figure 5.40 that as soon as the displacement has a value within the interval on the $c(t)$ axis given by

$$-\gamma \leqslant c(t) \leqslant \gamma, \quad (5.125)$$

all motion stops. This gives rise to the possibility of large, steady-state errors. For example, if the initial conditions are at point 1, the trajectory will be 1–2–3–4 and the system will not have any steady-state error. If the initial conditions are at point 5, however, the trajectory followed will be 5–6–7–8 and the system will have a steady-state error equal to γ.

We have already seen that dither is useful for eliminating the steady-state error due to Coulomb friction. Its effect on the steady-state error can easily be understood

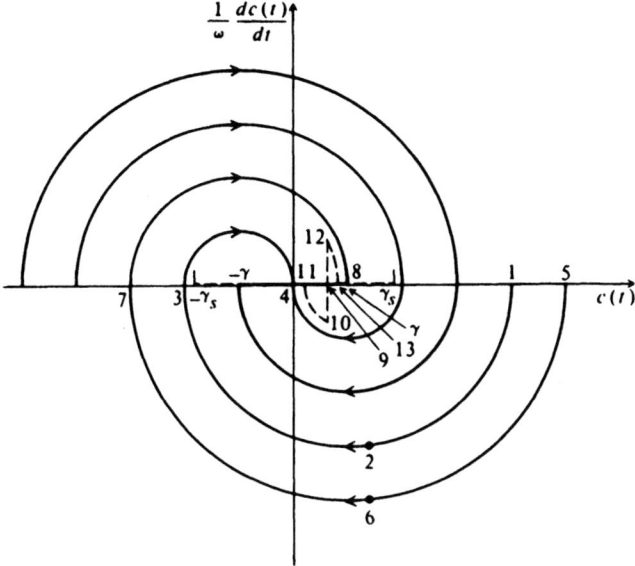

Figure 5.40 Phase portrait for a system containing Coulomb friction.

by studying the phase plane. For example, let us assume that a finite steady-state error exists corresponding to point 9 on Figure 5.40. With dither, the effect of a negative disturbance, segment 9–10, results in the system returning to point 11 on the $c(t)$ axis, while a positive disturbance, segment 9–12, results in the system returning to point 13 on the $c(t)$ axis. Because the projection 11–9 is greater than the projection 9–13, the system will tend to move towards the origin.

Before leaving this system, let us consider the required modification to the phase portrait in Figure 5.40 for stiction. Because stiction occurs only for zero velocity and is greater than Coulomb friction, its effect is to extend the termination line: $\pm \gamma$ to $\pm \gamma_s$. These extended terminations are illustrated in Figure 5.40.

B. Transformation of the Second-Order Differential Equation to a First-Order Equation

If the differential equation of the system cannot be easily integrated, a new differential equation in terms of the phase variables may be formed, from which the phase portrait can be obtained directly. This method can best be illustrated by considering the linear, second-order, undamped system discussed previously, namely,

$$\frac{d^2\theta(t)}{dt^2} + \omega_n^2 \theta(t) = 0. \quad (5.126)$$

Using the dot notation, we have

$$\ddot{\theta}(t) + \omega_n^2 \theta(t) = 0. \quad (5.127)$$

378 NONLINEAR CONTROL-SYSTEM DESIGN

Defining the state variables as

$$x_1(t) = \theta(t), \tag{5.128}$$
$$x_2(t) = \dot{\theta}(t), \tag{5.129}$$

we have

$$\dot{x}_1(t) = x_2(t), \tag{5.130}$$
$$\dot{x}_2(t) = -\omega_n^2 x_1(t). \tag{5.131}$$

Dividing Eq. (5.131) by Eq. (5.130), the following is obtained:

$$\frac{dx_2(t)}{dx_1(t)} = -\frac{\omega_n^2 x_1(t)}{x_2(t)}. \tag{5.132}$$

Equation (5.132) can be rewritten as

$$x_2(t)\, dx_2(t) + \omega_n^2 x_1(t)\, dx_1(t) = 0.$$

Integrating, we obtain

$$x_2^2(t) + \omega_n^2 x_1^2(t) = C,$$

where C is an arbitrary constant easily determined from the initial conditions. Solving for $x_2(t)$, we obtain

$$x_2(t) = \pm\sqrt{C - \omega_n^2 x_1^2(t)}.$$

In terms of the phase variable, this equation can be written as

$$\dot{\theta}(t) = \pm\sqrt{C - \omega_n^2 \theta^2(t)}. \tag{5.133}$$

The result of sketching various phase trajectories from Eq. (5.133) will be the phase portrait illustrated in Figure 5.38, where a normalized ordinate axis is used. The initial conditions determine the particular trajectory followed.

This techique can be easily extended to the systems whose phase portraits are illustrated in Figures 5.39 and 5.40. This method is by far the most useful analytic technique for obtaining the phase portrait of a system.

C. Method of Isoclines

The method of isoclines is a graphcial procedure for determining the phase portrait. It can be used even if the differential equation cannot be solved analytically. In practice, it is a very powerful method to use.

Isoclines are lines in the phase plane corresponding to constant slopes of the phase portrait. One starts with the differential equation in the form shown by Eq. (5.132).

5.15. CONSTRUCTION OF THE PHASE PORTRAIT

Here $dx_2(t)/dx_1(t)$, or $d\dot\theta(t)/d\theta(t)$, corresponds to the slope of the trajectories that form the phase portrait. Numerical values are assigned for the slopes $d\dot\theta(t)/d\theta(t)$, and Eq. (5.132) is used to find the corresponding points in the phase plane having those slopes. Once a set of isoclines is drawn, a trajectory may be drawn by starting at some point on an isocline and then proceeding to the next isocline along a straight line whose slope is the average of the slopes corresponding to the two isoclines. Because the procedure is a numerical approximation, closer spacing of the isoclines increases the accuracy of the resulting trajectory.

Let us illustrate the application of the isocline method to the lienar, second-order, undamped, and damped sytems whose portraits are given in Figures 5.38 and 5.39, respectively. For the undamped case, the family of isoclines can be drawn from Eq. (5.132). However, in order to plot the phase portrait on a normalized plane $[(1/\omega_n)\dot\theta(t)$ versus $\theta(t)]$, let

$$\dot\Theta(t) = \dot\theta(t)/\omega_n. \tag{5.134}$$

So we have

$$\frac{d\dot\Theta(t)}{d\theta(t)} = -\frac{\theta(t)}{\dot\Theta(t)} = m, \tag{5.135}$$

where m represents the slope of the trajectory.

Isoclines associated with slopes corresponding to Eq. (5.135) constitute a family of straight lines passing through the origin and are illustrated in Figure 5.41. Also shown is the construction of the phase trajectory starting with a point that lies on the

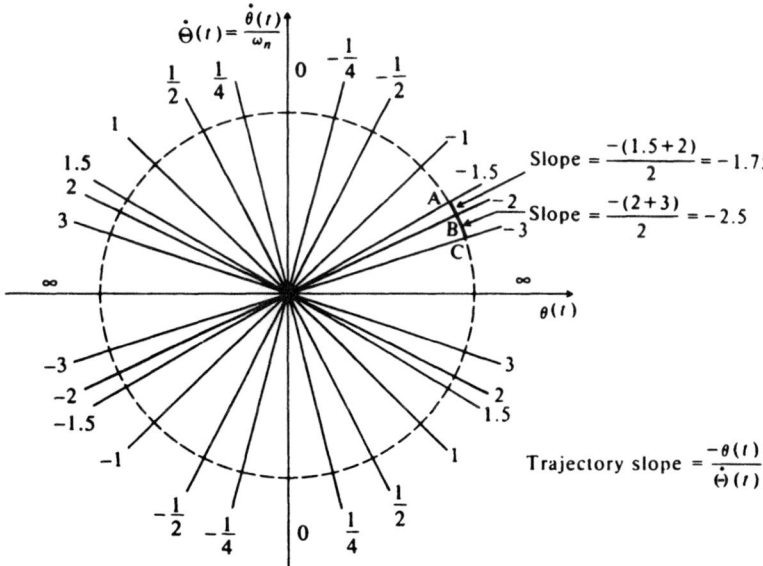

Figure 5.41 Construction of a phase trajectory for a linear, second-order, undamped system using the method of isoclines.

380 NONLINEAR CONTROL-SYSTEM DESIGN

isocline corresponding to a slope of -1.5. The motion of the trajectory drawn from point A on the -1.5 isocline, to the isocline whose slope is -2, has a slope in the phase plane that is the average of these two isoclines, or -1.75. This is indicated on Figure 5.41 as line segment AB. In addition, the following line segment, BC, whose slope is -2.5, is illustrated. The complete trajectory is shown dashed. It is obvious from this simple example that the accuracy of the isocline method depends on the number of isoclines drawn.

Let us next construct the phase portrait for the linear, second-order, damped system considered previously. From Eq. (5.112), the differential equation of the system is

$$\frac{d^2\theta(t)}{dt^2} + 2\zeta\omega_n \frac{d\theta(t)}{dt} + \omega_n^2\theta(t) = 0. \tag{5.136}$$

Let us assume that $\zeta = 0.5$ and using the dot notation, we have

$$\ddot{\theta}(t) + \omega_n\dot{\theta}(t) + \omega_n^2\theta(t) = 0. \tag{5.137}$$

Defining the state variables as

$$x_1(t) = \theta(t), \tag{5.138}$$
$$x_2(t) = \dot{\theta}(t), \tag{5.139}$$

then we have

$$\dot{x}_1(t) = x_2(t), \tag{5.140}$$
$$\dot{x}_2(t) = -\omega_n^2 x_1(t) - \omega_n x_2(t). \tag{5.141}$$

Dividing Eq. (5.141) by Eq. (5.140), we obtain the following:

$$\frac{dx_2(t)}{dx_1(t)} = -\omega_n^2 \frac{x_1(t)}{x_2(t)} - \omega_n. \tag{5.142}$$

Defining a normalized state variable

$$x_3(t) = \frac{x_2(t)}{\omega_n} = \frac{\dot{\theta}(t)}{\omega_n} = \dot{\Theta}(t),$$

we can write Eq. (5.142) as follows:

$$\frac{dx_3(t)}{dx_1(t)} = -1 - \frac{x_1(t)}{x_3(t)}, \tag{5.143}$$

or

$$\frac{d\dot{\Theta}(t)}{d\theta(t)} = -1 - \frac{\theta(t)}{\dot{\Theta}(t)}.$$

From this equation, the slope of the trajectories in the $\dot{\Theta}(t)$ versus $\theta(t)$ plane is given by

$$\text{trajectory slope} = -1 - \frac{\theta(t)}{\dot{\Theta}(t)}. \qquad (5.144)$$

Isoclines associated with slopes corresponding to Eq. (5.144) constitute a family of straight lines passing through the origin in the phase plane. This is illustrated in Figure 5.42. The construction of the trajectory whose initial condition is a point which lies on the isocline corresponding to a slope of 2 is illustrated. The segment of the trajectory drawn from point A on the isocline whose slope is 2 to that whose slope is 1 would be a straight line whose slope is the average of 2 and 1, or 1.5. This is indicated in Figure 5.42 as line segment AB. In addition, the following line segment, BC, whose slope is 0.5, is illustrated. The remaining trajectory is shown dashed. As for the undamped system, the accuracy depends greatly on the number of isoclines drawn. Observe that the motion of the trajectory is in the clockwise direction about the origin because for positive $x_3(t)$, $x_1(t)$ must be increasing, and for negative $x_3(t)$, $x_1(t)$ must be decreasing [see Eq. (5.143)].

5.16. CHARACTERISTICS OF THE PHASE PORTRAIT

Several properties of the phase portrait need to be singled out; their correct interpretation is important for the intelligent analysis of nonlinear control systems. We begin by defining and illustrating the notion of *singular points*. Then we illustrate limit cycles and follow this by showing how to determine time from the phase

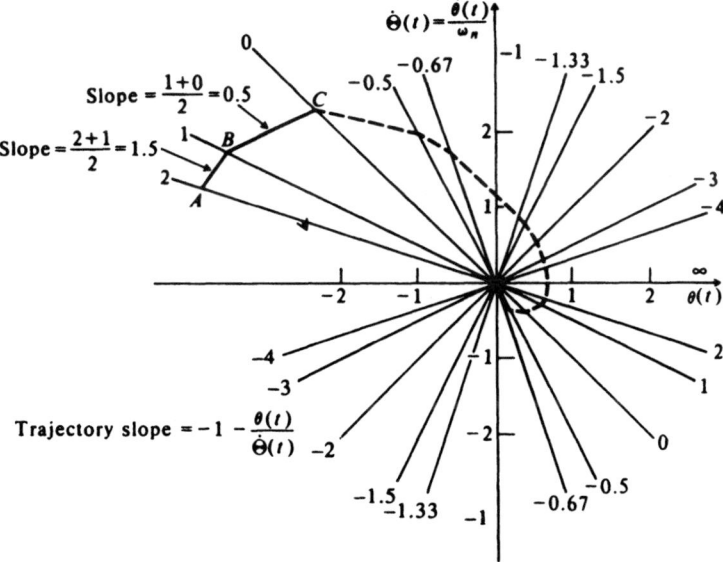

Figure 5.42 Construction of a phase trajectory for a linear, second-order, damped ($\zeta = 0.5$) system using the method of isoclines.

portrait. The section concludes with examples of several interesting and representative phase portraits.

A. Singular Points

Singular points are points of the phase plane where the system is in a state of equilibrium. At these points both velocity and acceleration of the system are zero. The origin is the only singular point in the phase portraits of the linear systems illustrated in Figures 5.38 and 5.39.

Consider the general second-order, nonlinear differential equation (5.110) which is unexcited from an external forcing function

$$\frac{d^2 x(t)}{dt^2} + A_1\left(x, \frac{dx}{dt}\right)\frac{dx(t)}{dt} + A_2\left(x, \frac{dx}{dt}\right)x(t) = 0.$$

If this equation is written in the state-variable from, the singular points are defined as the points that make the quantities in Eqs. (5.145) and (5.146) equal to zero:

$$\frac{dx(t)}{dt} = P(x, y), \tag{5.145}$$

$$\frac{dy(t)}{dt} = Q(x, y). \tag{5.146}$$

Here $x(t)$ and $y(t)$ are the state variables. The characteristics of singular points may vary greatly depending on the variations of the coefficients of the first-order differential equations given by Eqs. (5.145) and (5.146). The type of singular point may be found by means of a Taylor-series expansion of Eqs. (5.145) and (5.146) in the vicinity of the singular point. Assuming that the singularity occurs at $x(t) = A$ and $y(t) = B$, the result of expanding $P(x, y)$ and $Q(x, y)$ about these points is

$$\frac{dx(t)}{dt} = A_1(x - A) + A_2(y - B) + A_3(x - A)^2$$
$$+ A_4(x - A)(y - B) + A_5(y - B)^2 + \cdots. \tag{5.147}$$

$$\frac{dy(t)}{dt} = B_1(x - A) + B_2(y - B) + B_3(x - A)^2$$
$$B_4(x - A)(y - B) + B_5(y - B)^2 + \cdots. \tag{5.148}$$

We assume that the character of the singular points is determined entirely by the coefficients of the linear terms only: A_1, A_2, B_1, and B_2. This is certainly reasonable if a sufficiently small region is chosen in the vicinity of the singularity. What we have done is to characterize the system by its linear part in the vicinity of the singular point. Therefore, Eqs. (5.147) and (5.148) reduce to

$$\frac{dx(t)}{dt} = A_1(x(t) - A) + A_2(y(t) - B), \tag{5.149}$$

$$\frac{dy(t)}{dt} = B_1(x(t) - A) + B_2(y(t) - B). \tag{5.150}$$

We further simplify the problem by assuming that the singularity occurs at the origin. Therefore A and B are both zero. Equations (5.149) and (5.150) reduce to the form

$$\frac{dx(t)}{dt} = A_1 x(t) + A_2 y(t), \tag{5.151}$$

$$\frac{dy(t)}{dt} = B_1 x(t) + B_2 y(t). \tag{5.152}$$

We have not lost any generality in using this assumption, because the same result can always be obtained merely by changing the variables as follows:

$$x(t) - A = p, \quad y(t) - B = q. \tag{5.153}$$

The characteristics of the singular point can be determined by eliminating one of the two variables of Eqs. (5.151) and (5.152) and studying the resulting characteristic equation. Using the Laplace transform, the result is

$$X(s) = \frac{(s - B_2)x(0) + A_2 y(0)}{s^2 - (A_1 + B_2)s + (A_1 B_2 - A_2 B_1)}, \tag{5.154}$$

and the characteristic equation is given by

$$s^2 - (A_1 + B_2)s + (A_1 B_2 - A_2 B_1) = 0. \tag{5.155}$$

Assuming real coefficients, six different characteristic sets of roots of Eq. (5.155) are possible. These sets of roots result in singular points which can be classified as belonging to one of four types.

1. Node. A node is a point in the phase plane consisting of a family of trajectories which directly converge and approach it (stable node) or radiate from it (unstable node). A stable node occurs when the roots are both real and both lie in the left half of the s-plane. An unstable node occurs when the roots are both real and both lie in the right half of the s-plane. Figure 5.39 illustrates a stable node and Figure 5.43 an unstable node.

2. Focus. A focus is a point in the phase plane consisting of a family of spiral trajectories which either converge on the point (stable focus) or diverge from it (unstable focus). A stable focus occurs when the roots are complex conjugate and lie in the left half of the s-plane. An unstable focus occurs when the roots are complex conjugate and lie in the right half-plane. The origin of the phase portrait in Figure 5.44 is a stable focus. Figure 5.45 illustrates an unstable focus.

3. Center. A center is a point in the phase plane consisting of a family of closed curves encircling it. This occurs when the roots are complex conjugate and lie on the $j\omega$ axis. The origin of the phase portrait in Figure 5.38 is a center.

4. Saddle Point. A saddle point is a point in the phase plane that is characterized by the phase portrait illustrated in Figure 5.46. This occurs when the roots are

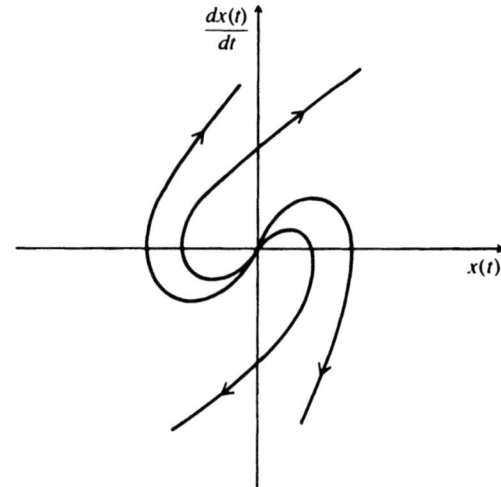

Figure 5.43 Phase portrait of an unstable node.

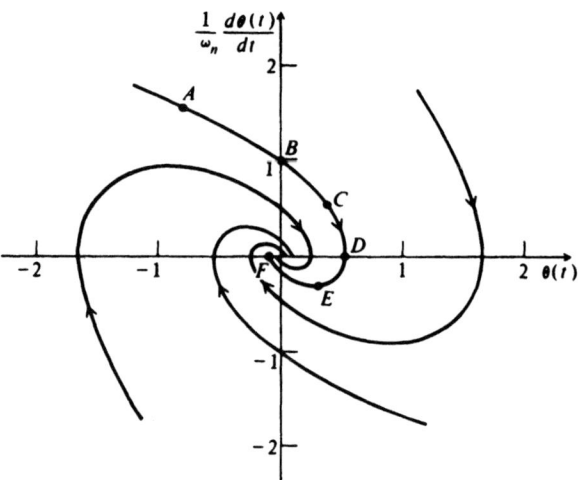

Figure 5.44 Phase portrait of a stable focus.

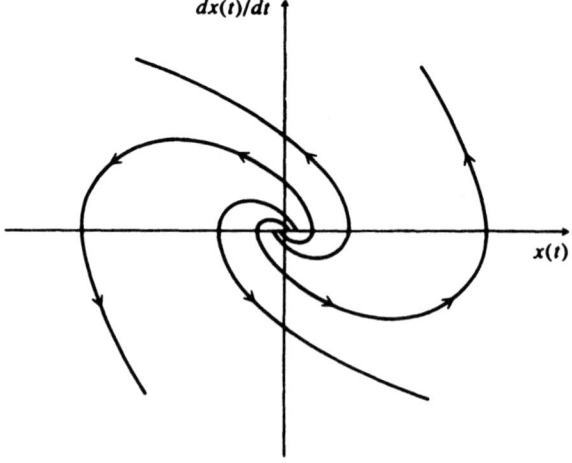

Figure 5.45 Phase portrait of an unstable focus.

real, with one in the right half-plane and the other in the left half-plane. Except for the case where the numerator of Eq. (5.154) contains a zero which exactly cancels the zero of the denominator lying in the right half-plane, this type of singularity always represents an unstable situation.

B. Limit Cycles

Limit cycles are defined as isolated closed paths of the phase portrait. The location and determination of the type of limit cycle, together with the singular points which exist in the phase plane, offer a complete description of the behavior of a nonlinear second-order system.

A limit cycle can be stable or unstable, depending on whether the paths in the neighborhood of the limit cycle converge toward it or diverge away from it. They can result from either soft or hard self-excitation. These situations can be portrayed on the phase plane. Figure 5.47 illustrates a system with a soft-excitation. Physically, this phase portrait may represent a system which has excessive gain for small signals and the output builds up in an unstable manner. With large signals the output also approaches the stable limit cycle from the outside as shown in Figure 5.47. A stable limit cycles exists between these two conditions and a sustained oscillation occurs. Figure 5.48 illustrates a system where a hard-self-excitation exists. The generation of an oscillation depends on the initial conditions. For example, let us assume that the initial state is at the stable node. For hard self-excitation to occur, a very large disturbance is required to change the state of the system to a region outside the unstable limit cycle. If this disturbance is sufficient for the operating point of the system to reach the stable limit cycle, a steady oscillation occurs.

Observe that all of the trajectories in Figures 5.38 through 5.48 are perpendicular to the x axis because the isocline slopes there are infinity (see Figure 5.41).

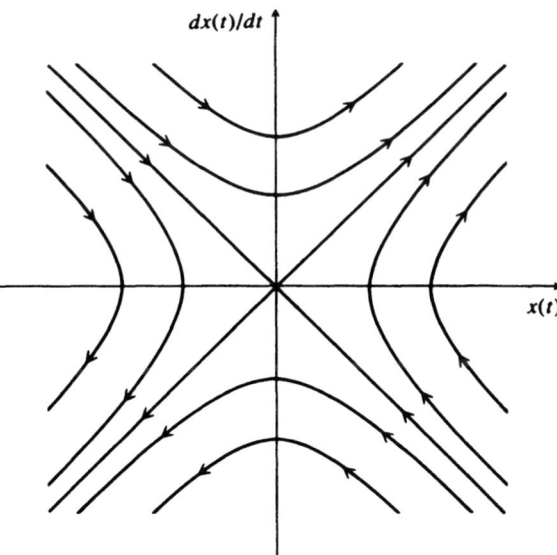

Figure 5.46 Phase portrait of a saddle point.

386 NONLINEAR CONTROL-SYSTEM DESIGN

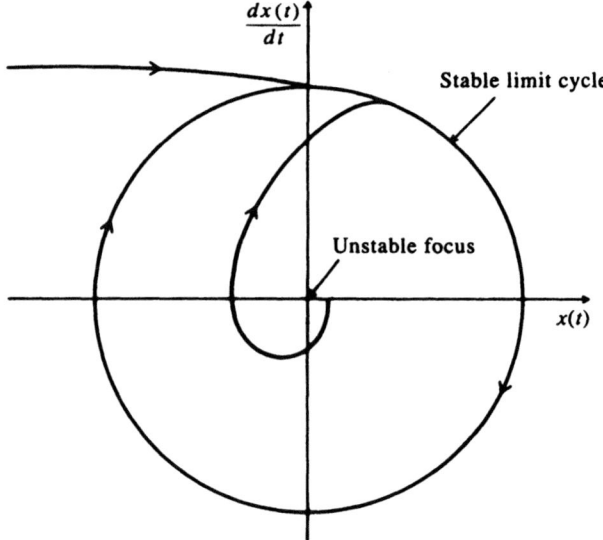

Figure 5.47 Phase portrait of a soft self-excitation.

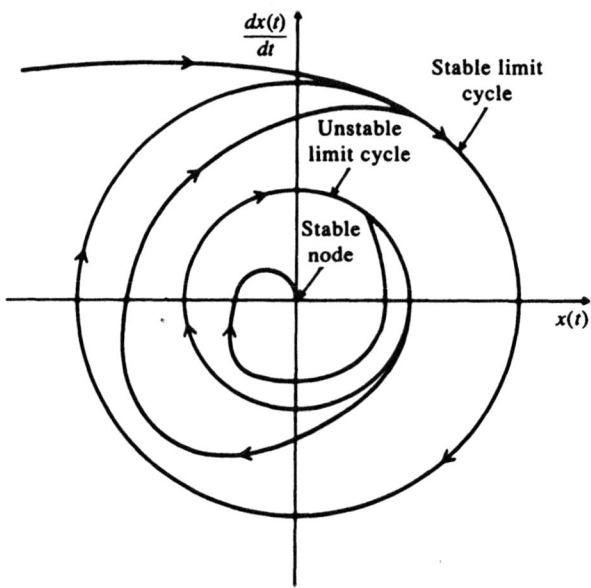

Figure 5.48 Phase portrait of hard self-excitation.

C. Determination of Time from the Phase Plane

Although we cannot solve for $x(t)$ and $dx(t)/dt$ as functions of time directly, it is possible to obtain time from the phase portrait. The variation of time can be easily found from the equation.

$$y(t) = dx(t)/dt. \tag{5.156}$$

Solving for t, we obtain the relationship

$$t = \int_A^B \frac{1}{y(t)} dx(t). \tag{5.157}$$

Equation (5.157) shows that if the phase portrait is replotted with $1/y(t)$ or $(dx(t)/dt)^{-1}$ as the ordinate and $x(t)$ as the abscissa, the area under the resulting curve represents time. This area can be evaluated with a planimeter, crudely by approximation using a series of rectangles, or with a computer if we know mathematical expressions for the segments of the phase trajectory. If we know mathematical expressions for the segments of the phase trajectory, in a piecewise-linear manner, then we can also find the time directly by integrating Eq. (5.157).

Let us use the phase trajectory labeled *ABCDEF* on the phase portrait of Figure 5.44 to illustrate the determination of time graphically. We assume that $\omega_n = 1$. Our first problem is to determine the time it takes for the motion to go from *A* to *C*. Figure 5.49 represents a plot of $[d\theta(t)/dt]^{-1}$ versus $\theta(t)$ for the interval *ABC*. Using a series of rectangles for area summation, we find that the time is approximately equal to 1.12 sec. If we attempt to find the time it takes for the trajectory to pass from *C* to *E*, we run into a problem, because $d\theta(t)/dt$ equals zero at point *D* and the reciprocal phase-plane plot goes to infinity. This certainly is not the true situation, and we must resort to an approximation in the vicinity of point *D*. The most practical approximation is to find the time it takes to go from point *C* to a small finite distance on the phase trajectory before point *D*, $D(-)$, and then the time it takes to go from a small finite distance past point *D*, $D(+)$, to point *E*. This technique is illustrated in Figure 5.50. Using the rectangular area summation technique, we find that it takes approximately 2.18 sec to go from point *C* to point *E*. Therefore, the total time it takes to traverse the segment *ABCDE* is approximately 3.3 sec.

An interesting observation to question is why does it take a longer time for the trajectory to go along segment *CDE* (2.18 sec) than segment *ABC* (1.12 sec), although the trajectory segment *CDE* is smaller than the trajectory segment *ABC* (see Figure 5.44)? The answer is that the control system is slowing down as the trajectory follows *A* to *B* to *C* to *D* to *E* to *F*, and ultimately comes to rest at point *F*. Therefore, it takes longer to pass from *C* to *D* to *E* than it does from *A* to *B* to *C*.

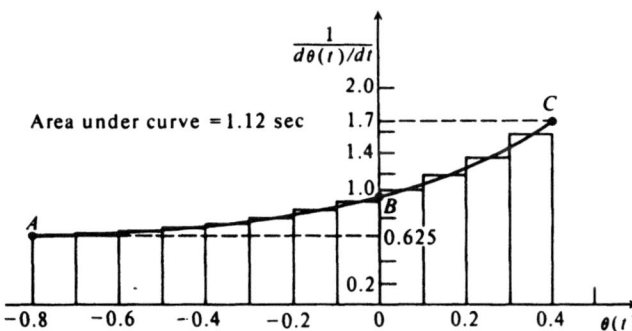

Figure 5.49 Finding time from a reciprocal phase-plane plot.

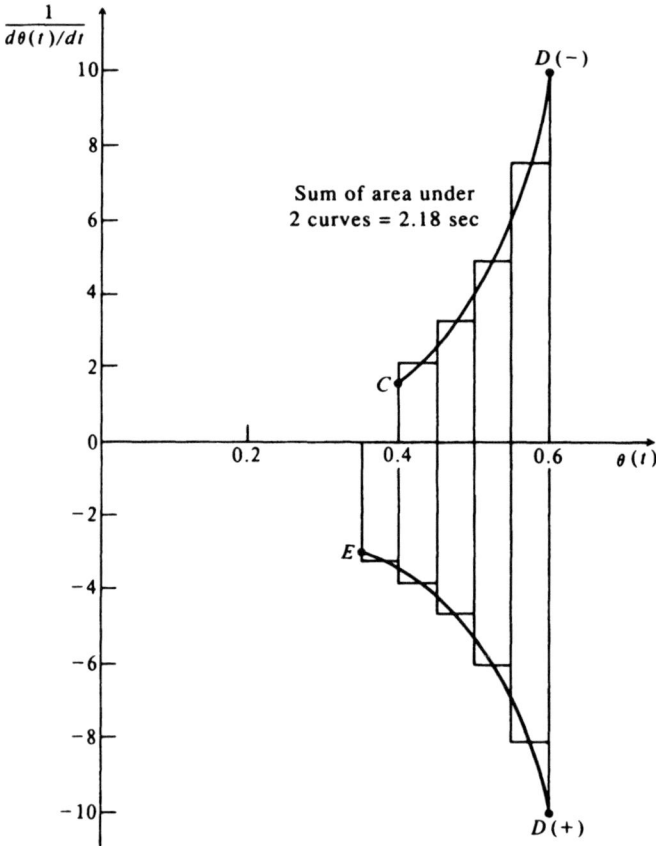

Figure 5.50 Approximating time from a reciprocal phase-plane plot when velocity passes through zero.

As was pointed out earlier, the time can be determined by integrating Eq. (5.157) directly if we know mathematical expressions for the segments of the phase trajectory (in a piecewise-linear manner), rather than using a series of rectangles for area summation as was done in this problem.

D. Representative Phase Portraits

In order to develop a great facility for intelligently interpreting phase portraits, we next present additional representative phase portraits. These, with the portraits already discussed, should provide a good facility for interpreting phase portraits. Specifically, we consider the phase portraits of undamped, second-order control systems that have the nonlinear characteristics of rate and position limiting. We compare the results with the portrait illustrated in Figure 5.41.

1. Rate Limiting of an Undamped, Second-Order System. Consider the linear, undamped, second-order system whose phase trajectory was illustrated in Figure 5.41. We shall change this linear system to a nonlinear one by adding a governor to the servomotor driving the load so that the maximum rate is limited. The resulting phase portrait is illustrated in Figure 5.51. If the initial conditions are such that the resulting phase trajectory lies anywhere within the dashed trajectory, the system will oscillate as the undamped linear system did, as shown in Figure 5.41. However, if the initial conditions are such that they lie outside the dashed trajectory of Figure 5.51, the output rate is limited to the maximum value allowed by the governor, as shown in Figure 5.51 by the dashed trajectory. Thereafter the system will oscillate indefinitely following the dashed trajectory.

2. Position Limiting of an Undamped, Second-Order System. Next consider the linear, undamped, second-order system that is made nonlinear by adding limit stops to the output shaft, so that the maximum positions are limited. The resulting phase portrait is given in Figure 5.52. The interpretations of this phase portrait is analogous to that of Figure 5.51, except that it is shifted by 90°.

5.17. PHASE PLANE FOR SYSTEMS CONTAINING EXTERNAL FORCING FUNCTIONS

In the preceding analysis of the phase plane, we have always assumed that the extenral forcing functions were zero, and that the system excitation came from the initial conditions. We will reconsider the nonlinear differential equation (5.110) when there is an external forcing function $u(t)$.

To illustrate the modification to our previous analysis, let us consider the linear second-order system illustrated in Figure 5.53. We want to find the phase trajectory of the system in the $c(t)$ vs. $\dot{c}(t)$ phase plane upon the application of a unit step, $r(t) = U(t)$. We will assume that the system's initial conditions are $c(0) = \dot{c}(0) = 0$, and that the system is underdamped. This results in the characteristic exponentially damped sinusoid response illustrated in Figure B.4. Figure 5.54a illustrates the corresponding phase trajectory for this case. Observe from this figure that the

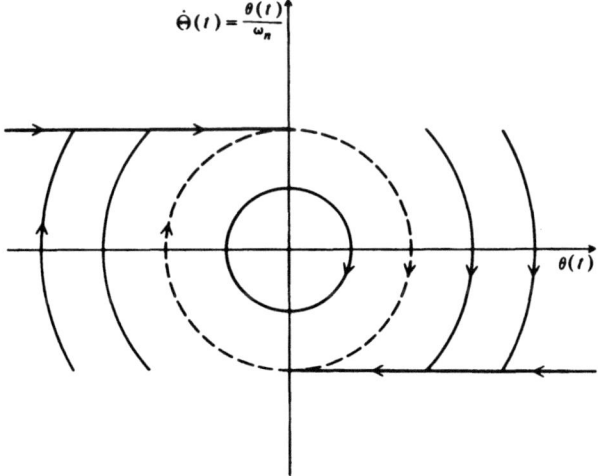

Figure 5.51 Phase portrait of a second-order, undamped system having rate limiting.

390 NONLINEAR CONTROL-SYSTEM DESIGN

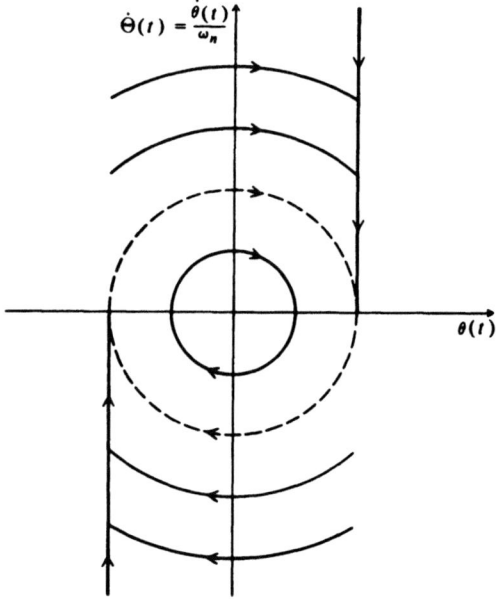

Figure 5.52 Phase portrait of a second-order, undamped system having position limiting.

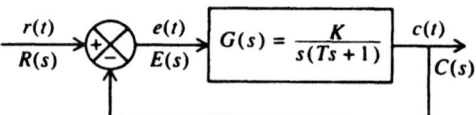

Figure 5.53 A linear, second-order system.

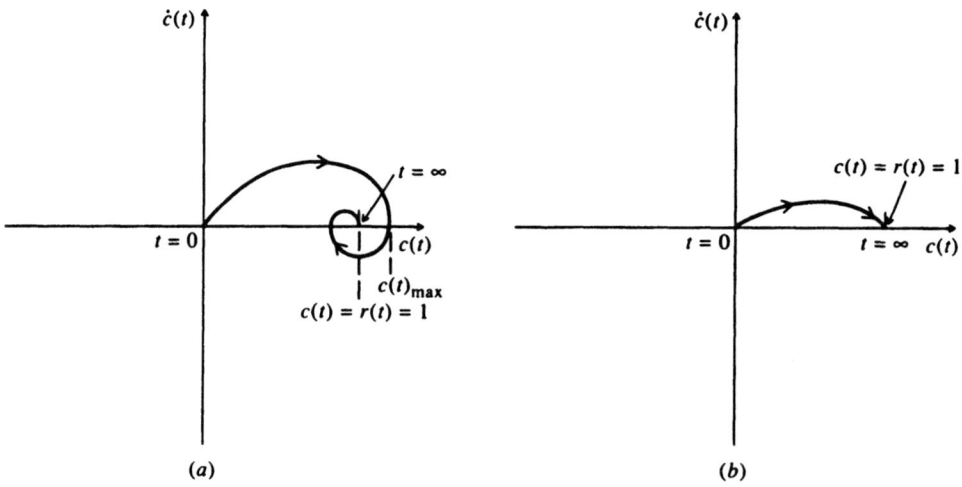

Figure 5.54 Phase plane for the output trajectory of a linear, second-order system having a step input for the cases when it is (a) underdamped and (b) overdamped.

5.17. PHASE PLANE FOR SYSTEMS CONTAINING EXTERNAL FORCING FUNCTIONS

value of the maximum overshoot of the system, $c(t)_{max}$, [see Eq. (B.32)] can be found from the phase trajectory. Figure 5.54b illustrates the phase trajectory when the system is critically damped.

We know from Figure C.4 the shape of the error's transient for a linear second-order system to a step input. How would the error's transient appear as a phase trajectory in the $e(t)$ vs. $\dot{e}(t)$ phase plane? We will first assume that the system is underdamped, and then we will assume critical damping. Figure 5.55a illustrates the corresponding phase trajectory for the underdamped case and Figure 5.55b illustrates the phase trajectory for the critical damping case.

As another interesting example, suppose that the input to the system of Figure 5.53 has an input that is a ramp, $r(t) = Vt$, and the system is assumed to be underdamped. We wish to find the phase trajectory of the system in the $e(t)$ vs $\dot{e}(t)$ phase plane. From Figure 5.53, we find that

$$C(s) = \frac{K}{s(Ts+1)} E(s). \tag{5.158}$$

Therefore, in the time domain,

$$T\ddot{c}(t) + \dot{c}(t) = Ke(t). \tag{5.159}$$

Because we know that

$$c(t) = r(t) - e(t), \tag{5.160}$$

substituting Eq. (5.160) into Eq. (5.159), we obtain

$$T(\ddot{r}(t) - \ddot{e}(t)) + (\dot{r}(t) - \dot{e}(t)) = Ke(t).$$

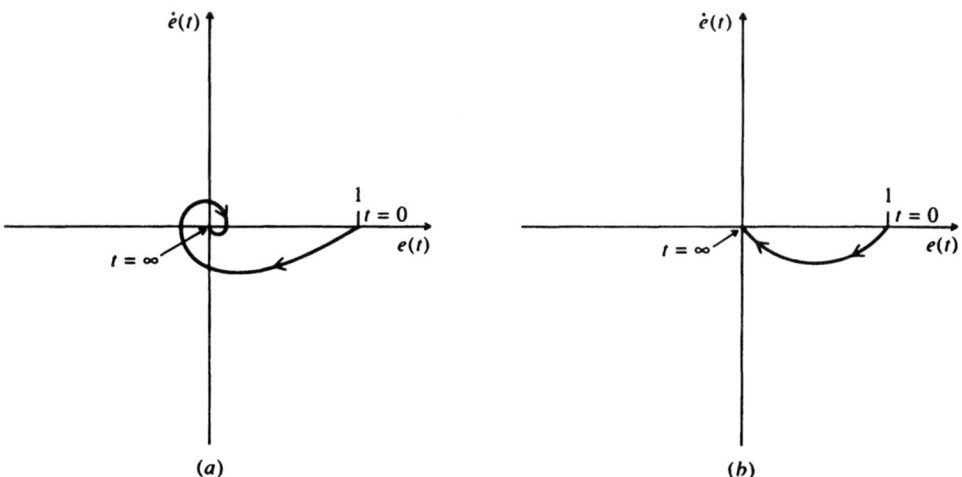

Figure 5.55 Phase plane for the error trajectory of a linear, second-order system having a unit step input for the case when it is (a) underdamped and (b) overdamped.

Simplifying,

$$T\ddot{e}(t) + \dot{e}(t) + Ke(t) = T\ddot{r}(t) + \dot{r}(t). \tag{5.161}$$

Because $\dot{r}(t)$ is given as

$$\dot{r}(t) = V, \tag{5.162}$$

then

$$\ddot{r}(t) = 0. \tag{5.163}$$

Substituting Eqs. (5.162) and (5.163) into Eq. (5.161),

$$T\ddot{e}(t) + \dot{e}(t) + Ke(t) = V. \tag{5.164}$$

Since $\dot{e}(t) = \ddot{e}(t) = 0$ in the steady state, then a singularity at

$$e(t) = V/K \tag{5.165}$$

exists, and the phase trajectory starts at its initial condition of $\dot{e}(0) = V$ and terminates at its singularity defined by Eq. (5.165) as illustrated in Figure 5.56.

5.18. DESIGN OF NONLINEAR FEEDBACK CONTROL SYSTEMS USING THE STATE-VARIABLE PHASE-PLANE METHOD

This section illustrates the design of a nonlinear feedback control system using the phase-plane method [41]. We use the analytic tools developed in Sections 5.14–5.17

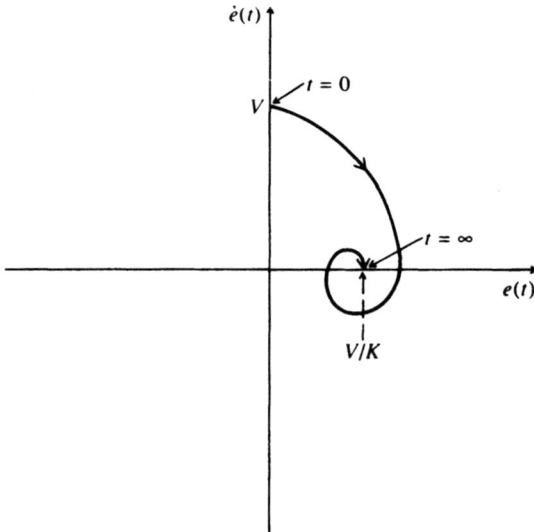

Figure 5.56 Phase plane for the error trajectory of a linear, second-order system having a unit ramp input Vt for an underdamped system.

to demonstrate the procedure to follow when designing a nonlinear system. Specifically, we consider an on-off control system. The primary function of this example is to use the transient response of the system as a guide for choosing the parameters. One of our main objectives will be to determine the existence of limit cycles as the parameters are varied. We will not, however, be able to obtain the margin of stability for the system from the phase plane. The following analysis should be compared with that obtained using the describing function for design in Section 5.10.

Let us consider the on-off control system illustrated in Figure 5.57. It has a two-position contactor which applies a corrective signal having the proper phase for control action. The problem is to determine a good polarity combination of the variable position feedback constant a and the variable velocity feedback constant b for stability and acceptable transient performance. We assume that the input $r(t)$ is zero and that the forcing function generated by the contactor is unity; therefore the general form of the differential equation for this system is given by

$$\frac{d^2c(t)}{dt^2} + 2\zeta\omega_n \frac{dc(t)}{dt} + \omega_n^2 c(t) = \text{sgn}\left(ac + b\frac{dc}{dt}\right), \qquad (5.166)$$

where $0 < \zeta < 1$. The sign of the unit forcing function is given by the sign of $ac + b(dc/dt)$. Assuming that $\zeta = 0.5$ and $\omega_n = 1$, the equation reduces to

$$\frac{d^2c(t)}{dt^2} + \frac{dc(t)}{dt} + c(t) = \text{sgn}\left(ac + b\frac{dc}{dt}\right). \qquad (5.167)$$

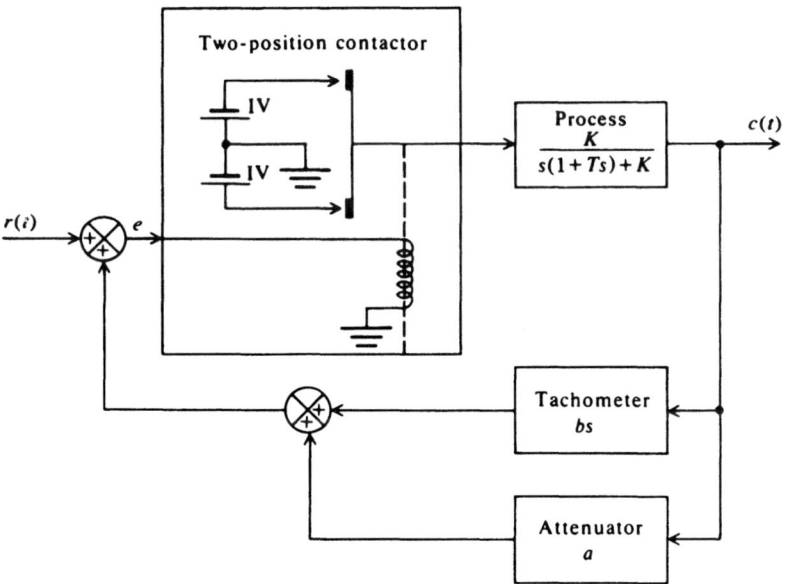

Figure 5.57 On-off control system having variable position and velocity feedback.

Specifically, we shall determine the transient response of this system for the following polarity combinations of a and b:

(A) $a > 0$, $\quad b > 0$,
(B) $a > 0$, $\quad b < 0$,
(C) $a < 0$, $\quad b > 0$,
(D) $a < 0$, $\quad b < 0$. (5.168)

In general the phase portrait for any of these cases may be obtained in a similar manner. We draw heavily on the results of our studies of singular points (see Section 5.16) in drawing the various phase portraits. The basic relation analyzed in our discussion of singular points was Eq. (5.110):

$$\frac{d^2 c(t)}{dt^2} + A_1\left(c, \frac{dc}{dt}\right)\frac{dc(t)}{dt} + A_2\left(c, \frac{dc}{dt}\right)c(t) = u(t). \tag{5.169}$$

If we are considering an underdamped, linear second-order system which is excited only by its initial conditions, we know that the differential equation (5.113) results in a phase portrait similar to that in Figure 5.39, where the origin acts as a stable node. In the case of an on-off system, however, we find that it is represented by a differential equation which is given by Eq. (5.167). The right-hand side of this equation can be thought of as either a unit positive or negative forcing function. The phase portrait of a system having a unit positive forcing function will have spirals that converge towards a stable focus at $c = 1$, $dc/dt = 0$. This was discussed in Section 5.17, and illustrated in Figure 5.54. The phase portrait of a system having a unit negative forcing function will have spirals that converge towards a stable focus at $c = -1$, $dc/dt = 0$. The on-off system we are considering actually has two stable foci because of the action of the two-position contactor. A switching line, defined by

$$ac(t) + b\frac{dc(t)}{dt} = 0 \tag{5.170}$$

separates the regions where the phase trajectory spirals towards $c = 1$, $dc/dt = 0$ or $c = -1$, $dc/dt = 0$. Using this general approach, which is valid for all four cases, stability and the transient response can be readily determined.

Case A. $a > 0$, $b > 0$. The switching line is a straight line given by the following relationship:

$$\frac{dc(t)}{dt} = -\frac{a}{b}c(t). \tag{5.171}$$

It passes through the origin and lies in the second and fourth quadrants. The sign of $ac + b(dc/dt)$ is positive in the region to the right of it and negative to the left of it. The phase portrait for this system, which is given in Figure 5.58, can be constructed using the method of isoclines or by transforming the second-order differential equation to a first-order equation or by use of a digital computer (see Section 5.19). Observe that three trajectories are possible, depending on the initial conditions.

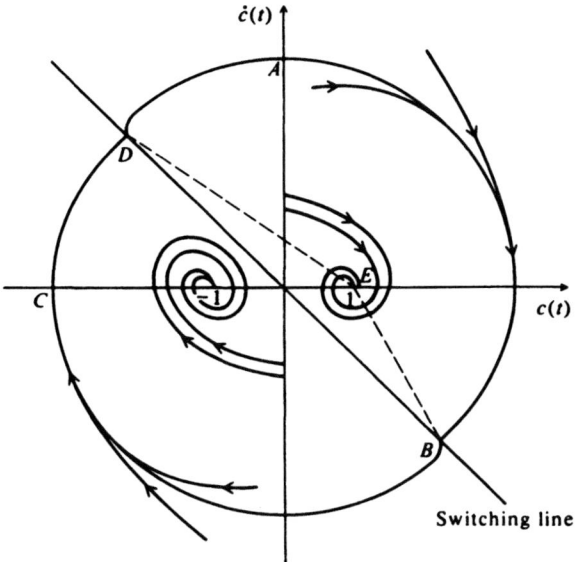

Figure 5.58 Phase portrait for an on-off control system where $a > 0$ and $b > 0$.

Two represent convergent motion where the phase portrait terminates on the foci at -1 or 1, depending on the final switching input. The third possible trajectory represents a limit cycle.

Any initial condition occurring beyond the limit cycle or very close to it in the enclosed area will eventually result in the trajectory reaching some part of the limit cycle, $ABCDA$. For a limit cycle to occur, the distance DE from the switching line to the corresponding focus must be greater than the distance BE, which represents the distance from the subsequent crossing of the line to the focus. This must be true with respect to the other focus as well. Trajectories that start sufficiently inside the limit cycle will spiral into one of the foci.

Case B. $a > 0$, $b < 0$. The switching line is a straight line given by

$$\frac{dc(t)}{dt} = \frac{a}{b} c(t). \tag{5.172}$$

It passes through the origin and lies in the first and third quadrants. In a manner similar to the first case, we can show the phase portrait to be as shown in Figure 5.59. Observe that any set of initial conditions which results in a trajectory intersecting the switching line inside the interval AB results in a motion spiraling towards one of the two stable foci. However, if the intersection occurs outside the interval AB, the trajectory theoretically ends on the switching line. Figure 5.59 indicates these points by C and D. Points E and F represents points of tangency with the switching lines for limiting trajectories. In reality, however, the system cannot just end at these points. This inconsistency is resolved by the fact that switching action of a contactor always has a certain time lag due to its dynamics. Therefore, when a solution reaches the switching line it actually proceeds for some small distance past it, before there is a

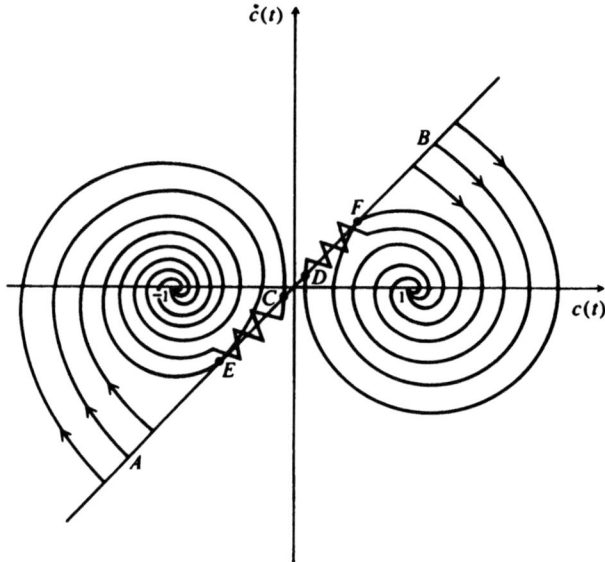

Figure 5.59 Phase portrait for an on-off control system where $a > 0, b < 0$.

change of sign in the forcing function. This results in a zigzag action along the switching line. Eventually, the trajectory spirals into one of the focis as shown in Figure 5.59. Physically, this is audible as a "chattering" of the contactor.

Case C. $a < 0, b > 0$. The switching line has the same form as that of case B. The phase portrait is given in Figure 5.60. For this case a stable limit of cycle always prevails. Regardless of the initial conditions, the final solution always winds up as a stable limit cycle.

Case D. $a < 0, b < 0$. The switching line has the same form as that of case A. The phase portrait, given in Figure 5.61, shows that no periodic solution exists. The solutions tend to end on the switching line. Because of the time lag of the relay, however, the trajectory zigzags towards the origin of the phase plane. The system finally oscillates at very high frequency and small amplitude around the origin.

It is interesting to observe that of the four cases considered, the control-system engineer would prefer the phase portrait of case D, the only configuration which resulted in a stable equilibrium state occurring around the origin. [Remember that $r(t) = 0$ and, therefore, $c(t) = 0$ in the steady-state as t approaches infinity.] However, we would have to tolerate some chattering around the origin with this linear switching system. To eliminate the chattering, one would have to use nonlinear switching techniques.

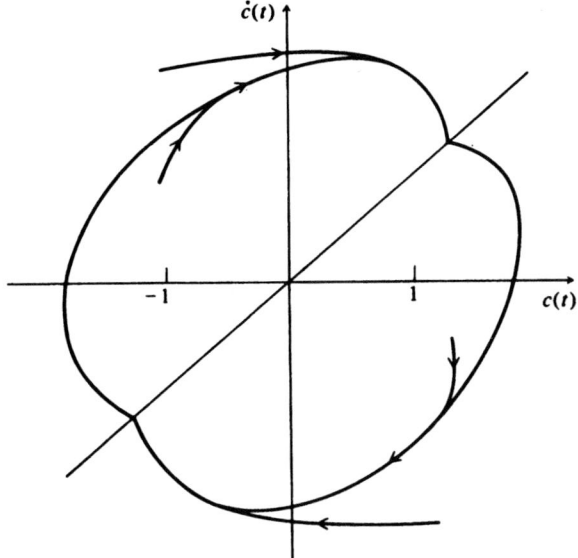

Figure 5.60 Phase portrait for an on-off control system where $a < 0, b > 0$.

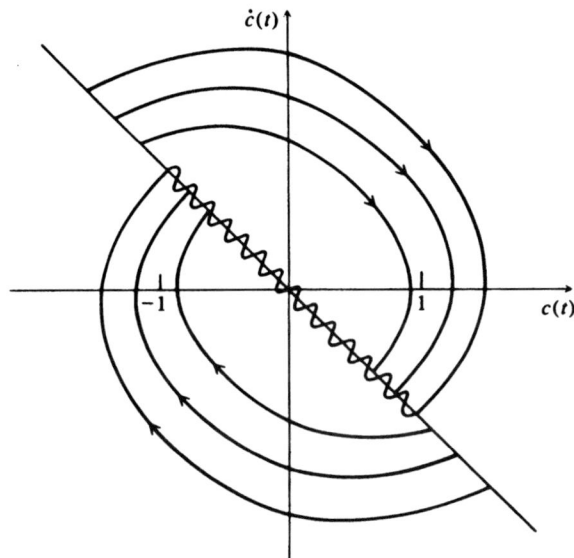

Figure 5.61 Phase portrait for an on-off control system where $a < 0, b < 0$.

5.19. DIGITAL COMPUTER PROGRAM FOR OBTAINING THE PHASE PLANE

This section presents a working program for obtaining a phase trajectory on the phase plane, and it applies the program to a practical system. The Fortran language will be used for the program.

Let us consider the linear, second-order system illustrated in Figure 5.62. It is desired to determine the phase trajectory of this system for the following set of initial conditions:

$$c(0) = 2.5 \text{ rad},$$
$$\dot{c}(0) = 0. \qquad (5.173)$$

It is assumed in this problem that the system does not have any external forcing function. The program used will determine the phase trajectory using the method of isoclines.

The isocline for this system is obtained as follows. The closed-loop transfer function of the system in Figure 5.62 is given by

$$\frac{C(s)}{R(s)} = \frac{4}{s^2 + s + 4}. \qquad (5.174)$$

In the time domain, the differential equation describing this equation is given by

$$\ddot{c}(t) + \dot{c}(t) + 4c(t) = 4r(t). \qquad (5.175)$$

Because the system does not have any external forcing function, $r(t) = 0$, and Eq. (5.175) reduces to

$$\ddot{c}(t) + \dot{c}(t) + 4c(t) = 0. \qquad (5.176)$$

Defining the state variables as

$$x_1(t) = c(t),$$
$$x_2(t) = \dot{c}(t), \qquad (5.177)$$

we obtain the following state equations to represent this system:

$$\dot{x}_1(t) = x_2(t), \qquad (5.178)$$
$$\dot{x}_2(t) = -4x_1(t) - x_2(t). \qquad (5.179)$$

Dividing Eq. (5.179) by (5.178), we obtain the following:

$$\frac{dx_2(t)}{dx_1(t)} = -\frac{4x_1(t)}{x_2(t)} - 1.$$

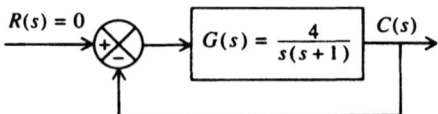

Figure 5.62 Linear, second-order system whose phase trajectory is to be obtained for initial conditions of $c(0) = 2.5$ rad and $\dot{c}(0) = 0$.

Simplifying this equation, we obtain the following:

$$0.5 \frac{dx_2(t)}{dx_1(t)} = -\frac{x_1(t)}{0.5x_2(t)} - 0.5. \tag{5.180}$$

Defining a normalized state variable

$$x_3(t) = 0.5x_2(t), \tag{5.181}$$

we can rewrite Eq. (5.180) as follows:

$$\frac{dx_3(t)}{dx_1(t)} = -\frac{x_1(t)}{x_3(t)} - 0.5. \tag{5.182}$$

Defining

$$\dot{\theta}(t) = 0.5\dot{c}(t), \tag{5.183}$$

Eq. (5.182) can be rewritten as

$$\frac{d\dot{\theta}(t)}{dc(t)} = -0.5 - \frac{c(t)}{\dot{\theta}(t)}. \tag{5.184}$$

Therefore, the slope of the trajectories in the $\dot{\theta}(t)$ vs $c(t)$ plane is given by

$$\text{trajectory slope} = -0.5 - \frac{c(t)}{\dot{\theta}(t)}. \tag{5.185}$$

Figure 5.63 illustrates the logic flow diagram for developing the program for computing the isocline lines and the phase trajectory. Table 5.7 illustrates the actual Fortran program for determining the isoclines and the phase trajectory. Table 5.8 illustrates the program output for determining the phase plane for the system of Figure 5.62 with the initial conditions given by Eq. (5.173). A plot of the results is given in Figure 5.64. The plot also compares the phase trajectory obtained using the method of isoclines with the theoretically obtained phase trajectory. An analysis of the resulting phase trajectory in the phase plane of Figure 5.64 indicates that the resulting trajectory has a stable node at the origin, and it is similar to the phase plane of Figure 5.42.

5.20. LIAPUNOV'S STABILITY CRITERIA

A. M. Liapunov [2] developed a fundamental method of determining the stability of a dynamic system based on generalization of energy considerations. This section presents the first and second methods of Liapunov and illustrates their application to nonlinear control systems [3,37].

400 NONLINEAR CONTROL-SYSTEM DESIGN

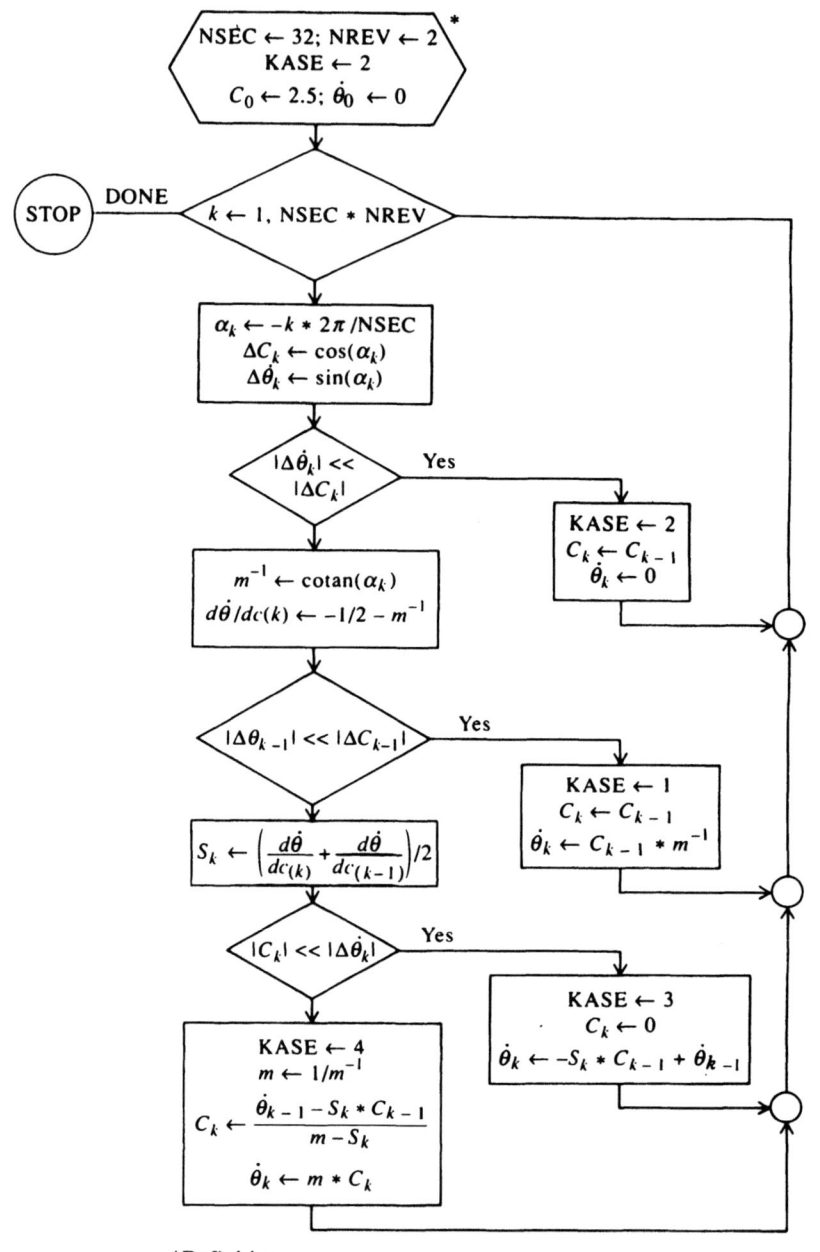

*Definitions

NSEC = Number of isocline lines (= division of 360°)
NREV = Number of times going around circle
KASE 1 = Case where the phase trajectory leaves the horizontal axis
  (isocline slope = ∞)
KASE 2 = Case where the phase trajectory enters the horizontal axis
  (isocline slope = ∞)
KASE 3 = Case where the phase trajectory enters the vertical axis
  (isocline slope = 0)
KASE 4 = Any other condition than KASE 1, 2, or 3

Figure 5.63 Logic flow diagram for phase-plane analysis.

Table 5.7. Computer Program for Phase-Plane Analysis (Fortran Program)

```
      data NSEC/32/,NREV/2/,C1/2.5/,DTH1/0.0/,KASE/2/,PI/3.141593/
      print55, C1, DTH1
      do 100 K=1, NSEC*NREV
        ANGLE=-K*2*PI/NSEC
        YDEL=sin(ANGLE)
        XDEL=cos(ANGLE)
c       see if 1/SLOPE → infinity (occurs at 0 or 180 degrees)
        if (ABS(XDEL).ge.1E-35) then
          if (ABS(YDEL/XDEL).le.1E-5) then
            KASE=2
            C=C1
            DTH=0
            goto80
          endif
        endif
c       otherwise, it's OK to compute 1/SLOPE
        SLPINV=cotan(ANGLE)
        DDTH=-0.5-SLPINV
c       see if 1/SLOPE → infinity on the previous loop iteration
        if (KASE.eq.2) then
          KASE=1
          C=C1
          DTH=C/SLPINV
        else
c         otherwise, calculate isocline intersection point
          S=(DDTH+DDTH1)/2
c         see if SLOPE → infinity occurs at 90 or 270 degrees)
          if (ABS(YDEL).ge.1E-35) then
            if (ABS(XDEL/YDEL).le.1E-5) then
              KASE=3
              C=0
              DTH=-S*C1+DTH1
              goto80
            endif
          end if
c         otherwise, both SLOPE & 1/SLOPE are finite → solve 2 equations
          KASE =4
          SLOPE =1/SLPINV
          C=(DTH1-S*C1)/(SLOPE-S)
          DTH=SLOPE *C
        end if
   80   continue @ come here to escape more testing
        print60, C, DTH, DDTH, DDTH1, S, KASE
        C1=C
        DTH1=DTH
        DDTH1=DDTH
  100 continue
      print65
   55 format(7x,'C    THETA-DOT   I.S.1   I.S.2  AVG.SLOPE',
     $' *CASE'/7x,'= ========   ======   ====   ',
     $ ========   ===='//2e12.4, '---> (INITIAL CONDITIONS)')
   60 format(5e12.4, I5)
   65 format(/' *NOTE: CASE=1 or 2 --> ISOCLINE SLOPES 1 OR 2 DIVERGE'/
     $       '        CASE=3 OR 4 --> ISOCLINE SLOPES VALID')
      end
```

Table 5.8. Program Output for Phase Plane of System in Figure 5.62

C	Theta-dot	I.S.1	I.S.2	Avg.slope	*Case
===	=========	======	======	=========	====
.2500 + 001	.0000	→ (INITIAL CONDITIONS) .4527 + 001	.0000	.0000	1
.2500 + 001	−.4973 + 000	.4527 + 001	.4527 + 001	.3221 + 001	4
.2352 + 001	−.9742 + 000	.1914 + 001	.1914 + 001	.1455 + 001	4
.2071 + 001	−.1384 + 001	.9966 + 000	.9966 + 000	.7483 + 000	4
.1678 + 001	−.1678 + 001	.5000 + 000	.5000 + 000	.3341 + 000	4
.1223 + 001	−.1830 + 001	.1682 + 000	.1682 + 000	.4120 − 001	4
.7657 + 000	−.1849 + 001	−.8579 + 001	−.8579 − 001	−.1934 + 000	4
.3518 + 000	−.1768 + 001	−.3011 + 000	−.3011 + 000	−.4005 + 000	3
.0000	−.1628 + 001	−.5000 + 000	−.5000 + 000	−.5995 + 000	4
−.2892 + 000	−.1454 + 001	−.6989 + 000	−.6989 + 000	−.8066 + 000	4
−.5239 + 000	−.1265 + 001	−.9142 + 000	−.9142 + 000	−.1041 + 001	4
−.7134 + 000	−.1068 + 001	−.1168 + 001	−.1168 + 001	−.1334 + 001	4
−.8651 + 000	−.8651 + 000	−.1500 + 001	−.1500 + 001	−.1748 + 001	4
−.9839 + 000	−.6574 + 000	−.1997 + 001	−.1997 + 001	−.2455 + 001	4
−.1071 + 001	−.4436 + 000	−.2914 + 001	−.2914 + 001	−.4221 + 001	4
−.1123 + 001	−.2234 + 000	−.5527 + 001	−.5527 + 001	−.4221 + 001	4
−.1123 + 001	.0000	−.5527 + 001	−.5527 + 001	−.4221 + 001	2
−.1123 + 001	.2234 + 000	.4527 + 001	−.4527 + 001	.3221 + 001	1
−.1057 + 001	.4377 + 000	.1914 + 001	−.4527 + 001	.1455 + 001	4
−.9303 + 000	.6216 + 000	.9966 + 000	−.1914 + 001	.7483 + 000	4
−.7537 + 000	.7537 + 000	.5000 + 000	−.9966 + 000	.3341 + 000	4
−.5493 + 000	.8220 + 000	.1682 + 000	−.5000 + 000	.4120 − 001	4
−.3440 + 000	.8305 + 000	−.8579 − 001	−.1682 + 000	−.1934 + 000	4
−.1580 + 000	.7945 + 000	−.3011 + 000	.8579 − 001	−.4005 + 000	4
.0000	.7312 + 000	−.5000 + 000	.3011 + 000	−.5995 + 000	3
.1300 + 000	.6533 + 000	−.6989 + 000	.5000 + 000	−.8066 + 000	4
.2354 + 000	.5683 + 000	−.9142 + 000	.6989 + 000	−.1041 + 001	4
.3205 + 000	.4797 + 000	−.1168 + 001	.9142 + 000	−.1334 + 001	4
.3887 + 000	.3887 + 000	−.1500 + 001	.1168 + 001	−.1748 + 001	4
.4421 + 000	.2954 + 000	−.1997 + 001	.1500 + 001	−.2455 + 001	4
.4812 + 000	.1993 + 000	−.2914 + 001	.1997 + 001		4

C	Theta-dot	I.S.1	I.S.2	Avg.slope	*Case
===	===	===	===	===	===
.5046 + 000	.1004 + 000	−.5527 + 001	−.2914 + 001	−.4221 + 001	4
.5046 + 000	.0000	−.5527 + 001	−.5527 + 001	−.4221 + 001	2
.5046 + 000	−.1004 + 000	.4527 + 001	−.5527 + 001	−.4221 + 001	1
.4747 + 000	−.1966 + 000	.1914 + 001	.4527 + 001	.3221 + 001	4
.4180 + 000	−.2793 + 000	.9966 + 001	.1914 + 001	.1455 + 001	4
.3386 + 000	−.3386 + 000	.5000 + 000	.9966 + 000	.7843 + 000	4
.2468 + 000	−.3693 + 000	.1682 + 000	.5000 + 000	.3341 + 000	4
.1546 + 000	−.3731 + 000	−.8579 − 001	.1682 + 000	.4120 − 001	4
.7100 − 001	−.3570 + 000	−.3011 + 000	−.8579 − 001	−.1934 + 000	4
.0000	−.3285 + 000	−.5000 + 000	−.3011 + 000	−.4005 + 000	3
−.5839 − 001	−.2935 + 000	−.6989 + 000	−.5000 + 000	−.5995 + 000	4
−.1058 + 000	−.2553 + 000	−.9142 + 000	−.6989 + 000	−.8066 + 000	4
−.1440 + 000	−.2155 + 000	−.1168 + 001	−.9142 + 000	−.1041 + 001	4
−.1746 + 000	−.1746 + 000	−.1500 + 001	−.1168 + 001	−.1334 + 001	4
−.1986 + 000	−.1327 + 000	−.1997 + 001	−.1500 + 001	−.1748 + 001	4
−.2162 + 000	−.8955 − 001	−.2914 + 001	−.1997 + 001	−.2455 + 001	4
−.2267 + 000	−.4510 − 001	−.5527 + 001	−.2914 + 001	−.4221 + 001	4
−.2267 + 000	.0000	−.5527 + 001	−.5527 + 001	−.4221 + 001	2
−.2267 + 000	.4510 − 001	.4527 + 001	−.5527 + 001	−.4221 + 001	1
−.2133 + 000	.8835 − 001	.1914 + 001	.4527 + 001	.3221 + 001	4
−.1878 + 000	.1255 + 000	.9966 + 000	.1914 + 001	.1455 + 001	4
−.1521 + 000	.1521 + 000	.5000 + 000	.9966 + 000	.7843 + 000	4
−.1109 + 000	.1659 + 000	.1682 + 000	.5000 + 000	.3341 + 001	4
−.6944 − 001	.1676 + 000	−.8579 − 001	.1682 + 000	.4119 − 001	4
−.3190 − 001	.1604 + 000	−.3011 + 000	−.8579 − 001	−.1934 + 000	4
.0000 000	.1476 + 000	−.5000 + 000	−.3011 + 000	−.4005 + 000	3
.2623 − 001	.1319 + 000	−.6989 + 000	−.5000 + 000	−.5995 + 000	4
.4751 − 001	.1147 + 000	−.9142 + 000	−.6989 + 000	−.8066 + 000	4
.6469 − 001	.9682 − 001	−.1168 + 001	−.9142 + 000	−.1041 + 001	4
.7846 − 001	.7846 − 001	−.1500 + 001	−.1168 + 001	−.1334 + 001	4
.8923 − 001	.5962 − 001	−.1997 + 001	−.1500 + 001	−.1748 + 001	4
.9713 − 001	.4023 − 001	−.2914 + 001	−.1997 + 001	−.2455 + 001	4
.1019 + 000	.2026 − 001	−.5527 + 001	−.2914 + 001	−.4221 + 001	4
.1019 + 000	.0000	−.5527 + 001	−.5527 + 001	−.4221 + 001	2

*NOTE: CASE = 1 or 2 → ISOCLINE SLOPES 1 OR 2 DIVERGE
  CASE = 3 OR 4→ ISOCLINE SLOPES VALID

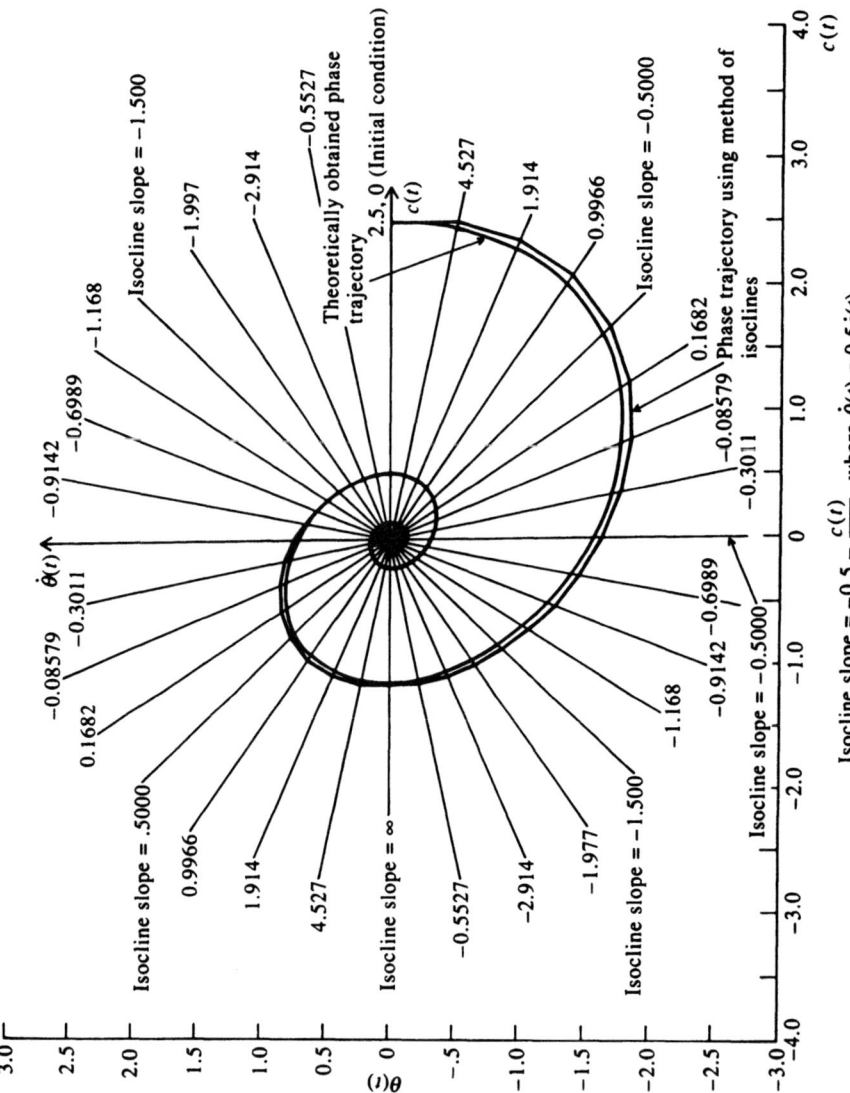

Figure 5.64 Phase trajectory for system of Figure 5.62.

A. Liapunov's First Method

Liapunov divided the general problem of analyzing the stability of nonlinear control systems into two classes. The first class consists of all those methods in which the differential equation of the system can be solved. System stability or instability is determined from this solution. This approach, which is known as *Liapunov's first method*, does not say anything of particular importance concerning the solution of the nonlinear differential equations. However, Liapunov did point out in his first method that the solution may be obtained in the form of a series from which stability can be determined using his second method. In addition, he proved that approximate solutions of nonlinear differential equations often yield useful stability information.

In order to illustrate Liapunov's first method, let us assume that the nonlinearity is single valued (has no hysteresis present) and has derivatives of every order in the vicinity of a point A. The nonlinear function, $y = f(x)$, can be expanded into a Taylor series as follows:

$$y = f(x) = y(A) + (x - A)\left(\frac{dy}{dx}\right)_A + \frac{1}{2!}(x - A)^2 \left(\frac{d^2y}{dx^2}\right)_A$$
$$+ \frac{1}{3!}(x - A)^3 \left(\frac{d^3y}{dx^3}\right)_A + \cdots + \frac{1}{n!}(x - A)^n \left(\frac{d^ny}{dx^n}\right)_A + \cdots. \quad (5.186)$$

Note that the first two terms of the series represent the linear approximation about the operating point of the actual nonlinearity. Liapunov proved that if the real parts of the roots of the characteristic equation corresponding to the differential equation of the linear approximation are different from zero, the equations of the linear approximation always give a correct answer to the question of staiblity of a nonlinear system [3,37]. This theorem means that we can use a linear approximation of the nonlinear equation and determine stability from it. If the roots of the linearized characteristic equation are negative, the motion is stable about the point in question. However, if any of the real parts of the roots of the characteristic equation of the linear approximation are positive, the motion is unstable about the operating point. In the special case of roots of the linearized characteristic equation having zero real parts, no conclusion may be drawn.

To illustrate the application of Liapunov's first method, consider the Van der Pol equation which describes the voltage buildup of an oscillator,

$$\ddot{v}(t) + u(v^2(t) - 1)\dot{v}(t) + Kv(t) = Q, \quad (5.187)$$

where

$$u < 0, \quad K > 0.$$

The equilibrium point of this system is determined when the velocity and acceleration are zero. Then,

$$Kv(t) = Q, \quad (5.188)$$

or
$$v(t) = \frac{Q}{K} = V = \text{equilibrium point.} \quad (5.189)$$

In order to determine behavior in the area of the equilibrium point V, let
$$v(t) = V + v_i(t), \quad (5.190)$$

where
$$\ddot{V} = \dot{V} = 0.$$

Substituting Eq. (5.190) into Eq. (5.187) we obtain the following expression:
$$\ddot{v}_i(t) + u[(V^2 + 2Vv_i(t) + v_i^2(t)) - 1]\dot{v}_i(t) + K(V + v_i(t)) = Q. \quad (5.191)$$

The linear approximation to this equation is given by
$$\ddot{v}_i(t) + u[V^2 - 1]\dot{v}_i(t) + Kv_i(t) = 0. \quad (5.192)$$

The characteristic equation is given by
$$s^2 + u[V^2 - 1]s + K = 0. \quad (5.193)$$

Applying the Routh–Hurwitz stability criterion, we find that, for stability,
$$u[V^2 - 1] > 0. \quad (5.194)$$

Equation (5.194) states that when $u < 0$, the system is stable for all $|V| < 1$. Liapunov's first method is not applicable when $V = 1$, because this condition results in zero real parts for the roots of the characteristic equation.

It is important to emphasize that Liapunov's first method determines stability in the immediate vicinity of the equilibrium point.

B. Liapunov's Second Method

This determines stability without actually having to solve the differential equation. In this method a function of the state variables having special properties is formed that can be compared to the sum of the kinetic and potential energy, and the derivative of the function with respect to time is taken. If this derivative is negative along the trajectories of the system, it can be shown that the system is asymptotically stable. The remainder of this section is devoted to the details and application of the method.

We introduce Liapunov's second method by first considering a linear system. Reconsider the simple mass-spring-damper mechanical system considered in Section 5.2. It was shown there that this system can be represented by the following differential equation:
$$M \frac{d^2 y(t)}{dt^2} + B \frac{dy(t)}{dt} + Ky(t) = f(t). \quad (5.195)$$

Assume that $M = B = K = 1$ and that $f(t) = 0$. Then we have

$$\ddot{y}(t) + \dot{y}(t) + y(t) = 0. \quad (5.196)$$

Defining the state variables as

$$x_1(t) = y(t), \quad (5.197)$$
$$x_2(t) = \dot{y}(t), \quad (5.198)$$

the system can be described by the following two first-order differential equations:

$$\dot{x}_1(t) = x_2(t) = \dot{y}(t), \quad (5.199)$$
$$\dot{x}_2(t) = -x_1(t) - x_2(t). \quad (5.200)$$

This simple linear system can easily be solved. Assuming the initial conditions are

$$x_1(0) = 1, \quad (5.201)$$
$$x_2(0) = 0, \quad (5.202)$$

then we obtain the following solutions:

$$x_1(t) = 1.15e^{-t/2}\sin(0.866t + \pi/3), \quad (5.203)$$
$$x_2(t) = -1.15e^{-t/2}\sin(0.866t). \quad (5.204)$$

Equations (5.203) and (5.204) are plotted in the time domain in Figure 5.65, and in the phase plane in Figure 5.66. These two figures completely determine the dynamics and stability of this simple mechanical system. The system is stable and the states $x_1(t)$ and $x_2(t)$ behave as indicated.

Now, let us look at this simple system from the viewpoint of energy. The total stored energy is given by

$$V(t) = \tfrac{1}{2}Kx_1^2(t) + \tfrac{1}{2}Mx_2^2(t). \quad (5.205)$$

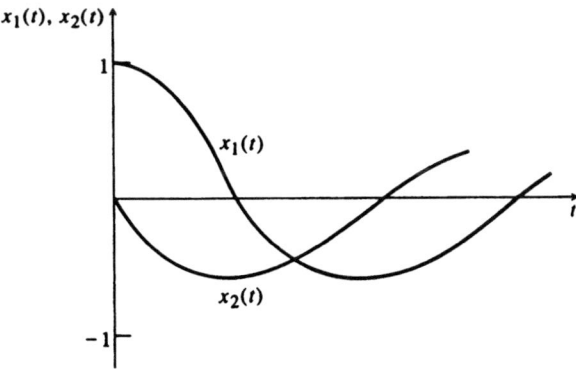

Figure 5.65 Time-domain response of a simple mechanical system.

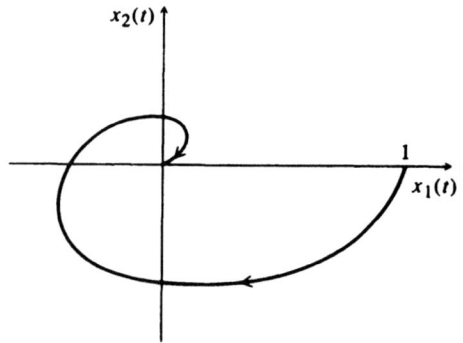

Figure 5.66 Phase trajectory of a simple mechanical system.

Because $K = M = 1$ in our simple example,

$$V(t) = \tfrac{1}{2}x_1^2(t) + \tfrac{1}{2}x_2^2(t). \tag{5.206}$$

This total energy is dissipated as heat in the damper at the rate of

$$\dot{V}(t) = -B\dot{x}_1(t)x_2(t) = -Bx_2^2(t). \tag{5.207}$$

Because $B = 1$, we obtain

$$\dot{V}(t) = -x_2^2(t). \tag{5.208}$$

Equation (5.206) determines the loci of constant stored energy in the $x_1(t)x_2(t)$ plane. Clearly they are circular for this simple example. Another important observation from Eq. (5.208) is that the energy rate is always negative and, therefore, these circles must get smaller and smaller with time. Figure 5.67 illustrates this characteristic in the phase plane. For this simple example, we can determine the time variation of $V(t)$ and $\dot{V}(t)$ explicitly by substituting Eqs. (5.203) and (5.204) into Eqs. (5.206) and (5.208). The results are as follows:

$$V(t) = 0.667e^{-t}[\sin^2(0.866t) + \sin^2(0.866t + \pi/3)], \tag{5.209}$$

$$\dot{V}(t) = -1.333e^{-t}\sin^2(0.866t). \tag{5.210}$$

Figure 5.68 illustrates the time variation of $V(t)$ and $\dot{V}(t)$. Comparing Figures 5.67 and 5.68, we conclude that the total stsored energy approaches zero as time approaches infinity. This implies that the system is asymptotically stable. By this is meant that the state will return to the origin from any point $\mathbf{x}(t)$ within a region R enclosing the origin. Asymptotic stability is the type of stability preferred by control engineers because it excludes a stable limit cycle.

The stability of nonlinear control systems depends on the particular state space in which the state vector ranges in addition to the type and magnitude of the input. Therefore, the stability of nonlinear control systems can also be classified on a regional basis as follows [6,16]:

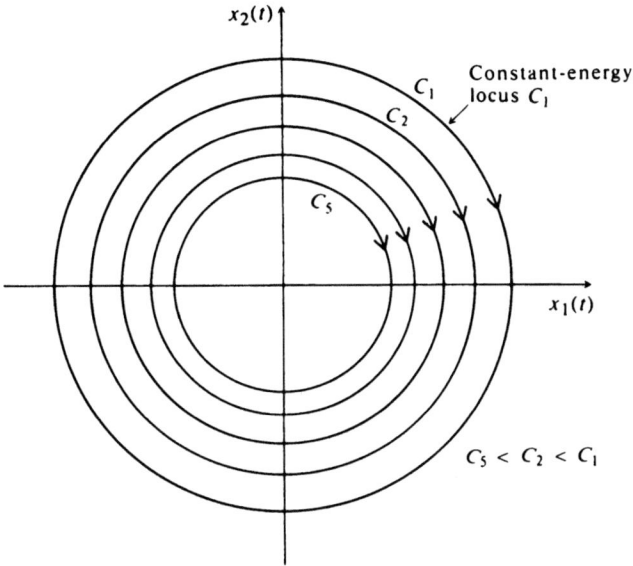

Figure 5.67 Constant-energy loci on the phase plane illustrating a decrease of energy with time.

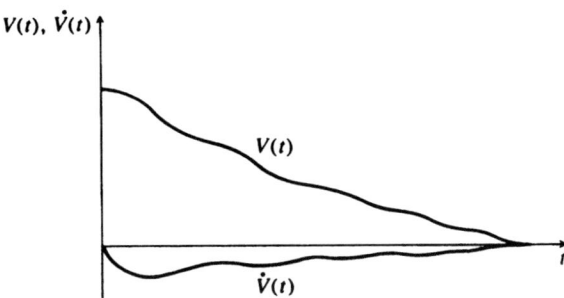

Figure 5.68 Variation of energy and energy rate with time.

(a) local stability, or stability in the small,
(b) finite stability,
(c) global stability, or stability in the large.

A nonlinear control system is denoted as being *locally stable* if it remains within an infinitesimal region about a singular point when subjected to a small perturbation. *Finite stability* refers to a system which returns to a singular point from any point $\mathbf{x}(t)$ within a region R of finite dimensions surrounding it. The system is said to be *globally stable* if the region R inclues the entire finite state space. Stability of either the local, finite, or global variety does not exclude limit cycles, but rather only excludes the possibility of the state point tending to travel to infinity. If the state point approaches the singularity as time approaches infinity, for any initial conditions within the region under consideration, then the system is described as being *asymptotically stable*. Asymptotic stability excludes a stable limit cycle as a possible

410 NONLINEAR CONTROL-SYSTEM DESIGN

dynamic equilibrium condition. The strongest possible condition that can be placed on a nonlinear control system, with parameters that do not vary with time, is global asymptotic stability.*

The formal definitions of positive definiteness of a scalar function $V(\mathbf{x})$ and similar functions are as follows.

1. Positive Definiteness of Scalar Functions. The scalar function $V(\mathbf{x})$ is positive definite in a region which includes the origin of the state space if $V(\mathbf{x}) > 0$ for all nonzero states $\mathbf{x}$ in the region, and $V(0) = 0$. An example of a positive definite scalar function is

$$V(\mathbf{x}) = 2x_1^2 + 4x_2^2. \qquad (5.211)$$

2. Negative Definiteness of Scalar Functions. The scalar function $V(\mathbf{x})$ is negative definite if $-V(\mathbf{x})$ is positive definite. An example of a negative definite scalar function is

$$V(\mathbf{x}) = -2x_1^2 - (4x_1 + 6x_2)^3. \qquad (5.212)$$

3. Indefiniteness of Scalar Functions. The scalar function $V(\mathbf{x})$ is indefinite in a region if it contains both positive and negative values. An example of an indefinite function is

$$V(\mathbf{x}) = 2x_1 x_2 + 3x_2^2. \qquad (5.213)$$

4. Quadratic Form of Scalar Function. The following quadratic form of the scalar function is also used very frequently in the analysis of the second method of Liapunov:

$$V(\mathbf{x}) = \mathbf{x}^T \mathbf{A} \mathbf{x} \qquad (5.214)$$

where $\mathbf{x}$ is a real vector and $\mathbf{A}$ is a real symmetric matrix.

The positive definiteness of the quadratic form of $V(\mathbf{x})$ is determined using Sylvester's criterion, which states that the necessary and sufficient conditions for the quadratic form of $V(\mathbf{x})$ to be positive definite are that all of the successive principal minors of the symmetric matrix $\mathbf{A}$ be positive. For example, let us assume that $\mathbf{A}$ is given by:

---

*There are about 30 different classes of stability currently in use. However, the types defined in this book are the important ones most frequently used by the control-system engineer.

5.20. LIAPUNOV'S STABILITY CRITERIA

$$\mathbf{A} = \begin{bmatrix} a_{11} & a_{12} & \cdots & a_{1n} \\ a_{12} & a_{22} & \cdots & a_{2n} \\ \vdots & \vdots & & \vdots \\ a_{1n} & a_{2n} & \cdots & a_{nn} \end{bmatrix}. \quad (5.215)$$

Therefore, the necessary and sufficient condition are that all the successive principal minors of **A** are positive as follows

$$a_{11} > 0; \quad \begin{vmatrix} a_{11} & a_{12} \\ a_{12} & a_{22} \end{vmatrix} > 0; \quad \ldots, \quad \begin{vmatrix} a_{11} & a_{12} & \cdots & a_{1n} \\ a_{12} & a_{22} & \cdots & a_{2n} \\ \vdots & \vdots & & \vdots \\ a_{1n} & a_{2n} & \cdots & a_{nn} \end{vmatrix} > 0. \quad (5.216)$$

As an example to illustrate the application of the quadratic form and Sylvester's criterion to test the positive definiteness of the energy function $V(\mathbf{x})$, consider the following scalar function $V(\mathbf{x})$:

$$V(\mathbf{x}) = 20x_1^2 + 8x_2^2 + 2x_3^2 + 4x_1x_2 - 4x_2x_3 - 8x_1x_3. \quad (5.217)$$

Applying the quadratic form of $V(\mathbf{x})$, we can express this in terms of the unknown elements of **A** as follows:

$$V(\mathbf{x}) = \mathbf{x}^T \mathbf{A} \mathbf{x} = [x_1 \ x_2 \ x_3] \begin{bmatrix} a_{11} & a_{12} & a_{13} \\ a_{12} & a_{22} & a_{23} \\ a_{13} & a_{23} & a_{33} \end{bmatrix} \begin{bmatrix} x_1 \\ x_2 \\ x_3 \end{bmatrix}. \quad (5.218)$$

We can multiply the matrices $\mathbf{x}^T$, **A**, and **x** and obtain the following in terms of the unknown elements of **A** as follows:

$$V(\mathbf{x}) = a_{11}x_1^2 + a_{12}x_1x_2 + a_{13}x_1x_3 + a_{12}x_1x_2 + a_{22}x_2^2 + a_{23}x_2x_3 + a_{13}x_1x_3 \\ + a_{23}x_2x_3 + a_{33}x_3^2. \quad (5.219)$$

Setting like coefficients of Eqs. (5.217) and (5.219) equal to each other, we find the values of the elements of the **A** matrix. The resulting **A** matrix is as follows:

$$\mathbf{A} = \begin{bmatrix} 20 & 2 & -4 \\ 2 & 8 & -2 \\ -4 & -2 & 2 \end{bmatrix}. \quad (5.220)$$

Applying Sylvester's criterion we find that

$$20 > 0; \quad \begin{vmatrix} 20 & 2 \\ 2 & 8 \end{vmatrix} = 156 > 0; \quad \begin{vmatrix} 20 & 2 & -4 \\ 2 & 8 & -2 \\ -4 & -2 & 2 \end{vmatrix} = 136 > 0. \quad (5.221)$$

In conclusion, since all of the successive principal minors of the matrix **A** are positive, then $V(\mathbf{x})$ is a positive definite scalar function.

A major factor in this analysis has been the choice of the energy function $V(t)$,

$$V(t) = \tfrac{1}{2}x_1^2(t) + \tfrac{1}{2}x_2^2(t). \tag{5.222}$$

This function has two very interesting properties. First it is positive for all nonzero values of $x_1(t)$ and $x_2(t)$. Secondly, it equals zero when $x_1(t) = x_2(t) = 0$. A scalar function having these properties is called a *positive-definite* function. By adding $V(t)$ as a third dimension to the $x_1(t)x_2(t)$ plane, the positive-definite function $V(x_1, x_2)$ appears as a cup-shaped three-dimensional surface as illustrated in Figure 5.69.

Liapunov's stability theorem can now be summarized for n-dimensional state space. A dynamic system of nth order is asymptotically stable if a positive-definite function $V(t)$ is found whose derivative with respect to time is negative along the trajectories of the system. In practice, it is fairly easy to find a function which is positive definite, but usually it is very hard to find a function where, in addition, $\dot{V}(t) < 0$ along the trajectories.

A justification of Liaponov's second method can best be presented by considering the phase plane of Figure 5.70. Contours of constant $V(t)$ are shown by the curves of C_1, C_2, and C_3. Assume that the phase trajectory of this second-order system, whose initial state is the point p, is described by the following state equations:

$$\dot{x}_1(t) = F_1(x_1, x_2), \tag{5.223}$$
$$\dot{x}_2(t) = F_2(x_1, x_2). \tag{5.224}$$

The positive-definite function $V(t)$ is assumed to be given by

$$V(t) = a^2 x_1^2(t) + b^2 x_2^2(t), \tag{5.225}$$

where a and b are unknown coefficients. The quantity $V(t)$ is permitted to take on successively larger constant values,

$$V(t) = 0, C_1, C_2, C_3, \ldots, \quad \text{where} \quad 0 < C_1 < C_2 < C_3 < \cdots. \tag{5.226}$$

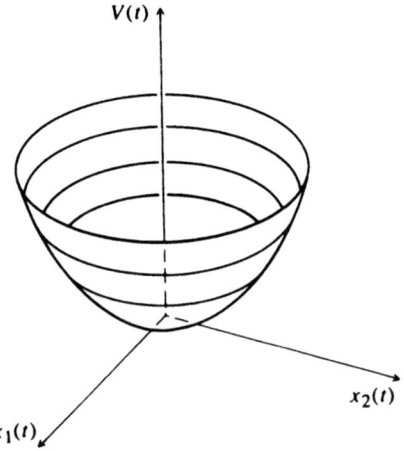

Figure 5.69 A positive-definite function.

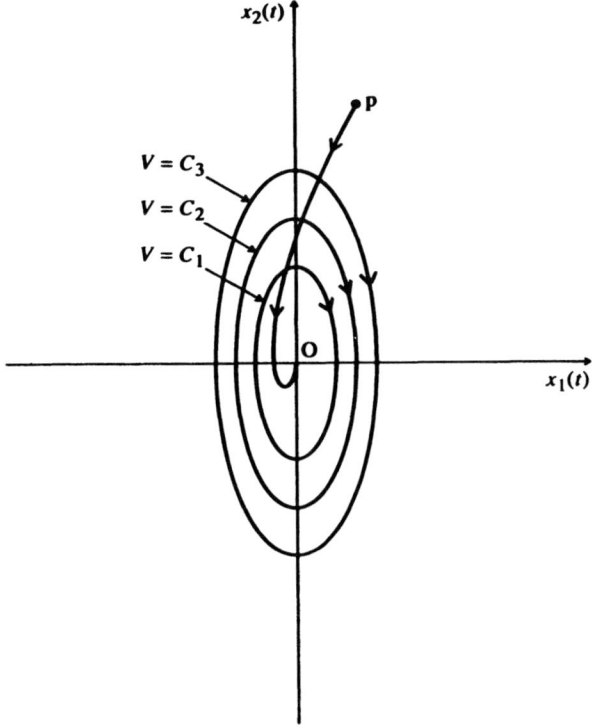

Figure 5.70 Justification of Liapunov's stability criterion.

Therefore, Eq. (5.225) results in a set of equations:

$$\begin{aligned} a^2 x_1^2 + b^2 x_2^2 &= 0, \\ a^2 x_1^2 + b^2 x_2^2 &= C_1, \\ a^2 x_1^2 + b^2 x_2^2 &= C_2, \\ &\cdots. \end{aligned} \quad (5.227)$$

When $V(t) = 0$, Eq. (5.227) describes the origin of the phase plane; for other values, the resulting equations describe ellipses in the phase plane. As shown in Figure 5.70, each succeeding ellipse contains within itself all of the preceding ellipses. The time derivative of the $V(t)$ function is given by

$$\frac{dV(t)}{dt} = \frac{\partial V(t)}{\partial x_1} \frac{dx_1}{dt} + \frac{\partial V(t)}{\partial x_2} \frac{dx_2}{dt}, \quad (5.228)$$

or

$$\frac{dV(t)}{dt} = \frac{\partial V(t)}{\partial x_1} \dot{x}_1(t) + \frac{\partial V(t)}{\partial x_2} \dot{x}_2(t). \quad (5.229)$$

Substituting Eqs. (5.223) and (5.224) into Eq. (5.229), we obtain

$$\frac{dV(t)}{dt} = \frac{\partial V(t)}{\partial x_1} F_1(x_1, x_2) + \frac{\partial V(t)}{\partial x_2} F_2(x_1, x_2). \tag{5.230}$$

Taking the partial derivatives of Eq. (5.225) as indicated, Eq. (5.230) can be rewritten as

$$\frac{dV(t)}{dt} = 2a^2 x_1 F_1(x_1, x_2) + 2b^2 x_2 F_2(x_1, x_2). \tag{5.231}$$

If $dV(t)/dt$ is negative, then the state must move from its initial state point p, in the direction of smaller values of $V(t)$ and toward the origin. This system would then be asymptotically stable.

C. Example 5.20-1

To illustrate the application of Liapunov's second method we consider the second-order differential equation

$$\frac{d^2 x(t)}{dt^2} + \left[K_1 + K_2 \left(\frac{dx(t)}{dt} \right)^2 \right] \frac{dx(t)}{dt} + x(t) = 0. \tag{5.232}$$

The state equations are

$$\frac{dx(t)}{dt} = y(t), \tag{5.233}$$

$$\frac{dy(t)}{dt} = -K_1 y(t) - K_2 y^3(t) - x(t), \tag{5.234}$$

where K_1 and K_2 are not both zero. Depending on the values of K_1 and K_2, stability or instability can result. For example, if $K_1 > 0$ and $K_2 > 0$, and $V(t)$ is given by

$$V(t) = x^2(t) + y^2(t), \tag{5.235}$$

this system is stable, because the derivative of $V(t)$ with respect to time t, $dV(t)/dt$, is negative, because

$$\frac{dV(t)}{dt} = 2\left(x(t) \frac{dx(t)}{dt} + y(t) \frac{dy(t)}{dt} \right) = -2(K_1 y^2(t) + K_2 y^4(t)) < 0. \tag{5.236}$$

Equilibrium occurs at the singularity located at the origin, and the equilibrium point is asymptotically stable.

It is left as an exercise to the reader to prove that the condition $K_1 < 0$ and $K_2 < 0$ corresponds to an unstable equilibrium; the condition $K_1 > 0$ and $K_2 < 0$ corresponds to a stable equilibrium only if $0 < y^2(t) < |K_1/K_2|$.

It is important to recognize that the stability conditions obtained from a particular $V(t)$ function are usually sufficient, but not necessary. In addition, a Liapunov function for any particular system is not unique. Therefore, if a particular $V(t)$ function should fail to demonstrate whether a particular system is stable or not, there is no assurance that another function could not be found that does determine stability or instability. There is also no assurance that exceeding the limits based on a particular $V(t)$ function will actually cause the equilibrium to be unstable. The Liapunov stability criterion is a conservative one.

As can be seen from this presentation, the primary use of Liapunov's second method is not the determination of stability where the answer may be found by other means, but is rather in the study of problems of stability which are not readily determined by other methods.

5.21. POPOV'S METHOD

An interesting and very powerful stability criterion for nonlinear control systems that are time invariant was introduced in 1959 by V. M. Popov, who obtained a frequency-domain criterion as a sufficient condition for asymptotic stability of single-loop control systems [13–20]. The method, as originally developed by Popov, is applicable to single-loop feedback systems containing time-invariant linear elements and time-invariant nonlinearities. An important feature of Popov's approach is that it is applicable to systems of high order. Once the frequency response of the linear element is known, very little additional calculation is required for determining stability of the nonlinear control system. It is an extension of the Nyquist diagram (see Section 1.7) to nonlinear systems.

This section presents Popov's stability criterion in terms of inequality constraints on the nonlinear element in conjunction with a modified frequency plot of the linear element. It will be shown that the most important and appealing feature of the Popov criterion is that it shares all of the desirable characteristics of the Nyquist method.

In order to introduce Popov's method, let us consider the nonlinear control system illustrated in Figure 5.71. It is composed of a linear, time-invariant process $G(s)$ and a nonlinear, time-invariant element $N[e(t)]$. The reference input $r(t)$ is assumed to be zero. Therefore, the response of this system can be expressed as

$$e(t) = e_0(t) - \int_0^t g(t-\tau)u(\tau)\,d\tau, \tag{5.237}$$

where

$$g(t) - L^{-1}[G(s)] = \text{unit impulse response},$$
$$-e_0(t) = \text{initial condition response}.$$

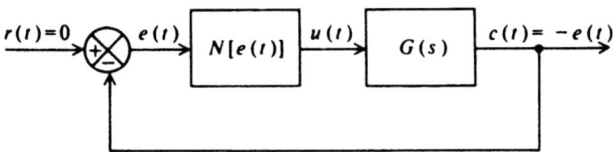

Figure 5.71 Nonlinear control system considered by Popov [14].

In this analysis, special restrictions are placed on the nonlinear and linear elements. For the nonlinear element $n[e(t)]$, it is assumed that the input-output relationship is restricted to lie within the region illustrated in Figure 5.72 where

$$0 \leqslant N[e(t)] \leqslant K, \tag{5.238}$$

and

$$u(t) = N[e(t)]e(t). \tag{5.239}$$

Furthermore, it is assumed that for all t, and for every finite value e_m, there is a finite value u_m such that

$$|u(t)| \leqslant u_m < \infty \quad \text{if} \quad |e(t)| \leqslant e_m. \tag{5.240}$$

The only assumption concerning the linear element $G(s)$ is that it is output stable of degree n for some value of n. By this we mean that if $n < 0$, the output response to an initial condition or an impulse may diverge, but when the output is multiplied by e^{nt} it will converge towards zero. For the case of $n > 0$, output stable means that the output response to an initial condition or an impulse will converge towards zero faster than the function e^{-nt}. In general, a linear element will be output stable of degree n if its transfer function $G(s)$ and initial condition response function $E_0(s)$ are rational functions of s and its poles all satisfy

$$\operatorname{Re} s < -n.$$

Therefore, n actually represents the settling rate of the linear element.

Popov's method is concerned with the asymptotic behavior of the control signal $u(t)$ and output $-e(t)$ of the linear element. Therefore, in addition to the definitions of asymptotic stability, local stability, finite stability, and global stability that were introduced in Section 5.20 in connection with the Liapunov stability criterion, we are

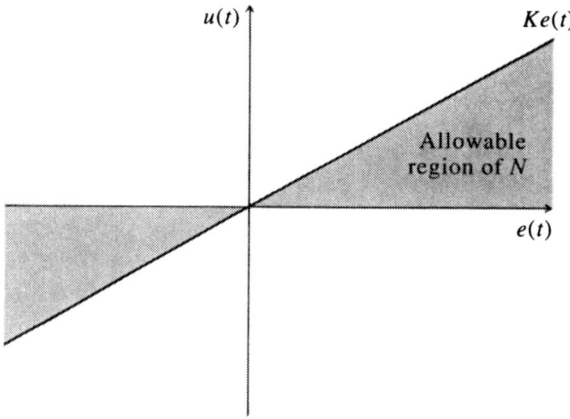

Figure 5.72 Restricted region of nonlinearity.

concerned here with control asymptoticity and output asymptoticity. *Control asymptoticity of degree n* exists if a real value n can be found such that for every set of initial conditions

$$\int_0^\infty [e^{nt} u(t)]^2 \, dt < \infty. \tag{5.241}$$

Output asymptoticity of degree n exists if a real value n can be found such that for every set of initial conditions

$$\int_0^\infty [e^{nt} e(t)]^2 \, dt < \infty. \tag{5.242}$$

These stability definitions can be clarified by considering the following lemma.

If the linear element $G(s)$ of Figure 5.71 is output stable of degree n, the input and output of the nonlinear element are bounded and satisfy Eq. (5.240), and the feedback system is control asymptotic of degree n, then

$$\lim_{t \to \infty} e^{nt} e(t) = 0. \tag{5.243}$$

Therefore, if this lemma is satisfied, $e(t)$ converges towards zero faster than e^{-nt} for $n > 0$.

How can we relate control asymptotic and output asymptoticity? Obviously there should be some relationship based on the properties of the linear element $G(s)$. Let us assume that the linear element is output stable of degree n. It can be shown that if the linear element is control asymptotic of degree n, then it is also output asymptotic of degree n. In addition, if for each set of initial conditions a number Q_0 exists such that

$$|e_0(t)| \leq Q_0 e^{-nt},$$

then there exists a number Q that is dependent on Q_0 such that

$$|e(t)| \leq Q e^{-nt}$$

for all values of t. This appears reasonable because a decaying control signal $u(t)$ that satisfies Eq. (5.241), when it is fed into the linear element whose unit-impulse response decays like $u(t)$, will produce an output $-e(t)$ that also decays in a similar manner.

Popov's fundamental theorem is based on the basic feedback control system illustrated in Figure 5.71. It is assumed that the linear system is output stable. The theorem states that for the feedback system to be absolutely control and output asymptotic for

$$0 \leq N[e(t)] \leq K,$$

it is sufficient that a real number q exists such that for all real $\omega \geq 0$ and an arbitrarily small $\delta > 0$, the following condition is satisfied:

$$\text{Re}[(1 + j\omega q) G(j\omega)] + 1/K \geq \delta > 0. \tag{5.244}$$

418 NONLINEAR CONTROL-SYSTEM DESIGN

The relation (5.244) is the Popov criterion. Depending on the type of nonlinearity present, the following restrictions on q and K are imposed:

(a) For a single-valued time-invariant nonlinearity:

$$-\infty < q < \infty \quad \text{if } 0 < K < \infty,$$
$$0 \leqslant q < \infty \quad \text{if } K = \infty.$$

(b) For a nonlinearity having passive hysteresis (see Figure 5.73):

$$-\infty < q \leqslant 0 \quad \text{and} \quad 0 < K < \infty.$$

(c) For a nonlinearity having active hysteresis (see Figure 5.74):

$$0 \leqslant q < \infty \quad \text{and} \quad 0 < K \leqslant \infty.$$

Examination of these three possible types of nonlinearities shows that the theorem allows for a tradeoff between the requirements on the nonlinear and linear elements.

Let us rewrite Eq. (5.244) as follows:

$$\operatorname{Re} G(j\omega) > -\frac{1}{K} + \omega q \operatorname{Im} G(j\omega). \tag{5.245}$$

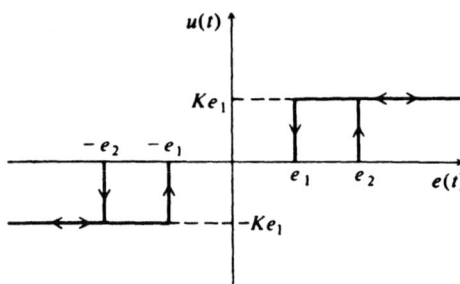

Figure 5.73 Passive hysteresis nonlinear characteristics.

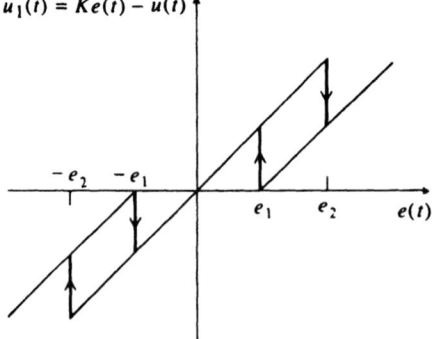

Figure 5.74 Active hysteresis nonlinear characteristics.

5.21. POPOV'S METHOD

Relation (5.245) states that for each frequency ω, the Nyquist plot of $G(j\omega)$ must lie to the right of a straight line given by

$$\operatorname{Re} G(j\omega) = -\frac{1}{K} + \omega q \operatorname{Im} G(j\omega). \tag{5.246}$$

This line is called the Popov line and is illustrated in Figure 5.75. The angles α and β are

$$\alpha = \tan^{-1} \omega q, \tag{5.247}$$

$$\beta = \tan^{-1} \frac{1}{\omega q}, \tag{5.248}$$

and it is clear that the slope of this line depends on the product ωq.

Stability depends on choosing a value of q such that, for each frequency ω, $G(j\omega)$ lies to the right of the Popov line. It is important to recognize from Eqs. (5.247) and (5.248) that the slope of this line is frequency dependent. A Popov line whose slope is not frequency dependent can be found in a modified frequency plane. In order to find the particular frequency-insensitive Popov line, a simple transformation is used. The modified frequency response function $G^*(s)$ is defined as

$$G^*(j\omega) \triangleq \operatorname{Re} G(j\omega) + j\omega \operatorname{Im} G(j\omega). \tag{5.249}$$

Therefore, Eq. (5.245) can be rewritten as

$$\operatorname{Re} G^*(j\omega) > -\frac{1}{K} + q \operatorname{Im} G^*(j\omega). \tag{5.250}$$

In the $G^*(j\omega)$ plane, the Popov line is defined by

$$\operatorname{Re} G^*(j\omega) = -\frac{1}{K} + q \operatorname{Im} G^*(j\omega),$$

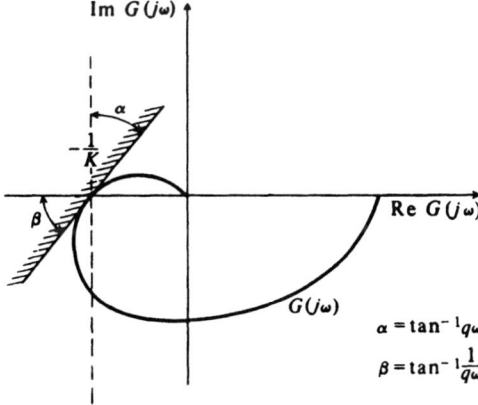

Figure 5.75 Popov's method when q is finite.

and is frequency insensitive. The Popov line in the $G^*(j\omega)$ plane is illustrated in Figures 5.76 and 5.77. The angle γ is defined as

$$\gamma = \tan^{-1} q.$$

Notice from Figures 5.76 and 5.77 that the $G^*(j\omega)$ locus passes to the right of a tangent to the locus at the point where $G^*(j\omega)$ intersects the negative real axis. These points are labeled $-1/K$. Therefore, K represents the maximum permissible gain for the system. For the case where $q = 0$, the Popov line expression reduces to

$$\operatorname{Re} G^*(j\omega) = -\frac{1}{K}$$

and the system is stable if it lies to the right of a vertical line passing through the point $-1/K_1$ as is illustrated in Figure 5.76. Notice that the case of $q = 0$ gives the most conservative value of gain permissible.

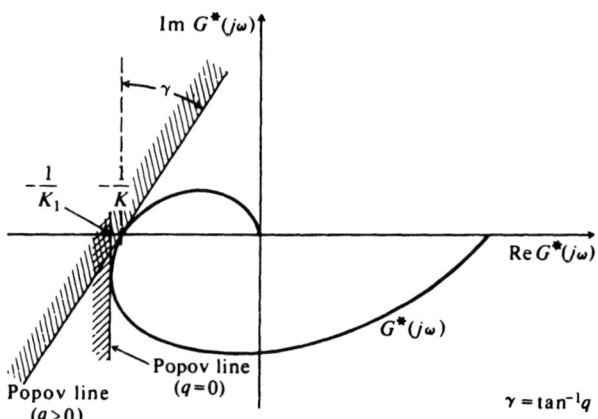

Figure 5.76 Popov line in the $G^*(j\omega)$ plane for the case where $q \geqslant 0$.

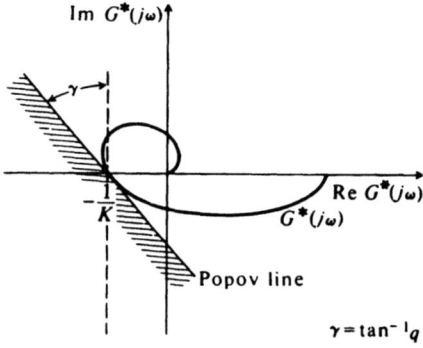

Figure 5.77 Popov line in the $G^*(j\omega)$ plane for the case where $q < 0$.

A. Example 5.21-1

As an example of applying Popov's method, consider the system illustrated in Figure 5.78. For the linear element, the initial condition response $e_0(t)$ is given by

$$e_0(t) = e_{10}e^{-t} + e_{20}e^{-2t} + e_{30}e^{-3t},$$

where e_{10}, e_{20}, and e_{30} depend on the initial conditions. The unit impulse response $g(t)$ is given by

$$g(t) = [0.5e^{-t} - e^{-2t} + 0.5e^{-3t}]U(t), \tag{5.251}$$

where $U(t)$ is a unit step function. Equation (5.251) indicates that the linear element is output stable and satisfies one of the necessary constraints in order to use Popov's method. The corresponding $G^*(j\omega)$-locus is illustrated in Figure 5.79. From this diagram, we can conclude that if the nonlinear element corresponds to a single-valued nonlinear element, and if $q = 0.5$, the Popov condition is satisfied when $0 < K \leqslant 60$.

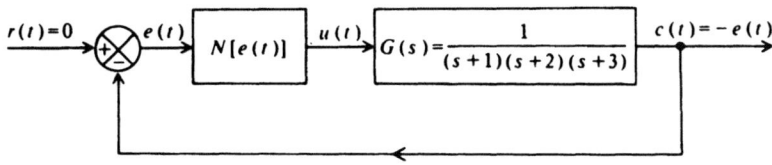

Figure 5.78 Nonlinear control-system example.

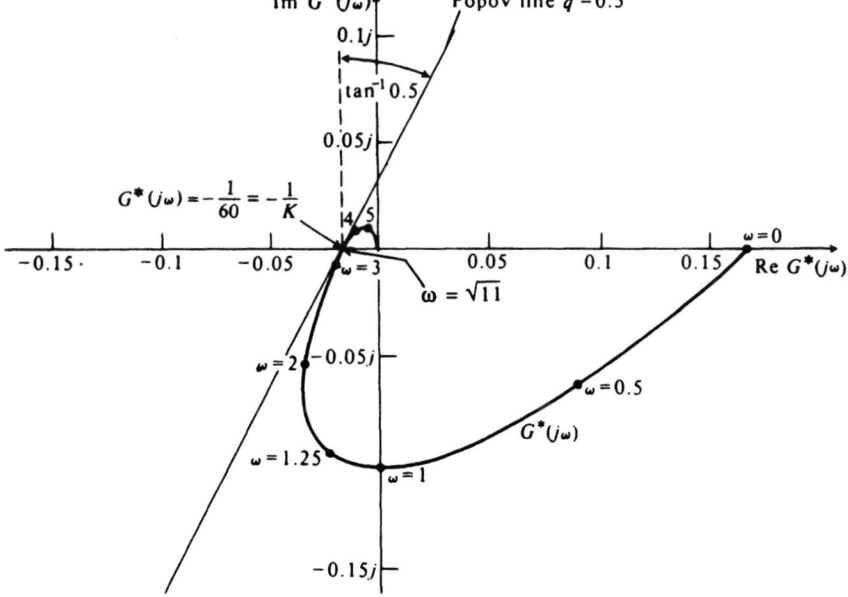

Figure 5.79 Popov analysis for system illustrated in Figure 5.78.

In conclusion, we see that the Popov method gives an exact and sufficient condition for determining the absolute stability of feedback systems having the configuration illustrated in Figure 5.71, with certain restrictions imposed. The inequality (5.244), which was given in terms of $G(j\omega)$ which makes the technique easily applicable to systems having high-order processes which are to be controlled. In addition, the method shares all of the desirable characteristics of the Nyquist method. In the following section, Popov's method is extended to other types of systems which are not necessarily restricted to systems in which the linear portion is output stable, and the nonlinearity is time invariant.

5.22. GENERALIZED CIRCLE CRITERION

The generalized circle criterion [17,18] enables ones to investigate the asymptotic behavior of a much wider class of systems than that for which Popov's theorem was originally intended. For example, this techique can be applied to systems having unstable or nonasymptotically stable plants, and time-variable nonlinearities. The generalized circle criterion presented in this section consists of modifying Popov's basic theorem in such a manner that the Popov condition can be applied directly to the original transfer function.*

Let us reconsider the basic nonlinear control system illustrated in Figure 5.71, and allow the nonlinearity to be time variable. It is assumed that the linear element of this system is not output stable. The generalized circle criterion is as follows: for the system of Figure 5.71 to be absolutely control and output asymptotic for

$$A < N[e(t), t] < B,$$

then it is sufficient that a real number q exist such that for all real ω, the following conditions are satisfied:

$$\left| G(j\omega) + \frac{(B+A) - j\omega q(B-A)}{2AB} \right|^2 - \left(\frac{B-A}{2AB} \right)^2 (1 + q^2\omega^2) \geq \delta > 0,$$

for $1/A > 1/B$ (5.252)

$$\left| G(j\omega) + \frac{(B+A) - j\omega q(B-A)}{2AB} \right|^2 - \left(\frac{B-A}{2AB} \right)^2 (1 + q^2\omega^2) \leq -\delta > 0,$$

for $1/A > 1/B$ (5.253)

The quantities $B - A$ and q are restricted as K and q were in Popov's method, discussed previously in Section 5.21.

Figure 5.80 illustrates the physical interpretation of relation (5.252), where it is assumed that $1/A > 1/B$. For each value of $\omega \geq 0$, the Nyquist plot of $G(j\omega)$ must lie outside the circle centered at[†]

[†]The circle criterion was originally developed only for the case of $q = 0$ [18]. However, the generalized circle criterion presented here is an extension which is valid for all values of q including zero [17].

$$-\frac{1}{2}\left(\frac{1}{A}+\frac{1}{B}\right)+\frac{1}{2}j\omega q\left(\frac{1}{A}-\frac{1}{B}\right), \tag{5.254}$$

which crosses the real axis at the points $-1/A$ and $-1/B$. It is interesting to note that if $1/B > 1/A$, then the Nyquist plot must lie inside the circle that is centered at the point given by Eq. (5.254).

Analysis of the generalized circle criterion is quite interesting. For example, if we let $A \to 0$ and $B \to K$, then relations (5.252) and (5.253) reduce to (5.245) which corresponds to the Popov condition (5.244). On the other hand, if we let $A \to C$ and $B \to C$, then we have a linear time-invariant system with gain C. For this case, the critical circle reduces to a point $-(1/C)$ in the $G(j\omega)$ plane, which is of course the critical point for the Nyquist diagram.

The generalized critical circle illustrated in Figure 5.80 and defined by Eqs. (5.252) and (5.253) is a function of frequency. Although all of the circles pass through the points $-1/A$ and $-1/B$, their centers move up with increasing values of $q\omega$. However, for the general nonlinearity case where the nonlinearity may contain hysteresis and is time variable, $q = 0$ and a set of circles results that are symmetrical about the real axis. Notice also from Figure 5.80 that tradeoffs can be made between the requirements on the linear and nonlinear elements. For example, by narrowing the sector

$$A < N[e(t), t] < B \tag{5.255}$$

the critical circles will be reduced and this will increase the permissible range of $G(j\omega)$.

Let us consider the application of the generalized circle criterion. Unlike the situation for the Popov line of Section 5.21, it is not advantageous to transform the critical circles from the $G(j\omega)$ plane into the $G^*(j\omega)$ plane, because this will only

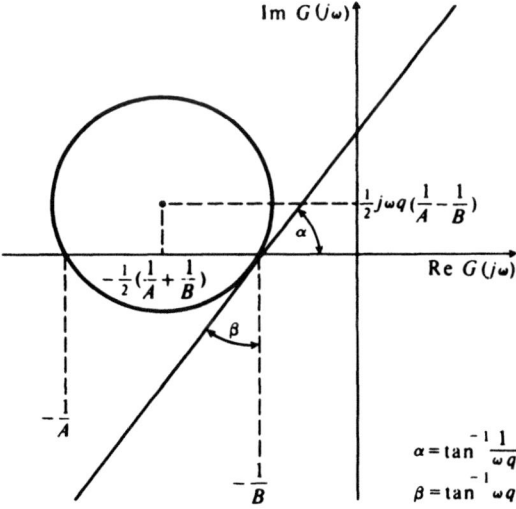

Figure 5.80 The generalized circle criterion for the case $1/A > 1/B$.

result in a family of curves that are not circles and whose shapes depend on both ω and q. However, it can easily be shown that if a tangent is drawn on the critical circle at the point $-1/B$, its angle α is given by

$$\alpha = \tan^{-1} \frac{1}{\omega q}. \qquad (5.256)$$

Figure 5.80 illustrates this angle and also the angle which is given by

$$\beta = \tan^{-1} \omega q. \qquad (5.257)$$

Comparing Figures 5.75 and 5.80, we observe that the tangent line on the critical circle has the same slope as the Popov line when $A = 0$. In addition, if

$$B = K$$

and

$$0 \leqslant A < B,$$

then the tangent line on the critical circle becomes identical to the Popov line shown in Figure 5.75. We also showed in Section 5.21 that the Popov line could be transformed into a frequency-independent line on the $G^*(j\omega)$ plane as was shown in Figures 5.76 and 5.77. Therefore, if $G^*(j\omega)$ lies to the right of the Popov line of Figures 5.76 and 5.77, then $G(j\omega)$ lies outside the critical circle illustrated in Figure 5.80 for all ω.

A. Example 5.22-1

As an example of the generalized circle criterion, let us consider a nonlinear control system that is time variable in the configuration illustrated in Figure 5.71, where the transfer function of the linear element is given by

$$G(s) = \frac{1}{(s-1)(s+2)(s+3)}, \qquad (5.258)$$

and the initial-condition response is

$$e_0(t) = e_{10}e^t + e_{20}e^{-2t} + e_{30}e^{-3t}, \qquad (5.259)$$

where e_{10}, e_{20}, and e_{30} are related to the initial conditions for a particular set of state variables. It is important to recognize that we are dealing with a nonlinear element that is time variable and a linear element that is not output stable, a problem which could not be solved using Popov's basic method. It is assumed that the nonlinear element corresponds to the general nonlinearity $N[e(t), t]$, and therefore q must be chosen equal to zero. The solution consists of plotting the frequency locus $G(j\omega)$ as is illustrated in Figure 5.81. The generalized circle criterion results in Popov sectors

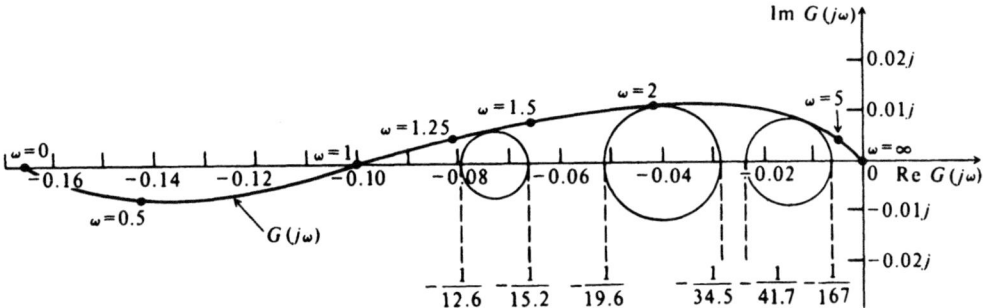

Figure 5.81 Generalized circle criterion example where

$$G(s) = \frac{1}{(s-1)(s+2)(s+3)}$$

and the nonlinearity corresponds to the general time variable nonlinearity case where $q = 0$.

$$A < N[e(t), t] < B \tag{5.260}$$

such that for each of these Popov sectors the $G(j\omega)$ locus lies on or outside a circle symmetrical about the real axis and passing through the points $-1/A$ and $-1/B$. For this particular example, the possible Popov sectors (illustrated in Figure 5.81) which result in stable systems are given by the following conditions:

$$\begin{aligned}
12.6 &< N[e(t), t] < 15.2, \\
19.6 &< N[e(t), t] < 34.5, \\
41.7 &< N[e(t), t] < 167.0.
\end{aligned} \tag{5.261}$$

Notice from Figure 5.81 that the size of the Popov sector, $B - A$, i.e., the difference between the lower bound A and upper bound B, depends on the size of the critical circle.

Therefore, we see that Popov's method can be extended to systems with unstable or nonasymptotically stable plants and time-varying nonlinearities by utilizing the generalized circle criterion. As shown in this section, the method permits the design of nonlinear systems whose linear elements are not output stable and whose nonlinearity can be time variable and correspond to any general nonlinearity. There are many practical situations where this is the very case involved, and the generalized circle criterion provides a very powerful method for solving this class of problems.

5.23. GUIDELINES FOR SELECTING THE "BEST" NONLINEAR CONTROL SYSTEM METHOD(S) PRESENTED FOR ANALYSIS AND DESIGN

Having learned the various methods available for analyzing nonlinear control systems in this chapter, how should the control-system engineer proceed for a specific problem? Which of the methods presented would you use? Unless guidelines are established for their use, uncertainty may exist as to the best approach(es) to be used in analyzing a nonlinear control system. This section provides guidelines for logically determining which analysis procedure(s) should be used in analyzing the stability of a particular nonlinear control system. These guidelines have been structured as a logic flow diagram, which is shown in Figure 5.82 [26]. This logic flow diagram should prove to be very valuable for integrating the various methods presented, and should prove useful to the student and the practicing control-system engineer.

In quasilinear systems, defined as those systems where the deviation from linearity is not too great, linear approximations may allow the use of conventional linear methods of analysis such as the Nyquist diagram, Bode diagram, or the root-locus method. This approach recognizes that certain control-system characteristics can change from one operating point to another, but it assumes that the control system is linear in the area of a specific operating point. For these control systems, the control-system engineer could use linear theory for analysis and design. This is the reason that linear theory has had such widespread use, although practical systems are never purely linear.

If the system cannot be approximated as a quasilinear system and linearizing approximations cannot be used, then we must use one or more of the nonlinear methods presented in this chapter for analysis and design.

If the linear and nonlinear portions of the system are time invariant, and the linear portion is stable (none of its roots are in the right half of the s-plane), then the describing function is recommended. Although it approximates a linearized transfer function by only considering the fundamental component of the output, it is a very powerful approach. Because of its approximation, it is essential that the method be checked by using an additional method. Simulation and the other nonlinear methods presented in this chapter can be used to check the describing function results.

If the nonlinear control system is second order, then the phase-plane and Liapunov methods are the most appropriate methods to use as a check. The Liapunov method could also be used as a check if the system were third order. If the system is third order or more, then Popov's method in the frequency domain can be used to determine asymptotic stability.

If the linear element is nonasymptotically stable, and the nonlinear element is time varying, then we can use the generalized circle criterion. This will determine allowable nonlinear gain ranges for system stability.

I always recommend that the nonlinear control system be simulated as a final check on system stability. It will assist in the check for factors which range from possible uncertainty regarding the validity of the assumptions, to analytic difficulties caused by system complexity. Simulation may also be necessary because of the failure of a technique to demonstrate stability conclusively. An example of this may be in the application of Liapunov's second method. The Liapunov condition for this method is sufficient, but not necessary, for stability. Therefore, failure to find a Liapunov function does not imply that the nonlinear control system is unstable. As noted in Figure 5.82, the simulation method is optional in some cases and, accordingly, it is shown as a dashed line for these cases.

5.23. GUIDELINES FOR SELECTING THE "BEST" NONLINEAR CONTROL SYSTEM METHOD

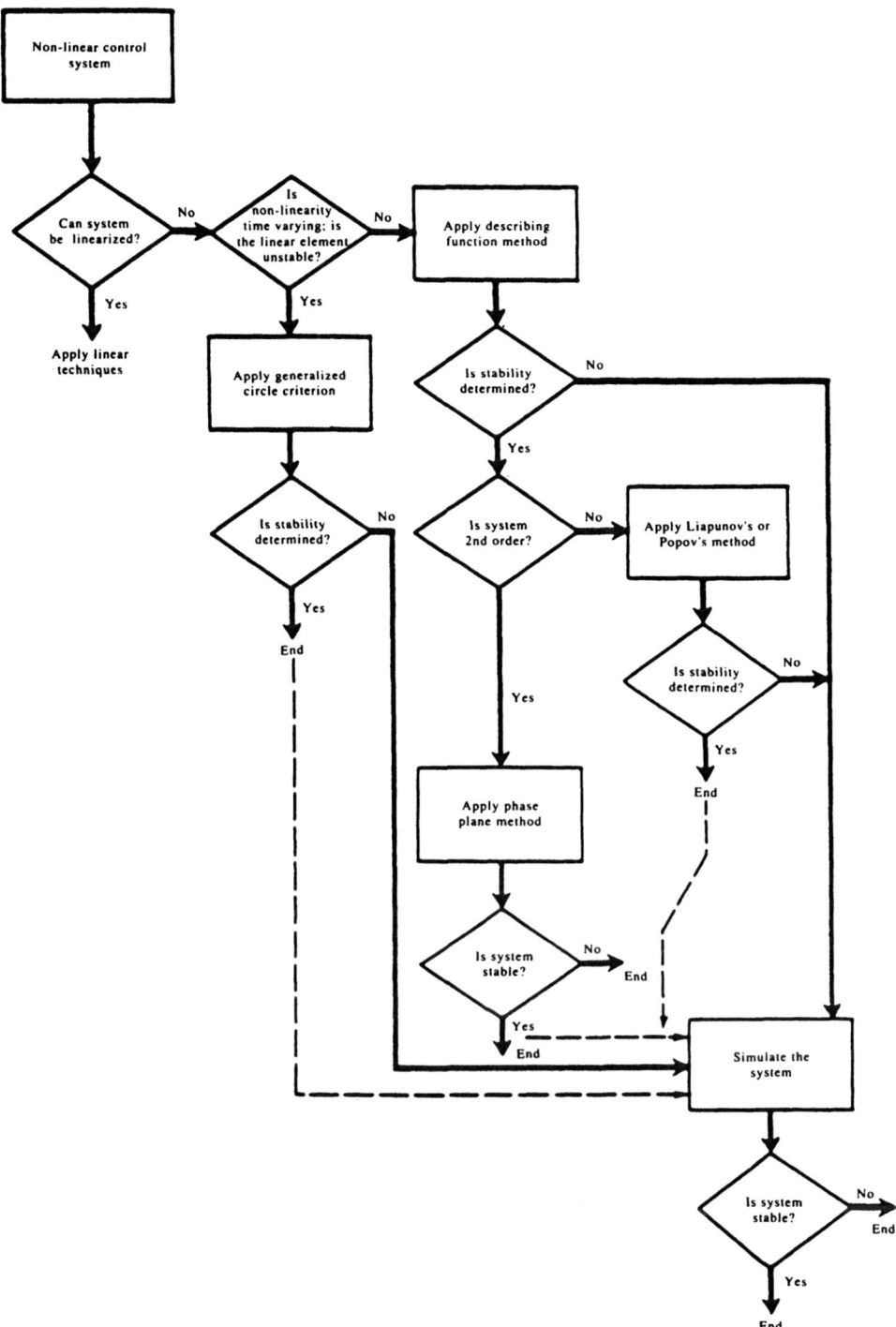

Figure 5.82 Logical approach for selecting method of analysis of nonlinear control systems [26]. (Reprinted from *Control Engineering*, October 1981, Copyright © 1981 Cahners Publishing Company).

428 NONLINEAR CONTROL-SYSTEM DESIGN

5.24. ILLUSTRATIVE PROBLEMS AND SOLUTIONS

This section provides a set of illustrative problems and their solutions to supplement the material presented in Chapter 5.

I5.1. A unity-feedback control system contains a linear element, $G(s)$, and a nonlinear element containing a dead zone. The transfer function of the linear element is given by

$$G(s) = \frac{3.56}{s(s+0.1)}$$

Assume that K_1 in Figure 5.6 equals 1 for the dead zone. Using the describing-function method on a gain-phase diagram, determine the existence of any limit cycles.

SOLUTION: The gain-phase diagram for this nonlinear control system is illustrated in Figure I5.1. It was obtained using MATLAB. The $-1/N$ curve was obtained from Figure 5.8. The gain-phase diagram shows that the system does not have a limit cycle, and it is stable. In addition, it can be concluded that the system is stable regardless of the size of the dead zone.

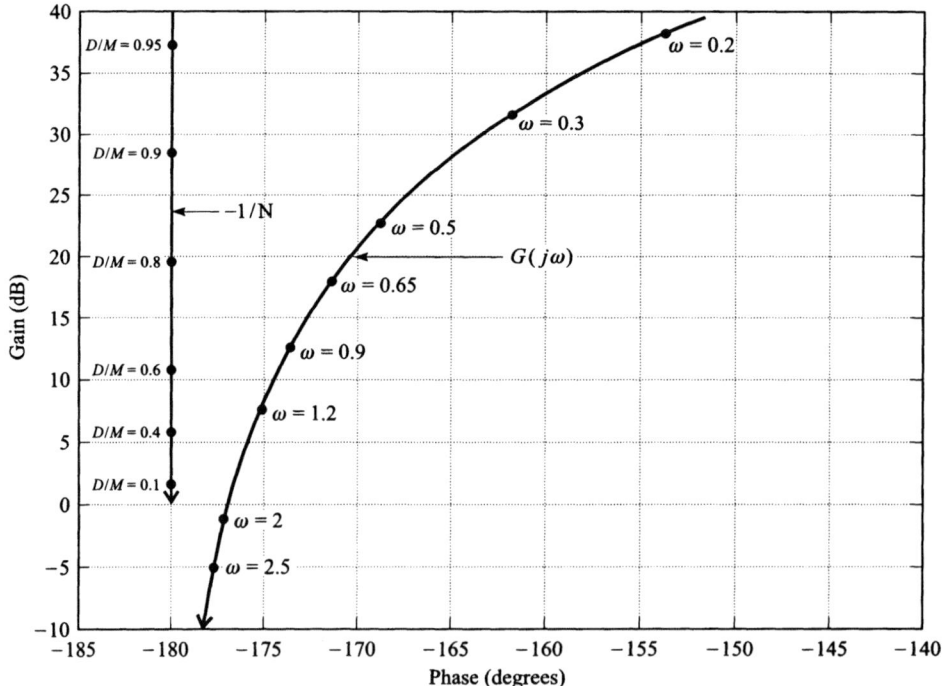

Figure I5.1

15.2. The following "bang-bang" control system represents the attitude-control system for an orbiting space satellite in which the reaction-jet moment $u(t)$ is applied in a "bang-bang" manner. Determine the state equations of this nonlinear system.

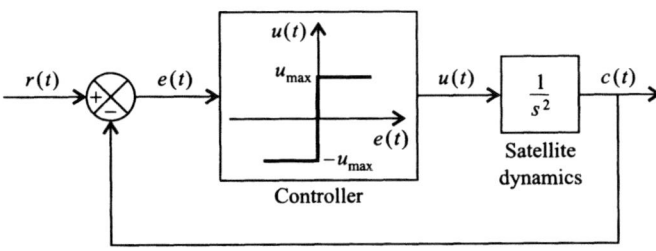

Figure I5.2

SOLUTION:
$$\ddot{c}(t) = u(t).$$

Let
$$x_1(t) = c(t),$$
$$x_2(t) = \dot{c}(t).$$

Therefore, the state equations are given by:
$$\dot{x}_1(t) = x_2(t),$$
$$\dot{x}_2(t) = u(t)$$

where the nonlinear relationship between $u(t)$ and $e(t)$ depends on the sign of $e(t) = r(t) - c(t)$ as follows:

$$u(t) = \frac{e(t)}{|e(t)|} u_{max} = \frac{r(t) - c(t)}{|r(t) - c(t)|} u_{max} = \frac{r(t) - x_1(t)}{|r(t) - x_1(t)|} u_{max}.$$

Therefore, the state equations are given by:
$$\dot{x}_1(t) = x_2(t), \qquad (15.2\text{-}1)$$
$$\dot{x}_2(t) = \frac{r(t) - x_1(t)}{|r(t) - x_1(t)|} u_{max}. \qquad (15.2\text{-}2)$$

15.3. The phase trajectory for the previous problem analyzed in Illustrative Problem I5.2 is to be determined.

(a) Determine the phase trajectory equation for this "bang-bang" control system by using the method of transforming the second-order differential equation to a first-order differential equation. Assume that $x_1(0)$ and

430 NONLINEAR CONTROL-SYSTEM DESIGN

$x_2(0)$ are the initial values of the states and the reference input is a constant equal to r_0.

(b) Draw the phase trajectory of this control system and determine what occurs when $x_1(t) < r_0$ and what occurs when $x_1(t) > r_0$.

SOLUTION: (a) Dividing Eq. (I5.2-2) by (I5.2-1) from Illustrative Problem I5.2, we obtain the following:

$$\frac{dx_2(t)}{dx_1(t)} = \left[\frac{r_0 - x_1(t)}{|r_0 - x_1(t)|} u_{\max}\right]\left[\frac{1}{x_2(t)}\right].$$

Separating the two variables, $x_1(t)$ and $x_2(t)$, and integrating we obtain the following:

$$\int_{x_2(0)}^{x_2(t)} x_2(t)\, dx_2(t) = \left[\frac{r_0 - x_1(t)}{|r_0 - x_1(t)|} u_{\max}\right]\int_{x_1(0)}^{x_1(t)} dx_1(t),$$

$$\left[\frac{x_2^2(t)}{2}\right]_{x_2(0)}^{x_2(t)} = \frac{r_0 - x_1(t)}{|r_0 - x_1(t)|} u_{\max}[x_1(t)]_{x_1(0)}^{x_1(t)}.$$

Therefore, the trajectory equation is given by:

$$\tfrac{1}{2}[x_2^2(t) - x_2^2(0)] = \left[\frac{r_0 - x_2(t)}{r_0 - x_1(t)} u_{\max}\right][x_1(t) - x_1(0)].$$

(b) When $x_1(t) < r_0$, then

$$x_2^2(t) = 2u_{\max} + \text{constant}.$$

When $x_1(t) > r_0$, then

$$x_2^2(t) = -2u_{\max} + \text{constant}.$$

Therefore, these two resulting trajectories represent two sets of parabolas. For the arbitrary initial conditions given, $x_1(0)$ and $x_2(0)$, the trajectories are sketched as follows and we observe that a limit cycle occurs:

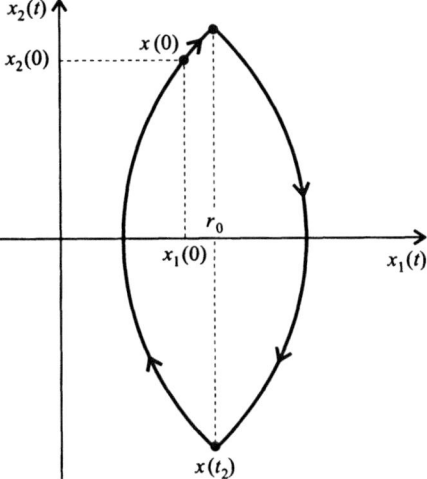

Figure I5.3

There is one additional important point to make in this illustrative problem. For the initial condition given at state $x(0)$, we note that $x_1(0)$ is less than r_0. The question arises as to whether the trajectory leaves the initial state, $x_1(0)$, and moves to the right or left. This question can be answered by reconsidering the state equation of this system given in Illustrative Problem I5.2 by Eq. (I5.2-1):

$$\frac{dx_1(t)}{dt} = x_2(t)$$

Because both $x_2(0)$ and dt are positive, then $dx_1(t)$ must also be positive and the motion along the trajectory must be to the right.

At $x(t_1)$, $x_1(t)$ now equals r_0 and as $x_1(t)$ becomes greater than r_0, the relay switches and we transfer to the new trajectory. The motion along this new trajectory continues until we reach state $x(t_2)$ and we again switch back to the original trajectory. The system then returns to its original state at $x(0)$. This periodic oscillation, or limit cycle, will continue unless the reference input r_0 changes.

I5.4. The phase plane for the nonlinear control system shown in Figure I5.4 containing a nonlinear element containing a dead zone and saturation is to be determined using the method of isoclines.

(a) Determine the general expression for the isocline equations. Assume the reference input, $r(t)$ is a constant r_0.

(b) Plot the phase trajectory for $e(0) = 2$ using the isocline method.

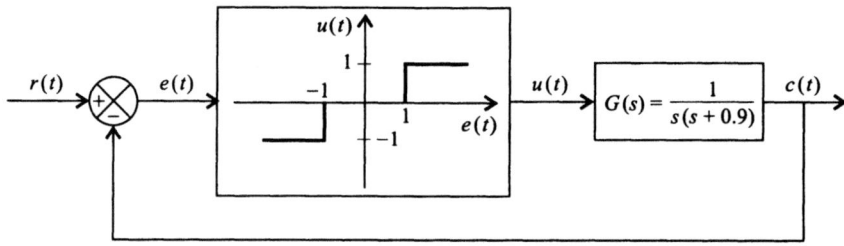

Figure 15.4(i)

SOLUTION: (a)
$$\frac{C(s)}{U(s)} = \frac{1}{s^2 + 0.9s}$$

Therefore,
$$\ddot{c}(t) + 0.9\dot{c}(t) = u(t)$$

where
$$u(t) = N(e(t)) = N[r_0(t) - c(t)]$$

where
$$N[e(t)] = +1 \quad \text{for } e(t) > 1,$$
$$N[e(t)] = 0 \quad \text{for } -1 < e(t) \leq 1,$$
$$N[e(t)] = -1 \quad \text{for } e(t) < -1.$$

Therefore,
$$\ddot{c}(t) + 0.9\dot{c}(t) = N[r_0 - c(t)].$$

Because $N(e(t))$ is a function of $e(t)$ rather than $c(t)$, we will define the states in terms of $e(t)$ and $\dot{e}(t)$ rather than $c(t)$ and $\dot{c}(t)$. Therefore, let the states be defined as:

$$x_1(t) = e(t) = [r_0(t) - c(t)],$$
$$x_2(t) = \dot{e}(t) = -\dot{c}(t).$$

Therefore, the state equations are given by:
$$\dot{x}_1(t) = x_2(t),$$
$$\dot{x}_2(t) = -0.9x_2(t) - N[x_1(t)].$$

Therefore, the isocline equations are given by:
$$\frac{dx_2(t)}{dx_1(t)} = \frac{-0.9x_2(t) - N[x_1(t)]}{x_2(t)} = -\frac{0.9x_2(t) + N[x_1(t)]}{x_2(t)}. \tag{15.4-1}$$

(b) *Region 1*: If $e(t) = x_1(t) > +1$, then $N[e(t)] = N[x_1(t)] = 1$. In this case the isocline equation, Eq. (I5.4-1) is given by

$$\frac{dx_2(t)}{dx_1(t)} = -\frac{0.9x_2(t) + 1}{x_2(t)}.$$

Because the isocline equation is independent of $x_1(t)$, the isoclines are all horizontal lines.

Region 2: If $-1 < e(t) < 1$, then $N[e(t)] = N[x_1(t)] = 0$. In this case the isocline equation, Eq. (I5.4-1), is given by

$$\frac{dx_2(t)}{dx_1(t)} = -\frac{[0.9x_2(t) + 0]}{x_2(t)} = -0.9.$$

In Region 2, all trajectories have a slope of -0.9.

Region 3: If $e(t) = x_1(t) < -1$, then $N[e(t)] = N[x_1(t)] = -1$. In this case the isocline equation, Eq. (I5.4-1), is given by

$$\frac{dx_2(t)}{dx_1(t)} = \frac{-0.9x_2(t) - 1}{x_2(t)}.$$

As in Region 1, since the isoclines are independent of $x_1(t)$, the isoclines are all horizontal lines [see Figure I5.4(ii)].

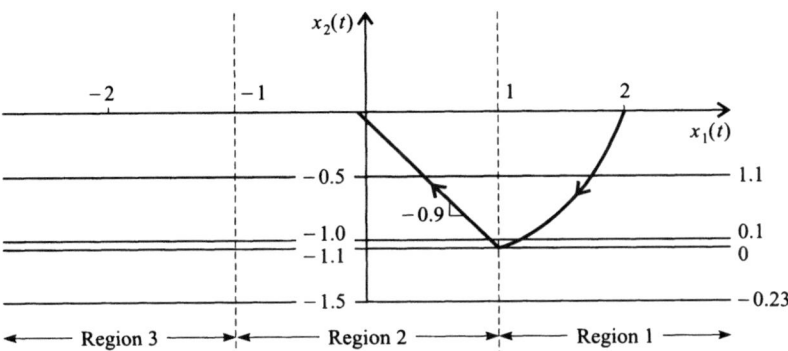

Figure I5.4(ii)

I5.5. In the following nonlinear control system, the control input, $u(t)$, is proportional to the error and it contains a nonlinear component $f(e)$ depending on the derivative of the error as follows:

$$u(t) = 2e(t) + f(e(t))\frac{de(t)}{dt}$$

434 NONLINEAR CONTROL-SYSTEM DESIGN

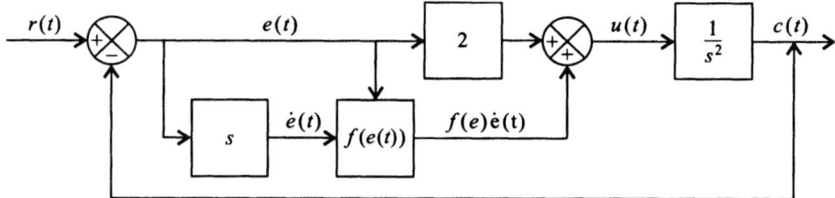

Figure 15.5

(a) Determine the state equations of this nonlinear control system. Assume that the reference input, $r(t)$, is constant. Define the state variables $x_1(t)$ and $x_2(t)$ in terms of $e(t)$ and $de(t)/dt$, respectively, in this problem.

(b) Determine whether the system is stable using Liapunov's second method.

SOLUTION: (a) For the linear portion of this control system,

$$2 \frac{d^2 c(t)}{dt^2} = u(t).$$

Substituting the value of $u(t)$ given in the problem, we obtain the following:

$$\frac{d^2 c(t)}{dt^2} = \frac{1}{2}\left[2e(t) + f(e(t)) \frac{de(t)}{dt}\right].$$

Since $e(t) = r(t) - c(t)$ and since $r(t)$ is a constant, we can rewrite this equation as:

$$\frac{d^2 e(t)}{dt^2} = -e(t) - \frac{1}{2} f(e(t)) \frac{de(t)}{dt}.$$

Defining the states as $x_1(t) = e(t)$ and $x_2(t) = de(t)/dt$, the state equations are given by:

$$\frac{dx_1(t)}{dt} = x_2(t) \qquad (15.5\text{-}1)$$

$$\frac{dx_2(t)}{dt} = -x(t) - \frac{1}{2} f(x_1(t)) x_2(t). \qquad (15.5\text{-}2)$$

(b) Let the energy function $V(t)$ be given by

$$V(t) = x_1^2(t) + x_2^2(t)$$

Therefore,

$$\frac{dV(t)}{dt} = 2x_1(t) \frac{dx_1(t)}{dt} + 2x_2(t) \frac{dx_2(t)}{dt}. \qquad (15.5\text{-}3)$$

Substituting the state equations (I5.5-1) and (I5.5-2) into the derivative of the energy function, Eq. (I5.5-3), we obtain the following:

$$\frac{dV(t)}{dt} = -f(x_1(t))x_2^2(t).$$

Because

$$x_2^2(t)$$

is always positive, then

$$\frac{dV(t)}{dt} < 0$$

and the system is asymptotically stable if

$$f(x_1(t)) > 0.$$

PROBLEMS

5.1. The torque speed characteristics of a two-phase, ac instrument servomotor are illustrated in Figure P5.1. Assume that the inertia of the rotor is 0.1 oz in^2 and that the load inertia and coefficient of viscous friction are negligible.

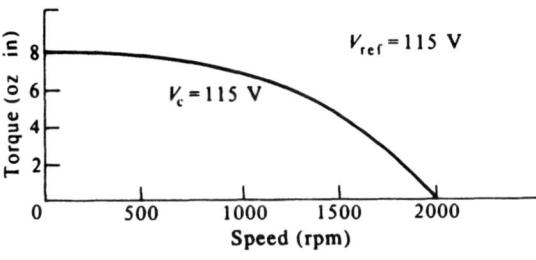

Figure P5.1

(a) Derive the transfer function of the motor, relating output position to control voltage using linearizing approximations. Approximate the characteristics by one straight line that is tangent to the exact characteristics at 1000 rpm, and by one straight line that goes through the two endpoints. Compare your results.

(b) How do the time constants derived in (a) change if the torque speed characteristics are approximated by two other straight lines: one at low speed that is tangent to the exact characteristics at 250 rpm, and one at high speed that is tangent to the exact characteristics at 1750 rpm.

436 NONLINEAR CONTROL-SYSTEM DESIGN

(c) Comparing your answers to (a) and (b), what conclusions can you reach?

5.2. Derive the describing function corresponding to the combined nonlinear characteristics of saturation S and dead zone D.

5.3. Derive the describing of a two-position contactor which does not exhibit any hysteresis effect.

5.4. An amplifying device in a feedback loop has the following nonlinear characteristics:

(a) No output signal for all inputs whose magnitude is less than E_1 V.

$$e_0(\omega t) = 0 \quad \text{when } |e_i(\omega t)| < |E_1|.$$

(b) Input signals whose magnitude is greater than E_1 V but less than E_2 V are amplified according to the relation

$$e_g(\omega t) = K[e_i(\omega t) - E_1],$$

where

$$|E_1| < |e_i(\omega t)| < |E_2|.$$

(c) For all input signals whose magnitude is greater than E_2 V, the output is given by

$$e_0(\omega t) = KE_2,$$

when

$$|e_i(\omega t)| > |E|_2.$$

Sketch the input–output characteristics and derive the describing function.

5.5. Derive the describing function for an amplifying device which has the nonlinear characteristics illustrated in Figure P5.5.

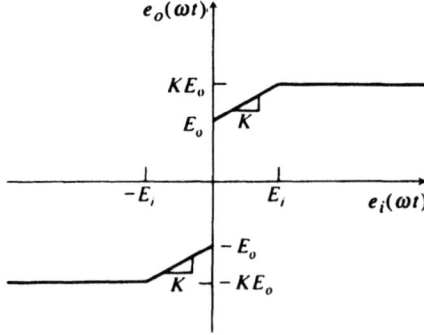

Figure P5.5

5.6. The characteristics of Figure P5.6 represent those of an on-off element (relay) having a dead zone D present. Determine the describing function for these relay characteristics assuming that the input to the nonlinear element is a sinusoid of $m(\omega t) = M \sin \omega t$. Reduce your final answer so that it is only in terms of the nonlinear relay characteristics D and K_1, and M.

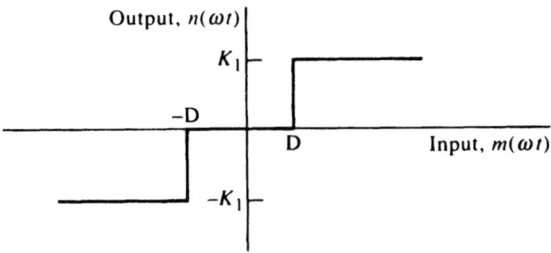

Figure P5.6

5.7. Determine the describing function for the nonlinear characteristic shown in Figure P5.7 consisting of a nonlinear dead zone and a linear gain, where the dead zone is represented by D and the gain is represented by K.

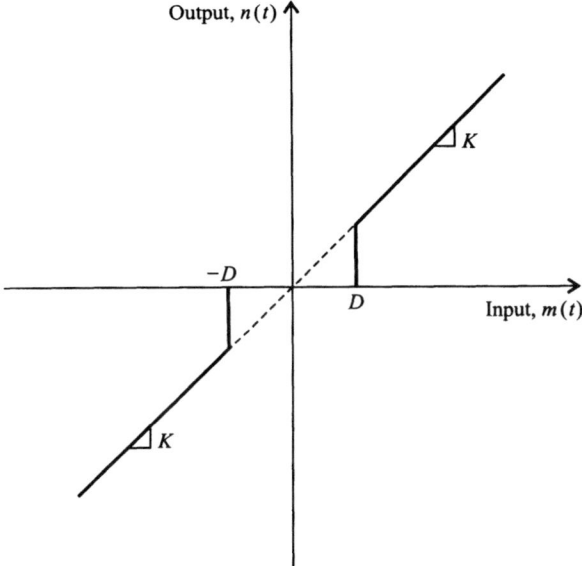

Figure P5.7

5.8. Determine the describing function for the nonlinear characteristic shown in Figure P5.8 consisting of a nonlinear Coulomb friction and linear viscous friction. The Coulomb friction level, which opposes velocity, is represented by F_c, and the linear vicous friction is represented by B. The input to this nonlinear characteristic is velocity, and the output is represented by friction (torque).

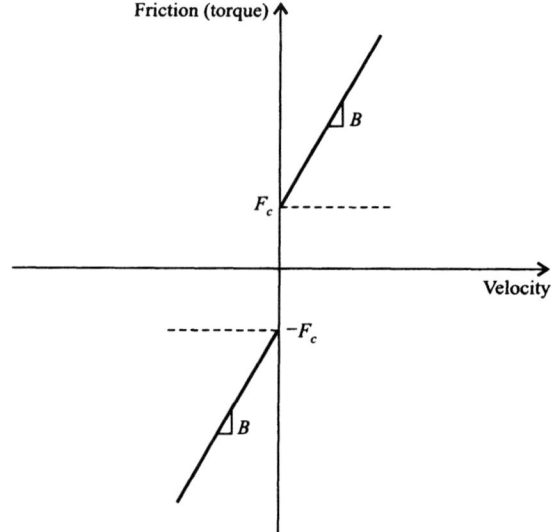

Figure P5.8

5.9. An instrument servo system used for positioning a load may be adequately represented by the block diagram shown in Figure P5.9, which indicates that the system contains backlash.

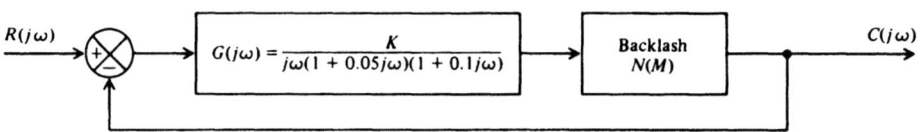

Figure P5.9

(a) Using the gain-phase diagram, determine the existence of any limit cycles if $K = 25$.

(b) What is the maximum gain that can be tolerated in $G(j\omega)$ before a limit cycle occurs?

5.10. In contrast to the on-off element having dead-zone and hysteresis characteristics present, as analyzed in Figure 5.17, an ideal relay has no dead-zone or hysteresis characteristics.

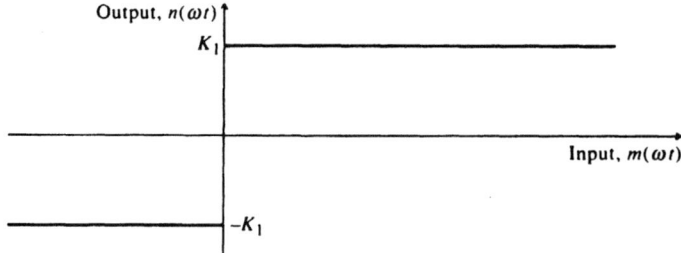

Figure P5.10

(a) For the ideal relay characteristics shown in Figure P5.10, determine its describing function analytically, assuming that the input to the nonlinear element is a sinusoid of $M \sin \omega t$.

(b) Check your answer by reducing Eq. (5.62) for the conditions of zero dead zone and hysteresis [A_1 of Eq. (5.61) now becomes zero]. Do your answers agree?

5.11. Utilizing the result of your derivation of the describing function for the ideal relay characteristics in Problem 5.10, consider the stability of the control system shown in Figure P5.11. For the value of $G(j\omega)$ indicated, determine the existence of any limit cycles and whether they are stable or unstable. Assume $K_1 = 1$.

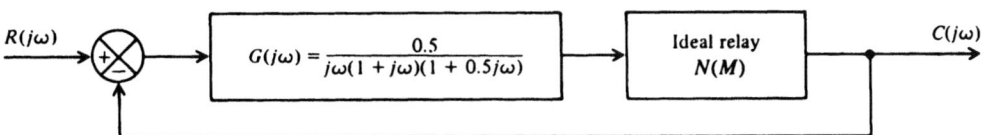

Figure P5.11

5.12. Repeat Problem 5.11 for

$$G(j\omega) = \frac{16.2}{j\omega(0.1 + j\omega)(0.385 + j\omega)}.$$

Assume $k_1 = 1$.

5.13. Assume that the block diagram of a load-positioning control system, which contains backlash, is represented as in Figure P5.9 and that the linear portion's transfer function is given by

$$G(j\omega) = \frac{20}{j\omega(1 + 0.5j\omega)^2}.$$

(a) Using the gain-phase diagram, determine the existence of any limit cycles. Are the limit cycles stable or unstable?

(b) Determine the gain which can permit a stability margin of 6 dB.

5.14. Repeat Problem 5.13 for

$$G(j\omega) = \frac{10}{j\omega(1 + 0.3j\omega)(1 + 0.1j\omega)}.$$

Use a stability margin of 3 dB in part (b).

5.15. Repeat Problem 5.13 for

$$G(j\omega) = \frac{15}{j\omega(1 + 0.3j\omega)(1 + 0.1j\omega)}.$$

5.16. A unity-feedback control system contains a nonlinear element containing backlash, $N(M)$, and a linear device whose transfer function is given by

$$G(s) = \frac{8}{s(1 + 0.4s)^2}.$$

(a) Using the gain-phase diagram, determine the existence of any limit cycles. If any limit cycles exist, determine the frequency ω and D/M at the intersection(s).

(b) Determine the gain which can permit a stability margin of 6 dB.

(c) It is desired to stabilize this control system by the introduction of rate feedback as shown in Figure 5.30. Determine the rate-feedback constant b which will permit a stability margin of 6°.

5.17. A unity-feedback system consists of cascaded elements which include a relay, pure integration, an amplifier which saturates, a two-phase ac servomotor, and a gear train containing backlash.

(a) Draw the block diagram and show the linear transfer functions and the describing functions for the nonlinear elements, symbolically.

(b) Qualitatively illustrate how you would predict the presence of an oscillation on a gain-phase diagram.

5.18. A unity-feedback instrument servo with a spring-loaded shaft has a dead zone of 2°. Assume that K_1 in Figure 5.6 equals 1. The transfer function for the linear portion of the system is given by

$$G(j\omega) = \frac{20}{j\omega(j\omega + 0.05)(j\omega + 0.1)}.$$

Utilizing the describing-function method on a gain-phase diagram, determine the conditions necessary for the existence of a limit cycle.

5.19. An instrument servo contains a three-position contactor which has a pull-in point and a drop-out point at errors of 0.02 and 0.01 rad, respectively. The transfer function of the linear portion of the unity-feedback control system is given by

$$G(j\omega) = \frac{10}{j\omega(j\omega + 0.1)}.$$

Determine whether a limit cycle exists using the describing-function analysis on a gain-phase diagram. Assume that K_1 in Figure 5.17 equals 1.

5.20. Repeat Problem 5.18 with the transfer function for the linear portion of the system given by

$$G(j\omega) = \frac{2}{j\omega(j\omega + 0.1)}.$$

5.21. A unity-feedback instrument servo is driven by an amplifier which saturates at 70% of rated voltage of the motor. Assume that the gain of the unsaturated amplifier is 40. The transfer function of the linear portion of the system, excluding the amplifier, is given by

$$G(j\omega) = \frac{0.25}{j\omega(j\omega + 2)}.$$

Utilizing the describing-function method on a gain-phase diagram, determine whether a limit cycle exists.

5.22. An instrument servo contains a three-position contactor which has a pull-in point and a drop-out point at errors of 0.03 and 0.01 rad, respectively. The transfer function of the linear portion is given by

$$G(j\omega) = \frac{10}{j\omega(j\omega + 4)}.$$

Determine whether a limit cycle exists utilizing the describing-function analysis on a gain-phase diagram. Assume K_1 in Figure 5.17 equals 1.

5.23. Repeat Problem 5.22 with the transfer function of the linear portion given by

$$G(j\omega) = \frac{1.4}{j\omega(j\omega + 4)}.$$

5.24. The rate-control system shown in Figure P5.24, using a tachometer for feedback, has a nonlinear element containing a dead zone of 2°. Assume that K_1 in Figure 5.6 equals 1. The transfer function of the linear element can be approximated as

$$G(s) = 10/s^2.$$

Using the describing-function method on a gain-phase diagram, determine whether any limit cycles occur.

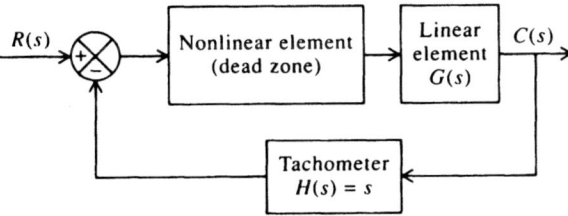

Figure P5.24

5.25. The Orbiting Astronomical Observatory (OAO), shown in Figure P5.25a, is designed to provide astronomers with a standardized stable platform in space readily adaptable to a variety of experiments [42]. Orbiting at 500 miles altitude, the OAO permitted scientists to examine any point in the heavens, unimpeded by the Earth's atmosphere. In addition, it has a precision and stability unsurpassed even by a stationary observatory. After injection into orbit and separation of the satellite from the second booster stage, the high-thrust gas jets stop any tumbling or rolling sensed by rate gyros and align the

(a)

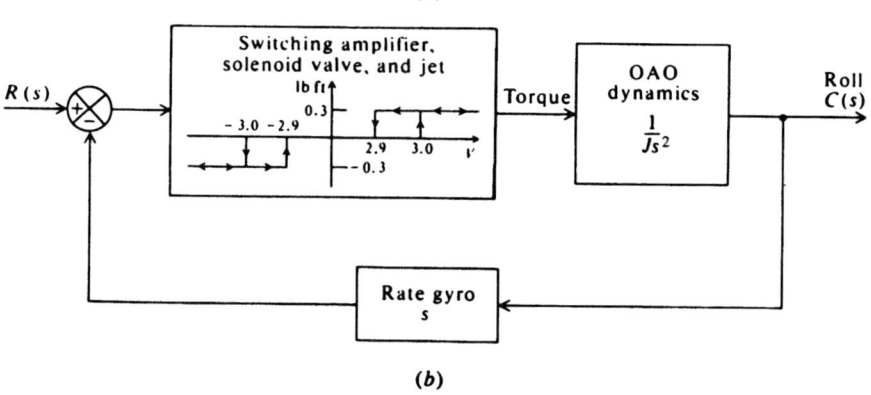

(b)

Figure P5.25 (a) NASA's OAO being checked out at Cape Kennedy. (Official NASA photo) (b) Equivalent block diagram of the roll coarse solar orientation control loop.

OAO's optical axis with the Earth-Sun line to $\pm\frac{1}{4}°$. The OAO utilizes coarse and fine solar orientation control systems. The coarse loop depends on gas jets firing and is a nonlinear control system. The fine loop depends on a momentum-exchange wheel and is a linear control system. Let us consider the nonlinear characteristics of the coarse solar orientation control system in this problem. Figure P5.25b illustrates an equivalent block diagram of the roll coarse solar orientation control loop. Basically, it consists of a switching amplifier which has very similar characteristics to the nonlinear on-off element having hysteresis that has been analyzed previously and which controls a solenoid valve and jet. As indicated in Figure P5.25b, the jet fires when the error reaches ± 3 V and stops firing when the error is reduced to ± 2.9 V. Assume that the resultant corrective torque produced by the jet is ± 0.3 lb ft, and the inertia in the roll axis is 1000 lb ft sec^2. The rate gyro which closes this rate loop, has a sensitivity of 10 V/(degree/sec). Using the describing-function analysis on a gain-phase diagram, determine the existence of any limit cycles.

5.26. An instrument servo system used for positioning a load may be adequately represented by the block diagram shown in Figure P5.26.

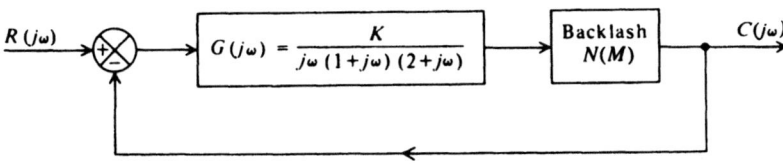

Figure P5.26

(a) Using the gain-phase diagram, prove that if $K = 4$ the system exhibits a limit cycle(s). Determine the range of frequencies and D/M ratios for which the limit cycle(s) exist. Are they stable or unstable?

(b) It is desired to stabilize this configuration by reducing the system gain. Determine the maximum value of gain K which will allow a minimum stability margin of 2 dB.

(c) It is desired to stabilize this configuration by adding a phase-lead network in cascade with $G(j\omega)$. Determine the time constants of the lead network

$$\frac{1 + j\omega\alpha T}{1 + j\omega t}$$

which will permit a minimum stability of 5°.

(d) It is desired to stabilize this configuration by the introduction of rate feedback as shown in Figure 5.30. Determine the rate-feedback constant b which will permit a minimum stability margin of 6°.

5.27. Repeat Problem 10.26 for

$$G(j\omega) = \frac{K(1+3j\omega)}{j\omega(1+2.1j\omega)^2}.$$

5.28. Repeat Problem 10.26 for

$$G(j\omega) = \frac{K(1+5j\omega)}{j\omega(1+1.67j\omega)^2}.$$

For part (d) to this problem, design for a minimum stability margin of 5°.

5.29. The control system shown in Figure P5.29 uses an ideal relay to apply a corrective signal of the proper phase to control the position of a servomotor. This type of nonlinear control system, which does not use an amplifier, is popularly known as a "bang-bang" control system.

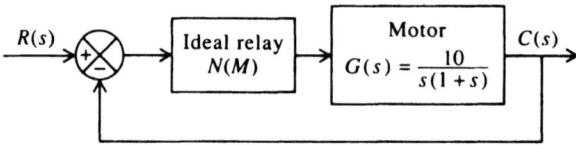

Figure P5.29

Using the result of your derivation of the describing function for the ideal characteristics in Problem 5.10 [or by reducing Eq. (5.62) of the textbook for the condition of zero dead zone and hysteresis], consider the stability of the control system shown. Determine the existence of any limit cycles, and whether they are stable or unstable. Assume $K_1 = 1$.

5.30. A positioning system consists of a major feedback path and three minor feedback paths as shown in Figure P5.30. Utilizing the signal-flow graph and describing-function approaches, determine whether any limit cycles exist if the nonlinearity corresponds to backlash and the system parameters correspond to the following values:

$$G_1(s) = \frac{2(1+10s)}{s(1+s)}, \qquad H_1(s) = 1.2,$$

$$G_2(s) = \frac{100(0.1s+1)(0.2s+1)}{(s+1)(0.5s+1)}, \qquad H_2(s) = 0.0282s,$$

$$G_3(s) = \frac{200(0.04s+1)}{s(0.02s+1)}, \qquad H_3(s) = 10s.$$

$$G_4(s) = \frac{4}{(0.1s+1)},$$

$$G_5(s) = \frac{5}{s},$$

$$G_6(s) = \frac{0.001}{s}.$$

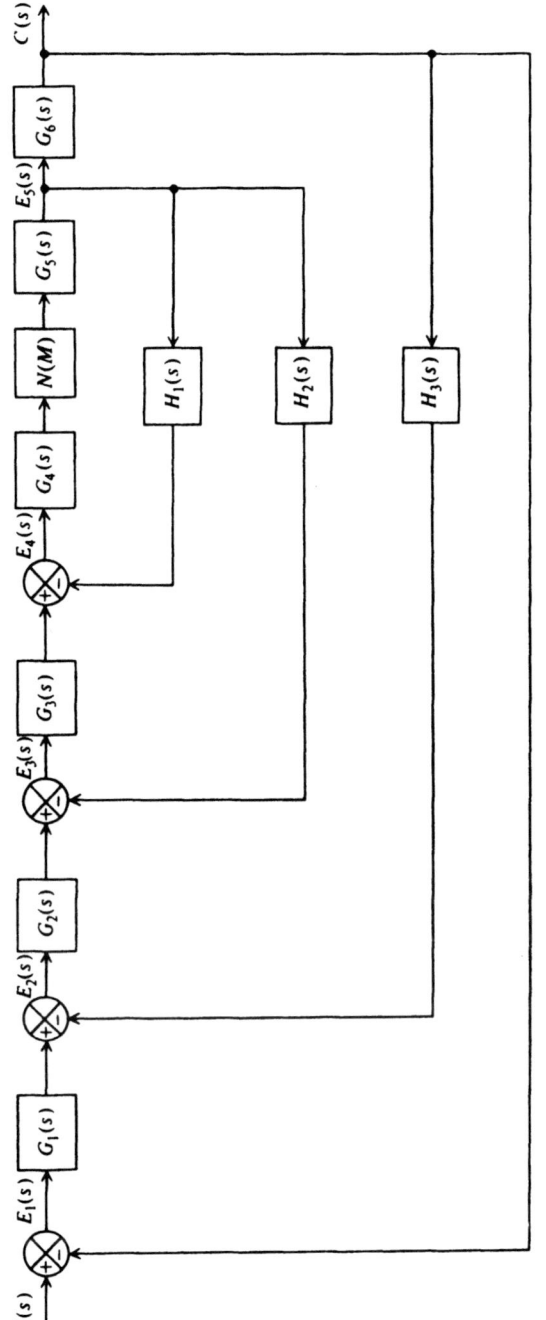

Figure P5.30

446 NONLINEAR CONTROL-SYSTEM DESIGN

5.31. Repeat Problem 5.30 with the nonlinearity corresponding to Coulomb friction.

5.32. In 1965 the Ranger unmanned space vehicle, shown in Figure P5.32a investigated the surface of the moon by means of a TV camera and instruments.

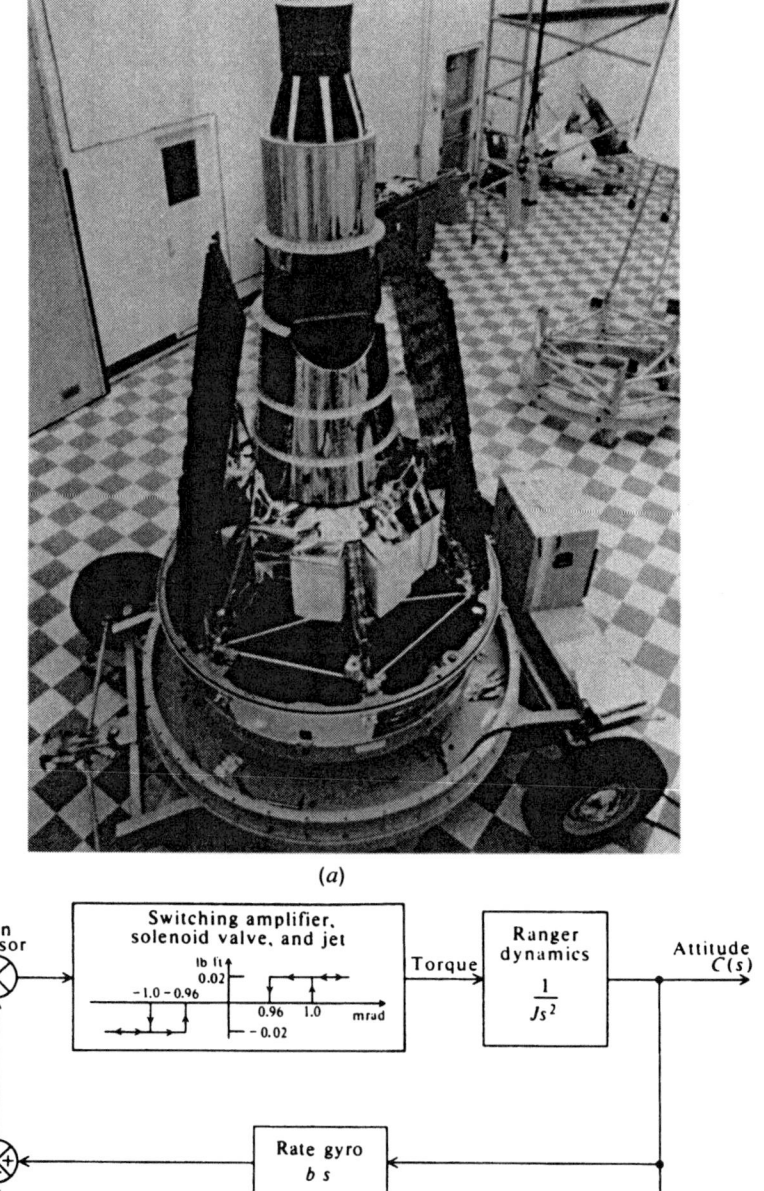

Figure P5.32 (a) Ranger spacecraft being checked out at Cape Kennedy. (Official NASA photo) (b) Equivalent block diagram of the Ranger attitude control system for the pitch axis.

From a control-system viewpoint, the Ranger attitude-control system must stabilize the vehicle from second-stage separation until lunar encounter [43]. The accuracy requirements are especially high, because the Ranger vehicle is an unmanned vehicle and its solar panels must point accurately at the Sun in order to obtain energy. In addition, the Ranger vehicle transmits data back to Earth by means of a narrow-beam antenna which requires very accurate pointing. The equivalent block diagram of the Ranger attitude-control system for the pitch axis is illustrated in Figure P5.32b. Error signals in position, generated by a sensor, are added to velocity error signals generated by a rate gyro. A switching amplifier, which has very similar characteristics to the nonlinear on-off element having hysteresis that has been analyzed previously, controls a solenoid valve and jet. As indicated in Figure P5.32b, the jet fires when the dead-zone error is equivalent to ± 1 mrad and stops firing when the error decreases to ± 0.96 mrad. Assume that the corrective torque produced by the jet is ± 0.02 lb ft, and the inertia of Ranger in the pitch axis is 110 lb ft sec^2. Utilizing the describing-function method, determine the existence of any limit cycles.

5.33. The block diagram of a load positioning control system containing backlash is shown in Figure P5.33a.

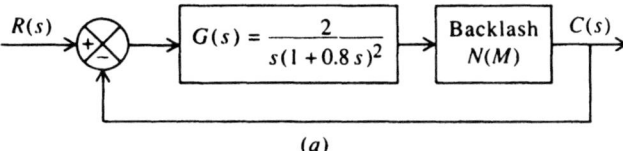

(a)

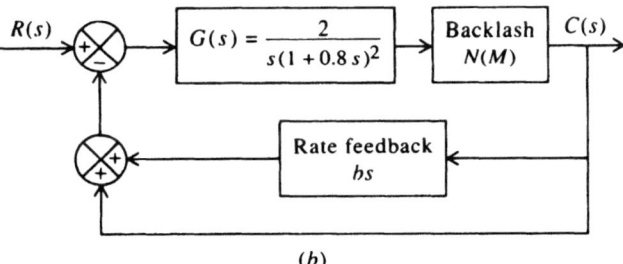

(b)

Figure P5.33

(a) Using the describing-function analysis (on a gain-phase diagram), determine the existence of any limit cycles.

(b) It is desired to stabilize this configuration by the introduction of rate feedback as shown in Figure P5.33b. Design the rate-feedback constant b for a minimum margin of stability of $20°$.

5.34. Repeat Problem 5.33 for

$$G(s) = \frac{3}{s(1+s)^2}.$$

In part b of this Problem, design the rate feedback constant b for a minimum margin of stability of 12°.

5.35. The control system of Figure P5.35 illustrates an undamped servo, where the motor torque is proportional to the error $e(t)$ and is not affected by velocity. Consider the composite gain of the forward part of the loop to be 800 dyne cm per radian of error.

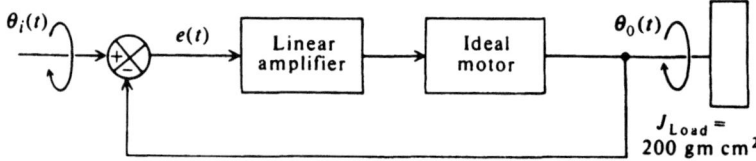

Figure P5.35

(a) Draw the phase trajectory for the following set of initial conditions:

$\theta_0(t) =$ initial output displacement (in rad) $= 1$ rad

$\dot{\theta}_0(t) =$ initial output velocity (in rad/sec) $= 0$.

(b) What conclusions can you reach from the phase trajectory?

5.36. The control system shown in Figure P5.36 illustrates a positioning servo system.

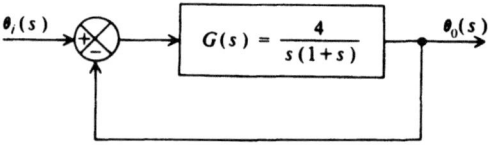

Figure P5.36

(a) Draw the phase trajectory for the following set of initial conditions:

$$\theta_0(t) = 2.5 \text{ rad}, \quad \dot{\theta}_0(t) = 0.$$

(b) What conclusions can you reach from the phase portrait?

5.37. Draw the phase portrait of a linear unity-feedback unexcited control system whose transfer function is given by

$$G(s) = \frac{10}{s(1+0.1s)}.$$

5.38. Repeat Problem 5.36 with a rate limiter at $-2\,\text{rad/sec}$ added to the system.

5.39. Repeat Problem 5.36 with a position limiter at $-0.5\,\text{rad}$ added to the system.

5.40. Repeat Problem 5.37 with the rate limiters at $\pm 1\,\text{rad/sec}$ added to the system.

5.41. Repeat Problem 5.37 with position limiters at $\pm 1\,\text{rad}$ added to the system.

5.42. Determine the time it takes for the phase trajectory illustrated in Figure P5.42 to transverse the following segments: AB, BC, CD, DE, AE.

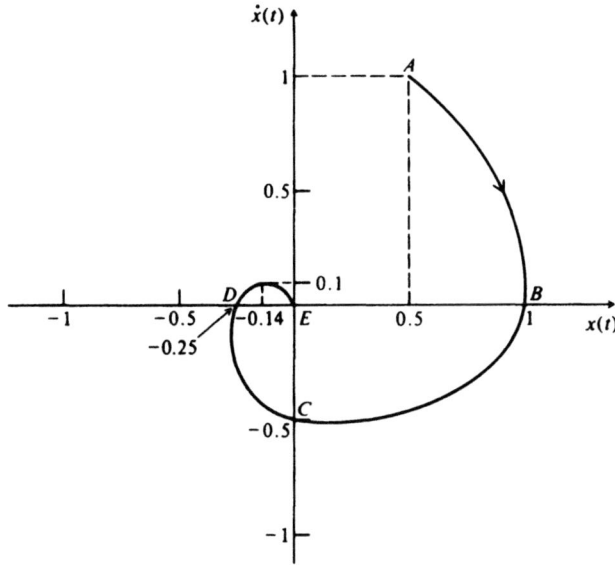

Figure P5.42

5.43. Manipulate the following equations into the form $d\dot{\theta}(t)/d\theta = f(\theta, t)$:

(a) $A\dfrac{d^2\theta(t)}{dt^2} + B\dfrac{d\theta(t)}{dt} + C\theta(t) = D$,

(b) $A\dfrac{d^2\theta(t)}{dt^2} + B\dfrac{d\theta(t)}{dt}\left|\dfrac{d\theta(t)}{dt}\right| + C\theta(t) = D$,

(c) $A\dfrac{d^2\theta(t)}{dt^2} + B\dfrac{d\theta(t)}{dt}|\theta(t)| + C\theta(t) = D$,

(d) $A\dfrac{d^2\theta(t)}{dt^2} + B\dfrac{d\theta(t)}{dt} + C\theta(t)|\theta(t)| = D$.

5.44. Derive the isocline equation for each of the following differential equations and draw the isoclines together with their slope markers:

(a) $\dfrac{d^2\theta(t)}{dt^2} + 0.8\dfrac{d\theta(t)}{dt} + 0.4\theta(t) = 0$,

(b) $\dfrac{d^2\theta(t)}{dt^2} + 0.8\dfrac{d\theta(t)}{dt} + 0.4\theta(t) = 0.6,$

(c) $\dfrac{d^2\theta(t)}{dt^2} + 0.8\dfrac{d\theta(t)}{dt}\left|\dfrac{d\theta(t)}{dt}\right| + 0.4\theta(t) = 0,$

(d) $\dfrac{d^2\theta(t)}{dt^2} + 0.8\dfrac{d\theta(t)}{dt}\left|\dfrac{d\theta(t)}{dt}\right| + 0.4\theta(t) = 0.6.$

5.45. Construct the phase portraits for each of the differential equations of Problem 5.44 using the method of isoclines.

5.46. Determine and construct the phase trajectory for a second-order system containing rate limiting where $\theta(0) = 0.7$ and $\dot\theta(0) = 0$. The differential equation for the system is given by

$$\dfrac{d^2\theta(t)}{dt^2} + 0.8\dfrac{d\theta(t)}{dt} + 0.4\theta(t) = 0$$

and

$$\left|\dfrac{d\theta(t)}{dt}\right|_{max} = 0.15.$$

5.47. Repeat Problem 5.46 for the same second-order differential equation, but with the initial conditions given by $\theta(0) = 1$ and $\dot\theta(0) = 0$ and rate limiting given by

$$\left|\dfrac{d\theta(t)}{dt}\right|_{max} = 0.2.$$

5.48. A linear unity-feedback control system contains a forward transfer function given by

$$G(s) = \dfrac{8}{s(1 + 0.5s)}.$$

(a) Draw the phase trajectory for the following set of initial conditions using the method of isoclines:

$$\dot c(0) = 2.0 \text{ rad/sec},$$
$$c(t) = 0.$$

(b) What kind of singular point does the origin of the phase plane represent? What conclusions can you reach from this phase trajectory?

(c) A position limiter is added to the output of this control system at ± 1.5 rad. How does this change the phase trajectory?

5.49. Determine and construct the phase trajectory of a nonlinear control system for which the describing equations are given by

$$\frac{d^2\theta(t)}{dt^2} + 0.8\frac{d\theta(t)}{dt} + \theta(t) = 0.5 \quad \text{where } \theta(t) > 0$$

and

$$\frac{d^2\theta(t)}{dt^2} + 0.8\frac{d\theta(t)}{dt} + \theta(t) = -0.5 \quad \text{where } \theta(t) < 0$$

Assume that the initial conditions are $\theta(0) = 0.5$ and $(d\theta(0)/dt) = 0$.

5.50. A phase trajectory of a nonlinear control system is illustrated in Figure P5.50.

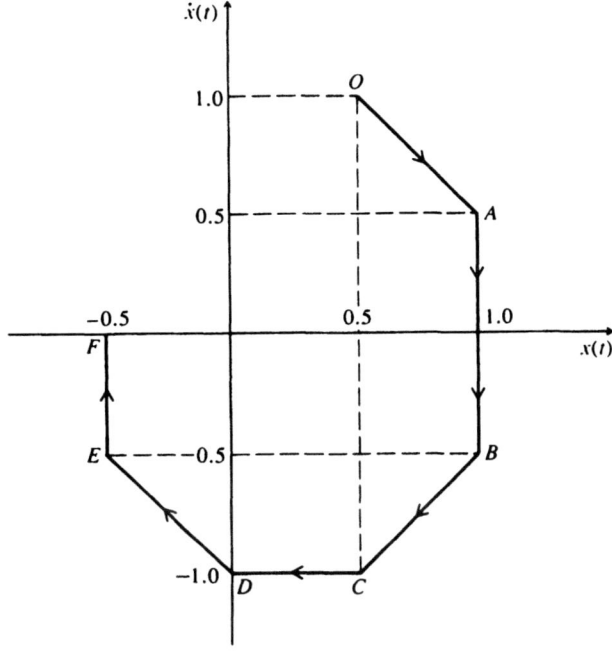

Figure P5.50

(a) Determine the time it takes to traverse the segment OA by direct integration.

(b) Determine the time it takes to traverse the segment DE by direct integration.

5.51. Using Liapunov's second method, determine the stability of the nonlinear control system described by the differential equation

$$\ddot{x}(t) + K_1(\dot{x}(t) + \dot{x}^2(t)) + K_2 x^2(t) + x(t) = 0$$

if

(a) $K_1 > 0, K_2 > 0$

(b) $K_1 < 0, K_2 < 0$

(c) $K_1 < 0, K_2 > 0$.

5.52. Using Liapunov's second method, determine the stability of the nonlinear second-order control system described by the differential equation

$$\ddot{x}(t) + [A + B(x(t))^2]\dot{x}(t) + x(t) = 0$$

if

(a) $A > 0, B > 0$

(b) $A < 0, B < 0$

(c) $A > 0, B < 0$.

5.53. Repeat Problem 5.52 for

$$\ddot{x}(t) + [A + Bx^4(t)]\dot{x}(t) + x(t) = 0.$$

5.54. Using Liapunov's second method, determine the stability of the control system described by the nonlinear differential equation

$$\frac{d^2 x(t)}{dt^2} + K_1 \frac{dx(t)}{dt} + K_2[x^8(t) + 2x(t)] = 0.$$

Assume that K_1 and K_2 are positive constants.

5.55. Utilizing Liapunov's second method, determine the stability of the control system described by the nonlinear differential equation

$$\ddot{x}(t) + K_1\dot{x}(t) + K_2\dot{x}^5(t) + x(t) = 0$$

if

(a) $K_1 > 0, K_2 > 0$

(b) $K_1 < 0, K_2 < 0$

(c) $K_1 > 0, K_2 < 0$

5.56. Repeat Problem 5.55 for the nonlinear differential equation

$$\ddot{x}(t) + [K_1 + K_2 x^4(t)]\dot{x}(t) + x(t) = 0.$$

5.57. A unity-feedback control system contains a linear and nonlinear element. The transfer function of the linear element is given by

$$G(s) = \frac{(s+3)}{s(s+2)^2(s+1.5)}.$$

The initial-condition response of the linear element is given by

$$e_0(t) = e_{10} + e_{20}e^{-2t} + e_{30}te^{-2t} + e_{40}e^{-1.5t},$$

where e_{10}, e_{20}, e_{30}, and e_{40} are related to the initial conditions and the unit-impulse response is given by

$$g(t) = [0.5 + 3.5e^{-2t} + te^{-2t} - 4e^{-1.5t}]U(t),$$

where $U(t)$ is the unit step input. Using Popov's method, determine the values of K which will result in a stable system, assuming that the nonlinear element is single valued with $q = 1.0$.

5.58. Repeat Problem 5.57 with $q = 0.75$ and the transfer function of the linear element given by

$$G(s) = \frac{(s+1)}{s(s+1.5)(s+2)^2}.$$

5.59. The control system shown in Figure P5.59 contains a nonlinear element $N[e(t)]$ and a linear element whose transfer function is $G(s)$. Assume that the nonlinearity corresponds to a general, single-valued, nonlinear element, with $q = 0$, and whose gain can vary between 0 and a maximum value of K. Using Popov's method, determine the values of K which will result in a stable control system.

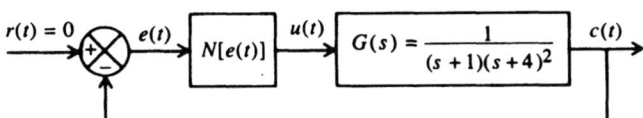

Figure P5.59

5.60. A unity-feedback control system containing a linear and nonlinear element is to be analyzed using Popov's method. The transfer function of the linear element is given by

$$G(s) = \frac{1}{s(s+1)^2}.$$

The nonlinear element corresponds to the general nonlinearity case ($q = 0$)

(a) Determine the frequency ω where the imaginary part of $G^*(j\omega)$ is zero.

(b) Using Popov's method, determine the values of K which will result in a stable system.

5.61 Repeat Problem 5.57 with $q = 1.11$ and the transfer function of the linear element given by

$$G(s) = \frac{(s+4)}{s(s+2)^2(s+1.5)}.$$

5.62. A unity-feedback control system contains a linear and nonlinear element. The transfer function of the linear element is given by

$$G(s) = \frac{(s+2)}{(s-1)(s+3)(s+4)}.$$

The initial condition response is

$$e_0(t) = e_{10}e^t + e_{20}e^{-3t} + e_{30}e^{-4t},$$

where e_{10}, e_{20}, and e_{30} arise from the initial conditions. Using the generalized circle criterion, determine possible values of $N[e(t), t]$ which will result in a stable system if the nonlinear element corresponds to the general nonlinearity case ($q = 0$).

5.63. Repeat Problem 5.62 with the transfer function of the linear element given by

$$G(s) = \frac{1}{(s-1)(s+3)(s+4)}.$$

5.64. Repeat Problem 5.62 with the transfer function of the linear element given by

$$G(s) = \frac{(s+6)}{(s-1)(s+3)(s+4)}.$$

REFERENCES

1. P. M. Lowitt and S. M. Shinners, "Type N—Integral space tracking configuration." *IEEE Trans. Mil. Electron.*, **MIL-9**, 88–98 (1965).
2. A. M. Liapunov, On the general problem of stability of motion. Ph.D. thesis, Kharkov 1892; reprinted (in French) in *Annals of Mathematics Studies*, Vol. 17. Princeton University Press, Princeton, NJ, 1949.
3. N. Minorsky, *Theory of Nonlinear Control Systems*. McGraw-Hill, New York, 1969.
4. E. Levinson, "Some saturation phenomena in servomechanisms." *Trans. Am. Inst. Electr. Eng.* **72**, 1 (1953).
5. C. A Ludeke, "The generation and extinction of subharmonics." In *Proceedings of the Symposium on Nonlinear Circuit Analysis*. Polytechnic Institute of Brooklyn, New York, 1953.
6. O. I. Elgerd, *Control Systems Theory*. McGraw-Hill, New York, 1967.

7. R. J. Kochenburger, "A frequency response method for analyzing and synthesizing contactor servo-mechanisms." *Trans. Am. Inst. Electr. Eng* **69**, 270 (1950).
8. E. C. Johnson, "Sinusoidal analysis of feedback-control systems containing nonlinear elements." *Trans. Am. Inst. Electr. Eng.* **71**, 169 (1952).
9. H. D. Grief, "Describing function method of servomechanism analysis applied to most commonly encountered nonlinearities." *Trans. Am. Inst. Electr. Eng.* **72**, 253 (1953).
10. J. G. Truxal, *Automatic Feedback Control System Synthesis*. McGraw-Hill, New York, 1955.
11. J. T. Leigh, *Essential of Nonlinear Control Theory*. Peter Peregrinus Ltd., London, 1983.
12. E. Levinson, "Phase-plane analysis." Electro-Technology **69**, 118 (1962).
13. C. A Desoer, "A generalization of the Popov criterion." *IEEE Trans. Autom. Control* **AC-10**, 182–185 (1965).
14. V. M. Popov, "Absolute stability of nonlinear systems of automatic control." *Autom. Remote Control* (USSR), **22**, 857–875 (1961).
15. V. A Jakubovic, "Frequency conditions for the absolute stability and dissipativity of control systems with a single differentiable nonlinearity." *Sov. Math. Engl. Transl.* **6**, 98–101 (1965).
16. S. Lefschetz, *Stability of Nonlinear Control Systems*. Academic Press, New York, 1965.
17. J. C. Hsu and A. U. Meyer, *Modern Control Principles and Applications*. McGraw-Hill, New York, 1968.
18. G. Zames, "On the input-output stability of time-varying nonlinear feedback systems. Part II. Condition involving circles in the frequency plane and sector nonlinearities." *IEEE Trans. Autom. Control*, **AC-11**, 465–476 (1966).
19. V. M. Popov and A. Halanay, "On the stability of nonlinear automatic control systems with lagging argument." *Autom. Remote Control* (USSR), **23**, 783–786 (1963).
20. A. V. Michailov, "Harmonic analysis in the theory of automatic control." *A. T. Moscow*, No. 3, p. 27, 1938.
21. B. O. Watkins, *Introduction to Control Systems*. Macmillan, New York, 1969.
22. H. H. Rosenbrick and C. Storey, *Computational Techniques for Chemical Engineers*. Pergamon, Oxford, 1966.
23. D. D. McCracken and W. S. Dorn, *Numerical Methods and FORTRAN Programming*. Wiley, New York, 1964.
24. R. W. Hamming, *Numerical Methods for Scientists and Engineers*. McGraw-Hill, New York, 1962.
25. S. M. Shinners, "Which computer—Analog, digital or hybrid?" *Mach. Des.*, **43**, 104–111 (1971).
26. S. M. Shinners, "*Guidelines for the analysis of nonlinear control systems*." Control Eng. **28**(11), 181–182 (1981).
27. D. P. Atherton, *Nonlinear Control Engineering*. Van Nostrand-Reinhold, London, 1982.
28. P. M. Lowitt and S. M. Shinners, "Integrated optimal synthesis for a radar tracker." In *Proceedings of the Seventh National Military Electronics Convention*, Washington, DC, 74–78, September 1963.
29. *MATLAB™ for MS-DOS Personal Computers User's Guide*, Control System Toolbox MathWorks, Inc., Natick, MA, 1997.
30. A. Grace, A. J. Laub, J. N. Little, and C. Thompson, *Control System Toolbox for Use with MATLAB™ User's Guide*. MathWorks, Inc., Natick, MA, 1990.
31. J. M. Boyle, M. P. Ford, and J. M. Maciejowski, "Multivariable toolbox for use with MATLAB." *IEEE Control Syst. Mag.* **9**(1), 59–65 (1989).

32. D. Graham and D. McRuer, *Analysis of Nonlinear Control Systems*. Wiley, New York, 1961.
33. R. Oldenburger and R. C. Boyer, "Effects of extra sinuosoidal inputs to nonlinear systems." In *Proceedings of the ASME Winter Annual Meeting*, New York, 1961.
34. S. M. Shinners, "Dual-input describing function." *Control Eng.* **18**, 53–55 (1971).
35. O. I. Elgerd, "Continuous control by high frequency signal injection." *Instrum. Control Syst.* **37**, 12 (1964).
36. R. C. Boyer, *Sinosoidal signal stabilization*. M. S. Thesis, Purdue University, Lafayette, IN, 1960.
37. J. E. Gibson, *Nonlinear Automatic Control*. McGraw-Hill, New York, 1963.
38. J. A. Aseltine, W. R. Beam, J. D. Palmer, and A. P. Sage, *Introduction to Computer Systems, Analysis, Design and Applications*. Wiley, New York, 1989.
39. N. Stern and R. A. Stern, *Introducing Quick BASIC 4.0 and 4.5: A Structured Approach*. Wiley, New York, 1986.
40. S. A Hovanessian and L. A. Pipes, *Digital Computer Methods in Engineering*. McGraw-Hill, New York, 1969.
41. C. Belove, ed., *Handbook of Modern Electronics and Electrical Engineering*. Wiley, New York, 1986.
42. O. Romaine, "OAO: NASA's biggest satellite yet." *Space/Aeronaut.* **40**, 54–58 (1962).
43. W. Turk, *Ranger Block III attitude control system*. Jet Propul. Lab. Tech. Rep. No. 32-663, 1964.

6

INTRODUCTION TO OPTIMAL CONTROL THEORY AND ITS APPLICATIONS

6.1. INTRODUCTION

From the presentation of linear and nonlinear systems in Chapters 2, 3, and 5, we recognize that the conventional design of feedback control systems has several disadvantages. The most serious disadvantage is that the control-system configurations we have analyzed and designed in the previous chapters were always assumed to be known in advance. But we never knew whether they were the best system to do the job. In contrast, optimal theory is concerned with obtaining a system which is the best possible with respect to a *standard* against which we can measure real performance. We denote this standard as the performance criterion (see Chapter 5† for a discussion of performance criteria). The task of designing control systems that are optimal, in some sense, is one of the most important and complex problems facing control engineers today.

Wiener [1], in the late 1940s, developed the concept of optimum design that was based on optimizing a performance criterion. McDonald [2] first applied the concept of optimization to control systems in 1950. His objective was to minimize the transient response time of a relay-type feedback control system to step inputs. In 1957, Draper and Li [3] wrote a booklet discussing the theoretical concepts of optimal control for an internal combustion engine. Their system attempted to minimize (or optimize) the consumption of fuel. Since that time many articles have been written on optimal control systems both in this country and abroad. Most important is the work of Belmann [4, 5], who developed the concept of dynamic programming, and Pontryagin, Boltynanskii, and Gamkrelidge [6, 7], who developed the maximum principle. Several good books have been written on the subject [8–11]. The purpose of this chapter is to introduce the principles of this theory and apply it to several important design problems associated with our time, such as the attitude-control system of space vehicles and the control system used to soft land a space vehicle on the Moon's surface. This chapter is not meant to compete with complete books that have been written on the subject [8–11]. Rather, its purpose is to extend to the reader (student or practicing control-system engineer) the methodology to synthesize a control system using optimal control theory as opposed to the analysis and design of a given control system, which has been the approach of the previous chapters. By synthesizing control systems, we mean to answer questions as to its configuration, whether it is linear or nonlinear, etc., because in this chapter we make no assumption as to the form of the system to start with.

6.2. CHARACTERISTICS OF THE OPTIMAL CONTROL PROBLEM

In this section, we consider the basic problem of optimal control theory [12]. It consists of choosing the input **u** to a control system so that the performance of the system is optimum with respect to some performance criterion. The basic goal of optimal control theory is to design control elements that meet a wide variety of requirements in the best possible manner. The structure of the optimal control problem can be described in the following way. One has:

1. A controlled process whose dynamics have the following form:

$$\dot{\mathbf{x}}(t) = f(\mathbf{x}(t), u(t), t) \tag{6.1}$$

2. Restrictions on the input **u**(t) and/or the plant state **x**(t);
3. A reference signal **r**(t) which signifies the desired output response;
4. A performance criterion having the following general form:

$$S = \int_{t_0}^{T} G(\mathbf{c}(t), \mathbf{u}(t), \mathbf{r}(t), t) dt. \tag{6.2}$$

In Eq. (6.2), the integrand G is referred to as the loss function and represents a measure of instantaneous change from ideal performance [13]. Therefore, the performance criterion is interpreted as the cumulative loss. The optimal control problem consists of the determination of the control input **u**(t) that minimizes the performance criterion S subject to certain constraints on **u**(t) and **x**(t).

Let us consider possible formulations of the performance criterion S to some practical problems.

A. Minimum-Time Problem

A very common problem that the control-system engineer encounters is to be able to drive a system's states from an initial value to its final state in the minimum amount of time. An example of this is a gun or missile launcher on a ship that is pointing in one direction. Suddenly, information is picked up from the ship's radar that a hostile target is heading toward the ship 180° away. What is the best strategy for the control system in travelling the 180° in the shortest amount of time? A similar problem is encountered when it is required that a satellite go from one point in space to another in the shortest amount of time for rendezvousing with another satellite. For this problem, the performance criterion is given by

$$S = \int_0^T dt. \tag{6.3}$$

This problem is solved in Section 6.6.

B. Minimum Fuel-Consumption Problem

A problem very common to the design of the attitude-control system of a manned space vehicle is to minimize its fuel consumption. Here the performance criterion is given by

$$S = \int_0^T |u(t)| dt, \tag{6.4}$$

where we show in Section 6.6 that $u(t)$ is the propellant rate of flow.

C. Minimum Energy-Consumption Problem

Another very common problem found in the design of the attitude-control system of an unmanned space vehicle (e.g., weather and communication satellites) is to minimize its solar energy consumption. The performance criterion for this problem is given by

$$S = \int_0^T u^2(t) dt. \tag{6.5}$$

This representation can be rationalized if we think of $u(t)$ as the current (or voltage) in the system. Because current and voltage are proportional to power, analogously, $u^2(t)$ represents power and its integral represents energy. This problem is also solved in Section 6.6.

D. Lunar Soft-Landing Problem

A very important occurrence in recent history was when the Apollo 11 mission soft-landed the lunar excursion module on the surface of the moon. The performance criterion for the lunar soft-landing problem is given by

$$S = \int_0^T \dot{m}(t) dt, \tag{6.6}$$

where $\dot{m}(t)$ represents the mass flow rate of the fuel, which we want to minimize. This problem is solved in Section 6.7.

E. Quadratic Optimal Control Problem

Section 3.11 presented the integral of the square of the error performance criterion as a useful performance index. From Eq. (3.228), this performance index is given by

$$S = \int_0^\infty e^2(t)\, dt. \tag{6.7}$$

Let us modify this in terms of the state variable $x_1(t)$:

$$S(t) = \int_0^\infty x_1^2(t)\, dt. \tag{6.8}$$

In practice, this performance index is more useful if we expand this to more than one state variable. For example, in terms of two state variables, this performance index would be as follows:

$$S = \int_0^\infty (x_1^2(t) + x_2^2(t))\, dt. \tag{6.9}$$

We can expand this to a more general representation for n state variables in terms of the following matrix operation which takes the product of the transpose of the state vector times the state vector:

$$\mathbf{x}^T(t)\mathbf{x}(t) = [x_1(t)\ x_2(t)\dots x_n(t)] \begin{bmatrix} x_1(t) \\ x_2(t) \\ \vdots \\ x_n(t) \end{bmatrix} = x_1^2(t) + x_2^2(t) + \dots + x_n^2(t). \tag{6.10}$$

Therefore, the performance index for n squared state variables is given by

$$S = \int_0^\infty [\mathbf{x}^T(t)\mathbf{x}(t)]\, dt. \tag{6.11}$$

The resulting performance index of Eq. (6.11) does not account for the control input $\mathbf{u}(t)$ which is necessary in practical problems. In general, we wish to account for the expenditure of energy as represented by the performance criterion of Eq. (6.5). Therefore, the following more general performance criterion is used in practice for quadratic optimal control formulation

$$S = \int_0^\infty (\mathbf{x}^T(t)\mathbf{Q}\mathbf{x}(t) + \mathbf{u}^T(t)\mathbf{R}\mathbf{u}(t))\, dt \tag{6.12}$$

where $\mathbf{Q}$ is a positive-definite real symmetric matrix, $\mathbf{R}$ is a positive-definite real symmetric matrix, and $\mathbf{u}(t)$ represents the input. Solution of Eq. (6.12) will allow for the solution of the optimal control vector $\mathbf{u}(t)$ for control systems where the state equation is given by Eq. (6.1).

The choice of input $\mathbf{u}(t)$, over the operating interval $t_0 = 0$ to $t = T$ sec, is denoted as the control law [14]. If the input $\mathbf{u}(t)$ minimizes the performance criterion S, then it is optimal. Under these conditions, the optimal control is denoted by $\mathbf{u}^0(t)$ and the optimal performance is denoted by S^0.

The optimal policy can be either an open-loop policy or a feedback (closed-loop) policy. The designation *open loop* is used if the controlled input is specified as a function of the input and only the initial state of the output. The optimal policy is designated *feedback* if the controlled input is specified as a function of the input and the current state of the plant. The nature of the policy is extremely important, because it indicates how $\mathbf{u}^0(t)$ is to be generated from $\mathbf{r}(t)$ and $\mathbf{c}(t)$. The representation of optimal control systems having open-loop and feedback policies is illustrated in Figure 6.1.

As the preceding chapters have shown, feedback has a great many advantages and, therefore, a feedback policy is preferred for optimal control. For example, feedback operation, in general, tends to make the system less sensitive to variations. In addition, feedback operation makes use of the most recent information on the state of the plant. As a result, if a disturbance within the system occurs in the feedback system, it operates optimally on the latest measurement of $\mathbf{c}(t)$. In an open-loop system, however, the entire input is preprogrammed only on the basis of the initial-state value of the output. In this case, any disturbances within the system destroy the optimality of operation.

The reader must be cautioned before continuing with this chapter. All problems do not have an optimal control solution. In some problems, the combination of performance criterion, process, desired state, and constraints on the input make an optimal control solution impossible because it requires performance beyond the physical constraints and capabilities of the control system. This question must be answered very early in the optimal control problem, because it may be necessary to compromise the performance criterion. In addition, the concepts of controllability and observability, which were presented in Sections 3.4 and 3.5, respectively, are very

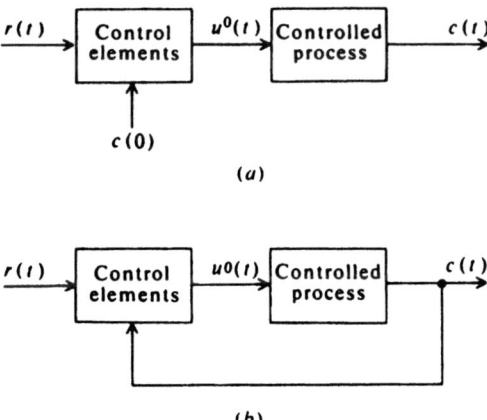

Figure 6.1 Open-loop and feedback optimal control systems; (a) open-loop optimal control system, (b) feedback optimal control system.

important in determining the existence of complete solutions to optimal control problems, [15–18]. Before we can design a system to be optimal, we must first determine whether the system is controllable and whether its states are observable, because the conditions on controllability and observability often govern the existence of a complete solution to an optimal control system.

6.3. CALCULUS OF VARIATIONS

The calculus of variations [13, 14, 19, 20] is concerned with obtaining the maxima and minima of entire functional expressions. It can be applied to a great many problems in the fields of classical mechanics, aerodynamics, optics, and control theory. We shall consider the calculus of variations only in the context of control theory. Three of its central problems are as follows:

1. The *Lagrange problem* is concerned with determining a function $\mathbf{u}(t)$ that minimizes a certain performance criterion S. Thus, for the given set of equations

$$\dot{\mathbf{x}}(t) = \mathbf{f}(\mathbf{x}(t), \mathbf{u}(t), t), \qquad (6.13)$$

$$\mathbf{c}(t) = \mathbf{L}\mathbf{x}(t), \qquad (6.14)$$

a set of initial conditions given by

$$\mathbf{x}(0) = \mathbf{p},$$

and a performance criterion given by

$$S = \int_0^T G(\mathbf{c}(t), \mathbf{u}(t), t)\, dt, \qquad (6.15)$$

this problem is concerned with determining the function $\mathbf{u}(t)$ that minimizes S.

2. The *Mayer problem* is concerned with determining a function $\mathbf{u}(t)$, which minimizes a certain performance criterion S evaluated at the end point and containing some variables whose final values are unspecified in advance. Thus, for the given set of equations

$$\dot{\mathbf{x}}(t) = \mathbf{f}(\mathbf{x}(t), \mathbf{u}(t), t), \qquad (6.16)$$

$$\mathbf{c}(t) = \mathbf{L}\mathbf{x}(t), \qquad (6.17)$$

a set of initial conditions given by

$$\mathbf{x}(0) = \mathbf{p}, \qquad (6.18)$$

a set of final conditions given by

$$\mathbf{x}(T) = \mathbf{q}, \qquad (6.19)$$

and a performance criterion given by

$$S = \int_0^T G(\mathbf{c}(t), \mathbf{u}(t), t)\, dt, \qquad (6.20)$$

this problem is concerned with determining the function $\mathbf{u}(t)$ that minimizes S.

3. The *Bolza problem* is concerned with determining a function $\mathbf{u}(t)$, that contains some variables that are unspecified in advance, and minimizing a certain performance criterion at the end point. Thus, for the given set of equations

$$\dot{\mathbf{x}}(t) = \mathbf{f}(\mathbf{x}(t), \mathbf{u}(t), t), \qquad (6.21)$$
$$\mathbf{c}(t) = \mathbf{L}\mathbf{x}(t),$$

a set of initial conditions given by

$$\mathbf{x}(0) = \mathbf{p},$$

a set of final conditions given by

$$\mathbf{x}(T) = \mathbf{q},$$

and a performance criterion given by

$$S = F[(\mathbf{c}(t), \mathbf{u}(t), t]_0^T + \int_0^T G(\mathbf{c}(t), \mathbf{u}(t), t)\, dt, \qquad (6.22)$$

the problem is concerned with determining the function $\mathbf{u}(t)$ that minimizes S.

The Bolza problem is the most general case of variational calculus. Most optimal control problems can be formulated as one of these three fundamental problems, and it is usually possible to introduce a mathematical substitution which can transform a Lagrange or Mayer problem into a Bolza problem.

In order to solve for $u^0(t)$ in a scalar problem using the calculus of variations, we must first solve the Euler–Lagrange equation. The derivation of this differential equation and demonstration of its computational difficulty can be shown by considering the Mayer scalar problem. It is assumed that $c(t) = x(t)$, and that Eq. (6.20) has been optimized and the optimum values $x^0(t)$ and $u^0(t)$ have been determined. For example, if $x^0(t)$ and $u^0(t)$ are substituted back into Eq. (6.20), then S^0 is optimized (let us assume that it is a minimum value).

What is the effect on S^0 if we now allow $x^0(t)$ and $u^0(t)$ to be perturbed by small arbitrary amounts $\Delta x(t)$ and $\Delta u(t)$, respectively? If S^0 is indeed optimum, then ΔS must be zero if we deviate slightly from it. We limit these perturbations, however, to $\Delta x(0) = \delta x(T) = 0$, in order to meet the constraints given by Eqs. (6.18) and (6.19). The resulting effect on S^0 in Eq. (6.20) can be determined by substituting

$$x(t) = x^0(t) + \Delta x(t), \qquad (6.23)$$
$$u(t) = u^0(t) + \Delta u(t) \qquad (6.24)$$

into the state equations of Eq. (6.16) as follows:

$$\frac{d}{dt}(x^0(t) + \Delta xz(t)) = f(x^0(t) + \Delta x(t), u^0(t) + \Delta u(t), t). \tag{6.25}$$

Using the linear terms of a Taylor-series expansion around the optimum values, we obtain

$$\dot{x}^0(t) + \Delta \dot{x}(t) = f(x^0(t), u^0(t), t) + \left(\frac{\partial f}{\partial x}\right)_0 \Delta x(t) + \left(\frac{\partial f}{\partial u}\right)_0 \Delta u(t). \tag{6.26}$$

Because

$$\dot{x}^0(t) = f(x^0(t), u^0(t), t), \tag{6.27}$$

we have

$$\Delta \dot{x}(t) = \left(\frac{\partial f}{\partial x}\right)_0 \Delta x(t) + \left(\frac{\partial f}{\partial u}\right)_0 \Delta u(t) \tag{6.28}$$

or

$$\Delta u(t) = \frac{\Delta \dot{x}(t) - (\partial f/\partial x)_0 \Delta x(t)}{(\partial f/\partial u)_0}. \tag{6.29}$$

Equations (6.28) and (6.20) indicate that if $\Delta x(t)$ is arbitrarily chosen, then $\Delta u(t)$ is restricted to values that obey this equation. Now, knowing the relationship between $x(t)$, $\Delta x(t)$, $u(t)$, and $\Delta u(t)$, can we say what effect $\Delta x(t)$ and $\Delta u(t)$ have on S^0? We can say that

$$\Delta S = S(x^0 + \Delta x, u^0 + \Delta u, t) - S(x^0, u^0, t). \tag{6.30}$$

Therefore,

$$\Delta S = \int_0^T G(x^0 + \Delta x, u^0 + \Delta u, t)dt - \int_0^T G(x^0, u^0, t)dt \tag{6.31}$$

Combining terms under the integral sign, we obtain

$$\Delta S = \int_0^T [G(x^0 + \Delta x, u^0 + \Delta u, t) - G(x^0, u^0, t)]dt. \tag{6.32}$$

By means of the Taylor series, we can show that

$$\Delta S = \int_0^T \left[\left(\frac{\partial G}{\partial x}\right)_0 \Delta x + \left(\frac{\partial G}{\partial u}\right)_0 \Delta u\right]dt. \tag{6.33}$$

Substituting Eq. (6.29) into Eq. (6.33), we obtain the following expression:

$$\Delta S = \int_0^T \left[\left(\frac{\partial G}{\partial x}\right)_0 \Delta x(t) + \left(\frac{\partial G}{\partial u}\right)_0 \frac{\Delta \dot{x}(t) - (\partial f/\partial x)_0 \Delta x(t)}{(\partial f/\partial u)_0} \right] dt. \quad (6.34)$$

Equation (6.34) can be simplified to

$$\Delta S = \int_0^T \left\{ \frac{(\partial G/\partial u)_0}{(\partial f/\partial u)_0} \Delta \dot{x}(t) + \left[\left(\frac{\partial G}{\partial x}\right)_0 - \left(\frac{\partial G}{\partial u}\right)_0 \frac{(\partial f/\partial x)_0}{(\partial f/\partial u)_0} \right] \Delta x(t) \right\} dt. \quad (6.35)$$

Integrating the first term of the integral by parts, we obtain

$$\int_0^T \frac{(\partial G/\partial u)_0}{(\partial f/\partial u)_0} \Delta \dot{x}(t) dt = \left[\frac{(\partial G/\partial u)_0}{(\partial f/\partial u)_0} \Delta x(t) \right]_0^T - \int_0^T \Delta x(t) \frac{d}{dt} \frac{(\partial G/\partial u)_0}{(\partial f/\partial u)_0} dt. \quad (6.36)$$

Because $x(0) = \Delta x(T) = 0$, the first term on the right-hand side of Eq. (6.36) is zero. Therefore, returning to Eq. (6.35), and substituting Eq. (6.36) into it, we obtain the following:

$$\Delta S = \int_0^T \left[\left(\frac{\partial G}{\partial x}\right)_0 - \left(\frac{\partial G}{\partial u}\right)_0 \frac{(\partial f/\partial x)_0}{(\partial f/\partial u)_0} - \frac{d}{dt} \frac{(\partial G/\partial u)_0}{(\partial f/\partial u)_0} \right] \Delta x(t) dt. \quad (6.37)$$

Equation (6.37) states the resulting effect on S for a perturbation of $\Delta x(t)$ and $\Delta u(t)$. However, if $x^0(t)$ and $u^0(t)$ do indeed result in an optimum S^0, then ΔS must be zero for a perturbation of $\Delta x(t)$. Therefore, it is necessary that during the interval $0 < t < T$,

$$\left(\frac{\partial G}{\partial x}\right)_0 - \left(\frac{\partial G}{\partial u}\right)_0 \frac{(\partial f/\partial x)_0}{(\partial f/\partial u)_0} - \frac{d}{dt} \frac{(\partial G/\partial u)_0}{(\partial f/\partial u)_0} = 0; \quad (6.38)$$

this is the well-known Euler–Lagrange differential equation.

The results of this exercise indicate that we have to solve the Euler–Lagrange differential equation, which cannot, in general, be integrated analytically. In addition, if numerical techniques are used, one finds that the Euler–Lagrange equations are unstable.

A. Example 6.3-1

Let us illustrate the application of the calculus of variations to a simple problem that is easily solvable [20]. Consider the open-loop system illustrated in Figure 6.2. It is desired to control the state between $x(0)$ and $x(T)$ in order that the performance criterion, which penalizes (in ISE sense) $u(t)$ and the deviation from the reference $x(t) = 0$,

$$S = \int_0^T (x^2(t) + u^2(t)) dt \quad (6.39)$$

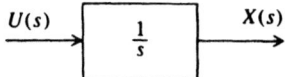

Figure 6.2 A simple integrating system.

is minimized. For this simple system, the state equation is given by

$$\dot{x}(t) = f(x(t), u(t), t) = u(t). \tag{6.40}$$

In addition, to obtain the Euler–Lagrange equation, we need to determine the following:

$$\left(\frac{\partial G}{\partial x}\right)_0 = 2x(t), \tag{6.41}$$

$$\left(\frac{\partial G}{\partial u}\right)_0 = 2u(t), \tag{6.42}$$

$$\frac{\partial f}{\partial x} = 0, \tag{6.43}$$

$$\frac{\partial f}{\partial u} = 1. \tag{6.44}$$

Substituting Eqs. (6.41)–(6.44) into Eq. (6.38), we obtain the following Euler–Lagrange equation:

$$x(t) - \dot{u}(t) = 0 \tag{6.45}$$

For this simple system, since Eqs. (6.40) and (6.45) are linear, the Laplace transformation can be used. We have

$$sX(s) - x(0) = U(s), \tag{6.46}$$

$$X(s) - [sU(s) - u(0)] = 0. \tag{6.47}$$

Observe from Eqs. (6.46) and (6.47) that we must obey the constraints given by $x(0)$ and $u(0)$. Because we know $x(0)$ and $x(T)$ but not $u(0)$, we will proceed by using only the one initial condition we know, $x(0)$. The value of $u(0)$ will be eliminated later on by attempting to meet the final condition $x(T)$. Solving for $X(s)$ from Eqs. (6.46) and (6.47), we obtain the expression

$$X(s) = \frac{sx(0) + u(0)}{s^2 - 1} \tag{6.48}$$

Taking the inverse transform, we have

$$x^0(t) = x(0)\cosh t + u(0)\sinh t. \tag{6.49}$$

The value of $u(0)$ must now be obtained in order that the state $x(T)$ is reached. Therefore,

$$x(T) = x(0)\cosh T + u(0)\sinh T. \tag{6.50}$$

Solving for $u(0)$, we obtain

$$u(0) = \frac{x(T) - x(0)\cosh T}{\sinh T}. \tag{6.51}$$

Substituting Eq. (6.51) into Eq. (6.49), we obtain the final expression for $x^0(t)$:

$$x^0(t) = x(0)\cosh t + \left[\frac{x(T) - x(0)\cosh T}{\sinh T}\right]\sinh t. \tag{6.52}$$

In addition, we can determine the corresponding optimal control from Eq. (6.40):

$$u^0(t) = \dot{x}^0(t) = x(0)\sinh t + \left[\frac{x(T) - x(0)\cosh T}{\sinh T}\right]\cosh t. \tag{6.53}$$

In this example, notice that the Euler–Lagrange equation turned out to be linear because the state dynamic equations were linear and the performance was quadratic. In addition, the Euler–Lagrange equation resulted in an open-loop optimal control solution and is based on a priori knowledge of the final state and the control interval.

In general, design of control systems by means of the calculus of variations usually leads to the solution of a two-point boundary-value problem. Usually, one is faced with a numerical trial-and-error solution to a resulting nonlinear differential equation. The calculus of variations is not commonly used by the control-system engineer for designing optimal control systems because it cannot easily handle "hard" constraints such as $|u| \leq u_{\max}$. Its use is limited to control systems that are linear, have no constraints on $x(t)$ and $u(t)$, and have a quadratic performance criterion.

6.4. DYNAMIC PROGRAMMING

The concept of dynamic programming [4, 5, 12, 21] as originally developed by Bellman, is based on the principle of optimality and the embedding approach. The principle of optimality states that regardless of the initial state and the initial decision, the remaining decision must form an optimal control policy with respect to the state resulting from the first decision. The embedding approach is one in which an optimal decision problem, is embedded in a series of smaller problems that are easier to solve.

468 INTRODUCTION TO OPTIMAL CONTROL THEORY AND ITS APPLICATIONS

A multistage decision process is an example of a problem that can be simplified considerably by applying the principle of dynamic programming. Utilizing the embedding approach and the principle of optimality, the total return of an N-stage decision process is reduced to the problem of solving a sequence of N single-stage decision processes. This permits a simple, systematic solution to the problem. In practice, dynamic programming is basically an optimization procedure that proceeds backwards in time. The solution is computed first over the last stage of the process, and successive solutions are computed for the remaining stages until the entire solution is obtained.

The principle of optimality results in a partial differential equation, known as the Hamilton–Jacobi equation can be derived from the definition of the performance criterion defined by Eq. (6.2):

$$S = \int_{t_0}^{T} G(\mathbf{c}(t), \mathbf{u}(t), \mathbf{r}(t), t)\, dt. \tag{6.54}$$

Let us assume that the optimal control input $\mathbf{u}^0(t)$ has been determined for a reference signal $\mathbf{r}(t)$ and a final time T. Therefore, the minimum value of this performance criterion S^0 is a function only of the initial state $\mathbf{c}(t_0)$ and the initial time t_0. From Eq. (6.54), we obtain S^0 as follows:

$$S^0 = S^0(\mathbf{c}(t_0), t_0) = \min_{\mathbf{u}(t)} \left[\int_{t_0}^{T} G(\mathbf{c}(t), \mathbf{u}(t), \mathbf{r}(t), t)\, dt \right]. \tag{6.55}$$

Applying the embedding approach and assuming that the last stage of the process occurs between t' and T. Eq. (6.55) may be rewritten as

$$S^0(\mathbf{c}(t_0), t_0) = \min_{\mathbf{u}(t)} \left[\int_{t_0}^{t'} G(\mathbf{c}(t), \mathbf{u}(t), \mathbf{r}(t), t)\, dt + \int_{t'}^{T} G(\mathbf{c}(t), \mathbf{u}(t), \mathbf{r}(t), t)\, dt \right]. \tag{6.56}$$

Applying the principle of optimality over the last stage, which occurs from t' to T, Eq. (6.56) can be rewritten as

$$S^0(\mathbf{c}(t_0), t_0) = \min_{\mathbf{u}(t)} \left[\int_{t_0}^{t'} G(\mathbf{c}(t), \mathbf{u}(t), \mathbf{r}(t), t)\, dt + S^0(\mathbf{c}(t'), t') \right]. \tag{6.57}$$

The first term of Eq. (6.57) may be approximated as

$$\int_{t_0}^{t'} G(\mathbf{c}(t), \mathbf{u}(t), \mathbf{r}(t), t)\, dt \approx [G(\mathbf{c}(t), \mathbf{u}(t), \mathbf{r}(t), t)]_{t=t_0} \Delta t, \tag{6.58}$$

by defining

$$t' = t_0 + \Delta t. \tag{6.59}$$

Expanding the second term of Eq. (6.57) by means of the linear terms of a Taylor series about the point $\mathbf{c}(t_0)$, t_0 yields

$$S^0(\mathbf{c}(t'), t') = S^0(\mathbf{c}(t_0), t_0) + \frac{\partial S^0}{\partial c_1}\Delta c_1(t) + \cdots + \frac{\partial S^0}{\partial c_n}\Delta c_n(t) + \frac{\partial S^0}{\partial t_0}\Delta t. \tag{6.60}$$

Substituting Eqs. (6.58) and (6.60) into (6.57), we obtain the following relationship:

$$S^0(\mathbf{c}(t_0), t_0) = \min_{\mathbf{u}(t_0)} \left[G(\mathbf{c}(t_0), \mathbf{u}(t_0), \mathbf{r}(t_0), t_0)\Delta t + S^0(\mathbf{c}(t_0), t_0) \right.$$
$$\left. + \frac{\partial S^0}{\partial c_1}\Delta c_1(t) + \cdots + \frac{\partial S^0}{\partial c_n}\Delta c_n(t) + \frac{\partial S^0}{\partial t_0}\Delta t \right]. \tag{6.61}$$

Taking $S^0(\mathbf{c}(t_0), t_0)$ outside the minimization operation [because it is not a function of $\mathbf{u}(t_0)$], dividing both sides of Eq. (6.61) by Δt, and letting $\Delta t \to 0$, we obtain

$$\min_{\mathbf{u}(t_0)} \left[G(\mathbf{c}(t_0), \mathbf{u}(t_0), \mathbf{r}(t_0), t_0) + \frac{\partial S^0}{\partial c_1}\dot{c}_1(t) + \cdots + \frac{\partial S^0}{\partial c_n}\dot{c}_n(t) + \frac{\partial S^0}{\partial t_0} \right] = 0. \tag{6.62}$$

Assuming that $\mathbf{u}^0(t_0)$ minimizes the bracketed term, Eq. (6.62) can be rewritten as

$$\frac{\partial S^0}{\partial t_0} + \sum_{i=1}^{n} \frac{\partial S^0}{\partial c_i}\dot{c}_i(t) + G(\mathbf{c}(t_0), \mathbf{u}^0(t_0), \mathbf{r}(t_0), t_0) = 0. \tag{6.63}$$

Because $\dot{c}_i(t)$ is a function of $\mathbf{c}(t)$ and $\mathbf{u}^0(t)$, we may write this relationship as

$$\dot{c}_i(t) = f_i(\mathbf{c}(t), \mathbf{u}^0(t)). \tag{6.64}$$

Substituting Eq. (6.64) into (6.63), we obtain the Hamilton–Jacobi equation:

$$\frac{\partial S^0}{\partial t_0} + \sum_{i=1}^{n} \frac{\partial S^0}{\partial c_i} f_i(\mathbf{c}(t), \mathbf{u}^0(t)) + G(\mathbf{c}(t), \mathbf{u}^0(t), \mathbf{r}(t), t_0) = 0. \tag{6.65}$$

The Hamilton–Jacobi equation is a partial differential equation whose boundary conditions can be obtained from the definition of the performance criterion [see Eq. (6.54)] and from the relationship

$$S^0(\mathbf{c}(t_0), t_0) = \int_{t_0}^{T} G(\mathbf{c}(t), \mathbf{u}^0(t), \mathbf{r}(t), t)\, dt. \tag{6.66}$$

If $\mathbf{c}(T)$ does not contain an impulse function at $t = T$, the boundary condition is given by

$$\lim_{t_o \to T} S^0(\mathbf{c}(t_0), t_0) = 0. \tag{6.67}$$

If $c(T)$ contains an impulse function at $t = T$, then the boundary condition is given by

$$\lim_{t_0 \to T} S^0(c(t_0), t_0) = [G(c(t), u^0(t), r(t), t]_{t=T}. \tag{6.68}$$

The solution to the Hamilton–Jacobi equation, with appropriate boundary conditions, results in the optimal control policy. However, because the Hamilton–Jacobi equation was derived from the embedding approach and the principle of optimality, it represents only a necessary condition for optimality. For example, if there are constraints on the inputs or outputs to the system, these constraints must also be considered as necessary conditions for optimality.

A. Example 6.4-1

As an example of solving an optimal control problem by means of dynamic programming, a simple regulator problem will be considered [22]. The objective in the regulator problem is to maintain the output at a fixed value. Therefore, the input $r(t) = 0$. For this problem consider the following.

1. The controlled process is a first-order linear plant whose dynamics are given by

$$\dot{c}(t) = Ac(t) + Bu(t). \tag{6.69}$$

2. The performance criterion is given by

$$S = \int_0^\infty c^2(t) \, dt. \tag{6.70}$$

3. The controlled input satisfies the following relationship:

$$-1 \leq u(t) \leq 1. \tag{6.71}$$

From Eq. (6.65), the scalar Hamilton–Jacobi equation is

$$\frac{\partial S^0}{\partial t_0} + \frac{\partial S^0}{\partial c} f(c(t), u^0(t)) + G(c(t), u^0(t), r(t), t_0) = 0. \tag{6.72}$$

From the statement of the problem, the terms of this equation are:

$$\frac{\partial S^0}{\partial t_0} = 0, \tag{6.73}$$

$$f(c(t), u^0(t)) = \dot{c}(t) = Ac(t) + Bu(t), \tag{6.74}$$

$$G(c(t), u^0(t), r(t), t_0) = c^2(t). \tag{6.75}$$

Substituting these equations into Eq. (6.72), we obtain

$$\frac{\partial S^0}{\partial c}(Ac(t) + Bu(t)) + c^2(t) = 0. \tag{6.76}$$

In order to minimize this expression with respect to $u(t)$, the constraint

$$-1 \leq u(t) \leq 1 \tag{6.77}$$

must be incorporated into the solution. Therefore,

$$\min_{u(t)} \left[\frac{\partial S^0}{\partial c}(Ac(t) + Bu(t)) + c^2(t) \right] = \min_{u(t)} \left[\frac{\partial S^0}{\partial c} Bu(t) \right]. \tag{6.78}$$

Clearly, the optimal control must satisfy

$$\begin{aligned} u^0(t) &= 1, \quad \text{for } \frac{\partial S^0}{\partial c} B < 0, \\ u^0(t) &= -1, \quad \text{for } \frac{\partial S^0}{\partial c} B > 0, \end{aligned} \tag{6.79}$$

or

$$u^0(t) = -\text{sgn}\left(B \frac{\partial S^0}{\partial c} \right). \tag{6.80}$$

Substituting Eq. (6.80) into (6.76), we obtain

$$\frac{\partial S^0}{\partial c}\left[Ac(t) - B\,\text{sgn}\left(B \frac{\partial S^0}{\partial c} \right) \right] + c^2(t) = 0. \tag{6.81}$$

Simplifying this equation, we obtain

$$Ac(t) \frac{\partial S^0}{\partial c} - \left| B \frac{\partial S^0}{\partial c} \right| + c^2(t) = 0, \tag{6.82}$$

or

$$c^2(t) + Ac(t) \frac{\partial S^0}{\partial c} - \left| B \frac{\partial S^0}{c} \right| = 0. \tag{6.83}$$

Equation (6.83) is a nonlinear partial differential equation. Except for some simple cases, the solution for S^0 requires the aid of a digital computer. The resulting solution for S^0 is then substituted into Eq. (6.80) in order to obtain the optimum $u^0(t)$. The conceptual block diagram of the resulting system is shown in Figure 6.3. It is

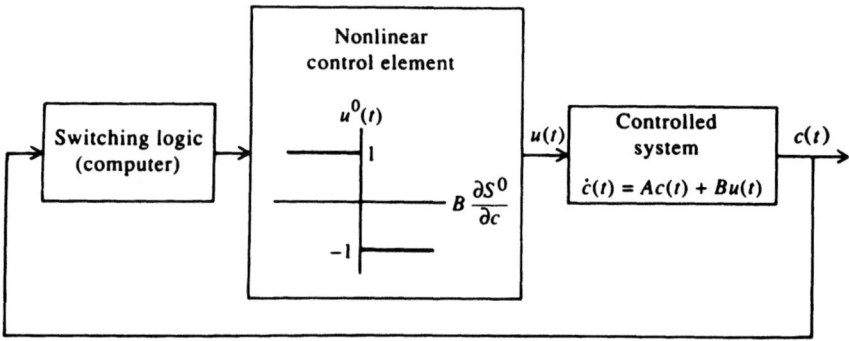

Figure 6.3 Conceptual block diagram for a scalar controlled system (containing a first-order linear plant) which is analyzed using dynamic programming.

important to recognize that this system is an intentional nonlinear system, as opposed to the unintentional nonlinear control systems that were analyzed and designed in Chapter 5. Because of the limits on the control input $u(t)$ optimal control-system designs are invariably nonlinear. That is the reason why nonlinear control systems were presented in this book in Chapter 5 prior to the presentation of optimal control systems in this chapter.

B. Example 6.4-2

As a second example of solving an optimal control problem by means of dynamic programming, consider the following problem.

1. The controlled process is a second-order linear plant whose dynamics are given by

$$\ddot{c}(t) + 4\dot{c}(t) + c(t) = u(t). \tag{6.84}$$

2. It is desired to minimize the response time so that the performance criterion is given by

$$S = \int_0^T dt. \tag{6.85}$$

3. The controlled input satisfies

$$-1 \leqslant u(t) \leqslant 1. \tag{6.86}$$

By defining the states

$$x_1(t) = c(t), \quad x_2(t) = \dot{c}(t), \tag{6.87}$$

the state equations for this system are given by

$$\dot{x}_1(t) = x_2(t), \quad \dot{x}_2(t) = -x_1(t) - 4x_2(t) + u(t). \tag{6.88}$$

The terms of the Hamilton–Jacobi equation [see Eq. (6.65)] for this problem are

$$\left.\begin{array}{l}\dfrac{\partial S^0}{\partial t_0} = 0, \\[6pt] \displaystyle\sum_{i=1}^{2} \dfrac{\partial S^0}{\partial c_i} \dot{c}_i(t) = \dfrac{\partial S^0}{\partial c} \dot{c}(t) + \dfrac{\partial S^0}{\partial \dot{c}}(-c(t) - 4\dot{c}(t) + u(t)). \\[6pt] G(x, u^0, r, t_0) = 1.\end{array}\right\} \quad (6.89)$$

Substituting these equations into the Hamilton–Jacobi equation, Eq. (6.65), we obtain

$$\dfrac{\partial S^0}{\partial c} \dot{c}(t) + \dfrac{\partial S^0}{\partial \dot{c}}(-c(t) - 4\dot{c}(t) + u(t)) + 1 = 0. \quad (6.90)$$

In order to minimize this expression with respect to $u(t)$, the constraint

$$-1 \leqslant u(t) \leqslant 1$$

must be incorporated into the solution. Therefore,

$$\min_{u(t)} \left[\dfrac{\partial S^0}{\partial c}\dot{c}(t) + \dfrac{\partial S^0}{\partial \dot{c}}(-c(t) - 4\dot{c}(t) + u(t)) + 1 \right] = \min_{u(t)} \left[\dfrac{\partial S^0}{\partial \dot{c}}(u(t)) \right].$$

Clearly, the optimal control satisfies

$$u^0(t) = -\operatorname{sgn}\left(\dfrac{\partial S^0}{\partial \dot{c}}\right). \quad (6.91)$$

Substituting this equation into Eq. (6.90), we obtain

$$\dot{c}(t)\left[\dfrac{\partial S^0}{\partial c} - 4\dfrac{\partial S^0}{\partial \dot{c}}\right] - c(t)\left[\dfrac{\partial S^0}{\partial \dot{c}}\right] - \left[\left|\dfrac{\partial S^0}{\partial \dot{c}}\right| - 1\right] = 0. \quad (6.92)$$

Equation (6.92) is a nonlinear partial differential equation that requires a digital computer to find S^0. The resulting solution for S^0 is then substituted into Eq. (6.91) in order to obtain the optimum $u^0(t)$. The conceptual block diagram of the resulting control system is illustrated in Figure 6.4. Notice that this system is also an intentional nonlinear control system, as was the previous system illustrated in Figure 6.3.

6.5. PONTRYAGIN'S MAXIMUM PRINCIPLE

In 1956, the Russian mathematicians Pontryagin, Boltyanskii, and Gamkrelidge developed the maximum principle [4, 6, 7, 16]. According to Pontryagin, the maximum principle was derived originally from the calculus of variations (see Section

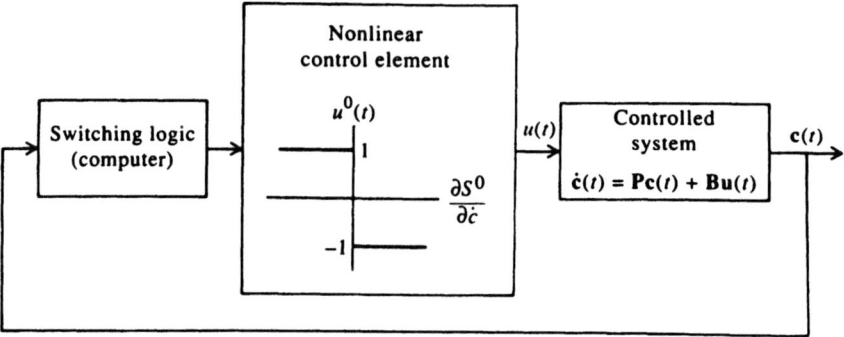

Figure 6.4 Conceptual block diagram for controlled system (containing a second-order linear plant) which is analyzed using dynamic programming.

6.3). Pontryagin's maximum principle is very similar to the calculus of variations, and is very closely related to dynamic programming. It is possible to obtain the maximum principle from dynamic programming by a simple change of variables.

In this section, Pontryagin's maximum principle and an illustrative example of its application are presented. In the following sections, it is applied to the space attitude-control problem and the lunar soft-landing problem.

Let us assume that we have a process whose dynamics are given by

$$\dot{c}_i(t) = f_i(\mathbf{c}(t), \mathbf{u}(t)), \quad i = 1, 2, \ldots, n \tag{6.93}$$

and a performance index

$$S = \int_0^T G(\mathbf{c}(t), \mathbf{u}(t), \mathbf{r}(t), t)dt, \tag{6.94}$$

which is to be minimized. The maximum principle requires that the optimal control input $\mathbf{u}^0(t)$ that minimizes S will maximize the scalar

$$H(t) = \sum_{i=1}^{n} p_i(t) f_i(\mathbf{c}(t), \mathbf{u}(t)) - G(\mathbf{c}(t), \mathbf{u}(t), \mathbf{r}(t), t_0), \tag{6.95}$$

where the process $\dot{p}_i(t)$ is defined as

$$\dot{p}_i(t) = -\frac{\partial H(t)}{\partial c_i(t)}, \quad i = 1, 2, \ldots, n. \tag{6.96}$$

The scalar H is called the Hamiltonian function; the vector $\mathbf{p}$ is known as the co-state. From Eqs. (6.93) and (6.95), $\dot{c}_i(t)$ can also be expressed in terms of H and p_i, as follows:

$$\dot{c}_i(t) = \frac{\partial H(t)}{\partial p_i(t)}, \quad i = 1, 2, \ldots, n. \tag{6.97}$$

The necessary conditions of the maximum principle can be obtained from the dynamic programming equations by a simple change of variables. Let us reconsider the Hamilton–Jacobi equation [see Eq. (6.65)]

$$\frac{\partial S^0}{\partial t_0} + \sum_{i=1}^{n} \frac{\partial S^0}{\partial c_i} f_i(\mathbf{c}(t), \mathbf{u}(t)) + G(\mathbf{c}(t), \mathbf{u}(t), \mathbf{r}(t), t_0) = 0. \tag{6.98}$$

Let us define

$$p_i(t) = -\frac{\partial S^0}{\partial c_i} \tag{6.99}$$

and

$$H(t) = \frac{\partial S^0}{\partial t_0}. \tag{6.100}$$

Substituting Eqs. (6.99) and (6.100) into (6.98), we obtain the following relationship:

$$H(t) = \sum_{i=1}^{N} p_i(t) f_i(\mathbf{c}(t), \mathbf{u}(t)) - G(\mathbf{c}(t), \mathbf{u}(t), \mathbf{r}(t), t_0), \tag{6.101}$$

which is the basic Hamiltonian function defined in Eq. (6.95). In addition, the relationship of Eq. (6.96) can be obtained from Eqs. (6.99) and (6.100) as follows:

$$\dot{p}_i(t) = \frac{dp_i(t)}{dt_0} = \frac{d}{dt_0}\left(-\frac{\partial S^0}{\partial c_i(t)}\right) = -\frac{\partial^2 S^0}{\partial t_0 \partial c_i(t)} = \frac{\partial H(t)}{\partial c_i(t)}. \tag{6.102}$$

Therefore, the basic maximum-principle relationships, as defined by Eqs. (6.95) and (6.96) have been justified from the dynamic programming relationship (the Hamilton–Jacobi equation) by a simple change of variables. However, it is important to note that this is not a proof of the maximum principle, because this derivation assumed the existance of the derivatives $\partial S^0/\partial c_i$ and $\partial S^0/\partial t_0$. On the contrary, there are many cases where they do not exist.

A. Example 6.5-1

As an example of an optimal control problem that can be solved by means of the maximum principle, let us reconsider the first-order problem in Section 6.4 which was solved by applying dynamic programming. The problem involves

1. A process whose dynamics are described by

$$\dot{c}(t) = Ac(t) + Bu(t). \tag{6.103}$$

476 INTRODUCTION TO OPTIMAL CONTROL THEORY AND ITS APPLICATIONS

2. A performance criterion given by

$$S = \int_0^\infty c^2(t)\, dt. \tag{6.104}$$

3. A control input constrained by the relationship

$$-1 \leqslant u(t) \leqslant 1. \tag{6.105}$$

From the above, the values of $f_i(\mathbf{c}(t), \mathbf{u}(t))$ and $G(\mathbf{c}(t), \mathbf{u}(t), \mathbf{r}(t), t_0)$ in the Hamiltonian equation are given by

$$f_i(\mathbf{c}(t), \mathbf{u}(t)) = Ac(t) + Bu(t), \tag{6.106}$$
$$G(\mathbf{c}(t), \mathbf{u}(t), \mathbf{r}(t), t_0) = c^2(t). \tag{6.107}$$

Substituting Eqs. (6.106) and (6.107) into the Hamiltonian equation [see Eq. (6.95)] we obtain the relationship

$$H(t) = p(t)(Ac(t) + Bu(t)) - c^2(t). \tag{6.108}$$

Applying the fundamental relationship of the maximum principle [see Eq. (6.96)]

$$\dot{p}_i(t) = -\frac{\partial H(t)}{\partial c_i(t)} \tag{6.109}$$

to Eq. (6.108), we obtain the first equation to be solved:

$$\dot{p}(t) = -\frac{\partial H(t)}{\partial c(t)} = -p(t)A + 2c(t). \tag{6.110}$$

A second equation can be obtained from the fact that the Hamiltonian is maximized when the performance criterion is minimized. Maximizing Eq. (6.108) with respect to $u(t)$:

$$\max_{u(t)} H[p(t)Bu(t)]. \tag{6.111}$$

Therefore,

$$u^0(t) = \text{sgn}(Bp(t)). \tag{6.112}$$

Notice that this is equivalent to the value found for $u(t)$ in Eq. (6.80) using dynamic programming. This can be shown by substituting Eq. (6.99) into Eq. (6.112). The result is Eq. (6.80), which we obtained using dynamic programming. Substituting Eq. (6.112) into (6.103), we obtain the second necessary condition

$$\dot{c}(t) = Ac(t) + B\,\text{sgn}(Bp(t)). \tag{6.113}$$

The solution to this optimal control problem, utilizing the maximum principle, has been reduced to the solution of the nonlinear ordinary differential equations given by Eqs. (6.110) and (6.113) for $p(t)$, which would then be substituted into Eq. (6.112) in order to obtain the optimum $u^0(t)$. In order to solve these equations, the boundary conditions must be utilized. Note that they define a two-point boundary-value problem.

It is interesting to compare the dynamic-programming solution with that obtained with the maximum principle for this problem. The dynamic-programming solution was reduced to the solution of one nonlinear partial differential equation [see Eq. (6.83)]. The maximum-principle solution of the same problem was reduced to the solution of two nonlinear ordinary differential equations [see Eqs. (6.110) and (6.113)]. Although both techniques result in equations that require digital computers for solution, the two nonlinear first-order ordinary differential equations (obtained using the maximum principle) are easier to solve than the nonlinear partial differential equation, which is a function of two variables (obtained using dynamic programming). In practice, however, the choice between these approaches will depend to a great extent on the particular problem. The application of Pontryagin's maximum principle to the synthesis of optimum attitude controllers for space vehicles is discussed in the following section, and its application to the lunar soft-landing problem is described in Section 6.7.

6.6. APPLICATION OF THE MAXIMUM PRINCIPLE TO THE SPACE ATTITUDE-CONTROL PROBLEM

Optimal control theory has been applied to a wide variety of important problems [4, 6, 8–10, 23–36]. It has been used to solve problems concerning the attitude control of space vehicles, lunar soft landing, control of traffic flow, the orbit transfer problem for interplanetary space vehicles, chemical process control problems, and problems concerned with communication systems. In this section, the use of the maximum principle for solving the attitude-control problem of various space vehicles is illustrated. The objective is to synthesize the optimum strategy for controlling the space vehicle to satisfy a given performance criterion.

Attitude control of a space vehicle encompasses a very wide variety of problems. During powered flight, the attitude-control system receives commands from a guidance system and controls the attitude of the vehicle. This causes the vehicle to pitch or yaw and results in changes in attitude and/or direction of the flight path. After the vehicle has attained the desired orbit, it is attitude stabilized with respect to some reference such as the Earth, Sun, or the stars. During reentry into the Earth's atmosphere, the vehicle is pitched over to the proper angle from the reference attitude by signals from a reference gyroscope. Then the firing of a retrorocket places the vehicle on a transfer orbit into the Earth's atmosphere. The attitude-control problem is even more complicated for manned space stations where orbit rendezvous is required for purposes of orbital refueling, crew changes, and/or satellite inspection.

This section is concerned with the attitude stabilization of a manned or unmanned vehicle in orbit about the Earth. It is to be stabilized perpendicular to the Earth's local vertical, as shown in Figure 6.5. Consider one plane of such a space vehicle orbiting the Earth that is slaved to the local vertical via horizon sensors and gyros. A

478 INTRODUCTION TO OPTIMAL CONTROL THEORY AND ITS APPLICATIONS

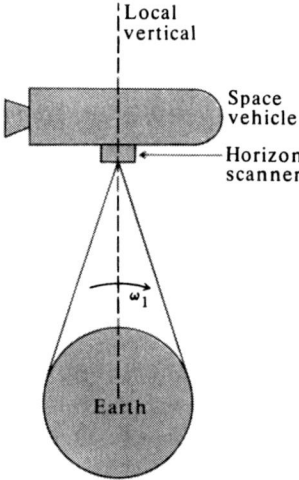

Figure 6.5 Attitude control via horizon scanners to establish the local vertical.

physical model of the problem is illustrated in Figure 6.6 and a block diagram of a typical attitude-control system is shown in Figure 6.7. The reference input position to the horizon tracker is denoted by $R(s)$ and the resultant output position of the attitude-control system is denoted by $C(s)$. The rate of change of the local vertical with respect to the Earth is denoted by ω_1, and the rate of the vehicle relative to the local vertical is denoted by ω_2. The vehicle inertial rate, which is the sum of ω_1 and ω_2, is denoted by ω_T.

Assuming that friction and disturbing forces are negligible, the motion of the space vehicle is given by the following simple second-order differential equation

$$\ddot{c}(t) = T(t)/J \qquad (6.114)$$

where $T(t)$ represents the control torque generated by the space vehicle to align its attitude with the reference. By defining

$$u(t) = \frac{T(t)}{J}, \quad \text{where } |u(t)| \leq 1, \qquad (6.115)$$

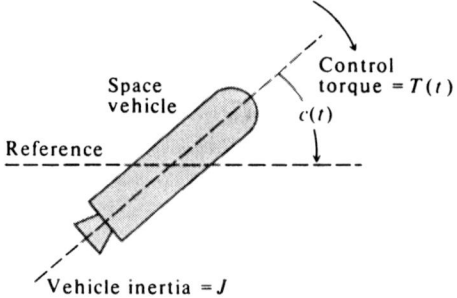

Figure 6.6 Attitude-control problem for one axis.

6.6. APPLICATION OF THE MAXIMUM PRINCIPLE TO THE SPACE ATTITUDE-CONTROL PROBLEM

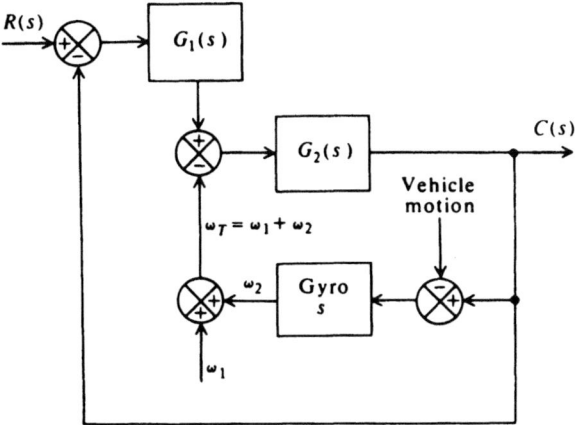

Figure 6.7 One axis of an attitude-control system.

Eq. (6.114) can be rewritten as

$$\ddot{c}(t) = u(t). \tag{6.116}$$

Utilizing state-variable notation, this second-order differential equation can be rewritten as two first-order differential equations. Let

$$c_1(t) = c(t), \quad c_2(t) = \dot{c}(t); \quad \dot{c}_1(t) = c_2(t), \quad \dot{c}_2(t) = u(t). \tag{6.117}$$

Then

$$\dot{\mathbf{c}}(t) = \mathbf{P}\mathbf{c}(t) + \mathbf{B}\mathbf{u}(t), \tag{6.118}$$

where

$$\mathbf{c}(t) = \begin{bmatrix} c_1(t) \\ c_2(t) \end{bmatrix}, \quad \dot{\mathbf{c}}(t) = \begin{bmatrix} \dot{c}_1(t) \\ \dot{c}_2(t) \end{bmatrix}, \quad \mathbf{P} = \begin{bmatrix} 0 & 1 \\ 0 & 0 \end{bmatrix},$$

$$\mathbf{B} = \begin{bmatrix} 0 & 0 \\ 0 & 1 \end{bmatrix}, \quad \mathbf{u}(t) = \begin{bmatrix} 0 \\ u(t) \end{bmatrix}. \tag{6.119}$$

The basic attitude-control problem is to maintain the vehicle at a referenced attitude. The desired equilibrium state for this problem is assumed to be the stable node at the origin of the $c_1 c_2$ plane.

In the following analysis the synthesis of the optimal attitude control system for several practical attitude-control problems is considered. For example, due to a disturbance torque or the command of a new reference attitude, what is the best strategy to minimize its response time? Other considerations may dictate that the amount of fuel or energy be minimized. These possibilities lead to the following problems:

1. the minimum-time problem,
2. the minimum fuel-consumption problem,
3. the minimum energy problem.

In all cases, it will be assumed that $u(t)$ is constrained to be $1 \leq u(t) \leq -1$.

1. The Minimum-Time Problem. Although the minimum-time problem can be solved with conventional techniques [34], it is synthesized here, utilizing Pontryagin's maximum principle as an introduction to the application of optimal control concepts. In addition, its solution is useful for comparison with the other problems that are subsequently considered.

Because it is desired to minimize the response time of the space vehicle, the performance criterion for the minimum-time problem is given by

$$S = \int_0^T dt = T. \tag{6.120}$$

Here the loss function is unity; that is,

$$G(\mathbf{c}(t), \mathbf{u}(t), \mathbf{r}(t), t) = 1. \tag{6.121}$$

Substituting Eqs. (6.117) and (6.121) into the expression for the Hamiltonian, Eq. (6.95), the following expression is obtained:

$$H(t) = p_1(t)c_2(t) + p_2(t)u(t) - 1. \tag{6.122}$$

The values of $p_1(t)$, $p_2(t)$, and $c_2(t)$ can be evaluated by applying Eqs. (6.96) and (6.97) to Eq. (6.122). The results are as follows:

$$\dot{p}_1(t) = -\frac{\partial H(t)}{\partial c_1(t)} = 0, \quad p_1(t) = K_1, \tag{6.123}$$

$$\dot{p}_2(t) = -\frac{\partial H(t)}{\partial c_2(t)} = -p_1(t), \quad p_2(t) = K_2 - p_1 t = K_2 - K_1 t, \tag{6.124}$$

where K_1 and K_2 represent constants of integration, and

$$\dot{c}_1(t) = \frac{\partial H(t)}{\partial p_1(t)} = c_2(t), \tag{6.125}$$

$$\dot{c}_2(t) = \frac{\partial H(t)}{\partial p_2(t)} = u(t). \tag{6.126}$$

Because the term $p_1(t)c_2(t) - 1$ in Eq. (6.122) is independent of the input $u(t)$, the maximization of the Hamiltonian function is concerned only with

$$\max_{u(t)} H[p_2(t)u(t)]. \tag{6.127}$$

6.6. APPLICATION OF THE MAXIMUM PRINCIPLE TO THE SPACE ATTITUDE-CONTROL PROBLEM

It is seen from Eq. (6.127) that the Hamiltonian function is maximized by choosing

$$u^0(t) = \text{sgn}[p_2(t)]. \tag{6.128}$$

Substituting Eq. (6.124) into Eq. (6.128), we obtain

$$u^0(t) = \text{sgn}[K_2 - K_1 t]. \tag{6.129}$$

The following conclusions can be drawn from this result:

1. The optimum input for minimization of the response time is piecewise constant.
2. The optimum input for minimum-time operation takes on only the values ± 1.
3. The sign of the optimum input for mimimum-time operation can change its value only once because $p_2(t)$ is a linear function of time [see Eq. (6.124)], and as t increases from zero to infinity, $u^0(t)$ will switch once when $K_2 - K_1 t = 0$. When $(K_2 - K_1 t) < 0$, $u^0(t) = -1$, and when $(K_2 - K_1 t) > 0$, $u^0(t) = 1$.

These conclusions clearly imply that the attitude-control system for the space vehicle should be bang-bang (on-off) when minimization of response time is the performance criterion. Physically this means that accelerating and then decelerating is the best that can be done in order to minimize the response time. The period of each action depends on the initial conditions of position and velocity.

The phase-plane representation and optimal switching curve can be formulated from consideration of Eqs. (6.116) and (6.117), when $u(t) = \pm 1$. When $u(t) = 1$, the following expressions are obtained:

$$\dot{c}_2(t) = 1, \tag{6.130}$$
$$c_2(t) = t + A_1, \tag{6.131}$$
$$\dot{c}_1(t) = t + A_1, \tag{6.132}$$
$$c_1(t) = \tfrac{1}{2}t^2 + A_1 t + A_2, \tag{6.133}$$

where A_1 and A_2 are constants of integration. By completing the square, Eq. (6.133) can be rearranged as follows:

$$c_1(t) = \tfrac{1}{2}(t + A_1)^2 + (A_2 - \tfrac{1}{2}A_1^2). \tag{6.134}$$

Substituting Eq. (6.132) into (6.134), we obtain the following relationship:

$$c_1(t) = \tfrac{1}{2}\dot{c}_1^2(t) + A_3, \tag{6.135}$$

where

$$A_3 = A_2 - \tfrac{1}{2}A_1^2. \tag{6.136}$$

Equation (6.135) defines the switching curve when $u = 1$. Similarly, the switching curve when $u = -1$ can be obtained as

$$c_1(t) = -\tfrac{1}{2}\dot{c}_1^2(t) + A_6, \tag{6.137}$$

where

$$A_6 = A_5 + \frac{1}{2}A_4^2. \tag{6.138}$$

The switching curves defined by Eqs. (6.135) and (6.137) define parabolas in the $c_1(t)\dot{c}_1(t)$ plane. The corresponding phase portrait and switching line for the minimum-time system is illustrated in Figure 6.8. The phase trajectory travels from its initial conditions to the switching line defined by Eq. (6.135) or Eq. (6.137). At the instant the state arrives at the switching curve, the system switches its control to the opposite phase and remains at this value until the state reaches the stable node at the origin. If the initial conditions are above the curve AOB, the system is under the control of $u(t) = -1$ until it reaches the arc BO. Then it switches to the control $u(t) = 1$, where it stays until the equilibrium state is reached. When the initial conditions are below AOB, the system is controlled by $u(t) = 1$ until the state reaches arc AO. At the instant that it arrives, the system switches to the control $u(t) = -1$ and remains at this value until the equilibrium state is reached.

Figure 6.9 shows the block diagram of the optimum control system just designed (A_3 and A_6 are assumed to equal zero). The basic control element required has ideal relay characteristics that can easily be implemented. The control system will accelerate and then decelerate in order to minimize the response time of the attitude-control system. It should be noted that the exact period of each phase is dependent upon the initial conditions relative to the switching curves.

2. The Minimum Fuel-Consumption Problem. In this problem it is assumed that the attitude-control system is powered by a reaction jet. It is desired to minimize the

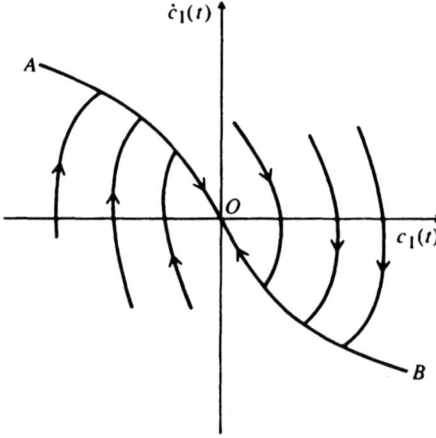

Figure 6.8 Phase-plane representation for the minimum-time problem.

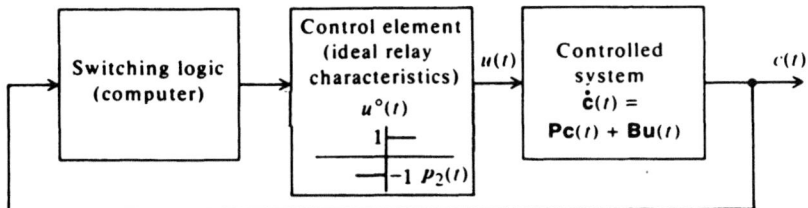

Figure 6.9 Synthesized system to minimize response time.

fuel consumption of the space vehicle. It is also assumed that a pure minimum fuel-consumption problem is inadequate from a practical viewpoint because it may result in an excessive response time. Therefore, the actual practical problem considered here is that of minimizing fuel consumption over a certain period of time. Although response time is of secondary importance in this problem, it must be considered. It is theoretically possible to bring the state of the system to the stable node with an arbitrarily small amount of fuel if the response time were not bounded. However, the response time must be considered fixed for practical systems. Therefore, it is assumed that the total response time is limited to T seconds.

In synthesizing a meaningful optimal control system, the reaction jet system must be understood. An important parameter used to characterize the performance of a reaction jet engine is its specific impulse $I_{sp}(t)$. It is defined as the ratio of the thrust F of a hypothetical engine to a propellant flow rate $\dot{\omega}(t)$ of one pound of propellant per second:

$$I_{sp}(t) = F/\dot{\omega}(t). \tag{6.139}$$

The units of I_{sp} from this relationship are pounds of thrust per pounds of propellant per second and are usually expressed in seconds. The specific impulse of an engine is indicative of how effectively each pound of propellant is utilized in producing a thrust force to the vehicle.

Because it is desired to minimize fuel consumption of the space vehicle over a period of T seconds, the performance criterion for this problem is given by

$$S = \int_0^T |u(t)|dt, \tag{6.140}$$

where $u(t)$ is defined as the propellant rate of flow and is constrained in magnitude by Eq. (6.115). For this problem, the loss function represents a measure of the total fuel flow rate:

$$G(\mathbf{c}(t), \mathbf{u}(t), \mathbf{r}(t), t) = |u(t)|. \tag{6.141}$$

Its time integral, as given by Eq. (6.140), represents a measure of the total fuel consumed in T seconds.

Substituting Eqs. (6.117) and (6.141) into the expression for the Hamiltonian function, Eq. (6.95), the following expression is obtained:

$$H(t) = p_1(t)c_2(t) + p_2(t)u(t) - |u(t)|. \tag{6.142}$$

The values of $p_1(t)$, $p_2(t)$, and $c_2(t)$ can be evaluated by applying Eqs. (6.96) and (6.97) to Eq. (6.142). The results are the same as Eqs. (6.123)–(6.126). Because the term $p_1(t)c_2(t)$ in Eq. (6.142) is independent of the input $u(t)$, the maximization of the Hamiltonian function is concerned only with

$$\max_{u(t)} H[p_2(t)u(t) - |u(t)|]. \tag{6.143}$$

It is obvious from Eq. (6.143) that the Hamiltonian function is maximized by choosing $u(t)$ as follows:

$$\begin{aligned} u^0(t) &= \text{sgn}[p_2(t)] \quad \text{for } |p_2(t)| \geqslant 1, \\ u^0 &= 0 \quad \text{for } |p_2(t)| < 1. \end{aligned} \tag{6.144}$$

Substituting Eq. (6.124) into Eq. (6.144), the following optimal control inputs are obtained:

$$\begin{aligned} u^0(t) &= \text{sgn}[K_2 - K_1 t], \quad \text{for } |p_2(t)| \geqslant 1, \\ u^0 &= 0 \quad \text{for } |p_2(t)| < 1. \end{aligned} \tag{6.145}$$

The following conclusions can be drawn from this result:

1. The optimum input is piecewise constant.
2. The optimum input can have values of only ± 1 and 0.

Although K_1 and K_2 are not known exactly, the fact that $p_2(t)$ is a linear function of time [see Eq. (6.124)] means that $u(t)$ must proceed in time as a nonrepeating sequence of values having the form $(\pm, 1, 0, \mp 1)$. These conclusions imply that the attitude-control system should be bang-bang and incorporate a dead zone. This function can be easily implemented. Physically, this means that in order to minimize the fuel consumption over a bounded period of time, the best one can do is to accelerate, coast at a constant velocity, and then decelerate. The periods for each action depend on the initial conditions of position and velocity.

The phase-plane representation and optimal switching curves can be formulated from considerations of Eqs. (6.116) and (6.117) when $u(t) = \pm 1$ and 0. When $u(t) = \pm 1$, results analogous to Eqs. (6.135) and (6.137) are obtained for the minimum-time problem. When $u(t) = 0$,

$$\dot{c}_2(t) = 0, \tag{6.146}$$
$$c_2(t) = \text{constant} \tag{6.147}$$

and the switching curve is a horizontal line parallel to the $c_1(t)$ axis. Typical phase trajectories for the minimum fuel-consumption problem when the response time is bounded are illustrated in Figure 6.10 as curves *lmno* and *l'm'n'o*. These results lead to the conclusion that if reaction jets are used for attitude control, then a nozzle that simply opens or closes without any intermediate settings provides the optimal system.

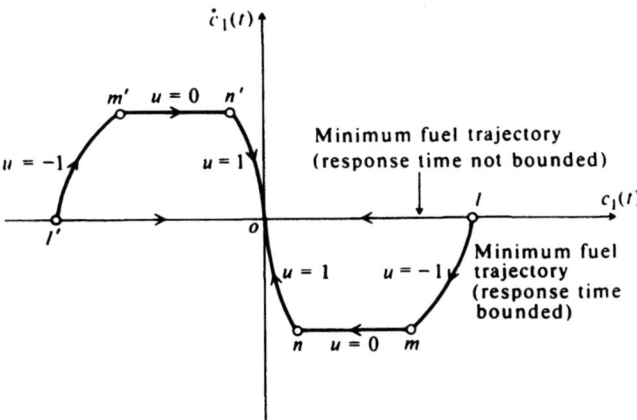

Figure 6.10 Typical phase trajectories for the minimum-fuel problem.

It is interesting to compare the phase trajectories of a pure minimum-fuel problem where the response time is not bounded and that of a minimum-fuel system where the response time is bounded. Typical phase trajectories for the pure minimum-fuel optimal problem are curves lo and $l'o$ in Figure 6.10. The corresponding time response of curves $lmno$ and lo can be obtained from an examination of the phase trajectories. It was shown in Section 5.16 that the variation of time along a phase trajectory can be obtained from [see Eq. (5.157)]

$$t = \int_l^o \frac{1}{\dot{c}_1(t)} dc_1(t). \tag{6.148}$$

The integral represents the area under the reciprocal phase-plane trajectory. A corresponding reciprocal plot for the phase trajectories $lmno$ and lo is illustrated in Figure 6.11. The points l and o area assumed to lie at infinity; this has a negligible effect on the area under curve $lmno$, because the area under these points approaches zero. However, the area under the rectangular curve lo is infinity. This illustrates that a pure minimum-fuel problem, where the response time is not bounded, results in an infinite response time and is not very practical.

Figure 6.12 shows the block diagram of the optimal control system. The basic control element required has the characteristics of a relay with a dead zone.

3. The Minimum-Energy Problem. In this problem it is assumed that the energy utilized in the attitude-control system of a space vehicle is to be minimized. This is a very practical problem for space vehicles utilizing momentum-exchanging devices such as control-moment gyros, inertia wheels, fluid flywheels, and magnetic moment devices in their attitude-control systems. All of these systems utilize electrical energy derived from such sources as batteries, fuel cells, and/or solar cells.

It is assumed that the input to the attitude-control system is an electrical signal $u(t)$. In addition, it is desired to utilize a minimum amount of electrical energy over a bounded period of time in order to accomplish a desired control. We know that the square of the signal utilized is proportional to power, and the time integral of power is

486 INTRODUCTION TO OPTIMAL CONTROL THEORY AND ITS APPLICATIONS

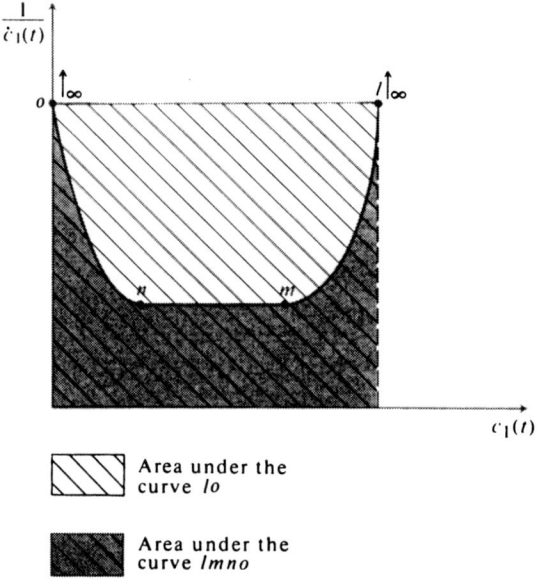

Area under the curve lo

Area under the curve $lmno$

Figure 6.11 Reciprocal phase-plane plot corresponding to Figure 6.10.

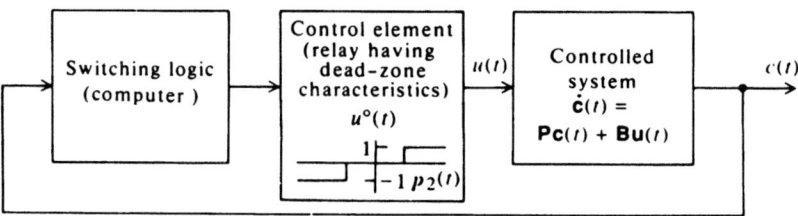

Figure 6.12 Synthesized system to minimize fuel consumption over a specified period of time.

energy. Because the square of the control signal is proportional to the power required for control, and the energy required is proportional to the integral of the square of the control signal, the minimum-energy problem can be formulated as a problem of minimum integral square control. The performance criterion is thus given by

$$S = \int_0^T u^2(t)dt. \tag{6.149}$$

Here the loss function represents power

$$G(\mathbf{c}(t), \mathbf{u}(t), \mathbf{r}(t), t) = u^2(t) \tag{6.150}$$

and its time integral, as given by Eq. (6.149), represents a measure of energy. Substituting Eqs. (6.117) and (6.150) into the expression for the Hamiltonian function. Eq. (6.95), we obtain

$$H(t) = p_1(t)c_2(t) + p_2(t)u(t) - u^2(t). \tag{6.151}$$

6.6. APPLICATION OF THE MAXIMUM PRINCIPLE TO THE SPACE ATTITUDE-CONTROL PROBLEM

The values of $p_1(t)$, $p_2(t)$, and $c_2(t)$, can be evaluated by applying Eqs. (6.96) and (6.97) to Eq. (6.151). Since the term $p_1(t)c_2(t)$ in Eq. (6.151) is independent of the input $u(t)$, the maximization of the Hamiltonian function is only concerned with the expression

$$\max_{u(t)} H[p_2(t)u(t) - u^2(t)]. \quad (6.152)$$

It is obvious from Eq. (6.152) that two cases exist for the Hamiltonian function to be maximized. To find the optimal value of $u^0(t)$, differentiation of Eq. (6.152) yields

$$\frac{\partial}{\partial u}[p_2(t)u(t) - u^2(t)] = 0,$$

$$u^0(t) = p_2(t)/2. \quad (6.153)$$

Therefore, if $|p_2(t)| \leqslant 2$, $u^0(t) = p_2(t)/2$. When $|p_2(t)| > 2$, then

$$u^0(t) = \text{sgn}[p_2(t)] \quad (6.154)$$

results in a maximum value of the Hamiltonian function. Substituting Eq. (6.124) into Eqs. (6.153) and (6.154), we obtain

$$u^0(t) = \frac{K_2}{2} - \frac{K_1}{2}t = K_2' - K_1't \quad \text{for } |p_2(t)| \leqslant 2,$$
$$u^0(t) = \text{sgn}[p_2(t)] = \text{sgn}[K_2 - K_1 t] \quad \text{for } |p_2(t)| > 2. \quad (6.155)$$

These results indicate that the control is linear over a fixed range and saturates whenever $|p_2(t)| > 2$. These conclusions imply that the attitude-control system for the space vehicle should contain a limiter whose output is linear for small signals and saturates for large signals in order to minimize energy.

Figure 6.13 shows the block diagram of the control system. The basic control element required is a limiter which will operate linearly until saturation is reached. The basic message of this solution is that the solar energy supplied to the attitude-control system is limited, and the power source cannot supply more power when the demand exceeds the supply. No doubt, you have experienced this with your local power company on occasion, but then we call it a "brownout."

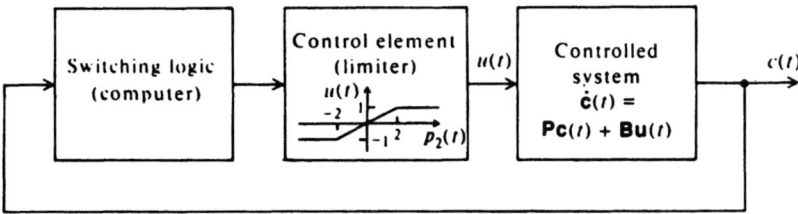

Figure 6.13 Synthesized system to minimize energy over a specified period of time.

488 INTRODUCTION TO OPTIMAL CONTROL THEORY AND ITS APPLICATIONS

The attitude-control problem has been used as a convenient illustration of the application of Pontryagin's maximum principle. The results of this analysis indicated that the resulting structure of the control element was in the form of a relay to minimize the response time, a relay with a dead zone to minimize the fuel consumption over a bounded period of time, and a limiter to minimize energy over a bounded period of time.

6.7. APPLICATION OF THE MAXIMUM PRINCIPLE TO THE LUNAR SOFT-LANDING PROBLEM

The Apollo 11 mission, in which astronauts Neil Armstrong and Edwin Aldrin successfully soft-landed the Lunar Excursion Module (LEM) on the lunar surface, was a historic event. Figure 6.14a is a photograph of the LEM vehicle, taken by astronaut Michael Collins from the Apollo Command Module window, after they separated. The problem of synthesizing an optimal control policy for minimal fuel thrust, during the terminal phase of a lunar soft-landing mission, is one that can best be solved using modern optimal control theory [37]. Let us first look at this problem from the basic physics involved in a lunar soft landing as shown in Figure 6.14b. It is assumed in this problem that the motion of the vehicle is vertical and subject to the following conditions:

(a) The only forces acting on the vehicle are its own weight and the thrust which acts as a braking force.
(b) The Moon is flat in the vicinity of the desired landing point.
(c) The propulsion system is capable of a mass-flow rate $\dot{m}(t)$ between zero and a lower fixed limit of α:

$$-\alpha \leqslant \dot{m}(t) \leqslant 0. \qquad (6.156)$$

On these assumptions, the motion of the vehicle is governed by the relation

$$\ddot{x}(t) = -\frac{K\dot{m}(t)}{m(t)} + g, \qquad (6.157)$$

where

$x(t)$ = altitude
$m(t)$ = total mass,
$\dot{m}(t)$ = mass flow rate ≤ 0,
g = acceleration of gravity at the surface of the moon,
K = velocity of exhaust gases = constant > 0.

In addition, system performance is based on the following criterion:

$$S = \int_0^T \dot{m}(t)dt. \qquad (6.158)$$

6.7. APPLICATION OF THE MAXIMUM PRINCIPLE TO THE LUNAR SOFT-LANDING PROBLEM

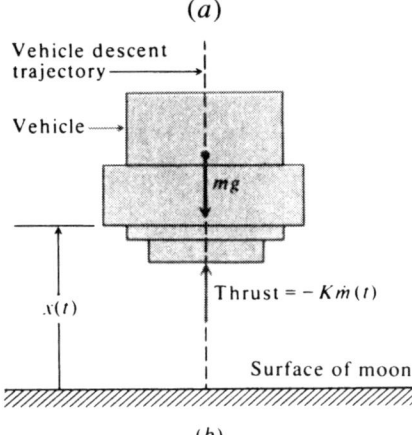

Figure 6.14 (a) Apollo II: Astronauts Neil Armstrong and Edwin Aldrin are inside the lunar module separated from the Apollo command module (Official NASA photograph) (b) Forces acting on vehicle.

Utilizing the maximum principle, we wish to determine the form of the optimal control policy.

Defining the states of this second-order system by

$$c_1(t) = x(t), \qquad (6.159)$$
$$c_2(t) = \dot{x}(t), \qquad (6.160)$$

and the input by

$$u(t) = \dot{m}(t), \tag{6.161}$$

the state equations of these systems are given by

$$\dot{c}_1(t) = c_2(t), \tag{6.162}$$

$$\dot{c}_2(t) = -\frac{K}{m(t)}u(t) + g. \tag{6.163}$$

The Hamiltonian function [see Eq. (6.95)] for this problem is given by

$$H(t) = \sum_{i=1}^{2} p_i(t)\dot{c}_i(t) - u(t) \tag{6.164}$$

or

$$H(t) = p_1(t)\dot{c}_1(t) + p_2(t)\dot{c}_2(t) - u(t). \tag{6.165}$$

Substituting Eqs. (6.162) and (6.163) into Eq. (6.165), we obtain the following:

$$H(t) = p_1(t)c_2(t) + p_2(t)\left(-\frac{K}{m(t)}u(t) + g\right) - u(t). \tag{6.166}$$

Simplifying this expression, we obtain the following:

$$H(t) = p_1(t)c_2(t) + p_2(t)g + \left[-p_2(t)\frac{K}{m(t)} - 1\right]u(t). \tag{6.167}$$

Applying the fundamental co-state relationship of the maximum principle [see Eq. (6.96)] to Eq. (6.167), we obtain the following:

$$\dot{p}_1(t) = -\frac{\partial H(t)}{\partial c_1(t)} = 0, \qquad p_1(t) = K_1, \tag{6.168}$$

$$\dot{p}_2(t) = -\frac{\partial H(t)}{\partial c_2(t)} = -p_1(t), \qquad p_2(t) = -K_1 t + K_2. \tag{6.169}$$

A third equation can be obtained from the determination of an admissible $u(t)$ which maximizes the Hamiltonian function. From Eq. (6.167), we wish to maximize

$$\max_{u(t)} H\left[\left(-p_2(t)\frac{K}{m(t)} - 1\right)u(t)\right]. \tag{6.170}$$

Defining

$$L = -p_2(t)\frac{K}{m(t)} - 1, \tag{6.171}$$

6.7. APPLICATION OF THE MAXIMUM PRINCIPLE TO THE LUNAR SOFT-LANDING PROBLEM

then an optimal control function $u^0(t)$ is obtained when

$$u^0(t) = \dot{m}(t) = \begin{cases} -\alpha, & \text{if } L < 0 \\ 0, & \text{if } L > 0. \end{cases} \quad (6.172)$$

In order to determine the optimal strategy for $u^0(t)$ in terms of $p_2(t)$, let us analyze L further.

$$L = -\frac{p_2(t)K}{m(t)} - 1 = -\left(\frac{p_2(t)K}{m(t)} + 1\right). \quad (6.173)$$

Therefore, note that

$$L < 0, \quad \text{when } \frac{p_2(t)K}{m(t)} + 1 > 0, \quad (6.174)$$

and

$$L > 0, \quad \text{when } \frac{p_2(t)K}{m(t)} + 1 < 0. \quad (6.175)$$

It is assumed that K is a positive constant, and because the mass $m(t)$ is a positive value, also, then for

$$L < 0, \quad p_2(t) > -\frac{m(t)}{K}, \quad (6.176)$$

and for

$$L > 0, \quad p_2(t) < -\frac{m(t)}{K}. \quad (6.177)$$

Therefore, the optimal control input, $u^0(t)$ is given by

$$u^0(t) = \dot{m}(t) = \begin{cases} -\alpha, & \text{if } p_2(t) > -\frac{m(t)}{K}, \\ 0, & \text{if } p_2(t) < -\frac{m(t)}{K}. \end{cases} \quad (6.178)$$

Therefore, the optimal thrust program in the vertical landing direction consists of either full thrust or a period of zero thrust (free-fall) for minimizing fuel consumption. Substituting Eq. (6.178) into the system dynamics given by Eq. (6.157), we obtain the third equation which must be solved:

$$\ddot{x}(t) = \begin{cases} \dfrac{K\alpha}{m(t)} + g, & \text{for } p_2 > -\dfrac{m(t)}{K}, \\ g, & \text{for } p_2(t) < -\dfrac{m(t)}{K}. \end{cases} \quad (6.179)$$

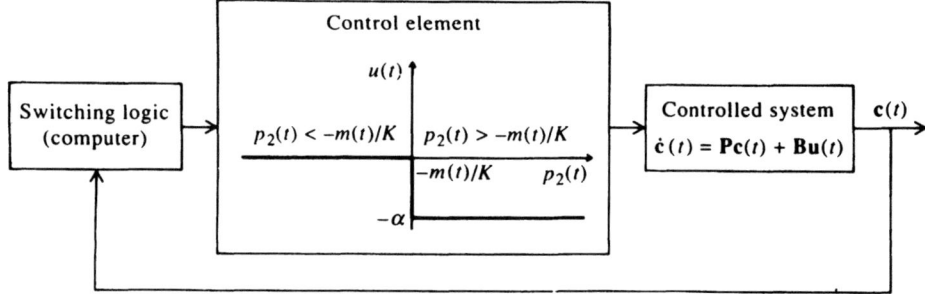

Figure 6.15 Synthesized system to minimize fuel consumption in the vertical axis of the lunar soft-landing problem.

Therefore, the solution to this problem using the maximum principle has been reduced to the solution of the nonlinear ordinary differential equations given by (6.168), (6.169), and (6.179).

An analysis of the switching strategy is very interesting. Equation (6.179) shows that switching occurs at the critical point $p_2(t) = -m(t)/K$. Because we note from Eq. (6.169) that $p_2(t)$ is the equation of a straight line, $p_2(t)$ will equal the critical switching value only once. [Note that the performance criterion of Eq. (6.158) is given over the time period from 0 to T]. In conclusion, the best optimal control input strategy in the vertical axis for soft landing on the Moon's surface is to fire the braking rockets just once for minimizing fuel consumption. The conceptual block diagram for the resulting optimal control system is illustrated in Figure 6.15.

6.8. ILLUSTRATIVE PROBLEMS AND SOLUTIONS

This section provides a set of illustrative problems and their solutions to supplement the material presented in Chapter 6.

I6.1. We wish to determine the optimal control policy using the maximum principle for a system whose controlled process is a first-order linear plant whose dynamics are given by

$$\dot{c}(t) = 2c(t) + 4u(t)$$

and whose performance criterion is given by

$$S = \int_0^T u^2(t)\,dt.$$

The controlled input must satisfy the following relationship:

$$-2 \leq u(t) \leq 2.$$

Determine the optimal control input, $u^0(t)$, which satisfies the requirements of this problem.

SOLUTION: From Eq. (6.95):

$$H(t) = \sum_{i=1}^{n} p_i(t) f_i[\mathbf{c}(t), \mathbf{u}(t)] - G[\mathbf{c}(t), \mathbf{u}(t), \mathbf{r}(t), t_0].$$

Substituting the problem parameters into this equation, we obtain the following:

$$H(t) = p(t)[2c(t) + 4u(t)] - u^2(t).$$

Maximizing this expression with respect to $u(t)$, we obtain:

$$\max_{u(t)} H[4p(t)u(t) - u^2(t)].$$

Taking the partial derivative of this expression with respect to $u(t)$, we obtain the following:

$$\frac{\partial}{\partial u}[4p(t)u(t) - u^2(t)] = 0$$

$$4p(t) - 2u(t) = 0$$

$$\therefore u^0(t) = 2p(t).$$

If $|p(t)| \leq 1$, $u^0(t) = 2p(t)$, and if $|p(t)| \geq 1$, $u^0(t) = 2 \text{ sgn } [p(t)]$.

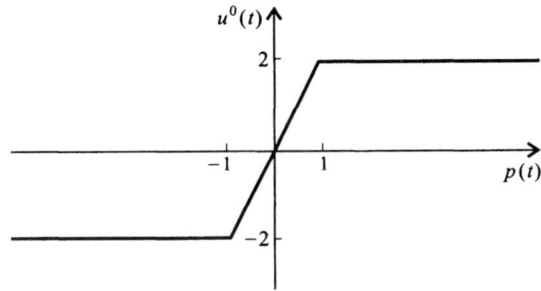

Figure I6.1

16.2. We wish to determine the optimal control policy for a control system described by the following differential equation:

$$\ddot{c}(t) + 6\dot{c}(t) + 5c(t) = 8u(t).$$

It is desired to minimize the performance criterion given by

$$S = \int_0^T u^2(t) \, dt$$

Assume that the input $u(t)$ must satisfy the following constraints:

$$-1 \leq u(t) \leq 1.$$

Determine the optimal control policy $u^0(t)$ using:

(a) Dynamic programming and

(b) Maximum principle.

(c) Show that the optimal control policy $u^0(t)$ obtained in parts (a) and (b) are equivalent.

SOLUTION: (a) Let $x_1(t) = c(t)$ and $x_2(t) = \dot{c}(t)$. Therefore,

$$\dot{x}_1(t) = x_2(t)$$

$$\dot{x}_2(t) = -5x_1(t) - 6x_2(t) + 8u(t).$$

From the Hamilton–Jacobi equation [see Eq. (6.65)],

$$\frac{\partial S^0}{\partial c(t)} \dot{c}(t) + \frac{\partial S^0}{\partial \dot{c}(t)} [-5c(t) - 6\dot{c}(t) + 8u(t)] + u^2(t) = 0.$$

To find the minimum of this function with respect to $u(t)$:

$$\frac{\partial}{\partial u(t)} \left[\frac{\partial S^0}{\partial \dot{c}(t)} (8u(t) + u^2(t)) \right] = 0,$$

$$8 \frac{\partial S^0}{\partial \dot{c}(t)} + 2u(t) = 0,$$

$$u(t) = -4 \frac{\partial S^0}{\partial \dot{c}(t)}.$$

Therefore,

$$u^0(t) = -4 \frac{\partial S^0}{\partial \dot{c}(t)} \quad \text{for} \quad \left| \frac{\partial S^0}{\partial \dot{c}(t)} \right| \leq \frac{1}{4},$$

$$u^0(t) = \text{sgn}\left(\frac{\partial S^0}{\partial \dot{c}(t)}\right) \quad \text{for} \quad \left| \frac{\partial S^0}{\partial \dot{c}(t)} \right| > \frac{1}{4}.$$

A sketch of this result is given by:

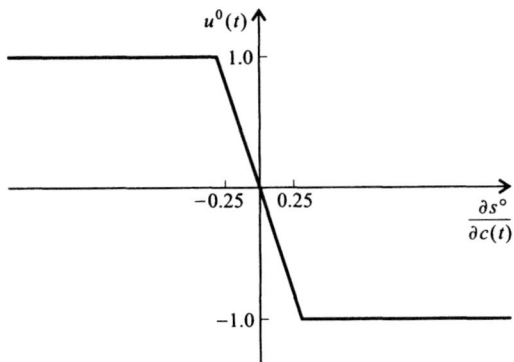

Figure I6.2(i)

(b) From the Hamiltonian function [see Eq. (6.101)].

$$H = p_1(t)x_2(t) + p_2(t)[-5x_1(t) - 6x_2(t) + 8u(t)] - u^2(t).$$

To find the maximum of this function with respect to $u(t)$:

$$\frac{\partial}{\partial u(t)}[8p_2(t)u(t) - u^2(t)] = 0,$$

$$8p_2(t) - 2u(t) = 0,$$

and

$$u(t) = 4p_2(t).$$

Therefore,

$$u^0(t) = 4p_2(t) \quad \text{for } |p_2(t)| \leq \frac{1}{4},$$
$$u^0(t) = \operatorname{sgn} p_2(t) \quad \text{for } |p_2(t)| > \frac{1}{4}.$$

A sketch of this result is given by:

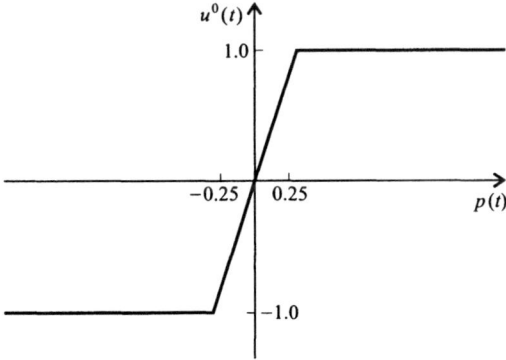

Figure I6.2(ii)

(c) Since

$$p_i = -\frac{\partial S^0}{\partial c_i}$$

[see Eq. 6.99)], then

$$u^0(t) = \operatorname{sgn} p_2(t) = \operatorname{sgn}\left(-\frac{\partial S^0}{\partial \dot{c}(t)}\right) = -\operatorname{sgn}\left(\frac{\partial S^0}{\partial \dot{c}(t)}\right).$$

PROBLEMS

6.1. Consider a simple open-loop system that consists of an integration and an amplifier having a gain of 2. It is desired to control the state of this system between $x(0)$ and $x(T)$ in order that the performance criterion

$$S = \int_0^T (4x^2(t) + 2u^2(t))dt$$

is minimized. Using the calculus of variations, determine the optimal control $u^0(t)$ that can achieve this.

6.2. Repeat Problem 6.1 with the performance criterion changed to

$$S = \int_0^T (2x^2(t) + 4u^2(t))dt.$$

What conclusions can you draw from your results?

6.3. Using dynamic programming, determine the optimal control policy for a second-order system where

$$\ddot{c}(t) + 4\dot{c}(t) + c(t) = u(t),$$

$$S = \int_0^T dt.$$

Assume the magnitude of the controlled input must be less than or equal to unity.

6.4. Repeat Problem 6.3 using the maximum principle.

6.5. Using dynamic programming, determine the optimal control policy for a third-order system where $-1 \leq u(t) \leq 1$ and

$$\dddot{c}(t) + 2\ddot{c}(t) + 6\dot{c}(t) = 4 + u(t),$$

$$S = \int_0^T dt.$$

6.6. Repeat Problem 6.5 with the dynamics given by

$$\dddot{c}(t) + 2\ddot{c}(t) + 6\dot{c}(t) + 4c(t) = u(t).$$

6.7. Consider a second-order system which is described by the following differential equation:

$$\ddot{c}(t) + 10\dot{c}(t) + 2c(t) = 4u(t).$$

It is desired to minimize the performance criterion given by

$$S = \int_0^T u(t)dt.$$

Assume that the input $u(t)$ must satisfy the following constraints:

$$-1 \leq u(t) \leq 1.$$

Determine the optimal control policy $u^0(t)$ using

(a) Dynamic programming and

(b) Maximum principle.

(c) Show that the optimal control policies $u^0(t)$ obtained in parts (a) and (b) are equivalent.

6.8. Consider a second-order system which is described by the following differential equation:

$$\ddot{c}(t) + 10\dot{c}(t) + 2c(t) = u(t).$$

It is desired to minimize the performance criterion given by

$$S = \int_0^T \sin u(t)dt.$$

Assume the input $u(t)$ must satisfy the following constraints:

$$-\pi/3 \leq u(t) \leq 0.$$

Determine the optimal control policy $u^0(t)$ using

(a) Dynamic programming and

(b) Maximum principle.

(c) Show that the optimal control policy $u^0(t)$ obtained in parts (a) and (b) are equivalent.

6.9. Consider a second-order control system which is described by the following differential equation:

$$\ddot{c}(t) + 4\dot{c}(t) + 3c(t) = 6u(t).$$

It is desired to minimize the performance criterion given by

$$S = \int_0^T u^2(t)dt.$$

Assume that the input $u(t)$ must satisfy the following constraints:

$$-1 \leqslant u(t) \leqslant 1$$

Determine the optimal control policy $u^0(t)$ using

(a) Dynamic programming and

(b) Maximum principle.

(c) Show that the optimal control policies $u^0(t)$ obtained in parts (a) and (b) are equivalent.

6.10. Repeat Problem 6.9 if the control system is described by the following differential equation:

$$\ddot{c}(t) + 7\dot{c}(t) + 2c(t) = 9u(t).$$

6.11. Using the maximum principle, determine the optimal control policy for a third-order system where $-1 \leqslant u(t) \leqslant 1$ and

$$\dddot{c}(t) + 2\ddot{c}(t) + 6\dot{c}(t) = 4 + u(t),$$

$$S = \int_0^T dt.$$

6.12. Repeat Problem 6.11 with the dynamics given by

$$\dddot{c}(t) + 2\ddot{c}(t) + 6\dot{c}(t) + 4c(t) = u(t).$$

6.13. Utilizing dynamic programming, determine the optimal control policy for a control system where

$$S = \int_0^\infty (Ac^2(t) + u^2(t))dt,$$

assuming that the plant dynamics are given by

$$\dot{c}(t) = Ku(t).$$

and that there are no constraints on the input.

6.14. A second-order space attitude-control system is characterized by the following state equations:

$$\dot{x}_1(t) = x_2(t),$$
$$\dot{x}_2(t) = -x_2(t) - x_1(t) + u(t).$$

Determine the control signal $u(t)$ such that the system is taken from the initial state $c(t_0)$ to the equilibrium state $c(t_f) = 0$ in the shortest time. Assume that

$$|u(t)| \leq U.$$

6.15. Repeat Problem 6.14 with the state equations given by

$$\dot{x}_1(t) = x_2(t),$$
$$\dot{x}_2(t) = -2x_2(t) - 4x_1(t) + u(t).$$

6.16. Repeat Problem 6.14 with the state equations given by

$$\dot{x}_1(t) = x_2(t),$$
$$\dot{x}_2(t) = -4x_2(t) + 4x_1(t) + u(t).$$

6.17. Determine the optimal control policy for a control system whose performance criterion is given by

$$S = \int_0^T (u(t) + c^2(t))dt,$$

and whose plant dynamics are given by

$$\dot{c}(t) = -c(t) + 4u(t).$$

Assume that the magnitude of the input ≤ 10. Solve using the maximum principle.

6.18. Aquaculture, the science of farming and husbandry of fresh water and marine organisms, is a field where optimal control theory has been applied for optimizing its operation [38]. An optimal policy for the raising of Maine lobsters has been developed. Experimental results indicate that Maine lobsters take from five to eight years to reach maturity when raised in their natural environment, whereas lobsters raised in water held at 70°F can reach maturity in two years. It has been shown that lobster growth is a function of temperature and lobster weight. When the costs of heating the water and maintaining the lobsters are included, the total expense for raising lobsters can be treated as an optimal control problem. The differential equations describing growth and cost of raising lobsters are given by the following:

$$\dot{W}(t) = K_0 W^\alpha (U(t) - U_0)^m \quad \text{for } U_0 \leq U(t) \leq U_{\max},$$
$$\dot{W}(t) = 0 \qquad\qquad\qquad \text{for all other temperatures,}$$
$$\dot{L} = K_1 W(t)(U(t) - U_0) + K_2 W(t) + K_3,$$

where

$W(t)$ = weight of the lobster (state variable),
$U(t)$ = water temperature for the lobster,
U_0 = empirically determined zero growth temperature,
U_{max} = maximum practical temperature for growing lobsters,
$L(t)$ = cost of raising lobsters (loss function).

The values K_0, α, and m are constants and K_1, K_2, and K_3 represent estimates of cost.

(a) Write the expressions for the Hamiltonian function and for the co-state vector for this aquaculture temperature control system.

(b) Assuming that m is greater than unity, determine the optimal control temperature, $U_0(t)$. What kind of temperature control system results?

6.19. Almost all agricultural crops suffer damage caused by certain prey (insects) that eat or otherwise destroy the crop. Nature keeps these pests in check by subjecting them to the role of prey, with respect to other insects that act as their predators. Predators usually cause little crop damage. Society, however, has tried to control pests by utilizing insecticides that are lethal not only to the prey, but to their predators as well. The unfortunate side effects of such control programs over the past are now being realized and analyzed by those concerned with the environment. Control engineers are entering the ecological domain in order to guide future pest-control programs. Any control attempt should start first with a model of the biological control system. By utilizing a biological model of the ecological environment, the effects of human control can be examined. Most prey–predator systems can be modeled by the following Lotka–Volterra equations [15]:

$$\frac{dN_1(t)}{dt} = AN_1(t) - BN_1(t)N_2(t),$$

$$\frac{dN_2(t)}{dt} = -CN_2(t) + DN_1(t)N_2(t),$$

where $N_1(t)$ and $N_2(t)$ are the instantaneous populations of the prey and predators, respectively, and A, B, C, and D are all positive constants. This model can be simplified by nondimensionalizing all the variables and eliminating all constants except one:

$$x_1(t) = \frac{N_1(t)}{N_{1r}} \quad N_{1r} = \frac{A}{D},$$

$$x_2(t) = \frac{N_2(t)}{N_{2r}}, \quad N_{2r} = \frac{A}{B},$$

$$\tau = \frac{t}{t_r}, \quad t_r = \frac{1}{A},$$

$$K = \frac{C}{A}.$$

The resulting equations are as follows:

$$\frac{dx_1(t)}{d\tau} = x_1(t)(1 - x_2(t)),$$

$$\frac{dx_2(t)}{d\tau} = x_2(t)(x_1(t) - K).$$

The result of human control on this biological control system is to decrease the growth rates of both prey and predators. Representing the human control effort as $u(t)$, which has the physical significance of using traps and chemical spray (insecticides), then

$$\frac{dx_1(t)}{d\tau} = x_1(t)(1 - x_2(t)) - u(t)x_1(t),$$
$$\frac{dx_2(t)}{d\tau} = x_2(t)(x_1(t) - K) - eu(t)x_2(t),$$

where e is a constant proportional to the relative effectiveness of the control used on the predators as compared with that used on the prey. In order to determine an optimal control strategy, the following loss function has been proposed:

$$G = (ax_1(t) + u(t)).$$

Its significance is that it assumes the two factors associated with the cost of the control program are those associated with the presence of the pests and those associated with using control—social and economic. Utilizing the maximum principle, determine the optimal control law for this biological system. Explain your result. Assume that $0 \leqslant u(t) \leqslant u_{\max}$.

6.20. Optimal control systems based on a quadratic performance criterion are very frequently used in practice by the control-system engineer. Consider a system with an initial displacement

$$\dot{\mathbf{x}}(t) = \mathbf{P}\mathbf{x}(t), \quad \mathbf{x}(0) = \mathbf{c},$$

in which it is desired to minimize the quadratic performance criterion

$$S = \int_0^\infty \mathbf{x}^T(t)\mathbf{Q}\mathbf{x}(t)dt,$$

where $\mathbf{Q}$ is a positive-definite (or positive-semidefinite) real matrix. If the eigenvalues of the companion matrix $\mathbf{P}$ have negative real parts ($\mathbf{P}$ is stable), then it can be shown that a matrix $\mathbf{A}$ exists which can be found from the following relationship that it satisfies:

$$\mathbf{P}^T\mathbf{A} + \mathbf{A}\mathbf{P} = -\mathbf{Q}.$$

In addition, because all of the eigenvalues of **P** have real parts, then $\mathbf{x}(\infty) \to 0$ and it can also be shown that the performance criterion can be obtained in terms of the initial condition $\mathbf{x}(0)$:

$$S = \mathbf{x}^T(0)\mathbf{A}\mathbf{x}(0).$$

It is desired to adjust the damping ratio ζ for the control system shown in Figure P6.20 in order that the integral of the square of the error (ISE) performance criterion is minimized. Using this approach, determine the value of ζ for this second-order system in order that the following performance criterion is minimized when the control system is subjected to a unit step input:

$$S = \int_0^\infty \mathbf{e}^T(t)\mathbf{Q}\mathbf{e}(t)dt = \int_0^\infty e^2(t)dt,$$

where

$$\mathbf{e}(t) = \begin{bmatrix} e_1(t) \\ e_2(t) \end{bmatrix} = \begin{bmatrix} e(t) \\ \dot{e}(t) \end{bmatrix}, \quad \mathbf{Q} = \begin{bmatrix} 1 & 0 \\ 0 & 0 \end{bmatrix}.$$

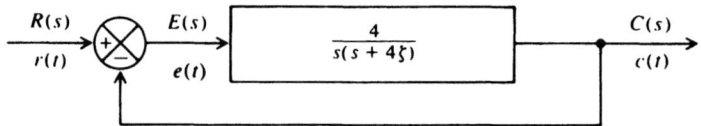

Figure P6.20

Assume that the system is initially at rest. Compare your results with the value obtained using classical techniques and discussed in Section 5.6‡.

6.21. Repeat Problem 6.8 if the performance criterion is modified to be

$$S = \int_0^T \cos u(t)dt,$$

and the input $u(t)$ must satisfy the following constraints:

$$2\pi/3 \leq u(t) \leq 4\pi/3.$$

REFERENCES

1. N. Weiner, *The Extrapolation, Interpolation and Smoothing of Stationary Time Series.* MIT Technology Press, Cambridge, MA, 1949.
2. D. McDonald, "Nonlinear techniques for improving servo performance." *Proc. Nat. Electron. Conf.* **6**, 400–421 (1950).

3. C. S Draper and V. T. Li, *Principles of Optimizing Control Systems and an Application to the Internal Combustion Engine.* American Society of Mechanical Engineers, New York, 1951.
4. R. E. Bellman, *Dynamic Programming.* Princeton University Press, Princeton, NJ, 1959.
5. R. E. Bellman, *Applied Dynamic Programming.* Princeton University Press, Princeton, NJ, 1962.
6. V. G. Boltyanskii, R. V. Gamkrelidge, and L. J. Pontryagin, "The theory of optimal proceses. I. The maximum principle." *Izv. Akad. Nauk SSR, Ser. Mat,* **24**, 3 (1960).
7. L. J. Pontryagin, "Optimal Control Processes." *Usp. Mat. Nauk* **14**, 3 (1959).
8. M. J. Grimble and A. Johanson, *Optimal Control and Stochastic Estimation: Theory and Application.* Wiley, New York, 1988.
9. M. Athans and P. L. Falb, *Optimal Control.* McGraw-Hill, New York. 1966.
10. A. P. Sage, *Optimum System Control.* Prentice-Hall, Englewood Cliffs, NJ, 1968.
11. B. D. O. Anderson and J. B. Moore, *Linear Optimal Control.* Prentice Hall, Englewod Cliffs, NJ, 1971.
12. R. E. Kalman, "Contributions to the theory of optimal control." *Bol. Soc. Mat. Mex.* **5**, 102–119 (1960).
13. G. Leitman, ed., *Optimization Techniques.* Academic Press, New York, 1962.
14. R. E. Kalman, *The Theory of Optimal Control and the Calculus of Variations*, Tech. Rep. No. 6–13. Research Institute of Applied Science, Baltimore, MD, 1961.
15. T. L. Vincent, "'Pest management programs via optimal control theory." In *Proceedings of the 1972 Joint Automatic Control Conference,* pp. 658–663.
16. R. E. Kalman, "On the general theory of control systems." In *Proceedings of the First International Congress of Automatic Control,* Moscow, 1960.
17. R. E. Kalman, Y. C. Ho, and K. S. Navendra, "Controllability of linear dynamical systems." *Contrib. Differ. Equa.* **1**, 189–213 (1961).
18. E. G. Gilbert, "Controllability and observability in multivariable control systems." *J. Control. Ser.* **A1**, 128–151 (1963).
19. J. T. Tou, *Modern Control Theory.* McGraw-Hill, New York, 1964.
20. O. I. Elgerd, *Control Systems Theory.* McGraw-Hill, New York, 1967.
21. R. E. Kalman, "When is a linear control system optimal?" *J. Basic Eng.* **86**, 51–60 (1964).
22. P. H. Dosik, "Synthesis of optimal control systems," *Electro-Technol.* **75**, 36–43 (1965).
23. E. L. Peterson, *Statistical Analysis and Optimization of Systems.* Wiley, New York, 1961.
24. B. Friedland, "The structure of optimum control systems." *J. Basic Eng.* **84**, 1–11 (1962).
25. E. B. Lee, "Mathematical aspects of the synthesis of linear minimum response time controllers." *IRE Trans. Autom. Control* **AC-5**, 283–290 (1960).
26. M. Athanassiades and O. J. Smith, "Theory and design of high order bang-bang control systems." *IRE Trans. Autom. Control* **AC-6**, 125–134 (1961).
27. L. W. Neustadt, "Time-optimal synthesis with position and integral limits." *J. Math. Anal. Appl.* **3**, 406–427 (1961).
28. I. Flügge-Lotz and H. Marback, "The optimal control of some attitude control systems for different performance criteria." In *Proceedings of the 1962 Joint Automatic Control Conference,* New York, pp. 12-1-1 to 12-1-12.
29. B. Friedland, "The design of optimum controllers for linear processes with energy limitations." In *Proceedings of the 1962 Joint Automatic Control Conference,* New York, pp. 12-4-1 to 12-4-12.

30. A. B. Pearson, "Synthesis of a minimum energy controller subject to an average power constraint." In *Proceedings of the 1962 Joint Automatic Control Conference*, New York, pp. 19-4-1 to 19-4-6.
31. M. Athanassiades, "Optimal control for linear time invariant plants with time, fuel, and energy constraints." *Trans. Am. Inst. Electr. Eng.* **81**, 321–325 (1962).
32. M. Athans, P. L. Falb, and R. I., Lacoss, "Time-, fuel-, and energy-optimal control of nonlinear norm-invariant systems." *IRE Trans. Autom. Control* **AC-8**, 196–202 (1963).
33. S. M. Shinners, "Optimal and adaptive control systems." *Electro-Technol.* **74**, 63–80 (1964).
34. T. M. Stout, "Effects of friction in an optimum relay servomechanism." *Trans. Am. Inst. Electr. Eng.* **72**, 329–335 (1953).
35. M. Athans, "The status of optimal control theory and applications for deterministic systems.' *IEEE Trans. Autom. Control* **AC-11**, 580–596 (1966).
36. S. Saelid, N. A. Jenssen, and J. G. Balchen, "Design and analysis of a dynamic positioning system based on Kalman filtering and optimal control." *IEEE Trans. Autom. Control* **AC-28**, 331–338 (1983).
37. J. S. Meditch, "On the problem of optimal thrust programming for a lunar soft landing." *IEEE Trans. Autom. Control* **AC-9**, 477–484 (1964).
38. L. W. Botsford, H. E. Rauch and R. A. Shlesser, "Optimal temperature control of a lobster plant," *IEEE Trans. Autom. Control* **AC-19**, 541–543 (1974).

7

CONTROL-SYSTEM DESIGN EXAMPLES: COMPLETE CASE STUDIES

7.1. INTRODUCTION

In the preceding chapters of this book, we have analyzed and designed control systems from specific viewpoints. For example, the Nyquist and Bode diagrams and the root-locus method were applied to linear control systems in Chapter 1, and Chapters 2 and 3, and extended to digital control systems in Chapter 4. The describing function, phase-plane, circle criterion, Liapunov's and Popov's methods were applied to the analysis and design of nonlinear control systems in Chapter 5. How do we take a global viewpoint of a control-system design problem and look at it from both linear and nonlinear viewpoints? We must also consider reliability, cost size, weight, and power consumption. We must design a working control system that meets all the specifications, that can be sold at a profit, that can be built on schedule, and that satisfies the customer's requirements.

In this chapter on complete case studies, we will employ the methods of the preceding chapters to design the following:

1. Design for the positioning system of a tracking radar which illustrates both linear and nonlinear design considerations jointly.
2. Design of the angular control system for a robot's joint.
3. State-variable design for the controller and full-order estimator for a space satellite.
4. Digital control system design for a microcomputer-controlled temperature control system.
5. Robust control system design for controlling the flaps of a hydrofoil.

These examples will illustrate the use of the appropriate methods presented in the book which are needed to design the control system for the intended applications.

The design examples will convey the overall approach and methodology used for designing control systems for a good cross section of applications.

7.2. OUTLINE OF PROCEDURE FOR DESIGNING A CONTROL SYSTEM [1, 2]

Due to the availability of a large number of techniques to solve the great variety of control-system problems present, the element of experience is very important to the approach used for the solution of a specific problem. Assuming that the control-system engineer has had some experience with the techniques described, then by logically considering the problem, the many methods described previously in this book provide a very powerful capability. An outline of a logical step-by-step procedure for designing a control system from its conception through the final hardware stage is illustrated in Figure 7.1 and described as follows:

1. Obtain a complete understanding of the job requirements with respect to
    (a) a general description of the problem;
    (b) the overall control-system performance and accuracy with respect to the steady-state and transient phases;
    (c) identification of the transfer function of the controlled process;
    (d) miscellaneous requirements as to reliability, schedule, cost, maintainability, size, weight, and available power.
2. Consider several alternative solutions, including electric and hydraulic power servo drives (see Sections 3.4[‡] and 3.5[‡], respectively), the use of continuous control or digital control, etc.
3. Choose the most desired approach based on the specifications, requirements, and elements fixed by the customer.
4. Interpret these requirements in terms of such closed-loop design characteristics as frequency and transient response.
5. Establish the approximate open-loop characteristics that will satisfy the closed-loop requirements.
6. Design the system and select the sensors, actuators, amplifier, and stabilization required (analog or digital) in order to satisfy step 5.
7. Review, refine, and simplify steps 5 and 6.
8. Simulate the system on a computer, including its linear and nonlinear characteristics, to check the design. Make any necessary changes to the design.
9. Build a prototype, and check the design experimentally. Make any necessary changes to the design.
10. Refine the design in order to optimize performance and minimize cost.

Observe from this approach that the procedure is an iterative one, and is itself a feedback process, as illustrated in Figure 7.1

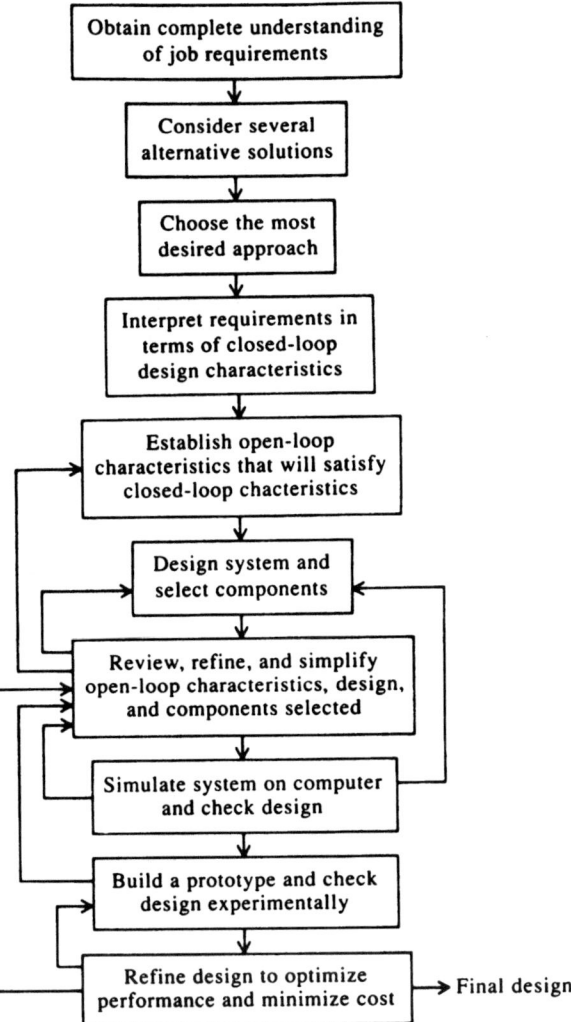

Figure 7.1 Procedure for designing a control system.

7.3. EXAMPLE 1: DESIGN FOR THE POSITIONING SYSTEM OF A TRACKING RADAR USING LINEAR AND NONLINEAR TECHNIQUES JOINTLY [1]

For our first case study, we will design the positioning system of a tracking radar using conventional linear and nonlinear techniques jointly. The demand for precise and smooth positioning of large loads, such as that of a tracking radar, has placed increased emphasis on synthesizing optimum positioning configurations. For example, low resonant frequencies, nonlinear friction, and large opposing wind torques are major problems associated with tracking radars using rotatable antennas (as opposed to tracking radars using electronic scanning) and other positioning systems [1, 3, 4].

508 CONTROL-SYSTEM DESIGN EXAMPLES: COMPLETE CASE STUDIES

Large antennas and associated supporting structures present obstacles to the design of an accurate and smooth tracker. Large masses associated with the load result in relatively low mechanical resonant frequencies, excessive frictional nonlinearities in the form of both Coulomb and static friction (see Section 5.8), and large opposing wind torques. The low resonant frequencies of the tracker necessitate low bandwidths and correspondingly small gain constants that adversely affect accuracy. In addition, the low mechanical resonant frequencies and nonlinear friction can also cause system instability. Assuming that a radome (sheltering structure for a radar antenna) is not used, the large opposing wind torques adversely affect accuracy and require high-power servo ratings. A radome may be undesirable for some applications because of its cost and resulting boresight shift.

For this application, we will assume that the structural resonance has its first peaking at 4 Hz of 13 dB, and there are higher resonance peaks at greater frequencies. The allowable resonances are shown on the Bode diagram in Figure 7.2. This type of structural resonance limits the dynamic accuracy of the friction-stabilization inner loop and, consequently, the auto-track outer loop. Assuming a minimum gain margin of 6 dB and a structural resonance peaking of 13 dB at 4 Hz, a system containing two pure integrations in the auto-track loop could result in an acceleration constant K_a equal to 1, as shown in Figure 7.2. This is insufficient for our design. Therefore, we want to go to an auto-track loop that has three pure integrations and provides for an infinite K_a.

In addition, we will assume that this tracking system must contend with three forms of friction. These are viscous, Coulomb, and static friction (stiction), which were analyzed in Section 5.8 from the describing-function method. We will analyze the effect of Coulomb and static friction on the system from a nonlinear, describing-function viewpoint.

Let us consider the possibility of using N pure integrations in the auto-tracking loop to track $d^{N-1}\theta/dt^{N-1}$ dynamics perfectly, and to overcome the structural reso-

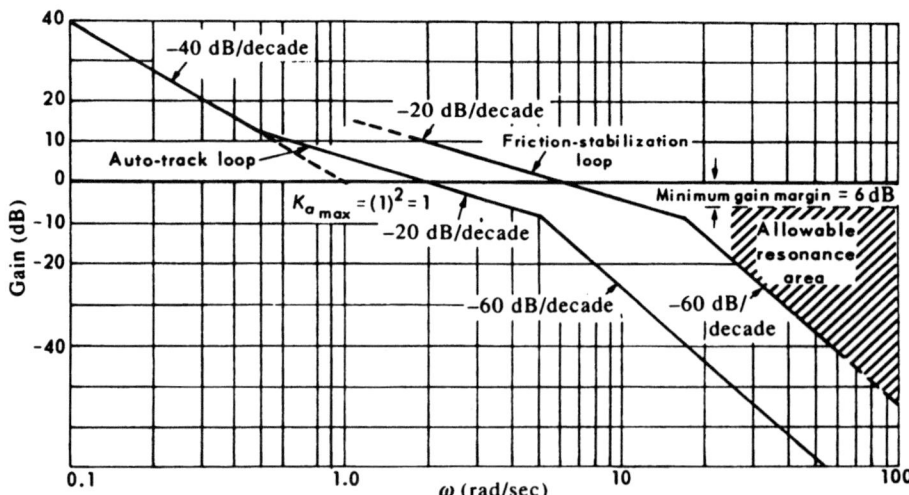

Figure 7.2 The resonance problem in a tracking radar using a rotatable antenna.

nance limitation [3, 4]. Conventional linear theory indicates that it is feasible to stabilize a positioning loop containing N pure integrations, where N can be equal to or greater than 3. These techniques also illustrate that the use of these systems in the presence of multiple nonlinearities does present nonlinear stability problems that must be solved. Let us examine the effects of five different auto-tracking loop types ranging from type 0 (contains zero pure integrations) to type-4 (contains four pure integrations) on the overall tracking-system stability and accuracy when operating in the presence of multiple frictional nonlinearities. A comprehensive analysis is performed on a single-loop tracking configuration and then extend to the case of the multiple-feedback loop. The concept is then extended to the design of a practical tracking radar system employing three pure integrations (type 3) in the auto-track loop. The tracking system, which is also analyzed from a nonlinear viewpoint, exhibits velocity and acceleration constants of infinity and, therefore, easily solves the accuracy problem introduced by low resonant frequencies. In addition, a highly damped rate-feedback loop further isolates the frictional nonlinearities, thereby preventing limit cycles due to nonlinear friction. It is shown that the practical tracking configuration utilizing multiple-feedback loops is completely stable from both a linear and nonlinear viewpoint, and exhibits very high accuracy.

To evaluate the effect of system type (the number of pure integrations in the positioning auto-track loop), consider the simple one-loop configuration containing a nonlinear element in Figure 7.3. Assuming that $H(s) = 1$, then Figure 7.4 shows how a positioning loop can be designed containing zero (type 0) to four pure integrations (type 4) in which the system is stable. All the systems have a 65° phase margin from purely linear considerations. Bode's first weighting-function theorem verifies the linear stabilization of type-N configurations (see Section 2.6).

To completely determine the stability of type-n positioning (tracking) loops, it is also necessary to analyze the system from a nonlinear viewpoint. All practical tracking and positioning systems must contend with the realistic nonlinear characteristics of Coulomb friction and stiction (see Section 5.8).

The characteristic equation of the single-loop configuration of Figure 7.4 can be written in the form of Eq. (7.1), from which stability can be readily determined:

$$1 + G(j\omega)H(j\omega)N(M, \omega) = 0. \tag{7.1}$$

In this equation, $G(j\omega)H(j\omega)$ represents the linear portion of the loop gain and $N(M, \omega)$ represents the describing function of the nonlinearity as a function of its

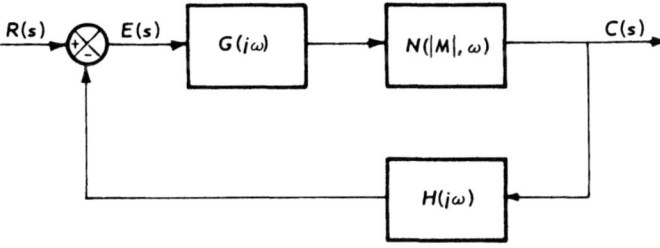

Figure 7.3 General nonlinear system.

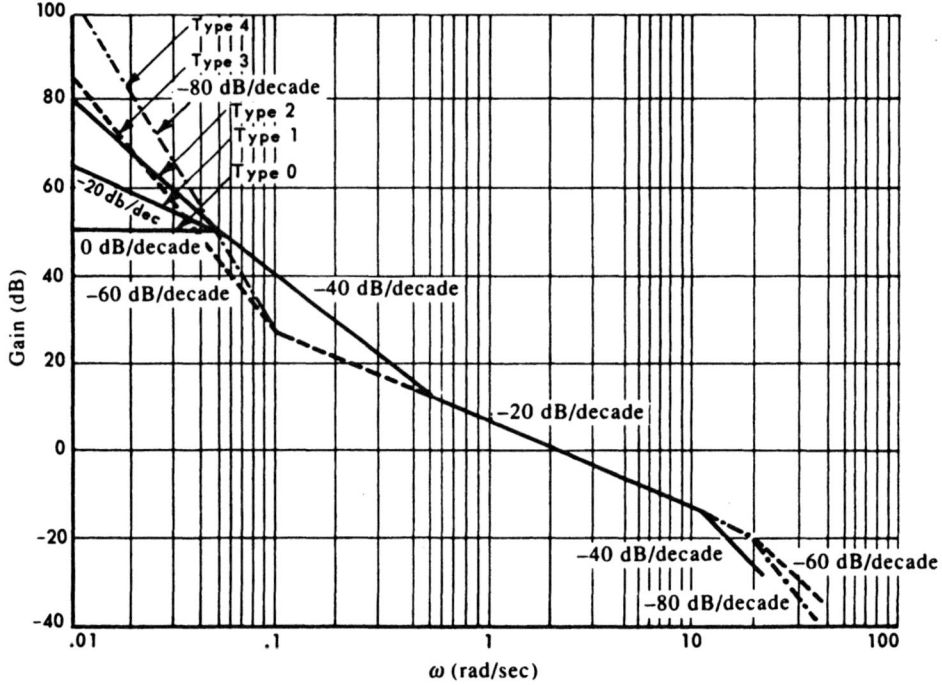

Figure 7.4 Bode diagrams of linearly stable type 0 through type 4 configurations.

input signal amplitude M and frequency ω. From Eq. (5.81), the criterion for nonlinear instability of the single-loop configuration is given by

$$G(j\omega)H(j\omega) = -\frac{1}{N(M,\omega)}. \qquad (7.2)$$

The gain-phase diagram will be used to perform this describing-function analysis as described in Sections 5.7–5.12. The describing function for the combined case of Coulomb friction and stiction [see Eq. (5.75)] has been used for analysis shown in Figure 7.5, where the effect of this combined nonlinearity on positioning loops having the type 0 through type 4 configurations illustrated in Figure 7.5 are shown. It is concluded that type 0, type 1, and type 2 single-loop configurations containing nonlinear friction are completely stable from nonlinear considerations. However, the type 3 and type 4 single-loop configurations are shown to exhibit limit cycles. The type 4 configuration clearly shows an intersection on the gain-phase diagram, whereas the type 3 configuration loci approach the describing-function loci at very low frequencies. However, the margin of stability of the type 3 system is negligible at moderate frequencies and borders on instability. Thus, for all practical purposes, it must be considered unstable.

Although single-loop systems greater than type 2 are unstable in single-loop configurations because of frictional nonlinearities, multiple-feedback loops can improve the situation. Therefore, the synthesis of a type 3 system using multiple loops, which is stable from both linear and nonlinear considerations, is described in this design.

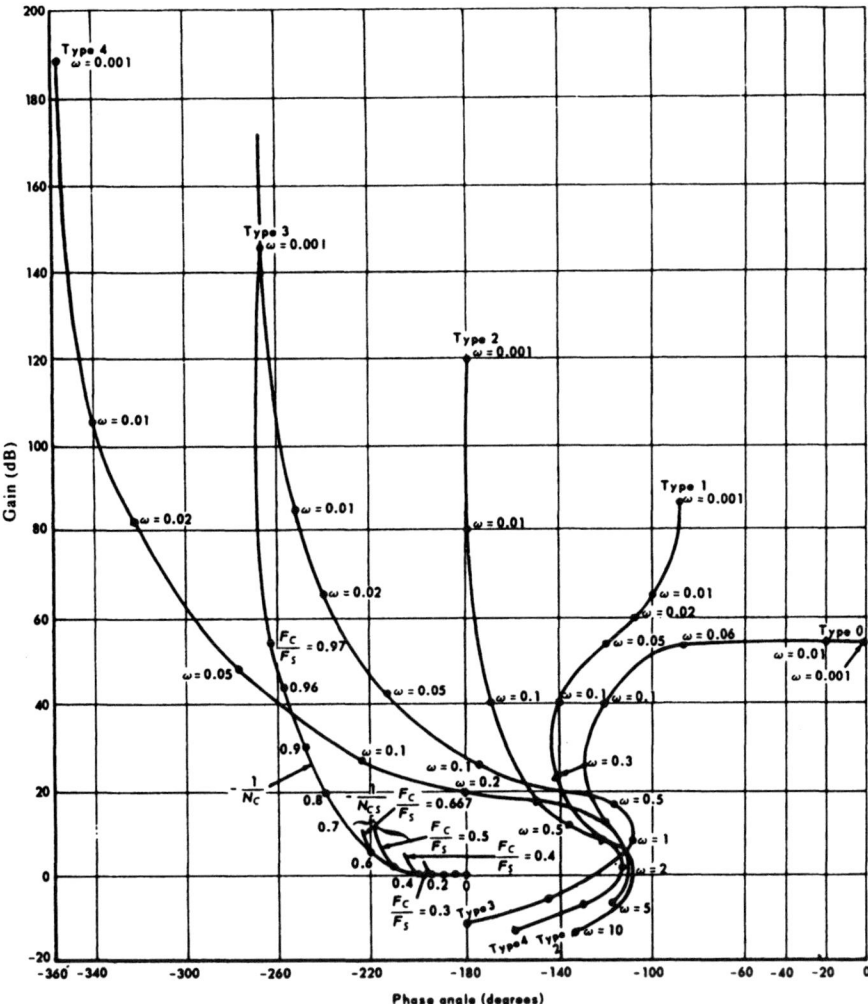

Figure 7.5 Describing-function analysis of type 0 through type 4 configurations.

The key to the successful synthesis of high-order systems is the proper design of the inner feedback loops. Great care must be exercised in isolating the nonlinear elements by a very heavily damped inner loop or loops. As an example of this approach, a type 3 tracking positioning system will be designed that contains the nonlinear frictional characteristics of Coulomb friction and stiction.

The basic multiple-loop configuration can be constructed around an analysis using describing-function and signal-flow graphs (see Sections 2.14–2.16[‡]). A method of accomplishing the design is illustrated with the aid of Figure 7.6, and the corresponding signal-flow graph is shown in Figure 7.7. Using Mason's theorem, given Eq. (2.135)[‡], the overall system transfer function is given by Eq. (7.3).

$$\frac{C(s)}{R(s)} = \frac{G_I \Delta_I}{\Delta}, \tag{7.3}$$

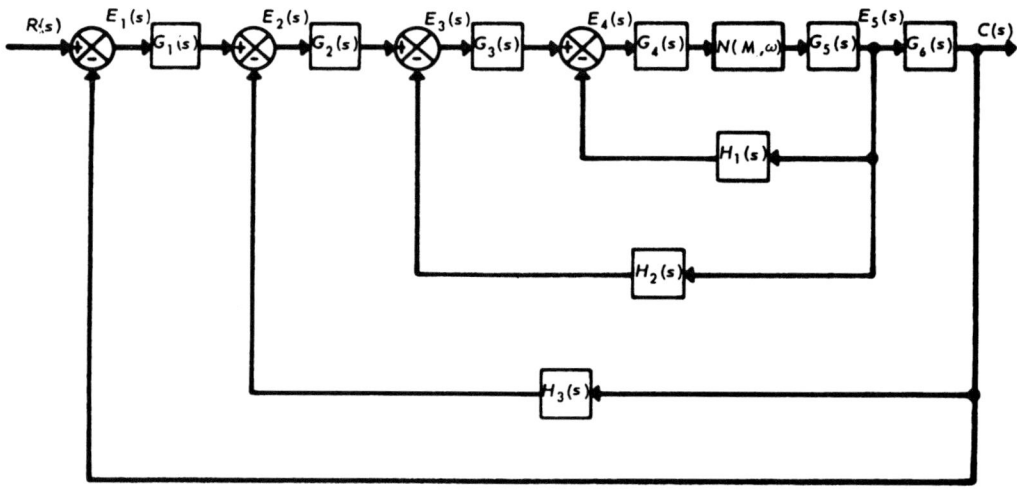

Figure 7.6 Representative multiple-feedback control system containing a nonlinearity common to all feedback paths.

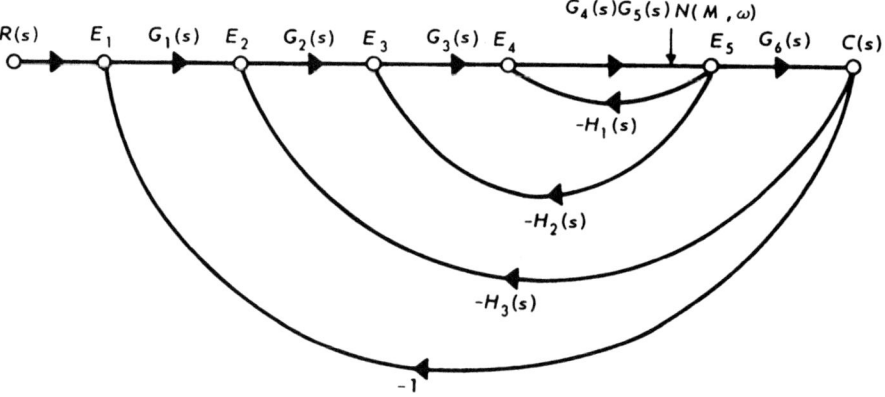

Figure 7.7 Signal-flow diagram for system shown in Figure 7.6.

where

$$\Delta = 1 + G_4(s)G_5(s)N(M, \omega)H_1(s) + G_3(s)G_4(s)G_5(s)N(M, \omega)H_2(s)$$
$$+ G_2(s)G_3(s)G_4(s)N(M, \omega)G_5(s)G_6(s)H_3(s)$$
$$+ G_1(s)G_2(s)G_3(s)G_4(s)N(M, \omega)G_5(s)G_6(s),$$
$$G_1 = G_1(s)G_2(s)G_3(s)G_4(s)N(M, \omega)G_5(s)G_6(s).$$
$$\Delta_I = 1.$$

Therefore,

$$\frac{C(s)}{R(s)} = \frac{G_1(s)G_2(s)G_3(s)G_4(s)N(M, \omega)G_5(s)G_6(s)}{1 + [G_4(s)G_5(s)H_1(s) + G_3(s)G_4(s)G_5(s)H_2(s)}.$$
$$+ G_2(s)G_3(s)G_4(s)G_5(s)G_6(s)H_3(s)$$
$$+ G_1(s)G_2(s)G_3(s)G_4(s)G_5(s)G_6(s)]N(M, \omega)$$

(7.4)

Stability for this system can be determined from the zeros of the characteristic equation:

$$1 + [G_4(s)G_5(s)H_1(s) + G_3(s)G_4(s)G_5(s)H_2(s)$$
$$+ G_2(s)G_3(s)G_4(s)G_5(s)G_6(s)H_3(s)$$
$$+ G_1(s)G_2(s)G_3(s)G_4(s)G_5(s)G_6(s)]N(M,\omega) = 0. \qquad (7.5)$$

From this step, stability can then be determined using the gain-phase diagram.

For the particular problem of a tracking radar containing a rotatable antenna, the general configuration in Fig. 6.6 applies. Therefore, we now have a technique readily applicable to the analysis of nonlinearities in a type 3 multiple-feedback system. The particular configuration to which this method will be applied consists of a tracking radar that uses a Ward–Leonard power drive system (see Section 3.4[‡]), having armature current and field current feedback to decrease the armature and field time constants (see Problem 3.7[‡]).

Figure 7.8 illustrates the equivalent configuration of the tracking radar control system. Notice that the hypothetical system considered in Figure 7.6 corresponds to the proposed structure. Therefore, the system stability can be studied from a gain-phase plot of Eq. (7.5). The values for $G_4(s)$, $G_5(s)$, $G_6(s)$, $H_1(s)$, $H_2(s)$, and $H_3(s)$ are dictated by practical considerations. It remains for the control-systems engineer to choose $G_1(s)$, $G_2(s)$ and $G_3(s)$.

The primary function of the rate feedback, $H_3(s)$, is to prevent oscillations due to the nonlinear friction characteristics of stiction and Coulomb friction. It is designed specifically with a very high damping ratio to overcome this problem.

The describing function analysis of this type 3 system is illustrated in Figure 7.9, which considers the case where Coulomb fraction F_c and stiction F_s are both present. It is extremely important for the control system to remain stable when the tracking loop is open and closed. When the tracking loop is open, the radar-system tracking is interrupted. This condition can be shown by removing $G_1(s)$ and its corresponding feedback path in Figure 7.8.

The values for $G_2(s)$ and $G_3(s)$ are dictated primarily from nonlinear considerations as shown on the gain-phase diagrams of Figure 7.9, while the value of $G_1(s)$ is dictated primarily from linear considerations as shown on the Bode diagram of Figure 7.10. However, there is some interdependence, and a trial-and-error solution is required. The resultant gain-phase diagram of Figure 7.9 in conjunction with the Bode diagram of Figure 7.10 indicate that the system is completely stable from a linear and nonlinear viewpoint when

$$G_1(s) = \frac{1.25(1+4s)^2}{s^2(1+0.0625s)^2}, \qquad (7.6)$$

$$G_2(s) = 5.4, \qquad (7.7)$$

$$G_3(s) = \frac{734(0.018s+1)}{s(0.01s+1)}. \qquad (7.8)$$

The open-loop transfer function for the Bode diagram of Figure 7.10 was obtained by assuming that the gain of the inner loop containing $H_3(s)$ as feedback in Figure

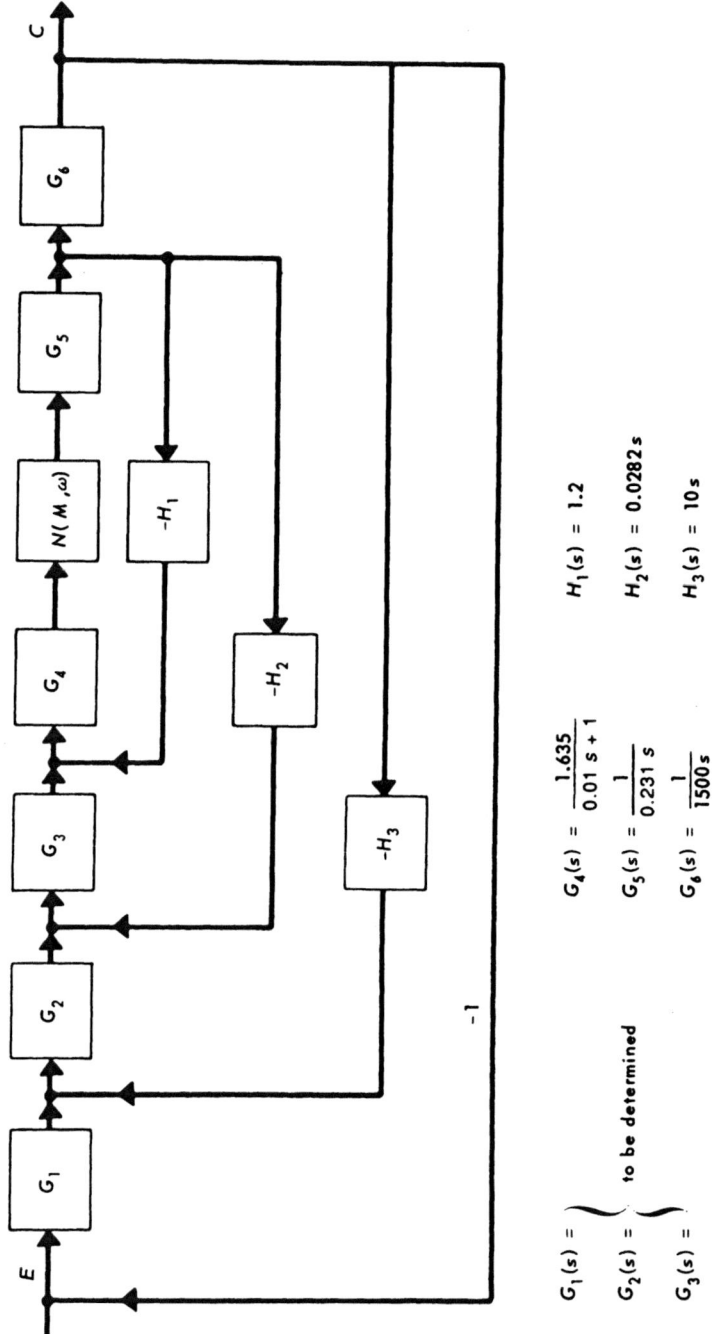

Figure 7.8 Equivalent representation of the tracking radar control system.

7.3. EXAMPLE 1: DESIGN FOR THE POSITIONING SYSTEM OF A TRACKING RADAR

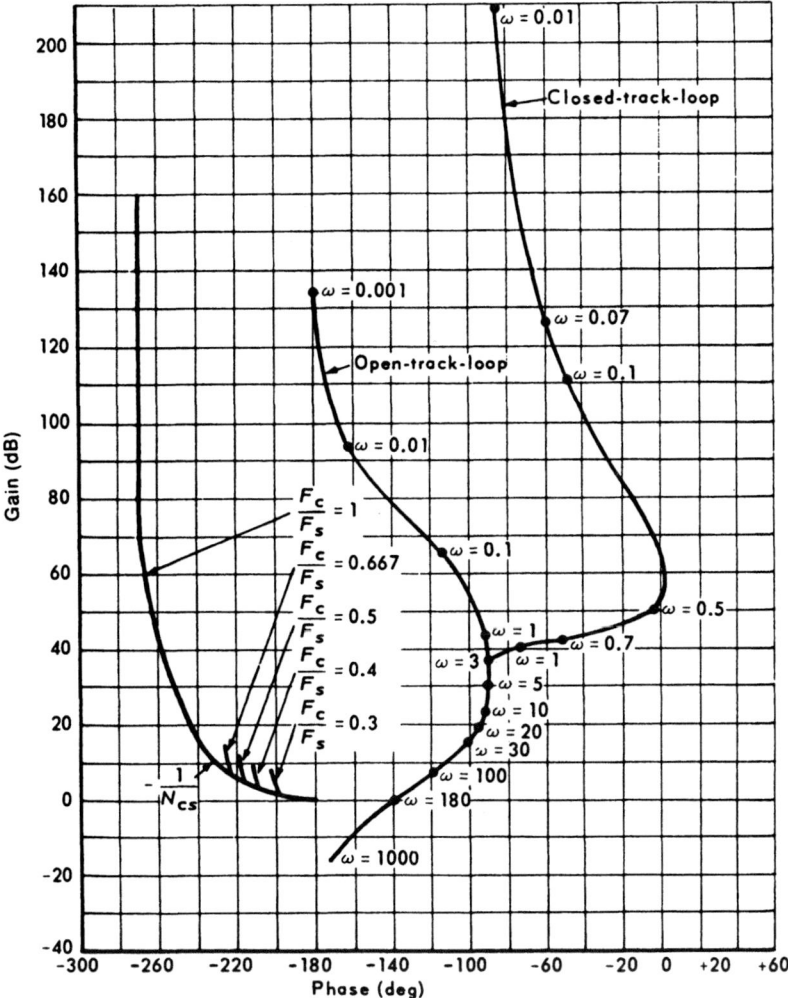

Figure 7.9 Describing-function analysis for Coulomb friction and static friction [1].

7.8 has a high gain over the frequencies of interest and, therefore, the auto-track loop sees its transfer function as [see Eq. (2.122)‡]

$$\frac{1}{H_3(s)} = \frac{1}{10s}. \tag{7.9}$$

Therefore, the open-loop transfer function of the auto-track loop is given by

$$\frac{0.125(1+4s)^2}{s^3(1+0.0625s)^2}. \tag{7.10}$$

This design has resulted in a type 3 ninth-order practical system that has zero steady-state error resulting from inputs of position, velocity, and acceleration.

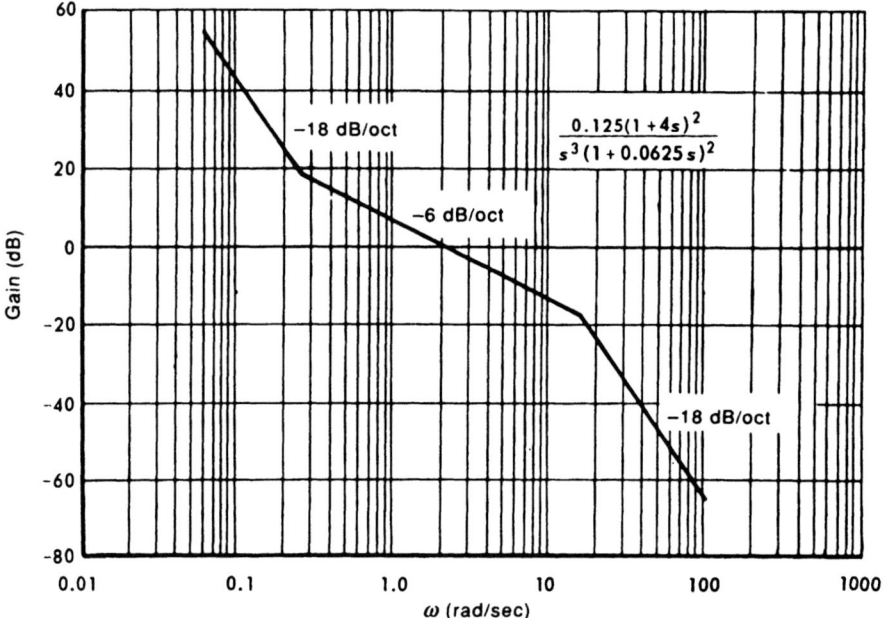

Figure 7.10 Tracking loop: open-loop frequency characteristics [1].

It is important to recognize that although this design problem focused on positioning a tracking radar containing a rotatable antenna, the approach is equally valid for the positioning control problem of any large load containing nonlinear friction.

7.4. EXAMPLE 2: DESIGN OF AN ANGULAR CONTROL SYSTEM FOR A ROBOT'S JOINT

Robots are playing an increasingly important role in manufacturing and other applications. Figure 7.11 shows an example of the use of robots to manufacture automobiles at the Ford Motor Company where a robot is used to automatically mate a hood inner panel to the outer panel for a Ford Taurus at its Woodhaven (MI) Stamping Plant.

For the second design example, the angular control system of a robot's joint will be designed. The specifications for this application are as follows:

(a) steady-state error due to a velocity input of 3.2 rad/sec should equal or be less than 0.1 rad; steady-state error of zero due to position inputs;
(b) phase margin greater than 60°;
(c) settling time of less than 1 sec;
(d) percent overshoot of less than 8%.

7.4. EXAMPLE 2: DESIGN OF AN ANGULAR CONTROL SYSTEM FOR A ROBOT'S JOINT

Figure 7.11 Robot being used to automatically mate a hood inner panel to the outer panel for a Ford Taurus at Ford Motor Company's Woodhaven (MI) Stamping Plant. (Courtesy of the Ford Motor Company)

The block diagram for controlling the robot's joint is illustrated in Figure 7.12. It will be assumed that the transfer function $\theta_c(s)/E_A(s)$ in this problem can be approximated by*

$$\frac{\theta_c(s)}{E_A(s)} = \frac{1}{s(s+10)}. \tag{7.11}$$

Because this transfer function has one pure integration, the specification requirement of zero steady-state error due to position inputs will be met (see Appendix C). Nonlinearities in this control system are assumed to be negligible.

As the first step in the design, the value of the amplifier gain K_A to satisfy the steady-state accuracy will be obtained. Assuming that $G(s) = 1$ for this part of the analysis, the forward function $\theta_c(s)/E(s)$ is found to be

$$\frac{\theta_c(s)}{E(s)} = \frac{0.1 K_A}{s(s+10)}. \tag{7.12}$$

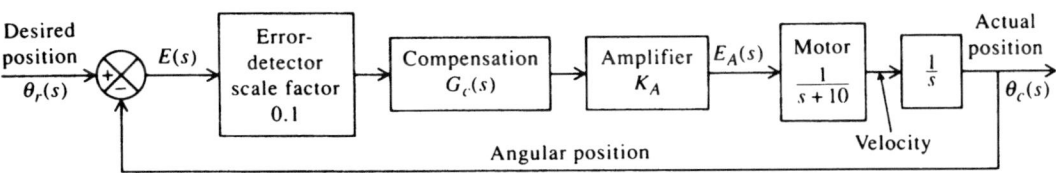

Figure 7.12 Block diagram for controlling a robot's joint.

*The actual transfer function of robot control systems are usually much more complex, but this approximation is adequate for determining the dominant roots of the system.

From the specifications, the value of the desired velocity constant K_v is given by

$$K_v = \frac{\omega}{\text{error}} = \frac{3.2 \text{ rad/sec}}{0.1 \text{ rad}} = 32/\text{sec}. \quad (7.13)$$

Therefore, the value of the amplifier gain K_A is given by (assuming $G_c(s) = 1$)

$$K_v = \lim_{s \to 0} sG(s)H(s) = \lim_{s \to 0} s \cdot \frac{0.1 K_A}{s(s+10)} = 32 \quad (7.14)$$

and

$$K_A = 3200. \quad (7.15)$$

With this value of amplifier gain and without any compensation, $G_c(s)$, the phase margin, percent overshoot, and settling time will be determined. The closed-loop transfer function of this second-order system is given by

$$\frac{\theta_c(s)}{\theta_r(s)} = \frac{G_0(s)}{1 + G_0(s)}, \quad (7.16)$$

where

$$G_0(s) = \frac{0.1(3200)}{s(s+10)} = \frac{320}{s(s+10)}. \quad (7.17)$$

Therefore,

$$\frac{\theta_c(s)}{\theta_r(s)} = \frac{320}{s^2 + 10s + 320}. \quad (7.18)$$

Using Eq. (B.2), the undamped natural frequency ω_n and damping ratio ζ are found to be

$$\omega_n = \sqrt{320} = 17.9 \text{ rad/sec}, \quad (7.19)$$

$$\zeta = \frac{10}{2(17.9)} = 0.28. \quad (7.20)$$

From Eq. (B.33), the maximum percent overshot for $\zeta = 0.28$ is 40.01%, much higher than the specification maximum of 8%. From Eq. (5.41)‡, the settling time for $\zeta = 0.28$ is 0.798 sec and is within the specification value of 1 sec. The Bode diagram of this uncompensated system is illustrated in Figures 7.13a and b, and shows a phase shift of $-148.9°$ at gain crossover frequency and a phase margin of only 31.1°, which is much less than the specified phase margin of 60° minimum. Figures 7.13a and b were obtained using MATLAB (see Section 1.7) and contained in the M-file which is part of the Advanced Modern Control System Theory and Design (AMCSTD) Toolbox and can be retrieved free from The MathWorks, Inc.

7.4. EXAMPLE 2: DESIGN OF AN ANGULAR CONTROL SYSTEM FOR A ROBOT'S JOINT

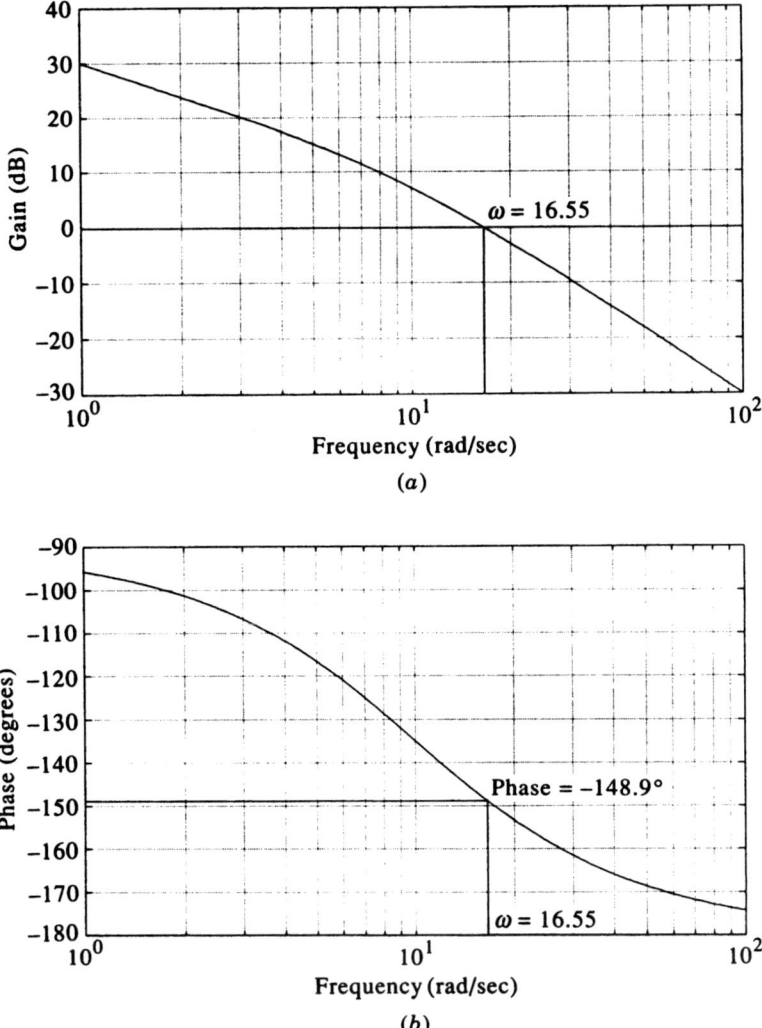

Figure 7.13 Bode diagram of uncompensated control system for robot.

anonymous FTP server at ftp://ftp.mathworths.com/pub/books/advshinners. Therefore, we must compensate this system to meet the specifications.

A phase-lead or phase-lag network can be selected for the design based on the specifications. However, from the tradeoffs between the two networks discussed in Section 2.10, a phase-lag network is selected, because it is less expensive (an increase in the amplifier gain is needed with a phase-lead network due to its dc attenuation), provides the compensated system with a smaller bandwidth and, therefore, the resulting system has less noise and the output response has less "jitter."

Using the approach discussed in Section 2.6, the following phase-lag network was selected:

$$G_c(s) = \frac{1 + 4.22s}{1 + 23.68s}. \tag{7.21}$$

Therefore, the open-loop transfer function is obtained from Eqs. (7.17) and (7.21) to be

$$G_0(s)G_c(s) = \frac{320}{s(s+10)} \frac{(1+4.22s)}{(1+23.68s)}, \qquad (7.22)$$

and its Bode diagram is drawn in Figures 7.14a and b. These figures were obtained using MATLAB (see Section 1.7) and are also contained in the M-file which is part of the AMCSTD Toolbox. The phase shift at gain crossover frequency of the compensated system is $-119.2°$, with a resulting phase margin of $60.8°$, and this satisfies the specification on phase margin ($60°$ minimum).

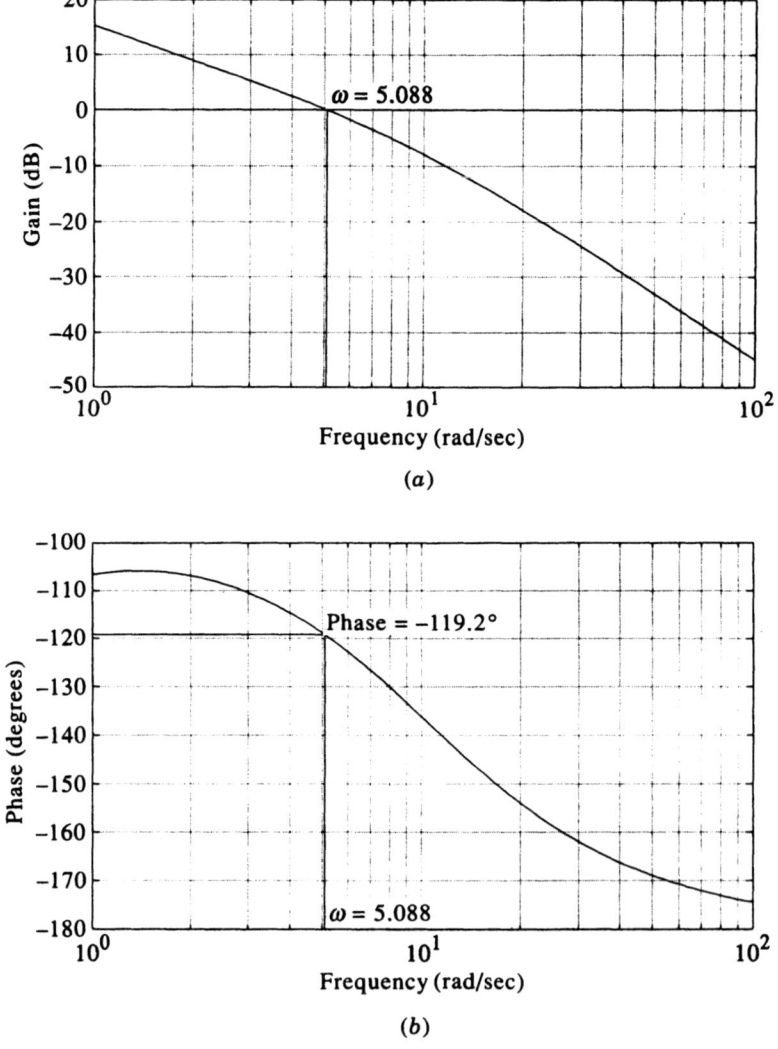

Figure 7.14 Bode diagram of compensated control system for robot.

Having met the stability and accuracy specification of the design, the resulting transient response must be checked. The closed-loop transfer function of the compensated system is

$$\frac{C(s)}{R(s)} = \frac{G_0(s)G_c(s)}{1 + G_0(s)G_c(s)} = \frac{320(1 + 4.22s)}{23.68s^3 + 237.8s^2 + 1360.4s + 320}. \tag{7.23}$$

Factoring the denominator, we obtain

$$\frac{C(s)}{R(s)} = \frac{320(1 + 4.22s)}{(s + 0.2455)(s + 4.90 + j5.57)(s + 4.90 - j5.57)}. \tag{7.24}$$

Thus, the dominant pair of complex-conjugate roots are given by $s = -4.90 \pm j5.57$. As shown in Figure B.3, the corresponding value of ω_n is given by 7.42. Therefore, the value of the damping ratio ζ for the compensated system can be obtained from Eq. (B.18) as

$$\cos \alpha = \frac{4.90}{7.41} = 0.66 = \zeta. \tag{7.25}$$

Therefore,

$$\zeta = 0.66. \tag{7.26}$$

Note that the damping ratio of the uncompensated system was 0.28 [see Eq. (7.20)]. From Eq. (B.33), the maximum percent overshoot for the compensated system for a damping ratio of 0.66 is reduced to 6.33%, and this satisfies the specification that it be less than 8 percent. From Eq. (5.41)‡, the settling time of the compensated system for a damping ratio of 0.66 is 0.82 sec, and this also satisfies the specification that it be less than 1 sec.

In conclusion, the phase-lag network of Eq. (7.21) in conjunction with the amplifier gain given by Eq. (7.15) meet all of the steady-state error, stability, and transient specifications for this design. Table 7.1 summarizes the specification requirements, and the uncompensated and compensated values for these parameters.

Table 7.1. Comparison of Design Parameters for Specification, Uncompensated System and Compensated System

Parameter	Specification	Uncompensated system	Compensated system
K_p	∞	∞	∞
K_v	32/sec	32/sec	32/sec
Phase margin	60° minimum	31.1°	60.8°
Settling time	< 1 sec	0.798 sec	0.82 sec
Maximum percent overshoot	< 8%	40.01%	6.33%

7.5. EXAMPLE 3: DESIGN OF THE CONTROLLER AND FULL-ORDER ESTIMATOR FOR A SPACE SATELLITE'S ATTITUDE-CONTROL SYSTEM WITH POLE PLACEMENT USING LINEAR-STATE-VARIABLE FEEDBACK

For the third design example, the controller and full-order estimator of a satellite's attitude control system will be designed with pole placement using linear-state-variable feedback.

The attitude-control problem is illustrated in Figure 7.15. The satellite is assumed to be rigid, operating in a frictionless environment, and disturbing forces are negligible. It is desired that the angle $\theta(t)$ be zero. However, the satellite will drift with time, and jets aboard the satellite will be fired so that $\theta(t)$ is driven to zero. The dynamics of the system are similar to the space attitude-control problem analyzed in Section 6.6. Assuming that the torque $T(t)$ due to the jet firings is the input to the system and the attitude angle $\theta(t)$ is the output, then the differential equation relating input and output is given by

$$T(t) = \frac{Jd^2\theta(t)}{dt^2}. \tag{7.27}$$

By defining

$$u(t) = \frac{T(t)}{J}, \tag{7.28}$$

then Eq. (7.27) can be written as

$$\frac{d^2\theta(t)}{dt^2} = u(t). \tag{7.29}$$

Using dot notation, Eq. (7.29) simplifies to

$$\ddot{\theta}(t) = u(t). \tag{7.30}$$

The design specifications for this attitude-control system are as follows:

(a) The system's controller must be critically damped.
(b) The settling time of the system's controller must be 1 sec or less.

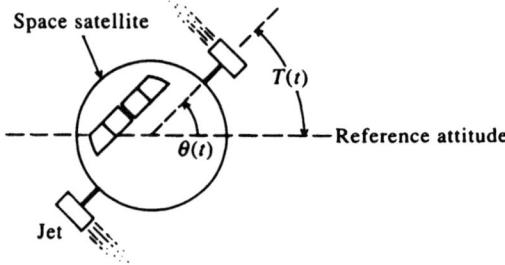

Figure 7.15 Control of space satellite's attitude.

7.5. EXAMPLE 3: A SPACE SATELLITE'S ATTITUDE-CONTROL SYSTEM

(c) The estimator must also be critically damped, an it must be two and one half times faster than the controller (e.g., $\omega_{n_{\text{estimator}}} = 2.5\omega_{n_{\text{controller}}}$).

The controller will be assumed to have the following form:

$$u(t) = -K_1 x_1(t) - K_2 x_2(t), \tag{7.31}$$

where $u(t)$ is the input torque, $x_1(t)$ is the attitude angle, and $x_2(t)$ is the attitude velocity.

The form of the controller's closed-loop transfer function is given by

$$\frac{\theta(s)}{U(s)} = \frac{\omega_n^2}{s^2 + 2\zeta\omega_n s + \omega_n^2}. \tag{7.32}$$

The value of ω_n can be determined by substituting the specification values of critical damping ($\zeta = 1$) and settling time (t_s) of 1 sec into Eq. (5.41)[‡]:

$$t_s = 4/\zeta\omega_n. \tag{7.33}$$

Solving for ω_n,

$$\omega_n = \frac{4}{t_s \zeta} = \frac{4}{(1)(1)} = 4 \text{ rad/sec.} \tag{7.34}$$

Therefore, the numerator and denominator's constant of Eq. (7.32) are given by

$$\omega_n^2 = (4)^2 = 16 \tag{7.35}$$

and Eq. (7.32) can be written as

$$\frac{\theta(s)}{U(s)} = \frac{16}{s^2 + 8s + 16}. \tag{7.36}$$

The resulting characteristic equation of this system is given by

$$\alpha_c(s) = s^2 + 8s + 16 = 0. \tag{7.37}$$

To find the controller gain **K**, Eq. (3.167) will be used:

$$|s\mathbf{I} - \mathbf{P} + \mathbf{bK}| = 0. \tag{7.38}$$

The state-equation representation of Eq. (7.30) is given by

$$x_1(t) = \theta(t),$$
$$x_2(t) = \dot{x}_1(t) = \dot{\theta}(t),$$

524 CONTROL-SYSTEM DESIGN EXAMPLES: COMPLETE CASE STUDIES

$$\begin{bmatrix} \dot{x}_1(t) \\ \dot{x}_2(t) \end{bmatrix} = \underbrace{\begin{bmatrix} 0 & 1 \\ 0 & 0 \end{bmatrix}}_{\mathbf{P}} \begin{bmatrix} x_1(t) \\ x_2(t) \end{bmatrix} + \underbrace{\begin{bmatrix} 0 \\ 1 \end{bmatrix}}_{\mathbf{b}} u(t), \qquad (7.39)$$

$$\theta(t) = \underbrace{[1 \quad 0]}_{\mathbf{L}} \begin{bmatrix} x_1(t) \\ x_2(t) \end{bmatrix}.$$

Substituting the values of **P** and **b** from Eq. (7.39) into Eq. (7.38), the following is obtained:

$$\left| \begin{bmatrix} s & 0 \\ 0 & s \end{bmatrix} - \begin{bmatrix} 0 & 1 \\ 0 & 0 \end{bmatrix} + \begin{bmatrix} 0 \\ 1 \end{bmatrix} [K_1 \quad K_2] \right| = 0. \qquad (7.40)$$

This simplifies to

$$\begin{bmatrix} s & -1 \\ K_1 & s+K_2 \end{bmatrix} = 0,$$

from which we obtain the characteristic equation of the controller:

$$\alpha_c(s) = s^2 + K_2 s + K_1 = 0. \qquad (7.41)$$

Comparing like coefficients in Eqs. (7.37) and (7.41), solutions of $K_1 = 16$ and $K_2 = 8$ are obtained. Therefore,

$$\mathbf{K} = [16 \quad 8], \qquad (7.42)$$

and the controller's characteristic equation is given by

$$\alpha_c(s) = s^2 + 8s + 16 = (s+4)^2 = 0. \qquad (7.43)$$

The controller is given by

$$u(t) = -\mathbf{K}\mathbf{x}(t) = -[16 \quad 8] \begin{bmatrix} x_1(t) \\ x_2(t) \end{bmatrix} = -16x_1(t) - 8x_2(t), \qquad (7.44)$$

which is the form of the controller specified [see Eq. (7.31)].

For the next part of the design, the full-order estimator will be designed. Because the estimator is specified to be 2.5 times faster than the controller,

$$\omega_{n_{\text{estimator}}} = 2.5 \omega_{n_{\text{controller}}}. \qquad (7.45)$$

Because $\omega_{n_{\text{controller}}} = 4 [see Eq. (7.34)]$,

$$\omega_{n_{\text{estimator}}} = 2.5(4) = 10 \text{ rad/sec.} \qquad (7.46)$$

7.5. EXAMPLE 3: A SPACE SATELLITE'S ATTITUDE-CONTROL SYSTEM

The estimator is specified to be critically damped. Therefore, the characteristic equation of the estimator is given by

$$\omega_e(s) = (s + 10)^2 = s^2 + 20s + 100 = 0. \tag{7.47}$$

To find the estimator, Eq. (8.173) will be used:

$$|s\mathbf{I} - (\mathbf{P} - \mathbf{ML})| = 0. \tag{7.48}$$

Substituting for **P** and **L** from Eq. (7.39) into Eq. (7.48), the following is obtained:

$$\left| \begin{bmatrix} s & 0 \\ 0 & s \end{bmatrix} - \begin{bmatrix} 0 & 1 \\ 0 & 0 \end{bmatrix} + \begin{bmatrix} m_1 \\ m_2 \end{bmatrix} [1 \ 0] \right| = 0, \tag{7.49}$$

which reduces to

$$\begin{vmatrix} s + m_1 & -1 \\ m_2 & s \end{vmatrix} = 0. \tag{7.50}$$

The resulting characteristic equation of the estimator in terns of m_1 and m_2 is given by

$$s^2 + m_1 s + m_2 = 0. \tag{7.51}$$

Setting like coefficients in Eqs. (7.47) and (7.51) equal to each other, the following is obtained: $m_1 = 20; m_2 = 100$. Therefore,

$$\mathbf{M} = \begin{bmatrix} 20 \\ 100 \end{bmatrix}, \tag{7.52}$$

and the characteristic equation of the estimator is given by

$$\alpha_e(s) = s^2 + 20s + 100 = (s + 10)^2 = 0. \tag{7.53}$$

Having designed the controller and estimator, the transfer function of the compensator will next be determined from Eq. (3.162) and Figure 3.19:

$$G_{\text{comp}} = \frac{U(s)}{\theta(s)} = -\mathbf{K}[s\mathbf{I} - \mathbf{P} + \mathbf{bK} + \mathbf{ML}]^{-1}\mathbf{M}. \tag{7.54}$$

Substituting Eqs. (7.39), (7.42), and (7.52) into Eq. (7.54), the following is obtained:

$$G_{\text{comp}}(s) = \frac{U(s)}{\theta(s)} = -[16 \ 8] \left[\begin{bmatrix} s & 0 \\ 0 & s \end{bmatrix} - \begin{bmatrix} 0 & 1 \\ 0 & 0 \end{bmatrix} + \begin{bmatrix} 0 \\ 1 \end{bmatrix} [16 \ 8] \right.$$
$$\left. + \begin{bmatrix} 20 \\ 100 \end{bmatrix} [1 \ 0] \right]^{-1} \begin{bmatrix} 20 \\ 200 \end{bmatrix}. \tag{7.55}$$

This reduces to

$$G_{comp}(s) = -\begin{bmatrix} 16 & 8 \end{bmatrix} \begin{bmatrix} s+20 & -1 \\ 116 & s+8 \end{bmatrix}^{-1} \begin{bmatrix} 20 \\ 100 \end{bmatrix}. \quad (7.56)$$

The inverse matrix portion of Eq. (7.56) is given by

$$\begin{bmatrix} s+20 & -1 \\ 116 & s+8 \end{bmatrix}^{-1} = \frac{\begin{bmatrix} s+8 & 1 \\ -116 & s+20 \end{bmatrix}}{s^2 + 28s + 276}. \quad (7.57)$$

Substitution of Eq. (7.57) into (7.56) results in the following:

$$G_{comp} = \frac{U(s)}{\theta(s)} = \frac{-1120(s + 1.429)}{s^2 + 28s + 276}. \quad (7.58)$$

Factoring the denominator, the following transfer function is obtained:

$$G_{comp}(s) = \frac{U(s)}{\theta(s)} = \frac{-1120(s + 1.429)}{(s + 14 \pm j8.944)^2}. \quad (7.59)$$

The transfer function of the satellite can be represented by [see Eq. (7.30)]

$$G(s) = \frac{\theta(s)}{U(s)} = \frac{1}{s^2}. \quad (7.60)$$

Therefore, the open-loop transfer function of the complete system is given by

$$G(s)G_{comp}(s) = \frac{-1120(s + 1.429)}{s^2(s + 14 \pm j8.944)^2}. \quad (7.61)$$

To check the resulting design, the root locus and Bode diagram will be used. For the root-locus design, the specific gain will be replaced with the variable gain K:

$$G(s)G_{comp}(s) = \frac{K(s + 1.429)}{s^2(s + 14 + j8.944)(s + 14 - j8.944)}, \quad (7.62)$$

which is shown in Figure 7.16. Observe that the root locus goes through the roots selected in Eqs. (7.43) and (7.53). These roots are shown in Figure 7.16 by asterisks. Note that $K = 1120$ at these roots. This figure was obtained using MATLAB (see Section 1.7), and is contained in the M-file which is part of the AMCSTD Toolbox and can be retrieved free from The MathWorks. Inc. anonymous FTP server at ftp://ftp.mathworks.com/pub/books/advshinners.

To draw the Bode diagram, the modified form of Eq. (7.62) is analyzed:

$$G(s)G_{comp}(s) = \frac{1120(s + 1.429)}{s^2(s^2 + 28s + 276)}. \quad (7.63)$$

7.5. EXAMPLE 3: A SPACE SATELLITE'S ATTITUDE-CONTROL SYSTEM

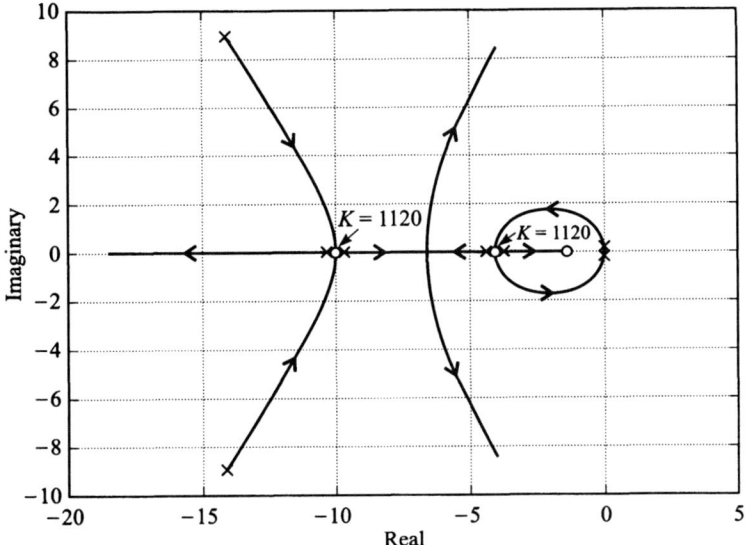

Figure 7.16 Root locus for compensated attitude-control system.

The resulting Bode diagram is drawn from the following modification to Eq. (7.63):

$$G(s)G_{comp}(s) = \frac{5.8(0.7s + 1)}{s^2} \frac{276}{s^2 + 28s + 276}. \tag{7.64}$$

The quadratic poles in the denominator have a natural resonance frequency ω_n and a damping ratio ζ given by

$$\omega_n = \sqrt{276} = 16.61 \text{ rad/sec}, \tag{7.65}$$

$$\zeta = \frac{28}{2(16.61)} = 0.84. \tag{7.66}$$

The resulting Bode diagram is shown in Figure 7.17. This figure was obtained using MATLAB (see Section 1.7), and is also contained in the M-file which is part of the AMCSTD Toolbox.

It is concluded that the uncompensated transfer function.

$$G(s) = 1/s^2, \tag{7.67}$$

has its phase margin increased from 0° to 46.78° (measured at gain crossover frequency of 4.174 rad/sec), and its gain margin is increased from $-\infty$ to 15.41 dB (measured at the phase crossover frequency of 15.36 rad/sec) for the compensator of Eq. (7.59). Notice that the gain crossover frequency of 4.174 rad/sec is approximately consistent with the controller's closed-loop roots of $\omega_n = 4$ and $\zeta = 1$. This is a reasonable result, as the slower roots of the controller are more dominant than the faster estimator roots on the system response.

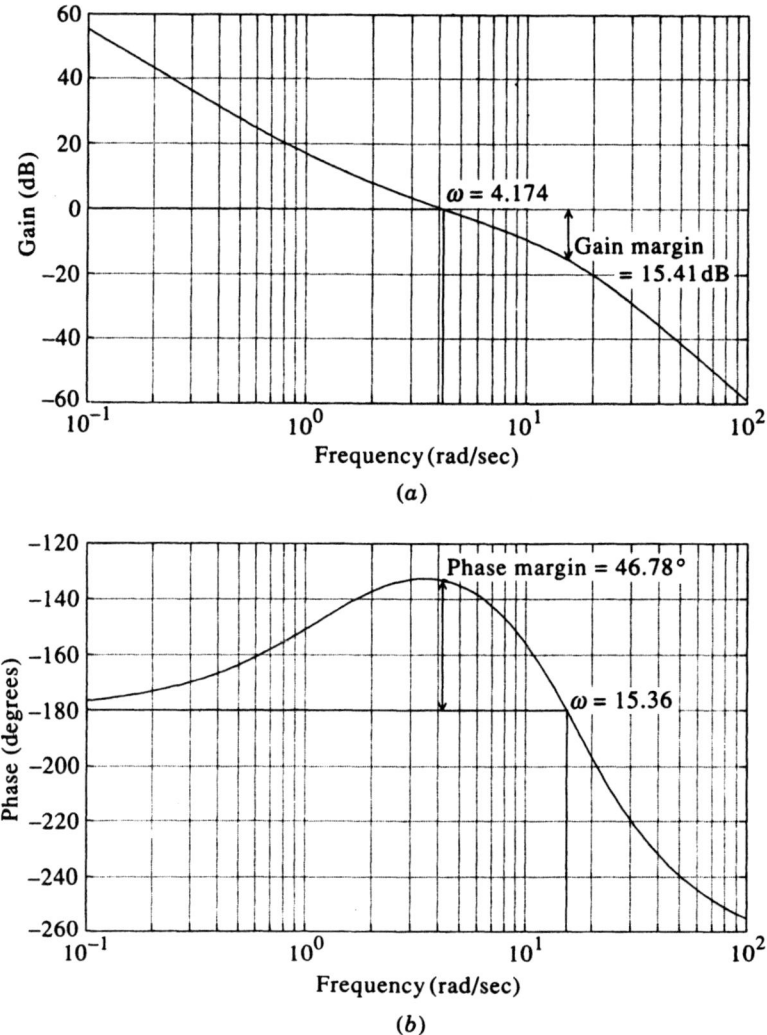

Figure 7.17 Bode diagram of compensated attitude-control system.

7.6. EXAMPLE 4: DESIGN OF A SAMPLED-DATA CONTROL SYSTEM FOR CONTROLLING THE TEMPERATURE OF A LIQUID IN A TANK

For the fourth design example in this chapter, the closed-loop temperature digital control system illustrated in Figure 7.18a will be designed. The microcomputer output controls the position of a solenoid valve, which then controls the quantity of steam into the tank coil. Feedback is obtained from a thermocouple in the tank, whose signal is amplified and then converted to a digital signal, for use by the microcomputer. In this manner, the microcomputer controls the temperature of the liquid contained in the tank in a closed-loop manner. The resulting block diagram of this thermal control system is illustrated in Figure 7.18b. The microcompu-

7.6. EXAMPLE 4: TEMPERATURE CONTROL SYSTEM FOR A LIQUID IN A TANK

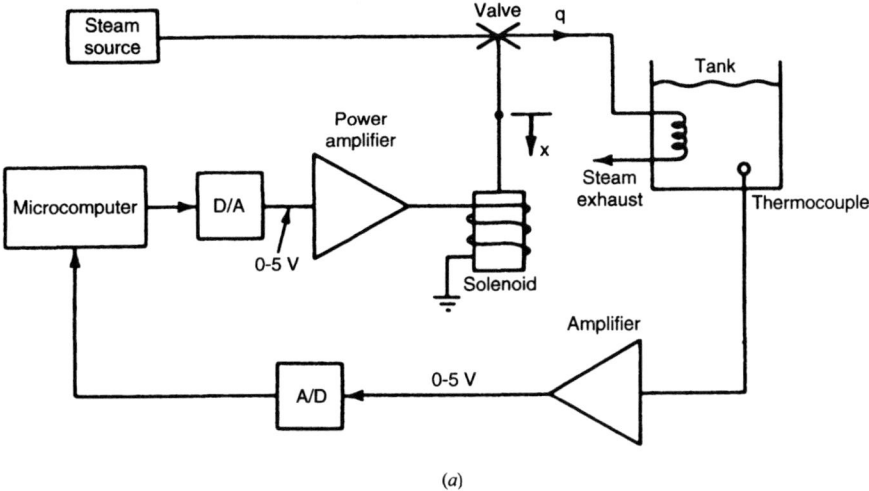

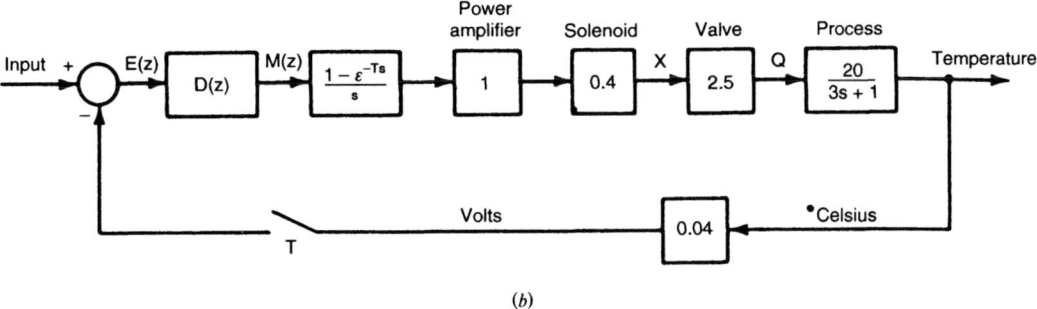

Figure 7.18 Temperature control system. (From Phillips and Nagle [5]. (Reprinted by permission of Prentice-Hall, Englewood Cliffs, NJ)

ter will be represented by $D(z)$, the zero-order hold by $G_H(s)$, and the transfer function of the thermal process for heating the liquid in the tank is given by [5]

$$G_p(s) = \frac{20}{3s+1}.$$

Observe that the transfer function of the thermal process $G_p(s)$ is of the same form as derived in Section 3.6‡, Eq. (3.152‡), for heating water in a tank.

The specifications for this design problem are as follows:

(a) The steady-state error for a step input of 10°C temperature should be less than or equal to 2%.
(b) The system should have a minimum phase margin of 60°.
(c) The sampling time is 0.5 sec.

The design problem is to determine the value of $D(z)$ to meet these specifications.

The system will first be analyzed without any compensation $[D(z) = 1]$. The closed-loop transfer function of the uncompensated system is given by the following:

$$\frac{T_c(z)}{T_r(z)} = \frac{(0.4)(2.5)G_H G_P(z)}{1 + (0.04)(0.4)(2.5)G_H G_P H(z)}, \quad (7.68)$$

where $T_r(z)$ and $T_c(z)$ represent the z transform of the desired input and actual output temperature, respectively, and

$$(0.4)(2.5)G_H G_P(z) = Z\left[\left(\frac{1-e^{-Ts}}{s}\right)\left(\frac{20}{3s+1}\right)\right]. \quad (7.69)$$

For the $T = 0.5$ sec, this reduces to the following:

$$G_H G_P(z) = \frac{3.07}{z - 0.8465}. \quad (7.70)$$

In addition, the feedback sensor's transfer function $H(z) = 0.04$. Therefore,

$$G_H G_P H(z) = 0.04\left(\frac{3.07}{z - 0.8465}\right). \quad (7.71)$$

Substituting Eqs. (7.70) and (7.71) into Eq. (7.68), the following closed-loop transfer function is obtained:

$$\frac{T_c(z)}{T_r(z)} = \frac{3.07}{z - 0.724}. \quad (7.72)$$

For a step input of 10°C,

$$T_r(z) = \frac{10z}{z - 1}. \quad (7.73)$$

The output response is obtained by substituting Eq. (7.73) into (7.72):

$$T_c(z) = \frac{10}{z - 1}\frac{3.07}{z - 0.724}. \quad (7.74)$$

Simplifying Eq. (7.74), the following is obtained:

$$T_c(z) = \frac{30.7z}{z^2 - 1.724z + 0.724}. \quad (7.75)$$

In terms of negative powers of z,

$$T_c(z) = \frac{30.7z^{-1}}{1 - 1.724z^{-1} + 0.724z^{-2}}. \quad (7.76)$$

To find the time-domain solution, the denominator of Eq. (7.76) is divided into its numerator and the following series is obtained:

7.6. EXAMPLE 4: TEMPERATURE CONTROL SYSTEM FOR A LIQUID IN A TANK

$$T_c(z) = 30.7z^{-1} + 52.92z^{-2} + 69.0z^{-3} + 80.6z^{-4} + \cdots + . \tag{7.77}$$

Taking the inverse z transform of Eq. (7.77), the following series is obtained in the discrete-time domain:

$$T_c^*(t) = 30.7\delta(t - 0.5) + 52.92\delta(t - 1) + 69.0\delta(t - 1.5) + 80.6\delta(t - 2) + \cdots. \tag{7.78}$$

The result is plotted in Figure 7.19 for terms taken out to 100 sec. From this plot, it is seen that the system is very overdamped and slow. It takes approximately 15 sec to reach its steady-state value of 111.112. To find the steady-state value, the final-value theorem of Eq. (4.61) will be applied to Eq. (7.74):

$$T_c(\infty) = \lim_{z \to 1}(1 - z^{-1})T_c(z). \tag{7.79}$$

Substituting Eq. (7.74) into Eq. (7.79), the following equation is obtained for the steady-state value:

$$T_c(\infty) = \lim_{z \to 1}\left(\frac{z - 1}{z}\right)\left(\frac{10z}{z - 1}\right)\left(\frac{3.07}{z - 0.724}\right) = 111.112. \tag{7.80}$$

The percent steady-state error for a step input of 10°C can be determined from

$$\frac{E(z)}{T_r(z)} = \frac{1}{1 + G_H G_P H(z)}. \tag{7.81}$$

Substituting Eq. (7.71) and

$$T_r(z) = \frac{10z}{z - 1} \tag{7.82}$$

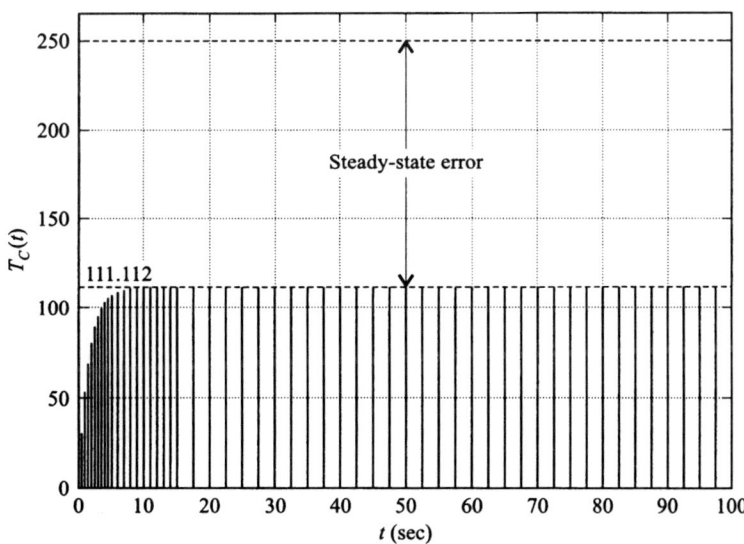

Figure 7.19 Transient response of uncompensated temperature control system.

532 CONTROL-SYSTEM DESIGN EXAMPLES: COMPLETE CASE STUDIES

into Eq. (7.81), the following is obtained:

$$E(z) = 10 \frac{z}{z-1} \frac{1}{1 + 0.04[3.07/(z - 0.8465)]}. \tag{7.83}$$

Therefore, applying the final-value theorem of Eq. (4.61) to Eq. (7.83), the following is obtained:

$$e_{ss} = \lim_{z \to 1} 10 \frac{(z-1)}{z} \left(\frac{z}{z-1}\right) \left(\frac{(z - 0.8465)}{z - 0.8465 + 0.1228}\right). \tag{7.84}$$

This reduces to

$$e_{ss} = 5.55. \tag{7.85}$$

Therefore, the steady-state error for the step input of 10°C is 55.5% which greatly exceeds the specification level of 2%. From the transient response of Figure 7.19, the final value of 111.112 is in error by 55.5% of the expected value of 250 (which is the reciprocal of the feedback transfer function of 0.04 times the 10° step input of temperature, or 250).

In the next stage of the design, the value of K will be determined, which results in a 2% error to a step input which is a specification requirement. With $D(z) = K$, the value of K can be determined as follows:

$$\frac{E(z)}{T_r(z)} = \frac{1}{1 + KG_H G_P H(z)}. \tag{7.86}$$

Substituting Eq. (7.71) into Eq. (7.86), the following is obtained:

$$\frac{E(z)}{T_r(z)} = \frac{z - 0.8465}{z - 0.8465 + 0.1228K}. \tag{7.87}$$

For a unit step input of temperature,

$$T_r(z) = \frac{z}{z - 1}. \tag{7.88}$$

Substituting Eq. (7.88) into Eq. (7.87), the following expression for error is obtained:

$$E(z) = \left(\frac{z}{z-1}\right)\left(\frac{z - 0.8465}{z - 0.8465 + 0.1228K}\right). \tag{7.89}$$

To satisfy a 2% steady-state error requirement for a unit step input, the final-value theorem is applied to Eq. (7.89) as follows (note that the percent error is the same for any constant input with $T = 0.5$ sec):

$$e_{ss} = 0.02 = \lim_{z \to 1} \frac{(z-1)}{z} \left(\frac{z}{z-1}\right)\left(\frac{z - 0.8465}{z - 0.8465 + 0.1228K}\right). \tag{7.90}$$

Solving for K,

$$K = 61.25. \quad (7.91)$$

Therefore, a gain of $D(z) = K = 61.25$ is needed to keep the steady-state error at 2%. It is interesting to analyze the stability of this system with $D(z) = K = 61.25$. The closed-loop transfer function is given by

$$\frac{T_c(z)}{T_r(z)} = \frac{KG_H G_P(z)}{1 + KG_H G_P H(z)}. \quad (7.92)$$

Substituting Eq. (7.71) into Eq. (7.92), the following is obtained:

$$\frac{T_c(z)}{T_r(z)} = \frac{3.07K}{z - (0.8465 - 0.1228K)}. \quad (7.93)$$

With $K = 61.25$, this simplifies to

$$\frac{T_c(z)}{T_r(z)} = \frac{188.04}{z + 6.675}. \quad (7.94)$$

Note that the characteristic equation obtained from Eq. (7.94) is given by

$$z + 6.675 = 0,$$

and indicates a root outside the unit circle at -6.675 in the z-plane. Therefore, this incompensated system will be unstable.

A digital controller will next be designed which has $K = 61.25$ and contains compensation to satisfy the specification requirements of a minimum phase margin of $60°$. The open-loop transfer function of the system with $K = 61.25$ is given by [see Eq. (7.71)]

$$KG_H G_P H(z) = (61.25)\frac{0.1228}{z - 0.8465} = \frac{7.522}{z - 0.8465}. \quad (7.95)$$

Transforming Eq. (7.95) into the w-plane, we use Eq. (4.175),

$$z = \frac{1 + (T/2)w}{1 - (T/2)w}. \quad (7.96)$$

Substituting Eq. (7.96) with $T = 0.5$ sec into Eq. (7.95), the following is obtained:

$$KG_H G_P H(w) = \frac{49(1 - w/4)}{(1 + w/0.3325)}. \quad (7.97)$$

To compensate the system, a zero will be added to cancel the pole term in Eq. (7.97). To achieve the desired phase margin, a pole term $(1 + w/0.039)$, was added as follows:

$$D(w) = \frac{1 + w/(0.3325)}{1 + w/(0.039)}. \quad (7.98)$$

534 CONTROL-SYSTEM DESIGN EXAMPLES: COMPLETE CASE STUDIES

Therefore, the open-loop transfer function of the compensated system is obtained by multiplying Eqs. (7.97) and (7.98). The result is as follows:

$$KG_H G_P H(w)D(w) = \frac{49(1 - w/4)}{1 + w/0.039}. \tag{7.99}$$

Figure 7.20 shows a Bode diagram of the compensated system, which has a phase margin of 62.32° and an infinite gain margin. Therefore, the specification on stability has been achieved. This Bode diagram was drawn using MATLAB (see Section 1.7), and is contained in the M-file that is part of the AMCSTD Toolbox which can be retrieved by the reader free from The MathWorks, Inc. anonymous FTP server at ftp://ftp.mathworks.com/pub/books/advshinners.

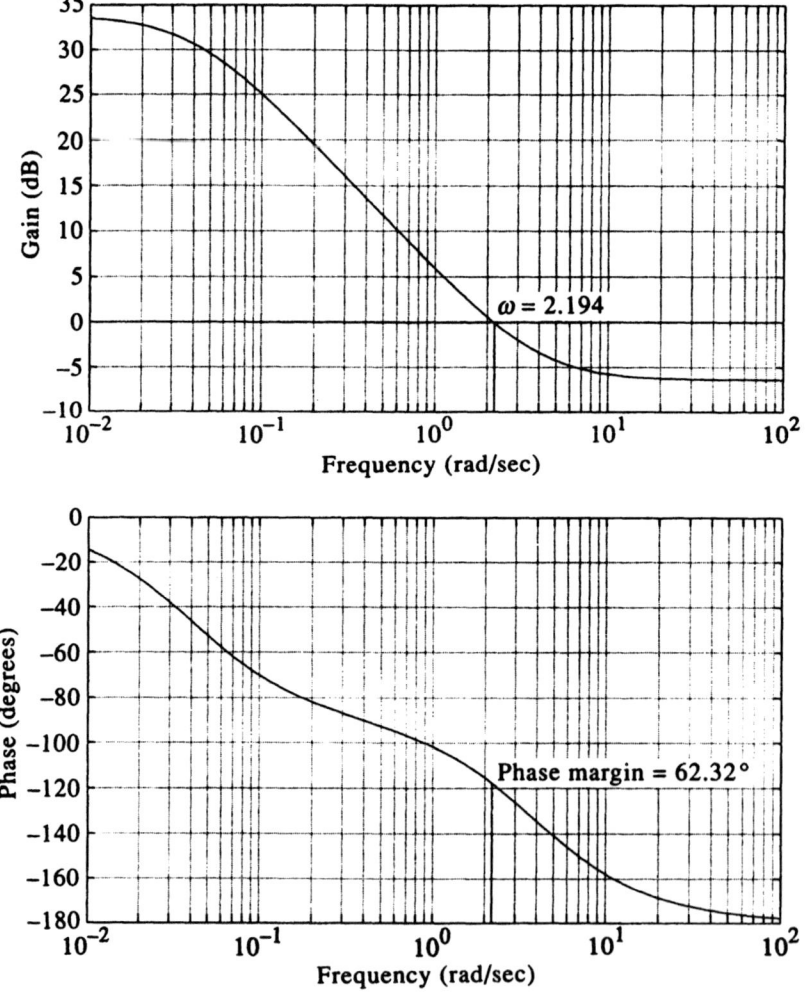

Figure 7.20 Bode diagram of compensated temperature control system.

7.6. EXAMPLE 4: TEMPERATURE CONTROL SYSTEM FOR A LIQUID IN A TANK

The compensating network of Eq. (7.98) will be converted back into the z plane using Eq. (4.167):

$$w = \frac{2}{T}\frac{z-1}{z+1}. \tag{7.100}$$

Because $T = 0.5$ sec, this reduces to

$$w = 4\frac{z-1}{z+1}. \tag{7.101}$$

Substituting Eq. (7.101) into Eq. (7.98), the following is obtained:

$$D(z) = 0.1258\frac{(z - 0.8465)}{(z - 0.9807)}. \tag{7.102}$$

Incorporating the gain of $K = 61.25$ [see Eq. (7.91) into the digital controller, the total transfer function of the digital controller is given by

$$KD(z) = (61.25)(0.1258)\frac{(z - 0.8465)}{(z - 0.9807)},$$

which reduces to

$$KD(z) = (7.705)\frac{(z - 0.8465)}{(z - 0.9807)}. \tag{7.103}$$

Therefore, the open-loop transfer function of the temperature control system can be obtained from Eq. (7.71) and (7.103) as follows:

$$KD(z)G_H G_P H(z) = 7.705\frac{(z - 0.8465)}{(z - 0.9807)}\frac{0.1228}{(z - 0.8465)},$$

which reduces to

$$KD(z)G_H G_P H(z) = \frac{0.9462}{z - 0.9807}. \tag{7.104}$$

The steady-state error of the resulting compensated system can be checked to determine that it meets the steady-state error specification as follows:

$$\frac{E(z)}{T_r(z)} = \frac{1}{1 + KD(z)G_H G_P H(z)}. \tag{7.105}$$

Substituting Eqs. (7.88) and (7.104) into Eq. (7.105), the following is obtained:

$$E(z) = \left(\frac{z}{z-1}\right)\left(\frac{1}{1 + 0.9462/(z - 0.9807)}\right). \tag{7.106}$$

Applying the final-value theorem to Eq. (7.106)

$$e_{ss} = \lim_{z \to 1} \frac{(z-1)}{z} \left(\frac{z}{z-1}\right)\left(\frac{1}{1+0.9462/(z-0.9807)}\right) = 0.02. \qquad (7.107)$$

Therefore, the compensated system has a 2% steady-state error to a unit step input, and satisfies the specification requirement on steady-state error.

To complete the design, the transient response of the compensated system will be determined. The closed-loop transfer function of the compensated system is given by

$$\frac{T_c(z)}{T_r(z)} = \frac{KD(z)G_H G_P(z)}{1 + KD(z)G_H G_P H(z)}. \qquad (7.108)$$

Substituting Eq. (7.104) into Eq. (7.108) and accounting for the fact that $H(z) = 0.04$, the following is obtained:

$$\frac{T_c(z)}{T_r(z)} = \frac{[(0.9462/0.04)/(z-0.9807)]}{1+0.9462/(z-0.9807)} = \frac{23.655}{z-0.0345}. \qquad (7.109)$$

Substituting a unit step input of temperature (the transient response can be determined using a unit step as well as an input of 10°C) for $T_r(z)$ in Eq. (7.109), the following expression of the output temperature is obtained:

$$T_c(z) = \left(\frac{z}{(z-1)}\right)\left(\frac{23.655}{(z-0.0345)}\right). \qquad (7.110)$$

From the following expression,

$$T_c(z) = \frac{23.655z}{z^2 - 1.0345z + 0.0345}. \qquad (7.111)$$

$T_c(z)$'s series output is obtained by dividing denominator into numerator:

$$T_c(z) = 23.655z^{-1} + 24.47z^{-2} + 24.49z^{-3} + 24.5z^{-4} + 24.5z^{-5} + \cdots. \qquad (7.112)$$

In the time domain, this series is given by

$$T_c^*(t) = 23.655\delta(t-T) + 24.47\delta(t-2T) + 24.49\delta(t-3T) \\ + 24.5\delta(t-4T) + 24.5\delta(t-5T) + 24.5\delta(t-6T) + \cdots, \qquad (7.113)$$

where $T = 0.5$ sec. The result is shown in Figure 7.21. From this graph, it can be seen that the system's response has improved dramatically compared to that of the uncompensated system shown in Figure 7.19. In approximately 2 sec, the system reaches its steady-state value of 24.5. This results in a 2% error with the ideal value of 25 (unit input divided by the feedback element's transfer function of 0.04).

In conclusion, this design has achieved all of its design specifications. The uncompensated system was stable, but it had a very large steady error and was very sluggish. Increasing its gain to meet the steady-state accuracy requirement made

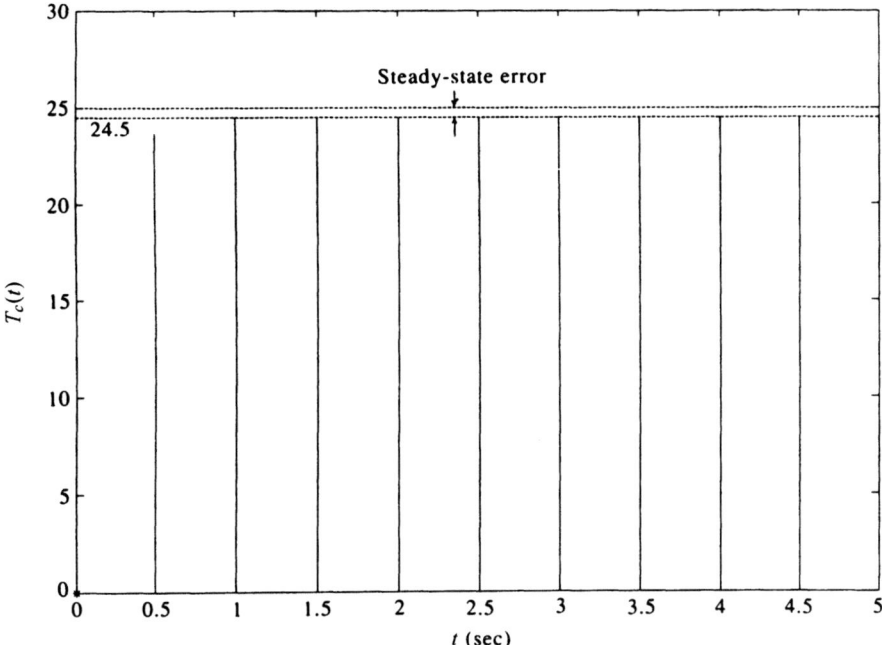

Figure 7.21 Transient response of compensated temperature control system.

the system unstable. By compensating it with the digital controller, it was stabilized, met its accuracy requirements, and its transient response was greatly improved.

7.7. EXAMPLE 5: DESIGN OF A ROBUST CONTROL SYSTEM FOR CONTROLLING THE FLAPS OF A HYDROFOIL [6–10]

In the fifth design example in this chapter, we wish to control the flaps of a hydrofoil. Consider the conceptual illustration in Figure 7.22 which illustrates sea waves hitting the flaps of the hydrofoil. This causes large resultant torques that try to turn the flaps from their desirable positions. These torques can be viewed as the torque distur-

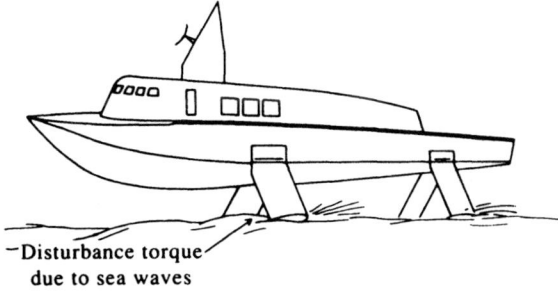

Figure 7.22 A hydrofoil boat illustrating sea waves causing disturbance torques.

bances $D(S)$ illustrated in the equivalent block diagram of the two-degrees-of-freedom robust control system for controlling the angular flap position shown in Figure 7.23. In the real world, sea waves and their resultant torque are a stochastic process. However, we will assume in this design example that the torque disturbances due to sea waves can be averaged for the environment in which the hydrofoil will operate and can be represented deterministically as a unit step input.

We wish to design the robust control system for controlling the angular position of the hydrofoil's flaps as illustrated in Figure 7.23 to minimize the effects of the disturbance torques $D(s)$ and with gain variations from $K = 40$ to 10 and 100.

The two-degrees-of-freedom control system illustrated in Figure 7.23 was analyzed thoroughly in Section 3.10 in the discussion of robust control systems. It was shown that since the transfer function $C(s)/D(s)$ [given by Eq. (3.198)] and the sensitivity of $H(s)$ with respect to K [given by Eq. (3.201)] are identical, then the same control system techniques can be used to suppress the effect of the disturbance $D(s)$ and achieve robustness (insensitivity) with respect to variations of K. The transfer function of the hydrofoil dynamics will be assumed in this case study to be given by the following transfer function:

$$G_0(s) = \frac{(1+0.2s)}{s(1+0.5s)(1+0.1s)}. \tag{7.114}$$

We will first assume that $G_{c1}(s) = G_{c2}(s) = 1$, and we will investigate the effect of the variation of K where $K = 10$, 40, and 100. Therefore,

$$KG_{c2}(s)G_0(s) = \frac{K(1+0.2s)}{s(1+0.5s)(1+0.1s)}. \tag{7.115}$$

Figure 7.24 illustrates the unit step response of the system when $K = 10$, 40, and 100. Table 7.2 lists the damping ratio of the unit step transient responses and the characteristic equation roots of this control system which were obtained using MATLAB. Observe that the variations in $K = 10$, 40, and 100 result in considerable variation in the transient responses of this control system. Figure 7.25 illustrates the root loci and location of the closed-loop, complex-conjugate roots of the cases being analyzed.

As discussed in Section 3.10, the design approach for this robust controller, $G_{c2}(s)$, is to place two zeros at (or near) the desired complex-conjugate loop poles at $-3.2262 \pm j11.5670$ for the nominal gain case of 40. Therefore,

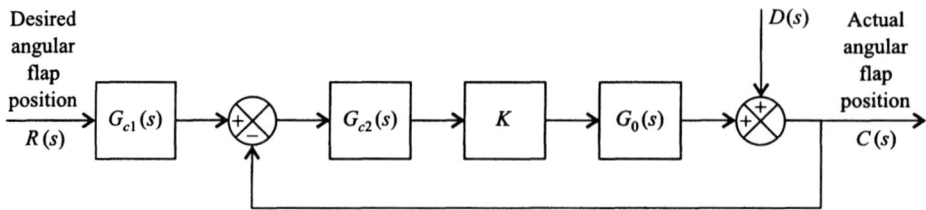

Figure 7.23 The equivalent block diagram for control of the hydrofoil boat of Figure 7.22 using a two-degrees-of-freedom robust controller.

7.7. EXAMPLE 5: ROBUST CONTROL SYSTEM DESIGN FOR CONTROLLING THE FLAPS OF A HYDROFOIL

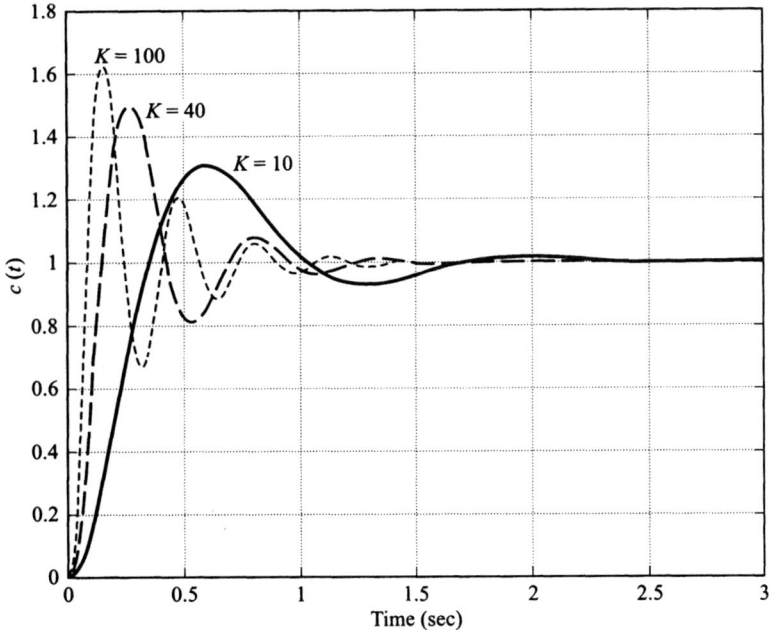

Figure 7.24 Unit step response of Figure 7.23 with $KG_{c2}(s)G_0(s)$ given by Eq. (7.115) and $G_{c1}(s) = 1$.

Table 7.2. Characteristics of the Control System Illustrated in Figure 7.23 where $KG_{c2}(s)G_0(s)$ is given by Eq. (7.115) and $G_{c1}(s) = 1$

K	Damping ratio	Roots of characteristic equation
100	0.173	$-5.1997, -3.4002 \pm j19.3153$
40	0.269	$-5.5477, -3.2262 \pm j11.5670$
10	0.431	$-7.5632, -2.2184 \pm j4.6392$

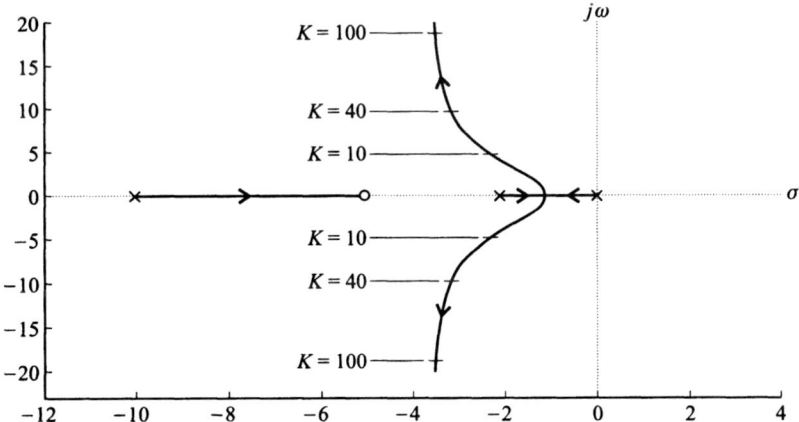

Figure 7.25 Root-locus for system of Figure 7.23 with $KG_{c2}(s)G_0(s)$ given by Eq. (7.115) and $G_{c1}(s) = 1$.

540 CONTROL-SYSTEM DESIGN EXAMPLES: COMPLETE CASE STUDIES

$$G_{c2}(s) = \frac{(s + 3.2262 + j11.5670)(s + 3.2262 - j11.5670)}{144.2}$$

$$= \frac{s^2 + 6.5423s + 144.2}{144.2}. \tag{7.116}$$

The forward-path transfer function of this control system with $KG_0(s)$ given by Eq. (7.115) and $G_{c2}(s)$ given by Eq. (7.116) is:

$$KG_{c2}(s)G_0(s) = \frac{K(1 + 0.2s)(s^2 + 6.4523s + 144.2)}{144.2s(1 + 0.5s)(1 + 0.1s)} \tag{7.117}$$

Table 7.3 lists the characteristics of the damping ratio and the characteristic equation roots of this control system, obtained using MATLAB, with the forward-loop transfer function given by Eq. (7.117). Observe that the damping ratios are much closer (0.317 to 0.483) than they were before the addition of the robust controller $G_{c2}(s)$ as shown in Table 7.2 (where the damping ratio previously varied from 0.173 to 0.431).

Figure 7.26 illustrates the root loci with the robust controller $G_{c2}(s)$ added, and the location of the closed-loop, complex-conjugate roots for the three cases. Observe from this root locus that by locating the two zeros of the forward-loop controller $G_{c2}(s)$ near the desired characteristic equation complex-conjugate roots, then the sensitivity of this control system is much better. The result is that the sensitivity of the system as K varies is much better than before.

Table 7.3. Characteristics of the Control System Illustrated in Figure 7.23 with the Forward-Loop Controller $G_{c2}(s)$ Added and where $KG_{c2}(s)G_0(s)$ is given by Eq. (7.117)

K	Damping ratio	Roots of characteristic equation
100	0.317	$-5.2097, -3.1939 \pm j9.5669$
40	0.372	$-5.5700, -3.0710 \pm j7.6586$
10	0.483	$-7.4587, -2.2112 \pm j4.0128$

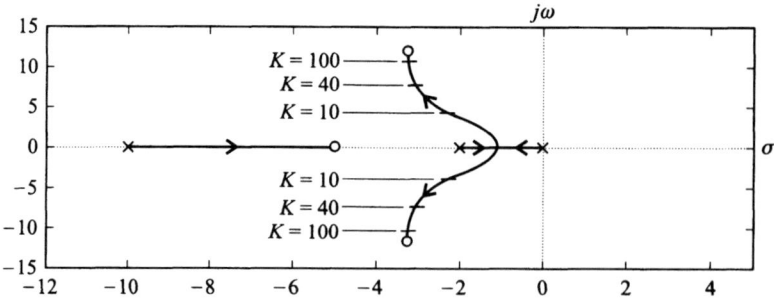

Figure 7.26 Root-locus for system shown in Figure 7.23 with $KG_{c2}(s)G_0(s)$ given by Eq. (7.117) and $G_{c1}(s) = 1$.

7.7. EXAMPLE 5: ROBUST CONTROL SYSTEM DESIGN FOR CONTROLLING THE FLAPS OF A HYDROFOIL

As was shown in Section 3.10 on robust control systems, it is necessary to add the series controller $G_{c1}(s)$ whose transfer function is the reciprocal of that given by $G_{c2}(s)$:

$$G_{c1}(s) = \frac{144.2}{s^2 + 6.5423s + 144.2}. \tag{7.118}$$

The unit step response of this control system with the forward-path transfer function of the control system given by Eq. (7.117), with $K = 10$, 40, and 100, and with the series controller transfer function given by Eq. (7.118), is illustrated in Figure 7.27. Comparing these unit step responses with those in Figure 7.24, we conclude that this control system has been made to be much less sensitive to variations in K. The maximum percent overshoots to a unit step input ranged from 30% to 61% in Figure 7.24 for the original system, and only 23% to 38% for the robust design in Figure 7.27. In addition, as was pointed out before, the robustness with respect to variations in K will also provide disturbance suppression using the same control techniques as the two-degrees-of-freedom control system shown in Figure 7.23.

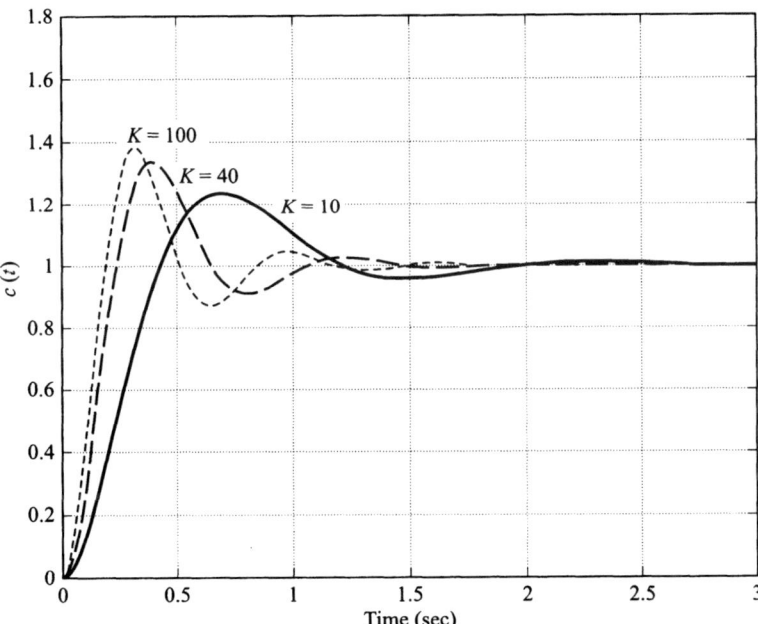

Figure 7.27 Unit step response for system of Figure 7.23 with $KG_{c2}(s)G_0(s)$ given by Eq. (7.117) and $G_{c1}(s)$ given by E1. (7.118).

PROBLEMS

7.1. We wish to design a hydrofoil ship, which is conceptually illustrated in Figure P7.1a, and whose block diagram is illustrated in Figure P7.1b. As illustrated, sea waves hitting the flaps cause large resultant torques that try to turn the flaps from their desirable positions. In effect, the sea waves can be viewed to be a torque disturbance $U_T(s)$, as shown in Figure P7.1b. Although sea waves and their resultant torque are a stochastic process, we will assume in this problem that the torque disturbance due to sea waves can be averaged for the environment in which this hydrofoil will operate to be a unit step input. The specifications for this hydrofoil control system are that the steady-state error due to a unit step input $U_T(s)$ shall not exceed a steady-state error at $E(s)$ of 0.1 and the stability requirement for the stabilized control system is a minimum phase margin of 50° and a minimum gain margin of 20 dB. In addition, a minimum gain crossover of 8 rad/sec is required to achieve the desired response time.

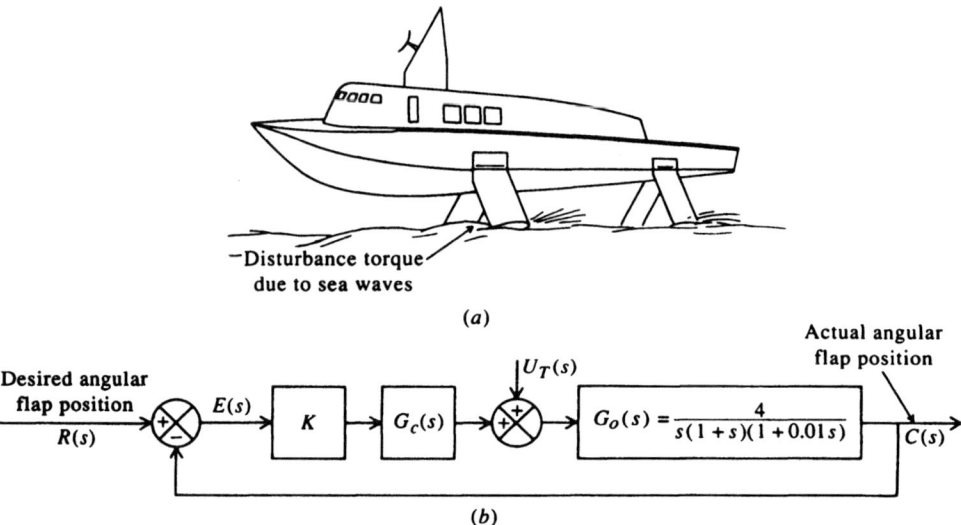

Figure P7.1 A hydrofoil boat illustrating sea waves causing disturbance torques (a), and its equivalent block diagram (b).

(a) With the compensation network $G_c(s)$ set equal to 1 and K set equal to 1, determine the phase and gain margins of this system.

(b) With the compensation network $G_c(s)$ set equal to 1, determine the gain K required to achieve the steady-state error requirement of 0.1 due to a unit torque disturbance at $U_T(s)$.

(c) With the gain K found in part (b), determine the resulting phase and gain margins of the systems.

(d) Design the compensation network $G_c(s)$ that will achieve the stability requirements.

7.2 It is desired to design the positioning system of the tracking radar illustrated in Figure P7.2a which does not have a radome (sheltering structure for a radar antenna) to prevent wind from causing torque disturbances. It has been shown in Reference 1 that winds having velocities of 50 knots can cause wind torque disturbances of approximately 4000 ft lb on a parabolic reflector of 16 ft in diameter. Wind and wind gust occurrence and its resultant wind torque disturbances on the positioning control system is a stochastic problem. However, for this problem, we will design for an average wind where this radar will be operating. It will be assumed that the average wind can be represented as a step of 10 units. The specifications for the design of this positioning loop require that the steady-state error of $E(s)$ due to a step input of 10 units at $U_T(s)$ should be 1.25. Stability requirements are that the compensated system shall have a minimum phase margin of 50° and a minimum gain margin of 20 dB. In addition, to minimize the effects of tracking jitter caused by noise in the tracking loop, we wish to limit the maximum gain crossover frequency to 1.5 rad/sec.

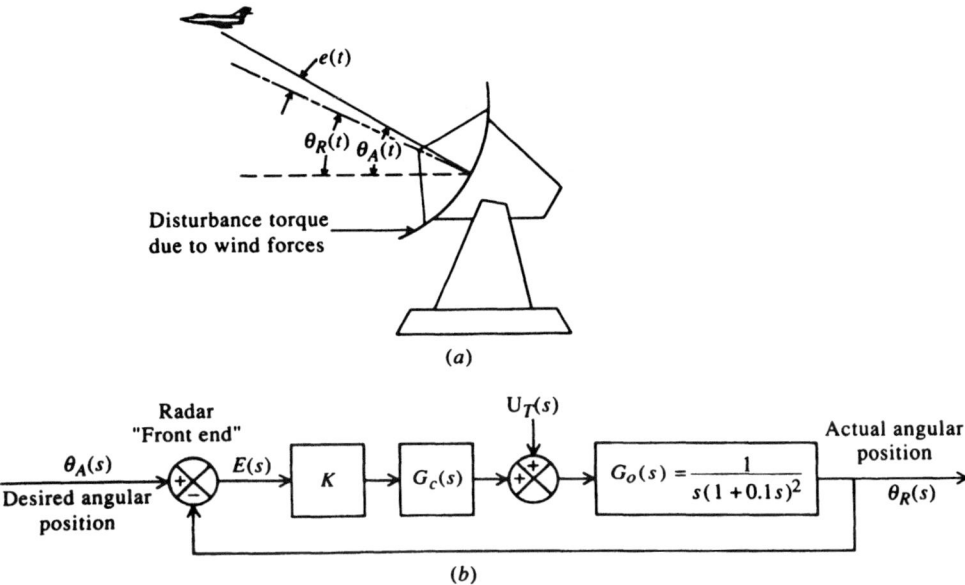

Figure P7.2 A tracking radar conceptually illustrating disturbance torques due to wind forces (a), and the equivalent block diagram of the tracking radar's positioning loop (b).

(a) With the compensation network $G_c(s)$ in Figure P7.2b, set equal to 1, find the gain K needed to achieve the steady-state error requirement of 1.25 due to the torque step disturbances at $U_T(s)$ of 10 units.

(b) With the value of K found in part (a), draw the Bode diagram of this system and find the resulting phase and gain margins.

(c) Design the compensation network $G_c(s)$ that will achieve the specified phase and gain margins.

(d) Determine the transient response of the uncompensated system of part (b) and the compensated system of part (c) to a unit step input.

7.3. Figure P7.3a shows a conceptual illustration of a robot performing a welding task in the manufacture of an automobile, which is a common application of robots. Suppose it is desired to design the angular positioning system for such a robot. Figure 7.3b illustrates the block diagram of the system. The specifications for the design of this system are that the positioning system have a damping ratio of 0.707, a position constant K_p of infinity, and a velocity constant K_v equal to 10.

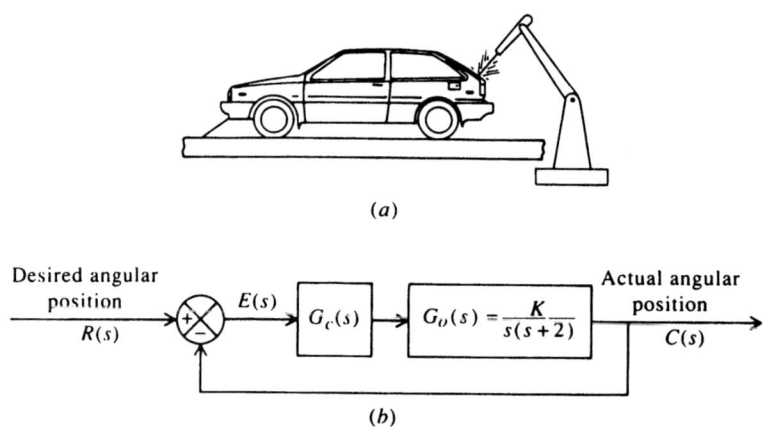

Figure P7.3 A conceptual illustration of a robot performing a welding task in the manufacture of an automobile (a), and an equivalent block diagram of the angular positioning system (b).

(a) With the compensation network $G_c(s)$ set equal to one, draw the root locus of the system.

(b) Find the value of gain K that will result in a damping ratio of 0.707, and indicate it on the root locus.

(c) Find the compensation $G_c(s)$ that will achieve a velocity constant of 10. Draw the resulting root locus.

(d) Find the transient response of the compensated system to a unit step input.

(e) How does the transient response of part (d) compare with the desired damping ratio of 0.707?

7.4. Repeat Problem 7.3 if the desired damping ratio is 0.5 instead of 0.707.

7.5. We wish to design an autopilot for controlling the roll of an airplane. In practice, the following three surfaces of an airplane are controlled in order to achieve the desired attitude of an aircraft: ailerons, elevators, and rudder, as illustrated in Figure P7.5a. The autopilot for controlling the roll of an aircraft can be controlled by adjusting the angle of the aileron's surface, which causes a torque to be developed because of the air pressure acting on it. This, in turn, causes the aircraft to roll. The autopilot acts to compare the actual roll angle

$\theta_{\text{actual}}(s)$ with the desired roll angle $\theta_{\text{reference}}(s)$. Any differences between these two quantities creates the actuating signal, which is amplified and drives the actuator until these two quantities are equal. Figure P7.5b is a simplified block diagram in which we assume that roll is independent of pitch and heading. (In practice, it is not.) Rate and attitude gyros are used to feed back rate and position information, respectively. The dynamics of the aircraft are approximated as

$$G(s) = \frac{1}{(s+4)^2}.$$

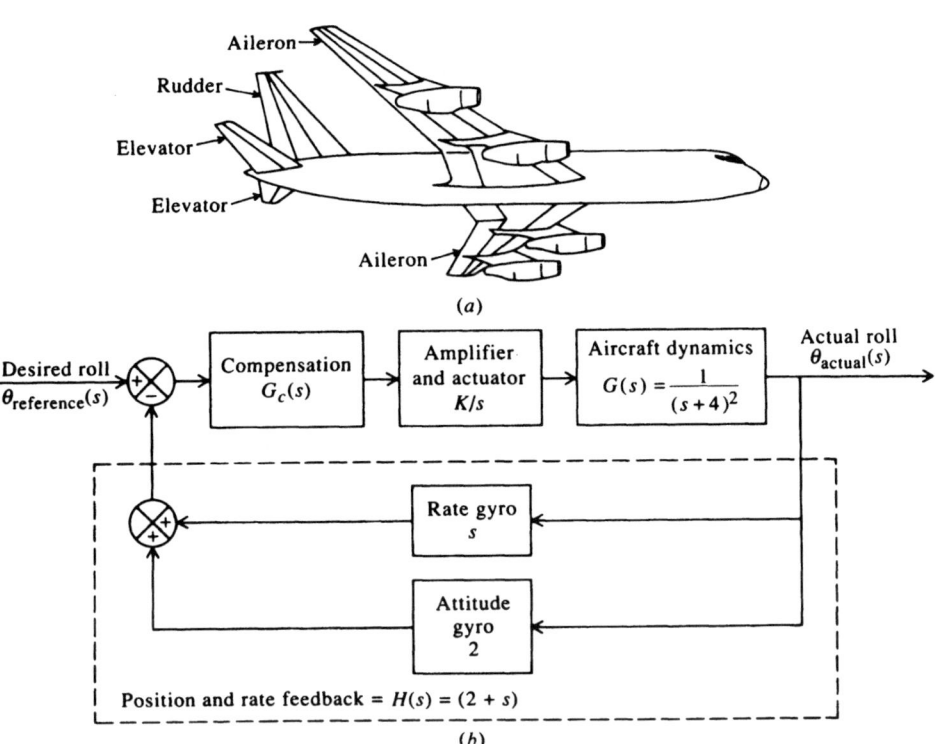

Figure P7.5 (a) Location of rudder, elevators, and ailerons on an airplane. (b) Simplified block diagram to control the roll of the airplane.

(a) Draw the root locus of this system assuming that the compensator's transfer function $G_c(s) = 1$.

(b) With $G_c(s) = 1$, determine the gain K from the root-locus diagram in order that the system has a damping ratio of 0.95. Assume that the transient behavior of the system is controlled by a pair of dominant complex-conjugate roots.

(c) It is desired that the system have a velocity constant of 40. Determine the increase in gain and the phase-lag network required to obtain a velocity constant of 40 and maintain a 0.95 damping ratio.

(d) Redraw the root locus for the compensated system. What is the new value of K for the compensated system? How does this affect the compensation $G_c(s)$ determined in part (c)?

7.6. We wish to design a combined compensator of a regulator that includes a controller and an estimator for a chemical process control system. Assume that the transfer function of the chemical process to be controlled is given by

$$\frac{C(s)}{U(s)} = G(s) = \frac{1}{(s+1)^2}.$$

Assume that the design specification of the controller is that it be critically damped with an $\omega_n = 2$ rad/sec, and that the estimator is also critically damped but with an $\omega_n = 10$ rad/sec.

(a) Find the controller's gain coefficients' matrix.

(b) Find the estimator's coefficients' vector.

(c) Design the compensator for the combined controller and estimator.

(d) Draw the root locus of the compensated system, and find the location of all the roots of the compensated system.

(e) Draw the Bode diagram for the compensated system,. Find its gain crossover frequency, phase margin, phase crossover frequency, and gain margin.

(f) How does the gain crossover frequency of part (e) compare with the controller's and estimator's closed-loop roots specified? Is this reasonable?

(g) Determine the transient response of the compensated system to a unit step input.

7.7. A student is experimenting with a simple sampled-data control system in the laboratory for the first time. The block diagram of the system he is experimenting with is shown in Figure P7.7. To determine the effect of the amplifier gain K on system stability, the student increases the gain K from zero to three. What kind of response to a unit step input will the student find for the following values of gain K:

(a) $K = 0.5$;

(b) $K = 1$;

(c) $K = 2$;

(d) $K = 3$.

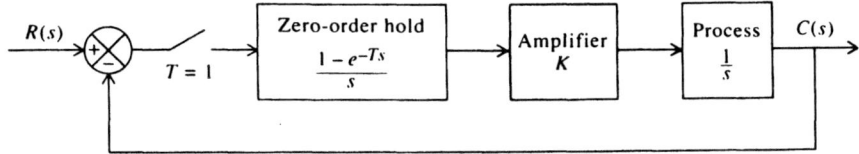

Figure P7.7

7.8. The design of the positioning system of a tracking radar is a very interesting control-system problem. Figure P7.8a illustrates a tracking radar which has wind torque disturbances acting on it, and the equivalent block diagram for one axis of the tracking radar's positioning loop is illustrated in Figure P7.8b. The block diagram illustrated represents a two-degrees-of-freedom robust control system which has the dual capability of minimizing the effects of the wind torque disturbance $D(s)$ while also being robust (insensitive) to variations in the gain K. The transfer function for the forward-loop transfer function for the tracking radar $G_0(s)$ is given by

$$G_0(s) = \frac{(1+0.5s)}{s(1+5s)(1+0.05s)}.$$

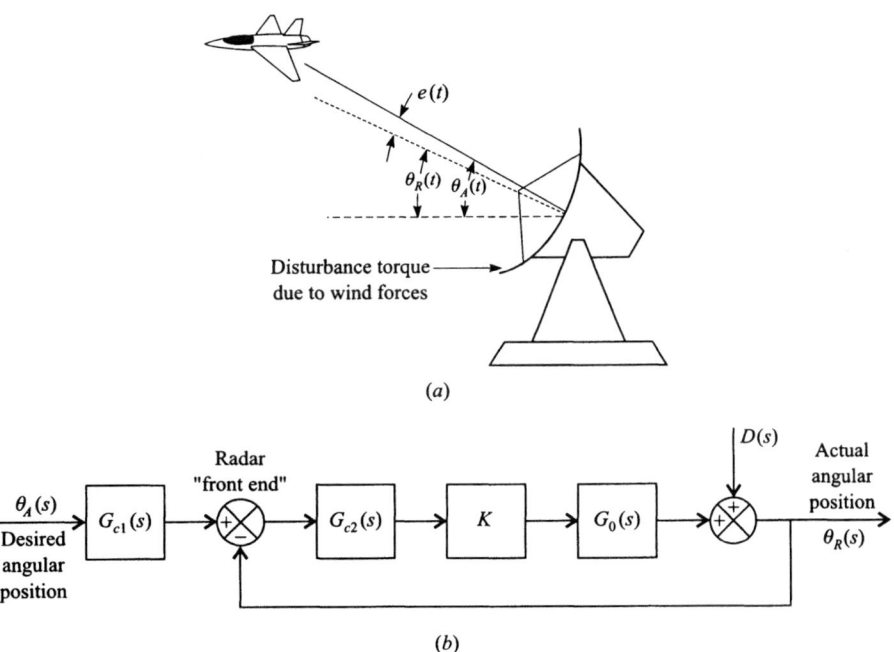

Figure P7.8 A tracking radar conceptually illustrating disturbance torques due to wind forces (a) and the equivalent block diagram of the tracking radar's positioning loop (b).

(a) Determine the transient response of this control system to a unit step input at $R(s)$ assuming that $G_{c1}(s) = G_{c2}(s) = 1$ and K has a nominal value of 130, but can also be as low as 20 or as high as 200.

(b) Draw the root locus for the conditions in part (a). Determine the location of the closed roots for $K = 20$, 130, and 200, and determine the damping ratio for the dominant complex-conjugate roots for these three cases.

(c) Design $G_{c2}(s)$ so that it contains the zeros which cancel the complex-conjugate roots for the case of nominal gain of 130.

(d) Draw the root locus for the conditions of part (c). Determine the location of the closed-loop roots for $K = 20$, 130, and 200, and determine the damping ratio for the dominant complex-conjugate roots for these three cases.

(e) Design the series controller $G_{c1}(s)$ and plot the resulting transient response of this control system containing $G_{c1}(s)$ and $G_{c2}(s)$ which was designed in part (c).

(f) How do the transient responses of part (e) compare with the transient responses of part (a)? Discuss the robustness of your resulting design to variations in the gain K.

REFERENCES

1. S. M. Shinners, *Techniques of System Engineering*, Chapter 8. McGraw-Hill, New York, 1967.
2. S.M. Shinners, *A Guide to Systems Engineering and Management*. Lexington Books, Lexington, MA, 1976.
3. P. M. Lowitt and S. M. Shinners, "Integrated optimal synthesis for a tracking radar." In *Proceedings of the 7th National Military Electronics Convention*, Washington, DC, 74–78 September, 1963.
4. P. M. Lowitt and S. M. Shinners, "'Type-N integral space tracking configurations." *IEEE Trans. Mil. Electron.* 416–424 (1965).
5. C. L. Phillips and H. T. Nagle, Jr., *Digital Control System Analysis and Design*. Prentice-Hall, Englewood Cliffs, NJ, 1984.
6. R. Y. Chiang and M. G. Safonou, "Modern CACSD Using the Robust Control Toolbox." *Proc. Conf. on Aerospace and Computational Control*, Oxnard, CA, August 28–30, 1989.
7. P. Dorato, *Robust Control*. IEEE Press, New York, 1987.
8. P. Dorato, "Case Studies in Robust Control Design." *IEEE Proc. Decision and Control Conf.*, December, 1990.
9. B. C. Kuo, *Automatic Control Systems*. Prentice Hall, Englewood Cliffs, NJ, 1995.
10. P. A. Ioannou and J. Sun, *Robust Adaptive Control*. Prentice Hall Professional Technical Reference, Des Moines, IA, 1995.

APPENDIX A

TUTORIAL FOR THE EFFECTIVE USE OF MATLAB

A.1. INTRODUCTION

This appendix presents the information necessary for the effective use of MATLAB which is used throughout this book. An important feature of this book is that most of the solutions shown in this book were generated using the commercially available software package called MATLAB which is available from the MathWorks, Inc. The *Advanced Modern Control System Theory and Design (AMCSTD) Toolbox*, which complements this book, and the M-files that were used to develop these solutions can be retrieved free from The Mathworks, Inc. anonymous FTP server at ftp://ftp.mathworks.com/pub/books/advshinners. These M-files are the *AMCSTD Toolbox* and were used to develop the graphical figures and problem solutions in this book. In this manner, the user of this book has available a toolbox of features/utilities created to enhance The Mathworks' Control System Toolbox, and the computer-generated solutions to the problems in this book. The *Modern Control System Theory and Design Toolbox* runs equally well with *The Student Edition of MATLAB* (authored by The Mathworks, Inc., and published by Prentice-Hall, Englewood Cliffs, NJ), as well as with or without the professional versions of MATLAB (with or without The MathWorks' Control System, Simulink, Nonlinear, and the Symbolic Toolboxes).

- MATLAB, an abbreviation for *MATrix LABoratory*, is a matrix-based system used for engineering calculations which has progressed to become the language used by most control-system engineers worldwide. This section focuses attention on the required background needed for designing control systems with MATLAB. MATLAB is a very useful language which operates interactively with the user, and will respond to all of the user's attempts at conversation.

It offers a variety of graphic output displays useful to control-system engineers such as system block diagram development (Simulink required), linear, log, semilog, polar, and contour plots. It is envisioned that this section will serve as a supplement to Reference 6, and enable students, professors, and practicing engineers to quickly learn the basics needed to use MATLAB with its associated toolboxes and the *Advanced Modern Control System Theory and Design Toolbox*.

This integrated textbook and software learning package is self-contained, and is designed for undergraduate courses on control systems and for the practicing engineer. This toolbox (software) contains extensive new features/utilities created to enhance MATLAB and several of The MathWorks' toolboxes. *Since this integrated learning package and the MATLAB software is self-contained, it is not necessary to purchase additional books/material to learn how to use MATLAB with this textbook.*

The software contained in the *Advanced Modern Control System Theory and Design Toolbox* contains the following features that makes this book self-contained for use with MATLAB:

- A *Tutorial File* has been created that contains the essentials necessary to understand and effectively utilize the MATLAB interface. This tutorial file aids the user in understanding the MATLAB interface where most other books require additional books for full comprehension. Features of this file are as follows:
  - MATLAB installation assistance
  - MATLAB performance improvement suggestions
  - MATLAB fundamentals
  - Understanding notations used by MATLAB
  - Control analysis using MATLAB
  - *Advanced Modern Control System Theory and Design Toolbox* use with MATLAB

- A *Demonstration m-file* gives the users a feel for the various utilities included in the *Advanced Modern Control System Theory and Design Toolbox*. Included are the following:
  - General purpose utilities
  - Linear, frequency domain, Bode diagrams (used starting in Chapter 1)
  - Linear, frequency domain, Nichols charts (used starting in Chapter 1)
  - Linear, frequency domain, Root locus (used starting in Chapter 1)
  - Nonlinear, frequency domain, Describing function (used in Chapter 5)
  - Linear, time domain
  - Conversions between discrete time domain and continuous time systems (used in Chapter 4)

  This demonstration helps the user learn how to use the MATLAB package easily with the *Advanced Modern Control System Theory and Design Toolbox* and, with the tutorial, makes this integrated package self-contained.

- Online HELP is available for all *Advanced Modern Control System Theory and Design Toolbox* utilities. Additionally, the "lookfor" command is fully supported in the online help for each *Advanced Modern Control System Theory and Design Toolbox* utility.

- A *Synopsis File* reviews and highlights the features of each chapter in *Advanced Modern Control System Theory and Design* in a concise manner which is helpful in guiding the professor on subjects to emphasize, and to the student and practicing engineer for reviewing and remembering important aspects of the coverage.

- The software is compatible with all editions the *The Student Edition of MATLAB* [31] and the Professional Versions of MATLAB [6], and it is compatible with The MathWorks' following software packages:
  - Control System Toolbox
  - Simulink Toolbox
  - Nonlinear Toolbox
  - Symbolic Toolbox

  I have attempted to make the presentation of MATLAB easy reading for the reader about to learn and use MATLAB. Most of this material can be read with little or no hands-on practice while getting a full grasp of most of the MATLAB capabilities.

A.2. First Time Usage—Software For Engineering

This section deals with the software. It does assume that you already have an understanding of the theoretical principals of the topics about to be presented. Any software that you may use now, or any time in the future, is only as good as your comprehension of the topics, and your ability to interpret the results presented to you by the computer. This cannot be stressed enough, because misinterpreted results due to whatever cause (incorrect inputs, computer round-offs, misinterpreted printouts, faulty programming code, etc.) can lead to any number of problems, depending on what you are doing. Computer literate people are all too aware of this, and have even given this a name: GIGO (garbage-in/garbage out).

With the advent of modern computers, engineers (as well as many other technical professionals) have had available a tool that can help them do their jobs quicker, more accurately, presentably, etc. The main question is what software are you going to use? Programming it all yourself is cumbersome, time-consuming, must be customized for each application, requires extensive understanding of all the topics encompassed, and on top of it all, is possibly ridden with errors. However, existing software packages have most of the features you require at a simple-to-use level ("user-friendly").

The major problem that most people have is which software package to buy! The correct answer has to depend on your current needs, and your expected near-term future needs. Far future needs probably should not be addressed due to the continually evolving computer software and hardware world, and should be re-evaluated when you get there. Some topics to consider are:

1. Ease of learning/use (user-friendly)
2. Analytic capability
3. Graphic capability
4. Expandability/modularity
5. Technical support/support groups
6. Hardware requirements
7. Price

Considering that this section is primarily prepared for a course in which it is beneficial for the student to own their own software (maximizing-capability/minimizing-costs), versus having a classroom licence (which usually ends up being a one-time experience), I suggest using "*The Student Edition of MATLAB*" by the MathWorks, Inc. It includes Software, Manual, User Groups, etc., and has most of the essentials that assist in learning a wide variety of topics at a very modest price.*

The Student Edition of MATLAB package gives you extensive mathematical/analytical/graphical capabilities, as well as a "taste" of two of the different professional-level toolboxes (Signal Toolbox and Control Toolbox) that they combine to call the "Signals and Systems Toolbox" (quite adequate for classroom training purposes of both topics). This is adequate for you to realize many control-theory applications, the programming potential of MATLAB, and the ease of dealing with toolbox add-ons (conceptually and mathematically). Another item of interest to all, is that there are available "FREE" toolboxes to all who desire. In their newsletters (that they willingly send to registered users) they list textbooks that use MATLAB and/or have toolboxes associated with the books. To top it off, they maintain a MATLAB User Group software library (nicknamed MUG) that contains all past newsletters, and FREE User software, as well as examples, and other useful information, available at no charge to you.

A.3. First Time Usage—MATLAB Installation

Installation is quite straightforward, just a little time-consuming. This step-by-step installation procedure focuses on the IBM (or compatible) PC. However, there are additional instructions in the AMCSTD Toolbox which apply equally as well to Macintosh and UNIX users. The user will need some basic knowledge of DOS and the system that you are working on. Certain typos in the first printing of *The Student Edition of MATLAB* for the IBM PC have been identified (subsequently corrected in later printings), and it is suggested that you correct them prior to reading and installing MATLAB:

1. p. 16: "three or more 360K" should read "two 360K"
2. p. 26: "\MATLABGEO" should read "\MATLAB\GEO"

---

*All references to pages or manuals herein are for *The Student Edition*, unless otherwise annotated.

Now simply follow the directions in Chapter 4. Once completed, you are ready to start using the student version of MATLAB. You may want to read the section on MATLAB fundamentals before trying to use MATLAB, but definitely read the following first.

If you have followed the installation section of MATLAB, all you have to do to start using MATLAB, is to type "MATLAB" and hit enter at the DOS prompt. After a moment to load, you should see the MATLAB copyright message on the top of the screen, followed by the line "HELP, DEMO, and INFO are available" (if proper installation has occurred). If this much does not come up whenever you activate MATLAB, something has gone wrong.

If you encounter any difficulties with the package:

1. Re-read the appropriate chapters in the textbook.
2. Check for obvious typographical errors.
3. For the *Student Edition*, consult the instructor (who has the available technical support). For the professional version, consult The MathWorks.

The first time (and only the first time) that you activate MATLAB, it will prompt you for registration information (serial number and your name) following its normal messages. The MATLAB book does not tell you anything about this step because it is a leftover feature from the professional version not described in any (student version or professional version) of their manuals (it is explained with a separate letter attached to the professional version). It is a formality, so that you can identify your copy of MATLAB without having to pull out your original diskettes when calling for technical support. Fill it in by answering the prompts that come on the screen. I do suggest entering your name, if nothing else but to personalize this copy as yours.

To add additional toolboxes to MATLAB, merely follow the directions on the bottom of page 26 (as if you are adding a library of your own). For installing the FREE *Advanced Modern Control System Theory and Design Toolbox* (henceforth called the AMCSTD toolbox), proceed as follows:

With the MCSTD Toolbox (referred to in the Second Edition of *Modern Control System Theory and Design*), there exists an install document file that will guide you (with a suggested approach) to integrating this Advanced Modern Control System Theory and Design Toolbox into your MATLAB installation. I have included all advanced functionality identified in the *Advanced Modern Control System Theory and Design* book into the MCSTD Toolbox (referred to in the Second Edition of *Modern Control System Theory and Design*). Both toolboxes are identical, however, the figures and problem solutions are different and appropriate for each book.

The problems, figures, Fortran and demo directories that come with that toolbox are non-essentials (but give a nice demo of the MCSTD features and a variety of worked examples from this book). They can be installed in a similar manner to the above. The readme.m, contents.m, tutorial, and synopsis files should be printed as reference material.

A.4. First Time Usage—Performance Tuning MATLAB

I wish to provide to the reader some of the things that I have learned in the proper use of MATLAB. For those of you who will spend hours at a time using MATLAB (student or professional version), this material will greatly benefit you. These are the suggestions that I make to my fellow colleagues using MATLAB, for significantly improving performance at an insignificant cost!

The first thing you notice when starting MATLAB, is how long it takes to start before you can type. Those of you with sharp eyes and/or minds might even notice a lot of hard disk activity occurring during this time. This is because MATLAB is an interpreter. That means for every command that you give, it must read the file with that command in it (and similarly the same for every command called inside the original command). With that in mind (and the fact that you can't change this), the faster the drive that has all these commands on them, the faster your performance.

The limiting factors for your choice, as you will see, is available computer memory and/or your desire to try something different. Buying a new and better drive is overkill and may not give you peak performance. Software that improves your current drive's speed is a step in the right direction; however, creating a drive out of computer memory (which is the fastest possible drive for your machine) is the ideal choice. The super-intelligent among you will even use a combination of these techniques (as I am about to explain). I have seen speed factor-improvements of 3 to 15 times using these techniques.

With the advantages come the disadvantages (and I feel I should point them out before going any farther). When creating memory drives, as the computer is turned off, all information on that drive is lost. My solution is to copy the standard toolboxes and software to that drive when I am going to use it (requires a short additional length of time), and keep my personal routines and data on my hard disk (the best of both worlds). Also, most PCs have limited memory which you use sparingly. My solution is: if you don't mind the speed, stick with what you have. Otherwise, merely buy and install more memory (the appropriate type: expanded or extended) for your computer.

The amount of memory that you desire for your memory drive (henceforth called ramdrive) depends on how much software you want to put there, the more you put the faster you go. Demo routines with their data do not have to be kept on the ramdrive after you have become familiar with them. But for now, you can add up the sizes of the files you want to move (add 10 percent for blank space) to see how big this ramdrive should be.

The commands that add this ramdrive to your system (provided that the memory is already there) are either vdisk, ramdrive (in DOS), or whatever appropriate software came with the memory that you purchased. An example of my computer setup (using DOS version 5.00 with high memory; your machine may different) for *The Student Edition of MATLAB* is:

1. Adding "DEVICE=C:\DOS\RAMDRIVE.SYS 2000/E" to the file "C:\CONFIG.SYS" makes a 2000KB (2MB) ramdrive. Most ramdrives make you add one or more lines to the file "C:\CONFIG.SYS." When your computer is rebooted, the ramdrive should be available for use.

2. Adding "DEVICE = C:\DOS\SMARTDRV.SYS" to the file "C:\CONFIG.SYS," adds disk caching upon reboot.
3. Adding "FASTOPEN C:" to the file "C:\AUTOEXEC.BAT," improves file access performance for that drive.

In step one, you may choose an appropriate size for your specific needs. I chose 2MB because it is adequate to hold all of *Student Edition of MATLAB* and the complete MCSTD *Toolbox*, serving my purpose quite well. You may choose not to put all that on your ramdrive, but only the parts that you are going to use (such as no demo files, compiler linking files, example files, etc.). On the other hand, you may have other toolboxes that you want to put on this ramdrive. This is a choice that you must make.

Now that you have your ramdrive set up (all changes have been made and you have rebooted), you want to take advantage of these changes. I will use "F:" as the letter of the ramdrive created—change according to your individual system. Your file that starts MATLAB up ("C:\MATLAB\BIN\MATLAB.BAT") has to reflect these changes. This is how I suggest altering that file:

1. Add to the top of the file a line for each directory that you want to copy which copies that directory from your hard drive to your ramdrive. One example is "XCOPY C:\MATLAB\MATLAB F:\MATLAB\MATLAB\". This would copy MATLAB from your hard disk C: to your ramdrive F:.
2. Alter the "SET MATLABPATH =" line by including the drive letter where the directory exists for each directory listed (you may mix between hard disk and ramdrive as you choose). In MATLAB version 4.0, the MATLABPATH is set in the file "MATLABRC.M".

The next performance suggestion concerns improving the speed of shelling to DOS with the MATLAB "!" command. If the computer uses the file "COMMAND.COM" from a ramdrive, it can load and execute the "!" command faster. Not only MATLAB will benefit from this tip, but also any program that shells to DOS. Further DOS performance tips I leave for the experts to suggest (but I do like this one in particular). This is my suggestion for altering the file "C:\AUTOEXEC.BAT":

1. Add "COPY C:\COMMAND.COM F:\" to the bottom of this file.
2. Add "SET COMSPEC = F:\COMMAND.COM" following the copy line. This tells DOS to use that file when shelling to DOS.

The last performance suggestion, as mentioned earlier, concerns demo files which take room but also make the computer search for each MATLAB command a little wider (since it checks all file names in the MATLABPATH until the file is found). I do not like losing anything, so what I do is move the demos/examples of each toolbox to separate directories (as I have already done in the AMCSTD toolbox). This way I can always add the demos/examples to the MATLABPATH in the MATLAB.BAT file. Personally, I like to do as little as possible, and just create a copy of the MATLAB.BAT file (I call

MATLABD.BAT) with those changes made to it. Also, I do not copy the demos/examples to the ramdrive (which would reduce the room on the ramdrive as well) unless I want them (as in MATLABD.BAT).

A.5. First Time Usage—MATLAB Fundamental Concepts

MATLAB is a very simple, but efficient, interpreter (versus compiler). This allows you to type in a line and have it execute immediately. The tremendous advantage is that you do not have to re-compile your program (as you would in Fortran, Pascal, or C) for every change you make before re-executing the code, as well as the ability to view the values of the variables without adding debugging write statements to your program. MATLAB does include an interface so that you can execute your Fortran or C code from inside MATLAB (see MATLAB manual section on MEX files). Like any other language, it has a small subset of commands (statements), from which other more sophisticated commands are developed by the users. There are four fundamental items in MATLAB: variables, functions, programming code and algebraic operators.

Variables, in MATLAB, are all treated as matrices. Text strings (which are stored as matrices), however, are automatically given a special attribute, so that the ASCII text value is printed (not the numeric value).

Functions in MATLAB operate slightly differently than most programming languages (with the possible exception of C++). Most MATLAB functions take none, one or more variables as parameters (within the parenthesis), and return none, one, or more variables as a result. An unusual aspect of this is that a single function can be used many different ways depending on how many parameters are being passed in, the type of data being passed in, and how many parameters are being retrieved.

Programming code, in MATLAB, is mainly for control flow, directory manipulation, data storage, and debugging purposes. Like most languages, this includes: if, else, end, for, while, ... etc. These will be discussed in more detail later in this text.

Algebraic operators, in MATLAB, are very similar to the math operators that you are used to, with only a few minor exceptions. The main consideration to be taken when using them is that they are (almost all) matrix math operators. If you want to use scalar mathematics, you must take care to use the scalar versions of the algebraic operators.

Another topic that should be mentioned in this section, is the concept of toolboxes. A toolbox, in MATLAB, is a collection of functions/utilities that work together on a specific topic (i.e., signal processing, control processing, system identification toolbox, ...) so that it (hopefully) meets all your needs on this topic, and you do not need to program any functions for your own needs. As your skills expand into other areas of expertise, there are other toolboxes (the MATLAB toolbox list is always growing) available for (in the professional world) a reasonable cost (versus you programming–debugging–testing your own code for each specific need). The MathWorks' desire to express this highly important fact has lead them to include the "Signal and Systems" toolbox with their student version. This toolbox gives the user a taste of the professional versions of the Signal and Systems Toolbox and the Control-System toolbox. A more detailed discussion on these and other toolboxes will follow later.

A.6. First Time Usage—Matrix Representations

Every variable of any sort in MATLAB is a matrix. The approach is to represent a wide variety of items as matrices. Once certain conventions (of which MATLAB has many predefined) are agreed upon, all that remains is to learn how to use them. All the types about to be described here have values which are either real or complex.

Scalar values start very simply, represented as a 1×1 matrix. They can take on any double precision value that the computer can represent, including some unusual ones: Infinity (Inf) and Not-A-Number (NaN).

Inf represents a number the representation of which is beyond the computer's ability. Almost any mathematical operation dealing with such a number yields the value Inf, as should be expected.

The use of NaN is difficult to illustrate; consider a set of data with one value in it that is totally absurd. If you use that value in your data set, your analysis results will be totally corrupted. Instead of removing that data from your set (which sometimes cannot be done) it would be nice to say that value is *N*ot really *a* valid *N*umber, hence NaN. Now we can agree to check for this value in our analysis routines, which is already done in many MATLAB routines.

Vector values become a little more sophisticated, represented as a $1 \times N$ matrix or a $N \times 1$ matrix (N being the length of the vector). The distinction between the two representations is used by MATLAB, and will be examined later. The typical uses of a vector are as a set of data, polynomial representations, etc. These uses will become more apparent to you as you practice with the topics presented later.

Matrix values are straightforward, as they represent a matrix. In MATLAB version 4.0, matrices have an added feature of being "sparse." Sparse matrix technology is a method for storing a large matrix filled with mostly zeros, in a much smaller storage space. This can aid in speed when dealing with these matrices. This feature does not change the features of a matrix, just the physical storage and speed of execution when used. Some typical matrix representation examples are sets of data samples, transition matrices, covariance matrices, etc. Once again, these uses will become more apparent to you as you practice with the topics presented later.

Strings are a little more confusing to understand. They are stored, like vectors, as a $1 \times N$ matrix. The difference between strings and vectors is that strings are given a special attribute (automatically set by MATLAB) so that they will be displayed properly on the screen. A non-vital detail (trivial and for reference purposes only) is that each letter in a string is stored as the numeric value of its ASCII representation.

A.7. First Time Usage—MATLAB Fundamentals

The MATLAB prompt that is listed/described in the manual is the "≫". At this prompt you will begin to learn how to use MATLAB. This practice session is created to give you the idea of how exit, find help, and inquire about variables in MATLAB.

MATLAB, like most modern software packages, tries to make the commands it uses very simple and logical. Most commands are like their English counterparts, making (hopefully) the commands simple for everyone to understand. One important fact that should be stated now is that MATLAB is case sensitive (this means that entering things in upper case is different than entering things in lower case). All commands that are to be entered in the following practice session should be in lower case unless specifically stated. After each command has been entered, to start the execution of the command press Enter or Return (this shall be assumed with each of the following commands, unless otherwise stated).

The first thing that any software should teach is how to exit the software and get back to your operating system (DOS). There are two wasy to do this: "exit" or "quit". Both are good examples of English-like commands. A good idea before exiting any software program is to save your work, but for now we shall not do this (especially since we have done no work, and you have not been shown how to save as of yet). After executing either of these commands, if you choose to continue using MATLAB, you must start MATLAB up because you have been returned to DOS.

The next thing that we are going to do is to "demo" some of the MATLAB features. Hopefully, you will have guessed that the command to do this is "demo", and you would be right. An extensive demo has been prepared by MATLAB, showing you a wide variety of its capabilities. You should take the time to try a few (or all) of these demos at your leisure.

The "help" feature of MATLAB is very useful to the user. This was designed so that you do not have to pull out the MATLAB manual (or toolbox manual(s)) in order to be able to understand the various commands currently available to you. There are several different ways to use the help command, depending on the version of MATLAB that you possess.

By typing "help" by itself, you will get either a list of all installed toolboxes (version 4.0 or later) or a list of all the currently available commands organized on different screen according to the toolboxes that you have installed (prior to version 4.0). As you add additional toolboxes to MATLAB, each toolbox will have its own screen listing the commands available in that toolbox. This is a nice feature which helps the individual think in a modular fashion when going to the professional world.

In version 4.0 or later, typing "help" followed by a space and the installed toolbox directory name, will list the commands in the toolbox with a oneline explanation of their functions (e.g., "help matlab").

Additionally, by typing "help" followed by a space and any listed command (in the manual or on the help screens), you will get an explanation of the command and how to use it (e.g., "help quit").

It is highly recommended that you review *The Student Edition of MATLAB*. I suggest getting some hands on practice while reading it. Functions worth reading about (or at least glancing at the help screens) are:

1. "!" allows you to execute DOS commands (from inside the MATLAB), without having to quit your MATLAB session (called shelling to DOS). When the DOS command that issues following the "!" has completed, you are returned to MATLAB without losing any of your MATLAB session.

   TSR warning: Do not use the "!" to run any TSR program. Start all TSR programs prior to entering MATLAB. TSRs take memory to add themselves to your environment (DOS). MATLAB is not aware of this memory usage, and will consequently use the memory, at some point in time, clobbering software and probably corrupting MATLAB and/or DOS (in memory). Do not be fooled if you do not see this happen, you may have been lucky in one of several ways this time. This time MATLAB may not have used the memory. This time what it clobbered may not have had any effect. Or you simply did not realize that the corrupted system is now returning corrupted results (MATLAB and/or DOS)!

2. "load" allows you to get data that has been stored. Two forms of data currently can be loaded: their own special form (called MAT files) and ASCII files that contain one numeric matrix (any size allowed) in it. This is useful to get data into MATLAB from other software packages.

3. "save" is the counterpart to load. This allows you to save data for future use. Like load, it saves as a MAT file or as an ASCII file (upon request). This is useful if you intend to export data to other programs after processing them in MATLAB.

4. "plot" is your first step into MATLAB's graphic world. There are too many different graphic commands to list, each with accompanying support functions. But as you may have guessed, "plot" plots data on the screen.

Getting printed copies of the graphs is done in several different ways (see Apendix C in *The Student Edition* manual): using a graphic screen dump to the printer utility (printed quality is only as good as the screen resolution), using the "meta" command with their "GPP" software (professional MATLAB prior to version 4.0 or later), and using the print command or print menu (professional MATLAB version 4.0 or later). Since this is for the student version, I shall discuss the screen dump utility.

DOS and MATLAB both come with screen dump TSRs (see prior TSR warning). If you intend to print any plots, you should load either one of these utilities prior to starting up MATLAB! The DOS utility is called "GRAPHICS." The MATLAB utility is "EGAEPSON" for Epson compatible printers, or "EGALASER" for laserjet compatible printers. If you have problems printing and have a VGA monitor, a suggestion is to restart MATLAB in the EGA mode.

Most functions are recognizable by their name and one-line descriptions. In *The Student Edition* manual (the only book that I reference by pages), you will find this section on pp. 185–194.

A.8. Control Systems—Data Representation

I will now show how to represent real systems as mathematical models. Any type of control system is going to be represented as matrices in MATLAB. The trick is to become fluent with their usage (which only really comes with practice). Each of the following subsections will make sense as you have been properly introduced to their topics by your instructor.

First and foremost, in most control-theory classes, is the *transfer function* representation of a system. This is where a system is represented by a series of polynomials, usually organized (as in most control-theory textbooks) as:

$$G(s) = \frac{C(s)}{R(s)} = \frac{A_m s^m + \ldots + A_1 s^1 + A_0 s^0}{B_n s^n + \ldots + B_1 s^1 + B_0 s^0}.$$

Here, a single-input/single-output (SISO) transfer function can be represented as two polynomials of decreasing powers (numerator polynomial divided by a demoninator polynomial). In MATLAB, a polynomial (numerator, denominator, or other) is represented as a row vector ($1 \times N$ matrix) containing the scale factors for each of the sequentially decreasing powers of s. This is a very important point to comprehend before reading the following examples:

1. "$17s^5 + 23s^3 + 8s^2 + 6$" converts to "[17 0 23 8 0 6]". The highest power of s is five, so there will be six numbers in this vector (s-powers 5 to 0). Powers of s that are not there have scale factors of zero (they hold the s-power sequence and are NOT to be forgotten).
2. "$5 + 12s + 4s^3 + 3$" converts to "[4 0 12 8]". If you cannot visually see this, then the best thing you can do for yourself is to rewrite the polynomial in "decreasing powers" of s as: "$4s^3 + 0s^2 + 12s^1 + (5+3)s^0$" or more simply "$4s^3 + 12s + 8$." You should now be able to see how to do this.

Next is the *zero-pole-gain method* that is similar to the transfer function in that the two methods represent the same equations, the equations have merely been manipulated into a different form. The difference is though that the polynomials have been factored into their roots, and "k" is the gain that makes the equations equal (namely the highest s-power numerator scale-factor divided by the highest s-power denominator scale-factor). This representation of the system is:

$$G(s) = \frac{Z(s)}{P(s)} = k\frac{(s - Z_1)(s - Z_2) \ldots (s - Z_N)}{(s - P_1)(s - P_2) \ldots (s - P_M)}$$

Here, a single-input/single-output (SISO) system (called the zero-pole-gain function) can be represented as two sets of roots (numerator and denominator) and a gain factor. When any of the numerator roots values goes to zero, the overall function value goes to zero; hence the numerator roots are called "Zeros." When any of the denominator root values goes to zero, the overall function value goes to infinity; hence, the donominator roots are called "Poles."

In MATLAB, a set of polynomial roots (numerator, denominator or other) is represented as a column vector ($N \times 1$ matrix) with each element representing one of these roots. *The Student Edition* attempts to show this "simple" transformation with one example followed by one example of the zero-pole-gain method. A few examples follow to challenge your understanding of "polynomial root" representation:

1. "$(s - 3)(s - 4.5)(s + 100)$" converts to "[3; 4.5; −100]". By now you should recognize the semicolon from MATLAB as meaning (among other things) a new line (new row). This makes the matrix listed above become a 3 rows by 1 column matrix (column matrix, three in length).
2. "$(s - 2)(s - 500)(s - 30)$" converts to "[500 ; 30 ; 2]". Your first guess would probably be different ([2 ; 500 ; 30] is also correct), because the order of the

elements in this column vector (roots) is irrelevant, since in multiplication the order in which you multiply is irrelevant (at least in scalar math). Therefore, any order, as long as all values are accounted for, is correct.

3. "$(s - 40)^3(s - 80)$" converts to "[40 ; 40 ; 40 ; 80]". Each root has to be accounted for. Multiple roots must be accounted for each time they are multiplied.
4. "$(s^3 - 11s^2 + 38s - 40)(s - 3)$" converts to "[2 ; 4 ; 5 ; 3]". You must always remember to factor to a single power of s (even if this makes complex values). $s^3 - 11s^2 + 38s - 40$ factors into $(s - 2)(s - 4)(s - 5)$.

Continuing, the easiest of the system representations to explain arises. State-variable has become popular in the professional industry, due largely to the ease of representing multi-input/multi-output systems. Several signals can be evaluated simultaneously in very sophisticated systems without having to trace through the transfer functions for each input to output. Several other subjects that may cross your path in your progression of control-theory knowledge (e.g. Kalman filter, sensitivity analysis), use this base structure to build upon. State-variable concepts are discussed starting in Section 1.7. For the present, let us consider the MATLAB representation of the following state and output equations, respectively:

$$\dot{x}(t) = Ax(t) + Bu(t)$$
$$\dot{y}(t) = Cx(t) + Du(t).$$

The "A" matrix represents the system dynamics, namely how the system is connected and where the integrators are located. "B" represents how inputs couple into the system. "C" represents how the outputs couple out of the system. "D" represents what portion of the inputs couples directly into the output. This representation may abstract the original design a little, but the advantages are quite significant. One such advantage is the improved precision of the representation, it suffers from fewer computer round-off-type problems.

Each of these three (transfer function, zero-pole-gain and state variable) representations have their advantages for visual inspection, but the conversion of them from one form to another can be cumbersome. MATLAB has realized this, and created several routines that allow you to convert from one representation to the next. A list of them, as well as their relations to each other, continuous and discrete versions, has been prepared and is stated under Model Conversions. A good exercise would be to pick a system, convert it to one of these representations, and then convert this representation to others with these routines. Practice in becoming familiar with these representations and converting between them is left to the readers' discretion.

Another aspect to consider is whether your system is a continous-time or discrete-time system. In reality, modern systems (digitally controlled) read the continous input (analog data) at discrete intervals making hybrid (sampled-data) systems. Breaking these systems apart, we can separately look at the continuous parts and discrete parts, or we can create "equivalent" models of these systems by using conversion routines. The concepts of hybrid systems are mind-boggling, and it is often preferable to convert their representations to one or the other (continuous or discrete) "equivalent" system for analysis.

A utility to convert from continuous to discrete, called "C2D" comes with the "Signal and Systems" toolbox. (The professional Control System Toolbox has many utilities that go back and forth with a variety of methods for converting.) The AMCSTD toolbox comes with the conceptual essentials (further elaborated in the demo m-file) to do almost all the conversions that the professional toolbox does. The AMCSTD toolbox even gives and uses examples of these conversions, since they are mostly polynomial substitutions (see polysbst in the AMCSTD toolbox). Once again, this is an excellent teaching tool for control-systems theory, while not giving you all of the professional Control System Toolbox features.

A visual implementation of the system modeling is also available. It utilizes the above data representations of the basic building blocks, and the capability to interconnect the blocks. The Simulink Toolbox, used for system simulations, allows building of the system models in the visual sense. The Simulink developed model can be "queried" for the states contained in the system model.

A.9. Summary of MATLAB and Advanced Modern Control System Theory and Design Toolbox Commands

An abbreviated list of the commonly predefined functions available from MATLAB, which are of prime interest to the control-system engineer, is listed in Table A.1. By entering these commands, MATLAB processes them immediately and displays the results determined.

A list of the additional commands available from the *Advanced Modern Control System Theory and Design Toolbox*, to supplement the MATLAB commands, which are also of prime interest to the control-system engineer, is given in Table A.2. By entering these commands, the data are processed immediately and the results are displayed. The control-system engineer will find these commands very useful in the analysis and design of control systems. For example, some practical control-system examples of the polynomial utilities are as follows (Bode plot and root locus are discussed starting in Chapter 1.):

- Polysbst   { go from s to $j\omega$ domain,
             { transform (scale/transport/rotate) axis in root locus plot
             { continuous to discrete transformations (substitutions)
- Polymag   { Bode plot: pick any gain frequency (0db = Phase margin (P.M.), 3db = Band Width (B.W.)
- Polyangl   { Bode plot: pick any phase frequency (-180 = Gain Margin (G.M.)
- Rootmag   { Root locus: pick a particular s/z magnitude (Unit Circle)
- Rootangl   { Root locus: pick a particular damping angle

Table A.1. Commands and Functions used by MATLAB [6]

Commands and Matrix Functions	Description
angle	Phase angle
ans	Answer when expression is not assigned
asin	Arcsine
atan	Arctangent
axis	Manual axis scaling on plots
bode	Plot Bode diagram (available only with Control System Toolbox, and the Student Edition of MATLAB)
clear	Remove items from memory and clear workspace
clg	Clear graph's screen
conj	Complex conjugate
conv	Multiplication; convolution
cos	Cosine
cosh	Hyperbolic cosine
cov	Covariance
deconv	Division; deconvolution
det	Determinant
disp	Display text or matrix
end	Terminates scope or "for," "while," and "if" statements
exit	Terminates program
exp	Exponential base e
eye	Identity matrix
figure	Opens new graphic window by creating figure object
grid	Draws grid lines for 2-D and 3-D plots
help	Online help for MATLAB functions and m-files
hold	Hold the current graph
home	Sends the cursor "home" to the upper left of the screen
imag	Imaginary part
inf	Infinity
inv	Matrix inverse
length	Length of vector
linspace	Generate linearly spaced vectors
log	Natural logarithm
loglog	Log–log scale plot
log10	Common logarithm (base 10)
logspace	Generate logarithmically spaced vectors
max	Maximum elements of a matrix
mean	Average or mean value of vectors and matrices
menu	Generate a menu of choices for user input
meshgrid	Generate X and Y arrays for 3-D plots
min	Minimum elements of a matrix
nyquist	Plot Nyquist frequency response (available only with the Control System Toolbox and, the Student Edition of MATLAB)
pi	Provide pi (π)
plot	Linear 2-D plot
plot3	Plots lines and point in 3-space
polar	Polar coordinate plot
poly	Characteristic polynomial

Table A.1. (*Continued*)

Commands and Matrix Functions	Description
polyfit	Polynomial curve fitting
prod	Product of the elements
quit	Terminate MATLAB
real	Real part
residue	Partial-fraction expansion
rlocus	Plot root locus (available only with the MCSTD Toolbox and with the Student Edition of MATLAB)
roots	Polynomial roots
save	Save workspace variables on disk
semilogx; semilogy	Semi-logarithmic 2-D plot (x-axis or y-axis logarithmic)
sin	Sine
sinh	Hyperbolic sine
size	Matrix dimensions
ss2tf	State-space to transfer function conversion
step	Plot unit step response (available only with the Control System Toolbox, and the Student Edition of MATLAB)
sum	Sum of the elements
tan	Tangent
tanh	Hyperbolic tangent
text	Add text to plot by creating text object
tf2ss	Transfer function to state-space conversion
title	Graph title
trace	Trace of a matrix
type	List file
what	Directory listing of m-files
who	List directory of variables in memory
xlabel	x-axis label
ylabel	y-axis label

Table A.2. Commands used by the Modern Control System Theory and Design Toolbox

Polynomial utilities	poly_add	— add two polynomials
	polysbst	— substitution of a polynomial variable with a polynomial
	polyder	— derivative of a polynomial (supplied with MATLAB)
	polyintg	— integral of a polynomial
	polymag	— locate roots of a polynomial that generate a given magnitude
	polyangl	— locate roots of a polynomial that generate a given angle
	rootmag	— locate roots of a polynomial that are at given magnitude
	rootangl	— locate roots of a polynomial that are at given angle
Interpolation utilities	crossing	— interpolates the index of specified values from a data set
	crosses	— interpolates the value of a data set at specified indices
	crosser	— iterates the solution of a function to a specified value
	margins	— analytic calculations of all the phase and gain margins
	rlpoba	— root locus point of break-away/break-in
Control utilities	rlaxis	— real axis portion of the root locus
	wpmp	— maximum feedback frequency location
	nichgrid	— Nichols grid at user requested inputs
Nonlinear Functions	back_lsh	— backlash response
(using the describing	dead_zn	— deadzone response
function implementation)	relays	— hysteresis response
Compatibility Functions	sbplot	— subplot replacement
	frz_axis	— axis replacement

APPENDIX B

CHARACTERISTIC RESPONSES OF SECOND-ORDER CONTROL SYSTEMS

The purpose of this appendix is to describe the transient response of a typical feedback control system which is referred to many times in this book. We consider a very common configuration in which a two-phase ac servomotor is enclosed by a simple unity feedback loop. Figure 4.1 illustrates the block diagram of this second-order system. For purposes of simplicity, the gain of the amplifier driving the motor is assumed to be unity.

The closed-loop transfer function of this system is given by

$$\frac{C(s)}{R(s)} = \frac{K_m/T_m}{s^2 + (1/T_m)s + K_m/T_m}. \tag{B.1}$$

By defining the undamped natural frequency ω_n and the dimensionless damping ratio ζ as

$$\omega_n^2 = \frac{K_m}{T_m} \text{ and } \zeta = \frac{1}{2\omega_n T_m}, \tag{B.2}$$

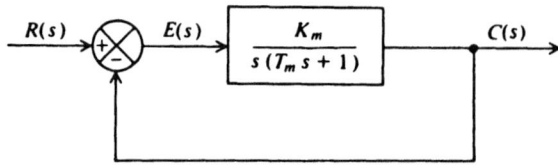

Figure B.1 Second-order feedback system containing a two-phase ac servomotor.

Eq. (B.1) can be rewritten as

$$\frac{C(s)}{R(s)} = \frac{\omega_n^2}{s^2 + 2\zeta\omega_n s + \omega_n^2}. \quad (B.3)$$

The parameters ω_n and ζ are very important for characterizing a system's response. Note from Eq. (B.3) that ω_n turns out to be the radian frequency of oscillation when $\zeta = 0$. As ζ increases from 0, the oscillation decays exponentially and becomes more damped. When $\zeta \geqslant 1$, an oscillation does not occur.

We assume that the initial conditions are zero and the input is a unit step. Therefore, $R(s) = 1/s$, and the Laplace transform of the output can be written as

$$C(s) = \frac{\omega_n^2}{s(s^2 + 2\zeta\omega_n s + \omega_n^2)}. \quad (B.4)$$

Factoring the denominator, we obtain

$$C(s) = \frac{\omega_n^2}{s(s + \zeta\omega_n - \omega_n\sqrt{\zeta^2 - 1})(s + \zeta\omega_n + \omega_n\sqrt{\zeta^2 - 1})}. \quad (B.5)$$

The exact solution for the output in the time domain is dependent on the value of ζ. When $\zeta \geqslant 1$, the second-order system has poles which lie along the negative real axis of the complex plane. When $\zeta < 1$, however, a pair of complex-conjugate poles result. We shall determine the output response to a step input for the three cases: where the damping ratio equals unity, is greater than unity, and is less than unity.

Case A. Damping Ratio Equals Unity

When $\zeta = 1$, Eq. (B.5) reduces to

$$C(s) = \frac{\omega_n^2}{s(s + \omega_n)^2}. \quad (B.6)$$

The time-domain response will next be obtained. The partial-fraction expansion of Eq. (B.6) is given by

$$C(s) = \frac{K_1}{(s + \omega_n)^2} + \frac{K_2}{s + \omega_n} + \frac{K_3}{s}. \quad (B.7)$$

We find that

$$K_1 = -\omega_n, \quad (B.8)$$
$$K_2 = -1, \quad (B,9)$$
$$K_3 = 1. \quad (B.10)$$

Substituting these constants into Eq. (B.7), we obtain

$$C(s) = \frac{-\omega_n}{(s + \omega_n)^2} - \frac{1}{s + \omega_n} + \frac{1}{s}. \quad (B.11)$$

The time-domain response of the output, $c(t)$, may be obtained by utilizing a table of Laplace transforms:

$$c(t) = -\omega_n t e^{-\omega_n t} - e^{-\omega_n t} + 1, \quad t \geq 0. \tag{B.12}$$

Figure B.2a illustrates the output response together with the unit step input. Notice that the output response exhibits no overshoots when $\zeta = 1$. The response is described as being critically damped.

Case B. Damping Ratio Greater than Unity

When $\zeta > 1$, the time-domain response can be obtained quite simply from its partial fraction expansion. This can be expressed as

$$C(s) = \frac{K_1}{s} + \frac{K_2}{s + \zeta\omega_n - \omega_n\sqrt{\zeta^2 - 1}} + \frac{K_3}{s + \zeta\omega_n + \omega_n\sqrt{\zeta^2 - 1}}, \tag{B.13}$$

where $\zeta\omega_n - \omega_n\sqrt{\zeta^2 - 1}$ and $\zeta\omega_n + \omega_n\sqrt{\zeta^2 - 1}$ are positive real numbers.

The constants K_1, K_2, and K_3 can be evaluated quite simply by the methods of partial fraction expansion. Their values are

$$K_1 = 1,$$
$$K_2 = \left[2\left(\zeta^2 - \zeta\sqrt{\zeta^2 - 1} - 1\right)\right]^{-1}, \tag{B.14}$$
$$K_3 = \left[2\left(\zeta^2 + \zeta\sqrt{\zeta^2 - 1} - 1\right)\right]^{-1}.$$

Therefore Eq. (B.13) can be written as

$$C(s) = s^{-1} + \left[2\left(\zeta^2 - \zeta\sqrt{\zeta^2 - 1} - 1\right)\right]^{-1}\left(s + \zeta\omega_n - \omega_n\sqrt{\zeta^2 - 1}\right)^{-1}$$
$$+ \left[2\left(\zeta^2 + \zeta\sqrt{\zeta^2 - 1} - 1\right)\right]^{-1}\left(s + \zeta\omega_n + \omega_n\sqrt{\zeta^2 - 1}\right)^{-1}. \tag{B.15}$$

The time-domain response of the output, $c(t)$, may be obtained by utilizing a table of Laplace transforms. It can be expressed as

$$c(t) = 1 + \left[2\left(\zeta^2 - \zeta\sqrt{\zeta^2 - 1} - 1\right)\right]^{-1} e^{-(\zeta - \sqrt{\zeta^2 - 1})\omega_n t}$$
$$+ \left[2\left(\zeta^2 + \zeta\sqrt{\zeta^2 - 1} - 1\right)\right]^{-1} e^{-(\zeta + \sqrt{\zeta^2 - 1})\omega_n t}, \quad t \geq 0. \tag{B.16}$$

Figure B.2b illustrates the output response together with the unit step input. Notice that when $\zeta > 1$, the output response exhibits no overshoots and takes longer to reach its final value than when $\zeta = 1$. This response is described as being overdamped.

APPENDIX B: CHARACTERISTIC RESPONSES OF SECOND-ORDER CONTROL SYSTEMS   **569**

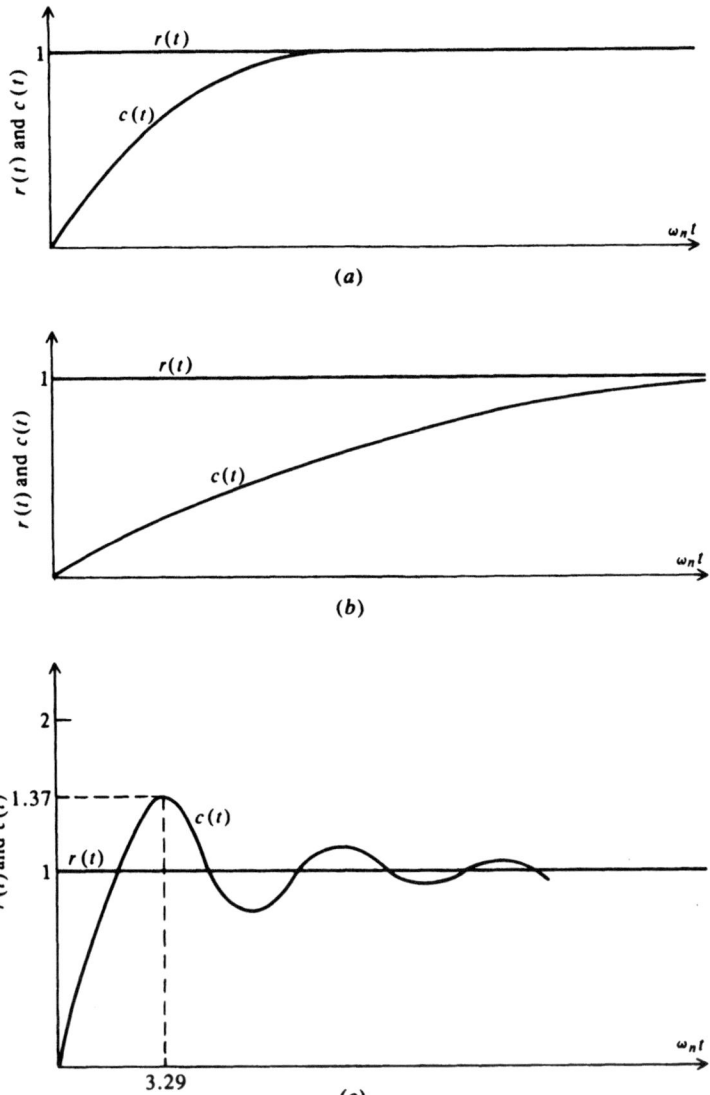

Figure B.2 (a) Input and output response for a critically damped second-order system; (b) input and output response for an overdamped second-order system; (c) input and output response for an underdamped second-order system ($\zeta = 0.3$).

Case C. Damping Ratio Less than Unity

When $\zeta < 1$, the time-domain response can be obained in an analogous manner. The solution is slightly more complex, however, because we now have a pair of complex-conjugate poles. The partial fraction expansion of Eq. (B.5) can be written as

$$C(s) = \frac{K_1}{s} + \frac{K_2}{s + \zeta\omega_n - j\omega_n\sqrt{1-\zeta^2}} + \frac{K_3}{s + \zeta\omega_n + j\omega_n\sqrt{1-\zeta^2}}. \qquad \text{(B.17)}$$

570 APPENDIX B: CHARACTERISTIC RESPONSES OF SECOND-ORDER CONTROL SYSTEMS

The constants K_1, K_2, and K_3 can be evaluated in an analogous manner by the method of partial fraction expansion, but the algebra becomes a little complicated. In order to simplify this situation somewhat, use is made of the relationship between the location of the complex-conjugate poles in the complex plane and the damping ratio ζ. The geometry of the configuration is illustrated in Figure B.3. Notice that the distance from the origin to either pole equals ω_n. In addition, the angle α has the following trigonometric properties:

$$\cos \alpha = \zeta, \qquad (B.18)$$

$$\sin \alpha = \sqrt{1 - \zeta^2}. \qquad (B.19)$$

Utilizing the relations given by Eqs. (B.18) and (B.19), the constants K_1, K_2, and K_3 can be expressed as

$$\begin{aligned} K_1 &= 1, \\ K_2 &= \frac{e^{-j\alpha}}{2j \sin \alpha}, \\ K_3 &= -\frac{e^{j\alpha}}{2j \sin \alpha}. \end{aligned} \qquad (B.20)$$

Therefore Eq. (B.17) can be written as

$$C(s) = \frac{1}{s} + \frac{e^{-j\alpha}}{2j \sin \alpha} \left(s + \zeta \omega_n - j\omega_n \sqrt{1 - \zeta^2} \right)^{-1} - \frac{e^{j\alpha}}{2j \sin \alpha} \left(s + \zeta \omega_n + j\omega_n \sqrt{1 - \zeta^2} \right)^{-1}. \qquad (B.21)$$

The time-domain response of the output, $c(t)$, may be obtained by utilizing a table of Laplace transforms. It can be expressed as

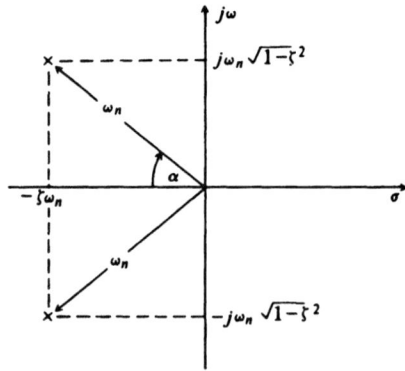

Figure B.3 Location of the complex-conjugate poles in the complex plane.

$$c(t) = 1 + \frac{e^{-j\alpha}}{2j \sin \alpha} e^{-(\zeta\omega_n - j\omega_n\sqrt{1-\zeta^2})t}$$
$$- \frac{e^{j\alpha}}{2j \sin \alpha} e^{-(\zeta\omega_n + j\omega_n\sqrt{1-\zeta^2})t}. \tag{B.22}$$

This can be simplified to

$$c(t) = 1 + \frac{e^{-\zeta\omega_n t}}{\sqrt{1-\zeta^2}} \frac{e^{j(\omega_n t\sqrt{1-\zeta^2}-\alpha)} - e^{-j(\omega_n t\sqrt{1-\zeta^2}-\alpha)}}{2j}, \tag{B.23}$$

or

$$c(t) = 1 - \frac{e^{-\zeta\omega_n t}}{\sqrt{1-\zeta^2}} \sin\left(\omega_n\sqrt{1-\zeta^2}\, t + \alpha\right), \quad t \geq 0. \tag{B.24}$$

Figure B.2c illustrates the output response for a value of ζ which has a value of 0.3, together with the unit step input. Notice that the output response exhibits several overshoots and undershoots before finally settling out. This response, which is characteristic of an exponentially damped sinusoid, is described as being underdamped.

Examination of Eq. (B.24) indicates that the term in the exponent, $\zeta\omega$, multiplies t and it controls the exponential decay or rise of the unit step response, $c(t)$. Therefore, $\zeta\omega_n$ determines the "damping" of the system and it is defined as the *damping factor* of the system. Note that the inverse of $\zeta\omega_n$ is proportional to the time constant of the system.

Observe from Eq. (B.12) that when the system is critically damped and $\zeta = 1$, the damping factor equals ω_n. Therefore, we view ζ as the damping ratio which is defined as follows:

$$\zeta = \text{damping ratio} = \frac{\zeta\omega_n}{\omega_n} = \frac{\text{damping factor}}{\text{damping factor at critical damping}}. \tag{B.25}$$

The time to the first overshoot and the value of the first overshoot are two interesting identifying characteristics for this type of response. We shall next derive these values in terms of the undamped natural frequency ω_n and the damping ratio ζ.

Equation (B.24) indicates that the damped natural radian frequency of oscillation of the system, ω_m, is

$$\omega_m = \omega_n\sqrt{1-\zeta^2}. \tag{B.26}$$

The damped natural cyclic frequency of oscillation of the system, f_m, is

$$f_m = \frac{\omega_m}{2\pi} = \frac{\omega_n\sqrt{1-\zeta^2}}{2\pi}. \tag{B.27}$$

The period of oscillation of the underdamped system, t_m, is

$$t_m = \frac{1}{f_m} = \frac{2\pi}{\omega_n\sqrt{1-\zeta^2}}. \tag{B.28}$$

APPENDIX B: CHARACTERISTIC RESPONSES OF SECOND-ORDER CONTROL SYSTEMS

The time at which the peak overshoot occurs, t_p, is found by differentiating $c(t)$, from Eq. (B.24), with respect to time and setting the derivative equal to zero:

$$\frac{dc(t)}{dt} = \zeta\omega_n e^{-\zeta\omega_n t}\left(\sqrt{1-\zeta^2}\right)^{-1}\sin\left(\omega_n\sqrt{1-\zeta^2}\,t + \alpha\right)$$
$$- \omega_n e^{-\zeta\omega_n t}\cos\left(\omega_n\sqrt{1-\zeta^2}\,t + \alpha\right) = 0$$

$$\frac{dc(t)}{dt} = \frac{\omega_n}{\sqrt{1-\zeta^2}}e^{-\zeta\omega_n t}\sin\left(\omega_n\sqrt{1-\zeta^2}\,t\right) = 0$$

This derivative is zero when

$$\omega_n\sqrt{1-\zeta^2}\,t = 0, \pi, 2\pi, \ldots$$

The peak overshoot occurs at the first value after zero, provided there are zero initial conditions. Therefore, the time to the first peak, t_p, and $\omega_n t_p$ are given by

$$t_p = \frac{\pi}{\omega_n\sqrt{1-\zeta^2}}, \tag{B.29}$$

$$\omega_n t_p = \frac{\pi}{\sqrt{1-\zeta^2}}. \tag{B.30}$$

For the case illustrated in Figure B.2c, where $\zeta = 0.3$, the time to the first overshoot is $3.29/\omega_n$.

Substituting Eq. (B.30) into Eq. (B.24) yields the value for the maximum instantaneous value of the output, $c(t)$:

$$c(t) = 1 - \frac{\exp\left(-\zeta\pi/\sqrt{1-\zeta^2}\right)}{\sqrt{1-\zeta^2}}\sin(\pi + \alpha). \tag{B.31}$$

This can be simplified by substituting

$$\sin(\pi + \alpha) = -\sin\alpha \quad \text{and} \quad \sin\alpha = \sqrt{1-\zeta^2}.$$

Therefore,

$$c(t)_{\max} = 1 + \exp\left(-\frac{\zeta\pi}{\sqrt{1-\zeta^2}}\right) \tag{B.32}$$

This is usually expressed as a percentage of the input. Therefore, for a unit step input,

$$\text{Maximum percent overshoot} = \exp\left(-\frac{\zeta\pi}{\sqrt{1-\zeta^2}}\right) \times 100\% \quad (B.33)$$

For the case illustrated in Figure B.2, where $\zeta = 0.3$, the maximum percent overshoot is 37%.

The second-order system is a very common and popular one. In order for the reader to become more familiar with its typical characteristic responses, Figure B.4 is shown to illustrate the resulting transient responses and percent maximum overshoots, rexpectively, for several values of damping ratios.

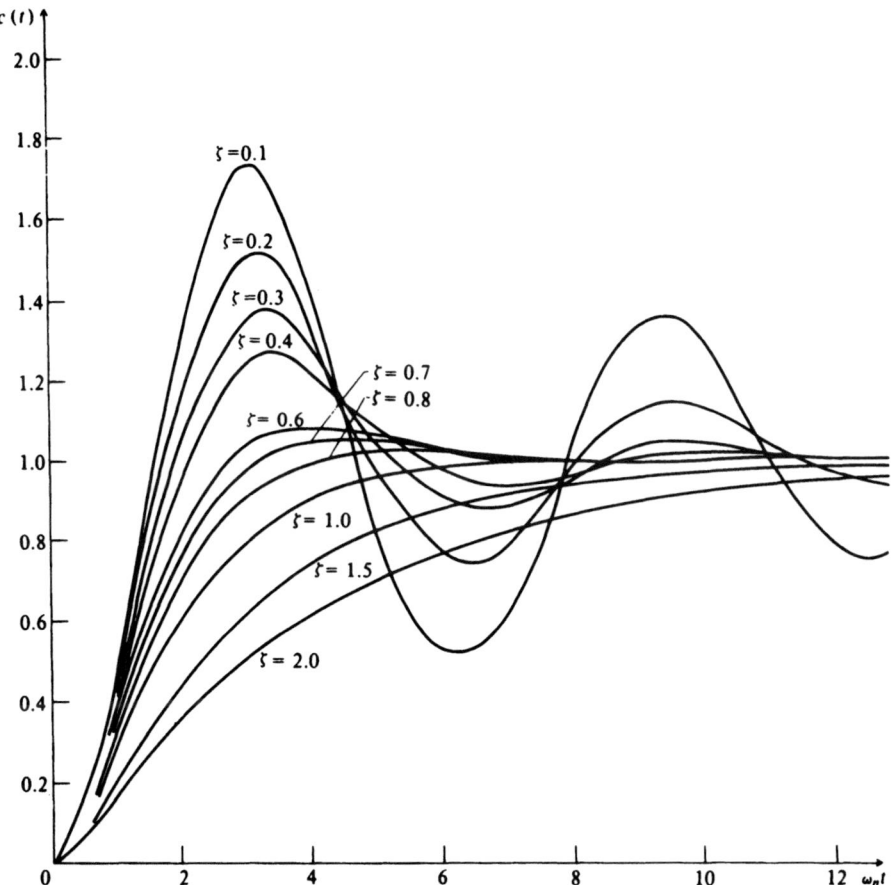

Figure B.4 Transient response curves of a second-order system to a unit step input.

APPENDIX C

STATIC ACCURACY

This appendix illustrates techniques that are available for determining the system accuracy. These are referred to several times in this book.

A method for determining the steady-state performance of any control system is to apply the final-value theorem of the Laplace transform. Let us consider the unity-feedback system shown in Figure C.1. The relation between the resulting system error, $E(s)$, for a given input $R(s)$ is given by

$$\frac{E(s)}{R(s)} = \frac{1}{1+G(s)}. \tag{C.1}$$

The steady-state error can be expressed as

$$e_{ss} = \lim_{t \to \infty} e(t) = \lim_{s \to 0} \frac{sR(s)}{1+G(s)}. \tag{C.2}$$

The control engineer is usually interested in test inputs of position, velocity, and acceleration. A step, ramp, and paraboloid are simple mathematical expressions which represent these physical quantities, respectively, and is illustrated in Figure C2. They are defined in Eqs. (C.3)–(C.5), where the notation $U(t)$ means a unit step for $t \geq 0$:

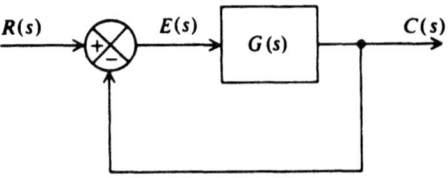

Figure C.1 A unity-feedback system.

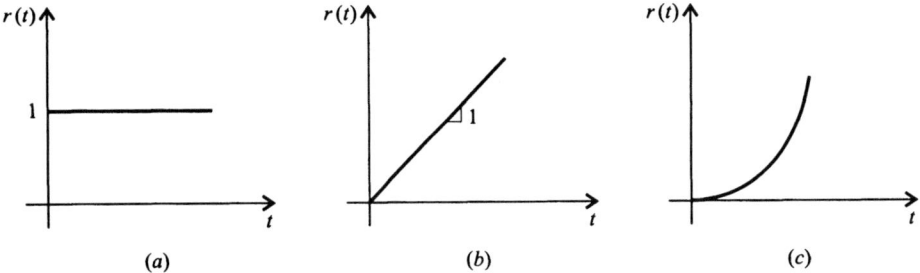

Figure C.2 Test inputs (a) Unit step representing a position input. (b) Unit ramp representing a velocity input. (c) Parabolic input representing an acceleration input.

$$\text{Position input: } r(t) = U(t), \qquad R(s) = 1/s, \qquad (C.3)$$
$$\text{Velocity input: } r(t) = tU(t), \qquad R(s) = 1/s^2, \qquad (C.4)$$
$$\text{Acceleration input: } r(t) = \frac{1}{2}t^2 U(t), \qquad R(s) = 1/s^3. \qquad (C.5)$$

We next determine the steady-state error of several types of systems for each of these three inputs: the unit step, unit ramp, and paraboloid. It is assumed that the loop transfer function $G(s)$ has the general form

$$G(s) = \frac{K(1 + T_1 s)(1 + T_2 s) \cdots (1 + T_m s)}{s^n[(T_o s)^2 + 2\zeta\omega_n s + 1](1 + T_a s)(1 + T_b s) \cdots (1 + T_q)}, \qquad (C.6)$$

where

s^n = a multiple pole at the origin of the complex plane,

K = gain factor of the expression.

The term s^n in the denominator is of particular significance. It represents the number of multiple poles in the denominator. Based on these number of pure integrations in the denominator of the open-loop transfer function, we classify the system by system "type". A control system is defined as type 0, type 1, type 2, type 3,..., if $n = 0$, $n = 1$, $n = 2$, $n = 3$, ..., respectively. We will show that as the system type is increased, the accuracy is improved. However, the stability problem becomes more difficult as the system type is increased. In practice, we usually never design a control system greater than type 2 because it is very difficult to stabilize a control system containing more than two pure integrations (although it can be done). Therefore, a tradeoff is required between steady-state accuracy and relative stability.

A. Unit Step (Position) Input

The steady-state error can be obtained by substituting $R(s) = 1/s$ into Eq. (C.2):

$$e_{ss} = \lim_{s \to 0} \frac{s(1/s)}{1 + G(s)} = \frac{1}{1 + \lim_{s \to 0} G(s)}. \qquad (C.7)$$

The quantity $\lim_{s \to 0} G(s)$ is defined as the position constant and is denoted by K_p:

$$K_p = \lim_{s \to 0} G(s). \qquad (C.8)$$

Therefore, the steady-state error in terms of the position constant is given by

$$e_{ss} = \frac{1}{1 + K_p}. \qquad (C.9)$$

Equation (C.9) states that the steady-state tracking error of a feedback control system having a unit step input equals $1/(1 +$ the position constant). Table C.1 summarizes the values of K_p and the resulting steady-state error as a function of the number of pure integrations of the open-loop transfer function $G(s)$.

Table C.1 indicates that the position constant is infinite for all systems which contian one or more pure integration(s) in the open-loop transfer function $G(s)$. Therefore, Eq. (C.9) implies that all systems containing at least one pure integration result in a theoretical steady-state positional response error of zero. Table C.1 indicates that the position constant is finite for a system containing no pure integrations and, therefore, the response error for a unit position input is $1/(1 + K)$.

B. Unit Ramp (Velocity) Input

The steady-state error can be obtained by substituting $R(s) = 1/s^2$ into Eq. (C.2):

$$e_{ss} = \lim_{s \to 0} \frac{s(1/s)^2}{1 + G(s)} = \frac{1}{\lim_{s \to 0} sG(s)}. \qquad (C.10)$$

The quantity $\lim_{s \to 0} sG(s)$ is defined as the velocity constant and is denoted by K_v:

$$K_v = \lim_{s \to 0} sG(s). \qquad (C.11)$$

Therefore, the steady-state error in terms of the velocity constant is given by

$$e_{ss} = 1/K_v. \qquad (C.12)$$

Table C.1.

Pure integrations of $G(s)$	K_p	e_{ss}
0	K	$1/(1 + K)$
1	∞	0
2	∞	0
.	.	.
.	.	.
.	.	.
n, where $n > 0$	∞	0

Equation (C.12) states that the steady-state response error of a feedback control system having a unit ramp equals the reciprocal of the velocity constant. Table C.2 summarizes the values of K_v and the resulting steady-state error as a function of the number of pure integrations of the open-loop transfer function $G(s)$.

Table C.2 indicates that the velocity constant is infinite for all systems that contain more than one pure integration in the open-loop transfer function $G(s)$. Therefore, Eq. (C.12) implies that all systems containing at least two pure integrations have a theoretical steady-state velocity response error of zero. Table C.2 indicates that a system containing no pure integrations cannot follow a velocity input. Table C.2 also indicates that a system containing one pure integration has a response of $1/K$ due to a unit ramp input.

C. Parabolic (Acceleration) Input

An expression for the steady-state error due to $r(t) = \frac{1}{2}t^2 U(t)$ can be obtained by substituting $R(s) = 1/s^3$ into Eq. (C.2):

$$e_{ss} = \lim_{s \to 0} \frac{s(1/s^3)}{1 + G(s)} = \frac{1}{\lim_{s \to 0} s^2 G(s)}. \tag{C.13}$$

The quantity $\lim_{s \to 0} s^2 G(s)$ is defined as the acceleration constant and is denoted by K_a:

$$K_a = \lim_{s \to 0} s^2 G(s). \tag{C.14}$$

Therefore, the steady-state error in terms of the acceleration constant is

$$e_{ss} = 1/K_a. \tag{C.15}$$

Equation (C.15) states that the steady-state response error of a feedback control system having this parabolic input equals the reciprocal of the acceleration constant. Table C.3 summarizes the values of K_a and the resulting steady-state error as a function of the number of pure integrations of the open-loop transfer function $G(s)$. Table C.3 indicates that the acceleration constant is infinite for all systems that contain three or more pure integrations in the open-loop transfer function $G(s)$, and

Table C.2.

Pure integrations of $G(s)$	K_v	e_{ss}
0	0	∞
1	K	$1/K$
2	∞	0
.	.	.
.	.	.
.	.	.
n, where $n > 1$	∞	0

Table C.3.

Pure integrations of $G(s)$	K_a	e_{ss}
0	0	∞
0	0	∞
2	K	$1/K$
3	∞	0
.	.	.
.	.	.
n, where $n > 2$	∞	0

the resulting steady-state error is zero. Therefore, Eq. (C.15) implies that all systems containing at last three pure integrations have a theoretical steady-state acceleration response error of zero. Table C.3 indicates that systems containing less than two pure integrations cannot follow an acceleration input. Table C.3 also indicates that a system containing two pure integrations has a response error of 1/K for this acceleration input.

A summary of the results derived appears in Table C.4. It is quite general and enables the reader to compare the capabilities of various types of systems. Notice from this table that the steady-state constants are zero, finite, or infinite. It is important to emphasize at this time that if the inputs are other than unit quantities the steady-state errors are proportionally increased because we are analyzing linear systems. For example, should the input to a system containing one pure integration be a ramp whose value is B position units (ft/sec, rad/sec, etc.), then the steady-state error as given by Eq. (C.12) would be modified to read

$$e_{ss} = B/K_v.$$

It should be noted that the unit of the velocity constant is 1/sec and that of the acceleration constant is 1/sec². The position constant K_p has no dimensions.

Let us now consider an input composed of position, velocity, and acceleration components which equal $AU(t)$ ft, $BU(t)$ ft/sec, and $C/2U(t)$ ft/sec², respectively. The form of the input can be represented as

$$r(t) = AU(t) + BtU(t) + \frac{1}{2}Ct^2U(t).$$

Table C.4. Summary of Steady-State Constants

Number of Pure Integrations	Constants		
	Position	Velocity	Acceleration
0	K_p	0	0
1	∞	K_v	0
2	∞	∞	K_a
3	∞	∞	∞

The steady-state response of the system may be obtained by considering each component of the input separately, and then adding the results by means of superposition. The resulting steady-state error is of the following form:

$$e_{ss} = \frac{A}{1+K_p} + \frac{B}{K_v} + \frac{C}{K_a}.$$

It is interesting to see how the various types of systems summarized in Table C.4 would respond to this input.

1. System Containing No Pure Integration. The steady-state error is

$$e_{ss} = \frac{A}{1+K_p} + \infty + \infty.$$

The result indicates that a system containing no pure integration will be able to follow the position input component of $AU(t)$ ft, but not velocity or acceleration inputs of $BU(t)$ ft/sec and $C/2U(t)$ ft/sec², respectively. Figure C.3 illustrates a typical step response for this system (when there are no velocity or acceleration components in the input).

2. System Containing One Pure Integration. The steady-state error is

$$e_{ss} = 0 + B/K_v + \infty.$$

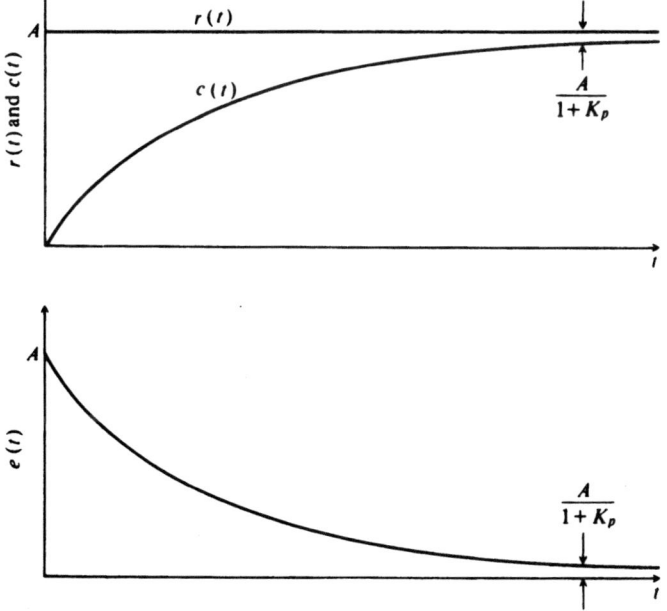

Figure C.3 Response of a system containing no pure integrations to a step input.

The result indicates that a system containing one pure integration will follow the position input component with zero error and the velocity input component with a finite error of B/K_v. This system will not, however, be able to follow the acceleration input component. The units of B/K_v are feet or radians. This should be interpreted to mean that there is a fixed positional error due to the constant velocity input component. Figure C.4 illustrates typical step and ramp responses of this system.

3. System Containing Two Pure Integrations.
The steady-state error is

$$e_{ss} = 0 + 0 + C/K_a.$$

The result indicates that a system containing two pure integrations will follow the position and velocity input components with zero error and the acceleration input component with a finite error of C/K_a. The units of C/K_a are feet or radians. This should be interpreted to mean that there is a fixed positional error due to the constant acceleration input component. Figure C.5 illustrates typical step, ramp, and acceleration responses of this system.

D. Relationships of Static Error Constants to Closed-Loop Poles and Zeros

It is often important to relate the position, velocity, and acceleration constants to the *closed-loop* poles and zeros. Let us consider a general single-loop, unity-feedback system having a forward transfer function, $G(s)$, where the closed-loop transfer function is given by

$$\frac{C(s)}{R(s)} = \frac{G(s)}{1 + G(s)}, \tag{C.16}$$

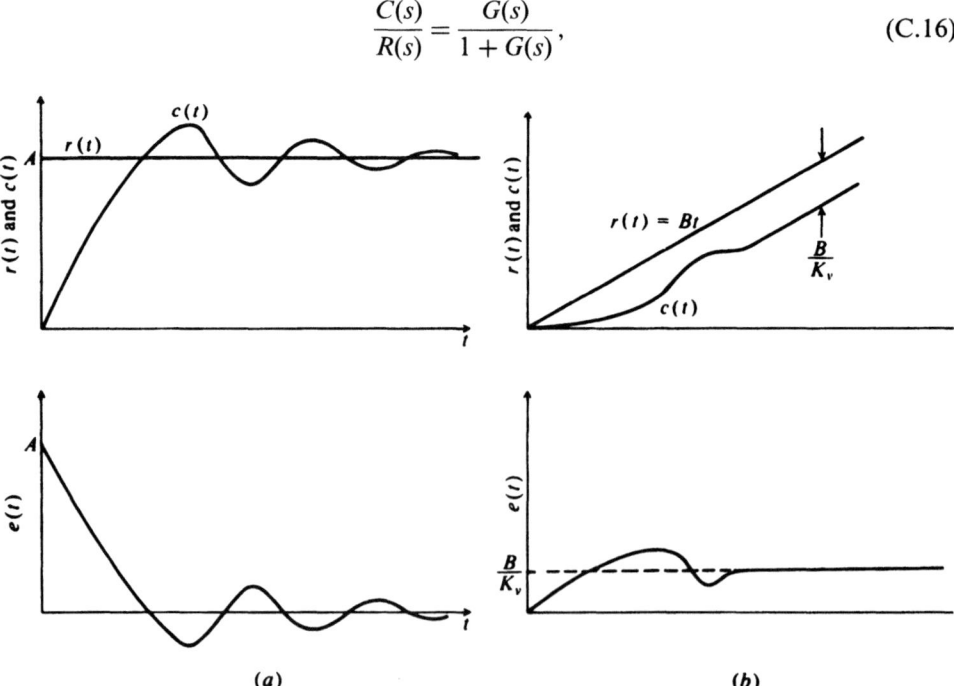

Figure C.4 Response of a system containing one pure integration to (a) step and (b) ramp inputs.

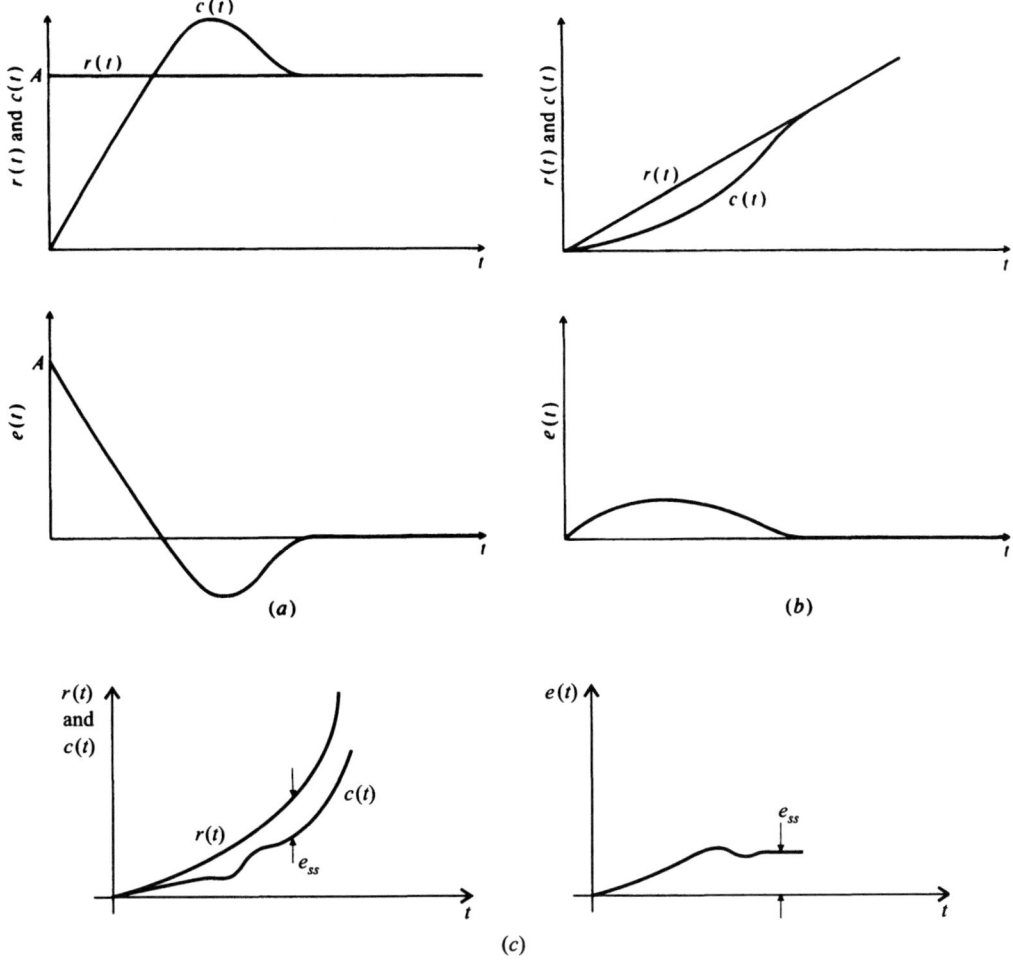

Figure C.5 Response of a system containing two pure integrations to (a) step, (b) ramp and (c) acceleration inputs.

and the relationship between input and error is given by

$$\frac{E(s)}{R(s)} = \frac{1}{1 + G(s)}. \tag{C.17}$$

Because $E(s) = R(s) - C(s)$, it is evident that

$$\frac{C(s)}{R(s)} = 1 - \frac{E(s)}{R(s)}. \tag{C.18}$$

It is assumed that $C(s)/R(s)$ can be represented by the rational function

$$\frac{C(s)}{R(s)} = \frac{K(s + z_1)(s + z_2) \cdots (s + z_m)}{(s + p_1)(s + p_2) \cdots (s + p_n)} = \frac{K \Pi_{j=1}^{m}(s + z_j)}{\Pi_{j=1}^{n}(s + p_j)}. \tag{C.19}$$

In addition, if $1/[1 + G(s)]$ in Eq. (C.17) is expanded as a power series in s, the error constants are defined in terms of the successive coefficients:

$$\frac{E(s)}{R(s)} = \frac{1}{1 + G(s)} = \frac{1}{1 + K_p} + \frac{1}{K_v}s + \frac{1}{K_a}s^2 + \cdots. \qquad \text{(C.20)}$$

Now the relationships between K_p, K_v, K_a, and the closed-loop poles and zeros can be determined.

1. Position Constant. Letting s approach zero in Eq. (C.17), we have

$$\frac{E(0)}{R(0)} = \lim_{s \to 0} \frac{1}{1 + G(s)}. \qquad \text{(C.21)}$$

Based on the definition of K_p in Eq. (C.8), we can rewrite Eq. (C.21) as

$$\frac{E(0)}{R(0)} = \frac{1}{1 + K_p}. \qquad \text{(C.22)}$$

Substituting Eq. (C.22) into Eq. (C.18) with s set at zero, we obtain the following:

$$\frac{C(0)}{R(0)} = \frac{K_p}{1 + K_p}.$$

Solving for K_p in terms of $C(0)/R(0)$, we have

$$K_p = \frac{C(0)/R(0)}{1 - C(0)/R(0)}. \qquad \text{(C.23)}$$

Using Eq. (C.19) to represent $C(s)/R(s)$ and letting s approach zero in Eq. (C.19), we have

$$\frac{C(0)}{R(0)} = \frac{K \Pi_{j=1}^{m} z_j}{\Pi_{j=1}^{n} p_j}, \qquad \text{(C.24)}$$

where

$$\prod_{j=1}^{m} z_j = \text{product of zeros,}$$

$$\prod_{j=1}^{n} p_j = \text{product of poles.}$$

Substituting Eq. (C.24) into (C.23), the following expression for K_p in terms of the closed-loop poles and zeros is obtained:

$$K_p = \frac{K \Pi_{j=1}^{m} z_j}{\Pi_{j=1}^{n} p_j - K \Pi_{j=1}^{m} z_j}. \qquad \text{(C.25)}$$

2. Velocity Constant.

In order to derive the velocity constant in terms of the closed-loop poles and zeros, let us substitute Eq. (C.20) into Eq. (C.18):

$$\frac{C(s)}{R(s)} = 1 - \frac{1}{1+K_p} - \frac{1}{K_v}s - \frac{1}{K_a}s^2 - \cdots.$$

Taking the derivative of this expression with respect to s, and then letting s equal zero, we obtain

$$\left[\frac{d}{ds}\left(\frac{C(s)}{R(s)}\right)\right]_{s=0} = -\frac{1}{K_v}. \tag{C.26}$$

In addition, we make use of the property that

$$\left[\frac{C(s)}{R(s)}\right]_{s=0} = \frac{C(0)}{R(0)} = 1 \tag{C.27}$$

in unity-feedback systems containing one or more pure integrations. Equation (C.27) says that a closed-loop unity-feedback system behaves with an ideal closed-loop transfer function of one at zero frequency. Dividing Eq. (C.26) by (C.27),

$$\frac{1}{K_v} = \frac{-\left[\frac{d}{ds}\left(\frac{C(s)}{R(s)}\right)\right]_{s=0}}{[C(s)/R(s)]_{s=0}} = -\left[\frac{d}{ds}\ln\frac{C(s)}{R(s)}\right]_{s=0}.$$

Substituting Eq. (C.19) into this equation, we have the following:

$$\frac{1}{K_v} = -\left\{\frac{d}{ds}[\ln K + \ln(s+z_1) + \cdots + \ln(s+z_m)\right.$$

$$\left. - \ln(s+p_1) - \cdots - \ln(s+p_n)]\right\}_{s=0}.$$

This can also be written as

$$\frac{1}{K_v} = -\left(\frac{1}{s+z_1} + \cdots + \frac{1}{s+z_m} - \frac{1}{s+p_1} - \cdots - \frac{1}{s+p_n}\right)_{s=0}, \tag{C.28}$$

or

$$\frac{1}{K_v} = \sum_{j=1}^{n}\frac{1}{p_j} - \sum_{j=1}^{m}\frac{1}{z_j}. \tag{C.29}$$

Therefore, $1/K_v$ equals the sum of the reciprocals of the closed-loop poles minus the sum of the reciprocals of the closed-loop zeros.

3. Acceleration Constant. The acceleration constant in terms of the closed-loop poles and zeros can be derived in a similar manner. We know from Eq. (C.20) that

$$\frac{E(s)}{R(s)} = \frac{1}{1+K_p} + \frac{1}{K_v}s + \frac{1}{K_a}s^2 + \cdots.$$

From Eq. (C.18), this can be rewritten as

$$\frac{C(s)}{R(s)} = 1 - \frac{1}{1+K_p} - \frac{1}{K_v}s - \frac{1}{K_a}s^2 - \cdots.$$

It is obvious from this equation that $-2/K_a$ equals the zero-frequency value of the second derivative of $C(s)/R(s)$. Writing this in terms of the logarithmic derivative:

$$\frac{d^2}{ds^2}\left[\ln\frac{C(s)}{R(s)}\right] = \frac{(C(s)/R(s))''}{C(s)/R(s)} - \left[\frac{(C(s)/R(s))'}{C(s)/R(s)}\right]^2.$$

Setting $C(0)/R(0)$ equal to one, then

$$-\frac{2}{K_a} = \left\{\frac{d^2}{ds^2}\left[\ln\frac{C(s)}{R(s)}\right]\right\}_{s=0} + \frac{1}{K_v^2}.$$

Differentiating the right-hand side of Eq. (C.28) and letting s equal zero yields the following expression:

$$-\frac{2}{K_a} = \frac{1}{K_v^2} + \sum_{j=1}^{n}\frac{1}{p_j^2} - \sum_{j=1}^{m}\frac{1}{z_j^2}, \qquad (C.30)$$

where K_v is defined by Eq. (C.29).

4. An Example. As an example, consider the second-order system illustrated in Figure B.1 of Appendix B whose characteristics are defined by Eqs. (B.1), (B.2), and (B.3):

$$\frac{C(s)}{R(s)} = \frac{K_m/T_m}{s^2 + (1/T_m)s + K_m/T_m},$$

or

$$\frac{C(s)}{R(s)} = \frac{\omega_n^2}{s^2 + 2\zeta\omega_n s + \omega_n^2},$$

where

$$\omega_n^2 = K_m/T_m \quad \text{and} \quad \zeta = 1/2\omega_n T_m.$$

This is representative of a wide class of control systems that was thoroughly analyzed in Appendix B. The problem is to determine K_v in terms of the parameters ζ and ω_n. K_m is the system gain and T_m is the time constant of the open-loop transfer function. The velocity constant of the simple system illustrated in Figure B.1 is

$$K_v = K_m$$

by inspection. Let us relate this velocity constant to ζ and ω_n from the basic definitions of K_m and T_m in terms of ζ and ω_n:

$$K_v = K_m = \omega_n^2 T_m = \omega_n^2 \left(\frac{1}{2\zeta\omega_n}\right) = \frac{\omega_n}{2\zeta}.$$

The fact that

$$K_v = \omega_n/2\zeta \tag{C.31}$$

is very important to remember for all second-order control systems that are characterized by a pair of complex-conjugate poles. Therefore, in order to obtain very accurate responses to velocity inputs, the damping ratio ζ is made very small. An example of this is in inertial navigation systems where accuracy is much more important than transient responses settling quickly.

ANSWERS TO SELECTED PROBLEMS

CHAPTER 1

1.4.
$$G(s)H(s) = \frac{100}{s(1 + 0.1s)}$$

1.6. (a) Crossover frequency = 4.483 rad/sec, phase margin = 32.44°, gain margin = 8.657 dB.

(b) $K = 9.9$

(c) Crossover frequency = 13.61 rad/sec, phase margin = −30.62°, gain margin = −11.26 dB.

CHAPTER 2

2.1. (a) The electrical network illustrated in Table 2.4‡, items 3, with
$$R_1 = 690 \, k\Omega, \quad R_2 = 51 \, k\Omega, \quad C_1 = 1.38 \, \mu F$$
and cascaded with an amplifier whose gain is 4.6 will satisfy the specifications

(b) The electrical network illustrated in Table 2.4‡, item 4, with
$$R_1 = 690 \, k\Omega, \quad R_2 = 51 \, k\Omega, \quad C_2 = 1.38 \, \mu F$$
and cascaded with an amplifier whose gain is 4.6 will satisfy the specifications.

(c) The electrical network illustrated in Table 2.4‡, item 5, with

$$R_1 = 595\,\text{k}\Omega, \quad R_2 = 25\,\text{k}\Omega, \quad C_1 = 1.68\,\mu\text{F}, \quad C_2 = 4\,\mu\text{F}$$

will satisfy the specifications.

2.2. (a) $\zeta = 0.5$; $\omega_n = 4$ rad/sec; maximum percent overshoot = 16.4%; steady-state error = 0.25.

(b) $b = 0.15$.

(c) Maximum percent overshoot = 1.5%; steady-state error = 0.4.

(d) Add a high-pass filter in cascade with the tachometer as illustrated in Figure 2.15.

2.5. (a) $G_c(s) = 1 + 0.382s$.

(b) $b = 0.382$.

2.7. $\omega_0 = \dfrac{1}{\sqrt{T_1 T_2}}$.

2.8. (a) $G_c(s) = \dfrac{0.5s + 1}{0.014s + 1} \dfrac{0.05s + 1}{0.01s + 1}$;

the attenuation of 1/178.6 due to the two lead networks is assumed to be compensated for by increasing the gain of the amplifier by 178.6. The result is a phase margin of 46.9° at a gain crossover frequency of 6.4 rad/sec, and a gain margin of 32.8 dB at a phase crossover frequency of 68 rad/sec.

(b) $\dfrac{12.5s + 1}{250s + 1}$.

2.11. A phase-lead network whose transfer function is

$$\dfrac{1}{15} \dfrac{1 + s}{1 + 0.067s}$$

The result is a phase margin of 44.8° at a gain crossover frequency of 2.35 rad/sec, and a gain margin of 11.2 dB at a phase crossover frequency of 11.2 rad/sec, assuming that the attenuation of the phase-lead network of $\tfrac{1}{15}$ is compensated for by boosting the gain of the amplifier by a factor of 15. Therefore, this phase-lead network achieves the design specification of a minimum phase margin of 60°.

2.12. Same answer as for Problem 2.11.

2.14. Two cascaded phase-lead networks are required to meet these specifications. Their transfer function, together with the gain given of 99, is given by

$$G_1(s) = 9.9 \dfrac{1 + 0.2s}{1 + 0.01s} \dfrac{1 + 0.1s}{1 + 0.005s}.$$

The constant attenuation of 400 due to the characteristics of these two phase-lead networks must be made up by an amplification increase of 400. The resulting phase margin is 46.09° at a gain crossover frequency of 55.58 rad/sec.

2.15.

(a) $\omega_c = 3$ rad/sec.

(b) $$G_c(s) = \frac{1}{40}\frac{(1+0.904s)}{(1+0.0226s)}.$$

It is assumed that the attenuation of 1/40 is compensated for by increasing the gain of the amplifier. This phase-lead network provides a phase margin of 71.86° at a new gain crossover frequency of 8.078 radians/second.

2.20. (a) $G_c(s) = \dfrac{1+0.416s}{1+0.384s}$

The constant attenuation of 1.08 due to the characteristics of this phase-lead network must be made up by an amplification increase of 1.08.

(b) For part (a), $\omega_p = 2.375$ rad/sec, $M_p = 1.154$ dB.

2.21. (a) $G_c(s) = \dfrac{0.5s+1}{0.001s+1}\dfrac{0.05s+1}{0.001s+1};$

The attenuation due to the phase-lead network is assumed to be compensated for by increasing the gain of the amplifiers.

(b) For the phase-lead network, $M_p = 0.7513$ dB, $\omega_p = 3.864$ rad/sec.

2.22. (a) $b = 0.25$.

(b) $e_{ss} = 0.3024$.

(c) The transfer function of the rate feedback path is given by

$$\frac{5s}{1+5s}0.25s.$$

(d) $e_{ss} = 0.0524$.

2.23. The phase-lag network is given by

$$\frac{s+0.0357}{s+0.001}.$$

2.28. $K = 4.22$.

2.38. $K = 2.226$.

2.40. (a) $K = 129.3$.

(b) $M_p = 4.24$ dB; $\omega_p = 10.67$ rad/sec.

2.41 $K = 25.57$.

CHAPTER 3

3.2.

(a) $x_1(t) = -4x_1(t) - h_1 x_1(t) + x_2(t)$

$x_2(t) = -Kx_1(t) + Kr(t)$

(b) $s^2 + 4s + h_1 s + K = 0$.

(c) $h_1 = 10$,

$K = 48$.

(d) $K_I = 64$,

$K_p = 10$.

(e) The proportional plus integral controller configuration of part (d) produces a zero term which would aid in the stability compensation of this control system. The controller configuration of part (a) does not have this feature, and it would be more difficult to stabilize.

3.6. The synthesized system is illustrated in Figure A3.6, where

$$K = 8, \quad h_1 = 1, \quad h_2 = \frac{5}{8}, \quad h_3 = \frac{5-\alpha}{8}.$$

A root locus analysis indicates that a good choice of α is approximately 2.5. Therefore, h_3 becomes $\frac{5}{16}$.

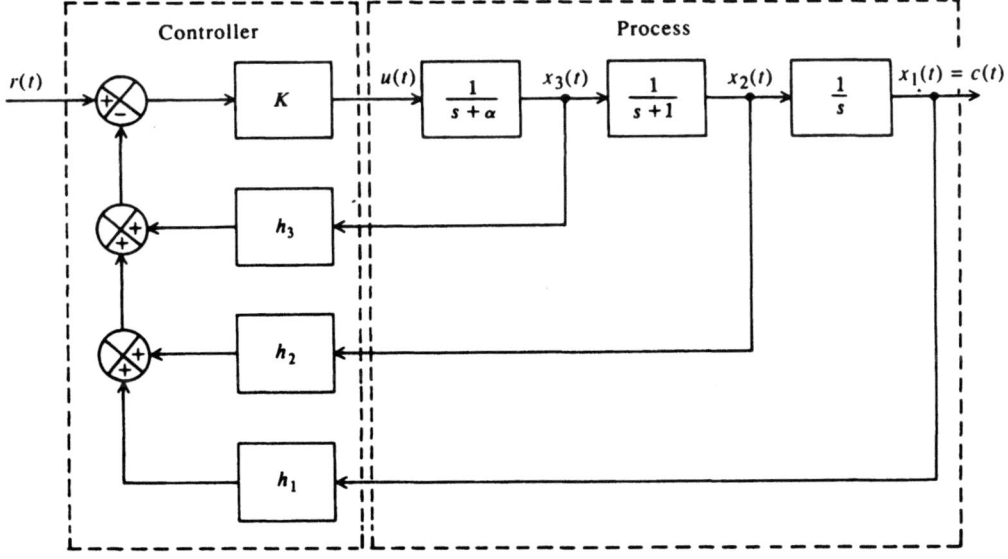

Figure A3.6

3.8. (a) Yes.

(b) see Figure A3.8.

(c) $K = 20,000$,

$h_1 = 1$,

$h_2 = 0.00995$.

(d) Root locus shows that the system is stable.

3.10. (a) $\mathbf{D} = \begin{bmatrix} 0 & 0 & 0 \\ 1 & -1 & 1 \\ 0 & 0 & 0 \end{bmatrix}$;

System is not controllable.

(b) $\mathbf{U} = \begin{bmatrix} 0 & 1 & -2 \\ 1 & -1 & 1 \\ 1 & 1 & 1 \end{bmatrix}$;

System is observable.

3.11. (a) $\mathbf{D} = \begin{bmatrix} 4 & -32 \\ 1 & -2 \end{bmatrix}$;

System is controllable.

(b) $\mathbf{U} = \begin{bmatrix} 1 & -8 \\ 0 & 0 \end{bmatrix}$;

System is unobservable.

3.13. $\mathbf{K} = \begin{bmatrix} K_1 \\ K_2 \end{bmatrix} = \begin{bmatrix} 12 \\ 48 \end{bmatrix}$.

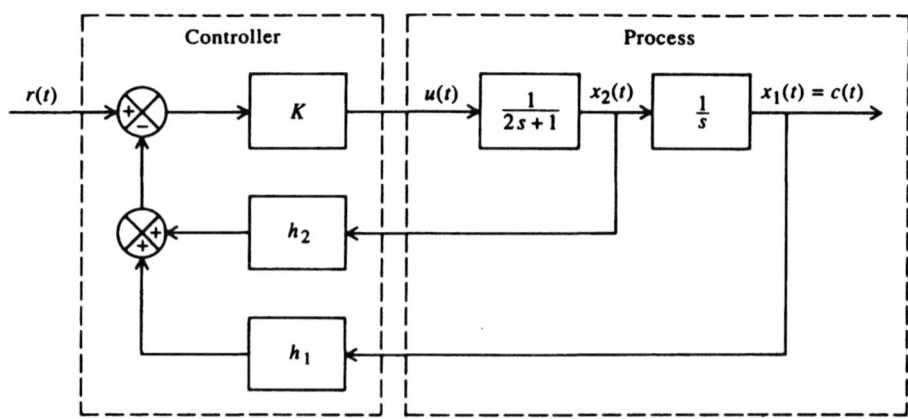

Figure A3.8

3.15. $\mathbf{M} = \begin{bmatrix} m_1 \\ m_2 \end{bmatrix} = \begin{bmatrix} 59 \\ 840 \end{bmatrix}.$

3.17. $\mathbf{M} = \begin{bmatrix} m_1 \\ m_2 \end{bmatrix} = \begin{bmatrix} 34 \\ 590 \end{bmatrix}.$

3.20. $H(s) = \dfrac{Y(s)}{U(s)} = \dfrac{1}{s^2 + 1.333s + 0.889}$

CHAPTER 4

4.3. $F(z) = 1 + 1/T\left[\tan^{-1}\left(\dfrac{\sin T}{z - \cos T}\right)\right].$

4.5. $R(z) = \dfrac{e^{aT}z}{e^{aT}z - 1}z^{-3}.$

4.7. (a) $e(k) = 2(1 - 0.5^k), \quad k \geq 0.$

(b) $e(k) = \delta(k-1) + 1.5\delta(k-2) + 1.75\delta(k-3) + 1.875\delta(k-4) + \cdots,$
$k = 0, 1, 2, \ldots.$

4.8. (a) $R(z) = 0.5z^{-1} + z^{-2} + 0.5z^{-3}.$

(b) $C(z) = 0.39935z^{-2} + 1.0256z^{-3} + 1.0155z^{-4} + 0.6159z^{-5}$
$+ 0.3736z^{-6} + 0.2266z^{-7} + 0.1374z^{-8} + 0.0834z^{-9}$
$+ 0.0506z^{-10} + 0.0307z^{-11} + 0.0186z^{-12} + \cdots,$

(c) $c^*(t) = 0.39935\delta(t-2) + 1.0256\delta(t-3) + 1.0155\delta(t-4)$
$+ 0.6159\delta(t-5) + 0.3736\delta(t-6) + 0.2266\delta(t-7)$
$+ 0.1374\delta(t-8) + 0.0834\delta(t-9) + 0.0506\delta(t-10)$
$+ 0.0307\delta(t-11) + 0.0186\delta(t-12) + \cdots,$

(d) $c(0) = 0,$
$c(\infty) = 0.$

4.10. $G(z) = (z/(z - e^{-T})) - (1/(z - e^P - 2T)).$

4.12. $c(k) = 2^k - k - 1, \quad k \geq 0.$

4.14. $\dfrac{C(z)}{R(z)} = G_1(z)G_2G_3G_6(z) + G_1G_4G_5G_6(z).$

4.18. (a) $G_{12}(z) = \dfrac{1.2142}{z - 0.3679}.$

(b) $C(z) = 1.26z^{-1} + 1.72z^{-2} + 1.89z^{-3} + 1.96z^{-4} + 1.98z^{-5}$
$+ 1.99z^{-6} + 1.99z^{-7} + \cdots.$

(c) $c^*(t) = 1.26\delta(t-1) + 1.72\delta(t-2) + 1.89\delta(t-3) + 1.96\delta(t-4)$

$+1.98\delta(t-5)+\cdots, \quad t \geq 0.$

(d) $c(0) = 0.$

(e) $c(\infty) = 2.$

4.20. (a) $G(z) = 0.5(z+1)/z(z-1).$

(b) $C(z) = 0.5z^{-1} + 1.5z^{-2} + 2.5z^{-3} + 3.5z^{-4} + 4.5z^{-5} + \cdots$

$c^*(t) = 0.5\delta(t-1) + 1.5\delta(t-2) + 2.5\delta(t-3)$
$+ 3.5\delta(t-4) + 4.5\delta(t-5) + \cdots.$

(c) $C(z) = 0.5z^{-2} + 2z^{-3} + 4.4z^{-4} + 8z^{-5} + 12.5z^{-6} + \cdots$

$c^*(t) = 0.5\delta(t-2) + 2\delta(t-3) + 4.4\delta(t-4) + 8\delta(t-5)$
$+ 12.5\delta(t-6) + \cdots.$

4.22. The system is unstable with two roots in the right half-plane.

4.25. The final-value theorem cannot be used in all three parts because the function is not analytic in the right half-plane. $f(0) = 0$ in all three parts.

4.28. (a) Unstable with one root outside the unit circle of z-plane.

(b) Stable.

(c) Unstable with two roots outside unit circle of z-plane.

4.29. (a) $c(\infty) = 4$

(b) $c(k) = 4[1 - k(0.5)^k - (0.5)^k].$

Therefore, $c(\infty) = 4.$

(c) Yes.

4.30. K must be less than 2.17.

4.41. (a) $x[(k+1)] = x(k) + Ty(k+1).$

(b) $\dfrac{Z(z)}{Y(z)} = \dfrac{Tz}{z-1}.$

4.42. (a) $x(k+1) = x(k) + \dfrac{T[y(k) + y(k+1)]}{2}.$

(b) $\dfrac{X(z)}{Y(z)} = \dfrac{T}{2}\dfrac{z+1}{z-1}.$

CHAPTER 5

5.1. (a) Transfer function of the motor with one straight line tangent to the exact characteristics at 1000 rev/min is

$$\frac{\theta(s)}{V_c(s)} = \frac{2.36}{s(0.0083s + 1)}.$$

Transfer function of the motor with one straight line going through the two endpoints is

$$\frac{\theta(s)}{V_c(s)} = \frac{1.79}{s(0.0068s + 1)}.$$

(b) Transfer function of the motor with one straight line tangent to the characteristics at 250 rev/min is

$$\frac{\theta(s)}{V_c(s)} = \frac{6.12}{s(0.0227s + 1)}.$$

Transfer function of the motor with one straight line tangent to the characteristics at 1750 rev/min is

$$\frac{\theta(s)}{V_c(s)} = \frac{2}{s(0.00334s + 1)}.$$

5.2.
$$N(M) = \frac{2K_1}{\pi}\left(-\frac{D}{M}\cos\sin^{-1}\frac{D}{M} - \sin^{-1}\frac{D}{M}\right.$$
$$\left. + \frac{S}{M}\cos\sin^{-1}\frac{S}{M} + \sin^{-1}\frac{S}{M}\right).$$

5.3.
$$N(M) = \frac{4K_1}{\pi M}\sqrt{1 - \left(\frac{D}{M}\right)^2}.$$

5.4.
$$N(M) = \frac{2K}{\pi}\left[\sin^{-1}\frac{E_2}{M} - \sin^{-1}\frac{E_1}{M} + \left(\frac{2E_1}{M} + \frac{E_2}{M}\right)\cos\sin^{-1}\frac{E_2}{M}\right.$$
$$\left. - \frac{E_1}{M}\cos\sin^{-1}\frac{E_1}{M}\right].$$

5.7. $N = 0$ for $M \le D$

$$N(M) = \frac{2K}{\pi}\left(\frac{\pi}{2} + \frac{\sin 2\theta}{2} - \sin^{-1}\frac{D}{M}\right)$$

where

$$\theta = \sin^{-1}\frac{D}{M}$$

ANSWERS TO SELECTED PROBLEMS

5.8. $N(M) = \dfrac{4 F_c}{\pi M} + B.$

5.17. (a) See Figure A5.17.

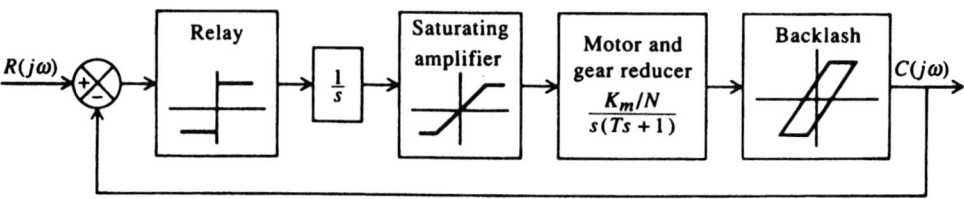

Figure A5.17

(b) For the linear elements, their transfer functions can be combined into a transfer function, $G(s)$. For the nonlinear elements, a new describing function which represents the combined nonlinear effects of saturation, the relay, and backlash would have to be derived. Then, $-1/N$ and $G(j\omega)$ could be plotted on the gain-phase diagram to determine the presence of an oscillation.

5.18. A stable limit cycle occurs at $D/M = 0.999$ and $\omega = 0.0707$ rad/sec.

5.19. A limit cycle does not exist.

5.21. System is stable and exhibits no limit cycle.

5.22. Unstable limit cycle occurs at $M/D = 0.045$ and $\omega = 15.6$ rad/sec.

5.23. System is stable and exhibits no limit cycle.

5.25. System does not exhibit the existence of any limit cycles.

5.26. (a) Limit cycles occur at $D/M = 0.8319$, $\omega = 0.3846$ rad/sec; $D/M = 0.2533$, $\omega = 1.033$ rad/sec.
The former limit cycle is unstable, the latter is stable.

(b) $K = 2.75.$

(c) $\alpha T = 0.9$ sec, $T = 0.4$ sec (assume increase in original gain by a factor of $\tfrac{9}{4}$).

(d) $b = 0.32.$

5.30. A stable limit cycle exists at $\omega = 30$ rad/sec and $D/M = 0.62$. An unstable limit cycle exists at $\omega = 3$ rad/sec and $D/M = 0.92$.

5.31. The describing-function analysis of the gain-phase diagram indicates that limit cycles do not exist for this system if the nonlinearity corresponds to Coulomb friction.

5.32. System does not exhibit the existence of any limit cycles.

5.35. (a) The phase trajectory is a circle through the point 1, 0 and whose center is the origin.

(b) The system is unstable, since no viscous damping is present.

5.43. (a) $\dfrac{d\dot{\theta}(t)}{d\theta(t)} = -\dfrac{B}{A} + \dfrac{D - C\theta(t)}{A\dot{\theta}(t)}$

(b) $\dfrac{d\dot{\theta}(t)}{d\theta(t)} = -\dfrac{B|\dot{\theta}(t)|}{A} + \dfrac{D - C\theta(t)}{A\dot{\theta}(t)}$

(c) $\dfrac{d\dot{\theta}(t)}{d\theta(t)} = -\dfrac{B|\theta(t)|}{A} + \dfrac{D - C\theta(t)}{A\dot{\theta}(t)}$

(d) $\dfrac{d\dot{\theta}(t)}{d\theta(t)} = -\dfrac{B}{A} + \dfrac{D - C\theta(t)|\theta(t)|}{A\dot{\theta}(t)}.$

5.44. (a) The isocline equation is given by

$$\dfrac{d\dot{\Theta}(t)}{d\theta(t)} = -2\zeta - \dfrac{\theta(t)}{\dot{\Theta}(t)},$$

where

$$\dot{\Theta}(t) = \dfrac{\dot{\theta}(t)}{\sqrt{0.4}}, \quad \zeta = \sqrt{0.4}.$$

(b) The isocline equation is given by

$$\dfrac{d\dot{\Theta}(t)}{d\theta(t)} = -2\zeta - \dfrac{\theta(t)}{\dot{\Theta}(t)},$$

where

$$\dot{\Theta}(t) = \dfrac{\dot{\theta}(t)}{\sqrt{0.4}}, \quad \zeta = \sqrt{0.4}.$$

(c) The isocline equation is given by

$$\dfrac{d\dot{\Theta}(t)}{d\theta(t)} = -2\zeta|\dot{\Theta}(t)| - \dfrac{\theta(t)}{\dot{\Theta}(t)},$$

where

$$\dot{\Theta}(t) = \dfrac{\dot{\theta}(t)}{\sqrt{0.4}}, \quad \zeta = \sqrt{0.4}.$$

(d) The isocline equation is given by

$$\frac{d\dot{\Theta}(t)}{d\theta(t)} = -\zeta|\dot{\Theta}(t)| - \frac{\theta(t)}{\dot{\Theta}(t)},$$

where

$$\dot{\Theta}(t) = \frac{\dot{\theta}(t)}{\sqrt{0.4}}, \quad \zeta = \sqrt{0.4}.$$

5.49. Defining

$$\theta_1(t) = \theta(t) - 0.5, \quad \text{for } \theta(t) > 0$$

and

$$\theta(t) = \theta(t) + 0.5, \quad \text{for } \theta(t) < 0,$$

the isocline equation is given by

$$\frac{d\dot{\Theta}_1(t)}{d\theta_1(t)} = -0.8 - \frac{\theta_1(t)}{\dot{\Theta}_1(t)},$$

where

$$\dot{\Theta}_1(t) = \frac{\dot{\theta}_1(t)}{\omega_n(t)}$$

and

$$\omega_n = 1.$$

5.51. (a) Asymptotically stable.

(b) Unstable.

(c) Asymptotically stable for

$$\left|\frac{K_2}{K_1}\right| > \left|\frac{\dot{x} + \dot{x}^2}{x^2}\right|.$$

5.57. Popov condition satisfied if $0 < K \leqslant 5.5$.

5.59. Popov condition satisfied if $0 < K \leqslant 89.14$.

5.63. Possible Popov sectors which result in stable systems are given by the following conditions:

$$6 < N[e(t), t] < 18.9, \quad 8.34 < N[e(t), t] < 47.6$$
$$33.3 < N[e(t), t] < 100.$$

CHAPTER 6

6.3. The optimal input is given by

$$u^0(t) = -\text{sgn}\left[\frac{\partial S^0}{\partial \dot{c}}\right]$$

and the nonlinear partial differential equation to be solved for S^0 is given by

$$\dot{c}(t)\left[\frac{\partial S^0}{\partial c} - 4\frac{\partial S^0}{\partial \dot{c}}\right] - c(t)\left[\frac{\partial S^0}{\partial \dot{c}}\right] - \left[\left|\frac{\partial S^0}{\partial \dot{c}}\right| - 1\right] = 0.$$

6.4. $u^0(t) = \text{sgn}[Ke^{2t}\cos\sqrt{3}t + \theta]$,

where K and θ are determined from the boundary conditions of the system.

6.7. (a) $u^0(t) = -\text{sgn}(4\frac{\partial S^0}{\partial \dot{c}} + 1)$,

(b) $u^0(t) = \text{sgn}(4p_2 - 1)$.

(c) Both results are equal because $p_2 = -\partial S/\partial \dot{c}$.

6.13. The optimal control policy is given by

$$u^0(t) = -\frac{K}{2}\frac{\partial S^0}{\partial c},$$

and the nonlinear partial differential equation to be solved for S^0 is given by

$$\frac{-K^2}{2}\left(\frac{\partial S^0}{\partial c}\right) + Ac^2 = 0.$$

6.14. $u^0(t) = U\text{sgn}[Ke^{t/2}\sin(0.866t + \theta)]$, where K and θ are determined from the boundary conditions of the system.

6.15. $u^0(t) = U\text{sgn}[Ke^t\sin(1.732t + \theta)]$, where K and θ are determined from the boundary conditions of the system.

6.16. $u^0(t) = U\text{sgn}[K_1 e^t + K_2 t e^t]$, where K_1, K_2, and θ are determined from the boundary conditions of the system.

CHAPTER 7

7.3. (a) Pertinent characteristics of the root locus are

1. Root locus occurs along the real axis between the origin and -2.
2. The asymptotes intersect the real axis at -1 with angles of $\pm 90°$.

598 ANSWERS TO SELECTED PROBLEMS

3. The point of breakaway from the real axis occurs at -1.

(b) $K = 2$.

(c) $G_c(s) = \dfrac{s + 0.1}{s + 0.01}$.

(d),(e) Maximum percent overshoot of the third-order compensated system is 11%, which compares with the maximum percent overshoot of a second-order system which has a damping ratio of 0.707 or 4.3%.

7.5. (b) Root-locus plot of Figure A7.5a indicates that $K = 2.927$.

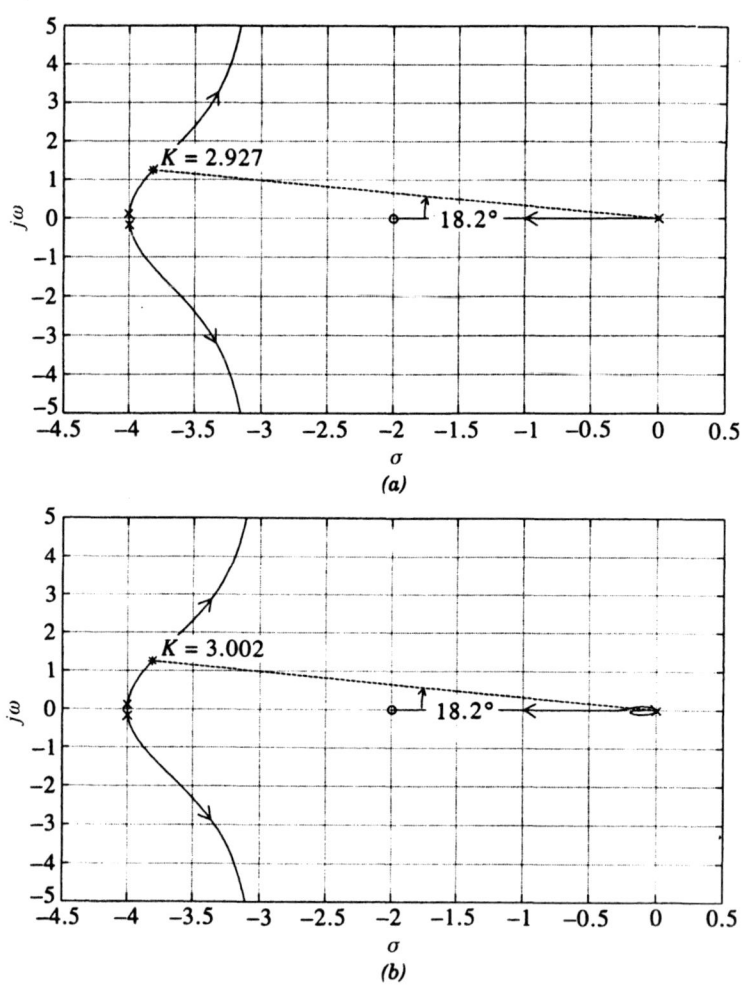

Figure A7.5

(c) $G_c(s) = \dfrac{s + 0.10922}{s + 0.001}$.

(d) Root-locus plot of Figure A7.5b indicates that

$$K = 3.002; \quad \text{therefore} \quad G_c(s) = \dfrac{s + 0.1066}{s + 0.001}.$$

7.7. (a) $c^*(t) = 0.5\delta(t-1) + 0.75\delta(t-2) + 0.875\delta(t-3)$
$+ 0.9375\delta(t-4) + 0.9688\delta(t-5) + 0.9844\delta(t-6)$
$+ 0.9922\delta(t-7) + 0.9961\delta(t-8) + 0.9980\delta(t-9) + \cdots$.

This corresponds to an overdamped response.

(b) $c^*(t) = \delta(t-1) + \delta(t-2) + \delta(t-3) + \delta(t-4)$
$+ \delta(t-5) + \delta(t-6) + \delta(t-7) + \delta(t-8) + \cdots$.

This corresponds to a critically damped system.

(c) $c^* = 2\delta(t-1) + 0\delta(t-2) + 2\delta(t-3) + 0\delta(t-4)$
$+ 2\delta(t-5) + 0\delta(t-6) + 2\delta(t-7) + 0\delta(t-8)$
$+ 2\delta(t-9) + \cdots$.

This corresponds to an oscillation having a stable amplitude of 2 (peak-to-peak) centered about 1.

(d) $c^*(t) = 3\delta(t-1) - 3\delta(t-2) + 9\delta(t-3) - 15\delta(t-4)$
$+ 33\delta(t-5) - 63\delta(t-6) + 129\delta(t-7)$
$- 255\delta(t-8) + 513\delta(t-9) + \cdots$.

This corresponds to a growing oscillation.

INDEX

Acceleration constant:
  definition of, 577
  determination from Bode diagram, 62
  relationship to poles and zeros, 584–585
Accuracy, static, 574–585
Ackermann's formula for design using pole placement, 137–143
  application example of, 139–143
Adaptive control systems, pitch flight, 116–117
Advanced z transform:
  characteristics of, 233–236
  table of, 235
Aircraft: adaptive pitch flight control of, 116–117
Analog-to-digital converters, 209–221
Antennas of tracking radars, mechanical resonance of, 508–510
APOLLO 11 mission, description of, 488–492
Aquaculture control system, optimal control application to, 499–500
Asymptotic stability, definition of, 408–414
Autopilots, control of airplane's roll using, 544–546

Backlash, definition of, 341–342
  describing function of, 342–344
back_lsh subroutine, Toolbox, 356
Bandwidth:
  effect of compensation on, 58, 93–96
  effect of noise on, 58
BASIC computer program:
  describing function from, 363–369
  Inverse z transform, 227–231
Bilinear transformation, 252
  modified, 257
Bode-diagram method, 16–21, 24–25
  approximate closed-loop response, 24–25
  break frequency of, 24–25
  case study design examples using, 507–516
  compensation using, 51–74
  digital system application of, 256–262, 528–537
  examples of, 54–67
  MATLAB used to obtain, 16–21
Bode's theorems:
  first theorem, 51–54
  second theorem, 54–55
Break frequency, definition of, 24–25

Calculus of variations, 462–467
  Bolza problem, 463
  Euler-Lagrange equation for, 465
  example of, 465–467
  Lagrange problem, 462
  Mayer problem, 462–463
Cascade compensation, methods of, 36–44
Cauchy's theorem:
  application to digital systems, 247
Center, definition of, 374
Characteristic response, second-order system, 566–573
Chemical control systems, 546
Circle criterion, *See* Generalized circle criterion
Closed-loop frequency response, 21–24
Compensation:
  Bode-diagram method of, 51–67
  cascade-compensation techniques, 36–44
  definition of, 33–34
  digital control application of, 257–261, 263–271
  linear continuous system use of, 33–36
  minor-loop feedback method of, 44–46
  Nichols chart method of, 21–24, 74–77
  root locus method of, 25–29, 74–93
  tradeoffs among methods of, 93–96
Complimentary sensitivity, definition of, 164
Computer programs:
  back_lsh, 356
  dead_zn, 356
  describing function using, 363–369
  inverse z transform, 230–231
  margins, 16
  MATLAB, 16–28, 272–273, 361–363, 549–565
  phase-plane obtained using, 367–401
  relays, 356
  rlpoba, 25
  root locus obtained using, 25
  rootangl, 25
  rootmag, 25
Controllability, definition of, 131–134
  example of, 134
  test for, 134

Damped frequency oscillation, definition of, 570–571
Damping factor, 571

Damping ratio:
  critical, 568
  definition of, 571
  overdamped, 570
  underdamped, 571
Data extrapolators, 215–221
  first-order hold, 217–221
  zero-order hold, 215–219
Dead zone, describing function of, 337–338
Dead zone nonlinearity, definition of, 337
dead_zn subroutine, Toolbox, 356
Describing functions, 336–369
  backlash, 342–344
  case study design example using, 507–516
  Coulomb friction, 347–348
  Dead zone, 337
  definition of, 334–336
  derivation of, 336–351
  design examples, 352–369
  guidelines for use of, 425–427
  hysteresis, 345–347
  MATLAB application to, 361–363
  prediction of oscillation using, 352–357
  saturation, 339–341
  software programs for obtaining, 361–369
  stiction, 347–351
Design:
  Bode-diagram method of, 51–74, 256–262, 507–516, 528–537
  cascade compensation techniques, 36–44
  general procedures for, 505–507
  linear continuous system, 33–177
  minor-loop feedback method of, 44–46
  Nichols chart method of, 21–24, 74–77
  robot's angular joint control, 516–522
  root-locus method of, 25–29, 74–93
  space vehicle attitude control, 477–492
  temperature control of liquid, 528–537
  tracking radar position loop, 507–516
Design procedure, general procedure for, 505–507
Differential equations, 326–327
  linear, 326
  nonlinear, 326–327
Digital control systems, 209–324, 528–537
  block-diagram algebra, 236–242
  Bode diagram applied to, 256-262, 528–537
  Case study of temperature control, 528–537
  digital filter, design of, 285–289
  examples of, 248–252, 256–285, 528–537
  Jury's criterion applied to, 255–256
  MATLAB applied to, 21–24, 74–77
  Nyquist diagram application to, 245–252
  Ragazzini's method applied to, 273–285
  root-locus diagram applied to, 263–271
  Routh-Hurwitz applied to, modified, 252–253
  sampling characteristics in, 211–215
  Schur-Cohn criterion applied to, 253–255
  signal-flow graphs applied to, 241–242
  stability analysis of, 245–271, 528–537
  temperature control, 528–537
Digital filters, 285–289
Digital-to-analog converters, 209–211
Digitization process, design of digital filters for, 285–289
Discrete-data control systems, See Digital control systems
Disturbance,
  effect on hydrofoil of sea wave, 537–542
  effect on tracking radar of wind, 507–516, 543–544
Dominant poles, 81–83
Dynamic programming, 467–473
  examples of, 470–473
  Hamilton-Jacobi equation for, 469
  relation to maximum principle, 477, 494–496

Ecological control systems, 500–501
Energy consumption minimization, use of optimal control for, 485–488
Error constants, definition of, 574–585
Estimator design:
  attitude system case study design, 477–492
  linear-state variable-feedback, 119–131, 143–156, 522–528
  pole placement, 119–131, 137–156
Estimator gain matrix M, definition of, 143
Euler-Lagrange equation, 465

Feedback compensation, 33–36, 44–45, 54–62, 64–67, 74–208
Final-value theorem: definition of z transform, 232–233
Finite stability, definition of, 409
First method of Liapunov, 405–406
First-order hold, 216–221
  amplitude characteristics of, 220
  phase characteristics of, 220
  transfer function of, 219
Focus, definition of, 383–384
FORTRAN Programs:
  phase-plane obtained using, 397–404
Fourier series:
  applied to describing function, 336–351
  applied to discrete systems, 213–215
Frequency spectrum, 213–215
Friction:
  Coulomb friction, 347–348
  describing function of, 347–351
  stiction, 347–348
Fuel consumption minimization, use of optimal control for, 482–485

Gain margin, definition of, 18
  multiple, 19

Gain-phase diagram, *See* Nichols chart
Generalized circle criterion, 422–425
    example of, 424–425
    guidelines for use of, 425–427
Global stability, definition of, 409

H^∞ control concepts, 162–175
    examples illustrating, 170–171, 174–175
    foundations of, 169–174
    MIMO case, 167–169
    SISO case, 174–175
    theoretical concepts of, 162–167
Hamilton-Jacobi equation, derivation of, 469
Hard self-excitation, derivation of, 328–329
Helicopter control system, 114–116
Human control systems, 114–116
Hydrofoils, effect of torque disturbances on, 537–541
    H.S. Denison, 104–105
    pitch-control system for, 104–105
Hysteresis:
    definition of, 345
    describing function of, 345–347

Ideal relay, characteristics of, 437
Initial-value theorem:
    definition of z transform, 232
Internal combustion engine:
    optimization of, 457
Inverse z transform, 227–230
    computer computational method, 229–230
    partial fraction expansion for, 228
    power-series expansion for, 228–229
Isocline method, phase-plane construction using, 372, 378–381

Jacobian matrices, definition of, 333
Jitter, due to noise, 5
Jump resonance, definition of, 328–329
Jury's stability criterion, 255–256

Lag network, 36–38
    Bode-diagram compensation using, 54, 56–58
    compensation using, 54–67
    Nichols chart compensation using, 74–75
    root-locus compensation using, 83–91
    tradeoffs of using, 93–96
Lag-lead network, 39–40
    tradeoffs of using, 93–96
Lead network, 38–39
    Bode diagram compensation using, 54, 56–58
    describing function compensation, 78–81
    digital control compensation, 256–261
    linear-state variable feedback, 130
    root locus compensation using, 78–81
    tradeoffs of using, 93–96

Liapunov's stability criteria, 399, 405–415
    design examples using, 414–415
    first method of, 405–406
    guidelines for use of, 425–427
    justification of, 412–414
    scalar function definiteness definition used for, 410–412
    second method of, 406–415
Limit cycles:
    definition of, 328
    phase-plane illustration of, 385–386
Linear algebraic techniques, 175–177
Linear differential equation, 326
    constant coefficient, 326
    homogeneous solution, 3
    particular solution, 3
    time-variable representation of, 326–327
Linear systems, 9–208
    compensation of, 33–208
    stability of, 3
    time varying, 326–327
Linearizing approximations, 330–334
Linear-state variable feedback, 119–131
    concept of, 119–124
    controller and estimator design, 143–154
    controller design using, 124–131
    estimator design using, 143–146
    general design procedure for, 119–124
    regulator design using, 146–154
Liquid-level control systems, 528–537
Local stability, definition of, 409
Logspace MATLAB command, definition of, 563
Lunar Excursion Module, equations for, 107–108, 488–492
Lunar landing, 488–492

Maintainability considerations, 5
margins subroutine, Toolbox, 16
Mason's theorem:
    digital system application of, 241–242
MATLAB, 549–565
    Bode diagram obtained using, 16–21
    commands, table of, 564–565
    conversion between continuous and discrete time, 15–16
    data representation of, 559–562
    Demonstration m-file, 550
    describing function analysis using, 361–362
    description of, 549–565
    fundamental concepts, 556
    Help, 550
    installation of, 552–553
    Modern Control System Theory and Design Toolbox, 549
    Nichols chart obtained using, 21–24
    Nonlinear Toolbox, 551
    Nyquist diagram obtained using, 12–16

MATLAB (*continued*)
  root-locus diagram obtained from, 25–28
  Simulink Toolbox, 551
  Student Edition of MATLAB, 549
  Symbolic Toolbox, 551
  Synopsis File, 550
  transformation from state-space to transfer function, 564
  transformation from transfer function to state-space form, 15
  retrieving from anonymous FTP server, 549
  tutorial, 550
  Tutorial File, 550
Matrix:
  closed-loop resolvent, 122
  controller gain K, 120
  estimator gain M, 143
  Jacobian, 333
Maximum percent overshoot:
  definition of, 573
  root-locus determination of, 81–83
Maximum principle, 473–492
  attitude control design using, 477–488
  co-state for, 474
  examples of, 475–492
  Hamiltonian function for, 474
  Lunar soft-landing system design, 488–492
  minimum-energy problem solution, 485–488
  minimum-fuel problem solution, 482–485
  minimum-time problem solution, 480–482
Maximum value of peaking, Nichols chart method for finding, 21–24
Method of isoclines, phase-plane construction using, 372, 378–381
Modern Control System Theory and Design Toolbox, 58
  retrieving from anonymous FTP server, 58
  table of commands used by, 74
Modified bilinear transformation, 257
Multivariable systems, 167–169

NASA programs:
  APOLLO 11, 488–489
  Lunar Excursion Module, 488–492
  Orbiting Astronomical Observer, 442–443
  Ranger unmanned space vehicle, 446–447
  Viking mission, 265–266
Negative-definite function, definition of, 410
Networks,
  phase-lag, 36–38
  phase-lag-lead, 39–40
  phase-lead, 38–39
Nichols chart:
  compensation and design using, 74–77
  MATLAB applied to obtain, 21–24
  maximum value of peaking from, 22–23
Node, phase-plane, 375

Noise:
  effect of bandwidth on, 58
  effect of compensation on, 58, 93–96
Nonlinear control systems, 325–456
  linearization of, 330–333
  methods available for analyzing, 329–330, 425–427
  properties of, 327–328
  relay servo, 392–397, 444
  simulation of, 426–427
Nonlinear differential equations, 326–327
Nonlinear system stability:
  describing functions for finding, 324–369
  generalized circle criterion for, 422–425
  guidelines for selecting method, 425–427
  Liapunov's criteria for finding, 399, 405–415
  phase-plane method for finding, 371–415
  piecewise-linear method for, 369–371
  Popov's method for finding, 415–422
Nonlinearities:
  backlash, 341–344
  Coulomb, 347–351
  dead zone, 337–338
  hysteresis, 345–347
  saturation, 339–341
  stiction, 347–351
Nuclear control systems, temperature-control system of, 103
Numerical methods, use of digital computers for, 332
Nyquist's stability criterion:
  digital control system application of, 245–252
  obtaining Nyquist diagram using MATLAB, 10–16

Observability, 135–137
On-off nonlinearity, describing function of, 345–347
Optimal control theory, 457–504
  aquaculture application of, 499–500
  attitude control application of, 477–488
  characteristics of, 458–462
  ecological system application of, 500–501
  lunar soft-landing problem, 459, 488–492
  minimum energy-problem, 459, 485–488
  minimum fuel-consumption problem, 459, 482–485
  minimum-time problem, 458–459, 480–482
  policy in, 461–462
  quadratic optimal control problem, 460, 501–502
Orbiting Astronomical Observatory, 442–443
Overdamped control system, definition of, 570
Overshoot:
  curves of, 574
  definition of, 572

Partial fraction expansion, z-transform inversion using, 228
Peak time, 572
Performance indices, quadratic, 460, 501–502
Phase portrait, definition of, 371–372
Phase trajectory, definition of, 372
Phase-lag network:
 Bode-diagram compensation using, 54, 56–58
 characteristics of, 36–38
 Nichols chart compensation using, 74–75
 robot joint case study use of, 516–521
 root-locus compensation using, 83–91
 tradeoffs of using, 93–96
Phase-lag-lead network:
 characteristics of, 39–40
 tradeoffs of using, 93–96
Phase-lead network:
 Bode-diagram compensation using, 54, 56–58
 characteristics of, 38–39
 describing function compensation, 78–81
 digital control compensation using, 256–261
 linear-state variable feedback using, 130
 root-locus compensation using, 78–81
 tradeoffs of using, 93–96
Phase-plane method, 371–404
 characteristics of, 381–389
 construction of, 372–381
 design examples using the, 392–397
 digital computer program for obtaining, 397–399
 energy determination from, 407–410
 guidelines for use of, 425–427
 representative phase portraits, 388–390
 software programs for obtaining, 397–399
 space-attitude control example, 481–485
 time determination from, 386–388
 with external forcing functions, 389–392
Phase-shift approximation, straight-line, 70–74
PD compensators, 47
PI compensators, 48, 103
PID compensators, 46–48
Piecewise-linear approximations, application to nonlinear systems, 330–334
Pole-placement design:
 using Ackermann's formula, 137–143
 using linear-state-variable feedback, 119–131
Pontryagin's maximum principle, 473–492
 attitude control design using, 477–492
 co-state for, 474
 examples of, 475–492
 Hamiltonian function for, 474
 Lunar soft-landing system design, 488–492
 minimum-energy problem solution, 485–488
 minimum-fuel problem solution, 482–485
 minimum-time problem solution, 480–482
Popov's method, 415–422
 design examples of, 421–422
 guidelines for use of, 425–427

Position constant:
 definition of, 576
 relationship to poles and zeros, 581–582
Position limiting, phase-plane illustration of, 389–390
Positive-definite function, definition of, 410
Power consumption, considerations of, 5
Power-series expansion, discrete system application of, 128–129
Procedures for design, 5–7
Programs:
 back_lsh, 356
 dead_zn, 356
 describing function using, 363–369
 discrete Bode and root locus, 272–273
 inverse z transform obtained from, 229–231
 margins, 16
 MATLAB, 15–28, 361–362, 549–565
 Nichols chart obtained from, 21–24
 phase-plane obtained using, 397–404
 relays, 356
 rlaxis, 25
 rlpoba, 25
 root locus obtained using, 25–28
 rootangl, 25
 rootmag, 25

Quadratic form of scalar function used for Liapunov's method, definition of, 410–411
Quadratic optimal control problem, 460, 501–502
Quasilinear control systems, definition of, 330

Ragazzini's method, 273–285
 application of, 276–282
 design rules, 275–276
 use of staleness factor with, 282–285
Ranger unmanned space vehicle, 446–447
Rate feedback:
 describing function compensation, 360–361
 linear system compensation using, 44–45
Rate limiting, phase-plane illustration of, 388–389
Regulator design, linear-state variable-feedback, 146–154
Relay servo:
 characteristics of, 392–397
 ideal, 444
 phase-plane analysis of, 392–397
 relays subroutine, Toolbox, 356
Reliability considerations, 5
rlaxis subroutine, Toolbox, 25
rlpoba subroutine, Toolbox, 25
Robotics:
 case study design of joint for, 516–521
 Ford Motor Company, 517
 welding of automobiles using, 544

Robust control systems, 156–162
  example illustrating, 158–162
  case study for controlling a hydrofoil, 537–541
  tracking radar positioning system robustness example, 547–548
Root-locus method, 409–454, 549–567
  case study design application of, 522–528
  compensation and design using, 75–93
  digital system application of, 528–537
  MATLAB applied to obtain, 25–28
rootangl subroutine, Toolbox, 25
rootmag subroutine, Toolbox, 25
Routh-Hurwitz stability criterion:
  application to root locus method, 87
  modified form for digital systems, 252–253

Saddle point, definition of, 383–385
Sampled-data control systems: 209–324
  case study of temperature control, 528–537
  definition of, 209
Sampled-data control system, See Digital control systems
Sampler-and-hold, 211–221
Sampling process:
  characteristics of, 211–221
  frequency-domain characteristics, 213–216
  ideal, 212
Saturation, 339–341
  definition of, 339
  describing function of, 340–341
Schedule considerations, 5
Schur-Cohn stability criterion, 253–255
  determinant, 253–254
Second-method of Liapunov, 406–415
Second-order control systems, characteristics of, 566–574
Sensitivity, definition of, 3–4
Sensitivity, 163–166
  complementary, 164
Shannon's sampling theorem, 214–215
Signal-flow graphs:
  case study design example using, 507–516
  digital system application of, 241–242
Simulation, nonlinear control system, 426–427
Singular points, definition of, 382–385
Size of equipment, considerations of, 5
Small-signal theory, 330–331
Soft self-excitation, definition of, 328–329
Space vehicles,
  APOLLO II, 488–492
  attitude control of, 477–488
  case study design examples, 522–528
  Lunar Excursion Module, 107–108, 488–492
  navigation equations for, 331–333
  Orbiting Astronomical Observer, 442–443
  Ranger Unmanned Space Vehicle, 446–447

Stability:
  asymptotic, 408–414
  Bode method for determining, 16–21, 24–25
  bounded input-bounded output, 3
  comparison of various methods, 93–96
  finite, 409
  global, 409
  guidelines for use of methods, 93–96
  Liapunov, 399, 405–415
  linear continuous system, 9–208
  local, 409
  Nichols method for determining, 21–24, 74–77
  Nyquist method for determining, 10–16, 245–252
  Routh-Hurwitz method, 87, 252–253
  summary of linear system methods, 93–96
Static accuracy, 574–585
Steady-state error, 575–586
  constants, 577–579
  from Bode diagram, 60–62
Stiction friction, 347–348
  definition of, 347
  describing function of, 347–351
  phase-plane application problem, 376–377
Subharmonic generation, definition of, 329
Subroutines from Advanced Modern Control System Theory and Design Toolbox:
  back_lsh, 356
  dead_zn, 356
  margins, 16
  relays, 356
  rlaxis, 25
  rlpoba, 25
  rootangl, 25
  rootmag, 25

Tables:
  acceleration constant and error, 578
  advanced z transform, 235
  compensation methods comparison, 94–95
  MATLAB commands and functions, 563–564
  MATLAB program listings for:
    Bode diagram obtained using, 18, 21
    describing function, 362–363
    Nichols chart obtained using, 24
    Nyquist diagram obtained using, 13, 16
    root locus method obtained using, 27–28
  position constant and error, 576
  steady-state constants summary, 578
  velocity constant and error, 577
  z transform, 227
Tank-level control system, design of, 528–537
Taylor series expansion, 332, 405, 469
Temperature control system:
  case study design of, 528–537
  nuclear power plant, 103
Thermal control systems, case study design of, 528–537

Time minimization, use of optimal control for, 458–459, 480–482
Toolbox subroutines, Advanced Modern Control System Theory and Design, *See* Subroutines for Advanced Modern Control System Theory and Design Toolbox
Tracking radar:
  design of positioning system for, 507–516
  example of, 543–544
Transfer-function concept:
  z transform, 221–236
Transformation between:
  state-space to transfer function form, 564
  transfer function to state-space form, 15
Transient response, 4
  determination from root locus, 75–93
  second-order systems, 566–574
Two-phase ac servomotor, nonlinear characteristics of, 330, 435–436
Type of control system, definition of, 575

Undamped natural frequency, definition of, 571
Underdamped control system, definition of, 571

Van der Pol equation, Liapunov's method applied to, 405
Velocity constant:
  definition of, 578
  determination from Bode diagram, 55
  relationship to poles and zeros, 583–584
w transforms:
  Bode diagrams using, 256–261
  case study design example using, 528–537
  definition of, 223–225
  modified, 257
Ward-Leonard system, case study example using, 513
Weight considerations, 5

z plane, relationship to s plane, 223–225
z transform: 221–236
  advanced, 233–236
  block diagram algebra, 236–242
  characteristics of, 242–245
  convergence of, 221–222
  exponential decay, 226
  final-value theorem, 232–233
  initial-value theorem, 232–233
  inverse, 227–230
  table of, 227
  transfer function using, 236–242
  unit ramp, 225–226
  unit step, 225
Zero-order hold, 215–219
  amplitude characteristics of, 218–219
  phase characteristics of, 218–219
  transfer function of, 218